TechRef

by

Thomas J. Glover
Millie M. Young

Third Edition

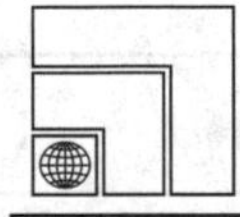

Sequoia Publishing, Inc.
Littleton, Colorado U.S.A.

This TechRef belongs to:

NAME: ______________________________

HOME ADDRESS: ______________________________

____

HOME PHONE: ______________________________

WORK PHONE: ______________________________

BUSINESS ADDRESS: ______________________________

____

In case of accident or serious illness, please notify:
Name:
Phone Number:

3rd Edition, 3rd Printing, June 1999

Sequoia Publishing, Inc.
9350 W. Cross Drive, Dept 101
Littleton, Colorado 80123
(303) 972-4167

ISBN 1–885071–15–9

Products by Sequoia Publishing, Inc.

Pocket Ref, 2nd Edition
by Thomas J. Glover
1989-1998, 544p, ISBN 1-885071-00-0

DeskRef, 2nd Edition
by Thomas J. Glover
6"x 9" version of Pocket Ref
1995-1998, 544p, ISBN 0-885071-06-X

MegaRef, Version 2
IBM PC and compatibles software version of the book *Pocket Ref.*

Pocket PC DIRectory, 1998 Edition
by Thomas J. Glover
November 1997, ISBN 1-885071-14-0

Pocket PCRef, 8th Edition
by Thomas J. Glover & Millie M. Young
November 1998, ISBN 1-885071-16-7

Preface

Sequoia Publishing, Inc. has made a serious effort to provide accurate information in this book. However, the probability exists that there are errors and misprints. Sequoia Publishing, Inc. and the authors do not represent the information as being exact and make no warranty of any kind with regard to the content of TechRef. Sequoia Publishing, Inc. and the authors shall not be held liable for any errors contained in TechRef or for incidental or consequential damages in connection with any use of the material herein.

The publishers would appreciate being notified of any errors, omissions, or misprints which may occur in this book. ***Your suggestions for future editions would also be greatly appreciated.***

The information in this manual was collected from numerous sources and if not properly acknowledged, Sequoia Publishing, Inc. and the authors would like to express their appreciation for those contributions. See page 6 for specific trade name, trade mark, and credit information.

Acknowledgments

TechRef would not have been possible without the efforts and endless patience of our families and many co-workers. Our deepest love and thanks to all of you.

Our deepest gratitude to Dave Derby, co-owner of Sequoia Publishing, for his technical editing, suggestions, and effort in tracking down the true meaning of Keyboard Scan Codes (a task no less difficult than tracking down the true meaning of life!).

Many thanks to Richard Young for his relentless pursuit of the perfect DOS Chapter. (Richard knows the true meaning of life and he has assured us that it has nothing to do with DOS!). Thanks to Liz Young, Trish Glover, Laurie Vendryes, Bob and Carrie Olson and Becky Tennessen for their help in compiling and verifying the Phone Book. Many thanks to Donna Baumgarten for her efforts in the never ending task of updating the hard and floppy drive sections.

Thank you never seems to be enough when you're saying it to the ones you care about the most! My family, Mary, Trish and Carrie, have supported and loved me through the whole monumental process of writing and publishing a book . . . Thank you and I love you. A very special thank you to my dear friend and co-author Millie, who has taught me the true meanings of courage, dedication and perseverance.

Thomas

It is amazing to me, what one person can accomplish when that accomplishment is based on the faith another person has in you. I share only in a small part of this book, the DOS Commands section, and though that may seem insignificant to some, it is a major accomplishment to this novice in the computer world. To the man I love, my gentle and patient husband Richard and our understanding offspring, Elizabeth, Christopher, and Stephanie, none of this would have been possible without you. And, especially to my mentor and friend Thomas, who doesn't know the meaning of limitations. To all of you who have had great faith in me and have allowed this humble sparrow to soar as an eagle, I give my sincerest thanks.

Millie

REFERENCES, TRADE NAMES and TRADE MARKS

The following are Registered Trademarks or Trade names:
ASCII – American Standard Code for Information Interchange
Commodore 64 – Commodore Computers
Diablo 630 – Xerox Corporation
Epson, FX–80 – Epson America Inc
Hayes – Hayes Microcomputer Products, Inc.
HP, HP-IB, Hewlett-Packard, Laserjet – Hewlett–Packard Company
IBM , AT, XT, PC, PS/2, PC Convertible, PC Jr., PC-DOS - International Business Machines Corporation
ISO – International Organization for Standardization
Macintosh, Apple IIc, Apple – Apple Computer, Inc.
Microsoft, MS–DOS, and Microsoft Windows – Microsoft Corporation
NEC, Pinwriter – NEC Corporation

The following books were used as references during the writing of TechRef. (They are all excellent references and should be added to any good reference library):

DOS Power Users Guide by Kris Jamsa
McGraw Hill, 1988, ISBN 0-07-881310-7
Hard Disk Handbook by Alfred Glossbrenner and Nick Anis
Osborne McGraw Hill, 1989, ISBN 0-07-881604-1
The Hard Disk Technical Guide by Douglas T. Anderson
PCS Publications, 1991
The Micro House Hard Drive Encyclopedia, Edited by Douglas T. Anderson, Micro House, 1992, 1993, 1994, 1995
Inside the IBM PC by Peter Norton
Brady Books, 1986, ISBN 0-89303-583-1
PC Magazine DOS Power Tools by Paul Somerson
Bantam Computer Books, 1988, ISBN 0-553-34526-5
Que's Computer User's Dictionary by Bryan Pfaffenberger
Que Corporation, 1990, ISBN 0-88022-540-8
Que's Upgrading & Repairing PCs by Scott Mueller
Que Corporation, 1994, ISBN 1-56529-736-9
MSDOS User's Guide and Reference, Ver 2.11, 3.0, 3.1, 3.2, 3.3, 4.01, 5.0, 6.0, & 6.2 by Microsoft Corporation.
Pocket Ref by Thomas J. Glover
Sequoia Publishing, Inc, 1989, ISBN 0-9622359-0-3
Supercharging MSDOS by Van Wolverton
Microsoft Press, 1986, ISBN 0-914845-95-0
The Winn Rosch Hardware Bible by Winn L. Rosch
Brady Books, 1989, ISBN 0-13-160979-3
PocketPOST by Data Depot
Clearwater, Florida (813) 446-3402
PC DOS Command Reference and Error Messages, Ver.6.0 and 6.3, by IBM Corporation.
Microsoft Windows User's Guide, Ver. 3.1, by Microsoft Corp.

NOTE: There are many more references, most of which are referenced on specific pages in TechRef. If we have omitted a reference, we apologize, please let us know and we will include it in the next printing of TechRef. See page 442 for additional hard drive references.

TABLE OF CONTENTS

NOTES

Chapter 1

ASCII and Numerics

COMPUTER ASCII CODES

The following ASCII (American Standard Code for Information Interchange) tables are used by most of the microcomputer industry. The codes occur in two sets: the "low–bit" set, from Dec 0 to Dec 127, and the "high–bit" set, from Dec 128 to Dec 255. The "low–bit" set is standard for almost all microcomputers but the "high–bit" set varies between the different computer brands. For instance, in the case of Apple computers and Epson printers, the "high–bit" set repeats the "low–bit" set except that the alphameric characters are italic. In the case of IBM and many other MSDOS systems, the "high–bit" set is composed of foreign language and box drawing characters and mathematic symbols.

Hex	Dec	Description	Abbr	Character	Control
00	0	Null	Null		Control @
01	1	Start Heading	SOH	☺	Control A
02	2	Start of Text	STX	☻	Control B
03	3	End of Text	ETX	♥	Control C
04	4	End Transmit	EOT	♦	Control D
05	5	Enquiry	ENQ	♣	Control E
06	6	Acknowledge	ACK	♠	Control F
07	7	Beep	BEL	•	Control G
08	8	Back space	BS	◘	Control H
09	9	Horizontal Tab	HT	○	Control I
0A	10	Line Feed	LF	◙	Control J
0B	11	Vertical Tab	VT	♂	Control K
0C	12	Form Feed	FF	♀	Control L
0D	13	Carriage Ret.	CR	♪	Control M
0E	14	Shift Out	SO	♫	Control N
0F	15	Shift In	SI	☼	Control O
10	16	Device Link Esc	DLE	►	Control P
11	17	Dev Cont 1 X-ON	DC1	◄	Control Q
12	18	Dev Control 2	DC2	↕	Control R
13	19	Dev Cont 3 X-OFF	DC3	‼	Control S
14	20	Dev Control 4	DC4	¶	Control T
15	21	Negative Ack	NAK	§	Control U
16	22	Synchronous Idle	SYN	▬	Control V
17	23	End Trans Block	ETB	↨	Control W
18	24	Cancel	CAN	↑	Control X
19	25	End Medium	EM	↓	Control Y
1A	26	Substitute	SUB	→	Control Z
1B	27	Escape	ESC	←	Control [

COMPUTER ASCII CODES

Hex	Dec	Description	Abbr	Character	Control
1C	28	Cursor Right	FS	└─	Control \
1D	29	Cursor Left	GS	↔	Control]
1E	30	Cursor Up	RS	▲	Control ^
1F	31	Cursor Down	US	▼	Control _

Hex	Dec	Character	Description
20	32		Space (SP)
21	33	!	Exclamation Point
22	34	“	Double Quote
23	35	#	Number sign
24	36	$	Dollar sign
25	37	%	Percent
26	38	&	Ampersand
27	39	’	Apostrophe
28	40	(	Left parenthesis
29	41	)	Right parenthesis
2A	42	*	Asterisk
2B	43	+	Plus sign
2C	44	,	Comma
2D	45	–	Minus sign
2E	46	.	Period
2F	47	/	Right or Front slash
30	48	0	Zero
31	49	1	One
32	50	2	Two
33	51	3	Three
34	52	4	Four
35	53	5	Five
36	54	6	Six
37	55	7	Seven
38	56	8	Eight
39	57	9	Nine
3A	58	:	Colon
3B	59	;	Semicolon
3C	60	<	Less than
3D	61	=	Equal sign
3E	62	>	Greater than
3F	63	?	Question mark
40	64	@	“at” symbol

COMPUTER ASCII CODES

Hex	Dec	Character	Description
41	65	A	Uppercase A
42	66	B	Uppercase B
43	67	C	Uppercase C
44	68	D	Uppercase D
45	69	E	Uppercase E
46	70	F	Uppercase F
47	71	G	Uppercase G
48	72	H	Uppercase H
49	73	I	Uppercase I
4A	74	J	Uppercase J
4B	75	K	Uppercase K
4C	76	L	Uppercase L
4D	77	M	Uppercase M
4E	78	N	Uppercase N
4F	79	O	Uppercase O
50	80	P	Uppercase P
51	81	Q	Uppercase Q
52	82	R	Uppercase R
53	83	S	Uppercase S
54	84	T	Uppercase T
55	85	U	Uppercase U
56	86	V	Uppercase V
57	87	W	Uppercase W
58	88	X	Uppercase X
59	89	Y	Uppercase Y
5A	90	Z	Uppercase Z
5B	91	[	Left bracket
5C	92	\	Left or Back Slash
5D	93	]	Right bracket
5E	94	^	Caret
5F	95	_	Underline
60	96	'	Accent
61	97	a	Lowercase a
62	98	b	Lowercase b
63	99	c	Lowercase c
64	100	d	Lowercase d
65	101	e	Lowercase e
66	102	f	Lowercase f
67	103	g	Lowercase g

COMPUTER ASCII CODES

Hex	Dec	Standard Character	Description
68	104	h	Lowercase h
69	105	i	Lowercase i
6A	106	j	Lowercase j
6B	107	k	Lowercase k
6C	108	l	Lowercase l
6D	109	m	Lowercase m
6E	110	n	Lowercase n
6F	111	o	Lowercase o
70	112	p	Lowercase p
71	113	q	Lowercase q
72	114	r	Lowercase r
73	115	s	Lowercase s
74	116	t	Lowercase t
75	117	u	Lowercase u
76	118	v	Lowercase v
77	119	w	Lowercase w
78	120	x	Lowercase x
79	121	y	Lowercase y
7A	122	z	Lowercase z
7B	123	{	Left brace
7C	124	|	Vertical line
7D	125	}	Right brace
7E	126	~	Tilde
7F	127	DEL	Delete

Hex	Dec	Standard Character	IBM Set	Standard Description
80	128	Null	Ç	Null
81	129	SOH	ü	Start Heading
82	130	STX	é	Start of Text
83	131	ETX	â	End of Text
84	132	EOT	ä	End Transmit
85	133	ENQ	à	Enquiry
86	134	ACK	å	Acknowledge
87	135	BEL	ç	Beep
88	136	BS	ê	Back Space
89	137	HT	ë	Horiz Tab
8A	138	LF	è	Line Feed

COMPUTER ASCII CODES

Hex	Dec	Standard Character	IBM Set	Standard Description
8B	139	VT	ï	Vertical Tab
8C	140	FF	î	Form Feed
8D	141	CR	ì	Carriage Return
8E	142	SO	Ä	Shift Out
8F	143	SI	Å	Shift In
90	144	DLE	É	Device Link Esc
91	145	DC1	æ	Device Cont 1 X–ON
92	146	DC2	Æ	Device Control 2
93	147	DC3	ô	Device Cont 3 X–OFF
94	148	DC4	ö	Device Control 4
95	149	NAK	ò	Negative Ack
96	150	SYN	û	Synchronous Idle
97	151	ETB	ù	End Transmit Block
98	152	CAN	ÿ	Cancel
99	153	EM	Ö	End Medium
9A	154	SUB	Ü	Substitute
9B	155	ESC	¢	Escape
9C	156	FS	£	Cursor Right
9D	157	GS	¥	Cursor Left
9E	158	RS	Pt	Cursor Up
9F	159	US	ƒ	Cursor Down
A0	160	Space	á	Space
A1	161	!	í	Italic Exclamation point
A2	162	“	ó	Italic Double quote
A3	163	#	ú	Italic Number sign
A4	164	$	ñ	Italic Dollar sign
A5	165	%	Ñ	Italic Percent
A6	166	&	ª	Italic Ampersand
A7	167	‘	º	Italic Apostrophe
A8	168	(	¿	Italic Left parenthesis
A9	169	)	⌐	Italic Right parenthesis
AA	170	*	¬	Italic asterisk
AB	171	+	½	Italic plus sign
AC	172	,	¼	Italic comma
AD	173	–	¡	Italic minus sign
AE	174	.	«	Italic period
AF	175	/	»	Italic right slash
B0	176	*0*	░	Italic Zero
B1	177	*1*	▒	Italic One

COMPUTER ASCII CODES

Hex	Dec	Standard Character	IBM Set	Standard Description
B2	178	2	▓	Italic Two
B3	179	3	│	Italic Three
B4	180	4	┤	Italic Four
B5	181	5	╡	Italic Five
B6	182	6	╢	Italic Six
B7	183	7	╖	Italic Seven
B8	184	8	╕	Italic Eight
B9	185	9	╣	Italic Nine
BA	186	:	║	Italic colon
BB	187	;	╗	Italic semicolon
BC	188	<	╝	Italic less than
BD	189	=	╜	Italic equal
BE	190	>	╛	Italic greater than
BF	191	?	┐	Italic question mark
C0	192	@	└	Italic "at" symbol
C1	193	A	┴	Italic A
C2	194	B	┬	Italic B
C3	195	C	├	Italic C
C4	196	D	─	Italic D
C5	197	E	┼	Italic E
C6	198	F	╞	Italic F
C7	199	G	╟	Italic G
C8	200	H	╚	Italic H
C9	201	I	╔	Italic I
CA	202	J	╩	Italic J
CB	203	K	╦	Italic K
CC	204	L	╠	Italic L
CD	205	M	═	Italic M
CE	206	N	╬	Italic N
CF	207	O	╧	Italic O
D0	208	P	╨	Italic P
D1	209	Q	╤	Italic Q
D2	210	R	╥	Italic R
D3	211	S	╙	Italic S
D4	212	T	╘	Italic T
D5	213	U	╒	Italic U
D6	214	V	╓	Italic V
D7	215	W	╫	Italic W
D8	216	X	╪	Italic X

COMPUTER ASCII CODES

Hex	Dec	Standard Character	IBM Set	Description
D9	217	*Y*	┘	Italic Y
DA	218	*Z*	┌	Italic Z
DB	219	*[*	█	Italic left bracket
DC	220	**	▄	Italic left or back slash
DD	221	*]*	▌	Italic right bracket
DE	222	*^*	▐	Italic caret
DF	223	_	▀	Italic underline
E0	224	'	α	Italic accent / alpha
E1	225	*a*	β	Italic a / beta
E2	226	*b*	Γ	Italic b / gamma
E3	227	*c*	π	Italic c / pi
E4	228	*d*	Σ	Italic d / sigma
E5	229	*e*	σ	Italic e / sigma
E6	230	*f*	μ	Italic f / mu
E7	231	*g*	γ	Italic g / gamma
E8	232	*h*	Φ	Italic h / phi
E9	233	*i*	θ	Italic i / theta
EA	234	*j*	Ω	Italic j / omega
EB	235	*k*	δ	Italic k / delta
EC	236	*l*	∞	Italic l / infinity
ED	237	*m*	Ø	Italic m / slashed zero
EE	238	*n*	∈	Italic n
EF	239	*o*	∩	Italic o
F0	240	*p*	≡	Italic p
F1	241	*q*	±	Italic q
F2	242	*r*	≥	Italic r
F3	243	*s*	≤	Italic s
F4	244	*t*	⌠	Italic t
F5	245	*u*	⌡	Italic u
F6	246	*v*	÷	Italic v
F7	247	*w*	≈	Italic w
F8	248	*x*	°	Italic x
F9	249	*y*	•	Italic y
FA	250	*z*	•	Italic z
FB	251	{	√	Italic left bracket
FC	252	|	^{n}	Italic vertical line
FD	253	}	^{2}	Italic right bracket
FE	254	~		Italic tilde
FF	255	Blank	Blank	Blank

NUMERIC PREFIXES

Prefix	Abbreviation	Pronounce	Multiplier
yocto	y	yok-to	10^{-24}
zepto	z	zep-to	10^{-21}
atto	a	at–to	10^{-18}
femto	f	fem–to	10^{-15}
pico	p	pe–ko	10^{-12}
nano	n	nan–o	10^{-9}
micro	m	mi–kro	10^{-6}
milli	m	mil – l	10^{-3}
centi	c	sent–ti	10^{-2}
deci	d	des – l	10^{-1}
deka	da	dek–a	10^{1}
hecto	h	hek–to	10^{2}
kilo	k	kil–o	10^{3}
mega	M	meg–a	10^{6}
giga	G	gig–a	10^{9}
tera	T	ter–a	10^{12}
peta	P	pe–ta	10^{15}
exa	E	ex–a	10^{18}
zetta	Z	za-ta	10^{21}
yotta	Y	yot-ta	10^{24}
		octillion	10^{27}
		nonillion	10^{30}

MEGABYTES AND KILOBYTES

1 kilobyte = 2^{10} bytes = exactly 1,024 bytes
1 megabyte = 2^{20} bytes = exactly 1,048,576 bytes
1 gigabyte = 2^{30} bytes = 1 billion bytes
1 terabyte = 2^{40} bytes = 1 trillion bytes
1 petabyte = 2^{50} bytes = 1 quadrillion bytes
1 byte = 8 bits (bit is short for binary digit)
8 bit computers (such as the 8088)
 move data in 1 byte chunks
16 bit computers (such as the 80286 and 80386SX)
 move data in 2 byte chunks
32 bit computers (80386DX,80486,Pentium, Power PC)
 move data in 4 byte chunks
64 bit computers (such as the Alpha AXP)
 move data in 8 byte chunks

POWERS OF 2

n	2^n	Hexadecimal
0	1	1
1	2	2
2	4	4
3	8	8
4	16	10
5	32	20
6	64	40
7	128	80
8	256	100
9	512	200
10	1024	400
11	2048	800
12	4096	1000
13	8192	2000
14	16384	4000
15	32768	8000
16	65536	10000
17	131072	20000
18	262144	40000
19	524288	80000
20	1048576	100000
21	2097152	200000
22	4194304	400000
23	8388608	800000
24	16777216	1000000
25	33554432	2000000
26	67108864	4000000
27	134217728	8000000
28	268435456	10000000
29	536870912	20000000
30	1073741824	40000000
31	2147483648	80000000
32	4294967296	100000000

POWERS OF 2

n	2^n	Hexadecimal
33	8589934592	200000000
34	17179869184	400000000
35	34359738368	800000000
36	68719476736	1000000000
37	137438953472	2000000000
38	274877906944	4000000000
39	549755813888	8000000000
40	1099511627776	10000000000
41	2199023255552	20000000000
42	4398046511104	40000000000
43	8796093022208	80000000000
44	17592186044416	100000000000
45	35184372088832	200000000000
46	70368744177664	400000000000
47	140737488355328	800000000000
48	281474976710656	1000000000000
49	562949953421312	2000000000000
50	1125899906842624	4000000000000
51	2251799813685248	8000000000000
52	4503599627370496	10000000000000
53	9007199254740992	20000000000000
54	18014398509481984	40000000000000
55	36028797018963968	80000000000000
56	72057594037927936	100000000000000
57	144115188075855872	200000000000000
58	288230376151711744	400000000000000
59	576460752303423488	800000000000000
60	1152921504606846976	1000000000000000
61	2305843009213693952	2000000000000000
62	4611686018427387904	4000000000000000
63	9223372036854775808	8000000000000000
64	18446744073709551616	10000000000000000

HEX to DECIMAL CONVERSION

Example: To convert the Hex number 1F7 to its decimal equivalent (Decimal 503), find 1F in the shaded left column of Hex numbers and follow the 1F row to the right, until it intersects the column with the *shaded* 7 at the top. The number at the intersection (503) is the decimal equivalent of Hex 1F7.

Standard Hex notation, using A through F to denote decimal values 10 through 15, is used in this table.

↓ Hex→	0	1	2	3	4	5	6	7
00	0	1	2	3	4	5	6	7
01	16	17	18	19	20	21	22	23
02	32	33	34	35	36	37	38	39
03	48	49	50	51	52	53	54	55
04	64	65	66	67	68	69	70	71
05	80	81	82	83	84	85	86	87
06	96	97	98	99	100	101	102	103
07	112	113	114	115	116	117	118	119
08	128	129	130	131	132	133	134	135
09	144	145	146	147	148	149	150	151
0A	160	161	162	163	164	165	166	167
0B	176	177	178	179	180	181	182	183
0C	192	193	194	195	196	197	198	199
0D	208	209	210	211	212	213	214	215
0E	224	225	226	227	228	229	230	231
0F	240	241	242	243	244	245	246	247
10	256	257	258	259	260	261	262	263
11	272	273	274	275	276	277	278	279
12	288	289	290	291	292	293	294	295
13	304	305	306	307	308	309	310	311
14	320	321	322	323	324	325	326	327
15	336	337	338	339	340	341	342	343
16	352	353	354	355	356	357	358	359
17	368	369	370	371	372	373	374	375
18	384	385	386	387	388	389	390	391
19	400	401	402	403	404	405	406	407
1A	416	417	418	419	420	421	422	423
1B	432	433	434	435	436	437	438	439
1C	448	449	450	451	452	453	454	455
1D	464	465	466	467	468	469	470	471
1E	480	481	482	483	484	485	486	487
1F	496	497	498	499	500	501	502	503
20	512	513	514	515	516	517	518	519
21	528	529	530	531	532	533	534	535
22	544	545	546	547	548	549	550	551
23	560	561	562	563	564	565	566	567
24	576	577	578	579	580	581	582	583
25	592	593	594	595	596	597	598	599

HEX to DECIMAL CONVERSION

Large number conversion: (Up to five Hexidecimal digits)
Find the fourth and fifth Hexidecimal significant digits in the following table and add their decimal equivalent to the value in the primary table. For example:

CB13F (Hex) = 786432 + 45056 + 319 = 831807 (Dec)

Hex	Dec	Hex	Dec	Hex	Dec	Hex	Dec
1000	4096	9000	36864	20000	131072	A0000	655360
2000	8192	A000	40960	30000	196608	B0000	720896
3000	12288	B000	45056	40000	262144	C0000	786432
4000	16384	C000	49152	50000	327680	D0000	851968
5000	20480	D000	53248	60000	393216	E0000	917504
6000	24576	E000	57344	70000	458752	F0000	983040
7000	28672	F000	61440	80000	524288		
8000	32768	10000	65536	90000	589824		

↓Hex→ 8	8	9	A	B	C	D	E	F
00	8	9	10	11	12	13	14	15
01	24	25	26	27	28	29	30	31
02	40	41	42	43	44	45	46	47
03	56	57	58	59	60	61	62	63
04	72	73	74	75	76	77	78	79
05	88	89	90	91	92	93	94	95
06	104	105	106	107	108	109	110	111
07	120	121	122	123	124	125	126	127
08	136	137	138	139	140	141	142	143
09	152	153	154	155	156	157	158	159
0A	168	169	170	171	172	173	174	175
0B	184	185	186	187	188	189	190	191
0C	200	201	202	203	204	205	206	207
0D	216	217	218	219	220	221	222	223
0E	232	233	234	235	236	237	238	239
0F	248	249	250	251	252	253	254	255
10	264	265	266	267	268	269	270	271
11	280	281	282	283	284	285	286	287
12	296	297	298	299	300	301	302	303
13	312	313	314	315	316	317	318	319
14	328	329	330	331	332	333	334	335
15	344	345	346	347	348	349	350	351
16	360	361	362	363	364	365	366	367
17	376	377	378	379	380	381	382	383
18	392	393	394	395	396	397	398	399
19	408	409	410	411	412	413	414	415
1A	424	425	426	427	428	429	430	431
1B	440	441	442	443	444	445	446	447
1C	456	457	458	459	460	461	462	463
1D	472	473	474	475	476	477	478	479
1E	488	489	490	491	492	493	494	495
1F	504	505	506	507	508	509	510	511
20	520	521	522	523	524	525	526	527
21	536	537	538	539	540	541	542	543
22	552	553	554	555	556	557	558	559
23	568	569	570	571	572	573	574	575
24	584	585	586	587	588	589	590	591
25	600	601	602	603	604	605	606	607

HEX to DECIMAL CONVERSION

↓Hex→	0	1	2	3	4	5	6	7
26	608	609	610	611	612	613	614	615
27	624	625	626	627	628	629	630	631
28	640	641	642	643	644	645	646	647
29	656	657	658	659	660	661	662	663
2A	672	673	674	675	676	677	678	679
2B	688	689	690	691	692	693	694	695
2C	704	705	706	707	708	709	710	711
2D	720	721	722	723	724	725	726	727
2E	736	737	738	739	740	741	742	743
2F	752	753	754	755	756	757	758	759
30	768	769	770	771	772	773	774	775
31	784	785	786	787	788	789	790	791
32	800	801	802	803	804	805	806	807
33	816	817	818	819	820	821	822	823
34	832	833	834	835	836	837	838	839
35	848	849	850	851	852	853	854	855
36	864	865	866	867	868	869	870	871
37	880	881	882	883	884	885	886	887
38	896	897	898	899	900	901	902	903
39	912	913	914	915	916	917	918	919
3A	928	929	930	931	932	933	934	935
3B	944	945	946	947	948	949	950	951
3C	960	961	962	963	964	965	966	967
3D	976	977	978	979	980	981	982	983
3E	992	993	994	995	996	997	998	999
3F	1008	1009	1010	1011	1012	1013	1014	1015
40	1024	1025	1026	1027	1028	1029	1030	1031
41	1040	1041	1042	1043	1044	1045	1046	1047
42	1056	1057	1058	1059	1060	1061	1062	1063
43	1072	1073	1074	1075	1076	1077	1078	1079
44	1088	1089	1090	1091	1092	1093	1094	1095
45	1104	1105	1106	1107	1108	1109	1110	1111
46	1120	1121	1122	1123	1124	1125	1126	1127
47	1136	1137	1138	1139	1140	1141	1142	1143
48	1152	1153	1154	1155	1156	1157	1158	1159
49	1168	1169	1170	1171	1172	1173	1174	1175
4A	1184	1185	1186	1187	1188	1189	1190	1191
4B	1200	1201	1202	1203	1204	1205	1206	1207
4C	1216	1217	1218	1219	1220	1221	1222	1223
4D	1232	1233	1234	1235	1236	1237	1238	1239
4E	1248	1249	1250	1251	1252	1253	1254	1255
4F	1264	1265	1266	1267	1268	1269	1270	1271
50	1280	1281	1282	1283	1284	1285	1286	1287
51	1296	1297	1298	1299	1300	1301	1302	1303
52	1312	1313	1314	1315	1316	1317	1318	1319
53	1328	1329	1330	1331	1332	1333	1334	1335
54	1344	1345	1346	1347	1348	1349	1350	1351
55	1360	1361	1362	1363	1364	1365	1366	1367
56	1376	1377	1378	1379	1380	1381	1382	1383
57	1392	1393	1394	1395	1396	1397	1398	1399
58	1408	1409	1410	1411	1412	1413	1414	1415
59	1424	1425	1426	1427	1428	1429	1430	1431
5A	1440	1441	1442	1443	1444	1445	1446	1447
5B	1456	1457	1458	1459	1460	1461	1462	1463
5C	1472	1473	1474	1475	1476	1477	1478	1479

HEX to DECIMAL CONVERSION

↓Hex→	8	9	A	B	C	D	E	F
26	616	617	618	619	620	621	622	623
27	632	633	634	635	636	637	638	639
28	648	649	650	651	652	653	654	655
29	664	665	666	667	668	669	670	671
2A	680	681	682	683	684	685	686	687
2B	696	697	698	699	700	701	702	703
2C	712	713	714	715	716	717	718	719
2D	728	729	730	731	732	733	734	735
2E	744	745	746	747	748	749	750	751
2F	760	761	762	763	764	765	766	767
30	776	777	778	779	780	781	782	783
31	792	793	794	795	796	797	798	799
32	808	809	810	811	812	813	814	815
33	824	825	826	827	828	829	830	831
34	840	841	842	843	844	845	846	847
35	856	857	858	859	860	861	862	863
36	872	873	874	875	876	877	878	879
37	888	889	890	891	892	893	894	895
38	904	905	906	907	908	909	910	911
39	920	921	922	923	924	925	926	927
3A	936	937	938	939	940	941	942	943
3B	952	953	954	955	956	957	958	959
3C	968	969	970	971	972	973	974	975
3D	984	985	986	987	988	989	990	991
3E	1000	1001	1002	1003	1004	1005	1006	1007
3F	1016	1017	1018	1019	1020	1021	1022	1023
40	1032	1033	1034	1035	1036	1037	1038	1039
41	1048	1049	1050	1051	1052	1053	1054	1055
42	1064	1065	1066	1067	1068	1069	1070	1071
43	1080	1081	1082	1083	1084	1085	1086	1087
44	1096	1097	1098	1099	1100	1101	1102	1103
45	1112	1113	1114	1115	1116	1117	1118	1119
46	1128	1129	1130	1131	1132	1133	1134	1135
47	1144	1145	1146	1147	1148	1149	1150	1151
48	1160	1161	1162	1163	1164	1165	1166	1167
49	1176	1177	1178	1179	1180	1181	1182	1183
4A	1192	1193	1194	1195	1196	1197	1198	1199
4B	1208	1209	1210	1211	1212	1213	1214	1215
4C	1224	1225	1226	1227	1228	1229	1230	1231
4D	1240	1241	1242	1243	1244	1245	1246	1247
4E	1256	1257	1258	1259	1260	1261	1262	1263
4F	1272	1273	1274	1275	1276	1277	1278	1279
50	1288	1289	1290	1291	1292	1293	1294	1295
51	1304	1305	1306	1307	1308	1309	1310	1311
52	1320	1321	1322	1323	1324	1325	1326	1327
53	1336	1337	1338	1339	1340	1341	1342	1343
54	1352	1353	1354	1355	1356	1357	1358	1359
55	1368	1369	1370	1371	1372	1373	1374	1375
56	1384	1385	1386	1387	1388	1389	1390	1391
57	1400	1401	1402	1403	1404	1405	1406	1407
58	1416	1417	1418	1419	1420	1421	1422	1423
59	1432	1433	1434	1435	1436	1437	1438	1439
5A	1448	1449	1450	1451	1452	1453	1454	1455
5B	1464	1465	1466	1467	1468	1469	1470	1471
5C	1480	1481	1482	1483	1484	1485	1486	1487

HEX to DECIMAL CONVERSION

↓ Hex→	0	1	2	3	4	5	6	7
5D	1488	1489	1490	1491	1492	1493	1494	1495
5E	1504	1505	1506	1507	1508	1509	1510	1511
5F	1520	1521	1522	1523	1524	1525	1526	1527
60	1536	1537	1538	1539	1540	1541	1542	1543
61	1552	1553	1554	1555	1556	1557	1558	1559
62	1568	1569	1570	1571	1572	1573	1574	1575
63	1584	1585	1586	1587	1588	1589	1590	1591
64	1600	1601	1602	1603	1604	1605	1606	1607
65	1616	1617	1618	1619	1620	1621	1622	1623
66	1632	1633	1634	1635	1636	1637	1638	1639
67	1648	1649	1650	1651	1652	1653	1654	1655
68	1664	1665	1666	1667	1668	1669	1670	1671
69	1680	1681	1682	1683	1684	1685	1686	1687
6A	1696	1697	1698	1699	1700	1701	1702	1703
6B	1712	1713	1714	1715	1716	1717	1718	1719
6C	1728	1729	1730	1731	1732	1733	1734	1735
6D	1744	1745	1746	1747	1748	1749	1750	1751
6E	1760	1761	1762	1763	1764	1765	1766	1767
6F	1776	1777	1778	1779	1780	1781	1782	1783
70	1792	1793	1794	1795	1796	1797	1798	1799
71	1808	1809	1810	1811	1812	1813	1814	1815
72	1824	1825	1826	1827	1828	1829	1830	1831
73	1840	1841	1842	1843	1844	1845	1846	1847
74	1856	1857	1858	1859	1860	1861	1862	1863
75	1872	1873	1874	1875	1876	1877	1878	1879
76	1888	1889	1890	1891	1892	1893	1894	1895
77	1904	1905	1906	1907	1908	1909	1910	1911
78	1920	1921	1922	1923	1924	1925	1926	1927
79	1936	1937	1938	1939	1940	1941	1942	1943
7A	1952	1953	1954	1955	1956	1957	1958	1959
7B	1968	1969	1970	1971	1972	1973	1974	1975
7C	1984	1985	1986	1987	1988	1989	1990	1991
7D	2000	2001	2002	2003	2004	2005	2006	2007
7E	2016	2017	2018	2019	2020	2021	2022	2023
7F	2032	2033	2034	2035	2036	2037	2038	2039
80	2048	2049	2050	2051	2052	2053	2054	2055
81	2064	2065	2066	2067	2068	2069	2070	2071
82	2080	2081	2082	2083	2084	2085	2086	2087
83	2096	2097	2098	2099	2100	2101	2102	2103
84	2112	2113	2114	2115	2116	2117	2118	2119
85	2128	2129	2130	2131	2132	2133	2134	2135
86	2144	2145	2146	2147	2148	2149	2150	2151
87	2160	2161	2162	2163	2164	2165	2166	2167
88	2176	2177	2178	2179	2180	2181	2182	2183
89	2192	2193	2194	2195	2196	2197	2198	2199
8A	2208	2209	2210	2211	2212	2213	2214	2215
8B	2224	2225	2226	2227	2228	2229	2230	2231
8C	2240	2241	2242	2243	2244	2245	2246	2247
8D	2256	2257	2258	2259	2260	2261	2262	2263
8E	2272	2273	2274	2275	2276	2277	2278	2279
8F	2288	2289	2290	2291	2292	2293	2294	2295
90	2304	2305	2306	2307	2308	2309	2310	2311
91	2320	2321	2322	2323	2324	2325	2326	2327
92	2336	2337	2338	2339	2340	2341	2342	2343
93	2352	2353	2354	2355	2356	2357	2358	2359

HEX to DECIMAL CONVERSION

↓Hex→	8	9	A	B	C	D	E	F
5D	1496	1497	1498	1499	1500	1501	1502	1503
5E	1512	1513	1514	1515	1516	1517	1518	1519
5F	1528	1529	1530	1531	1532	1533	1534	1535
60	1544	1545	1546	1547	1548	1549	1550	1551
61	1560	1561	1562	1563	1564	1565	1566	1567
62	1576	1577	1578	1579	1580	1581	1582	1583
63	1592	1593	1594	1595	1596	1597	1598	1599
64	1608	1609	1610	1611	1612	1613	1614	1615
65	1624	1625	1626	1627	1628	1629	1630	1631
66	1640	1641	1642	1643	1644	1645	1646	1647
67	1656	1657	1658	1659	1660	1661	1662	1663
68	1672	1673	1674	1675	1676	1677	1678	1679
69	1688	1689	1690	1691	1692	1693	1694	1695
6A	1704	1705	1706	1707	1708	1709	1710	1711
6B	1720	1721	1722	1723	1724	1725	1726	1727
6C	1736	1737	1738	1739	1740	1741	1742	1743
6D	1752	1753	1754	1755	1756	1757	1758	1759
6E	1768	1769	1770	1771	1772	1773	1774	1775
6F	1784	1785	1786	1787	1788	1789	1790	1791
70	1800	1801	1802	1803	1804	1805	1806	1807
71	1816	1817	1818	1819	1820	1821	1822	1823
72	1832	1833	1834	1835	1836	1837	1838	1839
73	1848	1849	1850	1851	1852	1853	1854	1855
74	1864	1865	1866	1867	1868	1869	1870	1871
75	1880	1881	1882	1883	1884	1885	1886	1887
76	1896	1897	1898	1899	1900	1901	1902	1903
77	1912	1913	1914	1915	1916	1917	1918	1919
78	1928	1929	1930	1931	1932	1933	1934	1935
79	1944	1945	1946	1947	1948	1949	1950	1951
7A	1960	1961	1962	1963	1964	1965	1966	1967
7B	1976	1977	1978	1979	1980	1981	1982	1983
7C	1992	1993	1994	1995	1996	1997	1998	1999
7D	2008	2009	2010	2011	2012	2013	2014	2015
7E	2024	2025	2026	2027	2028	2029	2030	2031
7F	2040	2041	2042	2043	2044	2045	2046	2047
80	2056	2057	2058	2059	2060	2061	2062	2063
81	2072	2073	2074	2075	2076	2077	2078	2079
82	2088	2089	2090	2091	2092	2093	2094	2095
83	2104	2105	2106	2107	2108	2109	2110	2111
84	2120	2121	2122	2123	2124	2125	2126	2127
85	2136	2137	2138	2139	2140	2141	2142	2143
86	2152	2153	2154	2155	2156	2157	2158	2159
87	2168	2169	2170	2171	2172	2173	2174	2175
88	2184	2185	2186	2187	2188	2189	2190	2191
89	2200	2201	2202	2203	2204	2205	2206	2207
8A	2216	2217	2218	2219	2220	2221	2222	2223
8B	2232	2233	2234	2235	2236	2237	2238	2239
8C	2248	2249	2250	2251	2252	2253	2254	2255
8D	2264	2265	2266	2267	2268	2269	2270	2271
8E	2280	2281	2282	2283	2284	2285	2286	2287
8F	2296	2297	2298	2299	2300	2301	2302	2303
90	2312	2313	2314	2315	2316	2317	2318	2319
91	2328	2329	2330	2331	2332	2333	2334	2335
92	2344	2345	2346	2347	2348	2349	2350	2351
93	2360	2361	2362	2363	2364	2365	2366	2367

HEX to DECIMAL CONVERSION

↓ Hex→	0	1	2	3	4	5	6	7
94	2368	2369	2370	2371	2372	2373	2374	2375
95	2384	2385	2386	2387	2388	2389	2390	2391
96	2400	2401	2402	2403	2404	2405	2406	2407
97	2416	2417	2418	2419	2420	2421	2422	2423
98	2432	2433	2434	2435	2436	2437	2438	2439
99	2448	2449	2450	2451	2452	2453	2454	2455
9A	2464	2465	2466	2467	2468	2469	2470	2471
9B	2480	2481	2482	2483	2484	2485	2486	2487
9C	2496	2497	2498	2499	2500	2501	2502	2503
9D	2512	2513	2514	2515	2516	2517	2518	2519
9E	2528	2529	2530	2531	2532	2533	2534	2535
9F	2544	2545	2546	2547	2548	2549	2550	2551
A0	2560	2561	2562	2563	2564	2565	2566	2567
A1	2576	2577	2578	2579	2580	2581	2582	2583
A2	2592	2593	2594	2595	2596	2597	2598	2599
A3	2608	2609	2610	2611	2612	2613	2614	2615
A4	2624	2625	2626	2627	2628	2629	2630	2631
A5	2640	2641	2642	2643	2644	2645	2646	2647
A6	2656	2657	2658	2659	2660	2661	2662	2663
A7	2672	2673	2674	2675	2676	2677	2678	2679
A8	2688	2689	2690	2691	2692	2693	2694	2695
A9	2704	2705	2706	2707	2708	2709	2710	2711
AA	2720	2721	2722	2723	2724	2725	2726	2727
AB	2736	2737	2738	2739	2740	2741	2742	2743
AC	2752	2753	2754	2755	2756	2757	2758	2759
AD	2768	2769	2770	2771	2772	2773	2774	2775
AE	2784	2785	2786	2787	2788	2789	2790	2791
AF	2800	2801	2802	2803	2804	2805	2806	2807
B0	2816	2817	2818	2819	2820	2821	2822	2823
B1	2832	2833	2834	2835	2836	2837	2838	2839
B2	2848	2849	2850	2851	2852	2853	2854	2855
B3	2864	2865	2866	2867	2868	2869	2870	2871
B4	2880	2881	2882	2883	2884	2885	2886	2887
B5	2896	2897	2898	2899	2900	2901	2902	2903
B6	2912	2913	2914	2915	2916	2917	2918	2919
B7	2928	2929	2930	2931	2932	2933	2934	2935
B8	2944	2945	2946	2947	2948	2949	2950	2951
B9	2960	2961	2962	2963	2964	2965	2966	2967
BA	2976	2977	2978	2979	2980	2981	2982	2983
BB	2992	2993	2994	2995	2996	2997	2998	2999
BC	3008	3009	3010	3011	3012	3013	3014	3015
BD	3024	3025	3026	3027	3028	3029	3030	3031
BE	3040	3041	3042	3043	3044	3045	3046	3047
BF	3056	3057	3058	3059	3060	3061	3062	3063
C0	3072	3073	3074	3075	3076	3077	3078	3079
C1	3088	3089	3090	3091	3092	3093	3094	3095
C2	3104	3105	3106	3107	3108	3109	3110	3111
C3	3120	3121	3122	3123	3124	3125	3126	3127
C4	3136	3137	3138	3139	3140	3141	3142	3143
C5	3152	3153	3154	3155	3156	3157	3158	3159
C6	3168	3169	3170	3171	3172	3173	3174	3175
C7	3184	3185	3186	3187	3188	3189	3190	3191
C8	3200	3201	3202	3203	3204	3205	3206	3207
C9	3216	3217	3218	3219	3220	3221	3222	3223
CA	3232	3233	3234	3235	3236	3237	3238	3239

HEX to DECIMAL CONVERSION

↓Hex→	8	9	A	B	C	D	E	F
94	2376	2377	2378	2379	2380	2381	2382	2383
95	2392	2393	2394	2395	2396	2397	2398	2399
96	2408	2409	2410	2411	2412	2413	2414	2415
97	2424	2425	2426	2427	2428	2429	2430	2431
98	2440	2441	2442	2443	2444	2445	2446	2447
99	2456	2457	2458	2459	2460	2461	2462	2463
9A	2472	2473	2474	2475	2476	2477	2478	2479
9B	2488	2489	2490	2491	2492	2493	2494	2495
9C	2504	2505	2506	2507	2508	2509	2510	2511
9D	2520	2521	2522	2523	2524	2525	2526	2527
9E	2536	2537	2538	2539	2540	2541	2542	2543
9F	2552	2553	2554	2555	2556	2557	2558	2559
A0	2568	2569	2570	2571	2572	2573	2574	2575
A1	2584	2585	2586	2587	2588	2589	2590	2591
A2	2600	2601	2602	2603	2604	2605	2606	2607
A3	2616	2617	2618	2619	2620	2621	2622	2623
A4	2632	2633	2634	2635	2636	2637	2638	2639
A5	2648	2649	2650	2651	2652	2653	2654	2655
A6	2664	2665	2666	2667	2668	2669	2670	2671
A7	2680	2681	2682	2683	2684	2685	2686	2687
A8	2696	2697	2698	2699	2700	2701	2702	2703
A9	2712	2713	2714	2715	2716	2717	2718	2719
AA	2728	2729	2730	2731	2732	2733	2734	2735
AB	2744	2745	2746	2747	2748	2749	2750	2751
AC	2760	2761	2762	2763	2764	2765	2766	2767
AD	2776	2777	2778	2779	2780	2781	2782	2783
AE	2792	2793	2794	2795	2796	2797	2798	2799
AF	2808	2809	2810	2811	2812	2813	2814	2815
B0	2824	2825	2826	2827	2828	2829	2830	2831
B1	2840	2841	2842	2843	2844	2845	2846	2847
B2	2856	2857	2858	2859	2860	2861	2862	2863
B3	2872	2873	2874	2875	2876	2877	2878	2879
B4	2888	2889	2890	2891	2892	2893	2894	2895
B5	2904	2905	2906	2907	2908	2909	2910	2911
B6	2920	2921	2922	2923	2924	2925	2926	2927
B7	2936	2937	2938	2939	2940	2941	2942	2943
B8	2952	2953	2954	2955	2956	2957	2958	2959
B9	2968	2969	2970	2971	2972	2973	2974	2975
BA	2984	2985	2986	2987	2988	2989	2990	2991
BB	3000	3001	3002	3003	3004	3005	3006	3007
BC	3016	3017	3018	3019	3020	3021	3022	3023
BD	3032	3033	3034	3035	3036	3037	3038	3039
BE	3048	3049	3050	3051	3052	3053	3054	3055
BF	3064	3065	3066	3067	3068	3069	3070	3071
C0	3080	3081	3082	3083	3084	3085	3086	3087
C1	3096	3097	3098	3099	3100	3101	3102	3103
C2	3112	3113	3114	3115	3116	3117	3118	3119
C3	3128	3129	3130	3131	3132	3133	3134	3135
C4	3144	3145	3146	3147	3148	3149	3150	3151
C5	3160	3161	3162	3163	3164	3165	3166	3167
C6	3176	3177	3178	3179	3180	3181	3182	3183
C7	3192	3193	3194	3195	3196	3197	3198	3199
C8	3208	3209	3210	3211	3212	3213	3214	3215
C9	3224	3225	3226	3227	3228	3229	3230	3231
CA	3240	3241	3242	3243	3244	3245	3246	3247

HEX to DECIMAL CONVERSION

↓Hex→	0	1	2	3	4	5	6	7
CB	3248	3249	3250	3251	3252	3253	3254	3255
CC	3264	3265	3266	3267	3268	3269	3270	3271
CD	3280	3281	3282	3283	3284	3285	3286	3287
CE	3296	3297	3298	3299	3300	3301	3302	3303
CF	3312	3313	3314	3315	3316	3317	3318	3319
D0	3328	3329	3330	3331	3332	3333	3334	3335
D1	3344	3345	3346	3347	3348	3349	3350	3351
D2	3360	3361	3362	3363	3364	3365	3366	3367
D3	3376	3377	3378	3379	3380	3381	3382	3383
D4	3392	3393	3394	3395	3396	3397	3398	3399
D5	3408	3409	3410	3411	3412	3413	3414	3415
D6	3424	3425	3426	3427	3428	3429	3430	3431
D7	3440	3441	3442	3443	3444	3445	3446	3447
D8	3456	3457	3458	3459	3460	3461	3462	3463
D9	3472	3473	3474	3475	3476	3477	3478	3479
DA	3488	3489	3490	3491	3492	3493	3494	3495
DB	3504	3505	3506	3507	3508	3509	3510	3511
DC	3520	3521	3522	3523	3524	3525	3526	3527
DD	3536	3537	3538	3539	3540	3541	3542	3543
DE	3552	3553	3554	3555	3556	3557	3558	3559
DF	3568	3569	3570	3571	3572	3573	3574	3575
E0	3584	3585	3586	3587	3588	3589	3590	3591
E1	3600	3601	3602	3603	3604	3605	3606	3607
E2	3616	3617	3618	3619	3620	3621	3622	3623
E3	3632	3633	3634	3635	3636	3637	3638	3639
E4	3648	3649	3650	3651	3652	3653	3654	3655
E5	3664	3665	3666	3667	3668	3669	3670	3671
E6	3680	3681	3682	3683	3684	3685	3686	3687
E7	3696	3697	3698	3699	3700	3701	3702	3703
E8	3712	3713	3714	3715	3716	3717	3718	3719
E9	3728	3729	3730	3731	3732	3733	3734	3735
EA	3744	3745	3746	3747	3748	3749	3750	3751
EB	3760	3761	3762	3763	3764	3765	3766	3767
EC	3776	3777	3778	3779	3780	3781	3782	3783
ED	3792	3793	3794	3795	3796	3797	3798	3799
EE	3808	3809	3810	3811	3812	3813	3814	3815
EF	3824	3825	3826	3827	3828	3829	3830	3831
F0	3840	3841	3842	3843	3844	3845	3846	3847
F1	3856	3857	3858	3859	3860	3861	3862	3863
F2	3872	3873	3874	3875	3876	3877	3878	3879
F3	3888	3889	3890	3891	3892	3893	3894	3895
F4	3904	3905	3906	3907	3908	3909	3910	3911
F5	3920	3921	3922	3923	3924	3925	3926	3927
F6	3936	3937	3938	3939	3940	3941	3942	3943
F7	3952	3953	3954	3955	3956	3957	3958	3959
F8	3968	3969	3970	3971	3972	3973	3974	3975
F9	3984	3985	3986	3987	3988	3989	3990	3991
FA	4000	4001	4002	4003	4004	4005	4006	4007
FB	4016	4017	4018	4019	4020	4021	4022	4023
FC	4032	4033	4034	4035	4036	4037	4038	4039
FD	4048	4049	4050	4051	4052	4053	4054	4055
FE	4064	4065	4066	4067	4068	4069	4070	4071
FF	4080	4081	4082	4083	4084	4085	4086	4087

HEX to DECIMAL CONVERSION

↓Hex→ 8		9	A	B	C	D	E	F
CB	3256	3257	3258	3259	3260	3261	3262	3263
CC	3272	3273	3274	3275	3276	3277	3278	3279
CD	3288	3289	3290	3291	3292	3293	3294	3295
CE	3304	3305	3306	3307	3308	3309	3310	3311
CF	3320	3321	3322	3323	3324	3325	3326	3327
D0	3336	3337	3338	3339	3340	3341	3342	3343
D1	3352	3353	3354	3355	3356	3357	3358	3359
D2	3368	3369	3370	3371	3372	3373	3374	3375
D3	3384	3385	3386	3387	3388	3389	3390	3391
D4	3400	3401	3402	3403	3404	3405	3406	3407
D5	3416	3417	3418	3419	3420	3421	3422	3423
D6	3432	3433	3434	3435	3436	3437	3438	3439
D7	3448	3449	3450	3451	3452	3453	3454	3455
D8	3464	3465	3466	3467	3468	3469	3470	3471
D9	3480	3481	3482	3483	3484	3485	3486	3487
DA	3496	3497	3498	3499	3500	3501	3502	3503
DB	3512	3513	3514	3515	3516	3517	3518	3519
DC	3528	3529	3530	3531	3532	3533	3534	3535
DD	3544	3545	3546	3547	3548	3549	3550	3551
DE	3560	3561	3562	3563	3564	3565	3566	3567
DF	3576	3577	3578	3579	3580	3581	3582	3583
E0	3592	3593	3594	3595	3596	3597	3598	3599
E1	3608	3609	3610	3611	3612	3613	3614	3615
E2	3624	3625	3626	3627	3628	3629	3630	3631
E3	3640	3641	3642	3643	3644	3645	3646	3647
E4	3656	3657	3658	3659	3660	3661	3662	3663
E5	3672	3673	3674	3675	3676	3677	3678	3679
E6	3688	3689	3690	3691	3692	3693	3694	3695
E7	3704	3705	3706	3707	3708	3709	3710	3711
E8	3720	3721	3722	3723	3724	3725	3726	3727
E9	3736	3737	3738	3739	3740	3741	3742	3743
EA	3752	3753	3754	3755	3756	3757	3758	3759
EB	3768	3769	3770	3771	3772	3773	3774	3775
EC	3784	3785	3786	3787	3788	3789	3790	3791
ED	3800	3801	3802	3803	3804	3805	3806	3807
EE	3816	3817	3818	3819	3820	3821	3822	3823
EF	3832	3833	3834	3835	3836	3837	3838	3839
F0	3848	3849	3850	3851	3852	3853	3854	3855
F1	3864	3865	3866	3867	3868	3869	3870	3871
F2	3880	3881	3882	3883	3884	3885	3886	3887
F3	3896	3897	3898	3899	3900	3901	3902	3903
F4	3912	3913	3914	3915	3916	3917	3918	3919
F5	3928	3929	3930	3931	3932	3933	3934	3935
F6	3944	3945	3946	3947	3948	3949	3950	3951
F7	3960	3961	3962	3963	3964	3965	3966	3967
F8	3976	3977	3978	3979	3980	3981	3982	3983
F9	3992	3993	3994	3995	3996	3997	3998	3999
FA	4008	4009	4010	4011	4012	4013	4014	4015
FB	4024	4025	4026	4027	4028	4029	4030	4031
FC	4040	4041	4042	4043	4044	4045	4046	4047
FD	4056	4057	4058	4059	4060	4061	4062	4063
FE	4072	4073	4074	4075	4076	4077	4078	4079
FF	4088	4089	4090	4091	4092	4093	4094	4095

ALPHABET-DEC-HEX-EBCDIC

Hex	Dec	Alph	EBCDIC
00	0	Null	00
01	1	SOH	01
02	2	STX	02
03	3	ETX	03
04	4	EOT	37
05	5	ENQ	2D
06	6	ACK	2E
07	7	BEL	2F
08	8	BS	16
09	9	HT	05
0A	10	LF	25
0B	11	VT	0B
0C	12	FF	0C
0D	13	CR	0D
0E	14	SO	0E
0F	15	SI	0F
10	16	DLE	10
11	17	DC1	11
12	18	DC2	12
13	19	DC3	13
14	20	DC4	3C
15	21	NAK	3D
16	22	SYN	32
17	23	ETB	11
18	24	CAN	18
19	25	EM	19
1A	26	SUB	3F
1B	27	ESC	27
1C	28	FS	22
1D	29	GS	–
1E	30	RS	35
1F	31	US	–
20	32	space	40
21	33	!	5A
22	34	“	7F
23	35	#	7B
24	36	$	5B
25	37	%	6C
26	38	&	50
27	39	‘	7D
28	40	(	4D
29	41	)	5D
2A	42	*	5C
2B	43	+	4E
2C	44	,	6B
2D	45	–	60
2E	46	.	4B
2F	47	/	61
30	48	0	F0
31	49	1	F1
32	50	2	F2
33	51	3	F3
34	52	4	F4
35	53	5	F5
36	54	6	F6
37	55	7	F7
38	56	8	F8
39	57	9	F9
3A	58	:	7A
3B	59	;	5E
3C	60	<	4C
3D	61	>	7E
3E	62	=	6E
3F	63	?	6F
40	64	@	7C
41	65	A	C1
42	66	B	C2
43	67	C	C3
44	68	D	C4
45	69	E	C5
46	70	F	C6
47	71	G	C7
48	72	H	C8
49	73	I	C9
4A	74	J	D1
4B	75	K	D2
4C	76	L	D3
4D	77	M	D4
4E	78	N	D5
4F	79	O	D6
50	80	P	D7
51	81	Q	D8
52	82	R	D9
53	83	S	E2
54	84	T	E3
55	85	U	E4
56	86	V	E5
57	87	W	E6
58	88	X	E7
59	89	Y	E8
5A	90	Z	E9
5B	91	[	–
5C	92	\	E0
5D	93	]	–
5E	94	^	–
5F	95	_	6D
60	96	‘	–
61	97	a	81
62	98	b	82
63	99	c	83
64	100	d	84
65	101	e	85
66	102	f	86
67	103	g	87
68	104	h	88
69	105	i	89
6A	106	j	91
6B	107	k	92
6C	108	l	93
6D	109	m	94
6E	110	n	95
6F	111	o	96
70	112	p	97
71	113	q	98
72	114	r	99
73	115	s	A2
74	116	t	A3
75	117	u	A4
76	118	v	A5
77	119	w	A6
78	120	x	A7
79	121	y	A8
7A	122	z	A9
7B	123	{	C0
7C	124	|	6A
7D	125	}	D0
7E	126	~	A1
7F	127	DEL	07

Chapter 2

PC Hardware

VIDEO STANDARDS

Video Standard (year)	Mode	Horz x Vert Resolution (pixels)	Simul-taneous Colors	Vert Freq Hz	Horz Freq kHz	Band Width MHz
MDA (1981)	Text	720x350	1	50Hz	18.43	16.257
HGC	Text	640x400	1	50	18.43	16.257
	Graph	720x348	1	50	"	"
CGA	Text	320x200	16	60	15.75	14.318
(1981)	Text	640x200	16	60	"	"
	Graph	320x200	4	60	"	"
	Graph	640x200	2	60	"	"
EGA Color	Text	640x350	16	60	15.75	14.318
(1985)	Graph	640x350	16	60	to	to
	Graph	320x200	16	60	21.85	16.257
	Graph	640x350	64	60	"	"
EGA Mono	Graph	640x350	1	50	"	"
MCGA	Text	320x400	16	70	31.50	25.175
(1987)	Text	640x400	16	70	"	"
	Graph	640x480	2	60	"	"
	Graph	320x200	256	70	"	"
VGA	Text	360x400	16	70	31.50	25.175
(1987)	Text	720x400	16	70	"	to
	Graph	640x350	16	70	"	28.322
	Graph	640x480	16	60	"	"
	Graph	640x480	2	60	"	"
	Graph	320x200	256	70	"	"
Super VGA	Graph	800x600	16	50,60	35,37	
(1989)	Graph	800x600	256	and	and	
	Graph	1024x768	16	72	60,80	
8514-A	Graph	1024x768	16	43.48	35.52	44.897
(1987)	Graph	640x480	256	60	31.5	"
	Graph	1024x768	256	43.48	35.52	"
XGA	Graph	640x480	256	43.48	35.52	
(1990)	Graph	1024x768	256	43.48	"	
	Graph	640x480	65536	60	31.5	
	Text	1056x400	16	70	"	

Note: Most video cards built around the standards listed above are downward compatible and will function in the modes of the earlier standards. For example, most VGA cards will operate in all of the MDA, CGA, and EGA modes.

VIDEO STANDARDS

Abbreviations for the graphics standards defined on the previous page are as follows:

MDA . . . . . . Monochrome Display Adapter
HGC . . . . . . Hercules Graphics Card
CGA . . . . . . Color Graphics Adapter
EGA. . . . . . . Enhanced Graphics Adapter
PGA. . . . . . . Professional Graphics Adapter
MCGA . . . . . Multi Color Graphics Array
VGA. . . . . . . Video Graphics Array - digital
8514-A. . . . . Video Graphics Array - analog
Super VGA . Super Video Graphics Array, VESA
XGA. . . . . . . Extended Graphics Array

Pixels are coded by assigning bits to the colors. 1 bit/pixel boards can only display 1 color, monochrome (the bit is either on or off). 2 bits/pixel boards can display 4 colors (CGA for example). 8 bits/pixel can display 256 colors (VGA for example). 24 bits/pixel can display 16,777,216 simultaneous colors. Video board memory limits the number of colors that a graphics adapter can store; for example, a 1024x768 adapter requires 786,432 bytes of memory in order to display 256 colors. Needless to say, future video memory requirements will continue to grow. Consider that a 4096x4096 image with 24 bit/pixel color will require nearly 50 Mb of video RAM.

KEYBOARD SCAN CODES

Generally, expanded PC/XT, AT and PS/2 keyboard scan codes are converted to PC/XT standard scan codes prior to ROM BIOS ASCII Code conversion. Notable exceptions are the F11 and F12 keys, which generate new scan codes (see table below). Extended ASCII characters and some special "characters" are achieved by combining 2 or more key presses.

Shaded areas in the table represent keys and scan codes of the standard 84 key PC/XT keyboard, however, the "Key #" listed in column 1 of the table is not the correct Key # for the XT class keyboard. See your computer's keyboard documentation for verification of the correct Key # to Key Name assignments. AT Scan Codes are only relevant to AT class and PS/2 (Models 50 and above) computers.

Key # for 101 Keybd	Key Name	XT scan codes Down • Up	AT hardware scan codes Down • Up
1	Esc	01 • 81	76 • F0 76
2	F1	3B • BB	05 • F0 05
3	F2	3C • BC	06 • F0 06
4	F3	3D • BD	04 • F0 04
5	F4	3E • BE	0C • F0 0C
6	F5	3F • BF	03 • F0 03
7	F6	40 • C0	0B • F0 0B
8	F7	41 • C1	83 • F0 83
9	F8	42 • C2	0A • F0 0A
10	F9	43 • C3	01 • F0 01
11	F10	44 • C4	09 • F0 09
12	F11	57 • D7	78 • F0 78
13	F12	58 • D8	07 • F0 07

Special Keys (expanded keyboards only)

14	***PrtScn / SysReq***		
14	–PRINT SCRN	E0 2A E0 37 •	E0 12 E0 7C •
14		E0 B7 E0 AA	E0 F0 7C E0 F0 12
14	–Sys Req (+ CTRL)	E0 37 • E0 B7	E0 7C • E0 F0 7C
14	–Sys Req (+ ALT)	54 •D4	84 • F0 84
15	ScrollLock	46 • C6	7E • F0 7E
16	***Pause / Break***		
16	–PAUSE (key alone)	E1 1D 45 E1 9D C5 •	E1 14 77 E1 F0 14 F0 77 •
16	(No Auto Repeat)	No Up Code	No Up Code
16	–BREAK (+ CTRL)	E0 46 E0 C6 •	E0 7E E0 F0 7E •
16	(No Auto Repeat)	No Up Code	No Up Code
31	***Insert Key***	E0 52 •E0 D2	E0 70 • E0 F0 70
31	–LEFT SHIFT case	E0 AA E0 52 •	E0 F0 12 E0 70 •
31		E0 D2 E0 2A	E0 F0 70 E0 12
31	–RIGHT SHIFT case	E0 B6 E0 52 •	E0 F0 59 E0 70 •
31		E0 D2 E0 36	E0 F0 70 E0 59
31	–NUM LOCK ON case	E0 2A E0 52 •	E0 12 E0 70 •
31		E0 D2 E0 AA	E0 F0 70 E0 F0 12

KEYBOARD SCAN CODES (cont.)

Key # for 101 Keybd	Key Name	XT scan codes Down • Up	AT hardware scan codes Down • Up
32	***Home***	E0 47 • E0 C7	E0 6C • E0 F0 6C
32	–LEFT SHIFT case	E0 AA E0 47 •	E0 F0 12 E0 6C •
32		E0 C7 E0 2A	E0 F0 6C E0 12
32	–RIGHT SHIFT case	E0 B6 E0 47 •	E0 F0 59 E0 6C •
32		E0 C7 E0 36	E0 F0 6C E0 59
32	–NUM LOCK ON case	E0 2A E0 47 •	E0 12 E0 6C •
32		E0 C7 E0 AA	E0 F0 6C E0 F0 12
33	***PageUp***	E0 49 • E0 C9	E0 7D • E0 F0 7D
33	–LEFT SHIFT case	E0 AA E0 49 •	E0 F0 12 E0 7D •
33		E0 C9 E0 2A	E0 F0 7D E0 12
33	–RIGHT SHIFT case	E0 B6 E0 49 •	E0 F0 59 E0 7D •
33		E0 C9 E0 36	E0 F0 7D E0 59
33	–NUM LOCK ON case	E0 2A E0 49 •	E0 12 E0 7D •
33		E0 C9 E0 AA	E0 F0 7D E0 F0 12
52	***Delete***	E0 53 • E0 D3	E0 71 • E0 F0 71
52	–LEFT SHIFT case	E0 AA E0 53 •	E0 F0 12 E0 71 •
52		E0 D3 E0 2A	E0 F0 71 E0 12
52	–RIGHT SHIFT case	E0 B6 E0 53 •	E0 F0 59 E0 71 •
52		E0 D3 E0 36	E0 F0 71 E0 59
52	–NUM LOCK ON case	E0 2A E0 53 •	E0 12 E0 71 •
52		E0 D3 E0 AA	E0 F0 71 E0 F0 12
53	***End***	E0 4F • E0 CF	E0 69 • E0 F0 69
53	–LEFT SHIFT case	E0 AA E0 4F •	E0 F0 12 E0 69 •
53		E0 CF E0 2A	E0 F0 69 E0 12
53	–RIGHT SHIFT case	E0 B6 E0 4F •	E0 F0 59 E0 69 •
53		E0 CF E0 36	E0 F0 69 E0 59
53	–NUM LOCK ON case	E0 2A E0 4F •	E0 12 E0 69 •
53		E0 CF E0 AA	E0 F0 69 E0 F0 12
54	***PageDown***	E0 51 • E0 D1	E0 7A • E0 F0 7A
54	–LEFT SHIFT case	E0 AA E0 51 •	E0 F0 12 E0 7A •
54		E0 D1 E0 2A	E0 F0 7A E0 12
54	–RIGHT SHIFT case	E0 B6 E0 51 •	E0 F0 59 E0 7A •
54		E0 D1 E0 36	E0 F0 7A E0 59
54	–NUM LOCK ON case	E0 2A E0 51 •	E0 12 E0 7A •
54		E0 D1 E0 AA	E0 F0 7A E0 F0 12
87	***UpArrow***	E0 48 • E0 C8	E0 75 • E0 F0 75
87	–LEFT SHIFT case	E0 AA E0 48 •	E0 F0 12 E0 75 •
87		E0 C8 E0 2A	E0 F0 75 E0 12
87	–RIGHT SHIFT case	E0 B6 E0 48 •	E0 F0 59 E0 75 •
87		E0 C8 E0 36	E0 F0 75 E0 59
87	–NUM LOCK ON case	E0 2A E0 48 •	E0 12 E0 75 •
87		E0 C8 E0 AA	E0 F0 75 E0 F0 12
97	***LeftArrow***	E0 4B • E0 CB	E0 6B • E0 F0 6B
97	–LEFT SHIFT case	E0 AA E0 4B •	E0 F0 12 E0 6B •
97		E0 CB E0 2A	E0 F0 6B E0 12
97	–RIGHT SHIFT case	E0 B6 E0 4B •	E0 F0 59 E0 6B •
97		E0 CB E0 36	E0 F0 6B E0 59
97	–NUM LOCK ON case	E0 2A E0 4B •	E0 12 E0 6B •
97		E0 CB E0 AA	E0 F0 6B E0 F0 12

KEYBOARD SCAN CODES (cont.)

Key # for 101 Keybd	Key Name	XT scan codes Down • Up	AT hardware scan codes Down • Up
98	*DownArrow*	E0 50 •E0 D0	E0 72 •E0 F0 72
98	–LEFT SHIFT case	E0 AA E0 50 •	E0 F0 12 E0 72 •
98		E0 D0 E0 2A	E0 F0 72 E0 12
98	–RIGHT SHIFT case	E0 B6 E0 50 •	E0 F0 59 E0 72 •
98		E0 D0 E0 36	E0 F0 72 E0 59
98	–NUM LOCK ON case	E0 2A E0 50 •	E0 12 E0 72 •
98		E0 D0 E0 AA	E0 F0 72 E0 F0 12
99	*RightArrow*	E0 4D •E0 CD	E0 74 •E0 F0 74
99	–LEFT SHIFT case	E0 AA E0 4D •	E0 F0 12 E0 74 •
99		E0 CD E0 2A	E0 F0 74 E0 12
99	–RIGHT SHIFT case	E0 B6 E0 4D •	E0 F0 59 E0 74 •
99		E0 CD E0 36	E0 F0 74 E0 59
99	–NUM LOCK ON case	E0 2A E0 4D •	E0 12 E0 74 •
99		E0 CD E0 AA	E0 F0 74 E0 F0 12

Alphanumeric Primary Keyboard Keys (includes expanded keys)

Key #	Key Name	XT scan codes	AT hardware scan codes
17	' ~ (accent, tilde)	29 • A9	0E • F0 0E
18	1 !	02 • 82	16 • F0 16
19	2 @	03 • 83	1E • F0 1E
20	3 #	04 • 84	26 • F0 26
21	4 $	05 • 85	25 • F0 25
22	5 %	06 • 86	2E • F0 2E
23	6 ^ (6, caret)	07 • 87	36 • F0 36
24	7 &	08 • 88	3D • F0 3D
25	8 * (8, asterisk)	09 • 89	3E • F0 3E
26	9 (	0A • 8A	46 • F0 46
27	0)	0B • 8B	45 • F0 45
28	– _ (dash,underline)	0C • 8C	4E • F0 4E
29	= + (equal, plus)	0D • 8D	55 • F0 55
30	Bkspace	0E • 8E	66 • F0 66
38	Tab	0F • 8F	0D • F0 0D
39	q Q	10 • 90	15 • F0 15
40	w W	11 • 91	1D • F0 1D
41	e E	12 • 92	24 • F0 24
42	r R	13 • 93	2D • F0 2D
43	t T	14 • 94	2C • F0 2C
44	y Y	15 • 95	35 • F0 35
45	u U	16 • 96	3C • F0 3C
46	i I	17 • 97	43 • F0 43
47	o O	18 • 98	44 • F0 44
48	p P	19 • 99	4D • F0 4D
49	[{	1A • 9A	54 • F0 54
50	] }	1B • 9B	5B • F0 5B
51	\ | (backslash,bar)	2B • AB	5D • F0 5D
59	CapsLock	3A • BA	58 • F0 58
60	a A	1E • 9E	1C • F0 1C
61	s S	1F • 9F	1B • F0 1B
62	d D	20 • A0	23 • F0 23
63	f F	21 • A1	2B • F0 2B
64	g G	22 • A2	34 • F0 34

KEYBOARD SCAN CODES (cont.)

Key # for 101 Keybd	Key Name	XT scan codes Down • Up	AT hardware scan codes Down • Up
65	h H	23 • A3	33 • F0 33
66	j J	24 • A4	3B • F0 3B
67	k K	25 • A5	42 • F0 42
68	l L	26 • A6	4B • F0 4B
69	; : (semicolon,colon)	27 • A7	4C • F0 4C
70	' " (single quote,double)	28 • A8	52 • F0 52
71	Enter	1C • 9C	5A • F0 5A
75	Shift(left)	2A • AA	12 • F0 12
76	z Z	2C • AC	1A • F0 1A
77	x X	2D • AD	22 • F0 22
78	c C	2E • AE	21 • F0 21
79	v V	2F • AF	2A • F0 2A
80	b B	30 • B0	32 • F0 32
81	n N	31 • B1	31 • F0 31
82	m M	32 • B2	3A • F0 3A
83	, < (comma,less than)	33 • B3	41 • F0 41
84	. > (period,greater than)	34 • B4	49 • F0 49
85	/ ? (forward slash, ?)	35 • B5	4A • F0 4A
86	Shift(right)	36 • B6	59 • F0 59
92	Ctrl(left)	1D • 9D	14 • F0 14
93	Alt(left)	38 • B8	11 • F0 11
94	Space	39 • B9	29 • F0 29
95	Alt(right)	E0 38 • E0 B8	E0 11 • E0 F0 11
96	Ctrl(right)	E0 1D • E0 9D	E0 14 • E0 F0 14

Keypad keys
(Includes expanded keyboard layout)

Key #	Key Name	XT scan codes Down • Up	AT hardware scan codes Down • Up
34	NumLock	45 • C5	77 • F0 77
35	/	E0 35 • E0 B5	E0 4A • E0 F0 4A
35	–LEFT SHIFT case	E0 AA E0 35 •	E0 F0 12 E0 4A •
35		E0 B5 E0 2A	E0 F0 4A E0 12
35	–RIGHT SHIFT case	E0 B6 E0 35 •	E0 F0 59 E0 4A •
35		E0 B5 E0 36	E0 F0 4A E0 59
36	* (PrtSc 84 key)	37 • B7	7C • F0 7C
37	–	4A • C4	7B • F0 7B
55	Home 7	47 • C7	6C • F0 6C
56	UpArrow 8	48 • C8	75 • F0 75
57	PageUp 9	49 • C9	7D • F0 7D
58	+	4E • CE	79 • F0 79
72	LeftArrow 4	4B • CB	6B • F0 6B
73	5	4C • CC	73 • F0 73
74	RightArrow 6	4D • CD	74 • F0 74
88	End 1	4F • CF	69 • F0 69
89	DownArrow 2	50 • D0	72 • F0 72
90	PageDown 3	51 • D1	7A • F0 7A
91	Enter	E0 1C • E0 9C	E0 5A • E0 F0 5A
100	Ins 0	52 • D2	70 • F0 70
101	Del .	53 • D3	71 • F0 71

IBM HARDWARE RELEASES

Date	Code	Hardware Release
04-24-81	FF	PC (the original!)
10-19-81	FF	PC (fixed bugs)
10-27-82	FF	PC hard drive support & 640k
11-08-82	FE	PC-XT
06-01-83	FD	PC jr
01-10-84	FC	AT
06-10-85	FC	AT revision 1
09-13-85	F9	PC Convertible
11-15-85	FC	AT w/speed control (30 meg HD)
01-10-86	FB	XT revision 1
04-21-86	FC	XT-286 model 2
05-09-86 to 02-05-87	FB	XT revision 2
09-02-86	FA	PS/2 Model 30
02-13-87 to 05-09-87	FC	PS/2 Model 50 model 4
02-13-87	FC	PS/2 Model 60 model 5
03-30-87	F8	PS/2 Model 80 16 MHz
06-26-87	FA	PS/2 Model 25
10-07-87	F8	PS/2 Model 80 20 MHz
01-28-88 to 04-18-88	FC	PS/2 Model 50Z
01-29-88 to 02-20-89	F8	PS/2 Model 70 - 386
08-25-88	FC	PS/2 Model 30 - 286
11-02-88	F8	PS/2 Model 55 -SX
01-18-89	F8	PS/2 Model P70 - 386
06-28-89	FC	PS/2 Model 25 - 286
06-28-89	FC	PS/2 Model 30 - 286
09-29-89 to 12-01-89	F8	PS/2 Model 70 - 486
11-21-89	F8	PS/2 Model 80 - 386-25
12-01-89	FC	PS/1 Model
02-08-90	F8	PS/2 Model 65 - SX
10-05-90	F8	PS/2 Model P75 - 486
02-27-91	F8	PS/2 Model L40 - SX
03-15-91 to 04-04-91	F8	PS/2 Model 35 - SX
03-15-91 to 04-04-91	F8	PS/2 Model 40 - SX
07-03-91	F8	PS/2 Model 57 - SX
?	F8	PS/2Model90-XP 486
?	F8	PS/2Model95-XP 486

IBM® PC/XT MOTHEROARD SWITCH 1 SETTINGS

Switch #	On/Off	Function
1	Off	Always off
2	On	Coprocessor NOT present in system
2	Off	Coprocessor present in system

Switch 3,4 System motherboard memory

Switch #	On/Off	Function
3,4	3 On, 4 On	PC=16K XT=64K
3,4	3 Off, 4 On	PC=32K XT=128K
3,4	3 On, 4 Off	PC=48K XT=192K
3,4	3 Off, 4 Off	PC=64K XT=256K
5,6	5 On, 6 On	EGA/VGA video adapter present
5,6	5 Off, 6 Off	Monochrome video adapter present
5,6	5 On, 6 Off	CGA video adapter present, 80x25 mode
5,6	5 Off, 6 On	CGA video adapter present, 40x25 mode
7,8	7 On, 8 On	One floppy disk drive present
7,8	7 Off, 8 On	Two floppy disk drives present
7,8	7 On, 8 Off	Three floppy disk drives present
7,8	7 Off, 8 Off	Four floppy disk drives present

IBM® PC MOTHEROARD SWITCH 2 SETTINGS (MEMORY)

System Memory Size	sw2-1	sw2-2	sw2-3	sw2-4	256K board sw2-5	64K board sw2-5
64K	On	On	On	On	On	Off
96K	Off	On	On	On	On	Off
128K	On	Off	On	On	On	Off
160K	Off	Off	On	On	On	Off
192K	On	On	Off	On	On	Off
224K	Off	On	Off	On	On	Off
256K	On	Off	Off	On	On	Off
288K	Off	Off	Off	On	On	Off
320K	On	On	On	Off	On	Off
352K	Off	On	On	Off	On	Off
384K	On	Off	On	Off	On	Off
416K	Off	Off	On	Off	On	Off
448K	On	On	Off	Off	On	Off
480K	Off	On	Off	Off	On	Off
512K	On	Off	Off	Off	On	Off
544K	Off	Off	Off	Off	On	Off
576K	On	On	On	On	Off	N/A
608K	Off	On	On	On	Off	N/A
640K	On	Off	On	On	Off	N/A
704K	On	On	Off	On	Off	N/A

Notes:

1. Switch 2 listed on this page is not used on an IBM® XT.

2. The 256K board listed at the head of column 6 is the PC2 motherboard. The 64K board at the head of column 7 is the PC1 motherboard.

3. Switch 1-3 and 1-4 on the previous page must both be OFF if the motherboard is fully populated with memory chips on either the 64K or 256K motherboard.

4. Switch 1 on the IBM® AT, is a single switch that selects whether the installed video adapter is color or monochrome.

RESISTOR COLOR CODES

Color	1st Digit(A)	2nd Digit(B)	Multiplier(C)	Tolerance(D)
Black	0	0	1	
Brown	1	1	10	1%
Red	2	2	100	2%
Orange	3	3	1,000	3%
Yellow	4	4	10,000	4%
Green	5	5	100,000	
Blue	6	6	1,000,000	
Violet	7	7	10,000,000	
Gray	8	8	100,000,000	
White	9	9	10^9	
Gold			0.1 (EIA)	5%
Silver			0.01 (EIA)	10%
No Color				20%

Example: Red–Red–Orange = 22,000 ohms, 20%

Additional information concerning the Axial Lead resistor can be obtained if Band A is a wide band. Case 1: If only Band A is wide, it indicates that the resistor is wirewound. Case 2: If Band A is wide and there is also a blue fifth band to the right of Band D on the Axial Lead Resistor, it indicates the resistor is wirewound and flame proof.

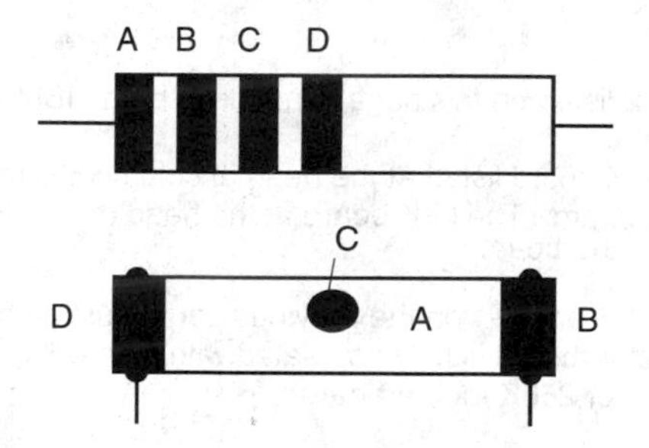

PAPER SIZES

Paper Size	Standard	Millimeters	Inches
Eight Crown	IMP	1461 x 1060	57-1/2 x 41-3/4
Antiquarian	IMP	1346 x 533	53 x 21
Quad Demy	IMP	1118 x 826	44 x 32-1/2
Double Princess	IMP	1118 x 711	44 x 28
Quad Crown	IMP	1016 x 762	40 x 30
Double Elephant	IMP	1016 x 686	40 x 27
B0	ISO	1000 x 1414	39.37 x 55.67
Arch-E	USA	914 x 1219	36 x 48
Double Demy	IMP	889 x 572	35 x 22-1/2
– **E**	ANSI	864 x 1118	34 x 44
A0	ISO	841 x 1189	33.11 x 46.81
Imperial	IMP	762 x 559	30 x 22
Princess	IMP	711 x 546	28 x 21-1/2
B1	ISO	707 x 1000	27.83 x 39.37
Arch-D	USA	610 x 914	24 x 36
A1	ISO	594 x 841	23.39 x 33.11
Demy	IMP	584 x 470	23 x 18-1/2
– **D**	ANSI	559 x 864	22 x 34
B2	ISO	500 x 707	19.68 x 27.83
Arch-C	USA	457 x 610	18 x 24
– **C**	ANSI	432 x 559	17 x 22
A2	ISO	420 x 594	16.54 x 23.39
B3	ISO	353 x 500	13.90 x 19.68
Brief	IMP	333 x 470	13-1/8 x 18-1/2
Foolscap folio	IMP	333 x 210	13-1/8 x 8-1/4
Arch-B	USA	305 x 457	12 x 18
A3	ISO	297 x 420	11.69 x 16.54
– **B**	ANSI	279 x 432	11 x 17
Demy quarto	IMP	273 x 216	10-3/4 x 8-1/2
B4	ISO	250 x 353	9.84 x 13.90
Crown quarto	IMP	241 x 184	9-1/2 x 7-1/4
Royal octavo	IMP	241 x 152	9-1/2 x 6
Arch-A	USA	229 x 305	9 x 12
Demy octavo	IMP	222 x 137	8-3/4 X 5-3/8
– **A**	ANSI	216 x 279	8.5 x 11
A4	ISO	210 x 297	8.27 x 11.69
Foolscap quarto	IMP	206 x 165	8-1/8 x 6-1/2
Crown Octavo	IMP	181 x 121	7-1/8 x 4-3/4
B5	ISO	176 x 250	6.93 x 9.84
A5	ISO	148 x 210	5.83 x 8.27
	USA	140 x 216	5.5 x 8.5
	USA	127 x 178	5 x 7
A6	ISO	105 x 148	4.13 x 5.83
	USA	102 x 127	4 x 5
	USA	76 x 102	3 x 5
A7	ISO	74 x 105	2.91 x 4.13
A8	ISO	52 x 74	2.05 x 2.91
A9	ISO	37 x 52	1.46 x 2.05
A10	ISO	26 x 37	1.02 x 1.46

Abbreviations for the above table are:

ISO . . . . International Organization for Standardization
ANSI . . . . American National Standards Institute
USA . . . . United States of America
IMP . . . . Imperial paper and plan sizes
Arch . . . . United States architectural standards

PARALLEL PRINTER INTERFACE

Printer Pin Number	Signal Description	Function	Signal Direction At Printer
1	STROBE	Reads in the data	Input
2	DATA Bit 0	Data line	Input
3	DATA Bit 1	Data line	Input
4	DATA Bit 2	Data line	Input
5	DATA Bit 3	Data line	Input
6	DATA Bit 4	Data line	Input
7	DATA Bit 5	Data line	Input
8	DATA Bit 6	Data line	Input
9	DATA Bit 7	Data line	Input
10	ACKNLG	Acknowledge receipt of data	Output
11	Busy	Printer is busy	Output
12	Paper Empty	Printer out of paper	Output
13	SLCT	Online mode indicator	Output
14	Auto Feed XT		Input
15	Not Used	Not Used	
16	Signal ground	Signal ground	
17	Frame ground	Frame ground	
18	+5 volts	+5 volts	
19-30	Ground	Return signals of pins 1–12, twisted pairs.	
31	Input Prime or INIT	Resets printer, clears buffer & initializes	Input
32	Fault or Error	Indicates offline mode.	Output
33	Signal ground	External ground	
34	Not Used	Not Used	
35	+5 Volts	+5 Volts (3.3 K-ohm)	
36	SLCT IN	TTL high level	Input

The above pinout is at the printer plug, computer side pinouts are on the next page. The "Parallel" or "Centronics" configuration for printer data transmission has become the de facto standard in the personal computer industry. This configuration was developed by a printer manufacturer (Centronics) as an alternative to serial data transmission. High data transfer rates are the main advantage of parallel and are attained by simultaneous transmission of all bits of a binary "word" (normally an ASCII code). Disadvantages of the parallel transfer are the requirement for 8 separate data lines and computer to printer cable lengths of less than 12 feet.

PARALLEL PINOUTS @ COMPUTER

DB25 Systems

Computer Pin Number	Signal Description	Function	Signal Direction At Computer
1	STROBE	Reads in the data	Output
2	DATA Bit 0	Data line	Output
3	DATA Bit 1	Data line	Output
4	DATA Bit 2	Data line	Output
5	DATA Bit 3	Data line	Output
6	DATA Bit 4	Data line	Output
7	DATA Bit 5	Data line	Output
8	DATA Bit 6	Data line	Output
9	DATA Bit 7	Data line	Output
10	ACKNLG	Acknowledge receipt of data	Input
11	Busy	Printer is busy	Input
12	Paper Empty	Printer out of paper	Input
13	SLCT	Online mode indicator	Input
14	Auto Feed XT		Input
15	Fault or Error	Indicates offline mode.	Input
16	Input Prime or INIT	Resets printer, clears buffer & initializes	Output
17	SLCT IN	TTL high level	Output
18-25	Ground	Return signals of pins 1–12, twisted pairs.	

LOOPBACK DIAGNOSTIC PLUGS

Parallel-IBM DB25	Parallel-Other DB25	Serial-IBM DB25	Serial-Other DB25
1 to 13	2 to 15	1 to 7	2 to 3
2 to 15	3 to 13	2 to 3	4 to 5
10 to 16	4 to 12	4 to 5 to 8	6 to 8 to 20 to 22
11 to 17	5 to 10	6 to 11 to 20 to 22	
12 to 14	6 to 11	15 to 17 to 23	
		18 to 25	

Loopback plugs work in conjunction with various software diagnostics programs and are used to determine whether or not a parallel or serial port is functioning correctly. The plugs labeled "IBM" will work with the IBM Corporation Advanced Diagnostics software and those labeled as "Other" will work with a variety of other programs such as Norton Diagnostics.

SERIAL I/O INTERFACES (RS232C)

Standard DB25 Pin Connector

Serial Pin Number	Signal Description	Function	Signal Direction At Device
1	FG	Frame ground	
2	TD	Transmit Data	Output
3	RD	Receive Data	Input
4	RTS	Request to Send	Output
5	CTS	Clear to Send	Input
6	DSR	Data Set Ready	Input
7	SG	Signal Ground	
8	DCD	Data Carrier Detect	Input
9	+V	+DC test voltage	Input
10	– V	– DC test voltage	Input
11	QM	Equalizer Mode	Input
12	(S)DCD	2nd Data Carrier Detect	Input
13	(S)CTS	2nd Clear to Send	Input
14	(S)TD	2nd Transmitted Data	Output
15	TC	Transmitter Clock	Input
16	(S)RD	2nd Received Data	Input
17	RC	Receiver Clock	Input
18	Not used	Not used	
19	(S)RTS	2nd Request to Send	Output
20	DTR	Data Terminal Ready	Output
21	SQ	Signal Quality Detect	Input
22	RI	Ring Indicator	Input
23		Data Rate Selector	Output
24	(TC)	External Transmitter Clk	Output
25	Not used	Not used	

IBM® Standard DB9 Pin Connector

Serial Pin Number	Signal Description	Function	Signal Direction At Device
1	DCD	Data Carrier Detect	Input
2	RD	Receive Data	Input
3	SD	Transmit Data	Output
4	DTR	Data Terminal Ready	Output
5	SG	Signal Ground	
6	DSR	Data Set Ready	Input
7	RTS	Request to Send	Output
8	CTS	Clear to Send	Input
9	RI	Ring Indicator	Input

NOTES ON SERIAL INTERFACING

Printers and asynchronous modems are relatively unsophisticated pieces of electronic equipment. Although all 25 pins of the Standard DB25 serial connector are listed 1 page back, only a few of the pins are needed for normal applications. The following list gives the necessary pins for each of the indicated applications.

1. "Dumb Terminals" – 1,2,3, & 7
2. Printers and asynchronous modems – 1,2,3,4,5,6,7,8, & 20
3. "Smart" and synchronous modems – 1,2,3,4,5,6,7,8,13,14, 15,17,20,22, & 24

Cable requirements also differ, depending on the particular hardware being used. The asynchronous modems normally use the 9 pin or 25 pin cables and are wired 1 to 1 (ie, pin 1 on one end of the cable goes to pin 1 on the other end of the cable.) Serial printers, however, have several wires switched in order to accommodate "handshaking" between computer and printer. The rewired junction is called a "Modem Eliminator". In the case of Standard DB25 the following are typical rewires:

Standard IBM PC

DB25 @ Computer	DB25 @ Printer
1	1
3	2
2	3
8	4
4	8
5 & 6	20
20	5 & 6
7	7

Second Standard PC

DB25 @ Computer	DB25 @ Printer
1	1
3	2
2	3
20	5, 6 & 8
7	7
5, 6 & 8	20

PC to Terminal

PC	Terminal
1	1
2	3
3	2
4	5
5	4
6 & 8	20
20	6 & 8
7	7

Std Hewlett-Packard

Std	Hewlett-Packard
1	1
2	3
3	2
4 & 5	8
8	4 & 5
6	20
7 & 22	7 & 22
17	15
11	12
12	11
15 & 24	17
20	6

GPIB I/O INTERFACE (IEEE-488)

The HPIB/GPIB/IEEE–488 standard is a very powerful interface developed originally by Hewlett–Packard (HP–IB). The interface has been adopted by a variety of groups, such as IEEE, and is known by names such as HP–IB, GPIB, IEEE–488 and IEC Standard 625–1 (outside the US). Worldwide use of this standard has come about due to its ease of use, handshaking protocol, and precisely defined function.

Information management is handled by three device types: Talkers, Listeners, and Controllers. Talkers send information, Listeners receive data, and Controllers manage the interactions. Up to 15 devices can be interconnected, but are usually located within 20 feet of the computer. Additional extenders can be used to access more than 15 devices.

GPIB 24 Line Bus

Pin Number	Signal Description	Function
1	DATA I/O 1	Data line I/O bus
2	DATA I/O 2	Data line I/O bus
3	DATA I/O 3	Data line I/O bus
4	DATA I/O 4	Data line I/O bus
5	EIO	End or Identify
6	DAV	Data valid
7	NRFD	Not Ready For Data
8	NDAC	Data Not Accepted
9	SRQ	Service Request
10	IFC	Interface Clear
11	ATN	Attention
12	Shield	or wire ground
13	DATA I/O 5	Data line I/O bus
14	DATA I/O 6	Data line I/O bus
15	DATA I/O 7	Data line I/O bus
16	DATA I/O 8	Data line I/O bus
17	REN	Remote Enable
18	Ground	Ground
19	Ground	Ground
20	Ground	Ground
21	Ground	Ground
22	Ground	Ground
23	Ground	Ground
24	Logic Ground	Logic Ground

Devices can be set up in star, linear or other combinations and are easily set up using male/female stackable connectors.

VIDEO CARD PINOUTS

Pin Number	Description
Monochrome Display Adapter (MDA and HGC)	
1 & 2	Ground
3, 4, & 5	Not Used
6	+ Intensity
7	+ Video
8	+ Horizontal Drive
9	– Vertical Drive
Color Graphics Display Adapter (CGA)	
1 & 2	Ground
3	Red
4	Green
5	Blue
6	+ Intensity
7	Reserved
8	+ Horizontal Drive
9	– Vertical Drive
CGA Composite Video (RCA phono jack)	
1 (pin)	1.5 volt DC video signal
2 (shell)	Ground
Enhanced Graphics Adapter (EGA)	
1	Ground
2	Secondary Red
3	Red
4	Green
5	Blue
6	Secondary Green / Intensity
7	Secondary Blue / Monochrome
8	Horizontal Drive
9	Vertical Drive

Video Graphics Array (VGA)

Pin	Color VGA	Monochrome VGA
1	Red (Output)	Not Used
2	Green (Output)	Monochrome Video
3	Blue (Output)	Not Used
4	Reserved	Not Used
5	Digital Ground	Ground
6	Red Return (Input)	Key
7	Green Return (Input)	Monochrome Ground
8	Blue Return (Input)	Not Used
9	Plug	No Connection
10	Digital Ground	Horizontal Sync Ground
11	Reserved	Not Used
12	Reserved	Vertical Sync Ground
13	Horizontal Sync (Output)	Horizontal Sync
14	Vertical Sync (Output)	Vertical Sync
15	Reserved	No Connection

KEYBOARD PLUG - 5 Pin Din

Pin #	Description
1	Clock (TTL signal)
2	Data (TTL signal)
3	Not used
4	Ground
5	Power (+5 volt)

KEYBOARD PLUG - 6 Pin MiniDin

Pin #	Description
1	Data (TTL signal)
2	Not used
3	Ground
4	Power (+5 volt)
5	Clock (TTL signal)
6	Not used

KEYBOARD PLUG - 6 Pin SDL

Pin #	Description
A	Not used
B	Data (TTL signal)
C	Ground
D	Clock (TTL signal)
E	Power (+5 volt)
F	Not used

MOUSE 9 Pin D-Shell

Pin #	Description
1	Not Used
2	Data
3	Clock
4	+5 Volt
5	Ground
6	Not Used
7	Enable Mouse
8	Mouse Ready
9	Not Used

MOUSE 6 Pin Mini DIN

Pin #	Description
1	Data
2	Not Used
3	Signal Ground
4	+5 Volt
5	Clock
6	Not Used
Shell	Shield Ground

MOUSE 9 Pin Microsoft Inport

Pin #	Description
1	+5 Volt
2	XA
3	XB
4	YA
5	YB

6	Switch 1
7	Switch 2
8	Switch 3
9	Signal Ground
Shell	Shield Ground

LIGHT PEN INTERFACE

Pin #	Description
1	– Light Pen Input
2	No connection
3	– Light Pen Switch
4	Chassis Ground
5	+5 Volts
6	+12 Volts

GAME CONTROL CABLE

Joystick Pin Number	Signal Description	Function	Signal Direction At Joystk
1	+5 Volts	Supply voltage	Input
2	Button 1	Push Button 1	Output
3	Position 0	X Coordinate	Output
4	Ground	Ground	
5	Ground	Ground	
6	Position 1	Y Coordinate	Output
7	Button 2	Push Button 2	Output
8	+5 Volts	Supply voltage	Input
9	+5 Volts	Supply voltage	Input
10	Button 3	Push Button 3	Output
11	Position 2	X Coordinate	Output
12	Ground	Ground	
13	Position 3	Y Coordinate	Output
14	Button 4	Push Button 4	Output
15	+5 Volts	Supply voltage	Input

286/386/486 BATTERY CONNECTOR

Pin #	Description
1	Ground
2	Not used
3	Not used, or alignment key
4	+6 volt

SPEAKER CONNECTOR

1	Audio
2	Alignment key
3	Ground
4	+5 volt

KEYBOARD LOCKOUT / POWER LED CONNECTOR-MOTHERBOARD

Pin #	Description
1	LED Power, +5 Volt
2	Alignment Key
3	Ground
4	Keyboard Lockout
5	Ground

PS-8 and 9 POWER CONNECTOR

Pin #	PS-8 (XT)	PS-8 (AT)	PS-9 (XT & AT)
1	Power ground	Power good	Ground
2	Align Key	+5 volt	Ground
3	+12 volt	+12 volt	–5 volt
4	–12 volt	–12 volt	+5 volt
5	Ground	Ground	+5 volt
6	Ground	Ground	+5 volt

ATX POWER CONNECTOR PLUG

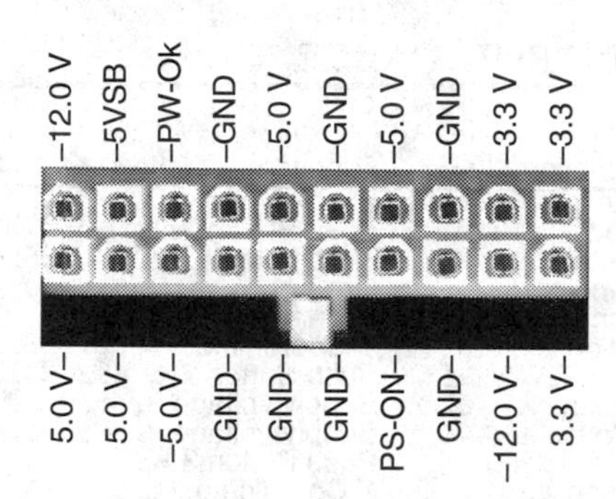

DISK DRIVE POWER CONNECTOR

Pin #	Description (4 pin molex)	Wire Color
1	+12 volt	Yellow
2	Ground	Black
3	Ground	Black
4	+5 volt	Red

UNIVERSAL SERIAL BUS CONNECTOR

Pin #	Description	Pin #	Description
1	+5 volt	6	+5 volt
2	Data –	7	Data –
3	Data +	8	Data +
4	Ground	9	Ground
5	No Connection	10	No Connection

CPU PROCESSOR TYPES

CPU Type	Date	MaxMem Phys/Virt	Bus Int/Ext	Number of Transistors	Speeds MHz
Advanced Micro Devices					
N80L286					10,12
AM386SX	7-91	4Gb	32/16	161k	25,33,40
AM386DX	3-91	4Gb	32/32	161k	25,33,40
AM486SX (doubler)	7-93	4Gb/64Tb	32/32	900k	33,40,25/50
AM486SXLV (3.3V)	7-93	4Gb/64Tb	32/32	900k	33
AM486DX	?	4Gb/64Tb	32/32	1,300k	33, 40
AM486DX2 (doubler)	?	4Gb/64Tb	32/32	1,300k	25/50,33/66,40/80
AM486DX4 (doubler, 3.3V)	?	4Gb/64Tb	32/32	1,300k	100
AM486DXLV (3.3V)	?	4Gb/64Tb	32/32	1,300k	33
AM486DXL2 (doubler)	?	4Gb/64Tb	32/32	1,300k	25/50,33/66,40/80
AM486DXL4 (doubler)	?	4Gb/64Tb	32/32	1,300k	50/100
AMD5x86 (3.3V)	12-95	4Gb/64Tb	32/32	1,300k	75
AMD-K5 (3.5V)	3-96	? / ?	64/64	4,300k	75,90,100,117,133
AMD-K6 (3.5V)	11-96	? / ?	64/64	8,800k	180,233,266,300
AMD-K6 MMX (2.9/3.2V)	4-97	? / ?	64/64	8,800k	166,200,233
AMD-K6 3D MMX (?)	1-98	? / ?	64/64	9,300k	300,350
AMD-K6-2 3D (?)	5-98	? / ?	64/64	9,300k	266,333,350,400,450
AMD-K6+ 3D MMX (?)	Q3-98	? / ?	64/64	21,300k	400
Advanced Micro Devices - Upgrade Chips					
AM186EM	9-94	? / ?	16/16	?	25,33,40
AM386EM	9-94	? / ?	32/32	?	25,33
AM486SE (3 or 5V)	9-94	? / ?	? / ?	?	25,33
Centaur Technology					
IDTWinChip C6 (3.3V)	5-97	? / ?	? / ?	5,400k	180,200,225,240
IDTWinChip C6+ (3.3V)	?	? / ?	? / ?	5,800k	266,300
IDTWinChip 2 (?)	?	? / ?	? / ?	?	266,300
IDTWinChip 2 3D (?)	?	? / ?	? / ?	?	?
IDTWinChip 2+ (?)	?	? / ?	? / ?	?	?
IDTWinChip 2+ ND (?)	?	? / ?	? / ?	?	?
IDTWinChip 3 (?)	?	? / ?	? / ?	?	400,600
Cyrix Corporation					
CX486SLC (3 or 5V)	4-92	16Mb	32/16	600k	20,25,33
CX486DLC	6-92	4Gb/64Tb	32/32	600k	25,33,40
CX486S	2-93	4Gb/64Tb	32/16	?	33,40
CX486DX	9-93	4Gb/64Tb	32/32	?	33,40,50
CX486DX2 (doubler)	9-93	4Gb/64Tb	32/32	?	25/50,33/66
CX486SLC2 (doubler)	10-93	4Gb/64Tb	32/16	?	25/50
CX486DXV (3V)	?	4Gb/64Tb	32/32	?	33
CX486DX2V (doubler, 3V)	?	4Gb/64Tb	32/32	?	25/50,33/66,40/80
5x86 (3.3V)	7-95	? / ?	32/64	2,000k	100,120
6x86 (3.3V)	10-95	? / ?	32/64	3,000k	100,110,120, 133, 150
6x86MX (3.3V)	5-97	? / ?	32/64	6,000k	150,166,188, 200,250,266
M3	Q2-98				
MediaGX (3.3V) (Multimedia accelerator chip)	2-97	? / ?	64/64	2,400k	120,133,150, 166,180,200,233
M II (3.3V)	4-98	? / ?	? / ?	?	300,333,350
MXi (?)	?	? / ?	? / ?	?	300,400

CPU PROCESSOR TYPES

CPU Type	Date	MaxMem Phys/Virt	Bus Int/Ext	Number of Transistors	Speeds MHz
Cyrix Corporation - Upgrade Chips					
CX486DRX2 (doubler)	8-93	4Gb/64Tb	32/32	?	16/32,20/40, 25/50,33/66
CX486SRX2 (doubler)	10-93	4Gb/64Tb	32/16	?	20/40,25/50
Digital Equipment Corporation					
Alpha 21064 (3.3V)	1992	16Gb/ ?	64/64	1,680k	150,300
Alpha 21064A (3.3V)	10-93	16Gb/ ?	64/64	2,800k	200,233,275,300
Alpha 21066 (3.3V)	?	16Gb/ ?	64/64	1,750k	166
Alpha 21066A (3.3V)	1-95	16Gb/ ?	64/64	1,750k	100,233
Digital Equipment Corporation (cont.)					
Alpha 21164 (3.3V)	1994	1Tb/8Tb	64/64	9,300k	266,300,333, 366,433,500,533,600
Alpha 21164A (3.3V)	?	? / ?	? / ?	9,000k	417
Alpha 21164PC (3.3V)	3-97	? / ?	64/64	3,400k	300,366,400, 466,533
EV6 (Alpha 21264)	H2-97	? / ?	64/64	15,200k	500 to 1,000
EV67 (Alpha 21264a)	1999			15,200k	800
EV7 (Alpha 21364)	2000			100,000k	500 to 750
EV78 (Alpha 21364a)	2002			100,000k	750
EV8 (Alpha 21464)	2003			250,000k	750 to 1,000
Evergreen Technologies Inc. - Upgrade Chips					
Rev To 486	?	? / ?	? / ?	?	25/50,25/75
Rev To DX4	9-94	? / ?	? / ?	?	25/75,33/100,50/100
Evergreen 586	?	? / ?	64/64	4,300k	133
Evergreen PR166	?	? / ?	64/64	3,000k	150
Evergreen MxPro	9-97	? / ?	64/64	8,800k	180,200
Exponential Technology, Inc.					
Exponential-X704	1998	? / ?	64/32	2,700k	410,533
(Out of business 5/97 after Apple declined use of the X704 in its Power Mac line)					
Fujitsu Microelectronics, Inc.					
TurboSPARC (3.3V)	9-96	? / ?	32/32	3,000k	160,170
Hewlett-Packard Company					
PA-7100LC (3.3V)	1994	4Gb/ ?	32/32	800k	60,80,100
PA-7150 (3.3V)	12-93	? / ?	32/32	850k	125
PA-7200 (3.3V)	1-95	? / ?	32/32	1,260k	120
PA-7300LC (3.3V)	10-95	? / ?	32/32	9,200k	160
PA-8000 (3.3V)	3-96	? / ?	64/64	?	180
PA-8200 (3.3V)	3-97	? / ?	64/64	3,800k	220
PA-8500 (3.3V)	10-98		64/64	120,000k	
Merced (with Intel)	1999				
Hitachi America, Ltd.					
SH7702 (3.3V)	?	4Gb/ ?	? / ?	?	45
SH7708 (3.3V)	?	4Gb/ ?	? / ?	?	60
IDT					
Win Chip C6	5-97			5,400k	200

CPU PROCESSOR TYPES

CPU Type	Date	MaxMem Phys/Virt	Bus Int/Ext	Number of Transistors	Speeds MHz
Intel Corporation					
8080	4-74	64Kb	8/8	6k	2
8086	6-78	1Mb	16/16	29k	5,8,10
8088	6-79	1Mb	16/8	29k	5,8
80286	2-82	16Mb/1Gb	16/16	134k	6,10,12
80386DX	10-85	4Gb/64Tb	32/32	275k	16,20,25,33
80386SX	6-88	4Gb/64Tb	32/16	275k	16,20,25,33
80486DX (3.3 or 5V)	4-89	4Gb/64Tb	32/32	1,200k	25,33,50
80386SL (3.3 or 5V)	10-90	4Gb/64Tb	32/16	855k	20,25
80486SX (3.3 or 5V)	4-91	4Gb/64Tb	32/32	1,185k	16,20,25,33
80486DX2(doubler,3.3or5V)	3-92	4Gb/64Tb	32/32	1,200k	25/50,33/66
80486SL (3.3 or 5V)	9-92	64Mb/64Tb	32/32	1,400k	20,25,33
Pentium (3.3 or 5V)	3-93	4Gb/64Tb	32/64	3,100k	60,66
80486DX4 (3.3 or 5V)	3-94	4Gb/64Tb	32/32	1,600k	75,100
Pentium SL (3.3 or 5V)	3-94	4Gb/64Tb	32/64	3,300k	75,100,120,133, 150,166,200
Pentium (3.3 or 5V)	10-94	4Gb/64Tb	32/64	3,200k	75,90,100,120
Pentium (3.3 or 5V)	6-95	4Gb/64Tb	32/64	3,300k	133,150,166,200
Pentium Pro (3.3V)	1-95	64Gb/64Tb	64/64	5,500k	150,166,180,200
Pent. w/ MMX (3.3 or 5V)	1-97	4Gb/64Tb	32/64	4,500k	166,200,233
Pentium II (?)	5-97	64Gb/64Tb	64/64	7,500k	233,266,300, 333,350,400,450
Celeron (?)	4-98	? / ?	? / ?	19,000k	266,300,333
Pentium II Xeon (?)	6-98	64Gb/64Tb	64/64	7,500k	400,450,500
P7 (Merced w/ HP)	2000				
Intel Corporation - Upgrade Chips					
SX2Overdrive (3.3 or 5V)	?	4Gb/64Tb	32/32	?	50
DX2Overdrive	?	4Gb/64Tb	32/32	900k	40,50,66
DX4Overdrive	10-94	4Gb/64Tb	32/32	1,600k	63,83
Pentium Overdrive	?	4Gb/64Tb	32/32	3,300k	120,133
Pent. Over. w/ MMX (3.3V)	?	? / ?	? / ?	?	125,150,166
Pentium II Overdrive	7-98	? / ?	? / ?	?	300,333
International Business Machines Corporation					
80386SLC	12-91	16Mb/ ?	32/16	800k	16,20,25
80486SLC2 (doubler)	9-92	16Mb/ ?	32/16	1,425k	20/40,25/50,33/66
80486SLC3 (tripler)	?	16Mb/ ?	32/16	1,425k	25/75
BL486DX (doubler)	?	4Gb/64Tb	32/32	~1,400k	25/50,33/66, 40/80,50/100
BL486DX2 (doubler)	?	4Gb/64Tb	32/32	~1,400k	25/5,33/66,40/80
BL486DXV (3.3V)	?	4Gb/64Tb	32/32	~1,400k	33,40
BL486DX2V (3.3V)	?	4Gb/64Tb	32/32	~1,400k	25/5,33/66,40/80
6x86 (3.3V)	?	? / ?	32/32	?	120,133,150
PowerPC 603 (3.3V)	?	? / ?	32/32	1,600k	66,80
PowerPC 603 (3.3V)	?	? / ?	32/32	2,500k	100
PowerPC 604 (3.3V)	10-94	? / ?	32/32	3,600k	100,120,133, 150,166,180
PowerPC 603e (3.3V)	10-95	? / ?	32/32	2,600k	150,160,166,180, 200,225,233,240,250
6x86MX (3.3V)	5-97	? / ?	32/64	?	150,165,188
PowerPC 604e (3.3V)	6-97	4Gb / ?	32/32	5,100k	160,180,200, 225,233,240,250,350

CPU Type	Date	MaxMem Phys/Virt	Bus Int/Ext	Number of Transistors	Speeds MHz
International Business Machines Corporation (cont.)					
PowerPC 750 (3.3V)	7-97	? / ?	32/32	6,350k	233,250,266,275,300 333,350,366,400
PowerPC 620 (3.3V)	?	? / ?	64/64	7,000k	
Power 3 (?)	10-97	? / ?	64/64	?	500
International Meta Systems, Inc.					
Meta 6000	1-98	? / ?	? / ?	?	225
Meta	1-98	? / ?	? / ?	?	150 to 200
Meta 6500	7-98	? / ?	? / ?	?	450
Meta 7000/	1-99	? / ?	? / ?	?	700
Kingston Technology Company - Upgrade Chips					
SLC/Now!-20 (386SLC-20)	?	? / ?	32/32	?	20
SLC/Now!-25 (386SLC-25)	?	? / ?	32/32	?	25
SLC/Now!-50 (486SLC-50)	?	? / ?	32/32	?	50
486/Now!					
(486/25CS3 or /33CD3)	?	? / ?	? / ?	?	25,33
(486/25PS3 or /33PS3)	?	? / ?	? / ?	?	25,33
Lightning 486					
(486/BL66 or /CLN66)	?	? / ?	? / ?	?	66
MCMaster					
(MC50PD or MC66PD)	?	? / ?	? / ?	?	50,66
TurboChip 486					
(TC486/100 or / 75)	?	? / ?	? / ?	?	75, 100
TurboChip 133 (TC5x86/133)	?	? / ?	? / ?	?	133
TurboChip 233 (?)	?	? / ?	? / ?	?	233
MIPS Technologies, Inc.					
R2000	1986	? / ?	32/32	110k	8
R3000 (5V)	1988	? / ?	32/32	?	40
R6000	1991	? / ?	32/32	?	66.7
R4000	1992	? / ?	64/64	1,100k	100
R4400 (3.3V)	1992	64Gb/ ?	64/64	2,300k	150,200,250
R4200 (3.3V)	1993	? / ?	64/64	1,400k	80
R4600 (3.3V)	1994	? / ?	64/64	1,850k	133,150
R8000 (3.3V)	1994	? / ?	64/64	3,430k	75,90
R4300i (3.3V)	1995	? / ?	64/64	1,700k	100,133
R4700 (3.3V)	1995	? / ?	64/64	1,800k	175
R5000 (3.3V)	1996	? / ?	64/64	3,600k	180,200,250
R10000 (3.3V)	1996	? / ?	64/64	6,800k	150,175,200,275
R5000A (3.3V)	1997	? / ?	64/64	3,600k	
RM7000	1997	? / ?	64/64	?	300
R12000 (3.3V)	1998		64/64		300
H1 Series	1999		64/64		
H2 Series	2000		64/64		
Motorola Communications and Electronics, Inc.					
68020 (5V)	1985	? / ?	? / ?	?	25
68030 (5V)	?	? / ?	? / ?	270k	50
68040 (5V)	1989	? / ?	? / ?	1,200k	25
68060 (3.3V)	1993	? / ?	? / ?	2,400k	50
PowerPC 601 (3.6V)	1993	? / ?	32/64	2,800k	66, 80, 100
PowerPC 603 (3.3V)	1994	4Gb/4Pb	32/64	1,600k	50, 66, 80

CPU PROCESSOR TYPES

CPU Type	Date	MaxMem Phys/Virt	Bus Int/Ext	Number of Transistors	Speeds MHz
PowerPC 604 (3.3V)	1994	4Gb/4Pb	32/64	3,600k	100, 133, 180
PowerPC 602 (3.3V)	1995	4Gb/4Pb	32/64	1,000k	66
PowerPC 603e (3.3V)	1995	4Gb/4Pb	32/64	2,600k	100,120,133,166, 180,200,250,275,300
PowerPC 620 (3.3V)	1995	1Tb/1Hb	64/64	7,000k	133, 200
PowerPC 604e (3.3V)	1996	4Gb/4Pb	32/64	5,100k	180,200,225,233, 250,300,350
PowerPC 630 (3.3V)	1997				600
PowerPC 740 (3.3V, G3 Series)	1998	4Gb/ ?	32/64	6,500k	200,233,266,300
PowerPC 750 (3.3V, G3 Series)	1998	4Gb/ ?	32/64	6,500k	200,233,266 300,333,366
PowerPC 604r (Mach 5, 3.3V)	1999				
PowerPC 770 (G3 Series)	1999			~30,000k	
G4 Series	2000			~50,000k	

NEC America, Inc.

CPU Type	Date	MaxMem Phys/Virt	Bus Int/Ext	Number of Transistors	Speeds MHz
V3	3-84	1Mb	16/16	63k	8,10
V20	3-84	1Mb	16/8	63k	8,10
VR4100 (3.3V)	?	? / ?	64/64	?	40
VR4101 (3.3V)	?	? / ?	64/64	?	33
VR4300 (3.3V)	?	? / ?	64/64	?	100,133
VR4400 (3.3V)	?	? / ?	64/64	?	200,250
VR4400MC (3.3V)	?	? / ?	64/64	?	200
VR5000 (3.3V)	?	? / ?	64/64	?	150,180,200
VR10000 (3.3V)	?	? / ?	64/64	?	200

NexGen, Inc.

CPU Type	Date	MaxMem Phys/Virt	Bus Int/Ext	Number of Transistors	Speeds MHz
Nx586	?	? / ?	32/32	3,500k	70,75,84,93

(Company acquired by Advanced Micro Devices in 1995.)

Philips Semiconductor

CPU Type	Date	MaxMem Phys/Virt	Bus Int/Ext	Number of Transistors	Speeds MHz
TriMedia TM-1000	7-97	? / ?	? / ?	5,500k	100,133,166

(Multimedia accelerator chip)

Ross Technology, Inc. - Upgrade Chips

CPU Type	Date	MaxMem Phys/Virt	Bus Int/Ext	Number of Transistors	Speeds MHz
hyperSPARC (5V)	1994	? / ?	32/32	1,500k	90,100,125, 142,150,180,200
Viper (?)	1999	? / ?	64/64	?	?

SGS-Thomson Microelectronics

CPU Type	Date	MaxMem Phys/Virt	Bus Int/Ext	Number of Transistors	Speeds MHz
ST486DX (5V)	?	4Gb/ ?	32/32	?	33,40,50
ST486DX2 (5V, doubler)	?	4Gb/ ?	32/32	?	50,66,80
ST6x86 (3.3V)	?	? / ?	64/32	?	80,100,110,120, 133,150
MPact R (3.3V)	4-97	? / ?	32	?	75

(Multimedia co-processor developed with Chromatic Research, Inc.)

Sun Microsystems, Inc.

CPU Type	Date	MaxMem Phys/Virt	Bus Int/Ext	Number of Transistors	Speeds MHz
Thunder I (5V)	1993	? / ?	32/32	6,000k	50
microSPARC-II	1994	? / ?	32/32	2,300k	85,100

Sun Microsystems, Inc. (cont.)

SuperSPARC-II (5V) . . . . 1995 . . . . . ? / ? . 32/32 . . . 3,100k . 75,90
UltraSPARC-I (3.3V) . . . . 10-95 . . . . . ? / ? . 64/64 . . . 5,200k . 143,167,182,200
UltraSPARC-II (3.3V) . . . 10-95 . . . . . ? / ? . 64/64 . . . 5,400k . 250,300,330
UltraSPARC-IIi (3.3V) . . . 10-96 . . . . . ? / ? . 64/64 . . . . ? . 266,300
microSPARC-IIep (3.3V) . . . . ? . . . . . ? / ? . 32/32 . . . . ? . 100 to 125
UltraSPARC-III (?) . . . . . . 1998 . . . . . ? / ? . ?/? . . . . ? . 600
UltraSPARC-IV (?) . . . . . . 2000 . . . . . ? / ? . ?/? . . . . ? . 1,000
UltraSPARC-V (?) . . . . . . 2002 . . . . . ? / ? . ?/? . . . . ? . 1,500

Texas Instruments, Inc.

TI486SXLC (3.3V) . . . . ? . . 16Mb/ ? . 32/16 . . . . ? . 25,33
TX486SXLC (5V) . . . . ? . . 16Mb/ ? . 32/16 . . . . ? . 33
TX486SXLC2 (5V, doubler) . . ? . . 16Mb/ ? . 32/16 . . . . ? . 25/50
TX486SL (3.3V) . . . . ? . . . 4Gb/ ? . 32/32 . . . . ? . 33
TX486SL (5V) . . . . ? . . . 4Gb/ ? . 32/32 . . . . ? . 40
TX486SXL2 (3.3V, doubler) . . ? . . . 4Gb/ ? . 32/32 . . . . ? . 20/40
TX486SXL2 (5V, doubler) . . . ? . . . 4Gb/ ? . 32/32 . . . . ? . 25/50

Texas Instruments, Inc. - Upgrade Chips

TI486SLC/E (5V) . . . . ? . . 16Mb/ ? . 32/16 . . . . ? . 25,33
TI486SLC/E-V (3.3V) . . . . . . ? . . 16Mb/ ? . 32/16 . . . . ? . 25
TI486DLC/E (5V) . . . . ? . . . 4Gb/ ? . 32/32 . . . . ? . 33,40
TI486DLC/E-V (3.3V) . . . . ? . . . 4Gb/ ? . 32/32 . . . . ? . 25,33

MATH CoPROCESSOR TYPES

CPU Type	CoProcessor
8086,8088,V20 & V30	8087
80286	80287XL
80386SX & SL	80387SX
80386DX	80387DX
80486SX	80487SX
80486DX	Built In
Pentium - all versions	Built In

CPU IDENTIFICATION

Intel 8088 - XT Class
5 and 8 MHz

Intel 80386DX
16-20-25-33 MHz

AMD 80L286
AT Class
6-10-12 MHz

CPU IDENTIFICATION

Intel 80486DX2
25/50, 33/66 MHz

Cyrix 486
25-80 MHz

Intel Pentium P54C
75-200 MHz

Intel Pentium P55C
MMX 150-233 MHz

AMD K6 MMX
PR166-233 MHz

AMD K6-2
266-450 MHz

AMD K5
PR75-PR133

Cyrix 586
100-120 MHz

CPU IDENTIFICATION

Cyrix MII
300-350 MHz

Cyrix 6x86 MX
150-266 MHz

Cyrix Media GX-MMX
120-233 MHz

Cyrix 6x86 M1
166-200 MHz

Intel Pentium II
233-450 MHz

I/O INTERFACES

Interface	Max Speed Megabytes/Sec	Max # of Devices per Channel	Max Cable Length
ESDI			
Hard drive	3	2	Internal
Fibre Channel			
FC-AL, half duplex	100-200	126	6 miles
FC-AL, full duplex	200-400	126	6 miles
IDE/EIDE			
IDE/EIDE/ATAPI			
PIO Mode 0	3.3	2	Internal
PIO Mode 1	5.2	2	Internal
PIO Mode 2	8.3	2	Internal
PIO Mode 3	11.1	2	Internal
PIO Mode 4	16.7	2	Internal
Ultra DMA	33	2	Internal
IEEE 1394			
1394-1995	12.5-50	63	14 ft
1394-FireWire & i.Link	12.5-50	63	14 ft
1394B	100-200	63	14 ft
Parallel			
Standard	0.15	1	6 ft
ECC (Extended Capability Port)	3	1	30 ft
EPP (Enhanced Parallel Port)	3	1	30 ft
SCSI			
SCSI-1	5	8	18 ft
Differential or LVD	5	8	75 ft
SCSI-2			
Fast, Fast Narrow	10	8	5 ft
Fast Wide, Wide	20	16	5 ft
If only 1 to 4 SCSI devices			15 ft
SCSI-3			
Ultra SCSI (Fast-20, Ultra Narrow)	20	8	5 ft
Ultra SCSI Differential	20	8	75 ft
Wide Ultra SCSI (Fast Wide 20)	40	16	5 ft
Wide Ultra SCSI Differential	40	16	75 ft
Ultra2 SCSI	40	8	36 ft
Wide Ultra2 SCSI	80	16	36 ft
Ultra3 SCSI	80	8	36 ft
Wide Ultra3 SCSI	160	16	36 ft
Serial			
Std w/ 16550 UART(115,200 bps)	0.015	1	

I/O INTERFACES (cont.)

ST506/412

Hard driveMFM & RLL · · · · · 0.6 to 0.94 · · · · · · · · · 2 · · Internal

SSA (Serial Storage Architecture)

Std 2 Channel System · · · · · · · 20-40 · · · · · · · 128· · · · · 75 ft

USB (Universal Serial Bus)

· 1.5 · · · · · · · · 27· · · · · 15 ft

PC MEMORY MAP

Address Range	Size	Description
ØØØØØ-ØØ3FF	1K	Interrupt Vectors
ØØ4ØØ-7FFFF	512K	Bios, DOS, 512K RAM Expansion
8ØØØØ-9FFFF	128K	128K RAM Expansion (Top of 640K)
AØØØØ-AFFFF	64K	EGA Video Buffer
BØØØØ-B7FFF	32K	Monochrome & other screen buffers
B8ØØØ-BFFFF	32K	CGA and EGA Buffers
AT LIM Expanded Memory 64K page is between 768K and 896K		
CØØØØ-C3FFF	16K	EGA Video Bios
C4ØØØ-C7FFF	16K	ROM Expansion Area
XT LIM Expanded Memory 64K page is between 800K and 960K		
C8ØØØ-CCFFF	20K	XT Hard Disk Controller Bios
CDØØØ-CFFFF	12K	User PROM, Memory mapped I/O
DØØØØ-DFFFF	64K	User PROM, normal LIM Location for Expanded Memory
EØØØØ-EFFFF	64K	ROM expansion, I/O for XT
FØØØØ-FDFFF	56K	ROM BASIC
FEØØØ-FFFD9	8K	BIOS
FFFFØ-FFFF4	4	1st Code run after system power on
FFFF5-FFFFC	8	BIOS Release Date
FFFFE-FFFFF	2	Machine ID (Top of 1 Meg RAM)
1ØØØØØ-FFFFFF	15Meg	AT Extended Memory

PC HARDWARE INTERRUPTS

NMI . . . . Non-Maskable Interrupt (Parity)

Interrupt Controller 1:

- IRQØ . . . . Timer Output
- IRQ1 . . . . Keyboard controller
- IRQ2 . . . . XT – Available
  AT – Route to Interrupt Controller 2, IRQ8 to 15
- IRQ3 . . . . Serial Port COM2: or SDLC (see page 62)
- IRQ4 . . . . Serial Port COM1: or SDLC (see page 62)
- IRQ5 . . . . XT – Hard Disk Controller
  AT – Parallel Printer Port 2
- IRQ6 . . . . Floppy Disk Controller
- IRQ7 . . . . Parallel Printer Port LPT1:

Interrupt Controller 2 (AT/Pentium Only):

- IRQ8 . . . . Real Time Clock
- IRQ9 . . . . Software redirect to IRQ2 (Int ØA Hex)
- IRQ10 . . . . . . Reserved
- IRQ11 . . . . . . Reserved
- IRQ12 . . . . . . Reserved
- IRQ13 . . . . . . 80287 Math Coprocessor
- IRQ14 . . . . . . Hard Disk Controller
- IRQ15 . . . . . . Some hard drive and SCSI controllers

DMA CHANNELS

XT 8 bit ISA Bus

Channel	Function
0	Dynamic memory refresh
1	Unassigned or SDLC
2	Floppy disk controller
3	Hard disk controller

16 bit ISA, EISA, and MCA Bus

Channel	Function
DMA Controller #1	
0	Dynamic memory refresh
1	Unassigned or SDLC
2	Floppy disk controller
3	Unassigned
DMA Controller #2	
4	First DMA Controller
5	Unassigned
6	Unassigned
7	Unassigned

SERIAL/COM: PORTS

Com: Port	PC / ISA IRQ / Address	PS2 / MCA IRQ / Address
1	4 / 03F8h	4 / 03F8h
2	3 / 02F8h	3 / 02F8h
3	4 / 03E8h*	3 / 3220h
4	3 / 02E8h*	3 / 3228h
5	not available	3 / 4220h
6	not available	3 / 4228h
7	not available	3 / 5220h
8	not available	3 / 5228h

* Note that some software and hardware products do not support the COM3: and COM4: addresses and interrupts

PC HARDWARE I/O MAP

8088 Class Systems

Address	Function
ØØØ–ØØF	DMA Controller (8237A)
Ø2Ø–Ø21	Interrupt controller (8259A)
Ø4Ø–Ø43	Timer (8253)
Ø6Ø–Ø63	(8255A)
Ø8Ø–Ø83	DMA page register (74LS612)
ØAØ–ØAF	NMI – Non Maskable Interrupt
2ØØ–2ØF	Game Port Joystick controller
21Ø–217	Expansion Unit
2E8–2EF	COM4: Serial Port (see page 62)
2F8–2FF	COM2: Serial Port
3ØØ–31F	Prototype Card
32Ø–32F	Hard Disk
378–37F	Parallel Printer Port 1
38Ø–38F	SDLC
3BØ–3BF	MDA – Monochrome Adapter and printer
3DØ–3D7	CGA – Color Graphics Adapter
3E8–3EF	COM3: Serial Port (see page 62)
3FØ–3F7	Floppy Diskette Controller
3F8–3FF	COM1: Serial Port

80286 /386/486 Class Systems

Address	Function
ØØØ–Ø1F	DMA Controller #1 (8237A–5)
Ø2Ø–Ø3F	Interrupt controller #1 (8259A)
Ø4Ø–Ø5F	Timer (8254)
Ø6Ø–Ø6F	Keyboard (8Ø42)
Ø7Ø–Ø7F	NMI – Non Maskable Interrupt & CMOS RAM
Ø8Ø–Ø9F	DMA page register (74LS612)
ØAØ–ØBF	Interrupt controller #2 (8259A)
ØCØ–ØDF	DMA Controller #2 (8237A)
ØFØ–ØFF	8Ø287 Math Coprocessor
1FØ–1F8	Hard Disk
2ØØ–2ØF	Game Port Joystick controller
258–25F	Intel Above Board
278–27F	Parallel Printer Port 2
2E8–2EF	COM4: Serial Port (see page 62)
2F8–2FF	COM2: Serial Port
3ØØ–31F	Prototype Card
378–37F	Parallel Printer Port 1
38Ø–38F	SDLC or Bisynchronous Comm Port 2
3AØ–3AF	Bisynchronous Comm Port 1
3BØ–3BF	MDA – Monochrome Adapter
3BC–3BE	Parallel Printer on Monochrome Adapter
3CØ–3CF	EGA – Reserved
3DØ–3D7	CGA – Color Graphics Adapter
3E8–3EF	COM3: Serial Port (see page 62)
3FØ–3F7	Floppy Diskette Controller
3F8–3FF	COM1: Serial Port

PC SOFTWARE INTERRUPTS

Address	Int #	Interrupt Name
ØØØ–ØØ3	Ø	Divide by zero
ØØ4–ØØ7	1	Single Step IRET
ØØ8–ØØB	2	NMI Non Maskable Interrupt
ØØC–ØØF	3	Breakpoint
Ø1Ø–Ø13	4	Overflow IRET
Ø14–Ø17	5	Print Screen
Ø18–Ø1F	6	Reserved Ø18–Ø1B and Ø1C–Ø1F
Ø2Ø–Ø23	8	Time of Day Ticker IRQØ
Ø24–Ø27	9	Keyboard IRQ1
Ø28–Ø2B	A	XT Reserved, AT IRQ2 direct to IRQ9
Ø2C–Ø2F	B	COM2 communications, IRQ3
Ø3Ø–Ø33	C	COM1 communications, IRQ4
Ø34–Ø37	D	XT Hard disk, AT Parallel Printer, IRQ5
Ø38–Ø3B	E	Floppy Diskette, IRQ6
Ø3C–Ø3F	F	Parallel Printer 1, IRQ7, slave 8259, IRET
Ø4Ø–Ø43	1Ø	ROM Handler – Video
Ø44–Ø47	11	ROM Handler – Equipment Check
Ø48–Ø4B	12	ROM Handler – Memory Check
Ø4C–Ø4F	13	ROM Handler – Diskette I/O
Ø5Ø–Ø53	14	ROM Handler – COMM I/O
Ø54–Ø57	15	XT Cassette, AT ROM Catchall Handlers
Ø58–Ø5B	16	ROM Handler – Keyboard I/O
Ø5C–Ø5F	17	ROM Handler – Printer I/O
Ø6Ø–Ø63	18	ROM Handler – Basic Startup
Ø64–Ø67	19	ROM Handler – Bootstrap
Ø68–Ø6B	1A	ROM Handler – Time of Day
Ø6C–Ø6F	1B	ROM Handler – Keyboard Break
Ø7Ø–Ø73	1C	ROM Handler – User Ticker
Ø74–Ø77	1D	ROM Pointer, Video Initialization
Ø78–Ø7B	1E	ROM Pointer, Diskette Parameters
Ø7C–Ø7F	1F	ROM Pointer, Graphics Characters Set 2
Ø8Ø–Ø83	2Ø	DOS – Terminate Program
Ø84–Ø87	21	DOS – Function Call
Ø88–Ø8B	22	DOS – Program's Terminate Address
Ø8C–Ø8F	23	DOS – Program's Control–Break Address
Ø9Ø–Ø93	24	DOS – Critical Error Handler
Ø94–Ø97	25	DOS – Absolute Disk Read
Ø98–Ø9B	26	DOS – Absolute Disk Write
Ø9C–Ø9F	27	DOS – TSR Terminate & Stay Ready
ØAØ–ØFF	28–3F	DOS – Idle Loop, IRET
1ØØ–1Ø3	4Ø	Hard Disk Pointer–Original Floppy Handler
1Ø4–1Ø7	41	ROM Pointer, XT Hard Disk Parameters
1Ø8–1ØB	42–45	Reserved
1ØC–1ØF	46	ROM Pointer, AT Hard Disk Parameters
11Ø–17F	47–5F	Reserved
18Ø–19F	6Ø–67	Reserved for User (67 is Expanded Mem)
1AØ–1BF	68–6F	Not Used
1CØ–1C3	7Ø	AT Real Time Clock, IRQ8
1C4–1C7	71	AT Redirect to IRQ2, IRQ9, LAN Adapter 1
1C8–1CB	72	AT Reserved, IRQ1Ø
1CC–1CF	73	AT Reserved, IRQ11
1DØ–1D3	74	AT Reserved, IRQ12
1D4–1D7	75	AT 8Ø287 Error to NMI, IRQ13
1D8–1DB	76	AT Hard Disk, IRQ14
1DC–1DF	77	AT Reserved, IRQ15
1EØ–1FF	78–7F	Not Used
2ØØ–217	8Ø–85	Reserved for BASIC
218–21B	86	NetBIOS, Relocated Interrupt 18H
218–3C3	87–FØ	Reserved for BASIC Interpreter
3C4–3FF	F1–FF	Not Used

AUDIO ERROR CODES

A variety of tests are executed automatically when computers are first turned on. Initially, the "Power-On Self Test" (POST) is run. It provides error or warning messages whenever a faulty component is encountered. Typically, two types of messages are issued: **Audio Beep Codes** and **Display Error Messages.**

Audio Beep Codes consist of a series of beeps that identify a faulty component. In the case of an IBM computer, if it is functioning normally, you will hear one short beep when the system is turned on. However, if a problem is detected, a series of beeps or no beeps will occur. The type and number of beeps define the problem. Audio Beep Codes for some of the major BIOS manufacturers are included below.

If the system has problems but completes the POST process, then additional errors may be reported in the form of **Display Error Messages**. The list of **Display Error Messages** is quite extensive and only the IBM PC/XT/PS2 messages are included in Pocket PCRef.

American Megatrends Bios (AMI)

Beeps	Error Description
Fatal Errors	
1	DRAM refresh failed
2	Parity circuit failed
3	Base 64K or CMOS RAM failed
4	System timer failed
5	Processor failed
6	Keyboard controller or gate A20 error
7	Virtual mode exception error
8	Display memory write/read test failed
9	ROM BIOS checksum failed
10	CMOS RAM shutdown register failed
11	Cache memory bad, do not enable cache
Nonfatal errors	
1 long, 3 short	Conventional/extended memory failed
1 long, 8 short	Display and retrace failed

Award Software Bios

Beeps	Error Description
1 long, 2 short	Video card failure

Any other beeps are probably RAM problems; no other audio errors

Dell Computer Corporation Bios

Beeps	Error Description
1-3	Video memory test failure
1-1-2	Testing CPU register
1-1-3	CMOS write/read test failed
1-1-4	ROM BIOS checksum bad
1-2-1	Programmable interval timer failed
1-2-2	DMA initialization failed
1-2-3	DMA page register write/read bad
1-3-1	RAM refresh verification failed
1-3-2	Testing first 64K RAM
1-3-3	First 64K RAM chip or data line bad, multi-bit
1-3-4	First 64K RAM odd/even logic bad
1-4-1	Address line fault in first 64K RAM
1-4-2	Parity error detected in first 64K RAM
2-1-1	Bit 0 fault in first 64K RAM
2-1-2	Bit 1 fault in first 64K RAM
2-1-3	Bit 2 fault in first 64K RAM
2-1-4	Bit 3 fault in first 64K RAM
2-2-1	Bit 4 fault in first 64K RAM
2-2-2	Bit 5 fault in first 64K RAM
2-2-3	Bit 6 fault in first 64K RAM
2-2-4	Bit 7 fault in first 64K RAM
2-3-1	Bit 8 fault in first 64K RAM
2-3-2	Bit 9 fault in first 64K RAM
2-3-3	Bit 10 fault in first 64K RAM
2-3-4	Bit 11 fault in first 64K RAM
2-4-1	Bit 12 fault in first 64K RAM
2-4-2	Bit 13 fault in first 64K RAM
2-4-3	Bit 14 fault in first 64K RAM
2-4-4	Bit 15 fault in first 64K RAM
3-1-1	Slave DMA register bad
3-1-2	Master DMA register bad
3-1-3	Master interrupt mask register bad
3-1-4	Slave interrupt mask register bad
3-2-2	Interrupt vector loading in progress
3-2-4	Keyboard controller test failed
3-3-1	CMOS RAM power bad; calculating checksum
3-3-2	CMOS configuration validation in progress
3-3-4	Video memory test failed
3-4-1	Video initialization failed
3-4-2	Video retrace failure
3-4-3	Search for video ROM in progress
none	Screen operable, running with video ROM
none	Monochrome monitor operable
none	Color monitor (40 column) operable
none	Color monitor (80 column) operable
4-2-1	Timer tick interrupt test in progress or bad

Dell Computer Corporation Bios (cont.)

Beeps	Error Description
4-2-2	Shutdown test in progress or bad
4-2-3	Gate A20 bad
4-2-4	Unexpected interrupt in protected mode
4-3-1	RAM test in progress or high address line bad FFFF
4-3-3	Interval timer channel 2 test or bad
4-3-4	Time-of-Day clock test or bad
4-4-1	Serial port test or bad
4-4-2	Parallel port test or bad
4-4-3	Math coprocessor test or bad
4-4-4	Cache test failure
5-1-2	BIOS update error; no RAM in system
5-1-3	BIOS update error; external video card detected
5-1-4	BIOS execution error
6-1-2	I/O controller failure
6-1-3	Keyboard controller failure
6-1-4	CMOS register test failure
6-2-1	BIOS shadowing failure
6-2-2	Pentium speed determination failure
6-2-3	No SIMM installed

IBM Corporation Bios

Beeps	Error Description
1 short	Successful Post, no errors
2 short	Initialization error - serial, parallel, floppy, ROM, or DMA
1 long, 1 short	System Board
1 long, 2 short	Video adapter or video memory failed
1 long, 3 short	Video adapter failed, EGA
3 long	3270 keyboard card failure
None	Power supply or system board
Continuous	Power supply or system board
Repeating short	Power supply or system board

Mylex and Eurosoft Bios

Beeps	Error Description
1	Always present to indicate start of beep coding
2	Video adapter bad or not detected
3	Keyboard controller error
4	Keyboard error
5	8259 Programmable Interrupt Controller (PIC) 1 Er
6	8259 PIC 2 error
7	DMA page register failure
8	RAM refresh error
9	RAM data test failed

Mylex and Eurosoft Bios (cont.)

Beeps	Error Description
10	RAM parity error occurred
11	8237 DMA controller 2 failed
12	CMOS RAM failure
13	8237 DMA controller 2 failed
14	CMOS RAM battery failure
15	CMOS RAM checksum error
16	RIOS ROM checksum error

Phoenix Bios

Beeps	Error Description
none/1-1-2	CPU reguster test in progress
1-1-3	CMOS write/read test failed
1-1-4	ROM BIOS checksum bad
1-2-1	Programmable interval timer failed
1-2-2	DMA initialization failed
1-2-3	DMA page register write/read bad
1-3-1	RAM refresh verification failed
none/1-3-2	Testing first 64K RAM
1-3-3	First 64K RAM chip or data line fault, multi-bit
1-3-4	First 64K RAM odd/even logic bad
1-4-1	Address line bad first 64K RAM
1-4-2	Parity error detected in first 64K RAM
1-4-3	EISA fail-safe timer test in progress
1-4-4	EISA s/w NMI port 462 test in progress
2-1-1	Bit 0 fault in first 64K RAM
2-1-2	Bit 1 fault in first 64K RAM
2-1-3	Bit 2 fault in first 64K RAM
2-1-4	Bit 3 fault in first 64K RAM
2-2-1	Bit 4 fault in first 64K RAM
2-2-2	Bit 5 fault in first 64K RAM
2-2-3	Bit 6 fault in first 64K RAM
2-2-4	Bit 7 fault in first 64K RAM
2-3-1	Bit 8 fault in first 64K RAM
2-3-2	Bit 9 fault in first 64K RAM
2-3-3	Bit 10 fault in first 64K RAM
2-3-4	Bit 11 fault in first 64K RAM
2-4-1	Bit 12 fault in first 64K RAM
2-4-2	Bit 13 fault in first 64K RAM
2-4-3	Bit 14 fault in first 64K RAM
2-4-4	Bit 15 fault in first 64K RAM
3-1-1	Slave DMA register bad
3-1-2	Master DMA register bad
3-1-3	Master interrupt mask register bad
3-1-4	Slave interrupt mask register bad
none/3-2-2	Interrupt vector loading in progress
3-2-4	Keyboard controller test failed

none/3-3-1 . . . . . CMOS RAM power bad; calculating checksum
none/3-3-2 . . . . . CMOS configuration validation in progress
3-3-4. . . . . Video memory test failed
3-4-1. . . . . Video initialization failed
3-4-2. . . . . Video retrace failure
none/3-4-3 . . . . . Search for video ROM in progress
none . . . . DDNIL bit scan failed
none . . . . Screen operable, running with video ROM
none . . . . Monochrome monitor operable
none . . . . Color monitor (40 column) operable
none . . . . Color monitor (80 column) operable
4-2-1. . . . . Timer tick interrupt test in progress or bad
4-2-2. . . . . Shutdown test in progress or bad
4-2-3. . . . . Gate A20 bad
4-2-4. . . . . Unexpected interrupt in protected mode
4-3-1. . . . . RAM test in progress or high address line bad FFFF
4-3-3. . . . . Interval timer channel 2 test or bad
4-3-4. . . . . Time-of-Day clock test or bad
4-4-1. . . . . Serial port test or bad
4-4-2. . . . . Parallel port test or bad
4-4-3. . . . . Math coprocessor test or bad
4-4-4. . . . . Cache test failure (Dell)
low-1-1-2 . . . . . . System board select bad (MCA only)
low-1-1-3 . . . . . . Extended CMOS RAM bad (MCA only)

Quadtel Bios

Beeps	Error Description
1 beep	POST ran okay and detected no error. System will now boot.
2 beeps.	POST detected a configuration error, or a CMOS RAM change since the last time you ran Setup. Check the CMOS battery and rerun Setup.
1 long, 2 short.	Faulty video configuration (no or bad video card installed), or bad ROM on a peripheral controller card (address range C0000 through FFFF)
1 long, 3+shorts	Faulty peripheral controller, such as VGA. Usually, the display shows a descriptive message. Check the setup of peripheral controllers.

Tandon Bios

Beeps	Error Description
long-short-long-short	8254 counter timer failure
short-long-short	RAM refresh failure
long-long-long	System RAM failure
short-short-short	BIOS RAM checksum failure
long-long	No video adapter is installed
long-long-long-long	Video adapter failure

IBM XT/AT CLASS ERROR CODES

Code	Description
01x	Undetermined problem errors
02x	Power supply errors
1xx	**System board error**
101	Interrupt failure
102	Timer failure
103	Timer interrupt failure
104	Protected mode failure
105	Last 8042 command not accepted
106	Converting logic test
107	Hot NMI test
108	Timer bus test
109	Direct memory access test error
121	Unexpected hardware interrupts occurred
131	Cassette wrap test failed
152	System board error: defective battery
161	System Options Error-(Run SETUP) [Battery failure]
162	System options not set correctly-(Run SETUP)
163	Time and date not set-(Run SETUP)
164	Memory size error-(Run SETUP)
165	Adaptor error
199	User indicated configuration not correct
2xx	**Memory (RAM) errors**
201	Memory test failed
202	Memory address error
203	Memory address error
3xx	**Keyboard errors**
301	Keyboard did not respond to software reset correctly or a stuck key failure was detected. If a stuck key was detected, the scan code for the key is displayed in hexadecimal. For example, the error code 49 301 indicates that key 73, the PgUp key failed (49Hex = 73Dec)
302	User indicated error from the keyboard test or AT system unit keylock is locked
303	Keyboard or system unit error
304	Keyboard or system unit error; CMOS does not match system
4xx	**Monochrome monitor errors**
401	Monochrome memory test, horizontal sync frequency test, or video test failed
408	User indicated display attributes failure
416	User indicated character set failure
424	User indicated 80X25 mode failure
432	Parallel port test failed (monochrome adapter)
5xx	**Color monitor errors**
501	Color memory test failed, horizontal sync frequency test, or video test failed
508	User indicated display attribute failure
516	User indicated character set failure
524	User indicated 80X25 mode failure
532	User indicated 40X25 mode failure
540	User indicated 320X200 graphics mode failure
548	User indicated 640X200 graphics mode failure
6xx	**Diskette drive errors**

IBM XT/AT CLASS ERROR CODES

Code	Description
601	Diskette power on diagnostics test failed
602	Diskette test failed; boot record is not valid
606	Diskette verify function failed
607	Write protected diskette
608	Bad command diskette status returned
610	Diskette initialization failed
611	Time-out - diskette status returned
612	Bad NEC - diskette status returned
613	Bad DMA - diskette status returned
621	Bad seek - diskette status returned
622	Bad CRC - diskette status returned
623	Record not found - diskette status returned
624	Bad address mark - diskette status returned
625	Bad NEC seek - diskette status returned
626	Diskette data compare error
7xx	**8087 or 80287 math coprocessor errors**
9xx	**Parallel printer adapter errors**
901	Parallel printer adapter test failed
10xx	**Reserved for parallel printer adapter**
11xx	**Asynchronous communications adapter errors**
1101	Async communications adapter test failed
12xx	**Alternate asynchronous communications adapter errors**
1201	Alternate asynchronous communications adapter test failed
13xx	**Game control adapter errors**
1301	Game control adapter test failed
1302	Joystick test failed
14xx	**Printer errors**
1401	Printer test failed
1404	Matrix printer failed
15xx	**Synchronous data link control (SDLC) comm adapter errors**
1510	8255 port B failure
1511	8255 port A failure
1512	8255 port C failure
1513	8253 timer 1 did not reach terminal count
1514	8253 timer 1 stuck on
1515	8253 timer 0 did not reach terminal count
1516	8253 timer 0 stuck on
1517	8253 timer 2 did not reach terminal count
1518	8253 timer 2 stuck on
1519	8273 port B error
1520	8273 port A error
1521	8273 command/read time-out
1522	Interrupt level 4 failure
1523	Ring Indicate stuck on
1524	Receive clock stuck on
1525	Transmit clock stuck on
1526	Test indicate stuck on
1527	Ring indicate not on
1528	Receive clock not on
1529	Transmit clock not on
1530	Test indicate not on
1531	Data set ready not on

IBM XT/AT CLASS ERROR CODES

Code	Description
1532	Carrier detect not on
1533	Clear to send not on
1534	Data set ready stuck on
1536	Clear to send stuck on
1537	Level 3 interrupt failure
1538	Receive interrupt results error
1539	Wrap data mis-compare
1540	DMA channel 1 error
1541	DMA channel 1 error
1542	Error in 8273 error checking or status reporting
1547	Stray interrupt level 4
1548	Stray interrupt level 3
1549	Interrupt presentation sequence time-out
16xx	**Display emulation errors (327x, 5520, 525x)**
17xx	**Fixed disk errors**
1701	Fixed disk POST error
1702	Fixed disk adapter error
1703	Fixed disk drive error
1704	Fixed disk adapter or drive error
1780	Fixed disk 0 failure
1781	Fixed disk 1 failure
1782	Fixed disk controller failure
1790	Fixed disk 0 error
1791	Fixed disk 1 error
18xx	**I/O expansion unit errors**
1801	I/O expansion unit POST error
1810	Enable/Disable failure
1811	Extender card wrap test failed (disabled)
1812	High order address lines failure (disabled)
1813	Wait state failure (disabled)
1814	Enable/Disable could not be set on
1815	Wait state failure (disabled)
1816	Extender card wrap test failed (enabled)
1817	High order address lines failure (enabled)
1818	Disable not functioning
1819	Wait request switch not set correctly
1820	Receiver card wrap test failure
1821	Receiver high order address lines failure
19xx	**3270 PC attachment card errors**
20xx	**Binary synchronous communications (BSC) adapter errors**
2010	8255 port A failure
2011	8255 port B failure
2012	8255 port C failure
2013	8253 timer 1 did not reach terminal count
2014	8253 timer 1 stuck on
2016	8253 timer 2 did not reach terminal count or timer 2 stuck on
2017	8251 Data set ready failed to come on
2018	8251 Clear to send not sensed
2019	8251 Data set ready stuck on
2020	8251 Clear to send stuck on
2021	8251 hardware reset failed
2022	8251 software reset failed
2023	8251 software "error reset" failed

Code	Description
2024	8251 transmit ready did not come on
2025	8251 receive ready did not come on
2026	8251 could not force "overrun" error status
2027	Interrupt failure - no timer interrupt
2028	Interrupt failure - transmit, replace card or planar
2029	Interrupt failure - transmit, replace card
2030	Interrupt failure - receive, replace card or planar
2031	Interrupt failure - receive, replace card
2033	Ring indicate stuck on
2034	Receive clock stuck on
2035	Transmit clock stuck on
2036	Test indicate stuck on
2037	Ring indicate stuck on
2038	Receive clock not on
2039	Transmit clock not on
2040	Test indicate not on
2041	Data set ready not on
2042	Carrier detect not on
2043	Clear to send not on
2044	Data set ready stuck on
2045	Carrier detect stuck on
2046	Clear to send stuck on
2047	Unexpected transmit interrupt
2048	Unexpected receive interrupt
2049	Transmit data did not equal receive data
2050	8251 detected overrun error
2051	Lost data set ready during data wrap
2052	Receive time-out during data wrap
21xx	**Alternate binary synchronous communications adapter errors**
2110	8255 port A failure
2111	8255 port B failure
2112	8255 port C failure
2113	8253 timer 1 did not reach terminal count
2114	8253 timer 1 stuck on
2115	8253 timer 2 did not reach terminal count or timer 2 stuck on
2116	8251 Data set ready failed to come on
2117	8251 Clear to send not sensed
2118	8251 Data set ready stuck on
2119	8251 Clear to send stuck on
2120	8251 hardware reset failed
2121	8251 software reset failed
2122	8251 software "error reset" failed
2123	8251 transmit ready did not come on
2124	8251 receive ready did not come on
2125	8251 could not force "overrun" error status
2126	Interrupt failure - no timer interrupt
2128	Interrupt failure - transmit, replace card or planar
2129	Interrupt failure - transmit, replace card
2130	Interrupt failure - receive, replace card or planar
2131	Interrupt failure - receive, replace card
2133	Ring indicate stuck on
2134	Receive clock stuck on
2135	Transmit clock stuck on

IBM XT/AT CLASS ERROR CODES

Code	Description
2136	Test indicate stuck on
2137	Ring indicate stuck on
2138	Receive clock not on
2139	Transmit clock not on
2140	Test indicate not on
2141	Data set ready not on
2142	Carrier detect not on
2143	Clear to send not on
2144	Data set ready stuck on
2145	Carrier detect stuck on
2146	Clear to send stuck on
2147	Unexpected transmit interrupt
2148	Unexpected receive interrupt
2149	Transmit data did not equal receive data
2150	8251 detected overrun error
2151	Lost data set ready during data wrap
2152	Receive time-out during data wrap
22xx	**Cluster adapter errors**
24xx	**Enhanced graphics adapter errors**
29xx	**Color matrix printer errors**
2901	
2902	
2904	
33xx	**Compact printer errors**

IBM is a registered trademark of the International Business Machine Corporation.

Chapter 3

Modems

See page 62 for information on Serial/COM: port addresses and interrupts.

MODEM STANDARDS

V.xx Standards are international data communication standards defined by CCITT (Consultative Committee for International Telephone and Telegraph).

Standard	Description
V.13	Simulated half-duplex for synchronous networks.
V.21	300 bps, compatible with Bell 103.
V.22	1200 bps, compatible with Bell 212A; full duplex; sync or async.
V.22bis	2400 bps with fall back to 1200 bps, compatible with Bell 212A and V.22; full duplex; sync or async.
V.23	1200 bps with 75 bps back channel for use in the United Kingdom.
V.25	Provides autodialing capabilities to sync or async dialup lines. Parallel interface.
V.25bis	Provides autodialing capabilities to sync or async dialup lines. Serial interface.
V.32	4800 and 9600 bps with fall back to 4800; full duplex, sync or async. The first universal standard for 9600 bps modems.
V.32bis	14,400 bps with fall back to 12000, 9600 , 7200 and 4800 bps. Sync or async; full duplex. V.32bis incorporates "fastrain" in which it can automatically increase or decrease modem speed during operation.
V.33	14,400 or 12,000 bps sync transmission over 4 wire leased lines. Used in very high speed super computer environments. V.32bis provides the same capability but over 2 wire dialup lines.
V.34	28,800 bps Standard approved in June 1994 and is the state-of-the-art protocol for high speed modem communications. It includes a 4-dimension 64 state trellis coding not found in the V.FC modems and also includes a V.8 high speed startup sequence.
V.42	LAP-M (Link Access Protocol) Error Correction and support for MNP levels 1 to 4; falls back to MNP 1-4 if LAP-M is not available.
V.42bis	V.42 with intelligent data compression and support for MNP5; compression up to 4:1.
V.FC or V.Fast	A class of modems incorporating some of the V.34 standards.

Bell Standards are USA data communication standards defined by Bell Labs and AT&T.

Standard	Description
Bell 103.	300 bps, async, full duplex over 2 wire dialup or leased lines. Comparable to V.21.
Bell 201B	2400 bps, sync, full duplex over 4 wire, half duplex over 2 wire dialup lines. Comparable to V.26.
Bell 201C	Same as 201B but dialup lines only.
Bell 208A	4800 bps, sync, full duplex over 4 wire leased line or half duplex over 2 wire leased line. Comparable to V.27
Bell 208B	Same as 208A but 2 wire dialup lines only
Bell 212A	1200 bps, sync or async, full duplex over 2 wire leased or dialup lines. Comparable to V.22.

MNP (Microcom Networking Protocol) Error Correction and Data Compression. In order to use MNP, the modems at both ends of the phone line must have the same MNP capability.

Standard	Description
MNP Level 1-4 .	Error correcting routines used to filter out line noise. It also reduces the size of data transferred by up to 20%, thereby speeding up transfers.
MNP Level 5 . . .	Conventional data compression of up to 2:1; useful for ASCII type files only not binary files like ZIP and ARC files. MNP 5 effectively doubles the baud rate of the transfer.

UART SERIAL CHIPS

The UART (Universal Asynchronous Receiver-Transmitter) is the heart of a computer's serial port and it provides a parallel to serial and serial to parallel translation link between computer and modem. The chips listed below are made by Intel (INS) and National Semiconductor (NS).

INS8250

The original UART used in IBM's first PC serial port. It has slow access cycle delays and requires extra NOPS between CPU read-write cycles. Several bugs (one of which was an interrupt enable problem) are present in the chip but are not serious. The 8250 was replaced by the 8250A. Chip will not work properly at 9600 bps.

INS8250A and INS82C50A

This chip is an upgrade to the original 8250 and fixes some of the original bugs. The "A" series chip was designed to correct the bugs in conjunction with the PC and XT BIOS and is therefore not compatible with many software packages and other computer's. Avoid using this chip! Chip will not work properly at 9600 bps.

INS8250B

The final upgrade of the 8250 chip series in which the bugs of the first two versions have been repaired. This chip will work in PC and XT class systems; however, it may or may not function correctly in 80286 and higher systems. Chip will not work properly at 9600 bps.

NS16450 and NS16C450

A higher speed version of the 8250 chip. It was designed for 80286 and higher systems and may not work correctly in PC and XT class computers. A scratch register (#7) has been included. The OS2 operating system requires the 16450 or higher in serial ports. Maximum data rate is 38,400 bps.

NS16550

This chip is an upgrade to the 16450. It provides higher baud rates and a DMA interface. It does not support FIFO (first in - first out). It works well in 80286 and higher systems and the maximum data rate is 115,200 bps.

NS16550A

A higher speed version of the 16550 chip. It was designed for 80286 and higher systems. It allows multiple DMA access and has a built-in 16 character transmit and receive FIFO (first in - first out) buffer. The 16550A is currently the recommended UART for high speed data communications. Maximum data rate is 115,200 bps.

HAYES COMPATIBLE MODEM COMMAND SETTINGS

Command Function

>>>>>*Note: all commands are not available on all modems!*<<<<<

+++ ....Default escape code, wait for modem to return state

A.............Force answer mode; Immediate answer on ring

A / ....Repeat last command line (Replaces AT)

AT ....Attention code

Cn ....n=Ø is Transmitter off, n=1 is on, (1=default)

Bn ....n=Ø is CCITT answer tone, n=1 is US/Canada Tone

Dn ....Dial telephone number

n= Ø to 9 for phone numbers

n= T is Touch Tone Dial, P is Pulse Dial

n= R is Originate Only, n= , is Pause

n= ! is xfer call to following extension

n= " is dial letters that follow

n= @ is Dial, Wait for answer, & continue

n= ; is Return to command mode after dialing

En ....n=Ø is no character echo in command state

n=1 is echo all characters in command state

Fn ....n=Ø is Half Duplex: n=1 if Full Duplex

Hn ....n=Ø is On Hook (Hang Up), n=1 is Off Hook

n=2 is Special Off Hook

In ....n=Ø is Display product code, n=1 show Check Sum

n=2 is show RAM test, n=3 is show call time length

n=4 is show current modem settings

Kn ....n=Ø at AT13 show last call length, n=1 show time

Ln ....Speaker volume control: n=Ø or 1 is low volume

n=2 is medium volume; n=3 is high volume

Mn ....n=Ø is Speaker always off, n=2 is always on

n=1 is Speaker on until carrier detected (default)

n=3 is Speaker on during CONNECT sequence only

Nn ....Auto data standard/speed adjust; n=Ø is connect at S37,

n=1 auto data standard and speed adjust to match

On ....n=Ø is return to on-line; n=1 is return to on-line & retain

Qn ....n=Ø is send Result Codes; n=1 is do not send code

n=2 is send result code only when originating call

SØ=n ......n=Ø to 255 rings before answer (see switch 5)

S1=n ....Counts rings from Ø to 255

S2=n ....Set escape code character, n=Ø to 127, 43 default

S3=n ....Set carriage return character, n=Ø to 127, 13 default

S4=n ....Set line feed character, n=Ø to 127, 10 default

S5=n ....Set backspace character, n=Ø to 127, 8 default

S6=n ....Wait time for dial tone, n=2 to 255 seconds

S7=n ....Wait time for carrier, n=2 to 255 seconds

S8=n ....Set duration of "," pause character, n=Ø to 255 sec.

S9=n ....Carrier detect response time, n=1 to 255 1/10 secs.

S10=n .....Delay time carrier loss to hang-up, n=1 to 255 1/10 s.

S11=n .....Duration & space of Touch Tones, n=50 to 255 ms.

S12=n .....Escape code guard time, n=50 to 255 1/50 seconds

S13=n .....UART Status Register Bit Mapped (reserved)

S14=n .....Option Register, Product code returned by AT1Ø

S15=n .....Flag Register (reserved)

Command	Function
S16=n	Self test mode. n=Ø is data mode (default), n=1 is Analog Loopback, n=2 is dial test, n=4 is Test Pattern, n=5 is Analog Loopback and Test Pattern
S18=n	Test timer for modem diagnostic tests
S37=n	Set line speed. Used in conjunction with Nn. n=Ø Attempt at speed of last AT command; n=1 to 3 attempt at 300bps; n=4 reserved; n=5 attempt 1200bps; n=6 attempt 2400bps; n=7 reserved; n=8 use 4800bps; n=9 use 9600; =10-12200bps; =11-14400bps; =12-7200
Sn ?	Send contents of Register n (Ø to 16) to Computer
Vn	n=Ø is send result codes as digits, n=1 is words
Wn	Protocol negotiation progress report; n=Ø is progress is not reported; n=1 is reported; n=2 is not reported but CONNECT XXXX message reports DCE speed
Xn	Send normal or extended result codes: n=Ø send basic set/blind dial; n=1 extended/blind dial; n=2 extended/dial tone; n=3 extended/blind & busy; n=4 extended/dial tone, busy
Yn	Long space disconnect: n=Ø is disabled; n=1 is enabled
Zn	Modem reset: n=Ø is power on; =1 to 3 user; =4 is factory
&Cn	n=Ø is DCD always active; n=1 active during connect
&Dn	n=Ø is DTR always ignored, =1 DTR causes return to command, =2 DTR disconnects, =3 disconnect/reset
&F	Get Factory Configuration
&Gn	n=Ø Disable Guard Tone, =1 is 550hz, =2 is 1800hz
&Kn	DTE: n=Ø is disable flow control, n=3 Enable RTS/CTS flow control; n=4 enable XON/XOFF flow control; n=5 enable transparent XON/XOFF flow control.
&Ln	n=Ø or 1 Speaker Volume Low, =2 medium, =3 high
&Mn	Communications mode (same as &Qn)
&Pn	n=Ø Pulse Make/Break Ratio USA 39% / 61% n=1 Pulse Make/Break Ratio UK 33% / 67%
&Qn	Communication mode: n=Ø is Async, Direct mode; n=4 modem issues OK result code; n=5 Error correction mode; n=6 Async, Normal mode; n=8 MNP; n=9 V.42 and V.42bis modes
&Rn	n=Ø is CTS tracks RTS, n=1 CTS always active
&Sn	n=Ø is DSR always active, n=1 DSR active at connect
&Tn	Test Commands: n=Ø end test, =1 local analog loopback, =3 local digital loopback, =4 enable Rmt digital loopback, =5 disable digital loopback, =6 request Rmt digital loop, =7 request Rmt dig loop & enter self test, =8 local analog loop & self test
&Vn	View current configuration
&W	Write Configuration to Memory
&Yn	n=Ø is Default is user configuration at NVRAM Ø ; n=1 default is user configuration at NVRAM location 1
&Zn=x	Store Phone Number “x” at location “n”. n=0,1,2, or 3

Chapter 4

DOS COMMANDS

Through MS-DOS® Version 6.22

This chapter is a concise general reference of DOS commands, listed in alphabetic order regardless of command type! In order to assist you in using the reference more effectively, a guide to conventions used in this chapter has been provided on page 82. A list of all DOS commands, grouped by command type, is located on page 84.

Editors Note: We strongly recommend that you upgrade your operating system with an official copy of MS-DOS 6.2x. Numerous functions and features that were not included in previous versions are now available and for the most part are bug free. The MS-DOS Users Guide and Technical Reference (order direct from Microsoft) are well written and are excellent resources. See page 6 for additional references. ***If you are using Version 6.0, it is strongly recommended that you do not use DBLSPACE or SMARTDRV. Both of these programs caused a variety of problems with hard drives and are considered not safe to use.*** MS-DOS 6.2x and several aftermarket programs are available which can safely provide the same features.

Command descriptions in this chapter are based on the following notations and syntax:

COMMAND NAME

Short Description: Long description

Syntax (shaded is optional):

COMMAND Drive: \Path /switches parameters

(Shaded areas indicate optional parameters and switches)

Examples: Samples of the syntax and command layout

Syntax Options:

Drive:\Path . . . . Drive & Directory containing command.
/switches. . . . . . *Switches* modify the way a command performs its particular function.
parameters . . . . Data (usually numeric) passed to the command when it's started.

Command Type and Version:

External command. . . . . DOS commands stored as files on a disk. All externals end in .EXE, .COM or .SYS.
Internal command . . . . . DOS commands contained in COMMAND.COM. These are loaded into the system on startup.
Batch command . . . . . . A script (text) file containing a sequence of commands to be run. The file always ends in .BAT
Config.sys command. . . Script (text) file containing start-up system configuration information and device drivers.
Network command. . . . . Will function on a network.
Introduced with Ver X.XX. . . The DOS version in which a command became available.

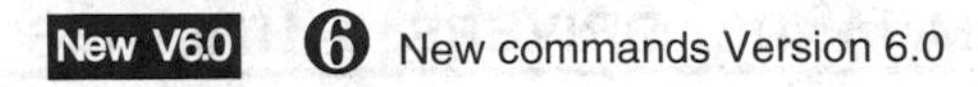

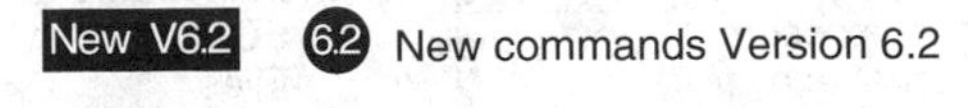

Danger V6.0 Dangerous Command Version 6.0

Removed V6.2 Command Removed Version 6.2

MS-DOS vs. PC-DOS

The following files contain the **D**isc **O**perating **S**ystem (DOS).

MS-DOS systems (most clones)

MSDOS.SYS
IO.SYS
COMMAND.COM

PC-DOS systems (IBM)

IBMBIO.COM
IBMDOS.COM
COMMAND.COM

These files (except COMMAND.COM) have attributes of "read only", "system" and "hidden" and are located in the root directory of the system's boot drive (hard drive or floppy drive). If any of these files are missing, the system will not start!

Despite the differences in these "operating system" files, most of the other commands prior to Version 6.0 have the same file names, e.g. both MS and PC use the FORMAT and FDISK programs to prepare a hard drive.

Due to space limitations, Sequoia Publishing is unable to provide information on commands for **PC**-DOS Versions 6.0, 6.1, and 6.3. Beginning with Version 6.0, Microsoft and IBM have taken radically different approaches to the commands supplied on the system disks, particularly the utility programs used for procedures such as disk repair and compression. We regret not being able to include these new PC-DOS commands, but we simply can't include the additional 100+ pages it would require. See page 85 for a list of the commands not covered.

MSDOS COMMANDS, DRIVERS & UTILITIES

External
Ados.com
Append.exe
Assign.com
Attrib.exe
Backinfo.exe
Backup.exe
Basic.exe
Basica.exe
Chkdsk.exe
Chkstate.sys
Command.com
Comp.exe
Country.sys
CV.com
Dblboot.bat
Dblspace.exe
Debug.exe
Defrag.exe
Deloldos.exe
Deltree.exe
Diskcomp.com
Diskcopy.com
Doskey.com
Dosshell.com
Dosshell.exe
Drvboot.bat
Drvspace.exe
Dvorak.sys
Edit.com
Edlin.exe
Emm386.exe
Exe2bin.exe
Expand.exe
Fasthelp.exe
Fastopen.exe
FC.exe
Fdisk.exe
Find.exe
Format.exe
Graftabl.com
Graphics.com
GW-Basic.exe
Help.com
Help.exe
Interlnk.exe
Intersvr.exe
Join.exe
Keyb.com
Keybxx.com
Label.exe
Link.exe
Loadfix.com
Mem.exe
Memmaker.exe
Mirror.com
Mode.com
More.com
Move.exe
Msav/Mwav.exe
Msbackup/
Mwbackup.exe
Mscdex.exe
Msd.com &.exe
Msherc.com
Nlsfunc.exe
Power.exe
Print.exe
Printfix.com
Qbasic.exe
Recover.exe
Replace.exe
Restore.exe
Scandisk.exe
Select.exe
Setup/
Busetup.exe
Setver.exe
Share.exe
Sizer.exe
Smartdrv.exe
Smartmon.exe
Sort.exe
Spatch.bat
Subst.exe
Sys.com
Tree.com
Truename.exe
Undelete/
Mwundel.exe
Unformat.com
Uninstal.exe
Vsafe.com
Wina20.386
Xcopy.exe

Internal
CD (Chdir)
Chcp
Chdir (CD)
Cls
Copy
Ctty
Date
Del (Erase)
Dir
Echo
Erase (Del)
Exit
For
LH(load high)
Loadhigh
MD (Mkdir)
Mkdir (MD)
Path
Prompt
RD (Rmdir)
Rem
Ren (Rename)
Rename (Ren)
Rmdir (RD)
Set
Time
Type
Ver
Verify
Vol

Config.sys
Ansi.sys
Break
Buffers
Command.com
Country.sys
Dblspace.sys
Device
Devicehigh
Display.sys
DOS
Driver.sys
Drivparm
Drvspace.sys
EGA.sys
Emm386.exe
Fastopen.exe
FCBS
Files
Himem.sys
Include
Install
Interlnk.exe
Kbdbuf.sys
Keyb.com/
Keyboard.sys
Lastdrive
Menucolor
Menudefault
Menuitem
Nlsfunc.exe
Numlock
Power.exe
Printer.sys
Ramdrive.sys/
Vdisk.sys
Rem
Setver.exe
Share.exe
Shell
Smartdrv.exe
Smartdrv.sys
Stacks
Submenu
Switchar
Switches

Batch
@
Break
Call
Choice.com
Echo
For
Goto
IF
Pause
Rem
Shift

Operating System

See also p. 83

Microsoft
MSDOS files:
Command.com
Io.sys
Msdos.sys

IBM
PCDOS files:
Command.com
Ibmbio.com
Ibmdos.com

Can Not Use on a Network

Chkdsk
Diskcomp
Diskcopy
Fastopen
Fdisk
Format
Join
Label
Recover
Scandisk
Subst
Sys
Unformat

Can Not Use While Running Windows

Append
Defrag
Emm386
Fastopen
Memmaker
Mscdex
Nlsfunc
Smartdrv
Subst
Vsafe

The Following PC-DOS Version 6.0, 6.1, and 6.3 Files are Not Described in this Edition of Pocket PCRef

See page 83

Cmosclk.sys
Cpbackup
Cpbdir
Cpsched
Datamon
Drvlock
E
Eject
Ibmavd
Ibmavw
Ibmavsp
Installhigh
Meutoini
Mouse
Pcformat
Pcmata
Pcmcs.sys
Pcmcs.exe
Pcmfdd
Pcmfdd.exe
Pcinfo
Pcmmtd
Pcmmtd.exe
Pcmscd.exe
Pcmscd
Pcmvcd.386
Pendos
Pendev.sys
Qconfig
Ramboost.exe
Ramsetup
Schedule
Setup
Umbcga.sys
Umbems.sys
Umbherc.sys
Umbmono.sys
Wnbackup
Wnschedl

DOS History

DOS Type	Release Date	Command COM (System File Sizes)	io and ibmbio (System File Sizes)	msdos & ibmdos (System File Sizes)	Loaded System (if High)
PC 1.0	8-4-81	3,231	1,920	6,400	13,312
MS 1.0	---	---	---	---	---
PC 1.1	5-7-82	4,959	1,920	6,400	14,336
MS 1.25	---	---	---	---	---
Zenith	4,986	1,713	6,138	---	---
PC 2.0	3-8-83	17,792	4,608	17,152	40,960
MS 2.0	---	---	---	---	---
Wang 2.01	12-22-83	15,877	30,482(Bios)	17,521	---
PC 2.1	10-20-83	17,792	4,736	17,024	40,960
MS 2.11	---	---	---	---	---
?mfg	11-17-83	15,957	6,836	17,176	25,680
PC 2.11	11-17-83	---	---	---	---
PC 2.11	5-30-84	18,272	5,120	17,408	---
PCAT&T 2.11	6-5-85	15,957	6,917	17,176	---
MSSanyo2.11	9-83-84	16,117	5,164	17,019	---
MS 2.25	---	---	---	---	---
PC 3.0	8-14-84	22,042	8,964	27,920	60,416
MS 3.0	---	---	---	---	---
PC 3.1	3-7-85	23,210	9,564	27,760	62,464
MS 3.1	---	---	---	---	---
PC 3.2	12-30-85	23,791	16,369	28,477	69,632
MS 3.2	7-7-86	23,612	16,138	28,480	55,568
MS 3.21	5-1-87	---	---	---	---
ZenithMS 3.21	9-28-87	23,948	18,501	28,480	---
PC 3.3	3-17-87	25,307	22,100	30,159	78,848
MS 3.3	7-24-87	25,276	22,357	30,128	55,440
MS 3.3a	2-2-88	25,308	22,398	30,128	---
MS 4.0	10-6-88	---	---	---	---
PC 4.01	3-89	---	---	---	---
MS 4.01	11-30-88	---	---	---	---
MS 4.01a	4-7-89	37,557	33,337	37,376	73,232
PC 5.0	5-9-91	47,987	33,430	37,378	---
MS 5.0	4-9-91	33,430	37,394	47,845	62,576 (21,776)
PC 5.00.1a	2-28-92	48,006	33,446	37,378	---
PC 5.02	9-1-92	47,990	33,718	37,362	---
MS 6.0	3-10-93	52,925	40,470	38,138	63,065 (17,197)
IBM 6.1	6-29-93	52,589	40,964	38,138	---
PC 6.1	9-30-93	52,797	40,964	38,138	---
MS 6.2R0	9-30-93	54,619	40,566	38,138	63,085 (22,093)
MS 6.22	5-31-94	54,645	40,774	38,138	63,085 (25,037)
PC 6.3	12-31-93	54,654	40,758	37,174	---

Windows History

Product	Release Date
Windows 1.0	11-85
Windows 2.0 (renamed Windows 286)	Fall 1987
Windows 386	Late 1987
Windows 3.0	5-22-90
Windows 3.1	4-6-92
Windows 3.11	
Windows for Workgroups 3.1	11-92
Windows for Workgroups 3.11	
Windows 95 Original	7-95
Service Pack 1 (Ver 4.00.95A)	12-31-95
OSR2 (Ver 4.00.95B)	9-96
OSR2.1	
OSR2.1 QFE	
OSR2.5 (Ver C) w/o USB support, IE4.0 req	2-98
OSR2.5 IE4.0 required	
Windows 98	6-98
Windows for Pen Computing 3.1	April 1992
Windows CE	
Windows NT 3.1 (1st NT Version)	1994
Windows NT 3.5	9-94
Windows NT 3.51	
Service Pack 1	
Service Pack 2	
Service Pack 3	
Service Pack 4	3-6-96
Service Pack 5	9-19-96
Windows NT 4.0	1996
Service Pack 1	
Service Pack 2	12-14-96
Service Pack 3	5-1-97
Windows NT 5.0	Estimated 1999

NOTE: *According to Microsoft, there were no official versions of MS-DOS prior to version 3.2. Prior to version 3.2, only OEM versions were sold with computers by the computer manufacturers. Slight variations in the sizes do occur, so use these as a general reference only.*

ADOS.COM

New V6.0

Starts AccessDOS: AccessDOS contains a set of public domain MS-DOS extensions developed for persons with motion and hearing disabilities by the University of Wisconsin.

Syntax (shaded is optional):

ADOS /a /c /L /m /x

Examples: ados /cos /c

Syntax options:

/a. . . . . Starts installation of AccessDOS.
/c. . . . . Runs in color mode.
/ L . . . . Runs in LCD mode.
/m . . . . Runs in monochrome mode.
/x. . . . . Runs in minimal mode.

Command Type and Version:

External command, Introduced with Ver 6.0.

Available in the MS-DOS 6.0, 6.21, and 6.22 Supplemental disks.

Notes:

1. See the ADOS.TXT and AREADME.TXT files in the Supplemental disks for user information.

ANSI.SYS

A device driver loaded through CONFIG.SYS that allows the user to control the computer's display and keyboard. Once the ANSI.SYS driver has been loaded, ANSI escape code sequences can be used to customize both the display and keyboard. This was developed by the American National Standards Institute (ANSI).

Syntax (shaded is optional):

DEVICE = Drive:\Path\ANSI.SYS /x /k /r

Examples: device=c:\dos\ansi.sys /x

If ANSI.SYS is loaded, try the following example for some enhancement of a color display:

PROMPT $e[35;44;1m$pge[33;44;1m

Syntax Options:

Drive: . . . . Letter of drive containing *\Path*.

\ *Path* . . . . Directory containing ANSI.SYS.

/x. . . . . Remaps 101-key keyboards so that the extended keys operate independently.

/k . . . . Extended keys on the 101-key keyboards will be ignored. This is particularly important on systems that do not accurately handle extended keyboard functions. Added in Version 5.0

/r (6.2) . . . . Used with screen-reading programs to adjust rate of line scrolling for easier reading.

Command Type and Version:

CONFIG.SYS command; Introduced with Ver 2.0

Notes:

1. The user has a lot of control over screen colors at the DOS level when the ANSI.SYS driver is loaded. See also PROMPT, p. 230.
2. The .SYS extension must be used in the syntax.
3. Using the Escape Code sequences is sometimes not an easy task. See PC Magazines book *DOS Power Tools, page 420,* for an example of how to write simple programs to send these codes.

ANSI escape sequences are a series of characters beginning with the ESCAPE (character 27) key, followed by open left bracket (**[**), followed by parameters sometimes, and ending with a letter or number. Note that the ending letter must be used in the correct upper or lower case format.

Parameters used in the escape sequences are as follows:

- *pl*. . . . . Line number (decimal value)
- *pc* . . . . Column number (decimal value)
- *pn* . . . . Specifies parameter is numeric
- *ps* . . . . Specific decimal number for a function. Multiple *ps* functions are separated with a **;**

ANSI escape sequences:

ESC [pl ; pc H. . . . Moves cursor to a specific line (*pl* parameter) and column (*pc* parameter). If no *pl* or *pc* is specified, the cursor goes to the Home position.

ESC [pl ; pc f . . . . Functions same as **ESC [*pl* ; *pc* H.**

ESC [*pn* A. . . . . . . Moves Cursor Up *pn* number of lines. If cursor is on top line, ANSI.SYS ignores sequence.

ESC [*pn* B. . . . . . . Moves Cursor Down *pn* number of lines. If the cursor is on the bottom line, ANSI.SYS ignores this sequence.

ESC [*pn* C. . . . . . . Moves Cursor Forward *pn* number of columns. If the cursor is at the farthermost right column, ANSI.SYS ignores this sequence.

ESC [*pn* D. . . . . . . Moves Cursor Backward *pn* number of lines. If the cursor is at the farthermost left column, ANSI.SYS ignores this sequence.

ESC [6n . . . . Reports status of selected device.

ESC [s . . . . Save Cursor Position. The cursor may be moved to the saved position by using the Restore Cursor sequence.

ESC [u . . . . Restore Cursor Position. Moves the cursor to the Save Cursor Position.

ESC [2 J. . . . . Erase Display. Erases the screen and returns the cursor to the home position.

ESC [K . . . . Erase Line. Erases all characters from the cursor to the end of the line.

ESC [*ps* ; . . ; *ps* m Sets graphics functions (text attributes and foreground and background colors). Note: These functions stay active until a new set of parameters is issued with this command.

Text Attributes: . All Attributes Off . . . 0
Bold On. 1
Faint On 2
Italic On. 3
Underscore . . . . 4 (Mono adapter only)
Blink On 5
Rapid Blink On 6
Reverse Video On. . 7
Concealed On. 8

Colors	Foreground	Background
Black	30	40
Red	31	41
Green	32	42
Yellow	33	43
Blue	34	44
Magenta	35	45
Cyan	36	46
White	37	47

Example: Try using the following PROMPT command if you have a color monitor and ANSI.SYS has been loaded in CONFIG.SYS.

PROMPT $e[35;44;1m$pge[33;44;1m

ESC [= *ps* h . . . . . Set Mode function. The active screen width and graphics mode type is changed with this sequence using the following values: ("mono" means monochrome).

ps	Mode (Graphics unless noted)	*ps*	Mode (Graphics unless noted)
0	40 x 25 mono (text)	13	320 x 200 color
1	40 x 25 mono (text)	14	640 x 200 color (16 color)
2	80 x 25 mono (text)	15	640 x 350 mono (2 color)
3	80 x 25 color (text)	16	640 x 350 color (16 color)
4	320 x 200 (4-color)	17	640 x 480 mono (2 color)
5	320 x 200 mono	18	640 x 480 (16 color)
6	640 x 200 mono	19	320 x 200 color (256 color)
7	Enables line wrapping		

ESC [= *ps* l (**l** in the sequence to the left is a lower case **L**) This sequence resets the Mode sequence described above. The *ps* parameter uses the same values as those shown in the Set Mode sequence above.

ESC [*code* **;** *string* **;...P** Redefine a specific keyboard key with a specific string of characters. *code* is one of the values in the ASCII Key Code table, on the next three pages, that represent keyboard keys or combinations of keys. Gray keys, keypad keys or codes shown in ( ) in the table may not function on some keyboards (try using the /x switch on the ANSI.SYS command line. *string* is either the decimal ASCII code for a single character (76 is the letter "C") or a string of characters in quotes ("<"). For example:

ESC ["< " ; "+" p ESC ["+" ; "< " p
ESC [60 ; 43 p ESC [43 ; 60 p

Both of the above sequences do the same task, they exchange the < and + keys.
Note that it is not possible to alter the ALT and Caps Lock keys.

NOTE: Some values listed in the ASCII Key Codes table below may not be valid for all computers! If in doubt, be sure to check the computer's documentation for verification.

ASCII Key Codes for ANSI.SYS

K means Key ➡ Key	K Code	SHIFT+K Code	CTRL+K Code	ALT+K Code
F1	0;59	0;84	0;94	0;104
F2	0;60	0;85	0;95	0;105
F3	0;61	0;86	0;96	0;106
F4	0;62	0;87	0;97	0;107
F5	0;63	0;88	0;98	0;108
F6	0;64	0;89	0;99	0;109
F7	0;65	0;90	0;100	0;110
F8	0;66	0;91	0;101	0;111
F9	0;67	0;92	0;102	0;112
F10	0;68	0;93	0;103	0;113
F11	0;133	0;135	0;137	0;139
F12	0;134	0;136	0;138	0;140
Home	0;71	55	0;119	—
Up Arrow	0;72	56	(0;141)	—
Page Up	0;73	57	0;132	—
Left Arrow	0;75	52	0;115	—
Right Arrow	0;77	54	0;116	—
End	0;79	49	0;117	—
Down Arrow	0;80	50	(0;145)	—
Page Down	0;81	51	0;118	—
Insert	0;82	48	(0;146)	—
Delete	0;83	46	(0;147)	—
Home (gray key)	224;71	224;71	224;119	224;151
Up Arrow (gray key)	224;72	224;72	224;141	224;152
Page Up (gray key)	224;73	224;73	224;132	224;153
Left Arrow (gray key)	224;75	224;75	224;115	224;155
Right Arrow (gray K)	224;77	224;77	224;116	224;157
End (gray key)	224;79	224;79	224;117	224;159
Down Arrow (gray key)	224;80	224;80	224;145	224;154
Page Down (gray key)	224;81	224;81	224;118	224;161
Insert (gray key)	224;82	224;82	224;146	224;162
Delete (gray key)	224;83	224;83	224;147	224;163
Print Screen	—	—	0;114	—
Pause/Break	—	—	0;0	—
Backspace	8	8	127	(0)

ASCII Key Codes for ANSI.SYS (cont.)

K means Key ➡ Key	K Code	SHIFT+K Code	CTRL+K Code	ALT+K Code
Tab	9	0;15	(0;148)	(0;165)
Null	0;3	—	—	—
A	97	65	1	0;30
B	98	66	2	0;48
C	99	66	3	0;46
D	100	68	4	0;32
Enter	13	—	10	(0;28)
E	101	69	5	0;18
F	102	70	6	0;33
G	103	71	7	0;34
H	104	72	8	0;35
I	105	73	9	0;23
J	106	74	10	0;36
K	107	75	11	0;37
L	108	76	12	0;38
M	109	77	13	0;50
N	110	78	14	0;49
O	111	79	15	0;24
P	112	80	16	0;25
Q	113	81	17	0;16
R	114	82	18	0;19
S	115	83	19	0;31
T	116	84	20	0;20
U	117	85	21	0;22
V	118	86	22	0;47
W	119	87	23	0;17
X	120	88	24	0;45
Y	121	89	25	0;21
Z	122	90	26	0;44
1	49	33	—	0;120
2	50	64	0	0;121
3	51	35	—	0;122
4	52	36	—	0;123
5	53	37	—	0;124
6	54	94	30	0;125
7	55	38	—	0;126
8	56	42	—	0;127
9	57	40	—	0;128
0	48	41	—	0;129
– (minus sign)	45	95	31	0;130
= (equal sign)	61	43	—	0;131

Key *(K means Key ➡)*	K Code	SHIFT+K Code	CTRL+K Code	ALT+K Code
[(left bracket)	91	123	27	0;26
] (right bracket)	93	125	29	0;27
\ (back slash).	92	124	28	0;43
; (semi-colon)	59	58	—	0;39
' (apostrophe)	39	34	—	0;40
, (comma)	44	60	—	0;51
. (period).	46	62	—	0;52
/ (forward slash) . . .	47	63	—	0;53
` (accent).	96	126	—	(0;41)
ENTER (on keypad)	13	—	10	(0;166)
/ (on keypad).	47	47	(0;142)	(0;74)
* (on keypad).	42	(0;144)	(0;78)	—
– (on keypad)	45	45	(0;149)	(0;164)
+ (on keypad)	43	43	(0;150)	(0;55)
5 (on keypad)	(0;76)	53	(0;143)	—

APPEND.EXE

Sets directory search order **:** Searches specified directories on specified drives to locate files outside of the current directory that have extensions other than .COM, .EXE, or .BAT. *Use Caution!*

Syntax (shaded is optional):

APPEND Drive: \Path /X /E /Path:on or off ;

Examples: APPEND /X /E
APPEND C:\WORDDATA; D:\PFS
APPEND ;

Syntax Options:

Drive: . . . . Letter of drive to be searched.
\Path . . . . Directory searched for data files.

/X :on or :off . . . Extends the DOS search path for specified files when executing programs. Processes SEARCH FIRST, FIND FIRST, and EXEC functions. :ON and :OFF, new to Version 5.0, toggles this switch on and off.

/Path :on or :off . .If path is already included for a program file, :on tells program to also search in appended directories. Default= :on; added in DOS Ver 5.0

/E . . . . Causes the appended path to be stored in the DOS environment and searched for there.

; . . . . Use ";" to separate multiple Drive:\Path statements on one line. APPEND ; by itself will cancel the APPEND list.

Command Type and Version:

External command; Network; Introduced with Ver 3.2

Notes:

1. /X and /E switches can only be used the first time you use Append. The line following the APPEND /X /E line contains the Drive:\Path.
2. You can not use any paths on the same command line as /X & /E.
3. :ON and :OFF switches are valid for Ver 5.0 and later.
4. Do not use APPEND with Windows.

ASSIGN.COM Removed V6.0

Assign disk drive: Instructs DOS to redirect disk operations on one drive to a different drive.

Syntax (shaded is optional):

ASSIGN Source = Target /status

Examples: ASSIGN A = B or ASSIGN A: = B: ❺
ASSIGN A = B B = C
ASSIGN
ASSIGN /status

Syntax Options:

ASSIGN . . . . . . ASSIGN with no switch cancels redirected drive assignments and sets them back to their original drives.

Source . . . . Letter(s) of source drive(s).

Target. . . . . Letter(s) of target drive(s). Starting with Version 5.0, a colon can be used with each assigned drive letter. For example; ASSIGN A: = B:

/Status . . . . Lists current drive assignments. Ver 5.0.

Command Type and Version:

External command; Network; Introduced with Ver 2.0 Removed from Version 6.0, considered too dangerous. Available in the MS-DOS 6.0, 6.21, 6.22 Supplemental Disks.

Notes:

1. DO NOT use a colon after a drive letter in versions prior to 5.0.
2. FORMAT, DISKCOPY, DISKCOMP, BACKUP, JOIN, LABEL, RESTORE, PRINT and SUBST cannot be used on ASSIGNed drives.
3. Be careful to reassign drives back to their original designations before running other programs.
4. If ASSIGN and APPEND are both used, the APPEND command must be used first.
5. See also the SUBST command.

ATTRIB.EXE Removed V6.0

Changes or displays file attributes: Sets, displays or clears a file's read-only, archive, system, and hidden attributes.

Syntax (shaded is optional):

ATTRIB +r-r +a-a +s-s +h-h Drive:\Path\Filename /s

Examples: ATTRIB wordfile.doc
ATTRIB +r wordfile.doc
ATTRIB +r d:\worddata*.* /s

Syntax Options:

Drive: . . . . Letter of drive containing *\path\filename*.

\Path . . . . Directory containing *filename*.

Filename. . . . . . Filename(s) of which attributes are to be displayed or changed. Wildcards (? and *) can be used for groups of files.

+ r. . . . . Sets file to read-only.

– r. . . . . Removes read-only attribute.

+ a. . . . . Sets the archive file attribute.

– a. . . . . Removes the archive file attribute.

+ s. . . . . Sets file as a system file. Ver. 5

– s. . . . . Removes system file attribute. Ver 5

+ h. . . . . Sets file as a hidden file. Ver 5

– h. . . . . Removes the hidden file attribute. Ver 5

/s. . . . . ATTRIB command processes files in the current directory and its subdirectories.

Command Type and Version:

External command; Network; Introduced with Ver 3.0

Notes:

1. When the system or hidden attribute is set , the read-only and *archive* attributes cannot be changed.
2. The *archive* attribute is used by the DOS BACKUP, RESTORE, and XCOPY commands when their **/m** switch is used and also the XCOPY command when the **/a** switch is used.

@ (at symbol)

Turns off the command echo function: In a batch file, placing the @ symbol at the start of a command line supresses the echoed display of the command on the screen.

Syntax (shaded is optional):

@ command

Examples: @xcopy a:*.* b:

@ECHO off

Syntax Options:

command . . . . Any DOS command.

Command Type and Version:

Batch command; Introduced with Ver 3.3

Notes:

1. Useful in preventing the words ECHO OFF from displaying on the screen when ECHO OFF is used in a Batch file. This command is useful if all screen echos need to be turned off in a Batch file.
2. See also ECHO.

BACKINFO.EXE Removed V6.0

MS-DOS utility: Allows viewing of files on a backup disk created by the DOS Version 3.3, 3.31, 4.0, 4.01, and 5.0 BACKUP command.

Syntax (shaded area optional):

BACKINFO drive1:

Example: backinfo b:

Syntax options:

drive1: . . . . Drive containing the BACKUP disk.

Command Type and Version:

External command, Introduced with Ver 3.3.
Removed from Ver 6.0

BACKUP.EXE

Removed V6.0

Back up files: Backs up files from one drive to another drive. Source and target drives may be either hard disks or floppy disks. DOSV6 use MSBACKUP.

Syntax (shaded is optional):

BACKUP Source:\Path\Filename Target: /s /m /a /d:date /t:time /f:size /L:LogDrive:\Path\Log

Examples: BACKUP C:*.* B: /s
BACKUP C:\DATA*.* B: /s /L:C:\LOG

Syntax Options:

Source:\Path. . . Source drive & directory to be backed up.

Filename. . . . . . Filename (s) to be backed up. Use of Wild cards (? and *) is allowed.

Target: . . . . Target drive for backed up files.

/s. . . . . Backs up all files in *Source:\Path* and subdirectories under *Source:\Path*

/m . . . . Backs up all files that have changed since the last backup (backup looks at the files archive attribute) and then turns off the files archive attribute.

/a. . . . . Adds new backup files to the existing backup disk (existing files are not deleted.) If a backup was made with DOS 3.2 or earlier, the /a switch is ignored.

/d:date. . . . . Only files created or modified after *date* are backed up. The way *date* is written depends on COUNTRY.SYS settings.

/t:time . . . . Only files created or modified after *time* are backed up. The way *time* is written depends on COUNTRY.SYS settings. Always use the */d:date* switch when */t:time* is used.

/f:size . . . . Format backup disk to the following *size*: (*size* can also be with k or kb, e.g. 160 can be 160k or 160kb; or 1200 can be 1200k, 1200kb, 1.2, 1.2m or 1.2mb, etc.)

size	**Disk size and type**
160	160k single sided DD 5.25" disk
180	180k single sided DD 5.25" disk
320	320k double sided DD 5.25" disk
360	360k double sided DD 5.25" disk
720	720k double sided DD, 3.5" disk
1200	1.2meg double sided HD, 5.25"
1440	1.44meg double sided HD, 3.5"
2880	2.88meg double sided, 3.5" disk

(DD=Double Density, HD=High Density)

/L: . . . . Creates a log file during a specific backup operation.

Logdrive:\Path . Drive & Directory where backup *Log* is to be sent.

Log . . . . Text file log of a backup operation.

Command Type and Version:

External command; Network; Introduced with Ver 2.0 Removed from Version 6.0, replaced with MSBACKUP.

Available in the MS-DOS 6.0 and 6.22 Supplemental Disks.

Notes:

1. See also RESTORE, COPY, XCOPY, DISKCOPY, IF
2. The sequence number of a backup disk can be checked by doing a DIR of the backup disk (Valid for version after DOS 3.3)
3. BACKUP does not backup the 3 system files, COMMAND.COM, MSDOS.SYS (or IBMDOS.SYS) , and IO.SYS (or IBMBIO.SYS).
4. BACKUP/RESTORE commands are not very compatible between pre DOS 5.0 version. DOS 5.0 will restore previous versions.
5. Do not use BACKUP when the ASSIGN, JOIN, or SUBST commands have been used.
6. When the IF ERRORLEVEL functions are used, BACKUP Exit Codes can be used to show why a backup failed (see IF):

Exit Code	Code Meaning
0	Successful backup
1	No files found to be backed up
2	File-sharing conflict, some files not backed up
3	BACKUP ter-minated by user with CTRL–C
4	Error termi-nated BACKUP procedure

7. Backup floppies are not readable by DOS, a special file format is used.

BASIC®.EXE and BASICA®.EXE

BASIC Computer Language: Depending on the system in use and version of DOS, it will run one of the BASIC interpreters (BASIC, BASICA, GW-BASIC, or QBASIC) and provide an environment for programming in the BASIC language. BASIC and BASICA are versions that were shipped with IBM® systems and were simply entry programs that started BASIC from the system's ROM. GW-BASIC is Microsoft's own version of BASIC that is shipped with MS-DOS versions through 4.01. For specifics on DOS 5.0/6.0 QBASIC, refer to page 231.

Syntax (shaded is optional):

BASIC Filename

Examples: BASIC Test.bas
BASICA

Syntax Options:

BASIC. . . . . BASIC without a filename just starts the BASIC Interpreter.

Filename. . . . . . A program written in BASIC that is loaded and run when the BASIC interpreter starts. The files normally end with .BAS

Command Type and Version:

External command; Network; Introduced with Ver 1.0

Notes:

1. See also QBASIC and GW-BASIC.

BREAK

Turns on/off the DOS check for Control-C or Control-Break: Determines when DOS looks for a Ctrl-C or Ctrl-Break more frequently in order to stop a program.

Syntax (shaded is optional):

BREAK on off

Examples: BREAK
BREAK = ON (syntax for CONFIG.SYS)
BREAK ON (syntax at DOS prompt)

Syntax Options:

BREAK . . . . BREAK, with no switches or options, displays the current setting of BREAK.

ON . . . . Tells DOS to check for Ctrl-C or Ctrl-Break from the keyboard, during disk reads and writes, and during screen and printer writes.

OFF . . . . Tells DOS to check for Ctrl-C or Ctrl-Break from the keyboard only during screen and printer writes.

Command Type and Version:

Internal command; CONFIG.SYS and Batch command; Introduced with Ver 2.0

Notes:

1. If BREAK is ON, your system will run slightly slower.
2. The default setting is BREAK=OFF.

BUFFERS

Sets number of disk buffers in memory: A disk buffer is a block of RAM memory that DOS uses to hold data while reading and writing data to a disk.

Syntax (shaded is optional):

BUFFERS = X ,Y

Examples: BUFFERS = 35
BUFFERS = 35,8

Syntax Options:

X . . . . The number of disk buffers allocated. The total may range from 1 to 99 for versions Ver 4.0 to 6.2x. Versions prior to 4.0 can be in the range from 2 to 255. Default values are as follows:

Buffers	Drive Configuration
2	. . . <128K RAM & 360k drive only
3	. . . <128K RAM & Disks over 360K
5	. . . 128K to 255K RAM
10	. . . 256K to 511K RAM
15	. . . 512K or more RAM

Y . . . . The number of secondary cache buffers. The total may range from 1 to 8, the default is 1.

Command Type and Version:

CONFIG.SYS command; Introduced with Ver 2.0

Notes:

1. Each buffer takes up approximately 532 bytes of RAM.
2. Standard buffer sizes should range from 20 to 30, unless more are required by a specific application (such as Dbase III Plus®)
3. If a disk cache program, such as SMARTDRV.SYS is used, the number of buffers can be set at 8 to 15 (sometimes lower).
4. In Ver 5.0, if DOS is in high memory, buffers are also in high mem.
5. The number of buffers (up to 35) significantly affects system speed; over 35, speed still increases but at much slower rate.
6. /X switch from earlier DOS versions is no longer available.

CALL

Calls a batch program: Starts one batch program from inside another batch program, without causing the initial batch program to stop.

Syntax (shaded is optional):

CALL Drive:\Path\ Filename Parameters

Examples: CALL C:\TEST %1

Syntax Options:

Drive: . . . . Letter of drive containing path.

\Path . . . . Path containing filename.

Filename. . . . . . Filename specifies name of the batch program to be called. *Filename* must have a .BAT extension.

Parameters . . . . Specifies command-line information required by the batch program, including switches, filenames, pass through parameters such as %1, and variables.

Command Type and Version:

Internal command; Batch; Introduced with Ver 3.3

Notes:

1. Any information that can be passed to a batch program can be contained in the *Batch-parameters*, including switches, filenames, replaceable parameters %1 through %9, and variables such as % Parity %
2. Pipes and redirection symbols cannot be used with CALL.
3. If a recursive call (a program that calls itself) is created, an exit condition must be provided or the two batch programs will loop endlessly.

CD or CHDIR

Change directory: Changes (moves) to another directory or shows the name of the current directory path.

Syntax (shaded is optional):

CD Drive:\Path

Examples: CD (displays current drive and directory)

CD D:\PFS (change to PFS directory on D: drive)

CD\ (changes to root directory)

Syntax Options:

Drive: . . . . Drive containing the subdirectory to be changed. CD does not move to Drive:, it remains on the current drive.

\Path . . . . Directory path name to be made current, if *Drive:* is the current drive. If *Drive:* is not the current drive, *\Path* is simply the active path on *Drive:* and the current drive and directory remain unchanged. Pathname can be no longer than 63 characters and (\) is to be used as the path's first character to move to the root directory.

Command Type and Version:

Internal command; Network; Introduced with Ver 2.0

Notes:

1. When a drive letter is not specified, the current drive is assumed.
2. **CD . .** specifies move up one directory level.

CHCP

Change code page: Displays or changes the number of the active code page for the command processor COMMAND.COM.

Syntax (shaded is optional):

CHCP *ccc*

Examples: CHCP (reports current *ccc* setting)
CHCP 863

Syntax Options:

ccc . . . . These are the numbers that represent the prepared system code pages defined by the COUNTRY.SYS command in the CONFIG.SYS file. Valid code page numbers are as follows:

437 . . . . United States
850 . . . . Multilingual (Latin I)
852 . . . . Slavic (Latin II)
860 . . . . Portuguese
863 . . . . Canadian-French
865 . . . . Nordic

Command Type and Version:

Internal command; Network; Introduced with Ver 3.3

Notes:

1. Once a specified code page has been selected, all programs that are started will use that new code page.
2. NLSFUNC (national language support functions) must be installed before a code page can be switched with CHCP.
3. MODE SELECT can also be used to change code pages.
4. See also DOS commands COUNTRY.SYS, NLSFUNC, DEVICE, and MODE.

CHKDSK.EXE

Checks disk: Scans the disk and reports size, disk memory available, RAM available and checks for and corrects logical errors. A status report is displayed on screen.

Syntax (shaded is optional):

CHKDSK Drive:\Path\Filename / f / v

Examples: CHKDSK C: / f
(If no Drive: is specified, the current drive is used.)

Syntax Options:

Drive: . . . . Drive letter of the disk to be checked.
\Path . . . . Directory path containing file to be checked.
Filename. . . . . . Name of file to be checked by CHKDSK for fragmentation. Wildcards * & ? are allowed.
/ f. . . . . Fixes logical errors on the disk.
/ v . . . . Verbose switch. Displays CHKDSK progress by listing each file in every directory as it is being checked.

Command Type and Version:

External command; Can NOT check a Network drive; Introduced with Ver 1.0

Notes:

1. CHKDSK analyzes a disk's File Allocation Table (FAT) and file system. / f must be specified in order to fix errors. If / f is not used, CHKDSK reports the error, but does not fix the error, even if you answer yes to fixing the error at the CHKDSK prompt.
2. When CHKDSK / f finds an error, it asks if you want to convert the "lost clusters" to files. If you answer Yes, files in the form FILE0001.CHK are created and the lost areas dumped into those files. You must then determine if any valuable information is in that file. If they don't contain useful information, delete them.
3. Do not use CHKDSK from inside any other program, especially Windows.
4. Only logical errors are repaired by CHKDSK, not physical errors.
5. CHKDSK will not work when SUBST, JOIN or ASSIGN has been used.

CHKSTATE.SYS

New V6.2

CHKSTATE is used only by MemMaker to track the memory optimization process: During the memory optimization process, MemMaker adds CHKSTATE.SYS to the beginning of the CONFIG.SYS file. When the memory optimization process is complete, MemMaker automatically removes CHKSTATE.SYS.

CHOICE.COM

Pauses the system and prompts the user to make a choice in a batch file: This command can only be used in batch programs.

Syntax (shaded is optional):

CHOICE /C:keys /N /S /T:c,nn text

Syntax Options:

/C:keys .... Defines which keys are allowed in the prompt. The : is optional. Displayed keys are separated by commas and will be enclosed in [ ] brackets. Multiple keystroke characters are allowed. Default is [YN] (yes/no).

/N .... Prevents display of prompt, but the specified keys are still valid.

/S .... Specifies that CHOICE is case sensitive.

/T:c,nn .... Forces CHOICE to pause for *nn* seconds before defaulting to a specified key *(c)*. nn can range from 0 to 99. The *c* key specified must be included in the */C:keys* definition.

text . . . . Defines what text is displayed before the prompt. Quotation marks (" ") must be used if a "/ " character is included in the prompt. Default for CHOICE is no text displayed.

Command Type and Version:

Internal Batch command; Network;
Introduced with Ver 6.0

Notes:

1. ERRORLEVEL 0 is returned if Control-C or Control-Break is pressed.

CLS

Clears or Erases Screen: All information is cleared from the DOS screen and the prompt and cursor is returned to the upper left corner of the screen.

Syntax:

CLS

Examples: CLS

Syntax Options:

None

Command Type and Version:

Internal command; Network; Introduced with Ver 2.0

Notes:

1. Screen colors set by ANSI.SYS will remain set.

2. If more than one video display is attached to the system, only the active display is cleared.
3. If ANSI.SYS is not loaded on the system, CLS will clear the screen to gray (or amber on an amber monitor, etc.) on black.

COMMAND.COM

Start a new DOS command processor: The command processor is responsible for displaying the prompt on the computer's display and contains all of the Internal DOS commands. It is also used to set variables such as environment size. Use the EXIT command to stop the new processor.

Syntax (shaded is optional):

COMMAND Drive:\Path\Device / e:xxxx / y / c text / k

In CONFIG.SYS use the following:

SHELL = Drive:\Path\ COMMAND.COM / e:xxxx / p / msg

Examples: COMMAND /e:1024
(use the following in CONFIG.SYS with SHELL)
SHELL = Drive:\Path\COMMAND.COM /e:512 /p

Syntax Options:

Drive:\Path . . . . Drive and \Path of the command device. Must be included if COMMAND.COM is not located in the root directory.

\Device . . . . Device for command input or outpur (see the CTTY command on page 116).

/ e:xxxx . . . . Set environment size in bytes (xxxx). Default for Ver 5.0,6.0, and 6.2x = 256 bytes; default for versions before 5.0 is 160 bytes.Range is 160 to 32768 bytes.

/p . . . . Makes the new command processor the permanent processor. Used only with SHELL command.

/c text . . . . Forces the command processor to perform the commands specified by *text*. On completion, it returns to the primary command processor. Must be last switch on command line.

/msg. . . . . Causes error messages to be stored in memory. The */p* switch must also be used when */msg* is used.

/k ❻. . . . . Execute a command, but after the command is executed, do not terminate the second COMMAND.COM that is running. Must be last switch on command line

/y 6.2. . . . . Tells COMMAND.COM to step through files specified by the /c text or /k switches

Command Type and Version:

External command;

CONFIG.SYS command when used with SHELL;

Introduced with Ver 1.0

Notes:

1. See also CTTY, EXIT and SHELL
2. Default environment sizes are commonly not large enough. Try setting the environment to 512 or 1024.
3. In Version 6.0, if DOS is unable to find COMMAND.COM, a warning message is issued that allows the user to "Enter correct name of Command Interpreter (e.g., C:\COMMAND.COM). This is a much improved error handling function and allows the system to complete the booting process.
4. Exercise caution when you are "messing around" with COMMAND.COM. It can get the user into some dangerous situations!
5. The SHELL command in CONFIG.SYS is the preferred method of increasing the environment size with the /e:xxxx switch.

COMP.EXE

Removed V6.0

Compare files: Compares the contents of two sets of disk files to see if they are the same or different. The comparison is made on a byte by byte basis. COMP displays filenames, locations and the differences found during the compare process .

Syntax (shaded is optional):

COMP Drive1:\Path1\File1 Drive2:\Path2\File2 /d /a /L /n=xx /c

Examples: COMP (prompts for file locations)
COMP C:\File1 D:\File2 /a

Syntax Options:

Drive1: Drive2: . Letters of drives containing the file (s) to be compared.

\Path1 \Path2 . . Paths of files to be compared.

File1 File2. . . . . Filenames to be compared. The names may be the same if they are in different locations. Wild cards (*?) are allowed.

/d. . . . . Displays file differences in decimal format, the default format is hexadecimal. Ver 5

/a. . . . . File differences displayed as characters. Ver 5.0

/L. . . . . Display Line numbers with different data instead of byte offsets. Ver 5.0

/n=xx. . . . . Compares the first number of lines (*xx*) in each file, even if files are different sizes. Ver 5.0

/c. . . . . Upper and lower case is ignored. Ver 5.0

Command Type and Version:

External command; Network; Introduced with Ver 1.0
Removed from Ver 6.0, replaced by FC.
Available on the MS-DOS 6.0, 6.21, and 6.22 Supplemental Disks.

Notes:

1. If the drive, path and filename information is not specific enough, COMP will prompt for the correct information
2. If more than 10 mismatches are found, COMP ends the compare.
3. See also DISKCOMP (for floppy disk comparisons) and FC.

COPY

Copies file(s) from one location to another: Files can also be combined (concatenated) using COPY.

Syntax (shaded is optional):

COPY /y / -y /a /b Source /a /b+Source /a /b +. .
Target /a /b /v

Examples: COPY C:\Test*.* D:\Test2
COPY Test1.txt + Text2.txt Test3.txt /a

Syntax Options:

Source . . . . Source Drive, Directory, and File(s) or Devices to be copied **from**.

Target . . . . Destination Drive, Directory, and File(s) or Devices being copied **to**.

/a . . . . Denotes an ASCII text file. If */a* preceeds a filename, that file and all following files are treated as ASCII files until a */b* switch is encountered, then files that follow are considered to be binary files. If */a* follows a filename, it applies to all files before and after the */a* until a */b* switch is encountered, then files that follow are considered to be binary files.

/b . . . . Denotes a Binary file. If */b* preceeds a filename, that file and all following files are treated as binary files until a */a* switch is encountered, then files that follow are considered to be ASCII files.

If */b* follows a filename, it applies to all files before and after the */b* until a */a* switch is encountered, then files that follow are considered to be ASCII files. /b forces copy to read exactly the number of bytes allocated to the file's size in the directory.

/ v . . . . Verifies files were copied correctly.

/ y 6.2 . . . . Directs COPY to replace existing file without confirmation prompt. Confirmation prompt is default.

/-y 6.2 . . . . Directs COPY to ask for confirmation prior to replacing existing files.

Command Type and Version:

Internal command; Network; Introduced with Ver 1.0

Notes:

1. COPY will only copy the contents of 1 directory. If a directory and its subdirectories need to be copied, use the XCOPY command.
2. COPY will not copy files 0 bytes in length, use XCOPY instead.
3. Both *Source* and *Destination* can be a device such as COM1: or LPT1:, however, when sending to *Destination*, if the */b* switch is used, all characters, including control codes, are sent to the device as data. If no switch is used, the data transfers as ASCII data and the transmitted control codes may perform their special function on the device. For example, if a Ctrl + L code is sent to a printer on LPT1:, the printer will form feed.
4. If Destination Filename is not specified, COPY will create a file with the same name and date and time of creation in the current directory (*Target*). If a file with the same name as *Filename* exists in the current directory, DOS will not copy the file and display an error message that says "File cannot be copied onto itself. 0 Files Copied".
5. If the + function is used to combine files, it is assumed that the files are ASCII files. Normally you should NOT combine binary files since the internal format of binary files may be different.
6. */v* slows down the copy process. If a verify error occurs, the message is displayed on the screen.
7. In order to change the date and time of a file during the copy process, use the following syntax:

   COPY /b Source + , ,
8. See also DISKCOPY and XCOPY.

COUNTRY and COUNTRY.SYS

Country dependent information: Enables DOS to use international time, date, currency, and case conversions.

Syntax (shaded is optional):

COUNTRY= ccc ppp Drive:\Path \Filename

Examples: COUNTRY = 002

Syntax Options:

ccc . . . . Country code number. Default 001, USA
ppp . . . . Code page number.
Drive:\Path . . . . Drive & subdirectory containing *Filename.*
Filename. . . . . . File containing country information.

Command Type and Version:

CONFIG.SYS; Introduced with Ver 3.0

Notes:

1. COUNTRY is put in CONFIG.SYS . If the *Drive:\Path\Filename* option is not used to specify which file contains country information, COUNTRY.SYS must be in the root directory of the system's boot drive so that COUNTRY can retrieve the country data.

Country Code	Country or Language	Code Page	Time Format	Date Format
001	United States	437, 850	2:35:00.00p	06-30-1991
002	Canadian-French	863, 850	14:35:00,00	1991-06-30
003	Latin America	850, 437	2:35:00.00p	30/06/1991
031	Netherlands	850, 437	14:35:00,00	30-06-1991
032	Belgium	850, 437	14:35:00,00	30/06/1991
033	France	850, 437	14:35:00,00	30.06.1991
034	Spain	850, 437	14:35:00,00	30/06/1991
036	Hungary	852, 850	14:35:00,00	1991-06-30
038	Croatia\Slovenia\ Yugoslavia\Serbia	852, 850	14:35:00,00	1991-06-30
039	Italy	850, 437	14.35.00,00	30/06/1991
041	Switzerland	850, 437	14,35,00.00	30.06.1991
042	Czech Rep\Slovakia	852, 850	14:35:00,00	1991-06-30
044	United Kingdom	437, 850	14:35:00.00	30/06/1991
045	Denmark	850, 865	14.35.00,00	30-06-1991
046	Sweden	850, 437	14.35.00,00	1991-06-30
047	Norway	850, 865	14:35:00,00	30.06.1991
048	Poland	852, 850	14:35:00,00	1991-06-30
049	Germany	850, 437	14:35:00,00	30.06.1991
055	Brazil	850, 437	14:35:00,00	30/06/1991

061	Intl. English	437, 850	14:35:00.00	30-06-1991
351	Portugal	850, 860	14:35:00,00	30-06-1991
358	Finland	850, 437	14.35.00,00	30.06.1991

CTTY

Change to a remote console: Allows you to choose the device from which you issue commands. USE WITH CAUTION, you could lose control of your system!

Syntax (shaded is optional):

CTTY Device

Examples: CTTY aux
CTTY com1
CTTY con

Syntax Options:

Device. . . . . Any valid DOS device for issuing commands. Examples include com1, com2, com3, com4, con, aux, prn (rare).

Command Type and Version:

Internal command; Network; Introduced with Ver 2.0

Notes:

1. *Device* refers to a character-oriented remote unit, or secondary terminal, that will be used for command input and *output*. This device name must be a valid MS/PC-DOS name, specifically, AUX, COM1, COM2, COM3, COM4, CON. The use of a colon after the device name is optional.
2. *ctty con* moves the input and output back to the main terminal (the local console screen and keyboard).
3. *When redirected, some programs that are designed to work with the video display's control codes may not function correctly.*
4. Other redirected IO or piping is not affected by CTTY.
5. CAUTION: the command CTTY NUL will disconnect the screen and keyboard !!!! Do not use unless the CTTY CON command is executed under some type of program control, such as a batch file.

CV.COM and CV.EXE **Removed V6.2**

CV starts the CodeView program: CodeView is a debugging utility for programs written in C.

Command Type and Version:

External command, Introduced with Ver 5.0.

Removed Ver. 6.2.

Available in the MS-DOS 6.0, 6.21, and 6.22 Supplemental Disks.

Notes:

1. CAUTION- Using CodeView CV.EXE Versions 3.0 to 3.13 with a 80386 memory manager such as EMM386 may cause loss of data. This problem has been fixed in Version 3.14 of CodeView. To start CodeView Versions 3.0 to 3.13 safely, use CV.COM.
2. Use HIMEM.SYS Version 2.77 or later with CodeView.

DATE

Date: Change and /or display the system date. (Note: This does not reset the computer's battery powered clock if DOS 3.21 or earlier is used.)

Syntax (shaded is optional):

DATE month-day-year

Examples: **date mm-dd-yy** (for North America)
Note: If COUNTRY in config.sys is set for a country other than a North American country, then the following syntax is used:

DATE dd-mm-yy for Europe
DATE yy-mm-dd for Far East

Syntax Description and Options:

month . . . . One or two digit number (1 to 12)

day . . . . One or two digit number (1 to 31). DOS knows the correct number of days in each month (28, 29, 30 or 31).

year. . . . . Two or four digit number (80 to 99 – The 19 is assumed for 1980 to 1999).

Command Type and Version:

Internal command; Network; Introduced with Ver 1.0

Notes:

1. You may separate the day, month and the year by the use of hyphens, periods or slashes.
2. If a system does not have an AUTOEXEC.BAT file in the root directory of the boot drive, the date and time functions are activated automatically when the system starts and the user is prompted for change or confirmation.
3. DOS has been programmed to change the year, month and day and adjusts the number of days in a month accordingly. DOS also knows which months have 28, 29, 30, or 31 days. DOS will issue errors if valid dates are not used.
4. Beginning with DOS 3.3, DATE and TIME both set the system's CMOS (battery powered) calendar (except in XT class systems)
5. See also TIME

DBLBOOT.BAT **Removed V6.22**

Creates a bootable DBLSPACE floppy disk:

Syntax:

DBLBOOT drive1:

Example: dblboot a:

Syntax options:

drive1:. . . . . Drive containing floppy disk to be compressed.

Command Type and Version:

External command, Introduced with MS-DOS Ver 6.0.

Available in the MS-DOS 6.0, 6.21, and 6.22 Supplemental Disks; Removed in version 6.22.

Notes:

1. DBLBOOT works only on high-density floppy disks (1.44 or 1.2 MB).
2. DBLSPACE must be installed prior to using DBLBOOT.

DBLSPACE.EXE

New V6.0 Danger V6.0 Removed V6.22

Utility to compress both hard and floppy disk drives so that there is more available storage space on the drive: Once the .EXE program has been run, DBLSPACE.SYS must be included in CONFIG.SYS. ***Many problems have been reported with the DOS 6.0 version of this program. USE WITH CAUTION or not at all, you could lose data on your drive!***

Syntax (shaded is optional):

DBLSPACE /Automount /Chkdsk /Compress /Convstac /Create /Defragment /Delete /Format /Info /List /Mount /Ratio /Size /Unmount

Syntax Options:

/Automount . . . . Automatically mount a compressed disk.

/ Chkdsk . . . . . . Check the validity of a compressed disk's directory and FAT and report the status of the drive.

/ Compress . . . . Start the compression process on a drive.

/ Convstac.	Removed V6.2 Converts a Stacker compressed drive to a DBLSPACE compressed drive.
/ Create.	Creates a new compressed drive in the free space of an existing drive.
/ Defragment. . .	Defragment the files on an existing drive.
/ Delete	Remove a compressed drive.
/ Format	Format a compressed drive.
/ Info	Display detailed information on a compressed drive.
/ List	Display a list of both compressed and uncompressed drives on a system. It does not report network drives.
/ Mount	Mount a compressed drive.
/ Ratio	Display and change the estimated compression ratio of a compressed drive.
/ Size.	Change the size of a compressed drive.
/ Uncompress 6.2	Uncompresses a drive compressed by DBLSPACE.
/ Unmount.	Unmount a compressed drive.

Command Type and Version:

External command; Introduced with Ver 6.0

Removed with Ver. 6.2, revision 2, and replaced by DRVSPACE.

Notes:

1. DBLSPACE can be run as a menu driven utility or with the command line switches listed under Syntax Options.
2. The maximum size of a DBLSPACE volume is 512 MB.
3. Default cluster size of a compressed volume is 8k.
4. When DBLSPACE.EXE is run, DBLSPACE.SYS is automatically placed in CONFIG.SYS as part of the installation process.
5. See Also DBLSPACE.SYS

DBLSPACE.SYS

New V6.0 **Danger V6.0** **Removed Ver 6.22**

Device driver that activates a compressed drive: DBLSPACE.SYS determines the final memory location of DBLSPACE.BIN, which provides access to the compressed drives. ***Many problems have been reported with the DOS 6.0 version of this program. USE WITH CAUTION or not at all, you could lose data on your drive!***

Syntax (shaded is optional):

DEVICE = Drive:\Path\ DBLSPACE.SYS
/ Move / Nohma

Examples: DEVICE = C:\DBLSPACE.SYS

It may also be loaded high using:
DEVICEHIGH = C:\DBLSPACE.SYS / Move

Syntax Options:

Drive:\ Path . . . Drive and Path of the DBLSPACE.SYS

/ Move. . . . . Moves the DBLSPACE.BIN file to a different location in memory. By default it is loaded at the top of conventional memory. /Move moves it to the bottom of conventional memory. Note that if DEVICEHIGH is used, it can be moved to upper memory, thereby freeing up conventional memory.

/ Nohma . . . . . . Tells DBLSPACE.SYS not to move DBLSPACE.BIN into high memory.

Command Type and Version:

CONFIG.SYS command; Introduced with Ver 6.0
Removed with Ver. 6.2, revision2, and replaced by DRVSPACE

Notes:

1. DBLSPACE can be run as a menu driven utility or with the command line switches listed under Syntax Options.
2. DBLSPACE.SYS is automatically inserted into CONFIG.SYS when the DBLSPACE.EXE installation program is run.
3. See also DBLSPACE.EXE and DEVICEHIGH.

DEBUG.EXE

Starts a debugging program: Debug is a program that provides a testing environment for binary and executable programs, i.e. all programs that have .EXE or .COM extensions . It is also commonly used to run executable programs that are in memory, such as a hard drive's setup program stored in ROM on a hard drive controller. The full use of DEBUG is beyond the scope of this book. Refer to books such as Microsoft's *DOS Manuals* or PC Magazine's *DOS Power Tools*.

Syntax (shaded is optional):

There are two methods of starting DEBUG.

Method 1:

DEBUG Drive:\Path\ Filename Parameter

Method 2:

DEBUG

Examples:

Method 1: DEBUG C:\test.exe

Method 2: DEBUG (run in command line mode)

Syntax Options:

Method 1:

Drive:\Path . . . . Drive and Path of the executable *Filename* to be tested.

Filename. . . . . . Name of executable file to be tested.

Parameter. . . . . Command line information needed by *Filename*.

Method 2:

Debug. . . . . Starts DEBUG in the command line mode where debug commands are given at the DEBUG hyphen prompt (**–**).

Command Type and Version:

External command; Introduced with Ver 1.0

Debug Commands for Method 2:

Case makes no difference; *address* and *range* is in hex

? . . . . Display list of all DEBUG commands.

A address . . . . . Assemble 8086/8087/8088 mnemonics directly into memory at *address* (hex).

C range address Compares contents of two memory blocks. *range* is the starting and ending address or starting address and length of Block 1 and *address* is the starting address of Block 2.

D range . . . . Dump (display) contents of memory with starting and ending addresses of *range.*

E address data . Enter data into memory starting at *address. data* is entered into successive bytes of memory.

F range data . . . Fill memory with *data* (hex or ASCII) in starting and ending addresses or starting address & length defined by *range*.

G=address bkp . Run program in memory starting at *address. bkp* defines 1 to 10 temporary breakpoints.

H hex1 hex2 . . . Does hexadecimal math on *hex1* & *hex2.* Two results are returned, first the sum of *hex1* and *hex2*; second, *hex1* minus *hex2*.

I port . . . . Read (input) & display 1 byte from *port.*

L address drive:start number . . . . .Load a file or specific drive sectors into memory. *address* is the memory location you want to load to. *drive* contains the sectors to be read.

start is the hex value of the first sector to be read. *number* is the number of consecutive sectors to load.

M range address. . . Copies memory contents from the starting and ending address or starting address and length of *range*. *address* is the starting address of the destination.

N d:\path\file parameters . . . . Name the *drive:\path\filename* of an executable file for Debug *L* or *W*. Also used to specify *parameters* for the executable file. *N* by itself clears the current specification.

O port data . . . . Output *data* to a *port* (by address).

P=address value Run a loop, string instruction,subroutine, or software interrupt starting at *address* and for *value* number of instructions.

Q. . . . . Stop DEBUG without saving the file being tested. Returns to DOS.

R register. . . . . . Display or alter CPU (central processing units) *register*. *R* by itself displays contents of all registers.

S range data. . . Search for *data* at the beginning and ending address of *range*.

T=address value Trace instructions starting at *address* and for *value* number of instructions.

U range . . . . Unassemble code at the start & end address or start address & length of *range*.

W address drive:start number . . . .Write a file or specific drive sectors into memory. *address* is the memory location you want to write to. *drive* contains the sectors to be written. *start* is the hex value of the first sector to be written. *number* is the number of consecutive sectors to write.

XA count . . . . . . Allocate count number of 16k expanded memory pages.

XD handle . . . . . Deallocate a handle to expanded memory

XM Lpage Ppage handle . . . . . . .Map a *Lpage* logical page of expanded memory belonging to *handle*, to a *Ppage* physical page of expanded memory.

XS. . . . . Display status information of expanded memory.

DEBUG ERROR MESSAGES: BF=Bad Flag; BP=Too many breakpoints; BR=Bad Register; DF=Double Flag

DEFRAG.EXE New V6.0

Reorganizes or defragments a disk in order to optimize disk drive performance.

Syntax (shaded is optional):

DEFRAG Drive: /F /U /S:order /B /Skiphigh /LCD /BW /GØ /A /H

Examples: DEFRAG C: /U /B

Syntax Options:

Drive:. . . . . Drive letter to be defragmented.

/ F . . . . Insures that no empty disk space remains between files.

/ U . . . . Leaves empty space, if any, between files.

/ S:order. . . Sort files in a specific sort *"order"*.
- N. . . In alphabetic order by name
- -N . . In reverse alphabetic name order
- E. . . In alphabetic order by extension
- -E . . In reverse alphabetic order by extension
- D. . . By date & time, earliest first
- -D . . By date & time, latest first
- S. . . By size, smallest first
- -S . . By size, largest first

/ B . . . . Reboot system after DEFRAG is done.

/ *Skiphigh*. . Load DEFRAG into conventional memory, instead of the default upper memory
/ *LCD*. . . . . Start DEFRAG in LCD color scheme mode.
/ *BW*. . . . . . Start DEFRAG in black & white color mode.
/ *GØ*. . . . . . Disable graphics mouse and character set.
/ *A* . . . . Start DEFRAG in Automatic mode.
/ *H* . . . . Moves hidden files.

Command Type and Version:

External command; Network; Introduced with Ver 6.0

Notes:

1. Do not use DEFRAG while Windows is running.
2. DEFRAG exit codes (ERRORLEVEL parameter) are:

0	Successful deframentation.
1	Internal error.
2	No free clusters, DEFRAG needs at least 1 free cluster.
3	Process aborted with CTRL+C by user.
4	General error.
5	Error occurred while reading a cluster.
6	Error occurred while writing a cluster.
7	Allocation error, correct using SCANDISK.
8	Memory error.
9	Insufficient memory for defragmentation.

DEL or ERASE

Delete or Erase: Deletes specified files from a directory.

Syntax (shaded is optional):

DEL Drive:\Path\ **Filename** /p

Examples: DEL *.*
DEL *.exe
DEL C:\budget\1990 /p
ERASE C:\Bin*.dbf

Syntax Options:

Drive: . . . . Drive letter containing *Path*
Path . . . . Subdirectory containing *Filename*
Filename . . . . . Filename(s) to be deleted.
/P . . . . Screen prompts user for confirmation of the file(s) to be deleted.

Command Type and Version:

Internal command; Network; Introduced with Ver 1.0

Notes:

1. Use of wildcards * and **?** is allowed. Use DEL *.* with caution, it will delete all files in the current directory. If you happen to be in the root directory of your boot drive when DEL *.* is used, COMMAND.COM, AUTOEXEC.BAT, CONFIG.SYS, etc will be deleted and the system will probably not start.
2. Files may be UNDELETED in DOS Versions 5.0, 6.0, and 6.2x..
3. See also RMDIR, MIRROR, and UNDELETE.

DELOLDOS.EXE

Removed V6.2

Directs DOS to delete the OLD_DOS directory: During setup (installation) DOS moves any previous DOS version files to a directory called OLD_DOS. The DELOLDOS command deletes the OLD_DOS directory and all contained files.

Syntax:

DELOLDOS

Examples: deloldos

Syntax options: None

Command Type and Version:

External command, Introduced Ver 6.0. Removed 6.2

Notes:

1. Deloldos should be the last step in the installation process for DOS Ver 6.0. When finished, DELOLDOS also deletes itself!

DELTREE.EXE New V6.0

Deletes a directory and all the files and subdirectories that are in it: Exercise caution when using this command.

Syntax (shaded is optional):

DELTREE /Y Drive:\Path\Filename

Examples: DELTREE /Y A:*.*
DELTREE /Y C:\DATA

Syntax Options:

Drive: . . . . Drive letter containing *\Path*
\ *Path* . . . . Subdirectory containing *\Filename*
\ *Filename*. . . . . Filename(s) to be deleted.
/Y. . . . . Completes DELTREE without first prompting for confirmation of the deletion. Don't use this switch if you can avoid it.

Command Type and Version:

External command; Network; Introduced with Ver 6.0

Notes:

1. If a filename is not specified, all files and subdirectories in the Drive:\Path are deleted.
2. Wild card are supported in the filenames.
3. Attributes such as read only, system and hidden are ignored when a filename is specified.
4. See also DEL and RMDIR.

DEVICE

Loads a device driver into memory: Device drivers are loaded by way of CONFIG.SYS.

Syntax (shaded is optional):

DEVICE = Drive:\Path\ Filename Parameters

Examples: DEVICE = C:\Dos\Himem.sys
DEVICE = Smartdrv.sys 1024 512

Syntax Options:

Drive:\Path . . . . Drive and directory(s) containing *Filename.*

\Filename . . . . . Driver to be loaded.

Parameters . . . . Switches and/or parameters needed by the device driver.

Command Type and Version:

CONFIG.SYS command; Introduced with Ver 2.0

Notes:

1. Standard installable device drivers are: ANSI.SYS, DISPLAY.SYS, DRIVER.SYS, EGA.SYS, PRINTER.SYS, RAMDRIVE.SYS, EMM386.EXE, HIMEM.SYS, and SMARTDRV.SYS. SMARTDRV.SYS is in DOS 5.0 only, SMARTDRV.EXE replaced it first in Windows and then in DOS 6. Other device drivers, such as SETVER and DBLSPACE or DRVSPACE may also be loaded.
2. COUNTRY.SYS and KEYBOARD.SYS are files, not device drivers. Do not try to load either of these files using the DEVICE command or your system will lock up and DOS will not be able to restart.
3. When new devices are purchased, such as a mouse or scanner, you will usually receive device driver software. Use DEVICE to install these drivers, making certain that the device driver is in the specified directory.
4. Install third party console drivers before DISPLAY.SYS.
5. See also DEVICEHIGH.

DEVICEHIGH

Load a device driver into upper memory: After DOS=umb and HIMEM.SYS have been loaded in CONFIG.SYS, DEVICEHIGH makes it possible to load device drivers into the upper memory area. Loading devices high will free up conventional memory for other programs.

Syntax (shaded is optional):

DEVICEHIGH = Drive:\Path\ Filename dswitch
or
DEVICEHIGH SIZE=hexsize Drive:\Path \Filename dswitch

DEVICEHIGH /L:(see below) / S Drive:\Path \Filename dswitch ❻

Examples: DEVICEHIGH = C:\Filename.sys
DEVICEHIGH SIZE=FF C:\Filename.sys

Syntax Options:

Drive:\Path . . . . Drive and Path of driver to be loaded high.

Filename. . . . . . Device driver to be loaded high.

dswitch . . . . Command line switches required by the device driver being loaded.

SIZE= *hexsize* . Minimum number of bytes (in hex) that must be available for DEVICEHIGH to try to load a driver in high memory. Ver 5

/ L:region1[,minsize1][;region2[,minsize2] ❻. . .
This switch specfies one or more memory regions into which to load a device driver. Normally, DOS loads the driver into the largest free UMB. / L allows a specific region to be selected. See your DOS manual for detailed information on using this switch.

/ S ❻. . . . . *Use / S only in conjunction with / L.* / S shrinks the UMB to its minimum size while a driver is loading and therefore makes the most efficient use of memory.

Command Type and Version:

CONFIG.SYS command; Introduced with Ver 5.0

Updated with different switches in Ver 6.0

Notes:

1. DOS=umb and HIMEM.SYS must be loaded before DEVICE-HIGH in order to function. The following is typical in CONFIG.SYS:

   DEVICE = C:\HIMEM.SYS

   DOS = umb

   DEVICE = C:\DOS\EMM386.EXE

   DEVICEHIGH = C:\Filename.sys

   As the example shows, EMM386.EXE or a comparable third-party product must be loaded before DEVICEHIGH will work. See DOS for more information.
2. If the driver being loaded high requires more high memory than is available, the system may lock-up. Use SIZE= to specify the memory required by the driver, after determining how much memory the driver normally takes by using MEM /DEBUG.
3. See also DOS, LOADHIGH, HIMEM.SYS and EMM386.
4. In MS-DOS Ver 6.0, see also MEMMAKER.

DIR

Directory: Displays the list of files and subdirectories within the current or a designated directory.

Syntax (shaded is optional):

DIR Drive:\Path\Filename /p /w /a:attrib / o:sort /s /b /L /c (hd)

Examples: DIR or DIR *.* (wild cards are allowed)
DIR *.exe /p

Syntax Options:

Drive:\Path . . . . Drive and subdirectory to be listed

\Filename . . . . . File name(s) and/or extension to display.

/p	Displays one screen of information, then pauses until any key is pressed.
/w.	Displays a wide screen list of files and subdirectories, but the file creation date & time, file size, and <DIR> subdirectory indicator are not shown.
/a : attrib	Displays only files with *attrib* attributes: h=hidden, –h=not hidden, s=system, –s=not system, d=directories, –d=files, a=files ready for archive, –a=files not changed, r=read only, –r=not read only. Introduced with Ver 5.0
/o : sort.	Displays by *sort* order: n=alphabetic by name, –n=reverse alphabetic, e=alphabetic by extension, –e=reverse extension alphabetic, d=earliest date/time 1st, –d=latest date/time 1st, s=smallest first, –s=largest 1st, g=group directories before files, –g=group directories after. Introduced with Ver 5.0 ❻*c=compression ratio (least compressed first), -c=compression ratio (most compressed first)*
/s	Show all occurrences in both the current directory and all subdirectories below it. Introduced with Ver 5.0
/b	Displays directory 1 line at a time. Ver 5
/L	Displays unsorted names in lowercase. Introduced with Ver 5.0
/c (hd) ❻	*Displays compression ratio. The optional* (hd) switch displays compression ratio of DBLSPACE files based on cluster size of host drive. If / w or / b switches are used, / c (hd) is ignored.

Command Type and Version:

Internal command; Network; Introduced with Ver 1.0

Notes:

1. The date and time formats displayed by the ***DIR*** command will vary, depending on which COUNTRY code is in CONFIG.SYS.
2. Place SET DIRCMD=/o /p in your Autoexec.bat file to force DIR to always sort and paginate the directory listing.

DISKCOMP.COM

Compares Disks: Compares the contents of the floppy disk in the Source drive to the contents of the floppy disk in the Target drive.

Syntax (shaded is optional):

DISKCOMP Source: Target: /1 /8

Examples:
DISKCOMP (first floppy disk drive is used)
DISKCOMP A: B: /1

Syntax Options:

Source: .... Source drive containing one of the floppy disks to be compared.
Target: .... Target drive containing the other disk to be compared.
/1 .... Compares only the first side of disks.
/8 .... Compares first 8 sectors per track.

Command Type and Version:

External command; Not for network.
Introduced with Ver 1.0

Notes:

1. DISKCOMP must be used with identical size floppy disks. It cannot be used with a hard drive.
2. If a target drive is not specified, DISKCOMP uses the current drive.
3. DISKCOMP prompts you when to swap disks as necessary.
4. DISKCOMP cannot compare double-sided disk with single-sided disk, or double-density disk with high-density disk.
5. Do not use DISKCOMP on a drive that is affected by the ASSIGN, JOIN, or SUBST commands or DISKCOMP will display an error message. Do not use DISKCOMP on a network drive.
6. When using DISKCOMP to compare a disk made with the COPY comand, although it is duplicate information, COPY may not put the information in the same location on the target disk and DISKCOMP will display an error message.

7. DISKCOMP exit codes are:

0	Disks are the same.
1	Disks are different.
2	Process aborted with CTRL+C by user.
3	Critical error.
4	Initialization error.

DISKCOPY.COM

Copies disks: Copies entire contents of the disk (including the DOS system files) in the source drive onto the disk in the target drive.

Syntax (shaded is optional):

DISKCOPY Source: Target: /1 / v /m

Examples:

DISKCOPY (current drive must be A: or B:)
DISKCOPY A: B: /1
DISKCOPY A: A: (prompts to change disks)

Syntax Options:

Source: . . . . The floppy disk to be copied.
Target: . . . . The floppy disk to be copied to.
/1. . . . . Copies one side of disk.
/ v . . . . Verifies that information is correctly copied. Introduced with Ver 5.0
/ m 6.2. . . . . Forces the use of only conventional memory for interim storage.

Command Type and Version:

External command; Not for networks;
Introduced Ver 1.0

Notes:

1. DISKCOPY must be used with identical size floppy disks only. It will not work with a hard disk.

2. If you do not enter a target drive, DOS uses the default drive as the target drive and DISKCOPY will overwrite all information that is on the target disk.
3. DISKCOPY will duplicate disk fragmentation from the source disk. Using the COPY command or the XCOPY command will give you a new disk that will be in sequential order and will not be fragmented.
4. DISKCOPY works only with removable (i.e. floppy) uncompressed disks.
5. DISKCOPY exit codes (ERRORLEVEL parameter) are:

0	Successful copy.
1	Nonfatal read/write error.
2	Process aborted with CTRL+C by user.
3	Critical error.
4	Initialization error.

DISPLAY.SYS

Driver that supports code page switching for the display: Supported types include Mono, CGA, EGA (includes VGA), and LCD.

Syntax (shaded is optional):

DEVICE = Drive:\Path\ DISPLAY.SYS
CON:= (type, hwcp, (n,m))

Examples:
DEVICE = DISPLAY.SYS con:=(ega,437,2)

Syntax Options:

Drive:\Path . . . . Drive & directory containing DISPLAY.SYS
type. . . . . Type of display adapter
hwcp . . . . The number assigned to a particular code page. Choices are as follows:
437 . . . . United States
850 . . . . Multilingual (Latin I)
852 . . . . Slavic (Latin II)
860 . . . . Portuguese
863 . . . . Canadian-French
865 . . . . Nordic

n . . . . Number of code pages supported by the hardware: Range is 0 through 6, max for EGA is 6, LCD is 1.

m. . . . . Number of subfonts supported by the hardware. Default=2 for EGA, 1 if LCD. If the *m* option is omitted, the parentheses around *n,m* can be omitted.

Command Type and Version:

CONFIG.SYS command; Introduced with Ver 3.3

Notes:

1. Code-page switching has no effect with monochrome and CGA display adapters.
2. If 3rd party console drivers are installed, make sure they are installed before DISPLAY.SYS.

DOS

Forces DOS to keep a link with the upper memory area or to load itself into high memory: HIMEM.SYS must be loaded before DOS= can be used. DOS is useful in that it is part of the program set that frees up conventional memory.

Syntax (shaded is optional):

DOS = high or low , umb or noumb

or

DOS = high or low, umb or noumb

Examples: DOS = high
DOS = umb
DOS = high, umb or DOS = umb, high

Syntax Options:

high. . . . . Loads a portion of DOS into high memory.

low . . . . Forces DOS to stay in conventional mem.

umb. . . . . Forces DOS to maintain a link between high (upper) memory and conventional memory.

noumb. . . . . Breaks the link between upper memory and conventional memory.

Command Type and Version:

CONFIG.SYS command: Introduced with Ver 5.0

Notes:

1. See also DEVICEHIGH and LOADHIGH.
2. UMB must be used in order to load either DOS or drivers into upper memory. EMM386.EXE or a comparable third party product must be loaded and configured in order to provide upper memory blocks from extended memory for DOS=UMB to work.
3. DOS can be placed anywhere in the CONFIG.SYS file.
4. UMB or NOUMB can be combined with HIGH or LOW in the same DOS = command line, see the example above.

DOSKEY.COM

Starts the DOSKEY program, which allows the user to edit command lines, create macros, and recall DOS commands:

Syntax (shaded is optional):

DOSKEY /reinstall /bufsize=nnn /macros /history /insert /overstrike /macroname=text

Examples: DOSKEY (start DOSKEY with defaults)
DOSKEY / history > special.bat

Syntax Options:

/ reinstall. . . . . . Installs DOSKEY again. If DOSKEY is currently running, this command clears the buffer.

/ bufsize=nnn . . Sets the size of the buffer where DOSKEY store commands. Default=512 bytes, minimum=256 bytes.

/ macros or */m*. . Displays the current list of DOSKEY macros.

/ history or */h*. . . Displays a list of all commands that were stored in memory.

/ insert. . . . . Sets typing to insert mode (text is not overwritten as typing occurs)

/ overstrike . . . . Sets typing to overstrike mode (text is overwritten as typing occurs)

/ macroname=. . Name of file created to hold *text* macro.

text . . . . The commands and text to be recorded in the file named *macroname.*

Command Type and Version:

External command; Network; Introduced with Ver 5.0

Notes:

1. */macros* and */history* can be used with DOS redirection to a file. e.g. DOSKEY */macros* > Macro.txt creates a text file list of macros.
2. DOSKEY is a very powerful program, see the Microsoft ***Users Guide and Reference*** for detailed comments and examples.

When DOSKEY is on, the following can be used to recall/edit commands from its command buffer:

Up Arrow. . . . . Recall command issued before currently displayed command.

Down Arrow. . . Recall command issued after the currently displayed command.

Page Up . . . . . Recall oldest command in current session.

Page Down . . . Recall most recent command in current session.

Left Arrow . . . . Moves cursor left one character.

Right Arrow . . . Moves cursor right one character.

Ctrl+Left Arrow Moves cursor left one word.

*Ctrl+Rght Arrow*Moves cursor right one word.

Home. . . . . Moves cursor to start of line.

End . . . . Moves cursor to end of line.

Esc . . . . Clears the display command line.

F1 . . . . Copy one character from last command buffer to the command line.

F2 .... Look forward for the next key typed after pressing F2.

F3 .... Copies the remainder of the current template line at the current cursor position to the command line.

F4 .... Delete all characters of the current temp late line, up to but not including the character pressed after F4 is pressed

F5 .... Copy current line to template and clear command line

F6 .... Put Ctrl+Z (end of line marker) at the end of the current line.

F7 .... Displays all commands and numbers, beginning with the oldest, currently stored in the command buffer.

Alt+F7 .... Delete all commands in command buffer.

F8 .... Locate the most recently used command in the buffer that begins with a specific character(s). At the DOS prompt, simply type those beginning characters and then press *F8*.

F9 .... Display the command associated with a specific command line number in buffer.

Alt+F10 ...... Delete all macros.

The following are special codes that can be used in creating macros. Code letters shown can be used in either upper or lower case.

$G .... Redirect output (same as >) to a device other than the screen. e.g. a printer.

GG........ Append output data (same as >>) to the end of a file instead of overwriting file.

$L .... Redirect input (same as <) to read from a device other than the keyboard.

$B .... Send output from macro to another command (same as |).

$T .... Used to separate commands in either a macro or at the DOSKEY command line.

$$ .... Used to specify the $ character

$1 to *$9* . . . . . . Batch parameters (similar to *%1*) for passing command line info to the macro when it is run.

*$ ** . . . . A replaceable parameter similar to *$1* to *$9*, except that everything that is typed on the command line after *macroname* is substituted for the *$ ** in the macro.

Macros are run by simply typing the *macroname* at the DOS prompt, followed by any parameter info such as *$1* or *$**. If a macro is created that has the same name as a normal DOS command, the DOS command is started by typing a space and then the command name, whereas with the macro, simply type the *macroname* without a space preceding it.

DOSSHELL.COM & EXE

Starts the DOS graphical user interface shell:

Syntax (shaded is optional):

DOSSHELL /t or /g :Res n /b

Examples: DOSSHELL / t
DOSSHELL / g:m
DOSSHELL / g /b

Syntax Options:

/ t. . . . . Directs DOSSHELL to start in text mode.

/ g . . . . Directs DOSSHELL to start in graphics mode.

:Res . . . . Screen resolution class. *l* (lowercase L) for Low, *m* for medium and *h* for high resolution.

n . . . . If there is more than one resolution available in the *Res* category, *n* provides additional information concerning which category to use.
n is hardware dependent.

/b . . . . Starts DOSSHELL in black & white mode or the state /t or /g is in.

Command Type and Version:

External command; Network; Introduced with Ver 4.0

Notes:

1. If DOSSHELL has already been started, the screen resolution can be changed from the options menu.
2. DOSSHELL is very useful for such tasks as renaming subdirectories.

DRIVER.SYS

Defines a logical drive from an existing physical drive: A logical drive is simply a drive letter used to point to the actual physical drive. The new drive letter established by DRIVER.SYS is the next highest drive letter above the system's highest current drive.

Syntax (shaded is optional):

DEVICE = Drive:\Path\ DRIVER.SYS /d:number /c / f:factor /h:heads /s:sectors / t:tracks

Examples:

DEVICE=C:\dos\driver.sys /d:1 /f:2 /h:2 /s:9 /t:80
(above configures a 3.5" 720k floppy drive, if the last hard drive was drive E:, then the 3.5 inch would be designated as drive F:)

Syntax Options:

Drive: . . . . Drive letter containing *\Path*

\Path . . . . Subdirectory containing *DRIVER.SYS*

/d: number . . . . Specifies physical drive number. Values must be in the range of 0 to 127. Normally, Drive A=0, Drive B=1, etc.

/c . . . . Specifies that the driver will be able to tell that the floppy disk drive door is open.

/ f: factor . . . . . . Specifies type of drive. Default value= 2

Factor .	Description
0	160kb/180kb or 320kb/360kb
1	1.2 megabyte (Mb)
2	720kb (3.5 in. disk)
7	1.44Mb (3.5 in. disk)
9	2.88Mb (3.5 in.disk)

/h: heads. . . . . . Specifies max. number of heads. Value for **heads** must be in the 1 to 99 range.

/s: sectors . . . . . Number of sectors per track, ranging in value from 1 to 99. The default varies according to the /f factor selected above. Normal values are 360kb and 720kb = 9 sectors, 1.44 meg = 18 sectors, 1.2 meg = 15 sectors and 2.8 meg = 36 sectors.

/t: tracks. . . . . . Number of tracks per side on the block device, ranging from 1 to 999. Default values vary according to the /f factor selected above. Normal values are 360kb = 40 tracks, 720kb, 1.44 meg, and 1.2 meg = 80 tracks.

Command Type and Version:

CONFIG.SYS command; Introduced with Ver 3.2

Notes:

1. DRIVER.SYS is commonly used to set up a 3.5 inch floppy drive on a system that does not support 3.5 inch drives directly. Setting up external 3.5 inch drives is also common.
2. See also the DRIVEPARM command, it is used to modify existing parameters of a physical device.
3. DRIVER.SYS can not be used to define hard drives. If hard drive logical drive assignments need to be changed, see the SUBST command.

4. If two DRIVER.SYS command lines are used for the same physical drive, then two logical drive letters will be assigned to the single physical drive.
5. XT class systems, with standard floppy controllers, will still need either a special driver or special controller in order to recognize a 1.44 or 2.8 Mb 3.5 inch floppy or 1.2 Mb 5-1/4 inch floppy.

DRIVPARM

Defines block device parameters: DRIVPARM allows the default or original device driver settings to be overridden when DOS is started.

Syntax (shaded is optional):

DRIVPARM=/d:number /c / f:factor /h:heads /i / n /s:sectors / t:tracks

Examples: DRIVPARM=/d:1 /c /f:2 /h:2 /s:9 /t:80
(above configures a 3.5" 720k floppy drive)

Syntax Options:

/d: number . . . . Specifies physical drive number. Numbers must be in the range of 0 to 255. Normally, Drive A=0, Drive B=1, etc.

/c . . . . Specifies that the driver will be able to tell that the floppy disk drive door is open.

/ f: factor . . . . . . Specifies type of drive. Default value= 2

Factor	Description
0	160K/180K or 320K/360
1	1.2 megabyte (MB)
2	720K (3.5 in. disk)
5	Hard disk
6	Tape
7	1.44MB (3.5 in. disk)
8	Read/write optical disk
9	2.88MB (3.5 in.disk)

/ h: heads . . . . . Specifies max. number of heads. Value for **heads** must be in the 1 to 99 range.

/ i ❹. . . . . Specifies an electronically-compatible 3.5 in. floppy disk drive. Use the / i switch if the ROM BIOS does not support 3.5 in. floppy disk drives.

/ n . . . . Non-removable block device.

/ s: sectors . . . . Number of sectors per track, ranging in value from 1 to 99. The default varies according to the /f factor selected above. Normal values are 360kb and 720kb = 9 sectors, 1.44 Mb = 18 sectors, 1.2 Mb = 15 sectors and 2.8 Mb = 36 sectors.

/ t: tracks . . . . . Number of tracks per side on the block device, ranging from 1 to 999. Default values vary according to the /f factor selected above. Normal values are 360kb = 40 tracks, 720kb, 1.44 Mb, and 1.2 Mb = 80 tracks.

Command Type and Version:

CONFIG.SYS command; Introduced with Ver 3.2

Notes:

1. DRIVPARM is particularly useful in configuring 3.5 inch floppy drives.
2. Settings in DRIVPARM will override any settings specified for a device prior to the DRIVPARM command line.
3. Although DRIVPARM is listed as an option in DOS Ver 3.3, the command will not function in that version.
4. DRIVPARM does not create new logical drives, it can only modify existing physical drive parameters.
5. See also DRIVER.SYS

DRVBOOT.BAT New Ver 6.22

Creates a bootable DRVSPACE floppy disk:

Syntax:

DRVBOOT drive1:

Example: drvboot a:

Syntax options:

drive1:. . . . . Drive containing floppy disk to be compressed.

Command Type and Version:

External command, Introduced with Ver 6.22.

Available in the MS-DOS 6.22 Supplemental Disks.

Notes:

1. DRVBOOT works only on high-density floppy disks (1.44 or 1.2 MB).
2. DRVSPACE must be installed prior to using DRVBOOT.

DRVSPACE.EXE

New Ver 6.22

Directs DOS to compress hard drives or floppy disks or configure compressed files:

Syntax (Shaded is optional):

DRVSPACE (starts the interactive DriveSpace program)

Examples: DRVSPACE

or

DRVSPACE (executes task command without starting the DriveSpace program)

Example: DRVSPACE /create c: /newdrive=d: /reserve=50

Syntax for Task Command Options:

/compress drive1: /newdrive=drive2 /reserve=size /f

Directs DOS to compress files on an existing disk (hard drive, floppy, or other removable media).

drive1:...........Specifies existing drive to compress.

/compress or /com..............Compresses the floppy disk or hard drive specified by drive1:.

/newdrive=drive2: or /new ..Identifies the drive letter for the uncompressed drive. After compression, the drive will contain an existing compressed drive (drive1:) and a new uncompressed drive (newdrive).

/reserve=size or /res....Size, in megabytes, of space to leave uncompressed. Space will be located on drive2:.

/f.....................Suppresses display of the final DriveSpace screen and returns to command prompt.

/create drive1: /newdrive=drive2 /reserve=size /size=size

Directs DOS to create a new compressed drive in free space on an uncompressed drive. The new compressed drive will provide more storage space than the amount of uncompressed storage it uses.

drive1:...........Specifies uncompressed existing drive containing space to create new drive.

/create...........or /cr Creates a new compressed drive in free space on the uncompressed drive specified by drive1:.

/newdrive=drive2: or /n ....Identifies the drive letter for the new compressed drive.

/reserve=size or /re .....Size, in megabytes, of space to leave uncompressed. Space will be located n drive2:. Can not use with /size=size.

/size=size or /si ....Total size, in megabytes, of the compressed .volume file. Can not use with /reserve=size.

/defragment /f drive1:

Directs DOS to defragment the specified compressed drive.

drive1:...........Specifies existing compressed drive to defragment.

/defragment or def.......Defragments specified compressed drive.

/f....................Specifies full defragmentation of specified drive.

/delete drive1:

Directs DOS to delete selected compressed drive and erase associated volume file.

drive1:...........Specifies drive to be deleted. Will not allow deletion of drive c:.

/delete or /del ....Deletes the specified drive.

/format drive1:

Directs DOS to format selected compressed drive. Caution-A compressed drive can not be unformatted after formatting using DRVSPACE /FORMAT.

drive1:...........Specifies drive to be formated. Will not allow formatting of drive c:.

/format or /f ...Formats the specified drive.

/info drive1:

Directs DOS to display information about selected compressed drive. Information includes free and unused space, name of compressed volume file, and estimated and actual compression ratios. Command may be used while Windows is running.

drive1:...........Specifies drive for which information is desired.

/format or /f ...Displays information for the specified drive.

/list

Directs DOS to list and describe, in brief terms, all available drives, except network and CD-ROM drives.

/list or /li.........Displays a list of all system drives, except CD-ROM or network drives.

/mount=nnn drive1: /newdrive=drive2

Directs DOS to create a reference between a compressed volume file (CVF) and a drive letter. DRVSPACE normally mounts compressed volume files automatically.

drive1:...........Specifies an existing drive containing the compressed volume file to be mounted. A drive must be specified.

/mount=ext or /mo=ext........Directs DOS to mount the compressed volume file with the filename extension specified by ext.

/newdrive=drive2: or /new ..Identifies the drive letter for the new drive.

/ratio=r.r drive1: /all

Directs DOS to change the estimated compression ratio of the specified compressed drive. DOS uses the ratio to estimate the amount of free space the drive contains.

drive1:...........Specifies existing compressed drive to defragment.

/ratio=r.r or /ra=r.r........Changes the ratio of specified compressed drive. Ratios are allowed in the range from 1.0 to 16.0. If not specified, DOS sets the ratio to the average compression ratio for all compressed files on the drive.

/all ....Specifies a change of all mounted compressed drives. Do not use if a drive is specified using drive1.

/size=size1 /reserve=size2 drive1:

Directs DOS to enlarge or reduce the current size of a compressed drive. The command is used to free-up space on a drive or enlarge a compressed drive if ample free space is available.

drive1:...........Specifies the drive containing to be resized.

/size=size1 or /si=size1 ......Changes the size of the drive specified by drive1: to size1 in megabytes. Can not be used with /reserve=size2.If neither switch is used, DOS will makes the compressed drive as small as possible.

/reserve=size2 or /res=size2......Size, in megabytes, of space to leave uncompressed. Can not use with /size=size1.

/uncompress drive1:

Directs DOS to uncompress files on an existing disk (hard drive, floppy, or other removable media). Uncompressing the last mounted drive also removes DRVSPACE.BIN from memory.

drive1:...........Specifies drive to uncompress.

/uncompress ....Uncompresses the floppy disk or hard drive specified by drive1:

/unmount drive1:

Directs DOS to eliminate a previous reference between a compressed volume file (CVF) and the specified drive. The unmounted drive is unavailable until again mounted. Drive c: can not be unmounted.

drive1:...........Specifies the drive to be unmounted. If no drive is specified DRVSPACE unmounts the current drive.

/unmount.......Directs DOS to unmount the specified drive.

Command Type and Version:

External command, Interactive
Introduced in MS-DOS Version 6.2, Revision 2.

Notes:

1. DRVSPACE is the Microsoft DOS Ver 6.2, Revision 2, replacement for DBLSPACE.
2. DRVSPACE requires 33Kb of memory to install.
3. DRVSPACE may slow down the speed of a system with a slow CPU.

DRVSPACE.SYS

New Ver 6.22

Device driver which directs DOS to move DRVSPACE.BIN to its final memory location: DRVSPACE.BIN provides DOS with access to compressed files. When the computer is started, DOS loads DRVSPACE.BIN at the top of conventional memory at the same time it loads other operating system functions; that is, prior to executing the CONFIG.SYS and AUTOEXEC.BAT files. After processing the CONFIG.SYS file, DOS moves DRVSPACE.BIN to the bottom of conventional memory. Running DRVSPACE.SETUP adds a command for DRVSPACE.SYS to the CONFIG.SYS file.

Syntax (shaded is optional):

DEVICE = DRVSPACE.SYS /move /nohma
or
DEVICEHIGH = DRVSPACE.SYS /move /nohma

Examples: DEVICE = DRVSPACE.SYS /move

Syntax Options:

/move . . . . Directs DOS to move DRVSPACE.BIN to its final memory location.
/nohma . . . . Tells DRVSPACE.SYS not to move DRVSPACE.BIN into high memory.

Command Type and Version:

External command, Introduced in Ver 6.2,revision 2.

DVORAK.SYS

Used with KEYB to provide an alternative to the standard QWERTY keyboard layout:

Syntax (shaded area optional):

KEYB nn,,drive1:\directory\DVORAK.SYS

Example: KEYB rh,,d:\dos\dvorak.sys

Syntax Options:

drive1:. . . . . Drive containing DVORAK.SYS.
directory. . . . . . Directory containing DVORAK.SYS.
nn . . . . Designates keyboard configuration.
. . . . dv = two-handed layout
. . . . rh = right-handed layout
. . . . lh = left-handed layout.

Command Type and Version:

External command, Introduced with MS-DOS Ver 6.0. Available in the MS-DOS 6.0, 6.21, and 6.22 Supplemental Disks.

Notes:

1. To return to the U. S. standard keyboard press CTRL+ALT+F1.
2. To return to the Dvorak keyboard layout press CTRL+ALT+F2.

ECHO

Display a message or turn command echo feature on or off: When batch files are run, DOS usually displays (echos) the name of the program being run to the display. This feature can be turned on or off with the ECHO command.

Syntax (shaded is optional):

ECHO Message | on | off

Examples: ECHO off
ECHO Enter program name to be run!
ECHO on

Syntax Options:

Message: . . . . . Text to be displayed on screen.
on . . . . Turn display echo on.
off . . . . Turn display echo off.

Command Type and Version:

Internal and Batch command; Introduced with Ver 2.0

Notes:

1. Use the @ symbol in front of a batch file command in order to turn the screen echo function off.
2. NOTE: in DOS 6.0, ECHO. (with the period) on a command line will output a blank line. ECHO by itself displays ECHO status.

EDIT.COM

Starts MS-DOS Editor: EDIT is a full-screen text editor which can create, save, edit and print ASCII text files.

Syntax (shaded is optional):

EDIT Drive:\Path \Filename /b /g /h /nohi

Examples: EDIT C:\Autoexec.bat
EDIT D:\Bin\Test.bat /h

Syntax Options:

Drive: *Path*. . . . Location of *Filename.*

Filename . . . . . Name of ASCII text file to be edited.

/b. . . . . Editor displayed in black and white.

/g. . . . . Provides CGA monitors with the fastest screen update.

/h. . . . . Allows monitor to display maximum number of lines on the screen.

/nohi . . . . Normally, DOS uses a 16 color mode for monitors. This switch enables the use of 8 color monitors.

Command Type and Version:

External command; Network; Introduced with Ver 5.0

Notes:

1. QBASIC.EXE must be in the same directory as EDIT or included in the DOS path. If it is not, EDIT will not function.
2. Shortcut keys that are shown on the bottom line of the screen may not display properly. If this occurs, use the */b* and */nohi* switches.

EDLIN.EXE Removed V6.0

Line oriented text editor: Edlin is an editor used to insert, change, copy, move and delete lines of text in an ASCII file. If a full screen editor is required, use EDIT (page 153). 24 lines of text can be displayed on the screen at one time.

Syntax (shaded is optional):

EDLIN Drive:\Path\ Filename /b

Examples: EDLIN Test
EDLIN C:\Autoexec.bat

Syntax Options:

Drive:\Path . . . . Drive and directory containg the file to be edited.

Filename. . . . . . File to be edited. If Edlin cannot find the file named *Filename*, it will automatically create the file in the specified *Drive:\Path* location.

/b . . . . Causes EDLIN to ignore Ctrl–Z (end of file character).

Command Type and Version:

External command; Network; Introduced with Ver 1.0
Removed from DOS Ver 6.0, use the EDIT command.
Available In the MS-DOS 6.0, 6.21, and 6.22 Supplemental Disks.

Notes:

1. Edlin can handle a maximum of 253 characters per line.
2. A full description of EDLIN is beyond the scope of this book. See a full DOS manual for additional details and instructions.
3. EDLIN uses an asterisk * prompt on a line by itself to ask for a command. If the * occurs after a line number, it indicates that that line number is the current line.

EDLIN Commands:(case doesn't matter)

? . . . . Displays the list of EDLIN commands.

Line. . . . . Just typing a number, at the prompt, displays the text contained in that line #.

Ctrl–C . . . . Exits user out of the insert (I) mode.

n A. . . . . Append *n* number of lines into memory from disk. Edlin will load till 75% of available memory is full.

L1,L2,L3,count C . . .Copy a block of lines. *L1*=first line to copy, *L2*=last line to copy, *L3*=line before which EDLIN is to insert the block, *count*=number of times to copy.

L1, L2 D. . . . . . Delete from line *L1* to line *L2*.

E. . . . . Write current file to disk and stop EDLIN.

L1 I. . . . . Insert lines before line *L1*. Ctrl-C stops.

L1, L2 L . . . . . . List (display) lines between *L1* and *L2*.

L1, L2, L3 M. . . Move a block of lines. *L1*=first line to
or *L1,+n,L3 M* . move, *L2*=last line to move, *L3*=line before which EDLIN is to move the block, *+n*=include the next *n* lines.

L1, L2 P. . . . . . Display all or part of the file one full screen of text at a time. *L1*=first line and *L2*=last line.

Q. . . . . Quit EDLIN without saving the current file to disk. Return to DOS.

L1,L2 ? R S1 S2 S3 . . .Replace a block of lines with a string. *L1*=first line to replace, *L2*=last line to replace, *?*=prompt user to confirm replacement, *S1*=string to be replaced, *S2*=Ctrl–Z separator, *S3*=string to replace S1.

L1,L2 ? S S1 . . Search between *L1* first line and *L2* last line for string *S1*. *?*=prompt user when string *S1* is located.

L1 T D:\Path\Filename . . .Transfer (merge) contents of a second file from disk into the current edited file. *L1*=line in current file before which user wants inserted file to be placed. *D:\Path\Filename*=name and directory location of file to be inserted into current file.

n W . . . . Write *n* number of lines, starting at the first line, to disk.

EGA.SYS

When usingTask Swapper with an EGA monitor, the EGA.SYS command saves and restores the display.

Syntax (shaded is optional):

DEVICE = Drive\path\ EGA.SYS

Examples: DEVICE=C:\Dos\EGA.SYS

Syntax Options:

Drive:\ Path. . . . Specifies the location of the EGA.SYS file.

Command Type and Version:

CONFIG.SYS command; Introduced with Ver 5.0

Notes:

1. To save memory when using a mouse on a system, install EGA.SYS before installing the mouse driver.

EMM386.EXE

Activates or deactivates expanded memory emulator for 80386 and higher systems: EMM386 is both a device driver loaded through CONFIG.SYS and an External command. It also enables or disables support of the Weitek coprocessor.

Syntax (shaded is optional):

To load EMM386 initially in CONFIG.SYS:

Device= Drive:\Path\ EMM386.EXE on ***or*** off ***or*** auto memory min=size w=on ***or*** w=off mx ***or*** frame = address ***or*** /pmmm pn=address x=mm–nn i=mm–nn b=address L=minXMS a=altregs h=handles d=nnn ram=mm-nn noems novcpi highscan verbose win=mm-nn nohi rom=mm-nn nomovexbda altboot

To use EMM386 as an External command:

EMM386 on ***or*** off ***or*** auto w=on ***or*** w=off /?

Examples: Device=C:\EMM386.EXE noems
EMM386 on (at DOS prompt)
EMM386 (at DOS prompt to show status)

Syntax Options:

Drive:\Path . . . . Drive and directory containing EMM386

EMM386 . . . . . . At the DOS prompt this displays the current status of EMM386.

on . . . . Activates EMM386 driver. (default)

off . . . . Deactivates EMM386 driver.

auto. . . . . Places EMM386 driver in auto mode, where expanded memory support is turned on when a program needs expanded memory.

memory. . . . . . . kbytes of memory allocated to EMM386. Default=256, Range=16 to 32768, use multiples of 16. This memory is in addition to low-memory backfilling.

w=on. . . . . Enable Weitek coprocessor support.

w=off. . . . . Disable Weitek coprocessor support.

m*x*. . . . . Address of page frame. Values for *x* can be 1 to 14 below. On systems with only 512k of memory, only 10 to 14 can be used.

1=C000 hex	8=DC00 hex
2=C400 hex	9=E000 hex
3=C800 hex	10=8000 hex
4=CC00 hex	11=8400 hex
5=D000 hex	12=8800 hex
6=D400 hex	13=8C00 hex
7=D800 hex	14=9000 hex

frame=*address* . Specific page-frame segment address for base page. *address* can be C000h to E000h and 8000h to 9000h, in increments of 400h.

/p*mmm* . . . . Address of page frame. *mmm* can range from C000h to E000h and 8000h to 9000h, in increments of 400h.

p*n*=*mmm*. . . . . . Specific segment address (*mmm*) of a specific page *n*. *n* can range from 0 to 255. *mmm* can range from 8000h to 9C00h and C000h to EC00h, in increments of 400h.

x=*mm–nn* . . . . . Excludes a range of segment addresses from EMS page use. *mm* and *nn* can both range from A000h to FFFFh, and are rounded off to the nearest 4k. x overrides i when two ranges overlap.

i=*mm–nn*. . . . . . Includes a range of segment addresses for EMS page or RAM use. *mm* and *nn* can both range from A000h to FFFFh, and are rounded off to the nearest 4k. x overrides i when two ranges overlap.

b=*address*. . . . . Lowest segment address that can be used for bank swapping of 16k EMS pages. Default=4000h, range=1000h to 4000h.

L=minXMS . . . . Specifies that *minXMS* kbytes of extended memory will remain after EMM386 has been loaded. Default=0

a=*altregs*. . . . . . *altregs* number of fast alternate register sets are allocated to EMM386. Default= 7, range=0 to 254. Each register uses an additional 400 bytes of memory.

h=*handles*. . . . . Number of handles EMM386 can have. Default=64, range=2 to 255.

d=*nnn* . . . . Kbytes of memory reserved for buffered DMA (direct memory access). Default= 16, range=16 to 256.

ram . . . . Upper memory and expanded memory access is provided.

noems. . . . . Upper memory access provided but not to expanded memory.

novcpi 6.2 . . . . Disables VCPI application support. Used with /noems.

highscan 6.2. . Directs EMM386 to check availability of upper memory for UMB or EMS windows.

verbose or v 6.2 Directs EMM386 to display error and/or status messages while loading.

Win=mm-nn . 6.2 Directs EMM386 to reserve the specified range of segment addresses for Windows. Values of mm and nn are in the range A000h through FFFh, rounded down to the nearest 4 Kb boundary. The /x switch takes precedence over /win if overlap occurs. The /win switch takes precedence over /ram, /rom, or /i switches if overlap occurs.

nohi 6.2. . . . . Forces EMM386 to load into convential memory thus increasing upper memory available for UMBs.

rom=mm-nn 6.2 Directs EMM386 to reserve the specified range of segment addresses for shadow RAM. Values of mm and nn are in the range A000h through FFFh, rounded down to the nearest 4 Kb boundary.

nomovexbda. 6.2 Directs EMM386 to keep extended BIOS data in conventional memory.

altboot . . . . *Provides an alternate boot sequence for* some computers with compatibility problems. Used if computer doesn't recognize Ctrl-Alt-Del.

/ ? . . . . Help with command line switches.

Command Type and Version:

External and CONFIG.SYS command;
Introduced with Ver 4.0

Notes:

1. HIMEM.SYS must be loaded before EMM386.EXE is loaded.
2. The .EXE extension of EMM386 must be used to load the driver.
3. The order of switches and parameters is not important.
4. Device=EMM386.EXE must precede DEVICEHIGH commands.
5. If enough memory is not available to set up a 64k page frame, the "Unable to set base address" error message will display.
6. DOS=umb must be used in CONFIG.SYS to provide access to the upper memory block.
7. See also DOS, HIMEM.SYS, DEVICEHIGH, and LOADHIGH.
8. Using EMM386.EXE and the Note 7 commands is a very complicated task. It is strongly recommended that the user spend a great deal of time with Microsoft's *MS-DOS 5.0 User's Guide and Reference* learning about memory management and system optimization.

EXE2BIN.EXE

Removed V6.0

Converts an executable file to a binary file: Converting executable files (.EXE extension) to files with a binary format, is only useful to software developers and is of no value to general users.

Syntax (shaded is optional):

EXE2BIN	Drive1:\Path1\	INfile
	Drive2:\Path2\	OUTfile

Examples: EXE2BIN C:\Test.exe C:\test.bin

Syntax Options:

Drive1:\Path1 . . Drive and directory of input .EXE file.
Drive2:\Path2 . . Drive and directory of output binary file.
INfile . . . . Input .EXE file to be converted.
OUTfile . . . . Output binary file.

Command Type and Version:

External command; Introduced with Ver 1.0
Removed from DOS Ver 6.0

Available in the MS-DOS 6.0, 6.21, and 6.22 Supplemental Disks.

Notes:

1. EXE2BIN is not for the general computer user, only programmers.
2. Default extensions for INfile is .EXE and for OUTfile is .BIN.
3. INfile must have been produced by LINK and must not be a packed file.
4. See also LINK

EXIT

Exits a secondary command processor and returns to the primary processor if one exists.

Syntax:

EXIT

Examples: EXIT

Syntax Options:

No options

Command Type and Version:

Internal; Network; Command processor function;
Introduced with Ver 2.0

Notes:

1. If a secondary command processor is not loaded (or /P is used with COMMAND.COM), the EXIT command will have no effect.
2. See Also COMMAND

EXPAND.EXE

Expands a compressed DOS file: Compressed files are not usable unless exanded. Use EXPAND to retrieve files from DOS installation or update disks.

Syntax (shaded is optional):

EXPAND Drive:\Path\ Filename Destination

Examples:
EXPAND B:\Dos\FIND.EX_ C:\Dos\FIND.EXE

Syntax Options:

Drive: \Path. . . . Specifies location and name of a compressed file to be expanded.

Filename. . . . . . File to be expanded.

Destination . . . . Target location where expanded files are to be placed. Destination can be a drive letter and colon, a filename, a directory name or a combinaton. A destination filename can only be used if a single compressed *Filename* is used.

Command Type and Version:

External command; Network; Introduced with Ver 5.0

Notes:

1. Wildcards (* and ?) **cannot** be used.
2. Compressed files, such as installation or update files, have a file extension which ends with an underscore character (_)
3. Although EXPAND is normally used by the DOS 5.0 Upgrade program to install all DOS 5.0 files, you can copy a single compressed file, such as FIND.EX_ , from an upgrade disk to the hard drive and EXPAND it for full use. A complete list of all files and what disk they are on is included in the file named PACKING.LST on upgrade disk 1 or 2.
4. One or more source filenames may be specified. Destination may include a filename only if a single source filename is specified. If no destination is specified, EXPAND prompts for it.

FASTHELP.EXE New V6.0

Displays a list and gives a brief description of all DOS 6.0 commands: This command is a direct replacement for the DOS Ver 5.0 HELP.
/? can be used in conjunction with other DOS commands to display the same help as FASTHELP would display for the same command.

Syntax (shaded is optional):

FASTHELP command

Examples: FASTHELP Chkdsk
FASTHELP
DISKCOPY /?

Syntax Options:

command . . . . . The particular DOS command that you want help about.

Command Type and Version:

External command; Network; Introduced with Ver 6.0

Notes:

1. FASTHELP without a command displays a list and brief description of all DOS 6.0 commands contained in the DOSHELP.HLP file.
2. Detailed information on DOS commands is available with the HELP command.
3. FASTHELP is a direct replacement for the DOS Ver 5.0 HELP command.

FASTOPEN.EXE

Fast opening of files: Decreases the amount of time to open frequently used files by keeping directory information in memory. FASTOPEN can be started at the DOS prompt or in either a Batch file or CONFIG.SYS. *DOS V4 is different, see manual.*

Syntax (shaded is optional):

To start in a Batch file or at the DOS Prompt:
FASTOPEN Drive1: = nnn Drive2:= nnn ... /x
To start in CONFIG.SYS use the following:
Install=Drive:\Path\FASTOPEN.EXE
Drive1: = nnn Drive2:=nnn ... /x

Examples: FASTOPEN C:=97 /x
Install=C:\DOS\FASTOPEN C:=97

Syntax Options:

Drive1: Drive2: . One or more drives FASTOPEN tracks.
nnn . . . . Number of files FASTOPEN can work with at the same time. The valid values are 10 through 999. 48 is the default.
/x. . . . . Creates the *name cache* in expanded memory rather than conventional memory. *name cache* is a buffer where names and locations of open files are stored.
Drive:\Path . . . . Drive and directory containing FASTOPEN.

Command Type and Version:

External and CONFIG.SYS command; NOT for Network
Introduced with Ver 3.3

Notes:

1. When placed in CONFIG.SYS, FASTOPEN.EXE must be used, not FASTOPEN without the extension.
2. FASTOPEN uses approximately 48 bytes of memory for each file that it tracks.
3. Deactivate FASTOPEN **BEFORE** disk compaction is used!!!!!
4. FASTOPEN works with hard drives only, not floppy drives.

FC.EXE

Compare two files and report the differences:
FC reports the differences it finds between two files and displays them on screen. The comparison can be of ASCII or binary files.

Syntax (shaded is optional):

FC /a /c /L /Lbx /n /t /w /nnn
Drive1:\Path\ File1 Drive2:\Path\ File2
or
FC /b Drive1:\Path\ File1 Drive2:\Path\ File2

Examples: FC /a C:\DATA\Test.txt D:\Master.txt
FC /b C:\DOS\MEM.EXE D:\UTIL\MEM2.EXE

Syntax Options:

Drive1:\Path . . . Drive and directory of first file *(File1)*.
Drive2:\Path . . . Drive and directory of second file *(File2)*.
File1 & File2 . . . The two files to be compared.
/ a . . . . Abbreviate ASCII comparison output, will only display first and last line of different block.
/ c . . . . Ignore upper/ lower case.
/ L . . . . Files compared in ASCII mode.
/ Lbx . . . . Set *x* lines of internal line buffer.
/ n . . . . During ASCII compare, displays line #s.
/ t. . . . . Do not expand tabs to spaces. Default is to treat tabs as spaces with stops at every 8th position.
/ w. . . . . During comparison, tabs and spaces are compressed. Also causes FC to ignore space that occurs at the beginning and end of lines.
/ nnn . . . . Set the number of consecutively matching lines before files are resynchronized.

/b . . . . Files compared in binary mode. This is the default for all files ending in .EXE, .COM, .SYS, .OBJ, .LIB and .BIN.

Command Type and Version:

External command; Network;
Introduced with Version 2.1

Notes:

1. See also COMP and DISKCOMP.
2. Use of wild cards (* or ?) is allowed.
3. For ASCII comparisons, the *File1* name is displayed, then the lines from *File1* that are different are displayed, then the first line to match in both files, then the *File2* name is displayed, then the lines from *File2* that are different, and finally, the first line to match in *File2*. FC uses a 100 line buffer to hold the lines being compared, if there are more than 100 lines of differences, FC cannot complete the comparison and issues a Resynch Failed error message.
4. For binary comparisons, the differences are reported on a single line as **xxxxxxxx: yy zz**, where xxxxxxxx is the hex address from the beginning of the file where the difference occurs. yy is the byte that is different in *File1* and zz is the byte that is different in *File2*. FC uses the same line buffer as Note 4 for binary comparisons, however if it runs out of memory, it will overlay portions of the memory until the comparison is completed.
5. FC is only available with MS-DOS®, not PC-DOS.

FCBS

Sets number of file control blocks that DOS can have open at the same time:

Syntax (shaded is optional):

FCBS = x

Examples: FCBS = 10

Syntax Options:

x . . . . File control blocks that DOS can have open at one time. Default = 4. Values can range from 1 through 255.

Command Type and Version:

CONFIG.SYS command; Introduced with Ver 3.0

Notes:

1. Normally, this command should only be used if a program specifically requires that FCBS be set to a specific value.
2. DOS may close a file opened earlier if there are not enough FCBs set aside.
3. The **,y** Syntax Option available in DOS Versions 4.01 and earlier, is no longer a valid option.

FDISK.EXE

Configures hard disk: After the low level format of a hard drive, FDISK is used to partition the drive for DOS. A series of menus are displayed to assist in the partitioning process. ***Caution:*** When a partition is deleted, all of the data stored on that partition is also deleted.

Syntax (shaded is optional):

FDISK /status

Examples: FDISK

Syntax Options:

/status ❻ . . . . *Display partition table info for hard* drives installed in the system.

/mbr . . . . Master boot record. Undocumented

Command Type and Version:

External command; Network, introduced with Ver 2.0

Notes:

1. Before DOS 3.3, FDISK did not create extended partitions or logical drives in the partitions. There could be only one DOS partition per drive. Until DOS 3.31 & 4.0, max size was 32Mb.
2. Using the FDISK command, you can accomplish the following:
   Create a primary DOS partition on a hard drive.
   Create an extended DOS partition on a hard drive.

Delete a partition on a hard drive.
Change the active partition on a hard drive.
Displays partition data for a hard drive.
Selects a different hard disk for partitioning.

3. Maximum partition size is 2 gigabytes.
4. In order to change the size of a partition, the partition must be deleted first, and a new partition created.
5. Drives formed by ASSIGN, SUBST, or JOIN cannot be partitioned with FDISK.
6. USE WITH CAUTION, backup hard drive data files before changing or deleting a partition.
7. The formatting of a hard drive for use by DOS is a three step process: Low level format, FDISK, then FORMAT. Note that IDE hard drives have been low level formatted at the factory, do not re-low level format these drives, only use FDISK then FORMAT.
8. See also FORMAT.

FILES

Sets the number of open files DOS can access.

Syntax:

FILES = nnn

Examples: FILES=20

Syntax Options:

nnn . . . . Number of files DOS can access, at one time, with valid values ranging from 8-255. The Default is 8.

Command Type and Version:

CONFIG.SYS command; Introduced with Ver 2.0

Notes:

1. The standard value for files is FILES=20, however, many software packages, such as database managers, will require values in the range of 35 to 40. See the documentation for each program you wish to run and verify that your FILES= statement is not smaller than that required by the program. It is all right if FILES= is larger than a program requires.

FIND.EXE

Looks for a text string in a file(s): Once the text string is located that FIND is searching for, it displays those lines of text containing the text string.

Syntax (shaded is optional):

FIND /v /c /n /i text Drive:\Path\ Filename

Examples: FIND /v /i "Dear Sir" C:\Test.doc
FIND "Dear Sir" Test.doc
FIND "Dear Sir" "Sincerely" "Help" C:\Test.doc

Syntax Options:

Drive:\Path . . . . Drive and directory containing *Filename*.
Filename. . . . . . File being searched for *Text*.
text . . . . Text string being searched for.
/v . . . . Display lines that do not contain *Text*.
/c . . . . Display line count of lines containing *Text*.
/n . . . . File's line number containing *Text*.
/i. . . . . Ignore upper/lower case during search. Ver 5.0

Command Type and Version:

External command; Network; Introduced with Ver 2.0

Notes:

1. Wild cards (* and ?) cannot be used in filenames being searched for by FIND. See the FOR command for help in this area.
2. FIND ignores carriage returns, so *Text* must be a string that does not contain any carriage returns.
3. If /c and /n are used together, the /n is disregarded.
4. If Filename is not specified, FIND will act as a filter for any standard device (keyboard, file, pipe, etc) and display those lines containing *Text*.
5. DOS provides three filter commands, FIND, MORE, and SORT.
6. /c /v used together will return a count of lines that do not contain *Text*.

FOR

A logical batch command that runs a specific command for each file in a group: FOR can be run from inside a batch file or at the DOS prompt.

Syntax (shaded is optional):

If used in a batch file, use the following:
FOR %%variable IN (set) DO command cpar
If used at the DOS prompt, use the following:
FOR %variable IN (set) DO command cpar

Examples:
FOR %T IN (*.doc, *.asc) DO DEL %T
(deletes all .doc and .asc files in current directory)

Syntax Options:

%variable . . . . . Replaceable variable for use at the DOS prompt. The *variable* name can be any character(s) except the numbers 0 to 9. FOR replaces *variable* with each text string contained in *(set)* and runs *command* over and over until all are processed.

%%variable. . . . Same as %*variable*, except for use in batch files only.

(set) . . . . One or more files or text strings on which *command* is to operate. ( ) is required

command . . . . . Any DOS command to be run on each item listed in *(set)*.

cpar. . . . . Parameters for *command*.

Command Type and Version:

Batch and Internal command; Introduced with Ver 2.0

Notes:

1. FOR..IN..DO commands cannot be nested on a single command line.
2. Wild cards (* and ?) are allowed in *(set)*.
3. Multiple %variable names are allowed.

FORMAT.EXE

Format a floppy or hard disk: A disk must be formatted before DOS can recognize it.

Syntax (shaded is optional):

There are 4 different syntax choices:

FORMAT Drive: /v:name /q /u /f:size /b /s /c
FORMAT Drive: /v:name /q /u /t:trak /n:sect /b /s /c
FORMAT Drive: /v:name /q /u /1 /4 /b /s /c
FORMAT Drive: /q /u /1 /4 /8 /b /s /c /autotest

Examples: FORMAT A: / s /autotest

Syntax Options:

Drive: . . . . Drive to be formatted. If no switches are used, the drive is formatted according to its system drive type.

/ v:name . . . . . . Assign the disk the volume label *name. name* can be up to 11 characters long. If /v is not used, DOS will automatically prompt the user for a volume name when the format process is finished. */v* is not compatible with */8.* See also the VOL, DIR, and LABEL commands.

/ q . . . . Quick format a disk by deleting the FAT (File Allocation Table) and root directory. Only use this on disks that have already been formatted. Ver 5.0

/ u . . . . Unconditional format. Destroys all data and UNFORMAT will not work. Use if read or write errors occur with this disk or when a new disk is to be formatted. Ver. 5.0

/ 1 . . . . Format 1 side of floppy only.

/ 4 . . . . Formats a DSDD (double-sided double-density) 5-1/4 inch, 360k floppy in a 1.2 m floppy drive. Warning: some 1.2m drives can not reliably do this format!

/ 8 . . . . Formats a 5-1/4 disk with 8 sectors per track. 8 sectors per track are necessary for use with pre DOS 2.0 operating systems.

/ f:size . . . . Floppy disk size. Use instead of /t and /n switches if possible:

160, 160k or 160kb . . . . 160k SSDD, 5-1/4"
180, 180k or 180kb . . . . 180k SSDD, 5-1/4"
320, 320k or 320kb . . . . 320k DSDD, 5-1/4"
360, 360k or 360kb . . . . 360k DSDD, 5-1/4"
720, 720k, or 720kb . . . . 720k DSDD, 3.5"
1200, k, kb, 1.2, 1.2m, 1.2mb . . . . 1.2m DSHD, 5-1/4"
1440, k, kb, 1.44, 1.44m, 1.44mb . 1.44m DSHD, 3.5"
2880, k, kb, 2.88, 2.88m, 2.88mb . 2.88m DSEHD, 3.5"

/ b . . . . Obsolete switch used to reserve space for the system files. No longer generally used, retained for compatibility only.

/ s . . . . Copies all 3 system files, [IO.SYS and MSDOS.SYS] or [IBMBIO.COM and IBMDOS.COM] and COMMAND.COM to the disk after formatting has finished. The DBLSPACE.BIN file is also copied to the target drive (if you are not using the DBLSPACE program, you can remove the hidden, system, read-only attributes from DBLSPACE.BIN on the target disk and then delete it.)

/ t:trak . . . . Number of tracks on disk, must be used with the */n* switch. Use */ f:size* switch if possible.

/ n:sect . . . . Number of sectors on disk, must be used with the */ t* switch. Use */ f:size* switch if possible.

/ autotest ❻ . . . Bypasses prompts during formatting. Note that this is an undocumented command.

/ c 6.2 . . . . Retests for bad cluster

Command Type and Version:

External command; Introduced with Ver 1.0

Notes:

1. New floppy disks need only be formatted in order to make the disk useable by DOS. Hard drives, however, require a 2 or 3 step format process which includes a low level format (Not on IDE drives), then partitioning with FDISK, and finally FORMAT.
2. If the / U switch is **not** used, UNFORMAT can unformat the disk. See also UNFORMAT
3. Format issues a warning when a hard drive is to be formatted.
4. Do not format Network drives or drives that have had ASSIGN, JOIN or SUBST used on the drive.
5. FORMAT / S and the DOS "SYS" command both copy the DBLSPACE.BIN file to the Target Disk.
6. FORMAT Exit codes are: 0 Successful FORMAT
   3 Aborted with Ctrl+C by user; 4 Fatal error other than 0,3, or 5
   5 No response to Proceed?

GOTO

Directs DOS to process commands starting with the line after a specified label: Within a Batch program, when DOS finds the specified label, it processes the commands beginning with the next line after that label.

Syntax (shaded is optional):

GOTO Label
:Label

Examples:	GOTO Start	
	Test.bat	(bypassed by GOTO)
	:Start	(must begin with :)

Syntax Options:

Label. . . . . Directs DOS to a specific line in a batch file. Valid values for *Label* can include spaces but cannot include other separators, such as equal signs and semicolons. GOTO will recognize only the first 8 characters of the Label name. *Label,* on the GOTO command line, does not begin with a colon and it must have a matching *Label* line in the batch program. The *Label* line in the batch program must begin with a colon. You can also substitute an environment variable enclosed in percent signs, e.g. %RETURN%, for *Label.*

Command Type and Version:

Internal command; only used in a Batch program; Introduced with Ver 2.0

Notes:

1. A batch-program line beginning with a colon (:) is a label line, and will not be processed as a command. When the line begins with a (:) colon, DOS ignores any commands on that line.

GRAFTABL.COM **Removed V6.0**

Allows a display to show extended characters in graphics mode from a specific code page: This command is required when a monitor is not able to display extended characters in graphics mode. (Most monitors do not need GRAFTABL.)

Syntax (shaded is optional):

GRAFTABL nnn
or
GRAFTABL /status

Examples:
GRAFTABL 860 (Portuguese code page)

Syntax Options:

nnn . . . . Code page used to define extended characters.
- 437 . . . . United States
- 850 . . . . Multilingual
- 852 . . . . Slavic
- 860 . . . . Portuguese
- 863 . . . . Canadian-French
- 865 . . . . Nordic

/status . . . . Identifies current country code page.

Command Type and Version:

External command; Network; Introduced with Ver 3.0
Beginning with MS-DOS Ver 6.0, GRAFTABL is only available on Microsoft's DOS Supplemental Disks.

Notes:

1. The active code page is not changed when GRAFTABL is run.
2. GRAFTABL uses approximately 1K of RAM.
3. GRAFTABL exit codes are as follows:
   - 0 . . . . Successful load of character set.
   - 1 . . . . Current character set replaced by new table.
   - 2 . . . . File error has occurred.
   - 3 . . . . Incorrect parameter, new table not loaded.
   - 4 . . . . Incorrect DOS version, 5.0 required.

GRAPHICS.COM

Configures DOS so that Print Screen (Shift+Print Scrn) can print a graphics screen to a printer. GRAPHICS supports CGA, EGA, and VGA display modes:

Syntax (shaded is optional):

GRAPHICS Type Drive:\Path\ Filename /r /b /Lcd /pb:std or /pb:Lcd

Examples: GRAPHICS color4 /b

Syntax Options:

Type	Printer type (HP=Hewlett-Packard)
color1	IBM Color Printer with black ribbon
color4	IBM Color Printer with RGB ribbon
color8	IBM Color Printer with CMY ribbon
hpdefault	Any HP PCL printer
deskjet	HP DeskJet printer
graphics	IBM Graphics, Proprinter or Quietwriter
graphicswide	IBM Graphics Printer with 11 inch carriage
laserjet	HP LaserJet printer
laserjetii	HP LaserJet II printer
paintjet	HP PaintJet printer
quietjet	HP QuietJet printer
quietjetplus	HP QuietJet Plus printer
ruggedwriter	HP Rugged Writer printer
ruggedwriterwide	HP Rugged Writerwide printer
thermal	IBM PC-convertible Thermal Printer
thinkjet	HP ThinkJet printer
Drive:\Path	Drive and directory containing *Filename.*
Filename	Printer profile where graphics screen is to be printed to. Default is GRAPHICS.PRO.
/r	Prints the image as white characters on a black background (black characters on a white background is the Default).

/b	Prints the background in color. (only color4 and color8 types are valid)
/Lcd	Prints image using an LCD screen aspect ratio instead of a CGA screen aspect ratio.
/pb:std	Sets printbox size. If this switch is used,
or /pb:Lcd . . .	you must check the GRAPHICS.PRO file and change each printbox line to *std* or *Lcd* so that it matches what you selected for */pb* :

Command Type and Version:

External command; Network; Introduced with Ver 2.0

Notes:

1. The GRAPHICS command does use a limited amount of conventional RAM when it is loaded.
2. Four shades of gray are printed if *color1* or *graphics* is in effect and the screen is in the 320x200 mode.
3. If a printer profile such as GRAPHICS.PRO is already loaded, and you wish to load a different .PRO file, the new .PRO must be smaller than the currently loaded .PRO. If it is larger, your system must be re-booted first in order for the larger profile to be loaded.
4. Use the Graphics or Graphicswide printer types if the printer you are using is an Epson.
5. Supported displays include EGA and VGA.
6. See also PRINT
7. Do not use the / b switch in conjunction with the / r switch or with a black and white printer.

GW-BASIC®.EXE

BASIC language intrepreter: GW-BASIC® is Microsoft's own version of BASIC that shipped with MS–DOS versions prior to Ver 5.0. Starting with Ver 5.0, QBASIC is shipped with DOS.

Syntax (shaded is optional):

GWBASIC Drive:\Path\Filename < Input
>> Output /f:n /i /s:n /c:n /m:n,n /d

Examples: GWBASIC (starts BASIC)
GWBASIC C:\BAS\test.bas /f:4 /d

Syntax Options:

Drive:\Path . . . . Drive and directory containing *Filename.*

Filename. . . . . . The BASIC program file to be run. The default file extension is **.**BAS

< *Input*. . . . . Standard input is read from *Input* file.

> *Output* . . . . . . Output is redirected to *Output* file or a device (screen, printer, etc)

>> . . . . Causes *Output* to be appended.

/ f:n . . . . Max number *n* of simultaneously open files while a BASIC program is running. Default is 3. */ i* must be used at the same time. Size requirement includes 194 bytes (File Control Block) plus 128 bytes (data buffer).

/ i. . . . . Forces static allocation of memory for file operations.

/ s:nn. . . . . Max record length *nn* for a file. Default is 128 bytes, maximum is 32,767 bytes.

/ c:nn. . . . . Allocates *nn* bytes of Receive buffer and 128 bytes of Transmit buffer for RS-232 (serial) communications. */c:0* disables support. Defaults are 256 byte receive buffer and 128 byte transmit buffer for each RS-232 card.

/ m:x,y. . . . . Sets the highest memory location *x* and the maximum block size *y* in bytes. Block size is in multiples of 16.

/ d . . . . Activates double-precision for the following functions: ATN, COS, EXP, LOG, SIN, SQR and TAN.

Command Type and Version:

External command; Network; Introduced with Ver 1.0

Notes:

1. See also BASIC, BASICA, and QBASIC.
2. Variables n, nn, x, and y listed above are all given in decimal values. If you wish to use hexadecimal values, precede the value with &H. If you wish to use octal values, precede the value with &O (O is the letter O, not zero).
3. A complete discussion of GW-BASIC is beyond the scope of this book. If you need information on GW-BASIC commands and how to program in BASIC, refer to Microsoft's manual on GW-BASIC or other texts on BASIC.
4. Different versions of GWBASIC were released and each needs to be run with its correct version of DOS.
5. Programs written in BASIC (IBM's version) may require small adjustments in order to run correctly under GW-BASIC

HELP.EXE - Version 5.0 only

Online information about MS-DOS version 5.0 commands:

Syntax (shaded is optional):

HELP command

Examples: HELP (brief description of commands)
HELP chkdsk
DISKCOPY / ? (see Note: 1 below)

Syntax Options:

Command . . . . . Any specific DOS version 5.0 command on which more information is desired.

Command Type and Version:

External command; Network; Introduced with Ver 5.0

FASTHELP in Ver 6.0 is the same as HELP in Ver 5.0

Notes:

1. You can get online HELP in two ways. Either specify the name of the command on the HELP command line or type the command name and the /? switch at the command prompt.

HELP - Version 6.0 and 6.2x New V6.0

Online information about MS-DOS Version 6.0 and 6.2x commands and a list of all DOS commands: The Ver 6.0 AND 6.2 information for HELP is much more detailed than FASTHELP or DOS Ver 5.0 HELP.

Syntax (shaded is optional):

HELP command /B /G /H /nohi

Examples: HELP (List of commands)
HELP chkdsk
DISKCOPY /? (see Note: 1 below)

Syntax Options:

Command..... Any specific DOS version 6.0 command on which more information is desired.

/B .... Display in black-and-white mode.

/G .... Display in CGA color mode.

/H .... Display HELP with the maximum number of lines that the display supports.

/nohi .... Turn high-intensity display off.

Command Type and Version:

External command; Network; Introduced with Ver 6.0

FASTHELP in Ver 6.0 and 6.2x is the same as HELP in Ver 5.0

Notes:

1. You can get online HELP in two ways. Either specify the name of the command on the HELP command line or type the command name and the /? switch at the command prompt.

Extended memory and HMA (high memory area) manager: HIMEM.SYS prevents programs from using the same memory locations at the same time.

Syntax (shaded is optional):

Device= Drive:\Path\ HIMEM.SYS /hmamin=m /numhandles=n /int15=xxx /machine:xxx /a20control:on or off /shadowram:on or off /cpuclock:on or off /EISA /verbose /test:on or off

Examples: Device=C:\Dos\HIMEM.SYS /test:off

Syntax Options:

Drive:\Path . . . . Drive and directory containing HIMEM.

/hmamin=m. . . . Minimum *m* kilobytes of memory a program must use before it can use the HMA. Default=0, Range=0 to 63. The most efficient use of HMA is accomplished by setting m to the amount of memory required by the program that uses the most HMA.

/numhandles=n. Maximum number (*n*) of EMB (extended memory block) handles that can be used at the same time. Each handle uses 6 bytes of RAM. Default=32, Range=1 to 128.

/int15=xxx . . . . . *xxx* kilobytes of memory are assigned to the Interrupt 15h interface. Programs must recognize VDisk headers in order to use this switch.

/machine:xxx. . . Defines a specific A20 handler *xxx* to be used. Normally, HIMEM automatically detects which A20 is to be used. Default=1. If the required handler is not

listed in the following table, see the README.TXT file in your DOS directory for additional information.

Number	Code	A20 handler
1	at	IBM PC/AT, Compuadd 386. JDR 386/33
2	ps2	IBM PS/2, Datamedia 386 /486, Unisys PowerPort
3	ptlcascade	Phoenix Cascade Bios
4	hpvectra	HP Vectra, A and A+
5	att6300plus	AT&T 6300 Plus
6	acer1100	Acer 1100
7	toshiba	Toshiba 1600, 1200XE and 5100
8	wyse	Wyse 12.5 MHz 286, Intel 361Z or 302, Hitachi HL500C,Compuadd 386
9	tulip	Tulip SX
10	zenith	Zenith ZBIOS
11	at1	IBM PC/AT
12	at2	IBM PC/AT (alt. delay)
12	css	CSS Labs
13	at3	IBM PC/AT (alt. delay)
13	philips	Philips
14	fasthp	HP Vectra
15 ❻	ibm7552	IBM 7552 Industrial Comp
16 ❻	bullmicral	Bull Micral 60
17 ❻	dell	Dell XBIOS

/a20control:on or */a20control:off* Off allows HIMEM.SYS to take control of the A20 line only if A20 was off when HIMEM.SYS was loaded. Default=*:on*

/shadowram:on. or */shadowram:off* If your system has Shadow RAM, *:off* switches the Shadow RAM off and returns control of that RAM to HIMEM. Default=*:off* if your system has less than 2 megabytes of RAM.

/cpuclock:on . . . If your system slows down when HIMEM.SYS is loaded, specifying *:on* might correct the problem. *:on* will slow down HIMEM.SYS.

/ EISA ❻. . . . . . Used only on EISA systems to specify that HIMEM allocates all available extended memory.

/ verbose or */ v* ❻ HIMEM displays status and error messages while loading. Hold ALT key down during system startup to disable /verbose.

/ test:on or *:off*. . Turns the HIMEM.SYS testing of all extended memory *:on* or *:off* during system startup.

Command Type and Version:

Config.sys command; Introduced with Ver 5.0

Notes:

1. Only one program at a time can use the high memory area.
2. HIMEM.SYS, or another XMS driver such as 386MAX or QEMM must be loaded before DOS can be loaded into HMA with the DOS=high command.
3. In most cases, command line switches do not need to be used. since the defaults are designed to work with most computer hardware.

IF

Performs a command based on the result of a condition in batch programs: If a conditional statement is true, DOS executes the command, if the condition is false, DOS ignores the command.

Syntax (shaded is optional):

Three syntax formats are valid:

IF not errorlevel nnn *command*
IF not string1==string2 *command*
IF not exist filename *command*

Examples: IF errorlevel 3 goto end

Syntax Options:

not. . . . . The command is to be carried out only if the statement is false.

errorlevel *nnn* . . True only if the previous program executed by COMMAND.COM had an exit code equal to or greater than *nnn*.

command . . . . . The specified command that DOS is to perform if the preceding condition is met.

string1==string2 True, only if *string1* and *string2* are the same. The values of *string1* and *string2* can be literal strings or batch variables.
Strings may not contain separators, such as commas, semicolons, spaces, etc.

exist *filename* . . True condition if *filename* exists.

Command Type and Version:

Internal command but only used in Batch programs;
Introduced with Ver 2.0

Notes:

1. The *errorlevel* parameter allows you to use exit codes as conditions. An exit code is returned to DOS whenever a programstops.
2. Use " " quotes around strings when comparing, it's safer.

INCLUDE

New V6.0

Includes the contents of one configuration block within another configuration block: This is one of five special CONFIG.SYS commands used to define multiple configurations.

Syntax:

INCLUDE=blockname

Syntax Options:

blockname . . . . The name of the configuration block to be included.

Command Type and Version:

CONFIG.SYS command; Introduced with Ver 6.0

Notes:

1. See also MENUITEM, MENUDEFAULT, MENUCOLOR, and SUBMENU. These are the other four special CONFIG.SYS commands used to define multiple configurations.
2. Refer to your DOS 6.0 manual for more information on setting up the special multiple configuration menus.

INSTALL

Loads a memory-resident program when DOS is started: Use the INSTALL command to load FASTOPEN, KEYB, NLSFUNC, or SHARE in CONFIG.SYS.

Syntax (shaded is optional):

INSTALL = Drive: \Path\ Filename parameters

Examples: INSTALL = C:\Dos\NLSFUNC

Syntax Options:

Drive:\Path . . . . Drive and directory containing *Filename.*

Filename . . . . . Name of memory-resident program that you want to run.

Parameters. . . . Command parameters, if any, required by *Filename.*

Command Type and Version:

Config.sys command; Network; Introduced with Ver 4.0

Notes:

1. Less memory is used when you load a program with INSTALL instead of loading from the AUTOEXEC.BAT file since an environment for a program is not created by INSTALL .
2. Do not use INSTALL to load programs that use shortcut keys, environment variables, or require COMMAND.COM for error handling.
3. Not all programs will function properly if loaded with INSTALL.
4. See also FASTOPEN, KEYB, NLSFUNC, SHARE, CONFIG.SYS.

INTERLNK

New V6.0

Link computers to share resources:

INTERLNK.EXE must be installed as a device driver in the CONFIG.SYS file before the INTERLNK and INTERSVR commands can be run.

Syntax (shaded is optional):

INTERLNK client : = server :

Examples: INTERLNK C: = F:

Syntax Options:

client : . . . . *The drive letter of the client drive that is* redirected to a drive on the server.

server : . . . . The drive letter on the server that will be redirected. If a letter is not specified, the client drive will no longer be redirected.

Command Type and Version:

External command; Network; Introduced with Ver 6.0

Notes:

1. See also INTERLNK.EXE and INTERSVR.
2. Note, the LASTDRIVE command may need to be used if drive letters greater than E are used.

INTERLNK.EXE New V6.0

Link computers to share resources:

INTERLNK.EXE must be installed as a device driver in the CONFIG.SYS file before the INTERLNK and INTERSVR commands can be run.

Syntax (shaded is optional):

Device= Drive: \Path\ INTERLNK.EXE /drives:n /noprinter /com:n|address /lpt:n|address /auto /noscan /low /baud:rate /v

Examples: Device=C:\ INTERLNK.EXE /drives:4

Syntax Options:

Drive:\Path . . . . Drive and directory containing the INTERLNK.EXE program.

/drives:n . . . . . . The number of redirected drives. Default is n=3. If n=0, only the printers are redirected.

/ noprinter . . . . . No printers are to be redirected. Default is INTERLNK redirects all ports.

/ com:n/address	Specifies that serial port *n* be used to transfer data. If *n* or the address is omitted, INTERLNK scans for the first available port. Default is INTERLNK redirects all ports.
/ lpt:n/address . .	Specifies that parallel port n be used to transfer data. If *n* or the address is omitted, INTERLNK scans for the first available port. Default is INTERLNK redirects all ports.
/ auto.	INTERLNK.EXE is installed in memory only if *client* can make a connection when the *server* starts up. Default is INTERLNK is installed whether or not *server* is there.
/ noscan	INTERLNK.EXE driver is installed, but a connection between *client* and *server* is prevented.
/ low	INTERLNK.EXE forces driver to be loaded into conventional memory. Default is driver loaded into upper memory if it is available.
/ baud:rate	Sets baud rate for com serial ports. Default=115200. Valid values are 9600, 19200, 38400, 57600, & 115200.
/ v	Used to resolve problems and conflicts between *com* and *lpt* ports and the computer's timer.

Command Type and Version:

CONFIG.SYS command; Network; Introduced Ver 6.0

Notes:

1. See also INTERSVR and INTERLNK the command.

INTERSVR.EXE New V6.0

Starts the INTERLNK server so that resources can be shared between linked computers: INTERLNK.EXE must be installed as a device driver in the CONFIG.SYS file before the INTERLNK and INTERSVR commands can be run.

Syntax (shaded is optional):

INTERSVR drive: /X=drive /lpt:n|address /com:n|address /baud:rate /b /v /rcopy

Examples: INTERSVR / rcopy

Syntax Options:

/ X=drive . . . . . . Specifies those drives that will not be redirected. Default is all drives are redirected.

/ lpt:n|address . . Specifies that serial port n be used to transfer data. If *n* or the address is omitted, INTERLNK scans for the first available port. Default is INTERSVR scans all ports.

/ com:n|address Specifies that serial port n be used to transfer data. If *n* or the address is omitted, INTERLNK scans for the first available port. Default is INTERSVR scans all ports.

/ baud:rate . . . . Sets baud rate for com serial ports. Default=115200. Valid values are 9600, 19200, 38400, 57600, & 115200.

/ b . . . . Display stat screen in black-and-white.

/ v . . . . Used to resolve problems and conflicts between *com* and *lpt* ports and the computer's timer.

/ rcopy. . . . . Copies all INTERLNK files from one computer to another. Note that a full 7 wire null-modem serial cable must be installed on the *com* port and the DOS MODE command must be available.

Command Type and Version:

External command; Network; Introduced with Ver 6.0

Notes:

1. See also INTERLNK.EXE and INTERLNK.
2. If port numbers for com and lpt are not specified, INTERLNK will scan and select the first port it finds.

JOIN.EXE

Removed V6.0

Joins a disk drive to a specific directory on another disk drive: Once joined, DOS treats the directories and files of the first drive as the contents of the second drive and path.

Syntax (shaded is optional):

Two syntax formats are valid:

JOIN Drive1: Drive2:\Path

JOIN Drive: /d

Examples: JOIN C: D:\Notes

JOIN C: D:\Notes\Bin (valid for DOS 5.0 only)

Syntax Options:

Drive1: . . . . Drive to be joined to *Drive2:\Path.*

Drive2:\Path . . . Drive and Path to which you want to JOIN *Drive1:*. Drive2:\Path must be empty and other than the root directory. With DOS Ver 5.0, you can JOIN to a subdirectory also, e.g. C:\Notes\Bin

Drive: . . . . Drive on which JOIN is to be canceled.

/ d. . . . . Cancels the JOIN command.

Command Type and Version:

External command; Introduced with Ver 3.0

Removed from MS DOS Version 6.0, however, it is available on Microsoft's MS-DOS 6.0 and 6.2x Supplemental Disks.

Considered too dangerous to use.

Notes:

1. Once you use the JOIN command, Drive1: becomes invalid.
2. If a specified path already exists before using JOIN, that directory cannot be used while JOIN is in effect. The specified directory must be empty or the JOIN operation will be incomplete and an error message will be displayed.
3. Commands that do not work with drives formed by JOIN are: ASSIGN, BACKUP, CHKDSK, DISKCOMP, DISKCOPY, FDISK, FORMAT, LABEL, MIRROR, RECOVER, RESTORE, SYS.
4. Use JOIN without parameters to show a list of the currently joined drives.

KBDBUF.SYS

New V6.0

A device driver that sets the number of key-strokes stored in the keyboard buffer.

Syntax (shaded is optional):

DEVICE = KBDBUF.SYS xxxx

Example: DEVICE = KBDBUF.SYS 200

Syntax Options (shaded is optional):

xxxx .... Designates the number of keystrokes held in the buffer. This number can range from 16 to 1024.

Command Type and Version:

CONFIG.SYS command

Introduced with MS-DOS Ver. 6.0

Available only on Microsoft's Supplemental Disks for MS-DOS Versions 6.0, 6.21 and 6.22.

KEYB.COM and KEYBOARD.SYS

Configures a keyboard for use with a specific language (installs alternate keyboard layout):

Syntax (shaded is optional):

If started in a batch file or at the DOS prompt:
KEYB xx,yyy,Drive:\Path\Filename /e / id:nn
If started in CONFIG.SYS:
install = Drive1:\Path1\KEYB.COM xx, yyy, Drive:\Path\Filename /e / id:nn

Examples: KEYB fr,850,437,C:\Dos\Keyboard.sys
install = C:\KEYB.COM fr , , C:\Dos\Keyboard.sys

Syntax Options:

xx . . . . Keyboard code. See table on next page.

yyy . . . . Code page. See table on next page.

Drive:\Path . . . . Drive and directory containing *Filename.*

Filename. . . . . . Keyboard definition file. Default=KEYBOARD.SYS

/e. . . . . Enhanced keyboard is being used.Ver5

/id:nn. . . . . Defines which keyboard is in use. See table on next page.

Drive1:\Path1 . . Drive and directory containing KEYB.COM

Command Type and Version:

External command; Network; Introduced with Ver 3.3

Notes:

1. When KEYB is installed through CONFIG.SYS, KEYB.COM with the .COM must be used. See also the CHCP command.
2. The Code Page specified with yyy must already be loaded on your system before KEYB is used.
3. You can switch from the default keyboard configuration to the KEYB configuration by pressing Ctrl+Alt+F2. To switch to the default keyboard configuration, press Ctrl+Alt+F1

4. The following are KEYB exit codes:
   - 0 .... KEYB definition file loaded successfully.
   - 1 .... Invalid Keyboard Code, Code Page, or syntax.
   - 2 .... Bad or missing keyboard definition file.
   - 4 .... Communication error with CON device.
   - 5 .... Requested Code Page has not been prepared.

The following table lists xx, yyy, and nnn values for different countries and languages.

Country or language	Keyboard Code *xx*	Code Page *yyy*	Keyboard ID *nnn*
Belgium	be	850,437	
Brazil	br	850,437	
Canadian-French	cf	850,863	
Czeck Republic	cz	852,850	
Denmark	dk	850,865	
Finland	su	850,437	
France	fr	850,437	120,189
Germany	gr	850,437	
Hungary	hu	852,850	
Italy	it	850,437	141,142
Latin America	la	850,437	
Netherlands	nl	850,437	
Norway	no	850,865	
Poland	pl	852,850	
Portugal	po	850,860	
Slovakia	sl	852,850	
Spain	sp	850,437	
Sweden	sv	850,437	
Switzerland (French)	sf	850,437	
Switzerland (German)	sg	850,437	
United Kingdom	uk	850,437	166,168
United States	us	850,437	
Yugoslavia	yu	852,850	

KEYBxx.COM

Loads a keyboard program for a specific country or keyboard type:

Syntax (shaded is optional):

KEYBxx

Examples: KEYBGR
KEYBUK

Syntax Options:

xx . . . . Code for a specific keyboard type:

KEYBdv . . . . Dvorak keyboard
KEYBfr. . . . . France
KEYBgr . . . . Germany
KEYBit. . . . . Italy
KEYBsp. . . . Spain
KEYBuk. . . . United Kingdom

Command Type and Version:

External command; Network; Introduced with Ver 3.0

Notes:

1. KEYBxx was discontinued after DOS version 3.2 and was replaced by KEYB.
2. Only one keyboard program can be loaded at a time.
3. You can switch from the default keyboard configuration to the KEYBxx configuration by pressing Ctrl+Alt+F2. To switch to the default keyboard configuration, press Ctrl+Alt+F1.
4. If you need to change from one keyboard type to another, restart the system after the changes have been made.

LABEL.EXE

Creates, changes or deletes the name or volume label of a disk: DOS displays the volume label and serial number, if it exists, as part of the directory listing.

Syntax (shaded is optional):

LABEL Drive: Label

Examples: LABEL
LABEL A: datadisc

Syntax Options:

Drive: . . . . Drive or diskette to be named.

Label. . . . . New volume label, up to 11 characters. A colon (**:**) must be included between the drive letter and label, but NO space.

Command Type and Version:

External command; Introduced with Ver 3.0

Notes:

1. Using the LABEL command without a label displays the following:

   Volume in Drive A is nnnnnnnnnnn
   Volume Serial Number is nnnn-nnnn
   Volume Label (11 characters, ENTER for none)?

2. The Volume label cannot include tabs. Spaces are allowed, but consecutive spaces may be treated as a single space.
3. **Do not** use the following characters in a volume label:

   * ? / \ | . , ; : + = [ ] ( ) & ^ < > “

4, LABEL is not case sensitive. (lower case is automatically converted to upper case.)

5. LABEL does not work on a drive created by ASSIGN, JOIN or SUBST.

LASTDRIVE

Number of drives installed: By default, the last drive is the one *after* the last drive used by your computer. DOS 4 and earlier it was E:

Syntax (shaded is optional):

LASTDRIVE = parameter

Examples: LASTDRIVE = F

Syntax Options:

parameter . . . . . A drive letter in the range of A through Z to correspond to the number of logical drives installed. Default is the drive after the last one used by the computer.

Command Type and Version:

CONFIG.SYS command; Introduced with Ver 3.0

Notes:

1. Memory is allocated by DOS for each drive specified by LASTDRIVE, therefore, don't specify more drives than are necessary.

LINK.EXE

Removed V5.0

8086 Object Linker that creates executable programs from Microsoft Macro Assembler (MASM) object files: LINK is for the experienced programmer and is not used by the general user.

Syntax (shaded is optional):

LINK (LINK prompts for file names, etc)
LINK object , execute , map , library options ;

Examples: LINK file /se:192 , , ;

Syntax Options:

object . . . . Object files to be linked together.
execute . . . . Name for created executable file.
map. . . . . Map listing file.
library . . . . Name(s) of library files to LINK.
options . . . . Options for the LINK program
; . . . . Terminates command line.

Command Type and Version:

External command; Introduced with Ver 1.0
Removed from Ver 5.0

Notes:

1. Further discussion of LINK is beyond the scope of POCKET PCRef.

LOADFIX.COM

Forces a program to load above the first 64k of conventional memory and then runs the program.

Syntax (shaded is optional):

LOADFIX Drive: \Path\ Filename parameters

Examples: LOADFIX C:\TEST.EXE

Syntax Options:

Drive:\Path. . . . Drive and directory containing *Filename.*
Filename. . . . . . Name of program that you want to run.
Parameters. . . . Command parameters, if any, required by *Filename.*

Command Type and Version:

External command; Introduced with Ver 5.0

Notes:

1. Use LOADFIX when the error message "Packed file corrupt" is reported during the execution of a program.

LOADHIGH or LH

***Loads programs into upper memory*:** Loading programs into upper memory frees up conventional memory for other programs. An upper memory manager such as EMM386 must be loaded first in order for LOADHIGH to function. LH and LOADHIGH are equivalent commands.

Syntax (shaded is optional):

LOADHIGH Drive:\Path\ Filename /L:region /s parameters

Examples: LOADHIGH C:\Dos\doskey.com
LH C:\Dos\doskey.com

Syntax Options:

Drive: \Path. . . . Drive and directory containing *Filename.*

Filename . . . . . Program to be loaded into high memory.

/ L:region. . . . . . Load the device driver into a specific upper memory region.

/s 6.2. . . . . Shrinks the upper-memory block (UMB) to minimum size while loading program. Used only with the */ L:region* switch. Typically used only by MEMMAKER.

parameters . . . . Command line parameters required by *Filename.*

Command Type and Version:

Internal command; Network; Introduced with Ver 5.0

Notes:

1. DOS=umb must be included in your CONFIG.SYS in order for LOADHIGH to function.
2. HIMEM.SYS and EMM386.EXE must be loaded in CONFIG.SYS on a 386/486 system in order to provide upper memory management for 386/486 systems. (Programs such as 386MAXand QEMM will provide the same capabilities.)
3. If there is not enough upper memory to load a program, DOS will load the program into conventional memory (no notice is given).
4. See also DEVICEHIGH, DOS, HIMEM.SYS, and EMM386.

5. When LOADHIGH is used, it is typically placed in the AUTOEXEC.BAT file.
6. Use MEM /c to see where programs are loaded.
7. Running MEMMAKER will automatically add all necessary LOADHIGH commands to AUTOEXEC.BAT

MD or MKDIR

Makes a Directory: Creates a new subdirectory under the current directory (if no Drive:\Path is specified). A new subdirectory on a different drive or under a different path can also be created. MD and MKDIR are equivalent commands.

Syntax (shaded is optional):

MD Drive:\Path\ subdirectory

Examples: MD contract
MKDIR contract
MD C:\contract\bin

Syntax Options:

Drive: . . . . Letter of drive for *subdirectory.*

Path . . . . Path where subdirectory is to be made. If no path is specified, e.g. C:\ only, the new directory is made a subdirectory under the root directory.

subdirectory . . . Name of the *subdirectory* being created.

Command Type and Version:

Internal command; Network; Introduced with Ver 2.0

Notes:

1. DOS will always assume that the MD command is on the current directory if no path is specified.
2. The maximum length of any path to the final subdirectory is 63 characters, including backslashes.

MEM.EXE

Display information about used and free system memory: Options are available that will display items such as which programs are loaded, the order of loaded programs, free memory, etc.

Syntax (shaded is optional):

MEM /program /page /a /c /d /f /m progname

Examples: MEM
MEM /classify

Syntax Options:

MEM . . . . Without any switches, the status of used and free memory is displayed.

/ program or */ p*. **DOS Version 4/5 only**: Displays the status of programs currently loaded into memory. This switch can not be used at the same time as */debug* and */classify*.

/ page or / p ❻ . **DOS Version 6 only.** Pauses display output after each screen.

/ a 6.2 . . . . Adds a line to the display stating the amount of memory available in HMA (High Memory Area)

/ c or / classify. . Displays the status of all programs and drivers currently loaded into conventional and upper memory. Other info, such as memory use and largest memory blocks available are also displayed. This switch can not be used at the same time as */program* and */debug*. Version 5.0

/ d or / debug . . Displays the status of programs and drivers currently loaded into memory. This switch can not be used at the same time as */program* and */classify*.

/ f or / free . . . . . Lists free regions in upper memory. */ free* can not be used with other switches, except */module*.

/m progname or */module progname* . . . Display info on a particular program loaded in memory. This switch can not be used with any other switches except */page*.

Command Type and Version:

External command; Network; Introduced with Ver 4.0

Notes:

1. Extended memory usage is displayed only if the installed system memory is 1 meg or greater. Only LIM 4.0 expanded memory use is displayed.
2. Total conventional memory=first 640k of RAM. Extended = mem above 1 meg. Expanded = bank switched LIM 4.0 memory.
3. If information is needed on hard drive available space, see the CHKDSK command.

MEMMAKER.EXE New V6.0

Optimizes computer memory by moving device drivers and memory-resident programs (TSR's) into upper memory: The system must be either a 386 or 486 and have extended memory available.

Syntax (shaded is optional):

MEMMAKER /b /batch /session /swap:drive /T /undo /w:size1,size2

Examples: MEMMAKER
MEMMAKER /undo

Syntax Options:

/b . . . . Display in black-and-white mode. Use if there are problems with your monochrome monitor.

/batch. . . . . Run MEMMAKER in unattended mode. This forces acceptance of defaults at all prompts. If an error occurs during the process, MEMMAKER restores the original AUTOEXEC.BAT,

CONFIG.SYS, and Windows SYSTEM.INI. Status messages and errors are reported in the MEMMAKER.STS file.

/ batch2. . . . . . . Completely automates the optimization process with absolutely no user intervention. No prompts, no pauses. Ver?

/ session . . . . . . This switch is only used by MEMMAKER during the optimizing process.

/ swap:drive . . . Specifies the drive letter of the system startup drive, if it has changed since the system started up. (encountered with some disk swapping programs)

/ T . . . . If problems are encountered between MEMMAKER and an IBM Token Ring network, use this switch. It disables the Token-Ring detection function.

/ undo . . . . Forces MEMMAKER to undo the most recent changes it has made to the system. This switch is normally used if problems are encountered after MEMMAKER has been run and you wish the system to be returned to its original confituration.

/ w:size1,size2 . Sets the upper memory size reserved for Windows translation buffers. Windows needs two separate areas of upper memory for the buffers. size1 is the size of the first area, size2 is the size of the second area. The default is no buffers are created (/ w:0,0).

Command Type and Version:

External command; Introduced with Ver 6.0

Notes:

1. See also DEVICEHIGH and LOADHIGH.
2. **WARNING: Do not run this program if Windows is running!**
3. CHKSTATE.SYS is a CONFIG.SYS command line that is automatically created by MEMMAKER during the optimization process. At the end of the process, it is automatically removed from CONFIG.SYS.

MENUCOLOR New V6.0

Command line to set text and background colors for the DOS startup menu in the CONFIG.SYS file: The startup menu is a list of system configuration choices that appear when your system is started. Each menu item is a set of CONFIG.SYS commands and is called a "configuration block." See your DOS manual for details of setting up and using the startup menu.

Syntax (shaded is optional):

MENUCOLOR = X , Y

Examples: MENUCOLOR 7, 9

Syntax Options:

X . . . . Sets menu text color. Valid values are 0 to 15.

, Y . . . . Sets screen background color. Valid values are 0 to 15. Default=0 (black).

Color Values

0=Black	8=Gray
1=Blue	9=Bright blue
2=Green	10=Bright green
3=Cyan	11=Bright cyan
4=Red	12=Bright red
5=Magenta	13=Bright magenta
6=Brown	14=Yellow
7=White	15=Bright white

Note: colors 8 to 15 blink on some displays.

Command Type and Version:

CONFIG.SYS command; Network; Introduced with Ver 6.0

Notes:

1. See also MENUDEFAULT, MENUITEM, NUMLOCK, INCLUDE and SUBMENU. All are used by the startup menu.
2. Don't make X and Y the same number, text won't show!

MENUDEFAULT New V6.0

Command line to set the default menu item for the DOS startup menu in CONFIG.SYS: The startup menu is a list of system configuration choices that appear when your system is started. Each menu item is a set of CONFIG.SYS commands and is called a "configuration block." See your DOS manual for details of setting up and using the startup menu.

Syntax (shaded is optional):

MENUDEFAULT = blockname , timeout

Examples: MENUDEFAULT = NET, 20

Syntax Options:

blockname . . . . Sets the default menu item. If no default is specified, item 1 is selected.

, *timeout* . . . . . . The number of seconds DOS waits before starting your computer with a default configuration.

Command Type and Version:

CONFIG.SYS command; Network;
Introduced with Ver 6.0

Notes:

1. See also MENUCOLOR, MENUITEM, NUMLOCK, INCLUDE and SUBMENU. All are used by the startup menu.

MENUITEM **New V6.0**

Command line to define a menu item for the DOS startup menu in CONFIG.SYS: The startup menu is a list of system configuration choices that appear when your system is started. Each menu item is a set of CONFIG.SYS commands and is called a "configuration block." See your DOS manual for details of setting up and using the startup menu.

Syntax (shaded is optional):

MENUITEM blockname , menutext

Examples: MENUITEM NET, Start your Network

Syntax Options:

blockname . . . . Defines a menu item on the startup menu. It is usable only within a menu block and there can be a maximum of nine menu items per menu. If DOS cannot find a specified name, the item will not appear on the startup menu. blockname can be up to 70 characters long but you cannot use spaces, \ (backslashes), / (forward slashes), commas, semicolons, equal signs or square brackets.

, menutext. . . . . Up to 70 characters of text to display for the menu item. If no text is given, DOS displays *blockname* as the menu item.

Command Type and Version:

CONFIG.SYS command; Network; Introduced with Ver 6.0

Notes:

1. See also MENUCOLOR, MENUDEFAULT, NUMLOCK, INCLUDE and SUBMENU. All are used by the startup menu.

MIRROR.COM **Removed V6.0**

Records information about 1 or more disks for use by UNFORMAT and UNDELETE ***commands***:

Syntax (shaded is optional):

Three syntax formats are valid:

MIRROR Drives: /1 /Tdrive – entries . . .
MIRROR /u
MIRROR /partn

Examples: MIRROR /u
MIRROR C: /Ta /Tc

Syntax Options:

Drives: . . . . The drive or drives to be MIRRORed.

/ 1 . . . . Instructs MIRROR to retain only the latest information about a disk. The default causes MIRROR to make a backup of existing information before new information is recorded.

/T*drive – entries* Loads a deletion–tracking program that maintains information so that the UNDELETE command can recover files. *drive* is required and is the drive to be MIRRORed. *entries* is optional and is the maximum number of entries in PCTRACKR.DEL (the deletion tracking file). *entries* can range from 1 to 999 and the *entries* defaults are as follows:

Disk Size	Default Entry	File Size
360k	25	5k
720k	50	9k
1.2 meg	75	14k
1.44 meg	75	14k
20 meg	101	18k
32 meg	202	36k
>32 meg	303	55k

/u . . . . Unload and disable the deletion tracking program. If other memory resident programs have been loaded after MIRROR, the */u* switch will not function.

/partn . . . . Save partitioning information for the UNFORMAT command. The information is saved on a floppy disk for use at a later time if partitions need to be rebuilt by UNFORMAT. The default drive to save the information to is A:, although a different drive can be specified at the prompt.

Command Type and Version:

External command; Network; Introduced with Ver 5.0
Removed from DOS Ver 6.0, functionally replaced by the UNDELETE / T command.
MIRROR is available on Microsoft's MS-DOS Ver. 6.0, 6.21, and 6.22 Supplemental Disks.

Notes:

1. If MIRROR is used without any switches, it saves information about the disk in the current drive.
2. Do not use MIRROR on any drive that has been redirected using the JOIN or SUBST commands. If ASSIGN is used, it must be used before MIRROR.
3. MIRROR saves a copy of a drive's FAT (file allocation table) and a copy of the drive's root directory. Since this information may change regularly, it is recommended that you use MIRROR regularly in order to maintain current information for UNFORMAT to use. It is recommended that MIRROR be placed in your AUTOEXEC.BAT file so that current information is saved every time your system is turned on or re-booted.
4. See also UNFORMAT and UNDELETE.
5. **DOS 6.0 Note:** MIRROR is still available from Microsoft as a supplemental disk, call them for details.

MODE.COM

Controls system devices such as display, serial ports, printer ports, and system settings:
NOTE: Since there are many functions that MODE addresses, they will each be treated separately in the following pages.

Command Type and Version:

External command; Network; Introduced with Ver 1.0

MODE to Display Device Status

Syntax (shaded is optional):

MODE device /status

Examples:
MODE (Display status of all system devices)
MODE con (Display console status)
MODE lpt1 /status

Syntax Options:

device . . . . Device for which status is requested.
/status or */sta* . . Displays status of redirected parallel printers.

Notes:

None

MODE to Configure Printer

Configures parallel port printers: Ports that can be addressed include PRN, LPT1, LPT2, and LPT3. Printer types that can be configured are IBM compatibles and Epson compatibles.

Syntax (shaded is optional):

```
MODE  Lptn  : c , L , r
MODE  Lptn  : cols=c  lines=L  retry=r
```

Examples: MODE Lpt2:132,6
MODE Lpt1 cols=132 lines=8

Syntax Options:

Lpt*n*. . . . . Parallel port to be configured. Valid numbers for *n* are 1, 2, and 3.

c or *cols=* . . . . . Number of character columns per line. Default=80, Values=80 or 132.

L or *lines=*. . . . . Number of vertical lines per inch. Default=6, Values=6 or 8.

r or *retry=* . . . . . Type of retry if time-out error occurs. This option leaves a memory resident piece of MODE in RAM. Valid *rs'* are:

- e . . . . Return busy port error from status check.
- b . . . . Return busy port "Busy" from status check.
- p . . . . Continue retry until printer accepts data.
- r. . . . . Return "Ready" from busy port status check.
- n . . . . Disable retry (Default). "none" is also valid.

Notes:

1. *retry=b* is equivalent to the "p" parameter in earlier DOS versions.
2. Ctrl+C will break out of a time-out loop.
3. PRN and LPT1 can be used interchangeably.
4. Do not use any *retry* options over a network.
5. The colon (:) with Lptn is optional.

MODE to Configure Serial Port

Configures a serial communications port: Ports that can be addressed include COM1, COM2, COM3, and COM4.

Syntax (shaded is optional):

MODE COMn :b , p , d , s , r
MODE COMn :baud=b parity=p data=d stop=s retry=r

Examples: MODE COM1:24,N,8,1

Syntax Options:

COM*n* . . . . Asynchronous serial port to be configured. Valid values are 1, 2, 3, and 4.

b or *baud=* . . . . Transmission rate in bits per second. Only the first 2 digits are required. Valid values are 11=110 baud, 15=150, 30=300, 60=600, 12=1200, 24=2400, 48=4800, 96=9600, & 19=19,200 baud.

p or *parity=* . . . . Parity check. N=none, E=even, O=odd, M=mark, S=space. Default=E

d or *data=* . . . . . Number of data bits in a character. Valid values are 5, 6, 7, 8. Default=7

s or *stop=* . . . . . Number of stop bits for end of character. Valid values are 1, 1.5 or 2. Default=1 (Default at 110 baud=2)

r or *retry=* . . . . . Type of retry if time-out error occurs. This option leaves a memory resident piece of MODE in RAM. Valid *r*s' are:

- e . . . . Return busy port error from status check.
- b . . . . Return busy port "Busy" from status check.
- p . . . . Continue retry until printer accepts data.
- r. . . . . Return "Ready" from busy port status check.
- n . . . . Disable retry (Default). "none" is also valid.

Notes:

1. If any parameters are omitted in the MODE statement, the most recent setting is used.
2. Do not use *retry* values over a network.
3. *retry=b* is equivalent to the "p" parameter in earlier DOS versions.

MODE to Redirect Printing

Redirects output from a parallel port to a serial port:

Syntax (shaded is optional):

MODE Lptm = COMn

Examples: MODE Lpt1: = COM1:
MODE Lpt1 = COM2

Syntax Options:

Lpt*m* . . . . The parallel port to be redirected. Valid *m* values are 1, 2, and 3.

COM*n* . . . . The serial port to be redirected to. Valid *n* values are 1, 2, 3, and 4

Notes:

1. Following a redirection, the original output direction can be restored by typing MODE lptm where m is the original printer port.

MODE to Set Device Code Pages

Selects, refreshes, prepares, or displays code page numbers for parallel printers and the console:

Syntax (shaded is optional):

MODE device codepage prepare= yyy
Drive:\Path\Filename
MODE device codepage select=yyy
MODE device codepage refresh
MODE device codepage /status

Examples:
MODE CON codepage prepare = 860
MODE LPT1 codepage /status

Syntax Options:

device . . . . Device to be affected. Valid values are CON, LPT1, LPT2, and LPT3.

codepage prepare or *cp prep*.. . . . Prepares the code page for the specific *device.* Use *codepage select* after this command.

Drive:\Path\Filename Drive, directory and file containing code page information (.CPI files) needed to prepare a code page.

- EGA.CPI . . . . . Enhanced graphics adapter or PS2
- EGA2.CPI . . . . Similar to EGA.CPI, but with more code pages.
- 4201.CPI . . . . . IBM Proprinters II and III, Model 4201 IBM Proprinters II & III, Model 4202
- 4208.CPI . . . . . IBM Proprinter X24E Model 4207 IBM Proprinter XL24E Model 4208
- 5202.CPI . . . . . IBM Quietwriter III Printer
- LCD.CPI. . . . . . IBM PC Convertible Liquid Crystal Disp.
- ISO.CPI . . . . . . Complies with Part 3 of ISO 9241 specification.

codepage select or *cp sel* Selects a code page for a specific device. *cp prep* above must be run first.

codepage refresh or *cp ref* If a code page is lost, this command reinstates it.

codepage . . . . . When used alone, codepage displays the numbers of the code pages that have been prepared for a specific device.

/status or */sta* . . Displays the current code page numbers

Notes:

1. See also NLSFUNC and CHCP.
2. EGA.CPI and EGA2.CPI are shipped with DOS. All others are supplied on Microsoft's MS-DOS Supplemental Disks.

MODE to Set Display Mode

Reconfigure or select active display adapter:

Syntax (shaded is optional):

```
MODE adapter , shift , t
MODE adapter , n
MODE CON : cols=c lines=n
```

Examples: MODE co80,r
MODE CON:cols=40 lines=43

Syntax Options:

adapter . . . . Display adapter category as follows:

- 40 or 80 . . . . Number of characters/line
- bw40 or bw80 . . . . CGA (color graphics with color disabled. Characters per line =40 or 80
- co40 or co80 . . . . . Color display with color enabled. Characters per line = 40 or 80.
- mono . . . . Monochrome display with 80 characters per line.

shift. . . . . Shift CGA screen left or right. Valid values are L for left, R for right.

t. . . . . Starts a test pattern for screen alignment.

n . . . . Vertical lines per screen. Valid values are 25, 43, and 50. ANSI.SYS must be loaded in CONFIG.SYS for this to work.

cols=. . . . . Characters or columns per line. Valid values are 40 and 80.

lines= . . . . Vertical lines per screen. Valid values are 25, 43, and 50. ANSI.SYS must be loaded in CONFIG.SYS for this to work.

Notes:

1. Some monitors do not support 43 and 50 vertical lines per screen.

MODE to Set Typematic Rate

Set the rate at which DOS repeats a character when a keyboard key is held down: Some keyboards do not recognize this command.

Syntax (shaded is optional):

MODE con : rate= r delay= d

Examples: MODE con : rate=20 delay=2

Syntax Options:

con or con: . . . . Keyboard

rate=*r* . . . . The rate that a character is repeated on the display when a key is held down. *r* Default=20 for AT keyboards, Default=21 for PS2 keyboards. *r* Range = 1 to 32, which is equivalent to the following: rate 1 = 2 characters per second (cps), 10 = 4.3 cps, 20 = 10 cps, 30 = 24 cps and 32 = 30 cps.

delay=*d*. . . . . . . The amount of time, after a key is held down, before the repeat function activates. *d* Default=2, *d* valid values are 1, 2, 3 and 4 (equivalent to 0.25, 0.50,0.75, and 1 second respectively). If a delay is specified, rate must also be specified.

Notes:

1. The keyboard must be an AT or PS/2 class or higher keyboard in order for this command to work.

MORE.COM

Displays output one screen at a time: MORE reads standard input from a pipe or redirected file and is typically used to view lengthy files. Each screen of information ends with the prompt -More- and you can press any key to view the next screen.

Syntax (shaded is optional):

MORE < Drive: \Path\ Filename

or

command | MORE

Examples: MORE < C:\Data.txt
DIR | MORE

Syntax Options:

Drive:\Path . . . . Drive and directory containing *Filename.*

Filename. . . . . . Name of file that supplies data to be displayed.

command . . . . . Name of command that supplies data to be displayed, for example, DIR

Command Type and Version:

External command; Network; Introduced with Ver 2.0

Notes:

1. When using the pipe (|) for redirection, you are able to use DOS commands, such as DIR, SORT, and TYPE with MORE, but the TEMP environment variable in AUTOEXEC.BAT file should be set first.
2. MORE saves input information in a temporary file on disk until the data is ready to be displayed. If there is no room on the disk, MORE will not work. Also, if the current drive is a write-protected drive, MORE will return an error.

MOVE.EXE New V6.0

Move files from one drive or directory to another: You can also move and rename complete directories, along with their files and subdirectories, to other drives or directories. **Warning:** DOS does not warn you if it is about to overwrite files with the same name.

Syntax (shaded is optional):

MOVE /Y /-Y Drive: \Path\ Filename
, Drive: \Path\ . . . Filename Destination

Examples: M

Syntax Options:

/Y 6.2 . . . . Directs MOVE to replace existing files without a confirmation prompt.

/-Y 6.2 . . . . Directs MOVE to ask for confirmation prior to replacing an existing file. (Default)

Drive:\Path. . . . Drive and directory containing *Filename.*

Filename. . . . . . Name of file(s) that you want to move.

Destination . . . . The new location of the file(s) being moved. This can be a drive, subdirectory, or combination of the two.

Command Type and Version:

External command; Network; Introduced with Ver 6.0

Notes:

1. If more than one file is being moved, the Destination must be a drive and subdirectory.

MSAV and MWAV.EXE 

Microsoft Anti-Virus scanners for DOS (MSAV) and Windows (MWAV).

Syntax (shaded is optional):

MSAV Drive: /S /C /R /A /L /N /P /F /ss /video /IN /BW /mono /LCD /FF /BF /NF /BT /NGM /LE /PS2

Examples: MSAV C: /A /N /F

Syntax Options:

Drive: .... Drive to be scanned. The Default is the current drive.

/S .... Scan but do not remove viruses.

/C .... Scan and remove viruses.

/R .... Create a MSAV.RPT report that lists the number of files scaned, the number of viruses found, and the number of viruses removed. Default=no report.

/A .... Scan all drives except A and B.

/L .... Scan all logical drives except networks.

/N .... Run in command mode, not graphical. Also, display contents of a MSAV.TXT file if it's present.

/P .... Run in command line mode w/ switches.

/F .... Do not display file names during scan.

/ss .... Set screen display size:
/25=25 lines, this is the default
/28=28 lines, use with VGA
/43=43 lines, use with EGA or VGA
/50=50 lines, use with VGA
/60=60 lines, use with VGA and Video7

/video .... Display list of valid video screen switches.

/IN .... Run MSAV using a color scheme.

/ BW . . . . Run MSAV in black-and-white mode.
/ mono . . . . Run MSAV in monochrome mode.
/ LCD . . . . Run MSAV in LCD mode.
/ FF . . . . Run MSAV in fast screen mode for CGA monitors. Screen quality is worse.
/ BF . . . . Use computer BIOS to display video.
/ NF . . . . Disable use of alternate screen fonts.
/ BT . . . . Enable graphics mouse in Windows.
/ NGM . . . . Use default mouse character instead of the graphics character.
/ LE . . . . Switch left and right mouse buttons.
/ PS2 . . . . Reset mouse if the mouse cursor locks up or disappears.

Command Type and Version:

External command; Network; Introduced with Ver 6.0

Notes:

1. MSAV is actually Central Point Software's Anti-Virus program which has been licensed to Microsoft.

MSBACKUP-MWBACKUP.EXE

New V6.0

Microsoft's menu driven program to backup and restore one or more files from one disk to another disk: This program is a replacement for BACKUP and RESTORE used in previous DOS versions. MSBACKUP is for DOS and MWBACKUP is for Windows.

Syntax (shaded is optional):

MSBACKUP setup_file /BW /LCD /MDA

Examples: MSBACKUP /BW

Syntax Options:

setup_file . . . . . Predefined setup that specifies which files to backup and the type of backup to be performed. MSBACKUP automatically creates this file if "save program settings". During the "save program" function, if no file name is specified, the file name DEFAULT.SET is used.

/BW . . . . Run screen in black-and-white mode.

/LCD . . . . Run screen in LCD mode.

/MDA . . . . Run screen in monochrome mode.

Command Type and Version:

External command; Network; Introduced with Ver 6.0

Notes:

1. MSBACKUP does not support the use of tape backups.
2. Backups and catalog files are compatible between MSBACKUP and MWBACKUP.

MSCDEX.EXE New V6.0

Microsoft's CD-ROM Extensions : MSCDEX is used in conjunction with the CD-ROM device driver that was shipped with the drive. It is normally executed in the AUTOEXEC.BAT file.

Syntax (shaded is optional):

MSCDEX /D:driver /D:driver2 . . . /E /K /S /V /L:letter /M:number

Examples: MSCDEX /D:1

Syntax Options:

/D:driver . . . . . . Drive signature for the first CD-ROM drive. Typically this is MSCDØØØØ. The drive signature must match that of the CD-ROM driver in CONFIG.SYS.

/D:driver2 . . . . . Drive signature of the second CD-ROM drive. Typically this is MSCDØØØ1.

/ E . . . . CD-ROM drive can use expanded memory, if available, to store sector buffers.

/ K . . . . Provide Kanji support for CD-ROM.

/ S . . . . Share CD-ROM on MS-NET network or Windows for workgroup servers.

/ V . . . . Display MSCDEX memory stats when the program starts.

/ L:letter . . . . Specifies drive letter for first CD-ROM. If more than one CD-ROM, DOS assigns the subsequent drive letters.

/ M:number . . . . Specifies the number of sector buffers.

Command Type and Version:

External command; Network; Introduced with Ver 6.0

Notes:

1. Do not start MSCDEX after Windows has been started.

MSD.COM & .EXE **New V6.0**

Microsoft's menu driven system diagnostics: This program provides detailed technical information about your system.

Syntax (shaded is optional):

MSD /I /B [/F drive:\path\filename]
[/P drive:\path\filename]
[/S drive:\path\filename]

Examples: MSD
MSD / B / I

Syntax Options:

/ I. . . . . Forces MSD to not initially detect hardware when it starts. This may be necessary if MSD is not running properly or locks up.

/ B. . . . . Run MSD in black-and-white mode.

drive:\path. . . . . Drive and path where a MSD report file is to be written.

/ F drive:\path\filename. . Prompts for a company, address, & phone to be written on the MSD report named *filename*.

/ P drive:\path\filename . Writes a complete MSD report to a file named *filename*.

/ S drive:\path\filename . Writes a summary MSD report to a file named *filename*.

Command Type and Version:

External command; Network; Introduced with Ver 6.0

Notes:

1. MSD has shipped with Windows for quite some time and is an excellent diagnostics tool.

MSHERC.COM

Installs support for Qbasic graphics programs using the Hercules graphics card:

Syntax (shaded is optional):

MSHERC /half

Examples: MSHERC /half

Syntax Options:

/half . . . . Use this switch if a color adapter card is also installed in the system.

Command Type and Version:

External command; Network; Introduced with Ver 5.0

NLSFUNC.EXE

National language support function, which loads country-specific information and code-page switching: Use NLSFUNC from either the command line or through **CONFIG.SYS.**

Syntax (shaded is optional):

At the DOS prompt:
NLSFUNC Drive:\Path\ Filename
If loaded through CONFIG.SYS:
INSTALL= Drive1:\Path1\ NLSFUNC.EXE
country

Examples: NLSFUNC C:\Bin\Newcode.sys

Syntax Options:

Drive:\Path . . . . Drive and directory containing *Filename.*
Filename. . . . . . File containing country-specific information.
Drive1:\Path1 . . Drive and directory containing NSLFUNC.
country . . . . Same as *Filename.*

Command Type and Version:

External & CONFIG.SYS command; Network; Introduced with Ver 3.3

Notes:

1. The COUNTRY command in CONFIG.SYS defines the default value for Drive:\Path \Filename. If there is no COUNTRY command in CONFIG.SYS, NLSFUNC looks for COUNTRY.SYS in the root directory of the start up drive.
2. See also CHCP and MODE.

NUMLOCK

New V6.0

Command line to set the NUM LOCK key to ON or OFF for the DOS startup menu in the CONFIG.SYS file: The startup menu is a list of system configuration choices that appear when your system is started. Each menu item is a set of CONFIG.SYS commands and is called a "configuration block." See your DOS manual for details of setting up and using the startup menu.

Syntax (shaded is optional):

NUMLOCK = ON or OFF

Examples: NUMLOCK = ON

Syntax Options:

ON .... Turns NUM LOCK key on.
OFF .... Turns NUM LOCK key off.

Command Type and Version:

CONFIG.SYS command; Network; Introduced with Ver 6.0

Notes:

1. See also MENUDEFAULT, MENUITEM, MENUCOLOR, INCLUDE and SUBMENU. All are used by the startup menu.

PATH

Sets a directory search path: DOS uses the path command to search for executable files in specified directories. The default is the current working directory.

Syntax (shaded is optional):

PATH Drive1: \Path1; Drive2: \Path2;...

Examples: PATH C:\ ;D:\ ;D:\Dos;D:\Utility\test
PATH (displays the current search path)
PATH ; (clears search-path settings other than default setting (current directory)

Syntax Options:

Drive1: Drive2: Specifies drive letters to be included in the search path

Path1 *Path 2* . Specifies directory (s) in the search path where DOS should look for files.

; . . . . Must be used to separate multiple *Drive:\Path* locations or if used as *Path ;* it clears search-path settings other than the default setting.

Command Type and Version:

Internal command; Network; Introduced with Ver 2.0

Notes:

1. The maximum number of characters allowed in the PATH statement is 127. See SUBST for ways to get around this limit. Also see the SET Path statement.
2. If files have the same name but different extensions, DOS searches for files in the following order: .COM, .EXE, .BAT.
3. If identical file names occur in different directories, DOS looks in the current directory first, then in locations specified in PATH in the order they are listed in the PATH statement.
4. A PATH command is usually included in the AUTOEXEC.BAT file so that it is issued at the time the system starts.

PAUSE

Pauses the processing of a batch file: Suspends processing of a batch file and prompts the user to press any key to continue.

Syntax (shaded is optional):

PAUSE

Examples: PAUSE

Syntax Options:

None

Command Type and Version:

Internal command; Only used in Batch Programs; Introduced with Ver 1.0

Notes:

1. Earlier versions of PAUSE indicated that a text comment could be inserted after PAUSE and the message would display when PAUSE ran, for example "PAUSE This is a test." This message function is not functional.
2. Ctrl+C or Ctrl Break will stop a Batch program while running or at pause

POWER

Reduces power consumption in a computer when applications and devices are idle: Once the POWER.EXE driver is loaded through the CONFIG.SYS file, POWER at the command line turns power on/off, reports status and sets conservation levels.

Syntax (shaded is optional):

POWER
ADV[:MAX or REG or MIN] or STD or OFF

Examples: POWER (displays current settings)
POWER OFF

Syntax Options:

ADV[:MAX or REG or MIN] . . . Conserves power when devices are idle. MAX=maximum power conservation, REG=default, balance conservation with device performance, MIN=higher device performance is needed.

STD . . . . If the computer supports APM, STD conserves power. If not supported, it turns off the power.

OFF . . . . Turns off power management.

Command Type and Version:

External command;
Network; Introduced with Ver 6.0

Notes:

1. See also POWER.EXE.
2. If the computer does not support APM, using STD will disable the power completely.

POWER.EXE

New V6.0

Reduces power consumption in a computer when applications and devices are idle:
This driver conforms to the Advanced Power Management (APM) specifications and is loaded through the CONFIG.SYS file.

Syntax (shaded is optional):

Device = Drive:\Path\ POWER.EXE
ADV[:MAX or REG or MIN] or STD or OFF /low

Examples: Device = POWER.EXE

Syntax Options:

Drive1\Path . . . Specifies the location of POWER.EXE

ADV[:MAX or REG or MIN] . . . Conserves power when devices are idle. MAX=maximum power conservation, REG=default, balance conservation with device performance, MIN=higher device performance is needed.

STD .... If the computer supports APM, STD conserves power. If not supported, it turns off the power.

OFF .... Turns off power management.

/ low .... Loads driver into conventional memory, even if upper memory is available. The default is load into upper memory.

Command Type and Version:

CONFIG.SYS command; Network; Introduced with Ver 6.0

Notes:

1. See also POWER.
2. If the computer does not support APM, using STD will disable the power completely.

PRINT.EXE

Prints a text file to a line printer, in the background. Other DOS commands can be executed at the same time PRINT is running:

Syntax (shaded is optional):

PRINT /d:device /b:size /u:ticks1 /m:ticks2 /s:ticks3 /q:qsize /t Drive:\Path\ Filename ... /c /p

Examples: PRINT C:\Test.txt /c C:\test2.txt /p
PRINT /d:Lpt1 /u:25

Syntax Options:

/d:device ...... Name of printer device.
Parallel Ports: Lpt1, Lpt2, Lpt3.
Serial Ports: com1,com2, com3, com4.
PRN and Lpt1 refer to the same parallel port. Default=PRN
/d must precede Filename.

/b:size . . . . Sets size (in bytes) of internal buffer. Default=512, Range=512 to 16384.

/u:ticks1 . . . . . . Maximum number of clock ticks PRINT is to wait for a printer to become available. Default=1, Value Range=1 to 255.

/m:ticks2 . . . . . . Maximum number of clock ticks PRINT can take to print a character on printer. Default=2, Value Range=1 to 255.

/s:ticks3. . . . . . . Maximum number of clock ticks allocated for background printing. Default=8, Value Range=1 to 255.

/q:qsize . . . . Max number of files allowed in print queue. Default=10 Value Range=4 to 32.

/ t. . . . . Removes files from the print queue.

Drive:\Path\Filename . Location & Filename of file to be printed.

/ c . . . . Removes files from the print queue. Both the */c* and */p* switches can be used on the same command line. When the */c* **precedes** the *Filenames* on the command line, it applies to all the files that follow until PRINTcomes to a */p,* in which case the */p* switch applies to the file preceding the */p*. When the */c* switch **follows** the *Filenames,* it applies to the file that precedes the */c* and all files that follow until PRINT comes to a */p* switch.

/ p . . . . Adds files to the print queue. Both the */c* and */p* switches can be used on the same command line. When the */p* **precedes** the *Filenames* on the command line, it applies to all the files that follow until PRINT comes to a */c,* in which case the */c* switch applies to the file preceding the */c*. When the */p* switch **follows** the *Filenames*, it applies to the file that precedes the */p* and all files that follow until PRINT comes to a */c* switch.

Command Type and Version:

External command; Introduced with Ver 2.0

Notes:

1. You can use the */d,/b,/u,/m,/s* and */q* switches only the first time you use PRINT. DOS must be restarted to use them again.
2. Use a program's own PRINT command to print files created with that program. PRINT only functions correctly with ASCII text.
3. Each queue entry includes a drive, directory and subdirectory and must not exceed 64 characters per entry.

PRINTER.SYS

Removed V6.0

Installable device driver that supports code-page switching for parallel ports PRN, LPT1, LPT2, AND LPT3:

Syntax (shaded is optional):

DEVICE = Drive:\Path\ PRINTER.SYS
LPTn = (type , hwcp , n)

Examples:
DEVICE=C:\Dos\PRINTER.SYS LPT1:=(4201,437,2)

Syntax Options:

Drive:\Path . . . . Drive and directory containing PRINTER.SYS

LPT*n*. . . . . LPT1, LPT2, or LPT3

type . . . . Type of printer in use. Valid values for *type* and the printer represented by each value are as follows:

4201 . . . IBM Proprinters II and III M.4201
IBM Proprinters II and III XL M.4202

4208 . . . IBM Proprinters X24E M.4207
IBM Proprinters XL24E M.4208

5202 . . . IBM Quietwriter III M.5202

hwcp. . . . . Code-page supported by your hardware. DOS supports the following code pages:

437 . . . . United States
850 . . . . Multilingual (Latin I)
852 . . . . Slavic (Latin II)
860 . . . . Portuguese
863 . . . . Canadian-French
865 . . . . Nordic

n . . . . Number of additional code-pages.

Command Type and Version:

CONFIG.SYS command; Introduced with Ver 3.3

Removed in Ver. 6.0, however it is available from Microsoft on the MS-DOS 6.0, 6.21, and 6.22 Supplemental Disks

PRINTFIX.COM New V6.0

Stops MS-DOS from Checking the status of the printer attached to the system:

Syntax:

PRINTFIX

Example: printfix

Syntax Options:

None

Command Type and Version:

External command, Introduced with Ver. 6.0

Available from Microsoft on the MS-DOS 6.0, 6.21, and 6.22 Supplemental Disks.

Notes:

1. Use only if printing problems occurred while installing MS-DOS 6.0, 6.21, or 6.22.

PROMPT

Change Prompt: Customizing prompt to display text or information and change color. Example: time or date, current directory or default drive.

Syntax (shaded is optional):

PROMPT Text $Characters

Examples: PROMPT pg (Most commonly used)
If ANSI.SYS is loaded and you have a color monitor, try the following for colors at the DOS level:
PROMPT $e[35;44;1m$pge[33;44;1m

Syntax Options:

PROMPT . . . . . PROMPT by itself resets to default prompt.

Text . . . . *Text* can be any typed message.

$Characters . . . Type in special characters from the table below to create special prompts.

Typed character **displayed prompt**

$q. . . . . The = character
$$. . . . . The $ sign
$t . . . . Current time
$d. . . . . Current date
$p. . . . . Current drive and path
$v. . . . . DOS version number
$n. . . . . Current drive
$g. . . . . >Greater-than symbol
$l . . . . <Less-than symbol
$b. . . . . (|) vertical bar
$_. . . . . Enter, first position of next line
$e. . . . . ASCII escape code (code 27)
$h. . . . . Backspace (deletes a **prompts** command line character)

Command Type and Version:

Internal command; Network; Introduced with Ver 2.0

Notes:

1. See also ANSI.SYS
2. The PROMPT command is typically inserted in AUTOEXEC.BAT

QBASIC®.EXE

Basic computer language: A program that reads instructions and interprets those instructions into executable computer code. A complete environment for programming in the Basic language is provided by the QBASIC program.

Syntax (shaded is optional):

QBASIC /b /editor /g /h /mbf /nohi /run Drive:\Path \Filename

Examples: QBASIC
QBASIC C:\Qb\Bin\Test

Syntax Options:

Drive:\Path . . . . Drive and directory containing *Filename.*
Filename . . . . . Name of file to load when QBASIC starts.
/b. . . . . QBASIC is displayed in black and white.
/editor . . . . Invokes EDIT, DOS full-screen text Editor.
/g. . . . . Fastest screen update of a CGA monitor.
/h. . . . . Displays max. number of display lines.
/mbf. . . . . Converts the resident functions MKS$, MKD$, CVS, and CVD to MKSMBF$, MKDMBF$, CVSMBF, and CVDMBF.
/nohi . . . . Allows use of monitor without high-intensity video support. COMPAQ laptop computers cannot use this switch.
/run . . . . The specified BASIC program is run before being displayed.

Command Type and Version:

External command; Network; Introduced with Ver 5.0

Notes:

1. QBASIC.EXE must be in the current directory, search path, or in same directory as EDIT.COM in order to use the DOS Editor.
2. Consecutive Basic programs can be run from a Batch file if the Basic system command and the /run switch is used.
3. If GW-BASIC programs need to be converted to QBASIC, read REMLINE.BAS in QBASIC's subdirectory.
4. If a monitor does not support shortcut keys, use */b* and */nohi.*

RAMDRIVE.SYS or VDISK.SYS

Creates a simulated hard disk from the system's RAM memory: RAM disks are much faster than hard disks but they are temporary (if the system shuts down, the data is lost).

Syntax (shaded is optional):

Device=Drive:\Path\ RAMDRIVE.SYS disksize sectorsize numentry /e /a

Examples:

Device=C:\Dos\RAMDRIVE.SYS 4096 /a

Syntax Options:

Drive:\Path .... Drive & directory containing RAMDRIVE.SYS

disksize....... Sets size of RAM disk in kilobytes. Valid sizes range from 4 to 32767.Default=64

sectorsize ..... Sets sector size in bytes. Valid sizes are 128, 256, and 512. Default=512. Do not change default if possible.

numentry...... Sets the number of files and directories that the RAM disk's root directory can hold. Default=64, range=2 to 1024. If this parameter is used, *disksize* and *sectorsize* must also be set.

/e .... RAM disk uses extended memory. 4Kb minimum extended memory is needed. Default=uses conventional memory.

/a .... RAM disk uses expanded memory. 4Kb minimum extended memory is needed. Default= uses conventional memory.

Command Type and Version:

CONFIG.SYS command; Introduced with Ver 3.1(Vdisk=3.0)

Notes:

1. Multiple RAM disks are allowed.
2. Always try to use */e* or */a* so that conventional RAM is not used.
3. A memory manager like HIMEM.SYS must be used if /e is used.
4. An expanded memory manager must be installed if /a is used.

RD or RMDIR

Removes a directory: You cannot delete a directory without first deleting its files and subdirectories. The directory must be empty except for the **"." and ". ."** symbols which represent the directory itself and the parent directory. RD and RMDIR are equivalent commands.

Syntax (shaded is optional):

RD Drive: \Path

Examples: RD \Data
RD \Data\Smith

Syntax Options:

Drive . . . . Drive containing *Path.*
Path . . . . Directory to be deleted.

Command Type and Version:

Internal command; Network; Introduced with Ver 2.0

Notes:

1. Use DIR to list hidden and system files and ATTRIB to remove hidden and system file attributes in order to empty directory.
2. When a backslash (\) is used before the first directory name in *Path*, DOS treats the directory as a subdirectory of the root directory. Omit the backslash (\) before the first directory name and DOS treats the directory as a subdirectory of the current directory.
3. The directory being deleted cannot be the current directory and must be an empty directory.

RECOVER.EXE Removed V6.0

Recovers readable information from a disk containing bad sectors: When CHKDSK reports bad sectors on a disk, use the RECOVER command to read a file, sector by sector, and recover data from the good sectors.

Syntax (shaded is optional):

RECOVER Drive:\Path\ Filename

Examples: RECOVER A:

Syntax Options:

Drive:\Path . . . . Drive and directory containing *Filename.*

\Filename . . . . . *Filename* to be recovered. If no *Filename* or *Path* is specified, the entire drive is recovered.

Command Type and Version:

External command; Introduced with Ver 2.0
Removed from Ver 6.0, deemed too dangerous.

Notes:

1. Wildcards (* and?) cannot be used with the RECOVER command.
2. When an entire disk is recovered, each file is placed in the root directory in a FILEnnnn.REC file. The 4 digit numbering sequence on each recovered file is as follows: FILE0001.REC, FILE0002, etc.
3. Since all data in bad sectors is lost when you recover a file, it is best to recover files one at a time, allowing you to edit each file and re-enter missing information.
4. If a drive was formed by the ASSIGN, JOIN or SUBST command, the RECOVER command will not work. It will not work with the BACKUP or RESTORE command since you must use RESTORE with backup files that you created with the BACKUP command.
5. RECOVER cannot recover files on a network drive.
6. If an entire drive is recovered, it is possible that some files will be lost, since the recovered files are written to the root directory and a limited number of files will fit in the root directory.
7. See also CHKDSK

REM

Allows use of remarks (comments) in a Batch file or in CONFIG.SYS: Any BATCH command or CONFIG.SYS line beginning with REM is ignored by DOS.

Syntax (shaded is optional):

REM Comment

Examples: REM begin files here

Syntax Options:

Comment . . . . . Line of text that you want to include as a comment.

Command Type and Version:

Internal command;

Batch command; Introduced with Ver 1.0

CONFIG.SYS command; Introduced with Ver 4.0

Notes:

1. ECHO ON must be used in the Batch or CONFIG.SYS file for a comment to be displayed.
2. REM can be used without a comment to add vertical spacing to a Batch file, but you can also use blank lines. Blank lines are ignored by DOS.
3. Do not use redirection characters **(>**or **<)** or pipe **(|)** in a Batch file comment.
4. a ";" can be used in place of REM in the WIN.INI file.

REN or RENAME

Renames a file(s): Changes the name(s) on all files matching a specified Filename. REN and RENAME are equivalent commands.

Syntax (shaded is optional):

REN Drive:\Path\ Filename1 Filename2

Examples: REN C:\ data*.dbf *.db2

Syntax Options:

Drive:\Path . . . . Drive and directory containing *Filename.*

Filename1. . . . . File(s) to be renamed.

Filename2. . . . . New name for file(s). You cannot rename Drive or Path.

Command Type and Version:

Internal command; Network; Introduced with Ver 1.0

Notes:

1. The use of Wildcards **(*** and **?)** are allowed.
2. You cannot duplicate a *Filename.*
3. See also LABEL, COPY and XCOPY.

REPLACE.EXE

Replaces files in the target drive with files from the source drive when the filenames are the same: If same name files are not on the target drive, the new files will be added to the target drive.

Syntax (shaded is optional):

REPLACE Source:\Path1\ Filename
Target:\Path2 /a /p /r /w

REPLACE Source:\Path1\ Filename
Target:\Path2 /p /r /s /w /u

Examples: REPLACE A:*.* C:\Test /a /s

Syntax Options:

Source:\Path1. . Source drive and directory containing *Filename.*

Filename. . . . . Name of source file.

Target:\Path2 . . Location of the destination file(s).

/a. . . . . Adds, instead of replacing, new files to the destination file. This switch **cannot** be used with */s* or */u*.

/p. . . . . Prompts for confirmation before adding a source file or replacing the destination file.

/r . . . . Replaces read-only and unprotected files.

/s. . . . . Searches subdirectories of the destination directory and replaces matching files with the source file. The */s* switch **cannot** be used with */a*.

/w . . . . Waits for a disk to be inserted before REPLACE starts copying. If */w* is not specified, REPLACE begins immediately.

/u. . . . . Updates or replaces files in the destination directory that are older than files in the source directory.

Command Type and Version:

External command; Network; Introduced with Ver 3.2

Notes:

1. REPLACE issues a message concerning the number of files that have been added or replaced when the operation is complete.
2. Use /w if you need to change disks during REPLACE.
3. REPLACE does not function on system or hidden files.
4. REPLACE returns the following exit codes: (see IF errorlevel)
   0 . . . . Files successfully added or replaced
   2 . . . . Source files could not be found
   3 . . . . Source or destination path could not be found
   5 . . . . User does not have access to files being replaced
   8 . . . . Insufficient system memory to complete command
   11 . . . Wrong command line syntax

RESTORE.EXE

***Restores files that were backed up using the BACKUP command*:** The "backed up" and "restored to" disk types do not have to be identical. In Ver 6.0, RESTORE will only restore backups made with previous versions of DOS. It will **NOT** restore backups made with the Ver 6.0 or 6.2x MSBACKUP program!

Syntax (shaded is optional):

RESTORE Drive1: Drive2: \Path\ Filename /s /p /b:date /a:date /e:time /L:time /m /n /d

Examples: RESTORE A: C:*.* /s
RESTORE B: D:\Data*.dbf /s /m

Syntax Options:

Drive1: . . . . Drive on which backed-up files are stored.
Drive2:\Path . . . Drive and directory to which backed-up files will be restored.
Filename. . . . . . Name(s) of backed-up file(s) to be restored.
/s . . . . Restores all subdirectories.
/p . . . . Prompts for permission to restore files that are read-only or files that have changed since last backup.
/b*:date* . . . . Restores files changed or modified on or before a specified *date.*
/a *:date*. . . . . . . Restores files changed or modified on or after a specified *date*.
/e*:time* . . . . Restores files changed or modified at or earlier than a specified *time*.
/L *:time*. . . . . . . Restores files changed or modified at or later than a specified *time*.
/m. . . . . Restores only files changed or modified since the last backup.
/n . . . . Restores files that no longer exist on the destination disk. (Drive2)

/*d* . . . . Without restoring, */d* displays a list of files on the backup disk that match names specified in *Filename.* Version 5.0

Command Type and Version:

External command; Network; Introduced with Ver 2.0

Notes:

1. RESTORE does not restore the system files (IO.SYS and MSDOS.SYS or IBMBIO.COM and IBMDOS.COM).
2. RESTORE will not function on drives that have been redirected with ASSIGN, JOIN, or SUBST.
3. MS-DOS RESTORE Version 5.0 will restore backups made with all previous versions of BACKUP.
4. RESTORE returns the following exit codes: (see IF errorlevel)
   - 0 . . . . Files successfully restored
   - 1 . . . . Files to be restored could not be found
   - 3 . . . . RESTORE stopped by user Ctrl+C
   - 4 . . . . RESTORE ended in error.
5. BACKUP is not included in DOS Ver 6.0, see the MSBACKUP utility program.

SCANDISK.EXE New V6.2

MS-DOS utility program to analyze and recover lost chains and lost clusters on hard or floppy disks to make more space available on these devices. SCANDISK also checks the surface of the disk for errors. Lost chains or lost clusters recovered by SCANDISK are saved in the root directory as files with a .CHK extension. The contents of each file can be examined using the MORE command or any text editor. The files can then be saved or deleted as needed. SCANDISK is an interactive program that steps the user through a series of options in order to scan and repair each selected drive.

Syntax (shaded is optional):

SCANDISK Drive1: Drive2: Volume_Name Drive:\Path\Filename /all /autofix /checkonly /custom /fragment /mono /nosave /nosummary /surface /undo Undo_Drive

Examples: SCANDISK C: /autofix
SCANDISK /all
SCANDISK /fragment C:\TEST\data

Syntax Options:

Drive: . . . . Identifies drive (disk) to scan.

Drive:\Path\Filename. . . . Identifies drive (disk), directory, and file to be checked for fragmentation

/all. . . . . Scan and repair all local drives.

/autofix . . . . Scan and repair without prompts.

/checkonly. . . . . Only scans the selected drives, no repairs are made. Can not be used with */custom* or */autofix*.

/custom. . . . . . . Scan and repair according to parameters set in SCANDISK.INI file. Can not be used with */autofix* or */checkonly*.

/fragment . . . . . Check for fragmentation of files on selected drives. Individual directories and files may be indicated and wildcards may be used.

/mono . . . . Execute in monochrome mode.

/nosave. . . . . . . Scans automatically and deletes any lost chain or cluster. Can be used only with */autofix*. If */nosave* is left off, all lost chains and clusters will automatically be saved as .CHK files in the root directory of the drive being scanned.

/nosummary . . . Disables full-screen summary display. Full-screen summary display is the default setting for SCANDISK.

/surface. . . . . . . Scans for physical errors on disk.

Volume_Name . Name of unmounted compressed volume (compressed using either DRVSPACE or DBLSPACE) to be scanned and repaired.

/undo. . . . . Undo any repairs made by SCANDISK. Use a blank disk as the undo disk.

Undo_Drive. . . . Drive containing the current undo disk.

Command Type and Version:

External command, Interactive, NOT for Network; Introduced with Version 6.2

Notes:

1. Do not use SCANDISK on CD-ROM drives, network drives, or drives created using ASSIGN, SUBST, JOIN, or INTERLNK.
2. Do not use SCANDISK on drives compressed using PC-DOS Ver 6.1.
3. All applications (including Windows) must be stopped before running SCANDISK or data may be lost.
4. Memory resident programs may need to be disabled in the AUTOEXEC.BAT and CONFIG.SYS files prior to running SCANDISK.
5. SCANDISK.INI file is a text file containing settings which determine how SCANDISK operates on start-up. Sections such as Environment and Custom contain the required settings. For more information see comments in the file.
6. SCANDISK is similar to CHKDSK but is more comprehensive in its analysis of a drive.
7. SCANDISK sets ERRORLEVEL to one of the following values upon return to the DOS prompt:
   0 - No problems detected.
   1 - Syntax error.
   2 - Unexpected termination due to an internal error or an out-of-memory error.
   3 - User exit prior to completion.
   4 - User exit during surface scan.
   254 - Disk problems found and all corrected.
   255 - Disk problems found but not all corrected.

SELECT.EXE **Removed V5.0**

Installs DOS on a new disk along with country specific information such as time and date formats and collating sequences: Select also formats the target disk, creates CONFIG.SYS and AUTOEXEC.BAT on a new disk and copies the source disk to the target disk.

Syntax (shaded is optional):

SELECT Source Target\Path yyy xx

Examples: SELECT B: A: 045 dk

Syntax Options:

Source . . . . Drive containing Information to be copied.
Target . . . . Drive containing disk onto which DOS is to be copied.
Path . . . . Name of directory containing information to be copied.
yyy . . . . Country code. See COUNTRY Command.
nn . . . . Keyboard code. See KEYB Command.

Command Type and Version:

External command; Introduced with Ver 3.0
Removed from Version 5.0

Notes:

1. WARNING: SELECT is used to install DOS for the first time. Everything on the *target* disk is erased. SELECT is not available for use on Version 5.0 and should be used with caution in earlier versions.
2. The *Source Drive* can be either Drive A:or Drive B:.
3. If a hard disk is used in the Target Drive, DOS will prompt for the correct internal label for that disk. If the wrong label is typed in, SELECT ends.

SET

Sets, removes or displays environment variables: SET is normally used in the AUTOEXEC.BAT file to set environment variables when the system starts. With DOS Ver 6.0, SET can be used in CONFIG.SYS. ❻

Syntax (shaded is optional):

SET variable = string

Examples:

SET (displays current environment settings)
SET TEMP=E:\Windows\Temp
SET variable =
(above clears *string* associated with *variable*)

Syntax Options:

variable....... The *variable* to be set or modified.
string......... Text *string* to be associated with *variable*.

Command Type and Version:

Internal command; Network; Introduced with Ver 2.0

Notes:

1. If SET is used to define values for both *variable* and *string*, DOS adds *variable* to the environment and associates *string* with it. If *variable* already existed, the new *variable* replaces the old one.
2. In a Batch file, SET can be used to create variables that can be used in the same way as %1 through %9. In order to use the new variable, it must be enclosed with %, e.g. %variable%
3. The SET command uses memory from the environment space. If the environment space is too small, DOS will issue the error message "Out of Environment Space". See the SHELL command and COMMAND.COM for ways to increase environment space.
4. See also PATH, PROMPT, SHELL and DIR for additional information on environment variables.

SETUP and BUSETUP.EXE

Programs which initially install MS-DOS.

Syntax(shaded is optional):

Initial installation from command prompt:

drive1:SETUP
Example: a:setup

Installation of certain utilities after initial installation from command line:

drive1:SETUP /e /f /u /i
Example: a:setup /e

Installation of certain utilities after initial installation by insertion of Setup disk and restart of computer:

drive1:BUSETUP /e /u
Example: a:busetup /e

Syntax options:

drive1:. . . . . . . Drive containing the SETUP program.

/ f. . . . . If the system drive A is not compatible with the Setup disk, this switch makes a minimal installation of DOS by copying essential command files on a floppy disk which is compatible with drive A.

/u. . . . . Used when installing MS-DOS 6 with certain third-party disk-partitioning software.

/e. . . . . Used to install Anti-Virus, Backup, or Undelete after initial installation.

/ i. . . . . Causes Setup to skip automatic hardware detection.

Command Type and Version:

External command, Introduced with Version 5.0.

Notes:

1. See the README.TXT files with MS-DOS Versions 5.0, 5 Upgrade, 6.0, and 6.2 for more information.
2. Press F3 twice to exit Setup.

SETVER.EXE

Sets the DOS version number that is reported to a program by MS-DOS® 5.0: If a program will not run under Ver 5.0 and issues the error "Incorrect DOS Version," adding the program to the SETVER file may allow the program to run.

Syntax (shaded is optional):

To initially load the SETVER table in CONFIG.SYS
Device = Drive:\Path\ SETVER.EXE
At DOS prompt or in Batch file:
SETVER Drive:\Path\ (Displays current table)
SETVER Drive:\Path Filename v.vv
SETVER Drive:\Path Filename /delete /d /quiet

Examples: Device=C:\DOS\SETVER.EXE

SETVER C:\DOS (Displays current ver. table)
SETVER C:\DOS TEST.EXE 3.30
(above adds TEST.EXE to the version table)
SETVER C:\DOS TEST.EXE /delete
(above deletes TEST.EXE from the version table)

Syntax Options:

Drive:\Path . . . . Drive and directory containing SETVER.

Filename. . . . . Program file to be added to version table. Must be a .EXE or .COM file. Wild cards are not allowed.

v.vv . . . . The DOS version number that should be reported to the program when it is run.

/delete or */ d* . . . Delete the version table entry for the *Filename* program.

/quiet. . . . . Hides the message normally displayed during the deletion process.

Command Type and Version:

External and CONFIG.SYS command;
Network; Introduced with Ver 5.0

Notes:

1. When loaded in CONFIG.SYS, the .EXE extension with SETVER.EXE must be used.
2. In order for SETVER to function at the DOS prompt or in a Batch file, it must first be loaded through CONFIG.SYS. SETVER is automatically added to CONFIG.SYS by the MS-DOS 5.0 setup program.
3. If you set a version number for your MS-DOS 5.0 COMMAND.COM, your system may not start.
4. If changes or additions or deletions are made to the SETVER table, your system must be restarted in order for the changes to take effect.
5. If a program starts correctly after it has been added to the SETVER table, the program may still not run correctly under Ver 5.0 if a compatibility problem exists.
6. If a program is added to the SETVER table and the program name is already in the table, the new entry and version number will replace the existing entry.
7. The following SETVER exit codes can be used in conjunction with the IF errorlevel command to report completion and error codes:

    0 . . . . SETVER function completed successfully
    1 . . . . Invalid command switch.
    2 . . . . Invalid *Filename*.
    3 . . . . Insufficient system memory to complete command.
    4 . . . . Invalid version number (*v.vv*) format specified.
    5 . . . . Specified entry not currently in version table.
    6 . . . . SETVER could not find the SETVER.EXE file.
    7 . . . . Invalid drive specified.
    8 . . . . Too many command line parameters specified by user.
    9 . . . . Missing command line parameter.
    10 . . . . . . Error while reading SETVER.EXE file.
    11 . . . . . . Corrupt SETVER.EXE file.
    12 . . . . . . Specified SETVER.EXE file does not support a version table.
    13 . . . . . . Insufficient space in version table to add a new entry.
    14 . . . . . . Error detected while writing to the SETVER.EXE file.

SHARE.EXE

Program that installs file-sharing and locking capabilities on hard disk: The share command is installed through AUTOEXEC.BAT or CONFIG.SYS and is used by networking, multitasking under Windows, DOSSHELL, and others.

Syntax (shaded is optional):

In a Batch file or at the DOS prompt:
SHARE /f:space /L:locks
In CONFIG.SYS:
INSTALL= Drive:\Path\ SHARE.EXE
/ f:space /L:locks

Examples: SHARE / f:4096 /L:40
INSTALL=C:\Dos\SHARE.EXE

Syntax Options:

Drive:\Path . . . . Drive and directory containing the SHARE.EXE file.

/f:space . . . . File space allocated in bytes for the DOS storage area used to record file-sharing information. Default=2048

/L:locks . . . . Number of files that are to be locked. Default=20

Command Type and Version:

External command; Network; Introduced with Ver 3.0

Notes:

1. In CONFIG.SYS, the .EXE extension must be included with SHARE.EXE
2. SHARE allows DOS to check and verify all read and write requests from programs.
3. The average length of a file name and its Path is 20 bytes. Use that value when calculating the */f:space* switch.
4. Beginning with Ver 5.0, SHARE is no longer required to support drive partitions >32mb.

SHELL

Specifies the name and location of a command interpreter, other than COMMAND.COM: Include the SHELL command to CONFIG.SYS to add a different Command Interpreter.

Syntax (shaded is optional):

SHELL = Drive:\Path\ Filename parameters

Examples:
SHELL=C:\COMMAND.COM /e:1024 /p

Syntax Options:

Drive:Path. . . . . Drive and directory containing *Filename.*

\Filename . . . . . Command Interpreter to be used.

Parameters. . . . Command-line parameters or switches to be used with Command Interpreter.

Command Type and Version:

CONFIG.SYS command; Introduced with Ver 2.0

Notes:

1. The SHELL command does not use or accept any switches, only the Command Interpreter uses switches .
2. The default Command Interpreter is COMMAND.COM.
3. SHELL must be used if the Command Interpreter is in a location other than the Root directory or if you need to change the environment size of COMMAND.COM.
4. **DOSSWAP** is the DOS Task Swapper and is used internally by the SHELL command. There are no switches for DOSSWAP and it should not be run from the DOS command line.

SHIFT

Allows a change in the position of replaceable command line parameters in a Batch file: Specifically, SHIFT copies the value of each replaceable parameter to the next lowest parameter (for example, %1 is copied to %0, %2 is copied to %1, etc).

Syntax:

SHIFT

Examples: SHIFT

Syntax Options:

None

Command Type and Version:

Internal command only used in Batch programs; Introduced with Ver 2.0

Notes:

1. Batch files, usually limited to ten parameters (%0 through %9) on the command line, can now use more than 10. This is made possible because if more than 10 parameters are used, those appearing after the 10th will be shifted one at a time into %9.
2. Once the parameters are shifted, they cannot be shifted back.

SIZER.EXE

SIZER is used only by MEMMAKER during the memory optimizing process. It is used to determine the size, in memory, of device drivers and memory resident programs. It is added automatically to AUTOEXEC.BAT or CONFIG.SYS in order to determine the memory size, and when MEMMAKER is finished, SIZER is automatically removed.

SMARTDRV.EXE New V6.0

Directs DOS to create a disk cache in extended memory or conventional memory: The cache effectively increases the speed of all disk functions. The SMARTDRV command allows management of the cache created by SMARTDRV.EXE.

Syntax (shaded is optional):

SMARTDRV /x Drive: + or - /b:buffer_size /c /e:element_size /f /initcachesize[wincachesize] / l /n /q /r /s /u / v

Example: SMARTDRV C-
SMARTDRV / r

DEVICE = SMARTDRV.EXE /x Drive: + or - /b:buffer_size /e:element_size /f /initcachesize[wincachesize] / l /n /q /r /s /u / v

Example: DEVICE = SMARTDRV.EXE C 1024 512
DEVICE = SMARTDRV.EXE /q

DEVICE = SMARTDRV.EXE /Double_buffer

Syntax options:

Drive: . . . . Identifies the drive (disk) that will use the cache. No specification allows all drives to use the cache. Ver 6.2 allows the caching of CD_ROM drives.

+ or - . . . . Cache-type (read or write) is enabled or disabled for identified drive. With Ver 6.0, "+" allows both read and write caching for the disk, "-" allows no caching, no specification for a floppy disk allows only read caching, and no specification for a hard drive allows both read and write caching. With Ver 6.2, "+" allows both read and write caching for the disk, "-" allows no cach-

ing, no specification for a floppy disk, CD-ROM drive, or drives created using INTERLINK allows only read caching, and no specification for a hard drive allows both read and write caching.

/b:buffer_size . . States the size of the read-ahead buffer. The buffer size can be set to any multiple of the element_size. The default size is 16 384 bytes (16K) which is twice the maximum (default) element_size of 8192 bytes (8K).

/c. . . . . Directs SMARTDRV to clear the buffer by writing all data in the cache to the cached disk. Use this switch before turning off the computer to save the cached data to the disk.

/e: element_size States the size of cache that SMARTDRV moves at one time. Element_size(bytes) can be one of the following: 1024, 2048, 4096, or 8192 (default).

/f 6.2 . . . . Directs SMARTDRV to write data in the cache to the disk after completion of each command. This is the default setting.

/initcachesize . . States size, in kilobytes, of the initial cache when SMARTDRV starts and Windows is not active. If not specified SMARTDRV sets initcachesize according to the amount of extended memory available as follows:

Extended Memory	Initcachesize
below 1MB	All extended
1MB to 2MB	1MB
2MB to 4MB	1MB
4MB to 6MB	2MB
above 6MB	2MB

/ L . . . . Limits SMARTDRV to only conventional (low) memory, even if extended memory (Upper Memory Blocks, UMB) is available.

/n 6.2 .... Directs SMARTDRV to write data in the cache to the disk only when the system is idle.

/q............. Directs SMARTDRV to load in the quiet mode with no messages on status or errors. Switch can not be used with /v.

/r .... Restarts SMARTDRV after clearing all data from the current cache to the cached disk.

/s............. Status of SMARTDRV is displayed.

/u 6.2 .... Disables the loading of CD-ROM caching.

/ v .... Directs SMARTDRV to display messages on status or errors when loading. Default is to not display messages unless error conditions are encountered. Switch can not be used with /q.

/ wincachesize . States, in kilobytes, the amount of cache that SMARTDRV will remove from initcachesize prior to starting Windows. If not specified SMARTDRV sets wincachesize according to the amount of extended memory available as follows:

Extended Memory	Wincachesize
below 1MB	0 (no cache)
from 1MB to 2MB	256KB
from 2MB to 4MB	512KB
from 4MB to 6MB	1MB
above 6MB	2MB

/ x 6.2 .... Directs SMARTDRV to disable write-behind caching for all drives.

/ Double_buffer. Directs SMARTDRV to perform double buffering which is needed for compatibility with some hard-disk controllers.

Command Type and Version:

External Command or Device Driver

Introduced with Ver 6.0, and some Windows before that.

Notes:

1. Do not start or load SMARTDRV while Windows is running.

2. For a CD-ROM drive to be cached SMARTDRV must load after MSCDEX.
3. MS-DOS LOADHIGH (LH) command can be used to load SMARTDRV high.
4. If the hard drive requires use of the double_buffer switch to perform properly, the double_buffer component of SMARTDRV must be loaded in conventional memory and the DEVICE command line for SMARTDRV must appear in the CONFIG.SYS file before the DEVICE command line for EMM386.
5. CONFIG.SYS must contain a DEVICE command which loads HIMEM.SYS or some other memory manager in order for SMARTDRV to use extended memory.
6. SMARTDRV is not an interactive program which steps the user through a series of screens.
7. If SMARTDRV is run without parameters being set, DOS will set up a disk cache using default parameters.

SMARTDRV.SYS Removed V6.0

Creates a disk cache in extended or expanded memory: A disk cache can significantly increase the speed of any disk operations.

Syntax (shaded is optional):

DEVICE = Drive:\Path\ SMARTDRV.SYS initsize minsize /a

Examples:
DEVICE=C:\DOS\SMARTDRV.SYS 1024 512

Syntax Options:

Drive:\Path . . . . Drive and directory containing SMARTDRV.SYS.

initsize. . . . . Initial size of disk cache in kilobytes. Default=256; Range=128 to 8192. Size is rounded off to 16k blocks.

minsize . . . . Minimum size of disk cache in kilobytes. Default=no minimum size. This option is important to programs such as Windows, which can reduce the cache size as required for its own use.

/a . . . . Specifies that the disk cache is to be set up in expanded memory. The Default places the cache in extended memory.

Command Type and Version:

CONFIG.SYS command; Introduced with Ver 4.0
Removed from MS-DOS Ver 6.0

Notes:

1. If no sizes are specified with SMARTDRV, then all available extended or expanded memory is allocated to the cache.
2. In order to use extended memory, HIMEM.SYS or another extended memory manager must be installed. HIMEM.SYS must precede SMARTDRV in CONFIG.SYS
3. On 80286 / 386 / 486 systems, extended memory is probably the best choice for SMARTDRV.
4. Do not use disk compaction programs while SMARTDRV is loaded.

SMARTMON.EXE New V6.0

Monitors SMARTDRV cache performance under Windows. Removed from DOS 6.2

Command Type and Version:

External command, Introduced with MS-DOS Ver 6.0.
Removed from DOS 6.2

SORT.EXE

A filtering program that reads the input, sorts the data and then writes the results to a screen, file or another device: The SORT command alphabetizes a file, rearranges in ascending or descending order by using a collating table based on Country Code and Code Page settings.

Syntax (shaded is optional):

SORT /r /+n < Drive1:\ Path1\ Filename1 > Drive2:\ Path2\ Filename2

command I SORT /r /+n > Drive2:\ Path2\ Filename2

Examples: SORT < C:\Data\Text.txt
DIR I SORT > C:\Sortdata.txt

Syntax Options:

Drive1:\Path1. . Drive and directory containing *Filename*.

Filename1. . . . . File containing data to be sorted.

Drive2:\Path2\\. Drive and directory containing *Filename2*.

Filename2. . . . . File in which to store sorted data.

Command. . . . . Specific command whose output is data to be sorted.

/r . . . . Reverses sorting order: Z to A and 9 to 0.

/+n. . . . . Sorts according to character in column *n*.

Command Type and Version:

External command; Network; Introduced with Ver 2.0

Notes:

1. Use the pipe (I) or the less-than (<) to direct data through SORT from a command or filename. Before using a pipe for redirection, set the TEMP environment variable in AUTOEXEC.BAT
2. Specify the MORE command to display information one screen at a time. You are prompted to continue after one screen is shown.
3. SORT is not case sensitive.
4. Files as large as 64K can be accommodated by SORT.
5. ASCII characters with codes higher than 127 are sorted based on the system's configuration with CONTRY.SYS.

SPATCH.BAT New V6.0

Batch file needed to maintain compatibility between MS-DOS 6.0, 6.21, or 6.22 and the permanent swap file established by Windows Ver 3.0.

Command Type and Version:

External command, Introduced with MS-DOS Ver 6.0. Available on the MS-DOS 6.0, 6.21, and 6.22 Supplemental Disks.

STACKS

***Supports the dynamic use of data stacks*: The STACKS command is used in CONFIG.SYS.**

Syntax:

STACKS = n,s

Examples: STACKS = 8, 512

Syntax Options:

n . . . . Defines the number of STACKS. Valid values for *n* are 0 and numbers in the range 8 to 64.

s . . . . Defines STACK size in bytes. Valid values for *s* are 0 and numbers in the range 32 to 512.

Command Type and Version:

CONFIG.SYS; Introduced with Ver 3.2

Notes:

1. Default setting for the STACKS command are as follows:

COMPUTER	STACKS
IBM PC, IBM PC/XT	0,0
IBM PC-PORTABLE.	0,0
OTHER	9, 128

2. When the values for *n* and *s* are specified at 0, DOS allocates no stacks. If your computer does not seem to function properly when STACKS are set to 0, return to the default values.

SUBMENU **New V6.0**

Command line to setup an item to display another set of choices for the DOS startup menu in CONFIG.SYS: The startup menu is a list of system configuration choices that appear when your system is started. Each menu item is a set of CONFIG.SYS commands and is called a "configuration block". See your DOS manual for details of setting up and using the startup menu.

Syntax (shaded is optional):

SUBMENU = blockname , menutext

Examples: SUBMENU = NET, Network Choices

Syntax Options:

blockname . . . . Sets the name of the associated menu block. The menu block must be defined somewhere else in the CONFIG.SYS file and can contain other menu definition commands. *Blockname* can be up to 70 characters but without spaces, backslashes, forward slashes, commas, semicolons, equal signs and square brackets.

, menutext. . . . . Text to be displayed for the menu item. If no text is defined, DOS displays the *blockname* as the menu item. menutext can be up to 70 characters long.

Command Type and Version:

CONFIG.SYS command; Network; Introduced with Ver 6.0

Notes:

1. See also MENUCOLOR, MENUITEM, NUMLOCK, INCLUDE and MENUDEFAULT. All are used by the startup menu.

SUBST.EXE

Substitutes a path with a drive letter: The SUBST command lets you use a drive letter (also known as a virtual drive) in commands as though it represents a physical drive.

Syntax (shaded is optional):

SUBST (Lists the virtual drives in effect)
SUBST Drive1: Drive2:\ Path
SUBST Drive1: /d (deletes virtual drive)

Examples: SUBST
SUBST R: B: \Data\Text.txt

Syntax Options:

Drive1: .... Virtual drive to which a path is assigned.

Drive2: .... Physical drive that contains the specified path.

Path .... Path to be assigned to the virtual drive named *Drive1:*

/d. .... Deletes the *Drive1:* virtual drive.

Command Type and Version:

External command; Introduced with Ver 3.1

Notes:

1. Commands that do not work on drives where SUBST has been used are as follows:

ASSIGN	DISKCOPY	RECOVER
BACKUP	FDISK	RESTORE
CHKDSK	FORMAT	SYS
DEFRAG	LABEL	UNDELETE /s
DISKCOMP	MIRROR	

2. A virtual drive letter must be included in the LASTDRIVE command in CONFIG.SYS.
3. Use SUBST rather than ASSIGN to ensure compatibility with future DOS versions.
4. If using drive letters higher than E, the LASTDRIVE command must also be used.
5. Do not use SUBST while Windows is running!

SWITCHAR

Changes the switch character: The forward slash, " / " is the standard switch character. SWITCHAR allows the user to choose another switch character.

Syntax (shaded is optional):

SWITCHAR= cc

Example: switchar = *

Syntax Options:

cc . . . . New switch character.

Command Type and Version:

CONFIG.SYS command, Introduced with Ver 2.0. Removed Version 3.0.

SWITCHES

***Forces enhanced keyboard to function like a conventional keyboard*:** This command is used in the CONFIG.SYS file.

Syntax (shaded is optional):

SWITCHES = /W /K /N /F

Examples: SWITCHES = / k

Syntax Options:

/ W . . . . If Windows 3.0 is used in enhanced mode and you have moved the WINA20.386 file, use this switch to tell DOS that the file has been moved.

/K ❻ . . . . *Ignores extended keys on 101-key* keyboards. It forces COMMAND.COM to use an older BIOS call to read the keyboard , making it possible to use certain older TSRs that depend on the older call. Actually, this switch was introduced in DOS V4.0, but was undocumented.

/N ❻ . . . . *Disables the F5 and F8 keys so that you* cannot bypass startup commands.

/F ❻ . . . . *Skips the 2 second system delay after* "Starting MS-DOS . . ." is displayed during startup.

Command Type and Version:

CONFIG.SYS command; Introduced with Ver 4.0

Notes:

1. Use the SWITCHES command when there is a program that does not properly interpret input from an enhanced keyboard. This command enables the enhanced keyboard to use conventional keyboard functions.
2. If SWITCHES=/k is used in a system that uses ANSI.SYS, be sure to also use the /k switch on the ANSI.SYS command.

SYS.COM

Copies the DOS system files (IO.SYS and MSDOS.SYS on MS-DOS systems or IBMBIO.COM and IBMDOS.COM on PC-DOS systems) and the Command Interpreter from one disk drive to another disk drive.

Syntax (shaded is optional):

SYS Drive1:\Path Drive2

Examples: SYS A: (current drive to drive A:)
SYS D:\ A: (copy from disk in D: to A:)

Syntax Options:

Drive1:\Path . . . Drive and directory where system files are located. If a path is not specified, DOS searches the root directory. If a drive is not specified, DOS uses the current drive as the system files source drive.

Drive2: . . . . Drive to which system files are to be copied. These files can be copied to a root directory only.

Command Type and Version:

External command; Introduced with Ver 1.0

Notes:

1. The order in which the SYS command files are copied is as follows: IO:SYS, MSDOS.SYS and COMMAND.COM.
2. The two system files no longer need to be "contiguous" in Ver 5.0. In simple terms, this means that pre DOS 3.3 disks do not need to be reformatted in order to install the Ver 5.0 operating system.
3. The SYS command will not work on drives redirected by ASSIGN, JOIN or SUBST.
4. The SYS command does not work on Network drives.
5. See also DISKCOPY, which duplicated disks of the same size (including transfer of the operating system). See also COPY and XCOPY for information on copying all files except system and hidden files.
6. ❻ With **DOS 6.0**, DBLSPACE.BIN is also copied to the target drive.
7. Pre DOS 5.0 can only be SYS Drive1:

TIME

Enter or change current system time: DOS uses the internal clock to update the directory with date and time when a file is created or changed.

Syntax (shaded is optional):

TIME Hours: Minutes: Seconds: Hundredths a *or* p

Examples: TIME
TIME 13:45 or TIME 1:45 p
TIME 11:28p

Syntax Options:

Hours:. . . . . Specifies the hour. One or two digit number with valid values from 0-23.

Minutes: . . . . . . Specifies the minute. One or two digit number with valid values from 0-59.

Seconds:. . . . . . Specifies the seconds. One or two digit number with valid values from 0-59.

Hundredths: . . . Specifies hundredths of a second. One or two digit number with valid values from 0-99.

a or *p* . . . . When a 12 hour time format is used instead of the 24 hour format, use **a** or **p** to specify A.M. or P.M. When a valid 12 hour time is entered and a parameter is not entered, *time* uses **a** (A.M.).

Command Type and Version:

Internal command; Network; Introduced with Ver 1.0

Notes:

1 Using *time* without parameters will display the current time and prompt you for a time change.

2. Use a colon (**:**) to separate hours, minutes, (seconds and hundred-ths of a second are optional), if as defined in COUNTRY, dependent information file for the United States.

3. With all versions of DOS 3.3 and later, the TIME command will update the system's battery powered clock (except XT-type systems.)

TREE.COM

Displays the directory structure of a path on a specific drive. See also DIR.

Syntax (shaded is optional):

TREE Drive:\ Path /f /a

Examples: TREE (all directories and subdirectories)

TREE \ (names of all subdirectories)

TREE D:\ /f | MORE

TREE D:\ /f > PRN

Syntax Options:

Drive:\Path . . . . Drive and directory containing disk for display of directory structure.

/f. . . . . Displays file names in each directory.

/a ❹ . . . . Text characters used for linking lines, instead of graphic characters. */a* is used with code pages that do not support graphic characters and to send output to printers that do not properly interpret graphic characters.

Command Type and Version:

External command; Network; Introduced with Ver 2.0

Notes:

1. The path structure displayed by the TREE command will depend upon the specified parameters on the command line.
2. The TREE command in MS-DOS 5.0 has been greatly enhanced.

TRUENAME

Displays the TRUENAME of directories and logical drives created with ASSIGN, JOIN, and SUBST.

Syntax (shaded if optional):

TRUENAME drive1: \ path \filename

Example: truename f:

Syntax options:

drive1:. . . . . Drive created by ASSIGN, JOIN, or SUBST.

\ path\ filename. Path and filename created by ASSIGN, JOIN, or SUBST.

Command Type and Version:

Internal command, Introduced with Ver 4.0.

TYPE

Screen display of a text file's contents: The TYPE command is used to view a text file without modifying it.

Syntax (shaded is optional):

TYPE Drive:\ Path\ Filename

Examples: TYPE C:\Act\Receivbl.dat
TYPE C:\Act\Receivbl.dat | MORE

Syntax Options:

Drive:\Path . . . . Drive and directory containing Filename.

\Filename . . . . . Name of text file to be viewed.

Command Type and Version:

Internal command; Network; Introduced with Ver 1.0

Notes:

1. Avoid using the TYPE command to display binary files or files created using a program as you may see strange characters on the screen which represent control codes used in binary files.
2. Use DIR to find the name of a file and EDLIN or EDIT to change its contents.
3. When using the pipe (|) for redirection, set the TEMP environment variable in AUTOEXEC.BAT.
4. See also DIR and MORE.

UNDELETE / MWUNDEL.EXE

Recovers files that have been deleted with the DEL command: UNDELETE is the DOS version and MWUNDEL is the Windows version.

Syntax (shaded is optional):

UNDELETE Drive:\Path\ Filename /List or /all /purge:drive /status /load [/dos or /dt or /ds] /sentry:drive /tracker:drive-entries /unload

Examples: UNDELETE /all
UNDELETE C:\Data*.*

Syntax Options:

Drive:\Path . . . . Drive and directory containing *Filename.*

Filename. . . . . . File to be undeleted. By default, all files in the current directory will be undeleted. Wild cards * and ? are allowed.

/ List . . . . Lists all deleted files in the *Drive:\Path* that can be undeleted, but does not undelete them.

/ all . . . . Recovers all deleted files without a confirmation prompt. If the deletion tracking file is present, it is used, otherwise deleted file information is taken from the DOS directory. See Note: 3.

/ purge:drive ❻ Deletes all files in the sentry directory on the specified *drive.*

/ status ❻. . . . . Displays the current UNDELETE protection level that is enabled.

/ load ❻ . . . . . . Load UNDELETE as memory resident, in order to track deleted files.

/ unload ❻ . . . . Unload the resident portion of the UNDELETE delete tracker.

/ dos . . . . Causes UNDELETE to ignore the deletion tracking file and recover only those files listed as deleted by DOS. A confirmation prompt occurs with each undelete.

/ dt. . . . . Causes UNDELETE to ignore the files listed as deleted by DOS and only recover those files listed in the deletion tracking file. A confirmation prompt occurs with each undelete.

/ ds ❻. . . . . UNDELETE only the files in the /Sentry directory.

/ Sentry:drive ❻ Specify the drive to be used for delete sentry files.

/ Tracker:drive-entries ❻ . . . Specify the drive to track deleted files on. The maximum number of deleted files to track can range from 1 to 999.

Command Type and Version:

External command; Introduced with Ver 5.0

Notes:

1. For best results, use MIRROR and the deletion tracking system.
2. When a file is recovered, it is assigned a # for the first character of its name, if a duplicate exists, another letter is selected, in order from the following list, until a unique filename is possible: #%&-1234567890ABCDEFGHIJKLMNOPQRSTUVWXYZ
3. If a switch is not specified with UNDELETE, the deletion tracking file is automatically used. If the deletion tracking file is not present, the DOS directory information is used. The deletion tracking system is much more accurate.
4. UNDELETE cannot undelete a directory.
5. UNDELETE cannot undelete a file if its directory has been deleted. A possible exception to this rule exists if the deleted directory was a main directory under the root directory and not a subdirectory of some other directory. If this is the case, see the UNFORMAT command. It is possible the directory and file can be saved. Use extreme caution with UNFORMAT and understand exactly what you are doing!!! If not used correctly, UNFORMAT can lose data and you might be worse off than when you started!
6. UNDELETE may not be able to recover a deleted file if data of any kind has been written to the disk since the file was deleted. If you accidentally delete a file, stop what you are doing immediately and run the UNDELETE program.
7. Some MIRROR commands from DOS 5.0 are included in the DOS 6.0 UNDELETE command.
8. See also the UNFORMAT command.

UNFORMAT.COM

Restores a disk that has been reformatted or restructured by the RECOVER command: UNFORMAT can also rebuild disk partition tables that have been corrupted. Do not use UNFORMAT on a network drive.

Syntax (shaded is optional):

UNFORMAT Drive: / J
UNFORMAT Drive: / U / L / test / P
UNFORMAT /partn / L

Examples: UNFORMAT C: / J
UNFORMAT A: / test

Syntax Options:

Drive: . . . . Drive containing disk to be unformatted.

/ J . Removed V6.0 . Check the file created by MIRROR for use with UNFORMAT to make sure it agrees with the system information. Use this switch only by itself.

/ U. Removed V6.0 . UNFORMAT a disk without using the MIRROR file.

/ L . . . . If */partn* is not used, */L* lists every file and directory found by UNFORMAT. Use if the MIRROR file is to be ignored. If */partn* is used also, */L* displays the complete partition table of the drive. Standard 512 byte sectors are assumed when the partition table size is displayed. ***Description for Version 5 ONLY.***

/ L . . . . Lists every file and subdirectory found by UNFORMAT. Default is to list only subdirectories and files that are fragmented. ***Description for Version 6.x ONLY.***

/ test . . . . Displays how UNFORMAT would rebuild information on the disk, but it does NOT unformat the disk. Use this

	switch only if you want UNFORMAT to ignore the MIRROR file. ***Description for Version 5 ONLY.***
/test	Displays how UNFORMAT would rebuild information on the disk, but it does NOT unformat the disk. ***Description for Version 6 ONLY.***
/P	Outputs messages to the LPT1 printer.
/partn .... Removed V6.0	Rebuilds and restores a corrupted partition table of a hard drive. This switch will only work if MIRROR was run previously and the PARTNSAV.FIL file is available to UNFORMAT.

Command Type and Version:

External command; Introduced with Ver 5.0

Notes:

1. Although UNFORMAT is a very powerful tool, it can also do a lot of damage if not used correctly. BE CAREFUL!
2. UNFORMAT normally restores a disk based on MIRROR information. If disk information has changed since MIRROR was run, UNFORMAT may not be able to recover it. Use MIRROR frequently in order to assure an accurate restoration of the disk.
3. If FORMAT with its */u* switch was used, UNFORMAT cannot restore the disk.
4. Per Microsoft's Ver 5.0 User's Guide: "The only case in which you would want to use a prior mirror file is the following: you use the MIRROR command, then the disk is corrupted, then you use the FORMAT command. If you use the MIRROR command and the FORMAT command after the disk is corrupted, the UNFORMAT command will not work. UNFORMAT searches the disk for the MIRROR file. Because UNFORMAT searches the disk directly, the disk does not have to be "readable" by MS-DOS for UNFORMAT to work. Do not use the FDISK command before using UNFORMAT; doing so can destroy information not saved by the MIRROR program."
5. If UNFORMAT does not use the MIRROR file, the restore will take much longer and be less reliable.
6. Without a MIRROR file, UNFORMAT cannot recover a file that is fragmented. It will recover what it can, then prompt for truncation of the file or delete the file.
7. If DOS displays the message "Invalid drive specification," the problem might be a corrupted disk partition table, which UNFORMAT can probably repair. In order to recover the disk partition table, the MIRROR file must be available.

8. When the */partn* switch is used, you are prompted to insert a system disk in drive A: and press ENTER to restart. The restart will allow DOS to read the new partition table data. Once the system has been restarted, use UNFORMAT without the */partn* switch to recover directories and the FAT (file allocation table).
9. See also UNDELETE, MIRROR, FORMAT, and FDISK.
10. In DOS Ver 5.0, the /p switch is not compatible with the /u switch.

UNINSTALL.EXE New V6.0

Restores the previous version of DOS after the MS-DOS 6 is installed: Used in conjunction with the Uninstall Disk to protect files while MS-DOS 6 is installed. If problems occurs during installation, UNINSTALL can be used to restore the previous version of DOS.

Command Type and Version:

External command, Introduced with Version 6.0.

VER

Displays DOS version number: Type **ver** and the version number will display on the screen.

Syntax (shaded is optional):

VER /R

Examples: VER

Syntax Options:

/ R ❻ . . . . Provides a more detailed report.

Command Type and Version:

Internal command; Network; Introduced with Ver 2.0

VERIFY

Disk verification: Verifies that the files are written correctly to a disk.

Syntax (shaded is optional):

VERIFY on / off

Examples: VERIFY on

Syntax Options:

Verify . . . . **Verify** without an option will state whether verification is turned on or off.

on . . . . Forces DOS to confirm that information is being written correctly. The verify command will function until the system is rebooted or **verify off** is used.

off . . . . Turns verification off once it is on.

Command Type and Version:

Internal command; Network; Introduced with Ver 2.0

Notes:

1. When the VERIFY command is used, DOS verifies data as it is written to a disk. This will slow writing speed slightly.
2. COPY / V or XCOPY / V can also be used to verify that files are being copied correctly but on a case by case basis.
3. Verify does not perform a physical disk to disk comparison.

VOL

Displays disk Volume label: The VOL command displays the name of volume label given to a disk when it was formatted. DOS Version 4.0 and greater will also display a volume serial number.

Syntax (shaded is optional):

VOL Drive:

Examples: VOL A: or VOL

Syntax Options:

VOL .... VOL, without options, displays the volume label and volume serial number of the disk in current drive.

Drive: .... Specifies the drive that contains the disk whose label is to be displayed.

Command Type and Version:

Internal command; Network; Introduced with Ver 2.0

Volume serial numbers introduced with DOS Ver 4.0

Notes:

1. See also FORMAT and LABEL.

VSAFE.COM **New V6.0**

Continuously monitors a system for viruses and displays a warning if it finds one:

VSAFE is a memory resident program that uses approximately 22k of memory. See Windows Note below.

Syntax (shaded is optional):

VSAFE /option + or - /NE /NX /A# /C# /N /D /U

Example: VSAFE / 2+ /NE /AV

Syntax Options:

/ option + or - . . Specifies how VSAFE looks for viruses. The + or - is used to either turn on or turn off the option. Options are as follows:

1 - Warn of a formatting request. Default=On

2 - Warn if a program tries to stay resident. Default=Off

3 - Disable all disk writes. Default=Off

4 - Check executable files that DOS opens. Default=On

5 - Check for boot sector viruses. Default=On
6 - Warns if a program tries to write to the boot sector or
partition table of a hard disk. Default=On
7 - Warns if a program tries to write to the boot sector of
a floppy disk. Default=Off
8 - Warns if an attempt is made to modify an executable file. Default=Off

/NE. . . . . Prevents VSAFE from loading into expanded memory.

/NX. . . . . Prevents VSAFE from loading into extended memory.

/A#. . . . . Sets the VSAFE hot key as Alt plus the key specified by #.

/C#. . . . . Sets the VSAFE hot key as Ctrl plus the key specified by #.

/N. . . . . Enable network drive monitoring.

/D. . . . . Disable CRC checksumming.

/U. . . . . Unloads VSAFE from memory.

Command Type and Version:

External command; Network; Introduced with Ver 6.0

Notes:

1. If VSAFE is to be used when Windows 3.1 is running, you must include " load=MWAVTSR.EXE" in the WIN.INI file.

WINA20.386

The WINA20.386 file must be located in the root directory in order for Microsoft Windows Ver. 3.0 to run in enhanced mode. It is automatically placed in the root directory by MS-DOS during the installation process:

If the file is not in the root directory, you will receive the message "You must have the file WINA20.386 in the root of your boot drive to run Windows in Enhanced Mode."

WINA20.386 must remain in the root directory unless the SWITCHES /W command is used to tell DOS that it has been moved. You must also add a DEVICE command under the [386Enh] section of your Windows SYSTEM.INI file, which specifies where WINA20.386 is now located.

Command Type and Version:

External command; Introduced with Ver 5.0

XCOPY.EXE

Copies files, directories, and subdirectories from one location to another location: XCOPY will not copy system or hidden files.

Syntax (shaded is optional):

XCOPY Source Destination /a /d:date /e /m /p /s /v /w /y /-y

Examples: XCOPY C:\Dos*.* D:\Dos2\ /s

Syntax Options:

Source . . . . Location and names of files to be copied.

Destination . . . . Destination of the files to be copied.

/a. . . . . Copies *Source* files that have their archive file attributes set **without** modifying it.

/ d:date . . . . Copies *Source* files that have been modified on or after a specific date.

/ e . . . . Copies subdirectories even if empty.

/m. . . . . Copies *Source* files that have their archive file attributes set and turns them off.
/p . . . . Prompts whether you want to create each destination file.
/s . . . . Copies directories and subdirectories, unless they are empty.
/v . . . . Verifies each file, as it is written, to confirm that the destination and source files are identical.
/w. . . . . Displays "Press any key to begin copying file (s)," and waits for response before starting to copy files.
/y 6.2 . . . . . . Directs XCOPY to replace existing files without a confirmation prompt.
/-y 6.2 . . . . . . Directs XCOPY to ask for confirmation prior to replacing an existing file. Default

Command Type and Version:

External command; Network; Introduced with Ver 3.2

Notes:

1. The default *Destination* is the current directory.
2. If the *Destination* subdirectory does not end with a " \ ", DOS will prompt you to find out if the subdirectory is a subdirectory or file.
3. XCOPY will not copy system or hidden files.
4. When a file is copied to *Destination*, the archive attribute is turned on, regardless of the file attribute in *Source*.
5. In order to copy between disks that are different formats, use XCOPY, not DISKCOPY, but remember that XCOPY does not copy the hidden or system files.
6. XCOPY exit codes are as follows: (see IF errorlevel)
   0 . . . . Files copied successfully
   1 . . . . Source files not found
   2 . . . . XCOPY stopped by user Ctrl+C
   4 . . . . One of the following errors ocurred:
   a. Initialization error
   b. Not enough disk space
   c. Insufficient memory available
   d. Invalid drive name
   e. Invalid syntax was used.
   5 . . . . Disk write error occurred.
7. When a files size is larger than 64k, use XCOPY instead of the COPY command.

Chapter 5

Microsoft Windows 3.1 Short Cut Keys

General Windows 3.1

ALT .... In an open Application: Use to activate the Menu Bar, same as F10.

ALT + BACKSPACE .... In a Text Box or Window: Use to undo last editing command, same as CTRL + Z.

ALT + DOWN ARROW .... In a Dialog Box: Use to close or open a selected List.

ALT + ENTER .... In 386 enhanced Mode: Moves an MS-DOS Application from a window to full screen and back.

ALT + ESC .... Moves immediately to the next open Application.

ALT + F4 .... (1) Use to exit Windows (2) In a Dialog Box: Use to cancel the Dialog Box (3) In any open Application: Use to Quit that Application.

ALT + HYPHEN .... In an open Application: Use to open Control Menu.

ALT + HYPHEN + N .... In an open Application: Use to minimize a second (or child) window.

ALT + HYPHEN + X .... In an open Application: Use to maximize a second (or child) window.

ALT + PRINT SCREEN ...... In an open Application: Copies an image of the active window to the Clipboard.

ALT + SHIFT + ESC .... Moves immediately to the previous open Application.

ALT + SPACEBAR .... In an open Application: Use to open Control Menu.

ALT + SPACEBAR + N .... In an open Application: Use to minimize a window.

ALT + SPACEBAR + M ...... In an open Application: Use to Move a window.

ALT + SPACEBAR + X .... In an open Application: Use to maximize a window.

ALT + TAB .... Displays and scrolls forward through a list of open Applications. Releasing the TAB key opens the selected Application.

ALT + TAB + TAB .... Displays only the Title Bar of open Applications and scrolls through the open Applications. Releasing the TAB key opens the selected Application.

ALT + Underlined Character .... (1) In an open Application: (a) Menu bar: Open Menu with Underlined Character (b) Menu: Select Menu Item with Underlined Character from Menu (2) In a Dialog

Box: Use to move to a Dialog Box item with Underlined Character.

ARROW KEYS ....Move between Menu commands, characters in a text box, or items in a list.

BACKSPACE.......................In a Text Box or Window: Use to delete character to the left of the cursor.

CTRL + ALT + Character....In 386 enhanced Mode: Use to assign Character as an Application Shortcut Key, Character is user selected and can be any letter, numeral, or special key.

CTRL + ALT + SHIFT + Character......................In 386 enhanced Mode: Use to assign Character as an Application Shortcut Key, Character is user selected and can be any letter, numeral, or special key.

CTRL + BACK SLASH (\) ...In a Dialog Box: Use to cancel all selected items from a list except the current item.

CTRL + C............................In a Text Box or Window: Use to copy selected text to the Clipboard, same as CTRL + INSERT.

CTRL + END........................In a Document: Use to move to end of document.

CTRL + ESC........................Opens the Task List window.

CTRL + F4...........................In an open Application: Use to close an active document or window.

CTRL + FORWARD SLASH (/)..........................In a Dialog Box: Use to select all items from a list.

CTRL + HOME ....In a Document: Use to move to the beginning of the document.

CTRL + INSERT ....In a Text Box or Window: Use to copy selected text to the Clipboard, same as CTRL + C.

CTRL + LEFT ARROW.......In a Dialog Box: Use to move left one word in a text box.

CTRL + RIGHT ARROW ....In a Dialog Box: Use to move right one word in a text box.

CTRL + SHIFT + Character ....In 386 enhanced Mode: Use to assign Character as an Application Shortcut Key, Character is user selected and can be any letter, numeral, or special key.

CTRL + SHIFT + ALT + Character.....................In 386 enhanced Mode: Use to assign Character as an Application Shortcut Key, Character is user selected and can be any letter, numeral, or special key.

CTRL + V............................In a Text Box or Window: Use to paste selected text from the Clipboard, same as SHIFT + INSERT.

CTRL + X............................In a Text Box or Window: Use to move selected text on to the Clipboard, same as SHIFT + DELETE.

CTRL + Z............................In a Text Box or Window: Use to undo last editing command, same as ALT + BACKSPACE.

DELETE..............................(1) Use to delete a group or program item (2) In a Text Box or Window: Use to delete character to the right of the cursor.

END...................................Move to the end of a line, screen, or list.

ENTER...............................In a Dialog Box: Use to close Dialog Box and initiate all highlighted commands.

ESC...................................In a Dialog Box: Use to cancel the Dialog Box.

F1......................................Starts the Help Program from within an open Application.

F10....................................In an open Application: Use to activate the Menu Bar, same as ALT.

HOME................................Move to beginning of a line, screen, or list.

PAGE DOWN......................Use to move down one screen.

PAGE UP............................Use to move up one screen.

PRINT SCREEN.................Use to copy an entire screen to the Clipboard.

SHIFT + ALT + ESC...........Moves immediately to the previous open application.

SHIFT + ALT + TAB...........Displays and scrolls backward through a list of open applications. Releasing the TAB key opens the selected Application.

SHIFT + CTRL + END........In a Document: Use to move to end of document.

SHIFT + CTRL + HOME.....In a Document: Use to move to beginning of document.

SHIFT + CTRL + LEFT ARROW........................In a Document: Use to move to previous word in document.

SHIFT + CTRL + RIGHT ARROW.........In a Document: Use to move to next word in document.

General Windows 3.1

SHIFT + DELETE ....In a Text Box or Window: Use to move selected text to the Clipboard, same as CTRL + X.

SHIFT + DOWN ARROW ...In a Document: Use to select whole line below the cursor location.

SHIFT + END......................In a Document: Use to move to end of a line.

SHIFT + F8 ....In a Dialog Box: Use to select nonconsecutive items from a list.

SHIFT + HOME ....In a Document: Use to move to beginning of a line.

SHIFT + INSERT ....In a Text Box or Window: Use to paste selected text from the Clipboard, same as CTRL + V.

SHIFT + LEFT ARROW......In a Document: Use to move one letter left in document.

SHIFT + RIGHT ARROW ...In a Document: Use to move one letter right in document.

SHIFT + TAB ....In a Dialog Box: Moves to previous command in the Dialog Box.

SHIFT + UP ARROW ....In a Document: Use to select whole line above the cursor location.

SPACEBAR ....In a Dialog Box: Use to choose a selected Command.

TAB....................................In a Dialog Box: Moves to next command in the Dialog Box.

Calendar Win 3.1

CTRL + END.......................In Day View: Use to move to 12 entries after the starting time.

CTRL + HOME ....In Day View: Use to move the starting time.

CTRL + INSERT ....In Day View: Use to move selection to the Clipboard.

CTRL + PAGE DOWN........In Day View: Use to move to next day. CTRL + PAGE UP ..In Day View: Use to move to previous day.

DOWN ARROW (1) ....In Month View: Use to move to next month (2) In Day View: Use to move to next time, same as ENTER.

ENTER ....(1) In Month View: Use to change day (2) In Day View: Use to move to next time, same as DOWN ARROW.

Calendar Win 3.1 (cont.)

PAGE DOWN ....(1) In Month View: Use to move to next month (2) In Day View: Use to move to next screen.
PAGE UP............................ (1) In Month View: Use to move to previous month (2) In Day View: Use to move to previous screen.
SHIFT + DELETE ....In Day View: Use to move a selection to the Clipboard.
SHIFT + INSERT ....In Day View: Use to paste a selection from the Clipboard to the appointment area or scratch pad.
TAB...................................... (1) In Month View: Use to move between date and scratch pad (2) In Day View: Use to move between appointment and scratch pad.
UP ARROW ....(1) In Month View: Use to move to previous week (2) In Day View: Use to move to previous time.

Cardfile Win 3.1

CTRL + END.........................Use to display the last card.
CTRL + HOME ....Use to display first card.
DOWN ARROW...................Use to scroll forward one card in list.
PAGE DOWN ....Use to scroll forward one card.
PAGE UP.............................Use to scroll backward one card.
SHIFT + CTRL + Character ....Use to display first card beginning with Character.
UP ARROW ....Use to scroll backward one card in list.

Clipboard Viewer Win 3.1

DELETE.............................Clear the contents of the Clipboard.

File Manager Win 3.1

ALT + ENTER......................Use to display properties of a file or directory.
ALT + F + N ....Use to rename a file.
ALT + F + U ....Use to undelete a file (MS-DOS 6.0 and 6.2 only)
ALT + V + A ....Use to display a file's date, file attributes, and size.
ALT + V + S ....Use to sort files by filename.
Character ....Go to directory or file where directory name or filename starts with Character.
CTRL + Drive Letter............Use to changed displayed drive.

File Manager Win 3.1 (cont.)

CTRL + * ....Use to expand all directories and subdirectories.
DELETE..............................Use to delete a directory or file
ENTER ....Use to display or hide a displayed directory's subdirectories, start an application, or open a file.
F2 ....Use to display drive list.
F5 ....Use to update the displayed file or directory.
F6 ....Use to scroll between the displayed drive, directory, and file, same as TAB.
F7 ....Use to move a displayed file or directory.
F8 ....Use to copy a displayed file or directory.
SHIFT + ENTER ....Use to open a new window and display contents of a directory.
TAB....................................Use to scroll between the displayed drive, directory, and file, same as F6.
+...Use to expand displayed directories one level to show subdirectories.
* ....Use to expand displayed subdirectory.
-..Use to collapse displayed subdirectory.

Help Program Win 3.1

ALT + F4.............................Use to quit the Help Program.
ALT + PRINT SCREEN ......Use to copy Help Screen to Clipboard.
CTRL + TAB ....Highlights all key words on a Help Screen.
SHIFT + TAB ....Use to move to previous Help Item.
TAB....................................Use to move to next Help Item.

Object Packager Win 3.1

TAB....................................Use to move between Content and Appearance windows.

Paintbrush Win 3.1

ARROW KEYS ....Use to move the cursor.
CTRL + S............................Use to save file.
CTRL + Z............................Use to undo everything drawn since selecting a tool.
DELETE..............................Use to simulate clicking the right mouse button.

Paintbrush Win 3.1 (cont.)

END ....Use to move to the bottom of the drawing area.
F9 + INSERT ....Use to simulate double-clicking the left mouse button.
HOME ....Use to move to the top of the drawing area.
INSERT ....Use to simulate clicking the left mouse button.
INSERT + ARROW KEYS..Use to simulate dragging the cursor.
PAGE DOWN ....Use to move down one screen.
PAGE UP............................Use to move up one screen.
SHIFT + DOWN ARROW...Use to move down one line.
SHIFT + END.......................Use to move to the right side of the drawing area.
SHIFT + HOME ....Use to move to the left side of the drawing area.
SHIFT + LEFT ARROW......Use to move left one space.
SHIFT + PAGE DOWN.......Use to move right one screen.
SHIFT + PAGE UP ....Use to move left one screen.
SHIFT + RIGHT ARROW ...Use to move right one space.
SHIFT + TAB ....Use to move among drawing area, palette, linesize box, and toolbox; same as TAB.
SHIFT + UP ARROW ....Use to move up one line.
TAB.......................................Use to move among drawing area, palette, linesize box, and toolbox; same as SHIFT + TAB.

Print Manager Win 3.1

CTRL + DOWN ARROW....Use to move selected document down in the queue.
CTRL + UP ARROW ....Use to move selected document up in the queue.
DOWN ARROW..................Use to move between queues or documents in a queue.
UP ARROW ....Use to move between queues or documents in a queue.

Program Manager Win 3.1

ALT + W..............................Use to move between groups, same as CTRL + TAB or CTRL + F6.
ARROW KEYS ....Use to move between items in a group window.
CTRL + F4 ....Use to close an active group window.

Program Manager Win 3.1 (cont.)

CTRL + F6 .... Use to move between groups, same as CTRL + TAB or ALT + W.
CTRL + TAB .... Use to move between groups, same as CTRL + F6 or ALT + W.
DELETE .... Use to delete a program item.
ENTER .... Use to open a selected Application.
SHIFT + F4 .... Use to tile the group windows.
SHIFT + F5 .... Use to cascade the group windows.

Sound Recorder & Media Player Win 3.1

END .... Use to move to the end of the sound when scroll bar is selected.
HOME .... Use to move to the beginning of the sound when scroll bar is selected.
LEFT ARROW .... Use to move backward when scroll bar is selected.
PAGE DOWN .... Use to move forward 1 second when scroll bar is selected.
PAGE UP .... Use to move backward 1 second when scroll bar is selected.
RIGHT ARROW .... Use to move forward when scroll bar is selected.

Write Win 3.1

ALT + BACKSPACE .... Use to undo last editing action.
ALT + F6 .... Use to switch between the document and the find/replace Dialog Box.
ARROW KEYS .... Use to move the picture size cursor.
CTRL + ENTER .... Use to insert manual page break.
CTRL + SHIFT + HYPHEN. Use to insert an invisible hyphen.
CTRL + Z .... Use to undo last typing action.
DOWN ARROW .... Use to select an object or picture, cursor must be above upper-left corner of object or picture
5 + DOWN ARROW .... Use to move to next paragraph, 5 is on the numeric key pad with the NUM LOCK key turned OFF.
5 + LEFT ARROW .... Use to move to next sentence, 5 is on the numeric key pad with the NUM LOCK key turned OFF.
5 + PAGE DOWN .... Use to move to next page, 5 is on the numeric key pad with the NUM LOCK key turned OFF.
5 + PAGE UP .... Use to move to previous page, 5 is on the numeric key pad with the NUM LOCK key turned OFF.

Write Win 3.1 (cont.)

5 + RIGHT ARROW............Use to move to previous sentence, 5 is on the numeric key pad with the NUM LOCK key turned OFF.

5 + UP ARROW..................Use to move to previous paragraph, 5 is on the numeric key pad with the NUM LOCK key turned OFF.

Chapter 6

Microsoft Windows 95 Shortcut Keys

General Windows 95 Shortcuts

ALT ....Activate Menu bar, same as F10
ALT+F4...............................Quit an application
ALT+S.................................Display Start Menu when no windows are open or applications selected on desktop. Use arrow keys to select menu commands.
ALT+TAB ....Switch to previous window
CTRL+C..............................Copy
CTRL+ESC.........................Open Start Menu
CTRL+ESC, then ALT+M ...Minimize all windows, return to desktop
CTRL+TAB ....Tab through pages in a properties dialog box
CTRL+V..............................Paste
CTRL+X..............................Cut
CTRL+Z..............................Undo last action
DELETE..............................Remove selected item
F1 ....Help
F10 ....Activate menu bar, same as ALT
SHIFT while inserting CD ROM
Bypass AutoPlay when inserting a compact disc
SHIFT+F10 ....View shortcut menu for a selected item, same as right mouse click

Windows 95 Desktop, My Computer and Windows Explorer

ALT+ENTER or ALT +
DOUBLE CLICK ....View item properties (same as right click, select "Properties") CTRL+SHIFT while dragging file ...Pull up menu to move, copy, or create shortcut.CTRL while dragging file ...Copy file
F2 ....Rename an object (some objects cannot be renamed)
F3 ....Find a folder or file
SHIFT+DELETE ....Delete an item immediately without placing it in the Recycle Bin.

Windows 95 My Computer and Windows Explorer

BACKSPACE......................View the folder one level up
CTRL+A.............................Select All
F5 ....Refresh a window
SHIFT while clicking
the Close (X) button.......Close the selected folder and all its parent folders

Windows 95 Explorer Only

ALT+ENTER.......................Display properties of a selected item
ALT+F+M............................Rename a selected file
ALT+F+S ....Create a shortcut to selected file
BACKSPACE.......................Go to parent folder
CTRL+Arrow Key................Scroll without changing the selected file
CTRL+G ....Go to
F6 ....Switch between panes
LEFT ARROW ....Collapse current selection if expanded; otherwise select parent folder
NUMLOCK + MINUS SIGN
(– on numeric keypad).Collapse the selected folder
NUMLOCK + PLUS SIGN
(+ on numeric keypad).Expand the selected folder
NUMLOCK + ASTERISK
(* on numeric keypad) Expand all subfolders under the selected folder
RIGHT ARROW..................Expand current selection if collapsed; otherwise select first subfolder
* on numeric keypad..........Expand everything under the selection
+ on numeric keypad ....Expand the selected item
– on numeric keypad ....Collapse the selected item

Windows 95 "Open" and "Save As" Dialogue Boxes

BACKSPACE......................Open folder one level up, if a folder is selected.
CTRL+SHIFT+TAB.............Move backward through category tabs
CTRL+TAB ....Move forward through category tabs
F4 ....Open the "Save In" or "Look In" List
F5 ....Refresh

Windows 95 Accessibility Options

To use Accessibility Options shortcut keys, the shortcut keys must be enabled.

LEFT ALT+ LEFT SHIFT+ NUMLOCK...................Toggle MouseKeys on and off
LEFT ALT+LEFT SHIFT + PRINT SCREEN..........Toggle High Contrast on and off
NUMLOCK for 5 seconds ...Toggle ToggleKeys on and off
RIGHT SHIFT for 8 seconds......Toggle FilterKeys on and off
SHIFT 5 times.....................Toggle StickyKeys on and off

Shortcuts Within an Open Windows 95 Application

ALT ....Activate Menu Bar, same as F10
ALT+BACKSPACE ....Undo last editing command, same as CTRL+Z.
ALT+ESC............................Open Start Menu, same as clicking on start button.
ALT+F4...............................Exit the application
ALT+HYPHEN ....Bring up Window sizing menu
ALT+HYPHEN +N ....Minimize a second (Child) window
ALT+PRINT SCREEN ....Copy an image of the active window to the clipboard
ALT+SHIFT+ESC ....Switch directly from one open application to another open application
ALT+SPACEBAR ....Open window sizing menu, same as ALT+HYPHEN.
ALT+SPACEBAR+N...........Minimize the current window
ALT+SPACEBAR+M ....Move a window
ALT+TAB ....Display open applications. Holding down tab key scrolls through the list; repeatedly pressing the TAB key displays open applications one at a time. Releasing the ALT key opens the selected application.
ALT + Underlined Character ....ALT activates the menu bar; underlined character pulls down corresponding menu. Within a menu, typing underlined character selects corresponding command. Within a dialogue box, typing underlined character moves cursor to corresponding field.
ARROW KEYS ....Move between menu items, characters in text box, or items in a list.
BACKSPACE......................Delete character to left of cursor

Within Open Win 95 Application (cont.)

CTRL+C .... Copy text to Clipboard
CTRL+END .... Move to end of document
CTRL+ESC .... Open Start Menu
CTRL+F4 .... Close an open application
CTRL+HOME .... Move to beginning of document
CTRL+LEFT ARROW .... Move cursor 1 item or word to the left
CTRL+RIGHT ARROW ...... Move cursor 1 item or word to the right
CTRL+V .... Paste
CTRL+X .... Cut (moves object to clipboard)
CTRL+Z .... Undo last editing command
DELETE .... Delete character to the right of the cursor.
END .... Move to the end of a line
ENTER .... In a dialog box, close the box and initiate all selected commands.
ESC .... Cancel a dialog box
F1 .... Access Help for the current application
F10 .... Activate Menu Bar, same as ALT.
HOME .... Move to beginning of current line
PAGE DOWN .... Move down one screen
PAGE UP .... Move up one screen
PRINT SCREEN .... Copy an entire screen to the clipboard
SHIFT+ALT+ESC .... Toggle to previous open application
SHIFT+ALT+TAB .... Bring up box to allow to to select an open application. Releasing tab key selects application
SHIFT+CTRL+END .... Select all text from cursor to end of document and moves cursor toend.
SHIFT+CTRL+HOME .... Select all text from cursor back to beginning of document and moves cursor to beginning
SHIFT+CTRL+LEFT ARROW .... Select word immediately left of cursor
SHIFT+CTRL+RIGHT ARROW .... Select word immediately right of cursor
SHIFT+DELETE .... Cut (moves text to clipboard, same as CTRL+X)
SHIFT+DOWN ARROW ..... Select one line of text just below or right of the cursor
SHIFT+END .... Select text from cursor to end of line
SHIFT+HOME .... Select text from the cursor back to the beginning of line
SHIFT+INSERT .... Paste text from clipboard at location of cursor
SHIFT+LEFT ARROW .... Select one character immediately left of cursor

Within Open Win 95 Application (cont.)

SHIFT+RIGHT ARROW	Select one character immediately right of cursor
SHIFT+TAB	Move one space backward (or left) in a dialog box
SHIFT+UP ARROW	Select one whole line above the cursor
TAB	Move one space forward (right) in a dialog box

Win 95 Calculator Shortcut Keys Standard Calculator

CTRL+C	Copy
CTRL+V	Paste
BACKSPACE	Delete last digit displayed, same as "back" button.
DELETE	Clear displayed number, same as "CE" button.
ESC	Clear current calculation, same as "C" button.
CTRL+L	Clear memory, same as "MC" button.
CTRL+M	Store displayed number in memory, same as "MS" button
CTRL+P	Add displayed # to memory, same as "M+" button.
CTRL+R	Recall number stored in memory, same as "MR" button.

Win 95 Scientific Calculator

Same as Standard Calculator above, Plus the Following Shortcuts

F9	+/-	Change sign of displayed number.
r	1/x	Calculate reciprocal of displayed number.
ENTER	=	Calculate last 2 numbers.
&	And	Calculate bitwise AND.
CTRL+A	Ave	Mean of numbers in Statistics Box.
BACKSPACE	Back	Delete last digit of displayed number.
F8	Bin	Convert to binary number system.
F4	Byte	Display lower 8 bits of number
o	cos	Calculate cosine of displayed number
INSERT	Dat	Enter number in Statistics Box.
F6	Dec	Convert to decimal number system.
F2	Deg	Set trigonometric input for degrees.
m	dms	Set degrees to degree-minute-second fmt.
F2	Dword	Show 32-bit representation of number.

Win 95 Scientific Calculator (cont.)

x ....exp ....Set to scientific notation.
v ....F-E ....Toggle scientific notation on/off.
F4 ....Grad........Set trigonometric input for gradient.
F5 ....Hex ....Convert to hexadecimal number system.
h ....hyp ....Enable hyperbolic function.
n ....In ....Calculate base e logarithm.
; ....Int............Show integer portion of a decimal value.
i ....Inv ....Enable inverse function.
L ....Log..........Calculate base 10 logarithm.
< ....Lsh ....Shift left, specify number of positions.
% ....Mod.........Display remainder of x/y.
! ....n! ....Compute factorial of displayed number.
~ ....Not ....Calculate bitwise inverse.
F7 ....Oct ....Convert to octal number system.
|(vertical bar)...Or............Calculate bitwise OR.
p ....PI ....Show value of pi (3.1415...)
F3 ....Rad ....Set trigonometric input for radians.
CTRL+D ....s ....Standard deviation when Population=n-1.
s ....sin ....Compute sine of displayed number.
i+@ ....sqrt (inv+x^2) ..Find square root of displayed number.
CTRL+S ....Sta ....Turn on statistics mode, open Statistics Box.
CTRL+T..........Sum ....Add values in Statistics Box.
t ....tan............Compute tangent of displayed number.
F3 ....Word ....Show lower 16 bits of current number.
^......................Xor ....Calculate bitwise exclusive OR.
@ ....x^2 ....Square the displayed number.
#......................x^3 ....Cube the displayed number.
y ....x^y ....Compute x to the yth power.

Windows 95 Control Panel/Printers

CTRL+A.............................Select All
CTRL+C.............................Copy
CTRL+V.............................Paste
CTRL+X.............................Cut
CTRL+Z.............................Undo

Windows 95 Note Pad

CTRL+C.............................Copy
CTRL+V.............................Paste
CTRL+X.............................Cut
CTRL+Z.............................Undo

Windows 95 Note Pad (cont.)

F1 .... Help
F3 .... Find Next
F5 .... Time/Date

Windows 95 Paint

ALT+F4 .... Exit application
CTRL+A .... View Color Box
CTRL+C .... Copy
CTRL+E .... Image Attributes
CTRL+F .... View bitmap
CTRL+I .... Invert Colors
CTRL+L .... Select All
CTRL+N .... Start new file
CTRL+O .... Open existing file
CTRL+P .... Print
CTRL+R .... Flip/Rotate image
CTRL+S .... Save
CTRL+T .... View Tool Box
CTRL+V .... Paste
CTRL+W .... Stretch or Skew image
CTRL+X .... Cut
CTRL+Z .... Undo
CTRL+SHIFT+N .... Clear Image
F4 .... Repeat last edit

Windows 95 WordPad

ALT+BACKSPACE .... Undo last edit, same as CTRL+Z
ALT+ENTER .... Object Properties
ALT+F6 .... Toggle between document and Find/Replace dialog box.
CTRL+A .... Select All
CTRL+C .... Copy
CTRL+F .... Find
CTRL+H .... Replace
CTRL+N .... Create New file
CTRL+O .... Open file
CTRL+P .... Print
CTRL+S .... Save
CTRL+V .... Paste
CTRL+X .... Cut
CTRL+Z .... Undo
F3 .... Find Next

Chapter 7

Hard Drive Specifications

STD 286/386/486 HARD DISK TYPES

Drive Type	# of Cylinders	# of Heads	Write Precomp	Land Zone	Size in Megabytes
1	306	4	128	305	10
2	615	4	300	615	21
3	615	6	300	615	31
4	940	8	512	940	63
5	940	6	512	940	47
6	615	4	65535	615	21
7	462	8	256	511	31
8	733	5	65535	733	31
9	900	15	65535	901	112
10	820	3	65535	820	21
11	855	5	65535	855	36
12	855	7	65535	855	50
13	306	8	128	319	21
14	733	7	65535	733	43
15	0	0	0	0	0
16	612	4	0	663	21
17	977	5	300	977	41
18	977	7	65535	977	57
19	1024	7	512	1023	60
20	733	5	300	732	31
21	733	7	300	732	43
22	733	5	300	733	31
23	306	4	0	336	10
24	698	7	300	732	42
25	615	4	0	615	21
26	1024	4	65535	1023	34
27	1024	5	65535	1023	43
28	1024	8	65535	1023	68
29	512	8	256	512	34
30	615	2	615	615	10
31	732	7	300	732	44
32	1023	5	65535	1023	44
33	306	4	0	340	10
34	976	5	488	977	42
35	1024	9	1024	1024	77
36	1024	5	512	1024	43
37	830	10	65535	830	69
38	823	10	256	824	68
39	615	4	128	664	21
40	615	8	128	664	41
41	917	15	65535	918	114
42	1023	15	65535	1024	127
43	823	10	512	823	68
44	820	6	65535	820	41
45	1024	8	65535	1024	68
46	925	9	65535	925	69
47	699	7	256	700	41

Note: Drive types over #24 vary between computer manufacturers

Hard Drive Table Syntax and Notations

See page 442 for comments on the hard drive data included in this chapter and a hard drive resource list. The following are descriptions of the information contained in the hard drive tables.

Telephone and BBS numbers for hard drive manufacturers are listed on the next page and in the Phone Book (page).

1. Format Size MB. . . . . Formatted drive size in megabytes (Mb)
2. Heads. . . . . . . Number of data heads
3. Cyl . . . . Number of cylinders
4. Sect/Trac . . . . Number of sectors per track, V=Variable
5. Translate . . . . Head-Cyl-Sector/Track Translation. *UNIV is a Universal Translation where any drive setup can be used as long as the total translated sectors is less than total drive sectors (Total drive sectors=physical heads x physical cylinders x physical sectors per track)
6. RWC. . . . . Start Reduced Write Current cylinder
7. WPC. . . . . Start Write Precompensation cylinder
8. Land zone. . . . Safe cylinder for parking drive heads
9. Seek Time . . . Avg. drive head access time, milliseconds
10. Interface . . . . Type of drive interface used ST412/506, ESDI, SCSI, IDE AT, IDE XT, EIDE
11. Encode . . . . . Data encoding method used on drive MFM, 2,7RLL, 1,7 RLL, RLL ZBR, ERLL
12. Form Factor. . Physical diameter and height of drive 5.25HH, 3.5HH, 3.5/3H, 2.5
13. Cache . . . . . . Read ahead cache/buffer, in kilobytes (kb)
14. mtbf . . . . Mean time between failures in kilohours (kh)
15. RPM . . . . Drive motor Revolutions Per Minute
16. Obselete? . . . Is the drive obsolete? Y=Yes

PLEASE NOTE: The density of information in the hard drive table has made it necessary to conserve space by abbreviating kilobytes "kb" as "k" and kilohours "kh" as "k".

Hard Drive Manufacturers Directory

The following table is a general summary of companies that have manufactured and/or are still manufacturing hard drives. The number of models shown is based on data contained in the Pocket PCRef Hard Drive Specifications table and Sequoia Publishing does not represent this summary as being exact. If you have information concerning the status of any of these companies, such as "XYZ Company went bankrupt in August, 1990" or "XYZ Company was bought by Q Company," please let us know so we can keep this section current. If a phone number is listed in the Status column, the company is in business.

Manufacturer	Status
Alps America	800-449-2577; No longer make hard drives.
Ampex	650-367-2011; No longer make hard drives.
Areal Technology, Inc	Out of Business
Atasi Technology, Inc	Out Of Business; Lipsig & Assoc provide support 408-733-1844
Aura Associates	781-290-4800; No longer make hard drives.
BASF	See Emtech Data Corp
Brand Technologies	Out of Business
Bull	508-294-6000; No longer makes hard drives.
C.itoh Electronics, Inc	800-347-2484; Doing business as Itochu Tech; sold hard drive division to Y-E Data.
Cardiff	760-752-5200; No longer make hard drives.
CDC	408-438-6550; See Seagate
Century Data	919-821-5696; Not a manufacturer.
CMI	Out of Business
CMS Enhancements, Inc	714-424-5520; Not a manufacturer. Ameriquest parent company
Cogito	Out of Business
Compaq	281-370-0670
Comport	Unknown
Conner Peripherals, Inc.	800-468-3472; Merged with Seagate Technology 2-5-96.
Core International	561-997-6033 Stopped Manufacturing hard drives August 1995. Split into 2 companies-Iowa Data Product Services (561-997-6033-old drive support) and Core Engineering (561-998-3800).

Manufacturer	Status
Data Direct Networks	310-247-0006
Digital Equipment Corp.	800-344-4825; Sold Storage Division to Quantum 1st Quarter 1995. Sold Direct Sales Division To PC Complete; OEM Hard Drives from Quantum & Seagate.
Disc Tec	407-671-5500; Maker of removable-hard drives.
Disctron (Otari)	Out Of Business
DMA	Out of Business
Eloch	Unknown
Emtech Data Corp	781-271-4000
Epson America	310-782-0770; No longer make hard drives.
Fuji Electric	510-438-9700; Do not manufacture hard drives in US.
Fujitsu America, Inc.	408-432-6333
Hewlett-Packard Co	Corporate: 415-857-1501; Most drives are OEM.
Hitachi America	914-332-5800
Hyosung	Unknown
IBM	408-256-1600
IBM Corp. (Storage Sys Div)	408-256-1600
IMI	Unknown
Integral Peripherals	303-449-8009
JCT	Unknown
JTS	888-587-0945
JVC Companies Of America	714-816-6500; No longer manufacture hard drives.
Kalok Corporation	Out of Business; see JTS at 888-587-0945.
Kyocera Electronics, Inc.	732-560-3400; No longer manufacture hard drives.
Lanstor	Unknown
Lapine	Unknown
Maxtor Corporation	408-432-1700; Sold XT product line to Sequel in 1992.
Mega Drive Systems	310-247-0006
Memorex	972-444-3500; No longer a manufacturer.
Micropolis Corp	Out of Business
Microscience Intl. Corp	Out of Business
Miniscribe Corporation	Out Of Business, Portions Bought By Maxtor Corporation
Mitsubishi Electronics	714-220-2500
Mitsumi Electronics Corp.	972-550-7300; No longer manufacture hard drives.

Manufacturer	Status
MMI	Unknown
NCL America	408-737-2496; No longer manufacture hard drives.
NCR Corp	800-531-2222; No longer manufacture hard drives; call AT&T Global Info.
NEC Technologies Inc	978-264-8000
NEI	Unknown
Newbury Data	Unknown
NPL	Unknown
Okidata	609-235-2600
Olivetti	509-927-5600; No longer make or support drives.
Optima Technology Corp	714-476-0515
Orca Technology Corp	Unknown
Otari	Out of Business
Pacific Magtron	408-956-8888; No longer make hard drives.
Panasonic	201-348-7000
Plus Development	408-894-4000; Bought Out By Quantum
Prairietek Corp	Unknown
Priam Corporation	Out Of Business; Lipsig & Assoc provide support 408-733-1844
Procom Technology	949-852-1000; Does Not manufacture drives, they Bundle
PTI (Peripheral Technology)	510-724-1486
Quantum Corporation	408-894-4000
Ricoh	800-955-3453; No longer manufacture drives.
RMS	212-840-8666; They say they have never manufactured drives.
Rodime Systems, Inc	Out of Business
Samsung	201-229-7000
Seagate Technologies	408-438-6550
Sequel, Inc	408-987-1000; Purchased XT model lines from Maxtor.
Shugart	520-294-0898
Siemens	Out of Business
Sony	408-432-1600
Storage Dimensions	408-954-0710; Do not manufacture hard drives, they bundle.
Syquest Technology	510-226-4000
Tandon Computer Corp	Out of Business; Filed Chapter 11 Bankruptcy March 1993.

Manufacturer	Status
Tandy Corp	817-390-3011; No longer manufacture hard drives.
Teac America, Inc.	213-726-0303
Texas Instruments	214-995-6611
Toshiba America, Inc.	714-583-3000
Tulin	978-283-2100; Not a manufacturer.
Vertex (see Priam)	Out Of Business; Lipsig & Assoc provide support 408-733-1844
Western Digital	714-932-5000
Xebec	Out of Business
Y-E Data America, Inc	847-855-0890; No longer manufacture drives, they make heads.
Zentec	Unknown

===

Total Number of Drives ··········4183

Drive Model	Format Size MB	Head	Cyl	Sect/ Trac	Translate H/C/S	RWC/ WPC	Land Zone
ALPS AMERICA							
DR311C	106	2	2108	V		Na/Na	Auto
DR311D	106	2	2108	V		Na/Na	Auto
DR312C	212	4	2108	V		Na/Na	Auto
DR312D	212	4	2108	V		Na/Na	Auto
DRND-10A	11	2	615	17		616/616	Auto
DRND-20A	21	4	615	17		616/616	Auto
DRPO-20A	16	2	615	26		616/616	Auto
DRPO-20D	16	2	615	26		616/616	
AMPEX							
PYXIS-13	11	4	320	17		132/132	
PYXIS-20	17	6	320	17		132/132	
PYXIS-27	22	8	320	17		132/132	
PYXIS-7	6	2	320	17		132/132	
AREAL TECHNOLOGY, INC							
A1080	1080					—/—	
A120	132	4	1070	63	10/535/50	NA/NA	None
A130	130	2	1438	V	5/856/60	—/—	
A180	183	4	1430	62	10/715/50	Na/Na	None
A260	260	4	1438	V	10/856/60	—/—	
A270	270					—/—	
A340	350	4	2120	V	12/950/60	—/—	
A520	526	6	2120	V	16/1020/63	—/—	
A540	540					—/—	
A810	810					—/—	
A85	86	2	1344	V		Auto/Auto	NA
A90	92	2	1430	62	10/715/25	Na/Na	None
BP100 (not made)	105	2	1720	V		Na/Na	Auto
BP200 (not made)						—/—	
BP50 (not made)						—/—	
MD2050 (not made)	49	2	819	V		—/—	
MD2060	62	2	1024	59	7/1024/17	Na/Na	None
MD2065	62	2	1024			—/—	
MD2080	81	2	1330	59	14/665/17	Na/Na	None
MD2085	86	2	1410	59	14/705/17	Na/Na	None
MD2100 (not made)	100	2	1638	V		—/—	
ATASI TECHNOLOGY, INC							
3020	17	3	645	17		320/320	
3033	28	5	645	17		320/320	
3046	39	7	645	17		320/320	644
3051	43	7	704	17		—/352	703
3051+	44	7	733	17		—/368	732
3053	44	7	733	17		350/368	
3075	67	8	1024	17		1025/1025	
3085	72	8	1024	17		—/512	1023
3128	128	8	1024	26		—/—	1023
519	159	15	1224	17		Na/Na	
519R	244	15	1224	26		Na/Na	
6120	1051	15	1925	71		Na/Na	Auto
638	338	15	1225	36		Na/Na	Auto
676	676	15	1632	54		Na/Na	Auto
7120	1034	15	1919	71		Na/Na	Auto
738	336	15	1225	36		Na/Na	Auto
776	668	15	1632	54		Na/Na	Auto
AURA ASSOCIATES							
AU126	125	4				—/—	
AU211	211					Na/Na	Auto

Drive Model	Seek Time	Interface	Encode	Form Factor	cache kb	mtbf	RPM	Obsolete?
ALPS AMERICA								
DR311C	13	IDE AT	1,7 RLL	3.5 3H		150k		Y
DR311D	13	SCSI-2	1,7 RLL	3.5 3H		150k		Y
DR312C	13	IDE AT	1,7 RLL	3.5 3H		150k		Y
DR312D	13	SCSI-2	1,7 RLL	3.5 3H		150k		Y
DRND-10A	60	ST412/506	MFM	3.5HH				Y
DRND-20A	60	ST412/506	MFM	3.5HH				Y
DRPO-20A	60	ST412/506	2,7 RLL	3.5HH				Y
DRPO-20D	60	ST412/506	2,7 RLL	3.5HH				Y
AMPEX								
PYXIS-13	90	ST412/506	MFM	5.25FH				Y
PYXIS-20	90	ST412/506	MFM	5.25FH				Y
PYXIS-27	90	ST412/506	MFM	5.25FH				Y
PYXIS-7	90	ST412/506	MFM	5.25FH				Y
AREAL TECHNOLOGY, INC								
A1080		ATA Fast		2.5 4H	128k	250k		Y
A120	15	IDE AT	2,7-1,7RLL	2.5 4H	32k	100k	2981	Y
A130	<15	IDE AT	1,7 RLL	2.5 4H		150k	2981	Y
A180	17	IDE XT-AT	2,7 RLL	2.5 4H	32k	100k	2981	Y
A260	<15	IDE AT	1,7 RLL	2.5 4H		150k	2981	Y
A270		ATA Fast		2.5 4H	128k	250k		Y
A340	13	IDE AT	1,7 RLL	2.5 4H		150k		Y
A520	13	IDE AT	1,7 RLL	2.5 4H		150k		Y
A540		ATA Fast		2.5 4H	128k	250k		Y
A810		ATA Fast		2.5 4H	128k	250k		Y
A85	15	IDE	2,7 RLL	2.5 4H		100k		Y
A90	15	IDE XT-AT	2,7 RLL	2.5 4H	32k	100k	2981	Y
BP100 (not made)	27	SCSI	2,7 RLL	2.5 4H				Y
BP200 (not made)								Y
BP50 (not made)								Y
MD2050 (not made)	28		2,7 RLL	2.5 4H				Y
MD2060	19	IDE AT	2,7 RLL	2.5 4H	32k	45k	1565	Y
MD2065	<16	IDE AT	RLL	2.5 4H		100k	2504	Y
MD2080	19	IDE AT	2,7 RLL	2.5 4H	32k	100k	1565	Y
MD2085	19	IDE AT	2,7 RLL	2.5 4H	32k	100k	2504	Y
MD2100 (not made)	29	SCSI	2,7 RLL	2.5 4H				Y
ATASI TECHNOLOGY, INC								
3020		ST412/506	MFM	5.25FH				Y
3033	30	ST412/506	MFM	5.25FH				Y
3046	30	ST412/506	MFM	5.25FH				Y
3051	33	ST412/506	MFM	5.25FH				Y
3051+		ST412/506	MFM	5.25FH				Y
3053	27	ST412/506	MFM	5.25FH				Y
3075	27	ST412/506	MFM	5.25FH				Y
3085	27	ST412/506	MFM	5.25FH				Y
3128		ST412/506	2,7 RLL	5.25FH				Y
519	22	ST412/506	MFM	5.25FH		40k		Y
519R	22	ST412/506	2,7 RLL	5.25FH		40k		Y
6120	14	ESDI	2,7 RLL	5.25FH		150k	3600	Y
638	18	ESDI		5.25FH		40k	3600	Y
676	16	ESDI	2,7 RLL	5.25FH		150k	3600	Y
7120	14	SCSI	2,7 RLL	5.25FH		150k	3600	Y
738	18	SCSI		5.25FH		40k	3600	Y
776	16	SCSI		5.25FH		150k	3600	Y
AURA ASSOCIATES								
AU126	17	PCMCIA-ATA	1,7 RLL	1.8 4H	32k	100k	5400	Y
AU211	13	ATA		1.8 4H	128k		3448	Y

Drive Model	Format Size MB	Head	Cyl	Sect/ Trac	Translate H/C/S	RWC/ WPC	Land Zone
AU211S	211					Na/Na	Auto
AU245	245					Na/Na	Auto
AU245S	245					Na/Na	Auto
AU43	42	2				—/—	
AU63	42	2				—/—	
AU85	85	4				—/—	

BASF

6185	23	6	440	17		220/220	
6186	15	4	440	17		220/220	
6187	8	2	440	17		220/220	
6188-R1	10	2	612	17		—/—	
6188-R3	21	4	612	17		—/—	

BRAND TECHNOLOGIES

9121A (not made)	107	5	1166	36	10/583/36	Na/Na	Auto
9121E (not made)	107	5	1166	36		Na/Na	Auto
9121S (not made)	107	5	1166	36		Na/Na	Auto
9170A	150	7	1165	36	14/583/36	Na/Na	Auto
9170E	150	7	1166	36		Na/Na	Auto
9170S	150	7	1166	36		Na/Na	Auto
9220A	200	9	1209	36	16/401/61	Na/Na	Auto
9220E	200	9	1210	36		Na/Na	Auto
9220S	200	9	1210	36		Na/Na	Auto
BT8085	71	8	1024	17		Na/Na	Auto
BT8128	109	8	1024	26		Na/Na	Auto
BT8170E	142	8	1024	34		Na/Na	Auto
BT8170S	142	8	1024	34		Na/Na	Auto
BT9400A (not made)	400	6	1800	36		Na/Na	Auto
BT9400S (not made)	400	6	1800	36	16/801/61	Na/Na	Auto
BT9650A (not made)	650	10	1800	36	16/1024/63	Na/Na	Auto
BT9650S (not made)	650	10	1800	36		Na/Na	Auto

BULL

D530	25	3	987	17		988/988	
D550	43	5	987	17		988/988	
D570	60	7	987	17		988/988	
D585	71	7	1166	17		1166/1166	

C.ITOH ELECTRONICS, INC

SEE YE-DATA						—/—	

CALLUNA TECHNOLOGY, INC.

105MB	105	4			4/828/31	—/—	
130MB	130	4			6/986/43	—/—	
170MB	170	4			8/923/45	—/—	
260MB	260	4			10/820/62	—/—	
520MB	520	4			16/1008/63	—/—	
85MB	85	4				—/—	
CT1040RM	1040	4				—/—	
CT260T2	260	1				—/—	

CARDIFF

F3053	44	5	1024	17		—/—	
F3080E	68	5	1024	26		Na/Na	
F3080S	68	5	1024	26		Na/Na	
F3127E	109	5	1024	35		Na/Na	
F3127S	109	5	1024	35		Na/Na	

Drive Model	Seek Time	Interface	Encode	Form Factor	cache kb	mtbf	RPM	Obsolete?
AU211S	13	SCSI-2		1.8 4H	128k		3448	Y
AU245	13	ATA		1.8 4H	128k		3448	Y
AU245S	13	SCSI-2		1.8 4H	128k		3448	Y
AU43	17	IDE AT	1,7 RLL	1.8 4H	32k	100k	5400	Y
AU63	17	PCMCIA-ATA	1,7 RLL	1.8 4H	32k	100k	5400	Y
AU85	17	IDE AT	1,7 RLL	1.8 4H	32k	100k	5400	Y

BASF

Drive Model	Seek Time	Interface	Encode	Form Factor	cache kb	mtbf	RPM	Obsolete?
6185	150/70?	ST412/506	MFM	5.25FH				Y
6186	70	ST412/506	MFM	5.25FH				Y
6187	70	ST412/506	MFM	5.25FH				Y
6188-R1	70	ST412/506	MFM	5.25FH				Y
6188-R3	70	ST412/506	MFM	5.25FH				Y

BRAND TECHNOLOGIES

Drive Model	Seek Time	Interface	Encode	Form Factor	cache kb	mtbf	RPM	Obsolete?
9121A (not made)	16.5	IDE AT	2,7 RLL	3.5HH		50k		Y
9121E (not made)	16.5	SCSI	2,7 RLL	3.5HH		50k		Y
9121S (not made)	16.5	SCSI	2,7 RLL	3.5HH		50k		Y
9170A	16.5	IDE AT	2,7 RLL	3.5HH	64k	50k		Y
9170E	16.5	ESDI	2,7 RLL	3.5HH		50k	3565	Y
9170S	16.5	SCSI	2,7 RLL	3.5HH	64k	50k		Y
9220A	16.5	IDE AT	2,7 RLL	3.5HH	64k	50k	3565	Y
9220E	16.5	ESDI	2,7 RLL	3.5HH		50k	3565	Y
9220S	16.5	SCSI	2,7 RLL	3.5HH	64k	50k		Y
BT8085	25	ST412/506	MFM	5.25FH		50k		Y
BT8128	25	ST412/506	2,7 RLL	5.25FH		50k		Y
BT8170E	25	ESDI	2,7 RLL	5.25FH		50k		Y
BT8170S	25	SCSI	2,7 RLL	5.25FH		50k		Y
BT9400A (not made)	12	IDE AT	1,7 RLL	5.25FH				Y
BT9400S (not made)	12	SCSI-2	1,7 RLL	5.25FH				Y
BT9650A (not made)	12	IDE AT	1,7 RLL	5.25FH				Y
BT9650S (not made)	12	SCSI-2	1,7 RLL	5.25FH				Y

BULL

Drive Model	Seek Time	Interface	Encode	Form Factor	cache kb	mtbf	RPM	Obsolete?
D530		ST412/506	MFM	5.25FH				Y
D550		ST412/506	MFM	5.25FH				Y
D570		ST412/506	MFM	5.25FH				Y
D585		ST412/506	2,7 RLL	5.25FH				Y

C.ITOH ELECTRONICS, INC

SEE YE-DATA

CALLUNA TECHNOLOGY, INC.

Drive Model	Seek Time	Interface	Encode	Form Factor	cache kb	mtbf	RPM	Obsolete?
105MB	18	PCMCIA-ATA	1,7 RLL	1.8	32k	150k		Y
130MB	16	PCMCIA-ATA	1,7 RLL	1.8	64k	150k		Y
170MB	16	PCMCIA-ATA	1,7 RLL	1.8	64k	150k		Y
260MB	16	PCMCIA-ATA	1,7 RLL	1.8	64k	150k		Y
520MB	12	PCMCIA-ATA	1,7 RLL	1.8	128k	150k		Y
85MB	18	PCMCIA-ATA	1,7 RLL	1.8	32k	150k		Y
CT1040RM	12	PCMCIA-ATA	1,7 RLL	1.8	128k	300k		Y
CT260T2	12	PCMCIA-ATA	1,7 RLL	1.8	128k	300k		Y

CARDIFF

Drive Model	Seek Time	Interface	Encode	Form Factor	cache kb	mtbf	RPM	Obsolete?
F3053	20	ST412/506	MFM	3.5HH				Y
F3080E	20	ESDI	2,7 RLL	3.5HH				Y
F3080S	20	SCSI	2,7 RLL	3.5HH				Y
F3127E	20	ESDI	2,7 RLL	3.5HH				Y
F3127S	20	SCSI	2,7 RLL	3.5HH				Y

Drive Model	Format Size MB	Head	Cyl	Sect/ Trac	Translate H/C/S	RWC/ WPC	Land Zone
CDC							
94151-25 WREN II	25	3	921	19		—/—	
94151-27 WREN II	26	3	921	19		—/—	
94151-42 Wren II	42	5	921	19		—/—	
94151-44 Wren II	44	5	921	19		—/—	
94151-59 Wren II	59	7	921	19		—/—	
94151-62 Wren II	62	7	921	19		—/—	
94151-76 Wren II	76	9	921	19		—/—	
94151-80 Wren II	80	9	921	19		—/—	
94151-80SA Wren II	72	9	921	17		—/—	
94151-80SC Wren II	70	9	921	17		—/—	
94151-86 Wren II	72	9	925	17		925/925	
94155-021 Wren I	18	3	697	17		697/697	
94155-025 Wren I	24	4	697	17		697/128	
94155-028 Wren I	24	3	697	17		698/128	
94155-029 Wren I	25	3	925	17		—/—	
94155-036 Wren I	31	5	733	17		697/128	

Conversion Chart: Part I

Old CDC/Imprimis model # to new Seagate model

CDC/Imprimis ➡	Seagate	Seagate ➡	CDC/Imprimis
94155-135	ST4135R	ST1090A	94354-090
94155-85	ST4085	ST1090N	94351-090
94155-86	ST4086	ST1100	94355-100
94155-96	ST4097	ST1111A	94354-111
94161-182	ST4182N	ST1111E	94356-111
94166-182	ST4182E	ST1111N	94351-111
94171-350	ST4350N	ST1126A	94354-126
94171-376	ST4376N	ST1126N	94351-126
94181-385H	ST4385N	ST1133A	94354-133
94181-702	ST4702N	ST1133NS	94351-133S
94186-383	ST4383E	ST1150R	94355-150
94186-383H	ST4384E	ST1156A	94354-155
94186-442	ST4442E	ST1156E	94356-155
94191-766	ST4766N	ST1156N	94351-155
94196-766	ST4766E	ST1156NS	94351-155S
94204-65	ST274A	ST1162A	94354-160
94204-71	ST280A	ST1162N	94351-160
94204-74	ST274A	ST1186A	94354-186
94204-81	ST280A	ST1186NS	94351-186S
94205-51	ST253	ST1201A	94354-200
94205-77	ST279R	ST1201E	94356-200
94211-106	ST2106N	ST1201N	94351-200
94216-106	ST2106E	ST1201NS	94351-200S
94221-125	ST2125N	ST1239A	94354-239
94241-502	ST2502N	ST1239NS	94351-230S
94244-274	ST2274A	ST2106E	94216-106
94244-383	ST2383A	ST2106N	94211-106
94246-182	ST2182E	ST2125N	94221-125
94246-383	ST2383E	ST2182E	94246-182
94351-090	ST1090N	ST2274A	94244-274
94351-111	ST1111N	ST2383A	94244-383
94351-126	ST1126N	ST2383E	94246-383
94351-133S	ST1133NS	ST2502N	94241-502
94351-155	ST1156N	ST253	94205-51
94351-155S	ST1156NS	ST274A	94204-74
94351-160	ST1162N	ST274A	94204-65
94351-186S	ST1186NS	ST279R	94205-77
94351-200	ST1201N	ST280A	94204-81
94351-200S	ST1201NS	ST280A	94204-71
94351-230S	ST1239NS	ST4085	94155-85
94354-090	ST1090A	ST4086	94155-86

Drive Model	Seek Time	Interface	Encode	Form Factor	cache kb	mtbf	RPM	Obsolete?
CDC								
94151-25 WREN II				5.25FH				Y
94151-27 WREN II				5.25FH				Y
94151-42 Wren II				5.25FH				Y
94151-44 Wren II				5.25FH				Y
94151-59 Wren II				5.25FH				Y
94151-62 Wren II				5.25FH				Y
94151-76 Wren II				5.25FH				Y
94151-80 Wren II				5.25FH				Y
94151-80SA Wren II	38	SCSI		5.25FH				Y
94151-80SC Wren II	38	SCSI		5.25FH				Y
94151-86 Wren II	38	ST412/506	MFM	5.25FH				Y
94155-021 Wren I		ST412/506	MFM	5.25FH				Y
94155-025 Wren I		ST412/506	MFM	5.25FH				Y
94155-028 Wren I	28	ST412/506	MFM	5.25FH				Y
94155-029 Wren I	28	ST412/506	MFM	5.25FH				Y
94155-036 Wren I		ST412/506	MFM	5.25FH				Y

Conversion Chart: Part II
Old CDC/Imprimis model # to new Seagate model

CDC/Imprimis ➡	Seagate	Seagate ➡	CDC/Imprimis
94354-111	ST1111A	ST4097	94155-96
94354-126	ST1126A	ST41200N	94601-12G/M
94354-133	ST1133A	ST41201J	97500-12G
94354-155	ST1156A	ST41201K	97509-12G
94354-160	ST1162A	ST4135R	94155-135
94354-186	ST1186A	ST41520N	97501-12G
94354-200	ST1201A	ST4182E	94166-182
94354-239	ST1239A	ST4182N	94161-182
94355-100	ST1100	ST4350N	94171-350
94355-150	ST1150R	ST4376N	94171-376
94356-111	ST1111E	ST4383E	94186-383
94356-155	ST1156E	ST4384E	94186-383H
94356-200	ST1201E	ST4385N	94181-385H
94601-12G/M	ST41200N	ST4442E	94186-442
94601-767H	ST4767N	ST4702N	94181-702
97100-80	ST683J	ST4766E	94196-766
97150-160	ST6165J	ST4766N	94191-766
97150-300	ST6315J	ST4767N	94601-767H
97150-340	ST6344J	ST6165J	97150-160
97150-500	ST6516J	ST6315J	97150-300
97200-1130	ST81123J	ST6344J	97150-340
97200-12G	ST81236J	ST6516J	97150-500
97200-23G	ST82272K	ST683J	97100-80
97200-25G	ST82500J	ST81123J	97200-1130
97200-368	ST8368J	ST81154K	97229-1150
97200-500	ST8500J	ST81236J	97200-12G
97200-736	ST8741J	ST81236K	97209-12G
97200-850	ST8851J	ST81236N	97201-12G
97201-12G	ST81236N	ST82105K	97289-21G
97201-25G	ST82500N	ST82272K	97200-23G
97201-368	ST8368N	ST82368K	97299-23G
97201-500	ST8500N	ST82500J	97200-25G
97201-736	ST8741N	ST82500K	97209-25G
97201-850	ST8851N	ST82500N	97201-25G
97209-12G	ST81236K	ST8368J	97200-368
97209-25G	ST82500K	ST8368N	97201-368
97229-1150	ST81154K	ST8500J	97200-500
97289-21G	ST82105K	ST8500N	97201-500
97299-23G	ST82368K	ST8741J	97200-736
97500-12G	ST41201J	ST8741N	97201-736
97501-12G	ST41520N	ST8851J	97200-850
97509-12G	ST41201K	ST8851N	97201-850

Drive Model	Format Size MB	Head	Cyl	Sect/ Trac	Translate H/C/S	RWC/ WPC	Land Zone
94155-037 Wren I	32	4	925	17		—-/—-	
94155-038 Wren I	31	5	733	17		734/0	
94155-048 Wren II	40	5	925	17		926/128	
94155-051 Wren II	43	5	989	17		990/128	
94155-057 Wren II	48	6	925	17		926/128	Auto
94155-057P Wren II	48	6	925	17		926/128	
94155-067 Wren II	56	7	925	17		926/128	Auto
94155-067P Wren II	56	7	925	17		926/128	
94155-077 Wren II	64	8	925	17		926/128	Auto
94155-085 Wren II	71	8	1024	17		1025/128	Auto
94155-085P Wren II	71	8	1024	17		1025/128	Auto
94155-086 Wren II	72	9	925	17		926/128	Auto
94155-087 Wren II	72	9	925	17		—-/—-	
94155-092 Wren II	77	9	989	17		—-/-1.0	
94155-092P Wren II	77	9	989	17		—-/128	
94155-096 Wren II	80	9	1024	17		—-/—-	Auto
94155-120 Wren II	120	8	960	26		961/128	Auto
94155-130 Wren II	122	9	1024	26		—-/128	
94155-135 Wren II	115	9	960	26		961/128	Auto
94156-048 Wren II	40	5	925	17		926/128	Auto
94156-067 Wren II	56	7	925	17		926/128	Auto
94156-086 Wren II	72	9	925	17		926/128	Auto
94156-48 Wren II	40					—-/—-	
94156-67 Wren II	56					—-/—-	
94156-86 Wren II	72					—-/—-	
94161-086 Wren III	86	5	969	35		Na/Na	Auto
94161-101 Wren III	84	5	969	34		Na/Na	Auto
94161-103 Wren III	104	6	969	35		Na/Na	Auto
94161-121 Wren III	121	7	969	35		Na/Na	Auto
94161-138 Wren III	138	8	969	35		Na/Na	Auto
94161-141 Wren III	118	7	969	35		Na/Na	Auto
94161-151 Wren III	151	9	969	34		Na/Na	Auto
94161-155 Wren III	132	9	969	35		—-/-1.0	
94161-156 Wren III	132	9	969	36		—-/-1.0	
94161-160 Wren III	160		969			—-/—-	
94161-182 Wren III	156	9	969	35		Na/Na	Auto
94161-182M Wren III	160	9	969			—-/—-	
94166-086 Wren III	86	5	969	35		—-/-1.0	
94166-101 Wren III	86	5	969	35		Na/Na	Auto
94166-103 Wren III	104	6	969	35		—-/-1.0	
94166-121 Wren III	107	6	969	36		Na/Na	Auto
94166-138 Wren III	138	8	969	35		—-/-1.0	
94166-141 Wren III	125	7	969	36		Na/Na	Auto
94166-161 COMPAQ	160	9	969	36		Na/Na	Auto
94166-161 Wren III	142	8	969	36		Na/Na	Auto
94166-182 Wren III	161	9	969	36		Na/Na	Auto
94171-300 Wren IV	300	9	1412			Na/Na	Auto
94171-307 Wren IV	300	9	1412			Na/Na	Auto
94171-327 Wren IV	300	9	1412			Na/Na	Auto
94171-330 Wren IV	330					—-/—-	
94171-344 Wren IV	323	9	1549	V		Na/Na	Auto
94171-350 Wren IV	307	9	1412	V		Na/Na	Auto
94171-375 Wren IV	330	9	1549	V		Na/Na	Auto
94171-376 Wren IV	330	9	1546	V		Na/Na	Auto
94171-376D Wren IV	323	9	1549	V		Na/Na	Auto
94181-383 Wren IV	330	15	1224			—-/—-	
94181-385D Wren V	337	15	791	V		Na/Na	Auto
94181-385H Wren V	337	15	791	V		Na/Na	Auto
94181-574 Wren V	574	15	1549	V		Na/Na	Auto
94181-702 Wren V	613	15	1546	V		Na/Na	Auto
94181-702D Wren V	601	15	1546	V		Na/Na	Auto
94181-702M Wren V	613	15	1549			—-/—-	
94186-265 Wren V	234	9	1412	36		Na/Na	Auto

Drive Model	Seek Time	Interface	Encode	Form Factor	cache kb	mtbf	RPM	Obsolete?
94155-037 Wren I	28	ST412/506	MFM	5.25FH				Y
94155-038 Wren I	28	ST412/506	MFM	5.25FH				Y
94155-048 Wren II	28	ST412/506	MFM	5.25FH				Y
94155-051 Wren II	28	ST412/506	MFM	5.25FH				Y
94155-057 Wren II	28	ST412/506	MFM	5.25FH		40k		Y
94155-057P Wren II	28	ST412/506	MFM	5.25FH				Y
94155-067 Wren II	28	ST412/506	MFM	5.25FH		40k		Y
94155-067P Wren II	38	ST412/506	MFM	5.25FH				Y
94155-077 Wren II	28	ST412/506	MFM	5.25FH		40k		Y
94155-085 Wren II	28	ST412/506	MFM	5.25FH		40k		Y
94155-085P Wren II	28	ST412/506	MFM	5.25FH		40k		Y
94155-086 Wren II	28	ST412/506	MFM	5.25FH		40k		Y
94155-087 Wren II	38	ESDI		5.25FH				Y
94155-092 Wren II	38	ST412/506	MFM	5.25FH				Y
94155-092P Wren II	38	ST412/506	MFM	5.25FH				Y
94155-096 Wren II	28	ST412/506	MFM	5.25FH		40k		Y
94155-120 Wren II	28	ST412/506	2,7 RLL	5.25FH		40k		Y
94155-130 Wren II	28	ST412/506	RLL	5.25FH				Y
94155-135 Wren II	28	ST412/506	2,7 RLL	5.25FH		40k		Y
94156-048 Wren II	28	ESDI	MFM	5.25FH		40k		Y
94156-067 Wren II	28	ESDI	MFM	5.25FH		40k		Y
94156-086 Wren II	28	ESDI	MFM	5.25FH		40k		Y
94156-48 Wren II		ESDI	ST412/506	5.25FH				Y
94156-67 Wren II		ESDI	ST412/506	5.25FH				Y
94156-86 Wren II		ESDI	ST412/506	5.25FH				Y
94161-086 Wren III	16.5	SCSI	2,7 RLL	5.25FH		100k		Y
94161-101 Wren III	16.5	SCSI	2,7 RLL	5.25FH		100k		Y
94161-103 Wren III	16.5	SCSI	2,7 RLL	5.25FH		100k		Y
94161-121 Wren III	16.5	SCSI	2,7 RLL	5.25FH		100k		Y
94161-138 Wren III	16.5	SCSI	2,7 RLL	5.25FH		100k		Y
94161-141 Wren III	16.5	SCSI	2,7 RLL	5.25FH		100k		Y
94161-151 Wren III	16.5	SCSI	2,7 RLL	5.25FH		100k		Y
94161-155 Wren III	17	SCSI	RLL	5.25FH				Y
94161-156 Wren III	17	SCSI	RLL	5.25FH				Y
94161-160 Wren III		SCSI	2,7 RLL	5.25FH				Y
94161-182 Wren III	16.5	SCSI	2,7 RLL	5.25FH		100k		Y
94161-182M Wren III	17	SCSI	ZBR	5.25FH				Y
94166-086 Wren III	25	ESDI	RLL	5.25FH				Y
94166-101 Wren III	16.5	ESDI	2,7 RLL	5.25FH		100k		Y
94166-103 Wren III	25	ESDI	RLL	5.25FH				Y
94166-121 Wren III	16.5	ESDI	2,7 RLL	5.25FH		100k		Y
94166-138 Wren III	25	ESDI	RLL	5.25FH				Y
94166-141 Wren III	16.5	ESDI	2,7 RLL	5.25FH		100k		Y
94166-161 COMPAQ		ESDI	2,7 RLL	5.25FH		100k		Y
94166-161 Wren III		ESDI	2,7 RLL	5.25FH		100k		Y
94166-182 Wren III	16.5	ESDI (10)	2,7 RLL	5.25FH		100k		Y
94171-300 Wren IV	17	SCSI	RLL ZBR	5.25FH				Y
94171-307 Wren IV	17	SCSI	RLL ZBR	5.25FH				Y
94171-327 Wren IV	17	SCSI	RLL ZBR	5.25FH				Y
94171-330 Wren IV		SCSI	RLL ZBR	5.25FH				Y
94171-344 Wren IV	18	SCSI	RLL ZBR	5.25FH				Y
94171-350 Wren IV	16.5	SCSI	RLL ZBR	5.25FH		100k		Y
94171-375 Wren IV	16	SCSI	RLL ZBR	5.25FH				Y
94171-376 Wren IV	17.5	SCSI	RLL ZBR	5.25FH		100k		Y
94171-376D Wren IV		SCSI	RLL ZBR	5.25HH		100k		Y
94181-383 Wren IV	18	SCSI	ZBR	5.25FH				Y
94181-385D Wren V		SCSI	RLL ZBR	5.25FH		100k		Y
94181-385H Wren V	10.7	SCSI	RLL ZBR	5.25FH		100k		Y
94181-574 Wren V	16	SCSI	RLL ZBR	5.25FH		100k		Y
94181-702 Wren V	16.5	SCSI	RLL ZBR	5.25FH		100k		Y
94181-702D Wren V		SCSI	RLL ZBR	5.25FH		100k		Y
94181-702M Wren V	17	SCSI	ZBR	5.25FH				Y
94186-265 Wren V		ESDI (10)	2,7 RLL	5.25FH		100k		Y

Drive Model	Format Size MB	Head	Cyl	Sect/ Trac	Translate H/C/S	RWC/ WPC	Land Zone
94186-324 Wren V	278	11	1412	35		Na/Na	Auto
94186-383 Wren V	338	7	1747	35		Na/Na	Auto
94186-383H Wren V	338	7	1747	35		Na/Na	Auto
94186-383S Wren V	338	13	1412	36		Na/Na	Auto
94186-442 Wren V	380	15	1412	35		Na/Na	Auto
94186-442S Wren V	390	15	1412	36		Na/Na	Auto
94191-766 Wren VI	677	15	1632	54		Na/Na	Auto
94191-766D Wren VI	677	15	1632	54		Na/Na	Auto
94196-383 Wren VI	338	7	1747	54		Na/Na	Auto
94196-766 Wren V	677	15	1632	54		Na/Na	Auto
94204-051 Wren II	43	5	989	26		Na/Na	Auto
94204-065 Wren II	63	5	948	26		Na/Na	Auto
94204-071 Wren II	63	5	1032	27		Na/Na	Auto
94204-074 Wren II	63	5	948	26		Na/Na	Auto
94204-081 Wren II	71	5	1032	27		Na/Na	Auto
94205-030 Wren II	26	3	989	17		989/—	Auto
94205-041 Wren II	43	4	989	17		990/128	Auto
94205-051 Wren II	43	5	989	17		990/128	Auto
94205-053 Wren II	43	5	1024	17		990/128	Auto
94205-071 Wren II	43	5	989	26		990/128	Auto
94205-075 Wren II	62	5	966	25		966/128	Auto
94205-077 Wren II	66	5	989	26		—/—	Auto
94208-062 Wren II	60	5	989	17		—/—	
94208-075 Wren II	66	5	989	26		Na/Na	Auto
94208-106 Wren II	91		989			—/—	
94208-51 WrenII	43		989			—/—	
94208-91 WrenII	80		989			—/—	
94208-951 Wren II	42	5	989	17		990/128	
94211-086 Wren III	72	5	1024			—/—	
94211-091 Wren III	77	5	1024	17		970/970	
94211-106 Wren III	92	5	1024	35		Na/Na	Auto
94211-106M Wren III	94	5	1024			1025/1025	
94211-209 Wren III	183	5	1547			1548/1548	
94216-106 Wren III	90	5	1024	34		Na/Na	Auto
94221-125 Wren V	110	3	1544	V		Na/Na	Auto
94221-169 Wren V	159	5	1310	V		Na/Na	Auto
94221-190 Wren V	190	5	1547	V		Na/Na	Auto
94221-209 Wren V	183	5	1544	V		Na/Na	Auto
94241-383 Wren VI	338	7	1400	V		Na/Na	Auto
94241-502 Wren VI		7	1765	V		Na/Na	Auto
94241-502M Wren VI		7	1765	V		Na/Na	Auto
94244-219 Wren VI	186	4	1747	54		1748/-1.0	
94244-274 Wren VI	233	5	1747	52		Na/Na	Auto
94244-383 Wren VI	338	7	1747	54		Na/Na	Auto
94246-182 Wren VI	161	4	1453	54		Na/Na	Auto
94246-383 Wren VI	338	17	1747			Na/Na	Auto
94311-136 Swift SL	120	5				Na/Na	Auto
94311-136S Swift SL	120	5	1247	36		Na/Na	Auto
94314-136 Swift SL	120	5				Na/Na	Auto
94316-111 Swift	98	5		36		Na/Na	Auto
94316-136 Swift SL	120	5		36		Na/Na	Auto
94316-155 Swift	138	7	1072	36		Na/Na	Auto
94316-200 Swift	177	5		36		Na/Na	Auto
94335-055 Swift SL	46	5				—/—	
94335-100 Swift	85	9	1072	17		—/—	
94335-150 Swift	128	9		26		—/—	
94351-090 Swift	80	5	1068			—/—	
94351-111 Swift	98	5	1068			Na/Na	Auto
94351-126 Swift	111	7	1068	29		Na/Na	Auto
94351-128 Swift	111	7	1068			Na/Na	Auto
94351-133S Swift	117	5	1268	36		Na/Na	Auto
94351-134 Swift	120	7	1268			—/—	
94351-135 Swift	121	6	1068			—/—	

Drive Model	Seek Time	Interface	Encode	Form Factor	cache kb	mtbf	RPM	Obsolete?
94186-324 Wren V		ESDI (10)	2,7 RLL	5.25FH				Y
94186-383 Wren V		ESDI (10)	2,7 RLL	5.25FH		100k		Y
94186-383H Wren V		ESDI (10)	2,7 RLL	5.25FH		100k		Y
94186-383S Wren V	19	SCSI	2,7 RLL	5.25FH		100k		Y
94186-442 Wren V		ESDI (10)	2,7 RLL	5.25FH		100k		Y
94186-442S Wren V	15	SCSI	2,7 RLL	5.25FH				Y
94191-766 Wren VI	15.5	SCSI	2,7 RLL	5.25FH		100k		Y
94191-766D Wren VI		SCSI	2,7 RLL	5.25FH		100k		Y
94196-383 Wren VI		ESDI (15)	2,7 RLL	5.25FH		100k		Y
94196-766 Wren V		ESDI (15)	2,7 RLL	5.25FH		100k		Y
94204-051 Wren II		IDE AT	2,7 RLL	5.25HH		40k		Y
94204-065 Wren II		IDE AT	2,7 RLL	5.25HH		40k		Y
94204-071 Wren II		IDE AT	2,7 RLL	5.25HH		40k		Y
94204-074 Wren II	28	IDE AT	2,7 RLL	5.25HH		40k		Y
94204-081 Wren II	28	IDE AT	2,7 RLL	5.25HH		40k		Y
94205-030 Wren II		ST412/506	MFM	5.25FH		40k		Y
94205-041 Wren II		ST412/506	MFM	5.25HH		40k		Y
94205-051 Wren II	28	ST412/506	MFM	5.25HH		40k		Y
94205-053 Wren II		ST412/506	MFM	5.25HH		40k		Y
94205-071 Wren II		ST412/506	RLL	5.25HH		40k		Y
94205-075 Wren II	28	ST412/506	RLL	5.25HH		40k		Y
94205-077 Wren II	28	ST412/506	2,7 RLL	5.25HH		40k		Y
94208-062 Wren II	28	COMPAQ	MFM	5.25HH				Y
94208-075 Wren II	30	IDE AT	2,7 RLL	5.25HH				Y
94208-106 Wren II		IDE AT		5.25HH				Y
94208-51 WrenII		IDE AT		5.25HH				Y
94208-91 WrenII		IDE AT		5.25HH				Y
94208-951 Wren II	28	COMPAQ	MFM	5.25FH				Y
94211-086 Wren III	18	SCSI	RLL	5.25HH				Y
94211-091 Wren III	18	SCSI	MFM	5.25HH				Y
94211-106 Wren III	18	SCSI	2,7 RLL	5.25HH		100k		Y
94211-106M Wren III	18	SCSI	ZBR	5.25FH				Y
94211-209 Wren III	18	SCSI	ZBR	3.5HH				Y
94216-106 Wren III	18	ESDI (10)	2,7 RLL	5.25HH		100k		Y
94221-125 Wren V	18	SCSI	RLL ZBR	5.25HH		100k		Y
94221-169 Wren V	18	SCSI	RLL ZBR	5.25HH		100k		Y
94221-190 Wren V	18	SCSI	RLL ZBR	5.25HH		100k		Y
94221-209 Wren V	18	SCSI	RLL ZBR	5.25HH		100k		Y
94241-383 Wren VI	14	SCSI	RLL ZBR	5.25HH		100k		Y
94241-502 Wren VI		SCSI	RLL ZBR	5.25HH		100k		Y
94241-502M Wren VI	16	SCSI(Mac)	RLL ZBR	5.25HH		100k		Y
94244-219 Wren VI	16	AT	RLL	5.25HH				Y
94244-274 Wren VI	16	IDE AT	2,7 RLL	5.25HH		100k		Y
94244-383 Wren VI	16	IDE AT	2,7 RLL	5.25HH		100k		Y
94246-182 Wren VI	16	ESDI (20)	2,7 RLL	5.25HH		100k		Y
94246-383 Wren VI	16	SCSI (20)	2,7 RLL	5.25HH		100k		Y
94311-136 Swift SL	15	SCSI	2,7 RLL	3.5 3H		70k		Y
94311-136S Swift SL	15	SCSI-2	2,7 RLL	3.5 3H		70k		Y
94314-136 Swift SL	15	IDE AT	2,7 RLL	3.5 3H		70k		Y
94316-111 Swift	23	ESDI	2,7 RLL	3.5HH		70k		Y
94316-136 Swift SL	15	ESDI	2,7 RLL	3.5 3H		70k		Y
94316-155 Swift	15	ESDI	2,7 RLL	3.5HH		70k		Y
94316-200 Swift	15	ESDI	2,7 RLL	3.5HH		70k		Y
94335-055 Swift SL	25	ST412/506	RLL	3.5HH				Y
94335-100 Swift	25	ST412/506	MFM	3.5HH				Y
94335-150 Swift	25	ST412/506	RLL	3.5HH				Y
94351-090 Swift	15	SCSI	RLL	3.5HH				Y
94351-111 Swift	15	SCSI	2,7 RLL	3.5HH		70k		Y
94351-126 Swift	15	SCSI	2,7 RLL	3.5HH		70k		Y
94351-128 Swift	15	SCSI	2,7 RLL	3.5HH		70k		Y
94351-133S Swift	15	SCSI-2	2,7 RLL	3.5HH		70k		Y
94351-134 Swift	15	SCSI	RLL	3.5HH				Y
94351-135 Swift	15	SCSI	RLL	3.5HH				Y

Drive Model	Format Size MB	Head	Cyl	Sect/ Trac	Translate H/C/S	RWC/ WPC	Land Zone
94351-155 Swift	138	7	1068	36		Na/Na	Auto
94351-155S Swift	138	7	1068	36		Na/Na	Auto
94351-160 Swift	143	9	1068	29		Na/Na	Auto
94351-172 Swift	177	9	1068	36		Na/Na	Auto
94351-186S Swift	164	7	1268	36		Na/Na	Auto
94351-200 Swift	178	7	1068	36		Na/Na	Auto
94351-200S Swift	177	9	1068	36		Na/Na	Auto
94351-230 Swift	210	9	1268	36		Na/Na	Auto
94351-230S Swift	210	9	1268	36		Na/Na	Auto
94354-090 Swift	80	5	102	29		—/-1.0	
94354-111 Swift	99	5	1072	36		Na/Na	Auto
94354-126 Swift	111	7	1072	29		Na/Na	Auto
94354-133 Swift	117	5	1272	36		Na/Na	Auto
94354-155 Swift	138	7	1072	36		Na/Na	Auto
94354-160 Swift	143	9	1072	29		Na/Na	Auto
94354-186 Swift	164	7	1272	36		Na/Na	Auto
94354-200 Swift	178	9	1072	36		Na/Na	Auto
94354-230 Swift	204					—/—	
94354-239 Swift	211	9	1272	36		Na/Na	Auto
94355-055 Swift II	46	5		17		—/—	Auto
94355-100 Swift	84	9	1072	17		1073/300	Auto
94355-150 Swift	128	9	1072	26		1073/300	Auto
94355-55 Swift	46					—/—	
94356-111 Swift	99	5	1072	36		Na/Na	Auto
94356-155 Swift	138	7	1072	36		Na/Na	Auto
94356-200 Swift	178	9	1072	36		Na/Na	Auto
94601-12D Wren VII	1035	15	1931	V		Na/Na	Auto
94601-12G Wren VII	1037	15	1937	V		Na/Na	Auto
94601-12GM Wren VII	1037	15	1937	V		Na/Na	Auto
94601-767H Wren VII	676	15	1356	V		Na/Na	Auto
97155-036	30			17		—/—	
9720-1123 Sabre	964	19				—/—	Auto
9720-1130 Sabre	1050	15	1635			—/—	Auto
9720-2270 Sabre	1948	19				—/—	Auto
9720-2500 Sabre	2145	19				—/—	Auto
9720-368 Sabre	368		1635			1218/1218	Auto
9720-500 Sabre	500	10	1217			1218/1218	Auto
9720-736 Sabre	736	15	1635			1636/1636	Auto
9720-850 Sabre	727	15	1381			1382/1382	Auto
97229-1150 Wren V	990	19				—/—	Auto
97501-15G Elite	1500	17				Na/Na	Auto
97509-12G Elite	1050	17				—/—	Auto
BJ7D5A/77731600	18	3	697	17		—/128	
BJ7D5A/77731601	18	3	697	17		—/128	
BJ7D5A/77731602	30	5	697	17		—/128	
BJ7D5A/77731603	30	5	697	17		—/128	
BJ7D5A/77731604	36	5	697			—/128	
BJ7D5A/77731605	30	5	697	17		—/128	
BJ7D5A/77731606	27			17		—/128	
BJ7D5A/77731607	18	3	697	17		—/128	
BJ7D5A/77731608	29	5	670	17		—/128	
BJ7D5A/77731609	30	5	697	17		—/128	
BJ7D5A/77731610	18	3	697	17		—/128	
BJ7D5A/77731611	30	5	697	17		—/128	
BJ7D5A/77731612	24	4	697	17		—/128	
BJ7D5A/77731613	31	5	733	17		—/128	
BJ7D5A/77731614	23	4	670	17		—/128	
BJ7D5A/77731615	24	4	697	17		—/128	
BJ7D5A/77731616	31	5	733	17		—/128	
BJ7D5A/77731617	30	5	697	17		—/128	
BJ7D5A/77731618	30	5	697	17		—/128	
BJ7D5A/77731619	30	5	697	17		—/128	
BJ7D5A/77731620	30	5	697	17		—/128	

Drive Model	Seek Time	Interface	Encode	Form Factor	cache kb	mtbf	RPM	Obsolete?
94351-155 Swift	15	SCSI	2,7 RLL	3.5HH		70k		Y
94351-155S Swift	15	SCSI-2	2,7 RLL	3.5HH		70k		Y
94351-160 Swift	15	SCSI	2,7 RLL	3.5HH		150k		Y
94351-172 Swift	15	SCSI	2,7 RLL	3.5HH		70k		Y
94351-186S Swift	15	SCSI-2	2,7 RLL	3.5HH		150k		Y
94351-200 Swift	15	SCSI	2,7 RLL	3.5HH		150k		Y
94351-200S Swift	15	SCSI-2	2,7 RLL	3.5HH		150k		Y
94351-230 Swift	15	SCSI		3.5HH		70k		Y
94351-230S Swift	15	SCSI-2	2,7 RLL	3.5HH		70k		Y
94354-090 Swift	15	AT	RLL	3.5HH				Y
94354-111 Swift	15	IDE AT	2,7 RLL	3.5HH		70k		Y
94354-126 Swift	15	IDE AT	2,7 RLL	3.5HH		150k		Y
94354-133 Swift	15	IDE AT	2,7 RLL	3.5HH		70k		Y
94354-155 Swift	15	IDE AT	2,7 RLL	3.5HH		70k		Y
94354-160 Swift	15	IDE AT	2,7 RLL	3.5HH		150k		Y
94354-186 Swift	15	IDE AT	2,7 RLL	3.5HH		150k		Y
94354-200 Swift	15	IDE AT	2,7 RLL	3.5HH		150k		Y
94354-230 Swift		IDE AT	2,7 RLL	3.5HH				Y
94354-239 Swift	15	IDE AT	2,7 RLL	3.5HH		70k		Y
94355-055 Swift II	25	ST412/506	MFM	3.5HH		70k		Y
94355-100 Swift	15	ST412/506	MFM	3.5HH		150k		Y
94355-150 Swift	15	ST412/506	2,7 RLL	3.5HH		150k		Y
94355-55 Swift		MFM		3.5HH				Y
94356-111 Swift	15	ESDI (10)	2,7 RLL	3.5HH		150k		Y
94356-155 Swift	15	ESDI (10)	2,7 RLL	3.5HH		70k		Y
94356-200 Swift	15	ESDI (10)	2,7 RLL	3.5HH		70k		Y
94601-12D Wren VII	15	SCSI	2,7 RLL	5.25FH		150k		Y
94601-12G Wren VII	15	SCSI	RLL ZBR	5.25FH		150k		Y
94601-12GM Wren VII	15	SCSI(Mac)	RLL ZBR	5.25FH		150k		Y
94601-767H Wren VII	15	SCSI(Mac)	RLL ZBR	5.25FH		100k		Y
97155-036		ST412/506	MFM	8.0 FH		70k		Y
9720-1123 Sabre	15	SMD	2,7 RLL	8.0 FH		70k		Y
9720-1130 Sabre	15	SMD/SCSI	2,7 RLL	8.0 FH		100k		Y
9720-2270 Sabre	12	SMD	2,7 RLL	8.0 FH		100k		Y
9720-2500 Sabre	12	SMD/SCSI	2,7 RLL	8.0 FH		100k		Y
9720-368 Sabre	18	SMD/SCSI	2,7 RLL	8.0 FH		30k		Y
9720-500 Sabre	18	SMD/SCSI	2,7 RLL	8.0 FH		30k		Y
9720-736 Sabre	15	SMD/SCSI	2,7 RLL	8.0 FH		50k		Y
9720-850 Sabre	15	SMD/SCSI	2,7 RLL	8.0 FH		50k		Y
97229-1150 Wren V	15	IPI-2		8.0 FH		100k		Y
97501-15G Elite	12	SCSI-2	RLL	5.25FH		100k		Y
97509-12G Elite	12	IPI-2		5.25FH		100k		Y
BJ7D5A/77731600		ST412/506	MFM	5.25FH				Y
BJ7D5A/77731601		ST412/506	MFM	5.25FH				Y
BJ7D5A/77731602		ST412/506	MFM	5.25FH				Y
BJ7D5A/77731603		ST412/506	MFM	5.25FH				Y
BJ7D5A/77731604		ST412/506	MFM	5.25FH				Y
BJ7D5A/77731605		ST412/506	MFM	5.25FH				Y
BJ7D5A/77731606		ST412/506	MFM	5.25FH				Y
BJ7D5A/77731607		ST412/506	MFM	5.25FH				Y
BJ7D5A/77731608		ST412/506	MFM	5.25FH				Y
BJ7D5A/77731609		ST412/506	MFM	5.25FH				Y
BJ7D5A/77731610		ST412/506	MFM	5.25FH				Y
BJ7D5A/77731611		ST412/506	MFM	5.25FH				Y
BJ7D5A/77731612		ST412/506	MFM	5.25FH				Y
BJ7D5A/77731613		ST412/506	MFM	5.25FH				Y
BJ7D5A/77731614		ST412/506	MFM	5.25FH				Y
BJ7D5A/77731615		ST412/506	MFM	5.25FH				Y
BJ7D5A/77731616		ST412/506	MFM	5.25FH				Y
BJ7D5A/77731617		ST412/506	MFM	5.25FH				Y
BJ7D5A/77731618		ST412/506	MFM	5.25FH				Y
BJ7D5A/77731619		ST412/506	MFM	5.25FH				Y
BJ7D5A/77731620		ST412/506	MFM	5.25FH				Y

Drive Model	Format Size MB	Head	Cyl	Sect/ Trac	Translate H/C/S	RWC/ WPC	Land Zone
Sabre 1123	964	19				—/—	Auto
Sabre 1150	990	19				—/—	Auto
Sabre 1230	1050	15	1635			—/—	Auto
Sabre 2270	1948	19				—/—	Auto
Sabre 2500	2145	19				—/—	Auto
Sabre 368	368	10	1635			—/—	Auto
Sabre 500	500	10	1217			—/—	Auto
Sabre 736	741	15	1217			—/—	Auto
Sabre 850	851	15	1635			—/—	Auto

CENTURY DATA

Drive Model	Format Size MB	Head	Cyl	Sect/ Trac	Translate H/C/S	RWC/ WPC	Land Zone
CAST-10203E	55	3	1050	35		Na/Na	Auto
CAST-10203S	55	3	1050	35		Na/Na	Auto
CAST-10304E	75	4	1050	35		Na/Na	Auto
CAST-10304S	75	4	1050	35		Na/Na	Auto
CAST-10305E	94	5	1050	35		Na/Na	Auto
CAST-10305S	94	5	1050	35		Na/Na	Auto
CAST-14404E	114	4	1590	35		Na/Na	Auto
CAST-14404S	114	4	1590	35		Na/Na	Auto
CAST-14405E	140	5	1590	35		Na/Na	Auto
CAST-14405S	140	5	1590	35		Na/Na	Auto
CAST-14406E	170	6	1590	35		Na/Na	Auto
CAST-14406S	170	6	1590	35		Na/Na	Auto
CAST-24509E	258	9	1599	35		Na/Na	Auto
CAST-24509S	258	9	1599	35		Na/Na	Auto
CAST-24611E	315	11	1599	35		Na/Na	Auto
CAST-24611S	315	11	1599	35		Na/Na	Auto
CAST-24713E	372	13	1599	35		Na/Na	Auto
CAST-24713S	372	13	1599	35		Na/Na	Auto

CMI

Drive Model	Format Size MB	Head	Cyl	Sect/ Trac	Translate H/C/S	RWC/ WPC	Land Zone
CM3412	10	4	306	17		306/256	
CM3426	20	4	615	17		616/256	
CM5018H	15	2		17		—/—	
CM5205	4	2	256	17		128/128	
CM5206	5	2	306	17		307/256	
CM5410	8	4	256	17		128/128	
CM5412	10	4	306	17		307/128	
CM5616	14	6	256	17		257/257	
CM5619	16	6	306	17		307/128	
CM5826	20	8	306	17		—/—	
CM6213	11	2	640	17		641/256	
CM6426	22	4	615	17		—/300	615
CM6426S	22	4	615	17		256/300	615
CM6640	33	6	615	17		616/300	615
CM7000	44	7	733	17		733/512	
CM7030	25	4	733	17		733/512	
CM7038	31	5	733	17		733/512	
CM7053	44	7	733	17		733/512	
CM7085	71	8	1024	17		1024/512	
CM7660	50	6	960	17		961/450	
CM7880	67	8	960	17		961/450	

CMS ENHANCEMENTS, INC

Drive Model	Format Size MB	Head	Cyl	Sect/ Trac	Translate H/C/S	RWC/ WPC	Land Zone
B1.0A1-U1	1281				16/2100/63	Na/Na	Auto
B340A4-U1	340				12/1010/55	Na/Na	Auto
B420A4-U1	425				16/1010/51	Na/Na	Auto
B540A4-U1	541				16/1023/63	Na/Na	Auto
B730A4-U1	731				16/1416/63	Na/Na	Auto
D20XT-OK	21	4	615	17		—/—	
D30XT-OK	32	4	615	26		—/—	
D40XT-OK	42	5	977	17		—/—	

Drive Model	Seek Time	Interface	Encode	Form Factor	cache kb	mtbf	RPM	Obsolete? ⇓
Sabre 1123	15					100k		Y
Sabre 1150	15					100k		Y
Sabre 1230	15					100k		Y
Sabre 2270	12					100k		Y
Sabre 2500	12					100k		Y
Sabre 368	18					30k		Y
Sabre 500	18					30k		Y
Sabre 736	15					50k		Y
Sabre 850	15					50k		Y
CENTURY DATA								
CAST-10203E	28	ESDI	2,7 RLL	5.25FH				Y
CAST-10203S	28	SCSI	2,7 RLL	5.25FH				Y
CAST-10304E	28	ESDI	2,7 RLL	5.25FH				Y
CAST-10304S	28	SCSI	2,7 RLL	5.25FH				Y
CAST-10305E	28	ESDI	2,7 RLL	5.25FH				Y
CAST-10305S	28	SCSI	2,7 RLL	5.25FH				Y
CAST-14404E	25	ESDI	2,7 RLL	5.25HH				Y
CAST-14404S	25	SCSI	2,7 RLL	5.25HH				Y
CAST-14405E	25	ESDI	2,7 RLL	5.25HH				Y
CAST-14405S	25	SCSI	2,7 RLL	5.25HH				Y
CAST-14406E	25	ESDI	2,7 RLL	5.25HH				Y
CAST-14406S	25	SCSI	2,7 RLL	5.25HH				Y
CAST-24509E	18	ESDI	2,7 RLL	5.25FH				Y
CAST-24509S	18	SCSI	2,7 RLL	5.25FH				Y
CAST-24611E	18	ESDI	2,7 RLL	5.25FH				Y
CAST-24611S	18	SCSI	2,7 RLL	5.25FH				Y
CAST-24713E	18	ESDI	2,7 RLL	5.25FH				Y
CAST-24713S	18	SCSI	2,7 RLL	5.25FH				Y
CMI								
CM3412		ST412/506	MFM	5.25FH				Y
CM3426	85	ST412/506	MFM	5.25FH				Y
CM5018H	85	ST412/506	MFM	5.25FH				Y
CM5205		ST412/506	MFM	5.25FH				Y
CM5206	102	ST412/506	MFM	5.25FH				Y
CM5410	102	ST412/506	MFM	5.25FH				Y
CM5412	85	ST412/506	MFM	5.25FH				Y
CM5616	102	ST412/506	MFM	5.25FH				Y
CM5619	85	ST412/506	MFM	5.25FH				Y
CM5826	102	ST412/506	MFM	5.25FH				Y
CM6213	48	ST412/506	MFM	5.25FH				Y
CM6426	39	ST412/506	MFM	5.25FH				Y
CM6426S	39	ST412/506	MFM	5.25FH				Y
CM6640	39	ST412/506	MFM	5.25FH				Y
CM7000	42	ST412/506	MFM	5.25FH				Y
CM7030	42	ST412/506	MFM	5.25FH				Y
CM7038	42	ST412/506	MFM	5.25FH				Y
CM7053	42	ST412/506	MFM	5.25FH				Y
CM7085	42	ST412/506	MFM	5.25FH				Y
CM7660	28	ST412/506	MFM	5.25FH				Y
CM7880	28	ST412/506	MFM	5.25FH				Y
CMS ENHANCEMENTS, INC								
B1.0A1-U1	10	IDE AT		3.5 3H		250k	4500	Y
B340A4-U1	13	IDE AT		3.5 3H		250k	3600	Y
B420A4-U1	13	IDE AT		3.5 3H		250k	3300	Y
B540A4-U1	14	IDE AT		3.5 3H		300k	3600	Y
B730A4-U1	11	IDE AT		3.5 3H		300k	4500	Y
D20XT-OK	62	ST412/506	MFM	3.5HH				Y
D30XT-OK	62	ST412/506	2,7 RLL	3.5HH				Y
D40XT-OK	24	ST412/506	MFM	3.5HH				Y

Drive Model	Format Size MB	Head	Cyl	Sect/ Trac	Translate H/C/S	RWC/ WPC	Land Zone
F115ESD1-T	115	7	915	35		—/—	Auto
F150AT-CA	150	9	969	34		—/—	
F150AT-WCA	151	9	969	34		—/—	Auto
F150EQ-WCA	151	9	969	34		—/—	Auto
F320AT-CA	320	15	1224	34		—/—	Auto
F70ESDI-T	73	7	583	35		—/—	Auto
H100286D-P	105	8	776	34		—/—	
H100386S-P	105	8	776	34		—/—	
H330E1 (PS Express)	330	7	1780	54		—/—	Auto
H340E1 (PS Express)	340	7	1780	54		—/—	Auto
H40M50-P	42	5	977	17		—/—	
H60286D-P	64	5	948	27		—/—	
H60SCSI-S	65	6	628	34		—/—	
H65M50-P	65	9	1024	17		—/—	
H80AT	84	9	1072	17		—/—	
H80SCSI-S	85	6	820	34		—/—	
HD20AT-S	21	4	615	17		—/—	
HD30AT-S	32	6	615	17		—/—	
HD40AT-S1	43	6	820	17		—/—	
K120M50Z-70P	125	8	925	33		—/—	
K20M25-WS	21	2	636	34		—/—	
K20M25/30-OK	21	4	615	17		—/—	
K20M25/30-WS	21	4	615	17		—/—	
K30M25/30-OK	32	6	615	17		—/—	
K30M25/30-WS	32	6	615	17		—/—	
K30M30E-P	31	4	615	25		—/—	
K40M25/30-WS	42	5	977	17		—/—	
K45M30286-ZS	48	6	615	26		—/—	
K50M50Z/70P	63	6	767	27		—/—	
K60M30286-ZS	61	5	921	26		—/—	
K80M25Z/30	84	9	1072	17		—/—	
K80M30286-WS	84	7	906	26		—/—	
LDSNECMS-20	20	4	575	32		—/—	
LDZE386-100	100	8	776	34		—/—	
PB340	340					Na/Na	Auto
PB520	520					Na/Na	Auto
PSEXPRESS 150	150					—/—	Auto
PSEXPRESS 320	320					—/—	Auto
SENTRY 180	180	5	1546			—/—	
SENTRY 300	290	9	1546			—/—	
SENTRY 600	600	15	1546			—/—	
SENTRY 90	90	5	1024			—/—	

COGITO

Drive Model	Format Size MB	Head	Cyl	Sect/ Trac	Translate H/C/S	RWC/ WPC	Land Zone
CG906	5	2	306	17		128/128	
CG912	10	4	306	17		128/128	
CG925	21	4	612	17		307/307	
PT912	11	2	612	17		307/307	
PT925	21	4	612	17		307/307	

COMPAQ

Drive Model	Format Size MB	Head	Cyl	Sect/ Trac	Translate H/C/S	RWC/ WPC	Land Zone
113640-001	43	2	1053	40		Na/Na	Auto
113641-001	112	8	832	33		Na/Na	Auto
115145-001	84	6	832	33		Na/Na	Auto
115147-001	325	7	1744	52		Na/Na	Auto
115158-001	651	15	1631	52		Na/Na	Auto
115627-001	112	8	832	33		Na/Na	Auto
115830-001	318	15	1220	34		Na/Na	Auto
116562-001	123	4	1552	39		Na/Na	Auto
116565-001	207	8	1336	38		Na/Na	Auto
122136-001	60	2	1520	39		Na/Na	Auto
131067-001	510	12	1806	46		Na/Na	Auto
131362-001	325	7	1744	52		Na/Na	Auto

Drive Model	Seek Time	Interface	Encode	Form Factor	cache kb	mtbf	RPM	Obsolete?
F115ESD1-T	30	ESDI	2,7 RLL	5.25FH		25k		Y
F150AT-CA	17	ESDI	2,7 RLL	5.25FH		40k		Y
F150AT-WCA	17	ESDI	2,7 RLL	5.25FH		40k		Y
F150EQ-WCA	17	ESDI	2,7 RLL	5.25FH		40k		Y
F320AT-CA	18	ESDI	2,7 RLL	5.25FH		40k		Y
F70ESDI-T	30	ESDI	2,7 RLL	5.25FH		25k		Y
H100286D-P	25	IDE AT		5.25HH		20k		Y
H100386S-P	25	IDE AT		5.25HH		20k		Y
H330E1 (PS Express)	14	ESDI	2,7 RLL	5.25HH		150k		Y
H340E1 (PS Express)	14	ESDI	2,7 RLL	5.25HH		150k		Y
H40M50-P	24	ST412/506	MFM	3.5HH		45k		Y
H60286D-P	29	IDE AT		5.25HH		40k		Y
H60SCSI-S	28	SCSI		5.25HH		45k		Y
H65M50-P	15	ST412/506	MFM	3.5HH		30k		Y
H80AT	15	SCSI		5.25HH		30k		Y
H80SCSI-S	28	SCSI		5.25HH		45k		Y
HD20AT-S	65	ST412/506	MFM	5.25HH		50k		Y
HD30AT-S	40	ST412/506	MFM	5.25HH		50k		Y
HD40AT-S1	28	ST412/506	MFM	5.25HH		50k		Y
K120M50Z-70P	23	MCA	2,7 RLL	3.5HH				Y
K20M25-WS	27	IDE AT		3.5HH		20k		Y
K20M25/30-OK	62	ST412/506	MFM	3.5HH		20k		Y
K20M25/30-WS	40	ST412/506	MFM	3.5HH		20k		Y
K30M25/30-OK	62	ST412/506	MFM	3.5HH		50k		Y
K30M25/30-WS	40	ST412/506	MFM	3.5HH		50k		Y
K30M30E-P	39	IDE AT		3.5HH		25k		Y
K40M25/30-WS	24	ST412/506	MFM	3.5HH		45k		Y
K45M30286-ZS	28	SCSI		3.5HH		45k		Y
K50M50Z/70P	27	MCA	2,7 RLL	3.5HH				Y
K60M30286-ZS	24	SCSI		3.5HH		40k		Y
K80M25Z/30	15	ST412/506	MFM	3.5HH				Y
K80M30286-WS	24	SCSI		3.5HH		40k		Y
LDSNECMS-20	28	IDE AT	2,7 RLL	3.5HH		20k		Y
LDZE386-100	25	IDE AT		3.5HH		20k		Y
PB340	12	SCSI-2	1,6 RLL		128k	150k	4200	Y
PB520	17	SCSI-2	1,7 RLL		128k	350k	4500	Y
PSEXPRESS 150	17	ESDI	2,7 RLL	5.25FH		40k		Y
PSEXPRESS 320	15	ESDI	2,7 RLL	5.25FH		40k		Y
SENTRY 180	18	SCSI		5.25FH		40k		Y
SENTRY 300	16.5	SCSI		5.25FH		30k		Y
SENTRY 600	16	SCSI		5.25FH		30k		Y
SENTRY 90	18	SCSI		5.25FH		40k		Y

COGITO

Drive Model	Seek Time	Interface	Encode	Form Factor	cache kb	mtbf	RPM	Obsolete?
CG906	93	ST412/506	MFM	5.25HH				Y
CG912	93	ST412/506	MFM	5.25HH				Y
CG925	93	ST412/506	MFM	5.25HH				Y
PT912	93	ST412/506	MFM	5.25HH				Y
PT925	93	ST412/506	MFM	5.25HH				Y

COMPAQ

Drive Model	Seek Time	Interface	Encode	Form Factor	cache kb	mtbf	RPM	Obsolete?
113640-001	29			3.5HH		40k		Y
113641-001	25			3.5HH		40k		Y
115145-001	25			3.5HH		40k		Y
115147-001	19	ESDI		5.25HH		60k		Y
115158-001	19	ESDI		5.25FH		40k		Y
115627-001	25			3.5HH		40k		Y
115830-001	18	ESDI		5.25FH		40k		Y
116562-001	19			3.5HH		40k		Y
116565-001	19			3.5HH		40k		Y
122136-001	19			3.5HH		40k		Y
131067-001	2			3.5HH		150k		Y
131362-001	18	ESDI		5.25HH		60k		Y

Drive Model	Format Size MB	Head	Cyl	Sect/ Trac	Translate H/C/S	RWC/ WPC	Land Zone
142018-001	1049	13	1974	56-96		—/—	
142216-001	2097	18		262668-108		—/—	
146742-001	2097	18		262668-108		—/—	
146742-003	1049	13	1974	56-96		—/—	
146742-005	4293	21		360682-135		—/—	
146742-006	4293	21		360682-135		—/—	
146742-007	2097	11		351186-135		—/—	
172492-002	421	4		251955-104	16/1010/51	—/—	
172493-001	1083	6		381161-117	16/2100/63	—/—	
172678-002	730	4		365864-128	16/1416/63	—/—	
172874-001	541	4		285358-118	9/1926/61	—/—	
196408-002	270	2		285358-118	14/944/40	—/—	
199580-001	4293	21		360682-135		—/—	
199597-001	4293	21		360682-135		—/—	
199642-001	2097	11		351186-135		—/—	
241139	6511	6		8895170-28		—/—	
262477	3227	6		5690114-23		—/—	
286118	4018	8		5690114-23		—/—	
286123	8038	4		11490403-26		—/—	

COMPORT

Drive Model	Format Size MB	Head	Cyl	Sect/ Trac	Translate H/C/S	RWC/ WPC	Land Zone
2040	44	4	820	26		—/—	
2041	44	4	820	26		—/—	
2082	86	6	820	34		—/—	

CONNER PERIPHERALS, INC.

Drive Model	Format Size MB	Head	Cyl	Sect/ Trac	Translate H/C/S	RWC/ WPC	Land Zone
CFA1080A	1080	8		72-114		—/—	
CFA1080S	1080	8		72-114		—/—	
CFA1275A	1278	6			16/2479/63	—/—	
CFA1275S	1278	6				—/—	
CFA170A	172	2	2111	V		Auto/Auto	NA
CFA170S	172	2	2111	67-91		—/—	
CFA2161A	2110	16	4095	63		—/—	
CFA270A	270	2		72-114		—/—	
CFA270S	270	2		72-114		—/—	
CFA340A	343	4		67-91		Na/Na	Auto
CFA340S	343	4		67-91		Na/Na	Auto
CFA425A	426					—/—	
CFA425S	426					—/—	
CFA540A	541	4		72-114		—/—	
CFA540S	541	4		72-114		—/—	
CFA810A	810	6		72-114		—/—	
CFA810S	810	6		72-114		—/—	
CFA850A	852	4				—/—	
CFA850S	852	4				—/—	
CFL350A	350	4	2225		12/905/63	—/—	
CFL420A	422	4	2393	V	16/818/63	—/—	
CFN170A	168	4		47-72		—/—	
CFN170S	168	4		47-72		—/—	
CFN250A	252	6		47-72	16/489/63	—/—	
CFN250S	252	6		47-72		—/—	
CFN340A	344	6		53-89	16/667/63	—/—	
CFN340S	344	6		53-89		—/—	
CFP1060D	1062	8				—/—	
CFP1060E	1062	8				—/—	
CFP1060S	1062	8				—/—	
CFP1060W	1062	8				—/—	
CFP1080E (Filepro)	1080	6		365866-120		—/—	
CFP1080S (Filepro)	1080	6		365866-120		—/—	
CFP2105E	2147	10		394867-139		—/—	
CFP2105S	2147	10		394867-139		—/—	
CFP2105W	2147	10		394867-139		—/—	
CFP2107E (Filepro)	2147	10		401669-124		—/—	

Drive Model	Seek Time	Interface	Encode	Form Factor	cache kb	mtbf	RPM	Obsolete?
142018-001	10	SCSI-2Fast		3.5HH			5400	Y
142216-001	9	SCSI-2Fast		3.5HH			6400	
146742-001	9	SCSI-2Fast		3.5HH			6400	
146742-003	10	SCSI-2Fast		3.5HH			5400	Y
146742-005	9	SCSI-2Fast		3.5HH			7200	
146742-006	9	SCSI-2FstW		3.5HH			7200	
146742-007	9	SCSI-2FstW		3.5HH			7200	
172492-002	14	IDE AT		3.5 3H	96k		3600	Y
172493-001	14	IDE AT		3.5 3H	128k		4495	
172678-002	11	IDE AT		3.5 3H	96k		4500	Y
172874-001	14	IDE AT		3.5 3H	96k		3600	Y
196408-002	14	IDE AT		3.5 3H	96k		3600	Y
199580-001	9	SCSI-2Fast		3.5HH			7200	
199597-001	9	SCSI-2FstW		3.5HH			7200	
199642-001	9	SCSI-2FstW		3.5HH			7200	
241139	13	SCSI-3ULTRA		5.25	512k		3600	
262477	12	SCSI-3ULTRA		3.5	128k		4500	
286118	12	SCSI-3ULTRA		3.5	128k		4500	
286123	12	SCSI-3ULTRA		3.5	128k		4000	

COMPORT

Drive Model	Seek Time	Interface	Encode	Form Factor	cache kb	mtbf	RPM	Obsolete?
2040	35	ST412/506	2,7 RLL	5.25HH		30k		Y
2041	29	IDE AT		5.25HH		30k		Y
2082	29	SCSI		5.25HH		30k		Y

CONNER PERIPHERALS, INC.

Drive Model	Seek Time	Interface	Encode	Form Factor	cache kb	mtbf	RPM	Obsolete?
CFA1080A	12	IDE AT	1,7 RLL	3.5 3H	256k	300k	4500	
CFA1080S	12	SCSI-2Fast	1,7 RLL	3.5 3H	256k	300k	4500	
CFA1275A	12	EIDE	1,7 RLL	3.5 3H	256k	300k	4500	
CFA1275S	12	SCSI-2	1,7 RLL	3.5 3H	256k	300k	4500	
CFA170A	13	IDE	1,7 RLL	3.5 3H	64k	250k		Y
CFA170S	13	SCSI-2	1,7 RLL	3.5 3H	64k	250k	4011	Y
CFA2161A		IDE AT		3.5 3H				
CFA270A	12	IDE AT	1,7 RLL	3.5 3H	256k	250k	4500	Y
CFA270S	12	SCSI-2	1,7 RLL	3.5 3H	256k	250k	4500	Y
CFA340A	13	IDE AT	1,7 RLL	3.5 3H	64k	300k	4011	Y
CFA340S	13	SCSI-2	1,7 RLL	3.5 3H	64k	300k	4011	Y
CFA425A	12	IDE AT	1,7 RLL	3.5 3H	64k	300k	4500	Y
CFA425S	12	SCSI-2	1,7 RLL	3.5 3H	64k	300k	4500	Y
CFA540A	12	IDE AT	1,7 RLL	3.5 3H	256k	300k	4500	Y
CFA540S	12	SCSI-2Fast	1,7 RLL	3.5 3H	256k	300k	4500	Y
CFA810A	12	IDE AT	1,7 RLL	3.5 3H	256k	300k	4500	Y
CFA810S	12	SCSI-2Fast	1,7 RLL	3.5 3H	256k	300k	4500	Y
CFA850A	12	IDE AT	1,7 RLL	3.5 3H	256k	300k	4500	Y
CFA850S	12	SCSI-2	1,7 RLL	3.5 3H	256k	300k	4500	Y
CFL350A	12	IDE AT	1,7 RLL	2.5 4H	32k	300k	3750	Y
CFL420A	12	IDE AT	1,7 RLL	2.5 4H	64k	300k	3600	Y
CFN170A	12	IDE AT	1,7 RLL	2.5 4H	32k	150k	4500	Y
CFN170S	12	SCSI	1,7 RLL	2.5 4H	32k	150k	4500	Y
CFN250A	12	IDE AT	1,7 RLL	2.5 4H	32k	150k	4500	Y
CFN250S	12	SCSI	1,7 RLL	2.5 4H	32k	150k	4500	Y
CFN340A	13	IDE AT	1,7 RLL	2.5 4H	32k	150k	4000	Y
CFN340S	13	SCSI	1,7 RLL	2.5 4H	32k	150k	4000	Y
CFP1060D	9	SCSI-2Fast	1,7 RLL	3.5 3H	512k	500k	5400	Y
CFP1060E	9	SCSI	1,7 RLL	3.5 3H	512k	500k	5400	Y
CFP1060S	9	SCSI-2Fast	1,7 RLL	3.5 3H	512k	500k	5400	Y
CFP1060W	9	SCSI-2FstW	1,7 RLL	3.5 3H	512k	500k	5400	Y
CFP1080E (Filepro)	11	SCSI-2FstW	1,7 RLL	3.5 3H	512k	1000k	5400	Y
CFP1080S (Filepro)	11	SCSI-2Fast	1,7 RLL	3.5 3H	256k	1000k	5400	Y
CFP2105E	9	SCSI-2FstW	1,7 RLL	3.5 3H	512k	1000k	5400	
CFP2105S	9	SCSI-2Fast	1,7 RLL	3.5 3H	512k	1000k	5400	
CFP2105W	9	SCSI-2FstW	1,7 RLL	3.5 3H	512k	1000k	5400	
CFP2107E (Filepro)	9	SCSI-2FstW	1,7 RLL	3.5 3H	512k	1000k	7200	

Drive Model	Format Size MB	Head	Cyl	Sect/ Trac	Translate H/C/S	RWC/ WPC	Land Zone
CFP2107S (Filepro)	2147	10	4016	69-124		—/—	
CFP2107W (Filepro)	2147	10	4016	69-124		—/—	
CFP4207E (Filepro)	4294	20	4016	69-124		—/—	
CFP4207S (Filepro)	4294	20	4016	69-124		—/—	
CFP4207W (Filepro)	4294	20	4016	69-124		—/—	
CFP4217C (Filepro)	4294		6028			Na/Na	Auto
CFP4217E (Filepro)	4294		6028			Na/Na	Auto
CFP4217S (Filepro)	4294		6028			Na/Na	Auto
CFP4217W (Filepro)	4294		6028			Na/Na	Auto
CFP4217WD (Filepro)	4294		6028			Na/Na	Auto
CFP9117C (Filepro)	9100		6028			Na/Na	Auto
CFP9117E (Filepro)	9100		6028			Na/Na	Auto
CFP9117S (Filepro)	9100		6028			Na/Na	Auto
CFP9117W (Filepro)	9100		6028			Na/Na	Auto
CFP9117WD (Filepro)	9100		6028			Na/Na	Auto
CFS1060A	1060	16	2064	63		—/—	
CFS1081A	1080	4	3930			—/—	
CFS1275A	1275	6	3640		16/2479/63	—/—	
CFS1276A	1275		4893			Na/Na	Auto
CFS1621A	1620	6	3930			—/—	
CFS2105S	2147	10	3948			—/—	
CFS210A	213	2		68-107		—/—	
CFS270A	270	2	2595		16/525/63	—/—	
CFS420A	426	4		68-107		—/—	
CFS425A	425	2	3687		16/826/63	—/—	
CFS540A	540	4	3517		16/1050/63	—/—	
CFS541A	540	2	3924			—/—	
CFS635A	635	3	3640			—/—	
CFS636A	635	2	4893			—/—	
CFS850A	850	4	3640		16/1652/63	—/—	
CP1044 (Derringer)	42.6	2				Na/Na	Auto
CP2020 (Kato)	21	2	653	32		Na/Na	Auto
CP2022	20	2	653	32	4/615/17	Na/Na	Auto
CP2024 (Kato)	21	2	653	32	4/615/17	Na/Na	Auto
CP2031	30	2			4/411/38	Na/Na	Auto
CP2034 (Pancho)	32	2	823	38	4/615/17	Na/Na	Auto
CP2040	43	4	548	38		Na/Na	Auto
CP2044 (Pancho)	42	4	552	38	5/977/17	Na/Na	Auto
CP2048 (Pancho)					4/548/38	Na/Na	Auto
CP2060	64	4	823	38		Na/Na	Auto
CP2064 (Pancho)	64	4	823	38	4/615/17	Na/Na	Auto
CP2084 (Pancho)	85	4	1096	38	8/548/38	Na/Na	Auto
CP2088	85	4		38	8/548/38	—/—	
CP2124 (Pancho)	120	4	1123	53	*UNIV T	Na/Na	Auto
CP2250	253					Na/Na	Auto
CP2254 (Trigger)	253					Na/Na	Auto
CP2304	209	8	1348	39	*UNIV T	Na/Na	Auto
CP3000	42	2	1045	40	5/980/17	Na/Na	Auto
CP30060	60	2	1524	39		Na/Na	Auto
CP30064 (Hopi)	60	2	1524	39	4/762/39	Na/Na	Auto
CP30064H (Hopi)	60	2	1524	39	4/762/39	Na/Na	Auto
CP30069 (Hopi)	60	2	1524	39		Na/Na	Auto
CP30080 (Hopi)	84	4	1053	39		Na/Na	Auto
CP30080E (Jaguar)	85	2	1806	46		Na/Na	Auto
CP30081	85	4	1058	39	8/526/39	Na/Na	Auto
CP30084 (Hopi)	84	4	1053	39	8/526/39	Na/Na	Auto
CP30084E (Jaguar)	85	2	1806	46	4/903/46	Na/Na	Auto
CP30100 (Hopi)	120	4	1522	39		Na/Na	Auto
CP30101	122	4	1524	9	8/762/39	—/—	761
CP30101 (Hopi)	121	8	761	39	*UNIV T	Na/Na	Auto
CP30101G	122	4	1524	9	8/762/39	—/—	761
CP30104 (Hopi)	121	4	1524	39	8/762/39	Na/Na	Auto
CP30104H (Hopi)	121	4	1524	39	8/762/39	Na/Na	Auto

Drive Model	Seek Time	Interface	Encode	Form Factor	cache kb	mtbf	RPM	Obsolete?
CFP2107S (Filepro)	9	SCSI-2Fast	1,7 RLL	3.5 3H	512k	1000k	7200	
CFP2107W (Filepro)	9	SCSI-2FstW	1,7 RLL	3.5 3H	512k	1000k	7200	
CFP4207E (Filepro)	9.5	SCSI-2FstW	1,7 RLL	3.5HH	512k	1000k	7200	
CFP4207S (Filepro)	9.5	SCSI-2Fast	1,7 RLL	3.5HH	512k	1000k	7200	
CFP4207W (Filepro)	9.5	SCSI-2FstW	1,7 RLL	3.5HH	512k	1000k	7200	
CFP4217C (Filepro)	9	SSA		3.5HH	512k	999k	7200	
CFP4217E (Filepro)	9	SCA		3.5HH	512k	999k	7200	
CFP4217S (Filepro)	9	SCSI-3		3.5HH	512k	999k	7200	
CFP4217W (Filepro)	9	SCSI-3Wide		3.5HH	512k	999k	7200	
CFP4217WD (Filepro)	9	SCSI-3Wide		3.5HH	512k	999k	7200	
CFP9117C (Filepro)	9	SSA	RLL 8,9	3.5HH	512k	999k	7200	
CFP9117E (Filepro)	9	SCA	RLL 8,9	3.5HH	512k	999k	7200	
CFP9117S (Filepro)	9	SCSI-3	RLL 8,9	3.5HH	512k	999k	7200	
CFP9117W (Filepro)	9	SCSI-3Wide	RLL 8,9	3.5HH	512k	999k	7200	
CFP9117WD (Filepro)	9	SCSI-3Wide	RLL 8,9	3.5HH	512k	999k	7200	
CFS1060A		IDE AT		3.5 3H				Y
CFS1081A	14	IDE AT	1,7 RLL	3.5 3H	64k	300k	3600	Y
CFS1275A	14	IDE	1,7 RLL	3.5 3H	64k	250k	3600	Y
CFS1276A	14	ATA-2	1,7 RLL	3.5 3H	64k	300k	4500	Y
CFS1621A	14	IDE AT	1,7 RLL	3.5 3H	64k	300k	3600	Y
CFS2105S	9	SCSI-2Fast	1,7 RLL	3.5 3H	512k	1000k	5400	
CFS210A	14	IDE AT	1,7 RLL	3.5 3H	32k	250k	3600	Y
CFS270A	14	IDE	1,7 RLL	3.5 3H	32k	250k	3400	Y
CFS420A	14	IDE AT	1,7 RLL	3.5 3H	32k	250k	3600	Y
CFS425A	14	IDE	1,7 RLL	3.5 3H	64k	250k	3600	Y
CFS540A	14	IDE	1,7 RLL	3.5 3H	64k	250k	3600	Y
CFS541A	14	IDE AT	1,7 RLL	3.5 3H	64k	300k	3600	Y
CFS635A	14	IDE AT	1,7 RLL	3.5 3H	64k	300k	3600	Y
CFS636A	13	ATA-2	1,7 RLL	3.5 3H	64k	300k	4500	Y
CFS850A	14	IDE	1,7 RLL	3.5 3H	64k	250k	3600	Y
CP1044 (Derringer)	19			2.5 4H	32k			Y
CP2020 (Kato)	23	SCSI	2,7 RLL	2.5 4H	8k	100k		Y
CP2022	23	IDE AT	2,7 RLL	3.5HH				Y
CP2024 (Kato)	23	IDE AT	2,7 RLL	2.5 4H	8k	100k	3433	Y
CP2031	19	ATA	2,7 RLL	2.5 4H	32k	100k		Y
CP2034 (Pancho)	19	IDE AT	2,7 RLL	2.5 4H	32k	100k	3433	Y
CP2040	17	SCSI	2,7 RLL	2.5 4H	32k	50k	3486	Y
CP2044 (Pancho)	19	IDE AT	2,7 RLL	2.5 4H	32k	100k	3486	Y
CP2048 (Pancho)	19	ATA	2,7 RLL	2.5 4H	32k	100k	3486	Y
CP2060	19	SCSI	2,7 RLL	2.5 4H	32k	50k	3486	Y
CP2064 (Pancho)	19	IDE AT	2,7 RLL	2.5 4H	32k	100k	3486	Y
CP2084 (Pancho)	19	IDE AT	1,7 RLL	2.5 4H	32k	150k	3486	Y
CP2088	19	IDE AT	1,7 RLL	2.5 4H	32k	100k	3486	Y
CP2124 (Pancho)	26	IDE AT	1,7 RLL	2.5 4H	32k	150k		Y
CP2250	12	SCSI		2.5 4H	32k			Y
CP2254 (Trigger)	12	ATA		2.5 4H	32k			Y
CP2304	19	IDE AT	RLL	3.5HH				Y
CP3000	28	IDE AT	2,7 RLL	3.5 3H	8k	150k	3557	Y
CP30060	19	SCSI	1,7 RLL	3.5 3H		150k		Y
CP30064 (Hopi)	19	IDE AT	1,7 RLL	3.5 3H	64k	100k	3400	Y
CP30064H (Hopi)	19	IDE AT	1,7 RLL	3.5 3H	32k	150k	3400	Y
CP30069 (Hopi)	19	MCA	1,7 RLL	3.5 3H	64k	100k	3399	Y
CP30080 (Hopi)	19	SCSI	1,7 RLL	3.5 3H	64k	100k	3400	Y
CP30080E (Jaguar)	17	SCSI	1,7 RLL	3.5 3H	32k	150k	3822	Y
CP30081	19	IDE AT	2,7 RLL	3.5 4H		150k		Y
CP30084 (Hopi)	19	IDE AT	1,7 RLL	3.5 3H	64k	100k	3400	Y
CP30084E (Jaguar)	17	IDE AT	1,7 RLL	3.5 3H	32k	150k	3822	Y
CP30100 (Hopi)	19	SCSI	2,7 RLL	3.5 3H	64k	150k	3400	Y
CP30101	19	IDE AT	2,7 RLL	3.5 3H				Y
CP30101 (Hopi)	10	IDE AT	2,7 RLL	3.5 3H				Y
CP30101G	19	IDE AT	2,7 RLL	3.5 3H				Y
CP30104 (Hopi)	19	IDE AT	1,7 RLL	3.5 3H	32k	100k	3400	Y
CP30104H (Hopi)	19	IDE AT	1,7 RLL	3.5 3H	32k	150k	3400	Y

Drive Model	Format Size MB	Head	Cyl	Sect/ Trac	Translate H/C/S	RWC/ WPC	Land Zone
CP30109 (Hopi)	120	4	1522	39		Na/Na	Auto
CP30124	126	2		62	5/895/55	—/—	
CP30170	172	2	2111	67-91		—/—	
CP30170E (Jaguar)	170	4	1806	46		Na/Na	Auto
CP30174	172	2	2111	67-91		—/—	
CP30174E (Jaguar)	170	4	1806	46	8/903/46	Na/Na	Auto
CP3020	21	2	636	33		Na/Na	Auto
CP30200 (Cougar)	212	4	2124	49		Na/Na	Auto
CP30201	212					—/—	
CP30204 (Cougar)	212	4		49	16/683/38	Na/Na	
CP3022	21	2	636	33	4/615/17	Na/Na	Auto
CP3023	21					—/—	
CP3024	22	2	636	33	4/615/17	Na/Na	Auto
CP30254	252	4	1985	62	10/895/55	Na/Na	Auto
CP30340	343	4		67-91		Na/Na	Auto
CP30344	343	4			16/665/63	Na/Na	Auto
CP3040	40	2	1026	40		Na/Na	Auto
CP3041	42	2	1047	40	5/977/17	Na/Na	Auto
CP3044	42	2	1047	40	5/977/17	Na/Na	Auto
CP3045	40					—/—	
CP30540	545	6	2243			—/—	
CP30544	545	6	2243		16/989/63	—/—	
CP3100	104	8	776	33		Na/Na	Auto
CP3101	104					—/—	
CP3102	104	8	776	33	*UNIV T	Na/Na	Auto
CP3104	104	8	776	33	13/925/17	Na/Na	Auto
CP3106	104					—/—	
CP3111	107	8	832	33	*UNIV T	Na/Na	Auto
CP3114	107	8	832	33	8/832/33	Na/Na	Auto
CP31370	1372	14	2386			—/—	
CP31374 Baja	1372	14				Na/Na	Auto
CP3150	52	4	776	33		Na/Na	Auto
CP3180	84	6	832	33		Na/Na	Auto
CP3181	84	6	832	33		Na/Na	Auto
CP3184	84	6	832	33	9/1024/17	Na/Na	Auto
CP320	20	2	752	26		Na/Na	Auto
CP3200	209	8	1366	38		Na/Na	Auto
CP3200F	212	8	1366	38		Na/Na	Auto
CP3201I	215	8	1348	39	*UNIV T	Na/Na	Auto
CP3204	209	8	1366	38	16/683/38	Na/Na	Auto
CP3204F	212	8	1366	38	16/683/38	Na/Na	Auto
CP3209F	212	8	1366	38	*UNIV T	Na/Na	Auto
CP321	20	2	752	26	4/615/17	Na/Na	Auto
CP323	20	2	752	26	4/615/17	Na/Na	Auto
CP324	20	2	752	26	4/615/17	Na/Na	Auto
CP3304 (Summit)	340	8	1806	46	16/659/63	Na/Na	Auto
CP3360 (Summit)	362	8	1807	49		Na/Na	Auto
CP3364 (Summit)	362	8	1808	49	16/702/63	Na/Na	Auto
CP340	42	4	788	26		Na/Na	Auto
CP341	42	4	805	26	5/977/17	Na/Na	Auto
CP341I	42	4	805	26	5/977/17	Na/Na	Auto
CP342	40	4	805	26	4/805/26	Na/Na	Auto
CP343 (Zenith)	43	4	805		5/977/17	Na/Na	Auto
CP344	43	4	805	26	5/977/17	Na/Na	Auto
CP346	42					—/—	
CP3500 (Summit)	510	12	1806	49		Na/Na	Auto
CP3501	510	12	1806	46		Auto/Auto	NA
CP3504 (Summit)	510	12		48	16/987/63	Na/Na	Auto
CP3505	510	12	1806	46		Na/Na	Auto
CP3540 (Summit)	543	12	1807	49		Na/Na	Auto
CP3544 (Summit)	544	12	1808	49	16/1023/63	Na/Na	Auto
CP4021	20					—/—	
CP4024 (Stubby)	21	2	627	34	4/615/17	Na/Na	Auto

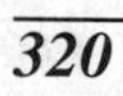

Drive Model	Seek Time	Interface	Encode	Form Factor	cache kb	mtbf	RPM	Obsolete?
CP30109 (Hopi)	19	MCA	2,7 RLL	3.5 3H	64k	150k	3400	Y
CP30124	14	IDE AT	1,7 RLL	3.5 3H	32k	250k	4542	Y
CP30170	13	SCSI-2	1,7 RLL	3.5 3H	64k	250k	4011	Y
CP30170E (Jaguar)	17	SCSI	1,7 RLL	3.5 3H	32k	150k	3833	Y
CP30174	13	IDE AT	1,7 RLL	3.5 3H	64k	250k	4011	Y
CP30174E (Jaguar)	17	IDE AT	1,7 RLL	3.5 3H	32k	150k	3833	Y
CP3020	27	SCSI	2,7 RLL	3.5 3H	8k	50k	3575	Y
CP30200 (Cougar)	12	SCSI-2	2,7 RLL	3.5 3H	256k	150k	4500	Y
CP30201		IDE AT	2,7 RLL	3.5 3H				Y
CP30204 (Cougar)	12	IDE AT	2,7 RLL	3.5 3H	256k	150k	4500	Y
CP3022	27	IDE AT	2,7 RLL	3.5 3H		50k		Y
CP3023		IDE AT	2,7 RLL	3.5 3H				Y
CP3024	27	IDE AT	2,7 RLL	3.5 3H	8k	50k	3575	Y
CP30254	14	IDE AT	1,7 RLL	3.5 3H	64k	250k	4542	Y
CP30340	13	SCSI-2	1,7 RLL	3.5 3H	64k	300k	4011	Y
CP30344	13	ATA		3.5 3H	64k	250k	4500	Y
CP3040	25	SCSI	2,7 RLL	3.5 3H	8k	50k	3557	Y
CP3041	25	IDE AT	2,7 RLL	3.5 3H		50k		Y
CP3044	25	IDE AT	2,7 RLL	3.5 3H	8k	50k	3557	Y
CP3045		IDE AT	2,7 RLL	3.5 3H				Y
CP30540	10	SCSI-2Fast	1,7 RLL	3.5 3H	256k	250k	5400	Y
CP30544	10	IDE AT	1,7 RLL	3.5 3H	256k	250k	5400	Y
CP3100	25	SCSI	2,7 RLL	3.5HH	32k	50k	3575	Y
CP3101		IDE AT	2,7 RLL	3.5HH				Y
CP3102	25	IDE AT	2,7 RLL	3.5HH	16k	50k		Y
CP3104	25	IDE AT	2,7 RLL	3.5HH	16k	30k	3575	Y
CP3106		IDE AT	2,7 RLL	3.5HH				Y
CP3111	25	IDE AT	2,7 RLL	3.5HH	16k	50k		Y
CP3114	25	IDE AT	2,7 RLL	3.5HH				Y
CP31370	10	SCSI-2Fast	1,7 RLL	3.5HH	256k	250k	5400	
CP31374 Baja	11	ATA			256k			
CP3150	25	SCSI	2,7 RLL	3.5HH		50k		Y
CP3180	25	SCSI	2,7 RLL	3.5HH	32k	50k	3575	Y
CP3181	25	IDE AT	2,7 RLL	3.5HH		50k		Y
CP3184	25	IDE AT	2,7 RLL	3.5HH	32k	50k	3575	Y
CP320		SCSI	2,7 RLL	3.5 3H				Y
CP3200	16	SCSI	2,7 RLL	3.5HH	64k	50k	3485	Y
CP3200F	16	SCSI	2,7 RLL	3.5HH	64k	50k	3485	Y
CP3201I	19	IDE AT	2,7 RLL	3.5HH		150k		Y
CP3204	19	IDE AT	2,7 RLL	3.5HH	64k	50k	3485	Y
CP3204F	16	IDE AT	2,7 RLL	3.5HH	64k	150k	3485	Y
CP3209F	16	IDE AT	2,7 RLL	3.5HH		50k		Y
CP321		IDE AT	2,7 RLL	3.5 3H				Y
CP323		ZENITH	2,7 RLL	3.5 3H				Y
CP324		IDE AT	2,7 RLL	3.5 3H				Y
CP3304 (Summit)		IDE AT	1,7 RLL	3.5HH		150k		Y
CP3360 (Summit)	12	SCSI-2	2,7 RLL	3.5HH	256k	150k	4500	Y
CP3364 (Summit)	12	IDE AT	2,7 RLL	3.5HH	256k	150k	4498	Y
CP340	29	SCSI	2,7 RLL	3.5HH	1k	20k	3600	Y
CP341	29	IDE AT	2,7 RLL	3.5HH				Y
CP341I	29	IDE AT	2,7 RLL	3.5HH				Y
CP342	29	IDE AT	2,7 RLL	3.5HH				Y
CP343 (Zenith)	29	ZENITH		3.5HH				Y
CP344	29	IDE AT	2,7 RLL	3.5HH	8k	20k	3600	Y
CP346		IDE AT	2,7 RLL	3.5HH				Y
CP3500 (Summit)	12	SCSI	2,7 RLL	3.5HH	256k	100k	3609	Y
CP3501	12	IDE AT	2,7 RLL	3.5HH		150k		Y
CP3504 (Summit)	12	IDE AT	2,7 RLL	3.5HH	256k	150k	3828	Y
CP3505	12	IDE AT	2,7 RLL	3.5HH				Y
CP3540 (Summit)	12	SCSI-2	2,7 RLL	3.5HH	256k	150k	4500	Y
CP3544 (Summit)	12	IDE AT	2,7 RLL	3.5HH	256k	150k	4498	Y
CP4021		IDE AT	2,7 RLL	3.5 4H				Y
CP4024 (Stubby)	<29	IDE AT	2,7 RLL	3.5 4H	8k	40k	2913	Y

Drive Model	Format Size MB	Head	Cyl	Sect/ Trac	Translate H/C/S	RWC/ WPC	Land Zone
CP4041	42					—/—	
CP4044 (Stubby)	43	2	1097	38	5/977/17	Na/Na	Auto
CP4084 (Gator)	85	2	1806	46		Na/Na	Auto
CP5500	510	20	2034	50		Na/Na	Auto

CORE INTERNATIONAL

Drive Model	Format Size MB	Head	Cyl	Sect/ Trac	Translate H/C/S	RWC/ WPC	Land Zone
3SHC230	230	5	1511	V		Na/Na	Auto
AT115	115	7	968	35		—/—	Auto
AT145	58	7	968			—/—	
AT150	156	9	968	35		—/—	Auto
AT20	20	4	615	17		—/—	Auto
AT26	26	3	988	17		—/—	
AT260	260	12	1212	35		—/—	Auto
AT30	32	5	733	17		—/—	
AT30R	49	5	733	26		—/—	
AT32	32	5	733	17		—/—	
AT32R	49	5	733	26		—/—	
AT40	40	5	924	17		—/—	
AT40F	40	4	564	35		—/—	Auto
AT40R	62	5	924	26		—/—	
ATPLUS20	21	4	615	17		—/—	
ATPLUS43	43	5	988	17		—/—	
ATPLUS43R	66	5	988	26		—/—	
ATPLUS44	44	7	733	17		—/—	
ATPLUS44R	68	7	733	26		—/—	
ATPLUS56	56	7	924	17		—/—	
ATPLUS63	42	5	988	17		—/—	
ATPLUS63R	65	65	988	26		—/—	
ATPLUS72	73	9	924	17		—/—	
ATPLUS72R	107	9	924	26		—/—	
ATPLUS80	80	9	1024			—/—	
ATPLUS80R	132	9	1024			—/—	
ATPLUS82	82	5	968	35		—/—	Auto
HC100	101	15	379	35		—/—	
HC1000	1056	15	1787	77		Na/Na	Auto
HC1000-20	1056	15	1787	77		—/—	Auto
HC1000S	1005	16	1918	64		—/—	Auto
HC150	150	7	1250	35		—/—	Auto
HC150FH	151	9	969	34		Na/Na	Auto
HC150S	155	9	969	35		—/—	Auto
HC175	177	9	1072	35		—/—	
HC200	200	8			12/986/33	—/—	
HC230	230	5				Na/Na	Auto
HC25	250					—/—	Auto
HC260	260	12	1212	35		Na/Na	
HC310	325	7	1747	52		Na/Na	Auto
HC310S	330	8	1447	56		—/—	Auto
HC315-20	340	8	1447	57		—/—	Auto
HC380	376	15	1412	35		—/—	
HC40	40	4	564	35		Na/Na	Auto
HC650	658	15	1661	53		—/—	Auto
HC650S	663	16	1447	56		—/—	Auto
HC655-20	680	16	1447	57		—/—	Auto
HC90	91	5	969	35		Na/Na	
MC120	120	8	920	32		Na/Na	Auto
MC60	60	4	928	32		Na/Na	Auto
OPTIMA 30	31	5	733	17		—/—	
OPTIMA 30R	48	5	733	26		—/—	
OPTIMA 40	41	5	963	17		—/—	
OPTIMA 40R	64	5	963	26		—/—	
OPTIMA 70	71	9	918	17		—/—	
OPTIMA 70R	109	9	918	17		—/—	
OPTIMA 80	80	9	1024	17		—/—	

Drive Model	Seek Time	Interface	Encode	Form Factor	cache kb	mtbf	RPM	Obsolete? ⇓
CP4041		IDE AT	2,7 RLL	3.5 4H				Y
CP4044 (Stubby)	<29	IDE AT	2,7 RLL	3.5 4H	8k	50k		Y
CP4084 (Gator)	19	IDE AT	2,7 RLL	3.5 4H	32k			Y
CP5500	12	SCSI-2	RLL		512k	150k	4498	Y
CORE INTERNATIONAL								
3SHC230	13	SCSI		3.5HH		150k		Y
AT115	16	ESDI		5.25FH		33k	3597	Y
AT145	17	ST412/506	MFM	5.25FH				Y
AT150	16	ESDI	2,7 RLL	5.25FH		33k	3597	Y
AT20	20	ST412/506	MFM	5.25FH		25k		Y
AT26	26	ST412/506	MFM	5.25HH		25k		Y
AT260	25	ESDI		5.25FH		25k	3524	Y
AT30	21	ST412/506	MFM	5.25FH		50k		Y
AT30R	21	ST412/506	2,7 RLL	5.25FH		50		Y
AT32	21	ST412/506	MFM	5.25HH		50k		Y
AT32R	21	ST412/506	2,7 RLL	5.25HH		50k		Y
AT40	26	ST412/506	MFM	5.25FH		50k		Y
AT40F	10	ESDI		5.25FH		33k	3597	Y
AT40R	26	ST412/506	2,7 RLL	5.25FH		50k		Y
ATPLUS20	26	ST412/506	MFM	5.25HH		50k		Y
ATPLUS43	26	ST412/506	MFM	5.25HH		50k		Y
ATPLUS43R	26	ST412/506	2,7 RLL	5.25HH		50k		Y
ATPLUS44	26	ST412/506	MFM	3.5HH		50k		Y
ATPLUS44R	26	ST412/506	2,7 RLL	3.5HH		50k		Y
ATPLUS56	26	ST412/506	MFM	5.25FH		33k		Y
ATPLUS63	26	ST412/506	MFM	5.25FH				Y
ATPLUS63R	26	ST412/506	2,7 RLL	5.25FH				Y
ATPLUS72	26	ST412/506	MFM	5.25FH		50k		Y
ATPLUS72R	26	ST412/506	2,7 RLL	5.25FH		50k		Y
ATPLUS80	15	ST412/506	MFM	3.5HH		50k		Y
ATPLUS80R	15	ST412/506	2,7 RLL	3.5HH		50k		Y
ATPLUS82	16	ESDI		5.25FH		33k	3597	Y
HC100	9	ESDI		5.25FH		50k		Y
HC1000	14	ESDI (24)	2,7 RLL	5.25FH		150k		Y
HC1000-20	14	ESDI	2,7 RLL	5.25FH		150k	3600	Y
HC1000S	15	SCSI	2,7 RLL	5.25FH		150k	4002	Y
HC150	17	ESDI	2,7 RLL	5.25HH		100k	3600	Y
HC150FH	16	ESDI (10)	2,7 RLL	5.25FH		100k		Y
HC150S	16.5	SCSI	2,7 RLL	5.25FH		150k	3597	Y
HC175	14	ESDI	2,7 RLL	5.25FH		50k		Y
HC200	16	IDE AT		5.25FH		150k		Y
HC230	13	SCSI		3.5FH		150k		Y
HC25		ESDI		5.25FH				Y
HC260	25	ESDI	2,7 RLL	5.25FH				Y
HC310	18	ESDI	2,7 RLL	5.25HH		100k	3600	Y
HC310S	16.5	SCSI	2,7 RLL	5.25FH		150k	4002	Y
HC315-20	17	ESDI	2,7 RLL	5.25FH		150k	4002	Y
HC380	16	ESDI	2,7 RLL	5.25FH		50k		Y
HC40	9	ESDI	2,7 RLL	5.25FH		50		Y
HC650	17	ESDI	2,7 RLL	5.25FH		100k	3600	Y
HC650S	16.5	SCSI		5.25FH		150k	4002	Y
HC655-20	17	ESDI	2,7 RLL	5.25FH		150k	4002	Y
HC90	16	ESDI	2,7 RLL	5.25HH		50k		Y
MC120	23	MCA		3.5HH		45k	3600	Y
MC60	23	MCA		3.5HH		45k	3600	Y
OPTIMA 30	21	ST412/506	MFM	5.25HH				Y
OPTIMA 30R	21	ST412/506	2,7 RLL	5.25HH				Y
OPTIMA 40	26	ST412/506	MFM	5.25HH		35k		Y
OPTIMA 40R	26	ST412/506	2,7 RLL	5.25HH		35k		Y
OPTIMA 70	26	ST412/506	MFM	5.25FH		35k		Y
OPTIMA 70R	26	ST412/506	2,7 RLL	5.25FH		35k		Y
OPTIMA 80	15	ST412/506	MFM	3.5HH		35k		Y

Drive Model	Format Size MB	Head	Cyl	Sect/ Trac	Translate H/C/S	RWC/ WPC	Land Zone
OPTIMA 80R	132	9	1024	26		—/—	

DIGITAL EQUIPMENT CORP.

Drive Model	Format Size MB	Head	Cyl	Sect/ Trac	Translate H/C/S	RWC/ WPC	Land Zone
CAPELLA 3055	550					Na/Na	Auto
CAPELLA 3110	1100					Na/Na	Auto
CAPELLA 3221	2200					Na/Na	Auto
DSP3053L	535	4	3117			Na/Na	Auto
DSP3080	852					Na/Na	Auto
DSP3085	852	14		57		—/—	
DSP3105	1050	14		57		—/—	
DSP3107L	1070	8	3117			Na/Na	Auto
DSP3133L	1337	10	3117			Na/Na	Auto
DSP3160	1600	16				—/—	
DSP3210	2148	16				Na/Na	Auto
DSP5300	3000	21				Na/Na	Auto
DSP5350	3572	25				—/—	
DSP5400	4000	26				Na/Na	Auto
DSRZ1BB-VW	2100					—/—	
DSRZ1CB-VW	4300					—/—	
DSRZ1CD-VW	4300					—/—	
DSRZ1DB-VW	9100					—/—	
DSRZ1DF-VA	9100					—/—	
DSRZ1DF-VW	9100					—/—	
DSRZ1EF-VA	18200					—/—	
DSRZ1EF-VW	18200					—/—	
DSRZ26N-VZ	1050					—/—	
DSRZ28L-VA	2100					—/—	
DSRZ28M-VZ	2100					—/—	
DSRZ29L-VA	4300					—/—	
DSRZ40-VA	9100					—/—	
RZ26N-VA	1050					—/—	
RZ26N-VW	1050					—/—	
RZ28D-VA	2100					—/—	
RZ28D-VW	2100					—/—	
RZ28M-VA	2100					—/—	
RZ28M-VW	2100					—/—	
RZ29B-VA	4300					—/—	
RZ29B-VW	4300					—/—	
SP3430	4300	20				Na/Na	Auto
VP3107	1075	5				Na/Na	Auto
VP3215	2150	10				Na/Na	Auto

DISC TEC

Drive Model	Format Size MB	Head	Cyl	Sect/ Trac	Translate H/C/S	RWC/ WPC	Land Zone
RHD-120	130					Na/Na	Auto
RHD-180	183					Na/Na	Auto
RHD-20 (Removable)	21	2	615	34		Na/Na	Auto
RHD-210	210					Na/Na	Auto
RHD-260	260					—/—	
RHD-340	340					—/—	
RHD-520	520					—/—	
RHD-60	62	2	1024	60		Na/Na	Auto
RHD-80	81					Na/Na	Auto

DISCTRON (OTARI)

Drive Model	Format Size MB	Head	Cyl	Sect/ Trac	Translate H/C/S	RWC/ WPC	Land Zone
D214	11	4	306	17		128/128	
D503	3	2	153	17		—/—	
D504	4	2	215	17		—/—	
D506	5	4	153	17		—/—	
D507	5	2	306	17		128/128	
D509	8	4	215	17		128/128	
D512	11	8	153	17		—/—	
D513	11	6	215	17		128/128	

Drive Model	Seek Time	Interface	Encode	Form Factor	cache kb	mtbf	RPM	Obsolete?
OPTIMA 80R	15	ST412/506	2,7 RLL	3.5HH		35k		Y

DIGITAL EQUIPMENT CORP.

Drive Model	Seek Time	Interface	Encode	Form Factor	cache kb	mtbf	RPM	Obsolete?
CAPELLA 3055	9	SCSI-2Fast		3H		700k	5400	Y
CAPELLA 3110	9	SCSI-2Fast		3H		700k	5400	Y
CAPELLA 3221	9	SCSI-2Fast		3H		700k	5400	
DSP3053L	9.5	SCSI-2Fast	1,7 RLL	3.5 3H	512k	500k	5400	Y
DSP3080	10	SCSI-2		3H	512k		5400	Y
DSP3085	9	SCSI-2Fast	1,7 RLL	3.5HH	512k	250k	5400	Y
DSP3105	9	SCSI-2Fast	1,7 RLL	3.5HH	512k	250k	5400	Y
DSP3107L	9.5	SCSI-2Diff	1,7 RLL	3.5 3H	512	500k	5400	Y
DSP3133L	9.5	SCSI-2Fast	1,7 RLL	3.5 3H	512k	500k	5400	Y
DSP3160	9.7	SCSI-2Fast	1,7 RLL	3.5HH	512k	350k	5400	Y
DSP3210	9.5	SCSI-2Fast	1,7 RLL	3.5HH	1024k	500k	5400	
DSP5300	12	SCSI-2Fast	1,7 RLL	5.25FH	512k	300k	5400	
DSP5350	12	SCSI-2Fast	1,7 RLL	5.25FH	512k	300k	5400	
DSP5400	12	SCSI-2Fast	1,7 RLL	5.25FH	1024k	300k	5400	
DSRZ1BB-VW	9	SCSI-2FstW		3.5FH	512k		7200	
DSRZ1CB-VW	9	Ultra SCSI		3.5FH	512k		7200	
DSRZ1CD-VW	8.5	Ultra SCSI		3.5 3H	512k		1000	
DSRZ1DB-VW	9	Ultra SCSI		3.5FH	512k		7200	
DSRZ1DF-VA	12	SCSI-2		3.5 3H	1000k		7200	
DSRZ1DF-VW	8	SCSI-2		3.5 3H	1000k		7200	
DSRZ1EF-VA	8	SCSI-2		3.5 3H	1000k		7200	
DSRZ1EF-VW	8	SCSI-2		3.5 3H	1000k		7200	
DSRZ26N-VZ	10	Ultra SCSI		3.5FH	480k		5400	Y
DSRZ28L-VA	9	Ultra SCSI		3.5FH	512k		7200	
DSRZ28M-VZ	10	Ultra SCSI		3.5FH	480k		5400	
DSRZ29L-VA	9	Ultra SCSI		3.5FH	512k		7200	
DSRZ40-VA	9	SCSI-2		3.5FH	512k		7200	
RZ26N-VA	14.5	SCSI-2Fast		3.5FH	480k		5400	Y
RZ26N-VW	14.5	SCSI-2FstW		3.5FH	480k		5400	Y
RZ28D-VA	12.2	SCSI-2Fast		3.5FH	480k		7200	Y
RZ28D-VW	12.2	SCSI-2FstW		3.5FH	480k		7200	Y
RZ28M-VA	14.5	SCSI-2Fast		3.5FH	480k		5400	
RZ28M-VW	14.5	SCSI-2FstW		3.5FH	480k		5400	Y
RZ29B-VA	12.2	SCSI-2Fast		3.5FH	1000k		7200	Y
RZ29B-VW	12.2	SCSI-2FstW		3.5FH	1000k		7200	Y
SP3430	9	SCSI-2Fast	1,7 RLL	3.5HH	2048k	800k	7200	
VP3107	9	SCSI-2Fast	1,7 RLL	3.5 3H	1024k	800k	7200	Y
VP3215	9	SCSI-2Fast	1,7 RLL	3.5 3H	1024k	800k	7200	

DISC TEC

Drive Model	Seek Time	Interface	Encode	Form Factor	cache kb	mtbf	RPM	Obsolete?
RHD-120	17	IDE AT	RLL	3.5 3H		100k		Y
RHD-180	15	IDE AT	RLL	3.5 3H		100k		Y
RHD-20 (Removable)	23	IDE AT	RLL	3.5 3H		20k		Y
RHD-210	19	IDE AT	RLL	3.5 3H		150k		Y
RHD-260	>14	IDE AT	RLL	3.5 3H		100k		Y
RHD-340	>14	IDE AT	RLL	3.5 3H		100k		Y
RHD-520	>14	IDE AT	RLL	3.5 3H		100k		
RHD-60	22	IDE AT	RLL	3.5 3H		45k		Y
RHD-80	16	IDE AT	RLL	3.5 3H		150k		Y

DISCTRON (OTARI)

Drive Model	Seek Time	Interface	Encode	Form Factor	cache kb	mtbf	RPM	Obsolete?
D214		ST412/506	MFM	5.25FH				Y
D503		ST412/506	MFM	5.25FH				Y
D504		ST412/506	MFM	5.25FH				Y
D506		ST412/506	MFM	5.25FH				Y
D507		ST412/506	MFM	5.25FH				Y
D509		ST412/506	MFM	5.25FH				Y
D512		ST412/506	MFM	5.25FH				Y
D513		ST412/506	MFM	5.25FH				Y

Drive Model	Format Size MB	Head	Cyl	Sect/ Trac	Translate H/C/S	RWC/ WPC	Land Zone
D514	11	4	306	17		128/128	
D518	15	8	215	17		128/128	
D519	16	6	306	17		128/128	
D526	21	8	306	17		128/128	

DMA

Drive Model	Format Size MB	Head	Cyl	Sect/ Trac	Translate H/C/S	RWC/ WPC	Land Zone
306	11	2	612	17		612/400	

ELOCH

Drive Model	Format Size MB	Head	Cyl	Sect/ Trac	Translate H/C/S	RWC/ WPC	Land Zone
DISCACHE10	10	4	320	17		321/321	
DISCACHE20	20	8	320	17		321/321	

EPSON

Drive Model	Format Size MB	Head	Cyl	Sect/ Trac	Translate H/C/S	RWC/ WPC	Land Zone
HD560	21	4	615	17		615/300	
HD830	10	2	612	17		—/—	
HD850	10	4	306	17		—/—	
HD860	21	4	612	17		—/—	
HMD710	10	2	615	17		—/—	
HMD720	21	4	615	17		—/—	
HMD726A	21	4	615	32		—/—	Auto
HMD755	21	2	615	34		—/—	
HMD765	42	4	615	34		—/—	
HMD976	69					—/—	

FUJI

Drive Model	Format Size MB	Head	Cyl	Sect/ Trac	Translate H/C/S	RWC/ WPC	Land Zone
FK301-13	10	4	306	17		307/128	
FK302-13	10	2	612	17		613/307	
FK302-26	21	4	612	17		613/307	
FK302-39	32	6	612	17		613/307	
FK303-52	40	8	615	17		—/616	
FK305-26	21	4	615	17		—/616	
FK305-26R	21		615	26		—/—	
FK305-39	32	6	615	17		—/616	
FK305-39R	32	4	615	26		—/616	
FK305-58	32	6	615	17		—/—	
FK305-58R	49	6	615	26		—/616	
FK308S-39R	45	6	615			—/—	
FK308S-58R	32	4	615	26		—/616	
FK309-26	21	4	615	17		—/616	
FK309-39R	32	4	615	26		—/616	
FK309S-50R	41	4	615			—/—	

FUJITSU AMERICA, INC.

Drive Model	Format Size MB	Head	Cyl	Sect/ Trac	Translate H/C/S	RWC/ WPC	Land Zone
M1603 SAU	540	3				—/—	
M1603 TAU	540	4				—/—	
M1606 SAU	1080	6	3457	94		—/—	
M1606 TAU	1080	6				—/—	
M1612 TAU	545	2	413385-153			—/—	
M1614 TAU	1090	4	413385-153			—/—	
M1623 TAU	1700	3				—/—	
M1624 TAU	2100	4				—/—	
M1636 TAU	1200	2				—/—	
M1638 TAU	2500	4				—/—	
M2225D	40	4	615	17		—/—	
M2225D2	20	4	615	17		—/—	
M2225DR	32	4	615	17		—/—	
M2226D	60	6	615	17		—/—	
M2226D2	30	6	615	17		—/—	
M2226DR	49	6	615	26		—/—	
M2227D	80	8	615	17		—/—	
M2227D2	42	8	615	17		—/—	

Drive Model	Seek Time	Interface	Encode	Form Factor	cache kb	mtbf	RPM	Obsolete?
D514		ST412/506	MFM	5.25FH				Y
D518		ST412/506	MFM	5.25FH				Y
D519		ST412/506	MFM	5.25FH				Y
D526		ST412/506	MFM	5.25FH				Y

DMA

Drive Model	Seek Time	Interface	Encode	Form Factor	cache kb	mtbf	RPM	Obsolete?
306	170?	ST412/506	MFM	5.25HH				Y

ELOCH

Drive Model	Seek Time	Interface	Encode	Form Factor	cache kb	mtbf	RPM	Obsolete?
DISCACHE10	65?	ST412/506	MFM	5.25FH				Y
DISCACHE20	65?	ST412/506	MFM	5.25FH				Y

EPSON

Drive Model	Seek Time	Interface	Encode	Form Factor	cache kb	mtbf	RPM	Obsolete?
HD560	78	ST412/506	MFM	5.25HH				Y
HD830	93	ST412/506	MFM	5.25HH				Y
HD850		ST412/506	MFM	5.25HH				Y
HD860		ST412/506	MFM	5.25HH				Y
HMD710	78	ST412/506	MFM	5.25HH				Y
HMD720	78	ST412/506	MFM	5.25HH				Y
HMD726A	80	SCSI	2,7 RLL	3.5HH		20k		Y
HMD755	80	ST412/506	2,7 RLL	5.25HH		20k		Y
HMD765	80	ST412/506	2,7 RLL	5.25HH		20k		Y
HMD976		SCSI		3.5HH				Y

FUJI

Drive Model	Seek Time	Interface	Encode	Form Factor	cache kb	mtbf	RPM	Obsolete?
FK301-13	65	ST412/506	MFM	3.5HH		45k		Y
FK302-13	65	ST412/506	MFM	3.5HH				Y
FK302-26	65	ST412/506	MFM	3.5HH				Y
FK302-39	65	ST412/506	MFM	3.5HH				Y
FK303-52	65?	ST412/506	MFM	3.5HH		20k		Y
FK305-26	65	ST412/506	MFM	3.5HH		20k	3350	Y
FK305-26R	65	ST412/506	2,7 RLL	3.5HH				Y
FK305-39	65	ST412/506	MFM	3.5HH		20k		Y
FK305-39R	65	ST412/506	2,7 RLL	3.5HH		20k	3350	Y
FK305-58	65	ST412/506	MFM	3.5HH		20k		Y
FK305-58R	65	ST412/506	2,7 RLL	3.5HH		20k	3350	Y
FK308S-39R	65	SCSI	2,7 RLL	3.5HH		20k		Y
FK308S-58R	65	ST412/506	2,7 RLL	3.5HH				Y
FK309-26	65	ST412/506	MFM	3.5HH		20k		Y
FK309-39R	65	ST412/506	2,7 RLL	3.5HH		20k		Y
FK309S-50R	45	SCSI	2,7 RLL	3.5HH		20k		Y

FUJITSU AMERICA, INC.

Drive Model	Seek Time	Interface	Encode	Form Factor	cache kb	mtbf	RPM	Obsolete?
M1603 SAU	10	SCSI-2Fast	1,7 RLL	3.5 3H	512k	800k	5400	Y
M1603 TAU	10	ATA-2	1,7 RLL	3.5 3H	256k	500k	5400	Y
M1606 SAU	10	SCSI-2Fast	1,7 RLL	3.5 3H	512k	800k	5400	Y
M1606 TAU	10	ATA-2	1,7 RLL	3.5 3H	256k	300k	5400	Y
M1612 TAU	11	ATA-2	PRML8,9	3.5 3H	64k	300k	4500	Y
M1614 TAU	11	ATA-2	PRML8,9	3.5 3H	64k	300k	4500	Y
M1623 TAU	10	ATA-2	PRML	3.5 3H	128k	500k	5400	Y
M1624 TAU	10	ATA-2	PRML	3.5 3H	128k	500k	5400	
M1636 TAU	10	ATA-2	PRML	3.5 3H	128k	500k	5400	Y
M1638 TAU	10	ATA-2	PRML	3.5 3H	128k	500k	5400	
M2225D	40	ST412/506	MFM	3.5HH		30k		Y
M2225D2	35	ST412/506	MFM	3.5HH				Y
M2225DR	35	ST412/506	2,7 RLL	3.5HH				Y
M2226D	40	ST412/506	MFM	3.5HH		30k		Y
M2226D2	35	ST412/506	MFM	3.5HH				Y
M2226DR	35	ST412/506	2,7 RLL	3.5HH				Y
M2227D	40	ST412/506	MFM	3.5HH		30k		Y
M2227D2	35	ST412/506	MFM	3.5HH				Y

Drive Model	Format Size MB	Head	Cyl	Sect/ Trac	Translate H/C/S	RWC/ WPC	Land Zone
M2227DR	65	8	615	26		—/—	
M2230	5	2	320	17		320/180	
M2230AS	5	2	320	17		320/320	
M2230AT	5	2	320	17		320/320	
M2231	5	2	306	17		—/—	
M2233	10	4	320	17		320/128	
M2233AS	10	4	320	17		320/320	
M2233AT	10	4	320	17		320/320	
M2234	15	6	320	17		320/128	
M2234AS	15	6	306	17		320/320	
M2235	21	8	320	17		320/128	
M2235AS	20	8	306	17		320/320	
M2241AS	26	4	754	17		—/375	754
M2241AS2	24	4	754	32		—/375	Auto
M2242AS	45	7	754	17		754/375	Auto
M2242AS2	43	7	754	17		—/—	Auto
M2243AS	72	11	754	17		754/375	Auto
M2243AS2	67	11	754	17		—/—	Auto
M2243R	110	7	1186	26		—/—	Auto
M2243T	68	7	1186	17		—/—	Auto
M2244E	73	5	823	35		Na/Na	Auto
M2244S	85U	5	823	65		Na/Na	Auto
M2244SA	85U	5	823	35		Na/Na	Auto
M2244SB	85U	5	823	19		Na/Na	Auto
M2245E	120	7	823	35		Na/Na	Auto
M2245S	120U	7	823	65		Na/Na	Auto
M2245SA	120U	7	823	35		Na/Na	Auto
M2245SB	120U	7	823	19		Na/Na	Auto
M2246E	138	10	823	35		Na/Na	Auto
M2246S	171U	10	823	65		Na/Na	Auto
M2246SA	171U	10	823	35		Na/Na	Auto
M2246SB	171U	10	823	19		Na/Na	Auto
M2247E	285	7	1243			Na/Na	Auto
M2247S	289	7	1243	65		Na/Na	Auto
M2247SA	160	7	1243	36		Na/Na	Auto
M2247SB	169	7	1243			Na/Na	Auto
M2248E	266	11	1243			Na/Na	Auto
M2248S	227	11	1243			Na/Na	Auto
M2248SA	252	11	1243	36		Na/Na	Auto
M2248SB	266	11	1243			Na/Na	Auto
M2249E	334	15	1243	35		Na/Na	Auto
M2249S	334	15	1243	35		Na/Na	Auto
M2249SA	334	15	1243	35		Na/Na	Auto
M2249SB	362	15	1243			Na/Na	Auto
M2261E	321	8	1658			Na/Na	Auto
M2261HA	357	8	1658	53		Na/Na	Auto
M2261S	321	8	1658			Na/Na	Auto
M2261SA	415U	8	1658	53		Na/Na	Auto
M2262E	448	11	1658			Na/Na	Auto
M2262HA	476	11	1658	51		Na/Na	Auto
M2262SA	476	11	1658	51		Na/Na	Auto
M2263E	688	15	1658	53		Na/Na	Auto
M2263HA	672	15	1658	53		Na/Na	Auto
M2263S	650	15	1658	53		Na/Na	Auto
M2266E	674	15	1658	53		Na/Na	Auto
M2266H	953	15	1658			Na/Na	Auto
M2266HA	1079	15	1658			Na/Na	Auto
M2266HB	1140	15	1658			Na/Na	Auto
M2266S	953	15	1658			Na/Na	Auto
M2266SA	1079	15	1658	65		Na/Na	Auto
M2266SB	1140	15	1658			Na/Na	Auto
M2344KS	690	27	624	NA		Na/Na	Auto
M2372K	823	27	745			—/—	

Drive Model	Seek Time	Interface	Encode	Form Factor	cache kb	mtbf	RPM	Obsolete?
M2227DR	35	ST412/506	2,7 RLL	3.5HH				Y
M2230	85	ST412/506	MFM	5.25FH				Y
M2230AS	27	ST412/506	MFM	5.25FH			3600	Y
M2230AT	8	ST412/506	MFM	5.25FH			3600	Y
M2231	85	ST412/506	MFM	5.25FH				Y
M2233	80	ST412/506	MFM	5.25FH				Y
M2233AS	27	ST412/506	MFM	5.25FH			3600	Y
M2233AT	8	ST412/506	MFM	5.25FH			3600	Y
M2234	8	ST412/506	MFM	5.25FH			3600	Y
M2234AS	27	ST412/506	MFM	5.25FH			3600	Y
M2235	85	ST412/506	MFM	5.25FH			3600	Y
M2235AS	27	ST412/506	MFM	5.25FH			3600	Y
M2241AS		ST412/506	MFM	5.25FH				Y
M2241AS2	30	ST412/506	MFM	5.25FH		20k		Y
M2242AS	30	ST412/506	MFM	5.25FH		30k		Y
M2242AS2	30	ST412/506	MFM	5.25FH				Y
M2243AS	30	ST412/506	MFM	5.25FH		30k		Y
M2243AS2	30	ST412/506	MFM	5.25FH				Y
M2243R	25	ST412/506	2,7 RLL	5.25HH				Y
M2243T	25	ST412/506	MFM	5.25HH				Y
M2244E	25	ESDI	2,7 RLL	5.25FH		30k		Y
M2244S	25	SCSI	2,7 RLL	5.25FH		35k	3600	Y
M2244SA	25	SCSI	2,7 RLL	5.25FH		35k	3600	Y
M2244SB	25	SCSI	2,7 RLL	5.25FH		35k	3600	Y
M2245E	25	ESDI	2,7 RLL	5.25FH				Y
M2245S	25	SCSI	2,7 RLL	5.25FH			3600	Y
M2245SA	25	SCSI	2,7 RLL	5.25FH			3600	Y
M2245SB	25	SCSI	2,7 RLL	5.25FH			3600	Y
M2246E	25	ESDI	2,7 RLL	5.25FH		30k		Y
M2246S	25	SCSI	2,7 RLL	5.25FH		30k	3600	Y
M2246SA	25	SCSI	2,7 RLL	5.25FH		30k	3600	Y
M2246SB	25	SCSI	2,7 RLL	5.25FH		30k	3600	Y
M2247E	18	ESDI	1,7 RLL	5.25FH		30k		Y
M2247S	18	SCSI	1,7 RLL	5.25FH		30k		Y
M2247SA	18	SCSI	1,7 RLL	5.25FH		30k		Y
M2247SB	18	SCSI	1,7 RLL	5.25FH		30k		Y
M2248E	18	ESDI	1,7 RLL	5.25FH		130k		Y
M2248S	18	SCSI	1,7 RLL	5.25FH		130k		Y
M2248SA	18	SCSI	1,7 RLL	5.25FH		130k		Y
M2248SB	18	SCSI	1,7 RLL	5.25FH		130k		Y
M2249E	18	ESDI	1,7 RLL	5.25FH		30k		Y
M2249S	18	SCSI	1,7 RLL	5.25FH		30k		Y
M2249SA	18	SCSI	1,7 RLL	5.25FH		30k		Y
M2249SB	18	SCSI	1,7 RLL	5.25FH		30k		Y
M2261E	16	ESDI	1,7 RLL	5.25FH		200k		Y
M2261HA	16	SCSI	1,7 RLL	5.25FH		200k		Y
M2261S	16	SCSI	2,7 RLL	5.25FH		200k		Y
M2261SA		SCSI		5.25FH				Y
M2262E	16	ESDI	1,7 RLL	5.25FH		200k		Y
M2262HA	16	SCSI	1,7 RLL	5.25FH		200k		Y
M2262SA	16	SCSI	1,7 RLL	5.25FH		200k		Y
M2263E	16	ESDI	1,7 RLL	5.25FH		30k	3600	Y
M2263HA	16	SCSI	1,7 RLL	5.25FH		200k		Y
M2263S	16	SCSI	1,7 RLL	5.25FH		30k		Y
M2266E	16	ESDI	1,7 RLL	5.25FH		200k		Y
M2266H	14.5	SCSI	1,7 RLL	5.25FH		200k	3600	Y
M2266HA	14.5	SCSI	1,7 RLL	5.25FH		200k	3600	Y
M2266HB	14.5	SCSI	1,7 RLL	5.25FH		200k	3600	Y
M2266S	14.5	SCSI	1,7 RLL	5.25FH		200k	3600	Y
M2266SA	14.5	SCSI	1,7 RLL	5.25FH	256k	200k	3600	Y
M2266SB	14.5	SCSI	1,7 RLL	5.25FH		200k	3600	Y
M2344KS	16	SCSI/SMD	RLL	8 FH				Y
M2372K	16	HSMD	2,7 RLL					Y

Drive Model	Format Size MB	Head	Cyl	Sect/ Trac	Translate H/C/S	RWC/ WPC	Land Zone
M2372KS	823	27	745			—/—	
M2382K	1000	27	745			—/—	
M2382P	1000	27	745			—/—	
M2392K	2020	21	1916			—/—	
M2511A	128	1	9952	25		—/—	
M2611H	46	2	1334	34		Na/Na	Auto
M2611S	46	2	1334	68		Na/Na	Auto
M2611SA	46	2	1334	34		Na/Na	Auto
M2611SB	46	2	1334	17		Na/Na	Auto
M2611T	45	2	1334	33	4/667/33	Na/Na	Auto
M2612ES	90	4	1334			Na/Na	Auto
M2612ESA	90	4	1334	34		Na/Na	Auto
M2612ESB	90	4	1334			Na/Na	Auto
M2612ET	90	4	1334	34	8/667/33	Na/Na	Auto
M2612S	92	4	1334	34		Na/Na	Auto
M2612SA	91	4	1334	33		Na/Na	Auto
M2612T	90	4	1334	33	8/667/33	Na/Na	Auto
M2613ES	139	6	1334			Na/Na	Auto
M2613ESA	137	6	1334	34		Na/Na	Auto
M2613ESB	139	6	1334			Na/Na	Auto
M2613ET	137	6	1334	34	12/667/33	Na/Na	Auto
M2613S	139	6	1334	34		Na/Na	Auto
M2613SA	137	6	1334	34		Na/Na	Auto
M2613SB	139	6	1334	17		Na/Na	Auto
M2613T	137	6	1334	34	12/667/33	Na/Na	Auto
M2614ES	185	8	1334			Na/Na	Auto
M2614ESA	182	8	1334	34		Na/Na	Auto
M2614ESB	185	8	1334			Na/Na	Auto
M2614ET	180	8	1334	34	16/667/33	Na/Na	Auto
M2614S	185	8	1334	34		Na/Na	Auto
M2614SA	182	8	1334	34		Na/Na	Auto
M2614SB	186	8	1334	17		Na/Na	Auto
M2614T	180	8	1334	34	16/667/33	Na/Na	Auto
M2616ESA	105	4	1542	34		Na/Na	Auto
M2616ET	105	4	1542	34	8/771/33	Na/Na	Auto
M2616SA	105	4	1542			Na/Na	Auto
M2616T	105	4	1542		8/771/33	Na/Na	Auto
M2621S	235	5	1435			Na/Na	Auto
M2622F	293	7	1435			—/—	
M2622FA	330	7	1435			—/—	
M2622S	330	7	1153	80		Na/Na	Auto
M2622SA	329	7	1429	56-70		Na/Na	Auto
M2622T	326	7	1435		10/1013/63	Na/Na	Auto
M2623F	377	9	1429	V		Na/Na	Auto
M2623FA	498	9	1435			—/—	
M2623S	425	9	1153	80		Na/Na	Auto
M2623SA	425	9	1429	64		Na/Na	Auto
M2623T	420	9	1435		13/002/63	Na/Na	Auto
M2624F	461	6	1435			—/—	
M2624FA	520	11	1435			—/—	
M2624S	520	11	1463	63		Na/Na	Auto
M2624SA	520	11	1429	64		Na/Na	Auto
M2624T	513	11	1429	63	16/995/63	Na/Na	Auto
M2635S	160	4	1569			—/—	
M2635T	160	4	1569		8/620/63	—/—	
M2637S	240	6	1574	49		—/—	
M2637SA	240	6	1574			—/—	
M2637T	240	6	1569		8/930/63	—/—	
M2651SA	1400	16	1944	88		—/—	
M2652H	1628	20	1893	84		Na/Na	Auto
M2652HA	1600	20	1944			Na/Na	Auto
M2652HD	1628	20	1893	84		Na/Na	Auto
M2652P	1600	20	1893			Na/Na	Auto

Drive Model	Seek Time	Interface	Encode	Form Factor	cache kb	mtbf	RPM	Obsolete? ⇓
M2372KS	16	SCSI	2,7 RLL					Y
M2382K	16	ESMD	1,7 RLL					Y
M2382P	16	IPI	1,7 RLL					Y
M2392K	12	ESMD	1,7 RLL					Y
M2511A	30	SCSI-2	1,7 RLL	3.5 3H	256k	30k	3600	Y
M2611H	25	SCSI	1,7 RLL	3.5HH		50k		Y
M2611S	25	SCSI	1,7 RLL	3.5HH		50k		Y
M2611SA	25	SCSI	1,7 RLL	3.5HH	24k	50k	3490	Y
M2611SB	25	SCSI	1,7 RLL	3.5HH		50k		Y
M2611T	25	IDE AT	1,7 RLL	3.5HH	64k	50k	3490	Y
M2612ES	20	SCSI	1,7 RLL	3.5HH				Y
M2612ESA	20	SCSI	1,7 RLL	3.5HH	24k	30k	3490	Y
M2612ESB	20	SCSI	1,7 RLL	3.5HH				Y
M2612ET	20	IDE AT	1,7 RLL	3.5HH	64k	50k	3490	Y
M2612S	20	SCSI	1,7 RLL	3.5HH		50k	3490	Y
M2612SA	25	SCSI	1,7 RLL	3.5HH	24k	30k	3490	Y
M2612T	25	IDE AT	1,7 RLL	3.5HH	64k	50k	3490	Y
M2613ES	20	SCSI	1,7 RLL	3.5HH				Y
M2613ESA	20	SCSI	1,7 RLL	3.5HH	24k	30k	3490	Y
M2613ESB	20	SCSI	1,7 RLL	3.5HH				Y
M2613ET	20	IDE AT	1,7 RLL	3.5HH	64k	50k	3490	Y
M2613S	20	SCSI	1,7 RLL	3.5HH		50k	3490	Y
M2613SA	25	SCSI	1,7 RLL	3.5HH	24k	30k	3490	Y
M2613SB	20	SCSI	1,7 RLL	3.5HH		50k		Y
M2613T	25	IDE AT	1,7 RLL	3.5HH	64k	50k	3490	Y
M2614ES	20	SCSI	1,7 RLL	3.5HH				Y
M2614ESA	20	SCSI	1,7 RLL	3.5HH	24k	30k	3490	Y
M2614ESB	20	SCSI	1,7 RLL	3.5HH				Y
M2614ET	20	IDE AT	1,7 RLL	3.5HH		50k		Y
M2614S	20	SCSI	1,7 RLL	3.5HH		50k	3490	Y
M2614SA	25	SCSI	1,7 RLL	3.5HH	24k	30k	3490	Y
M2614SB	20	SCSI	1,7 RLL	3.5HH		50k		Y
M2614T	20	IDE AT	1,7 RLL	3.5HH	64k	50k	3490	Y
M2616ESA	20	SCSI	1,7 RLL	3.5HH	24k	30k	3490	Y
M2616ET	20	IDE AT	1,7 RLL	3.5HH				Y
M2616SA	20	SCSI	1,7 RLL	3.5HH	24k	50k	3490	Y
M2616T	20	IDE AT	1,7 RLL	3.5HH	64k	50k	3490	Y
M2621S	12	SCSI-2	1,7 RLL	3.5HH				Y
M2622F	12	SCSI	1,7 RLL	3.5HH			4400	Y
M2622FA	12	SCSI-1/2	1,7 RLL	3.5HH	240k	200k	4400	Y
M2622S	12	SCSI-2	1,7 RLL	3.5HH			4400	Y
M2622SA	12	SCSI-2	1,7 RLL	3.5HH	240k	200k	4400	Y
M2622T	12	IDE AT	1,7 RLL	3.5HH	240k	200k	4400	Y
M2623F	12	SCSI 1/2	1,7 RLL	3.5HH		200k	4400	Y
M2623FA	12	SCSI-1/2	1,7 RLL	3.5HH	240k	200k	4400	Y
M2623S	12	SCSI-2	1,7 RLL	3.5HH	240k	200k	4400	Y
M2623SA	12	SCSI-2	1,7 RLL	3.5HH	240k	200k	4400	Y
M2623T	12	IDE AT	1,7 RLL	3.5HH	240k	200k	4400	Y
M2624F	12	SCSI	1,7 RLL	3.5HH			4400	Y
M2624FA	12	SCSI-1/2	1,7 RLL	3.5HH	240k	200k	4400	Y
M2624S	12	SCSI-2	1,7 RLL	3.5HH	240k	200k	4400	Y
M2624SA	12	SCSI-2	1,7 RLL	3.5HH	240k	200k	4400	Y
M2624T	12	IDE AT	1,7 RLL	3.5HH	240k	200k	4400	Y
M2635S	14	SCSI-2	1,7 RLL	2.5 4H	256k	>150k	4500	Y
M2635T	14	IDE AT	1,7 RLL	2.5 4H	256k	>150k	4500	Y
M2637S	14	SCSI-2	1,7 RLL	2.5 4H	256	150k	4500	Y
M2637SA	14.5	SCSI-2	1,7 RLL	FH	256k		4500	Y
M2637T	14	IDE AT	1,7 RLL	2.5 4H	256k	150k	4500	Y
M2651SA	12	SCSI-2	1,7 RLL	5.25FH	256k	300k	5400	Y
M2652H	11	SCSI-2	1,7 RLL	5.25HH		200k	5400	Y
M2652HA	11	SCSI-2Diff		FH		300k	5400	Y
M2652HD	11	SCSI-2	1,7 RLL	5.25FH		200k	5400	Y
M2652P	11	IPI-2		FH		300k	5400	Y

Drive Model	Format Size MB	Head	Cyl	Sect/ Trac	Translate H/C/S	RWC/ WPC	Land Zone
M2652S	1628	20	1893	84		Na/Na	Auto
M2652SA	1750	20	1944	88		Na/Na	Auto
M2653	1400	15	2078	88		—/—	
M2654HA	2000	21	2179			Na/Na	Auto
M2654SA	2061	21	2170	88		—/—	
M2671P	2640	15	2671			—/—	
M2681SAU	264	3	2379			—/—	
M2681TAU	264	3	2379		11/977/48	—/—	
M2682SAU	350	4	2379	64-90		—/—	
M2682TAU	352	4	2378	64-90	11/992/63	—/—	
M2684SAU	525	6	2379	74		—/—	
M2684TAU	525	6	2379		16/1024/63	—/—	
M2691EHA	645	9	1818	V		Na/Na	Auto
M2691EQ	756U	9	1831			—/—	
M2691ER	756U	9	1831			—/—	
M2691ESA	645	9	1818	V		Na/Na	Auto
M2692EQ	925U	11	1831			—/—	
M2692ER	925U	11	1831			—/—	
M2693EQ	1093U	13	1831			—/—	
M2693ER	1093U	13	1831			—/—	
M2694EHA	1080	15	1818	V		Na/Na	Auto
M2694EQ	1261U	15	1831			—/—	
M2694ER	1261U	15	1831			—/—	
M2694ESA	1080	15	1818	V		Na/Na	Auto
M2703S	260	3	2305			—/—	
M2703T	260	3	2305			—/—	
M2704	260	3				—/—	
M2704S	350	4	2305			—/—	
M2704T	350	4	2305			—/—	
M2705	350	4				—/—	
M2706	530	6				—/—	
M2706S	530	6	2305			—/—	
M2706T	530	6	2305			—/—	
M2712TAM	540	1				—/—	
M2713TAM	1080	2				—/—	
M2714TAM	1080	2				—/—	
M2723	1200	3				—/—	
M2724	1600	4				—/—	
M2903	2100	14	3139			—/—	
M2909	3100	20	3139			—/—	
M2914	2100	7				—/—	
M2915	2100	16	3012			—/—	
M2927	1100	4				—/—	
M2932	2170	10	3422			—/—	
M2934	4350	19	3422			—/—	
M2948S	8800	18	5751			—/—	
M2949S	9100	18	5772			—/—	
M2952S	2200	5	5565			—/—	
M2954S	4400	9	5565			—/—	
MAA3182	18200	19	9040			—/—	
MAB3045	4550	5	8490			—/—	
MAB3091 SB	9100	10				—/—	
MAC3045	4550	5	8690	156-24		—/—	
MAC3045	4550	5	8690	156-24		—/—	
MAC3045	4550	5	8690	156-24		—/—	
MAC3091	9100	10	8690	156-24		—/—	
MAC3091	9100	10	8690	156-24		—/—	
MAC3091	9100	10	8690	156-24		—/—	
MHA2021	2160	4				—/—	
MHA2032	3240	6				—/—	
MHB2021	2160	4				—/—	
MHB2032	3240	6				—/—	
MHC2020AT	4090	6	7229			—/—	

Drive Model	Seek Time	Interface	Encode	Form Factor	cache kb	mtbf	RPM	Obsolete?
M2652S	11	SCSI-2	1,7 RLL	5.25FH		200k	5400	Y
M2652SA	11	SCSI-2	1,7 RLL	5.25			5400	Y
M2653	12	SCSI-2Diff	1,7 RLL	5.25	256k		5400	Y
M2654HA	12	SCSI-2Diff		FH		300k	5400	Y
M2654SA	12	SCSI-2	1,7 RLL	5.25FH	256k	300k	5400	
M2671P	12	IPI-2	1,7 RLL	8 FH		200k	4340	
M2681SAU	12	SCSI-2	1,7 RLL	3.5 3H	256k	250k	4500	Y
M2681TAU	12	IDE AT	1,7 RLL	3.5 3H	256k	250k	4500	Y
M2682SAU	12	SCSI-2	1,7 RLL	3.5 3H	256k	250k	4500	Y
M2682TAU	12	IDE AT	1,7 RLL	3.5 3H	256k	250k	4500	Y
M2684SAU	12	SCSI-2	1,7 RLL	3.5 3H	256k	250k	4500	Y
M2684TAU	12	IDE AT	1,7 RLL	3.5 3H	256k	250k	4500	Y
M2691EHA	10	SCSI-2	1,7 RLL	3.5HH	256k	300k	5400	Y
M2691EQ	10	SCSI	1,7 RLL	3.5HH	512k		5400	Y
M2691ER	10	SCSI-2Diff	1,7 RLL	3.5HH	512k		5400	Y
M2691ESA	10	SCSI-2	1,7 RLL	3.5HH	256k	300k	5400	Y
M2692EQ	10	SCSI	1,7 RLL	3.5HH	512k		5400	Y
M2692ER	10	SCSI-2Diff	1,7 RLL	3.5HH	512k		5400	Y
M2693EQ	10	SCSI	1,7 RLL	3.5HH	512k		5400	Y
M2693ER	10	SCSI-2Diff	1,7 RLL	3.5HH	512k		5400	Y
M2694EHA	10	SCSI-2Diff	1,7 RLL	3.5HH	256k	300k	5400	Y
M2694EQ	10	SCSI	1,7 RLL	3.5HH	512k		5400	Y
M2694ER	10	SCSI-2Diff	1,7 RLL	3.5HH	512k		5400	Y
M2694ESA	10	SCSI-2	1,7 RLL	3.5HH	512k	300k	5400	Y
M2703S	12	SCSI-2Fast	RLL	2.5 4H	512k	300k	5400	Y
M2703T	12	ATA-2	RLL	2.5 4H	256k	300k	5400	Y
M2704	12	SCSI		2.5 4H	256	250k	5400	Y
M2704S	12	SCSI-2Fast	RLL	2.5 4H	512k	300k	5400	Y
M2704T	12	ATA-2	RLL	2.5 4H	256k	300k	5400	Y
M2705	12	SCSI		2.5 4H	256	250k	5400	Y
M2706	12	SCSI		2.5 4H	512k	300k	5400	Y
M2706S	12	SCSI-2Fast	RLL	2.5 4H	512k	300k	5400	Y
M2706T	12	ATA-2	RLL	2.5 4H	256k	300k	5400	Y
M2712TAM	12	ATA	PRML8,9	2.5 4H	128k	300k	3634	Y
M2713TAM	12	ATA	PRML8,9	2.5 4H	128k	300k	3634	Y
M2714TAM	12	ATA	PRML8,9	2.5 4H	128k	300k	3634	Y
M2723	12	ATA-3	PRML	3.5 3H	128k	300k	4000	Y
M2724	12	ATA-3	PRML	3.5 3H	128k	300k	4000	Y
M2903	10.5	SCSI-2FstW	RLL	3.5HH	512k	500k	5400	Y
M2909	10.5	SCSI-2FstW	RLL	3.5HH	512k	500k	5400	Y
M2914	9	SCSI-2FstW		3.5HH	512k	500k	7200	Y
M2915	9.8	SCSI-2FstW	RLL	3.5HH	512k	500k	7200	Y
M2927	10.5	SCSI-2FstW		3.5HH	512k	500k	5400	Y
M2932	11	SCSI-2Fast	RLL	3.5HH	510k	800k	7200	Y
M2934	11	SCSI-2Fast	RLL	3.5HH	510k	800k	7200	
M2948S	10	SCSI-2FstW	PR4ML	3.5HH	512k	1000k	7200	
M2949S	10	SCSI-2FstW	RLL 0,4,4	3.5HH	512k	1000k	7200	
M2952S	8	SCSI-2FstW	RLL 8,9	3.5 3H	512k	1000k	7200	Y
M2954S	8	SCSI-2FstW	RLL 8,9	3.5 3H	512k	1000k	7200	
MAA3182	9	Ultra2 LVD	PRML8,9	3.5HH	512k	1000k	7200	
MAB3045	8.5	SCSI-2	PRML8,9	3.5 3H	512k	1000k	7200	
MAB3091 SB	16	UlSCSI40Fst	PRML		512k		7200	
MAC3045	7.5	Ultra SCSI	PRML8,9	3.5 3H	512k	1000k	1003	
MAC3045	7.5	Ultra2 LVD	PRML8,9	3.5 3H	512k	1000k	1003	
MAC3045	7.5	FC-AL	PRML8,9	3.5 3H	512k	1000k	1003	
MAC3091	7.5	FC-AL	PRML8,9	3.5 3H	512k	1000k	1003	
MAC3091	7.5	Ultra2 LVD	PRML8,9	3.5 3H	512k	1000k	1003	
MAC3091	7.5	Ultra SCSI	PRML8,9	3.5 3H	512k	1000k	1003	
MHA2021	13	ATA-3	PRML	2.5 3H	128k	300k	4000	
MHA2032	13	ATA-3	PRML	2.5 3H	128k	300k	4000	
MHB2021	13	ATA-3	PRML	2.5 4H	128k	300k	4000	
MHB2032	13	ATA-3	PRML	2.5 4H	128k	300k	4000	
MHC2020AT	13	ATA-3	PRML	2.5 4H	512k	300k	4000	

Drive Model	Format Size MB	Head	Cyl	Sect/ Trac	Translate H/C/S	RWC/ WPC	Land Zone
MHD2021AT	2160	3	7289			—/—	
MHD2032AT	3250	4	7317			—/—	
MPA3017A	1750	2	8713	132-25		—/—	
MPA3026AT	2620	3	8713	132-25		—/—	
MPA3035AT	3500	4	8713	132-25		—/—	
MPA3043AT	4370	5	8713	132-25		—/—	
MPA3052AT	5250	6	8713	132-25		—/—	
MPB3021AT	2160	2	8983	168-30		—/—	
MPB3032AT	3240	3	8983	168-30		—/—	
MPB3043AT	4320	4	8983	168-30		—/—	
MPB3052AT	5240	5	8983	168-30		—/—	
MPB3064AT	6480	6	8983	168-30		—/—	
MPC3032AT	3240	2	11116	192-33		—/—	
MPC3043AT	4320	3	11116	192-33		—/—	
MPC3045AH	4500	4	10424	162-24		—/—	
MPC3064AT	6480	4	11116	192-33		—/—	
MPC3065AH	6500	6	10424	162-24		—/—	
MPC3084AT	8450	6	11116	192-33		—/—	
MPC3096AT	9740	6	11116	192-33		—/—	
MPC3102AT	10200	6	11116	192-33		—/—	

HEWLETT-PACKARD CO

Drive Model	Format Size MB	Head	Cyl	Sect/ Trac	Translate H/C/S	RWC/ WPC	Land Zone
HP97501A	10	2	698	28	8/142/17	—/—	
HP97501B	20	2	1400	28	8/288/17	—/—	
HP97530E	136	4				Na/Na	Auto
HP97530S	204	6				Na/Na	Auto
HP97532D	215	4	1643	64*V		Na/Na	Auto
HP97532E	215	4	1643	64		Na/Na	Auto
HP97532S	215	4	1643	64		Na/Na	Auto
HP97532T	215	4	1643	64		Na/Na	Auto
HP97533D	323	6	1643	64		Na/Na	Auto
HP97533E	323	6	1643	64		Na/Na	Auto
HP97533S	323	6	1643	64		Na/Na	Auto
HP97533T	323	6	1643	64		Na/Na	Auto
HP97536D	646	12	1643	64		Na/Na	Auto
HP97536E	646	12	1643	64		Na/Na	Auto
HP97536S	646	12	1643	64		Na/Na	Auto
HP97536SP	320					—/—	
HP97536SX	322					—/—	
HP97536T	646	12	1643	64		Na/Na	Auto
HP97536TA	320					—/—	
HP97544D	331	8	1447	56		Na/Na	Auto
HP97544E	337	8	1447	56		Na/Na	Auto
HP97544P	331	8	1447	56		Na/Na	Auto
HP97544S	331	8	1447	56		Na/Na	Auto
HP97544SA	331					—/—	
HP97544T	331	8	1447	56		Na/Na	Auto
HP97548D	663	16	1447	56		Na/Na	Auto
HP97548E	675	16	1447	56		Na/Na	Auto
HP97548P	663	16	1447	56		Na/Na	Auto
HP97548S	663	16	1447	56		Na/Na	Auto
HP97548SZ	663					—/—	
HP97548T	663	16	1447	56		Na/Na	Auto
HP97549P	1001	16	1911	64		Na/Na	Auto
HP97549T	1001	16	1911	69		Na/Na	Auto
HP97556	786					—/—	
HP97556E	688	11	1697	72		Na/Na	Auto
HP97556P	677	11	1670	72		Na/Na	Auto
HP97556T	677	11	1670	72		Na/Na	Auto
HP97558E	1084	15	1962	72		Na/Na	Auto
HP97558P	1069	15	1935	72		Na/Na	Auto
HP97558T	1069	15	1935	72		Na/Na	Auto
HP97560	1300					—/—	

Drive Model	Seek Time	Interface	Encode	Form Factor	cache kb	mtbf	RPM	Obsolete?
MHD2021AT	13	ATA-3	PRML	2.5 4H		300k	4000	
MHD2032AT	13	ATA-3	PRML	2.5 4H		300k	4000	
MPA3017A	10	ATA-3	PRML8,9	3.5 3H	128k	500k	5400	Y
MPA3026AT	10	ATA-3	PRML8,9	3.5 3H	128k	500k	5400	
MPA3035AT	10	ATA-3	PRML8,9	3.5 3H	128k	500k	5400	
MPA3043AT	10	ATA-3	PRML8,9	3.5 3H	128k	500k	5400	
MPA3052AT	10	ATA-3	PRML8,9	3.5 3H	128k	500k	5400	
MPB3021AT	10	ATA-3	PRML8,9	3.5 3H	256k	500k	5400	
MPB3032AT	10	ATA-3	PRML8,9	3.5 3H	256k	500k	5400	
MPB3043AT	10	ATA-3	PRML8,9	3.5 3H	256k	500k	5400	
MPB3052AT	10	ATA-3	PRML8,9	3.5 3H	256k	500k	5400	
MPB3064AT	10	ATA-3	PRML8,9	3.5 3H	256k	500k	5400	
MPC3032AT		ATA-3	Prml16,17	3.5 3H	256k	500k	5400	
MPC3043AT		ATA-3	Prml16,17	3.5 3H	256k	500k	5400	
MPC3045AH	9	ATA-3	PRML8,9	3.5 3H	512k	500k	7200	
MPC3064AT		ATA-3	Prml16,17	3.5 3H	256k	500k	5400	
MPC3065AH	9	ATA-3	PRML8,9	3.5 3H	512k	500k	7200	
MPC3084AT		ATA-3	Prml16,17	3.5 3H	256k	500k	5400	
MPC3096AT		ATA-3	Prml16,17	3.5 3H	256k	500k	5400	
MPC3102AT		ATA-3	Prml16,17	3.5 3H	256k	500k	5400	

HEWLETT-PACKARD CO

Drive Model	Seek Time	Interface	Encode	Form Factor	cache kb	mtbf	RPM	Obsolete?
HP97501A	75		MFM	3.5HH				Y
HP97501B			MFM	3.5HH				Y
HP97530E	18	ESDI	2,7 RLL	5.25FH				Y
HP97530S	18	SCSI	2,7 RLL	5.25FH				Y
HP97532D	17	SCSI	2,7 RLL	5.25FH	16k	99k	3348	Y
HP97532E	17	ESDI (10)	2,7 RLL	5.25FH	16k	99k	3348	Y
HP97532S	17	SCSI	2,7 RLL	5.25FH	16k	99k	3348	Y
HP97532T	17	SCSI	2,7 RLL	5.25FH	16k	99k	3348	Y
HP97533D	17	SCSI	2,7 RLL	5.25FH	16k	99k	3348	Y
HP97533E	17	ESDI	2,7 RLL	5.25FH	16k	99k	3348	Y
HP97533S	17	SCSI	2,7 RLL	5.25FH	16k	99k	3348	Y
HP97533T	17	SCSI	2,7 RLL	5.25FH	16k	99k	3348	Y
HP97536D	17	SCSI	2,7 RLL	5.25FH	16k	99k	3348	Y
HP97536E	17	ESDI	2,7 RLL	5.25FH	16k	99k	3348	Y
HP97536S	17	SCSI	2,7 RLL	5.25FH	16k	99k	3348	Y
HP97536SP		SCSI	2,7 RLL	5.25FH				Y
HP97536SX		SCSI	2,7 RLL	5.25FH				Y
HP97536T	17	SCSI	2,7 RLL	5.25FH	16k	99k	3348	Y
HP97536TA		SCSI	2,7 RLL	5.25FH				Y
HP97544D	16	SCSI	2,7 RLL	5.25HH	64k	150k	4002	Y
HP97544E	17	ESDI	2,7 RLL	5.25FH	64k	150k	4002	Y
HP97544P	17	SCSI-2	2,7 RLL	5.25FH	64k	150k	4002	Y
HP97544S	16	SCSI	2,7 RLL	5.25FH	64k	150k	4002	Y
HP97544SA		SCSI	2,7 RLL	5.25FH				Y
HP97544T	17	SCSI-2	2,7 RLL	5.25FH	64k	150k	4002	Y
HP97548D	16	SCSI	2,7 RLL	5.25FH	64k	150k	4002	Y
HP97548E	17	ESDI	2,7 RLL	5.25FH	64k	150k	4002	Y
HP97548P	17	SCSI-2	2,7 RLL	5.25FH	64k	150k	4002	Y
HP97548S	16	SCSI	2,7 RLL	5.25FH	64k	150k	4002	Y
HP97548SZ		SCSI	2,7 RLL	5.25FH				Y
HP97548T	17	SCSI-2	2,7 RLL	5.25FH	64k	150k	4002	Y
HP97549P	17	SCSI-2	2,7 RLL	5.25FH	128k	150k	4002	
HP97549T	17	SCSI-2	2,7 RLL	5.25FH	128k	150k	4002	
HP97556			2,7 RLL	5.25FH				Y
HP97556E	14	ESDI	2,7 RLL	5.25FH	128k	150k	4002	Y
HP97556P	14	SCSI-2	2,7 RLL	5.25FH	128k	150k	4002	Y
HP97556T	14	SCSI-2	2,7 RLL	5.25FH	128k	150k	4002	Y
HP97558E	14	ESDI	2,7 RLL	5.25FH	128k	150k	4002	
HP97558P	14	SCSI-2	2,7 RLL	5.25FH	128k	150k	4002	
HP97558T	14	SCSI-2	2,7 RLL	5.25FH	128k	150k	4002	
HP97560		SCSI-2	2,7 RLL	5.25FH		150k		Y

Drive Model	Format Size MB	Head	Cyl	Sect/ Trac	Translate H/C/S	RWC/ WPC	Land Zone
HP97560E	1374	19	1962	72		Na/Na	Auto
HP97560P	1355	19	1935	72		Na/Na	Auto
HP97560T	1355	19	1935	72		Na/Na	Auto
HPC2233 ATA	238	5	1546	V	16/462/63	Na/Na	Auto
HPC2233S	234	5	1546	V		Na/Na	Auto
HPC2234 ATA	334	7	1546	V	16/647/63	Na/Na	Auto
HPC2234S	328	7	1546	V		Na/Na	Auto
HPC2235A	429					Na/Na	Auto
HPC2235S	422	9	1546	V		Na/Na	Auto
HPC2244	566	7	2051	79		—/—	
HPC2245	728	9	2051	79		—/—	
HPC2246	890	11	2051	79		—/—	
HPC2247	1052	13	1981	56-96		Na/Na	Auto
HPC2247D	1052	13	1981	56-96		Na/Na	Auto
HPC2247SE	1052	13	1981	56-96		Na/Na	Auto
HPC2247W	1052	13	1981	56-96		Na/Na	Auto
HPC2270S	320					—/—	
HPC2271S	663					—/—	
HPC2490D	2100	18	258268-108			—/—	
HPC2490SE	2100	18	258268-108			—/—	
HPC2490W	2100	18	258268-108			—/—	
HPC3007	1370		2255			Na/Na	Auto
HPC3009	1792		2255			Na/Na	Auto
HPC3010	2003					Na/Na	Auto
HPC3013A	21	3	700		4/615/17	—/—	
HPC3014A	42	4	786			—/—	
HPC3031A	21	3				—/—	
HPC3323D	1050	7	291072-120			Na/Na	Auto
HPC3323SE	1050	7	291072-120			Na/Na	Auto
HPC3323W	1050	7	291072-120			Na/Na	Auto
HPC3324	1050	9	3703	100		—/—	
HPC3325A	2170	9	3610100-14			—/—	
HPC3335 ATA	429	9	1546	V		Na/Na	Auto
HPC3550	2000					—/—	
HPC3555	1000					—/—	
HPC3653A	8700	20	5371124-17			—/—	
HPC3724D	1200	5	3610100-14			Na/Na	Auto
HPC3724S	1200	5	3610100-14			Na/Na	Auto
HPC3724W	1200	5	3610100-14			Na/Na	Auto
HPC3725D	2170	9	3610100-14			—/—	
HPC3725S	2170	9	3610100-14			—/—	
HPC3725W	2170	9	3610100-14			—/—	
HPC5270A	1084	4		91-155		—/—	
HPC5271A	1626	6		91-155		—/—	
HPC5272A	1336	4		94-162		—/—	
HPC5273A	2004	6		94-162		—/—	
HPC5273AK	1336	4		94-162		—/—	
HPC5280A	1084	4		91-155		—/—	
HPC5281A	1626	6		91-155		—/—	
HPC5283A	2004	6		94-162		—/—	
HPC5421SK	8700	20	5371124-17			—/—	
HPC5421TK	8700	20	5371124-17			—/—	
HPC5435A	1336	4		94-162		—/—	
HPC5435AK	1300					—/—	
HPC5436AK	2004	6		94-162		—/—	
HPD1296A	21	4	615	17		0/300	670
HPD1297A	42	6	820	17		—/—	
HPD1660A	340	8	1457	57		Na/Na	Auto
HPD1661A	680	16	1457	57		Na/Na	Auto
HPD1674A	108	6	820	40		—/—	
HPD1675A	155	6	820	40		—/—	
HPD1676A	310	6	820	40		—/—	
HPD1697A	240	4	1800		8/930/63	—/—	

Drive Model	Seek Time	Interface	Encode	Form Factor	cache kb	mtbf	RPM	Obsolete?
HP97560E	14	ESDI	2,7 RLL	5.25FH	128k	150k	4002	
HP97560P	14	SCSI-2	2,7 RLL	5.25FH	128k	150k	4002	
HP97560T	14	SCSI-2	2,7 RLL	5.25FH	128k	150k	4002	
HPC2233 ATA	12.6	IDE AT	2,7 RLL	3.5HH	64k	150k	3600	Y
HPC2233S	12	SCSI-2	2,7 RLL	3.5HH	64k	150k	3600	Y
HPC2234 ATA	12.6	IDE AT	2,7 RLL	3.5HH	64k	150k	3600	Y
HPC2234S	12	SCSI-2	2,7 RLL	3.5HH	64k	150k	3600	Y
HPC2235A	13	ATA		FH		150k		Y
HPC2235S	12	SCSI-2	2,7 RLL	3.5HH	64k	150k	3600	Y
HPC2244	10	SCSI-2	1,7 RLL	3.5HH	256k	300k	5400	Y
HPC2245	10	SCSI-2	1,7 RLL	3.5HH	256k	300	5400	Y
HPC2246	10	SCSI-2	1,7 RLL	3.5HH	256k	300	5400	Y
HPC2247	10	SCSI-2	1,7 RLL	3.5HH	256k	300k	5400	Y
HPC2247D	10	SCSI-2Diff	1,7 RLL	3.5HH	256k	300k	5400	Y
HPC2247SE	10	SCSI-2	1,7 RLL	3.5HH	256k	300k	5400	Y
HPC2247W	10	SCSI-2FstW	1,7 RLL	3.5HH	256k	300k	5400	Y
HPC2270S		SCSI		5.25FH				Y
HPC2271S		SCSI		5.25FH				Y
HPC2490D	9	SCSI-2Diff		3.5HH		500k	6400	Y
HPC2490SE	9	SCSI		3.5HH		500k	6400	Y
HPC2490W	9	SCSI-2FstW		3.5HH		500k	6400	Y
HPC3007	12	SCSI-2		FH	256k	300k	5400	
HPC3009	12	SCSI-2		FH	256k	300k	5400	
HPC3010	12	SCSI-2		FH	256k	300k	5400	
HPC3013A	15	IDE AT	1,7 RLL	1.3 4H		300k		Y
HPC3014A	18	IDE		1.3 4H		300k	5310	Y
HPC3031A	18	IDE		1.3 4H		300k	5310	Y
HPC3323D	9.5	SCSI-2Diff		3.5 3H	512k	500k	5400	Y
HPC3323SE	9.5	SCSI-2		3.5 3H	512k	500k	5400	Y
HPC3323W	9.5	SCSI-2Diff		3.5 3H	512k	500k	5400	Y
HPC3324	9.5	SCSI-2	1,7 RLL	3.5 3H	512k	500	5400	
HPC3325A	10.5	SCSI-2	PRML	3.5 3H	512k		5400	
HPC3335 ATA	12.6	IDE AT	2,7 RLL	3.5HH	64k	150k	3600	Y
HPC3550		SCSI-2FstW		3.5HH				Y
HPC3555		SCSI-2FstW		3.5HH				Y
HPC3653A	9	SE SCSI	PRML	3.5HH	512k		7200	Y
HPC3724D	9.5	SCSI-2Diff		3.5 3H		800k	5400	Y
HPC3724S	9.5	SCSI-2		3.5 3H		800k	5400	Y
HPC3724W	9.5	SCSI-2FstW		3.5 3H		800k	5400	Y
HPC3725D	9.5	SCSI-2Diff		3.5 3H		800k	5400	Y
HPC3725S	9.5	SCSI-2		3.5 3H		800k	5400	Y
HPC3725W	9.5	SCSI-2Diff		3.5 3H		800k	5400	Y
HPC5270A	<12	EIDE/ATA-2		3.5HH	128k	300k	4480	
HPC5271A	<12	EIDE/ATA-2		3.5HH	128k	300k	4480	
HPC5272A	<12	EIDE/AT	1,7 RLL	3.5 3H	64k	300k	4480	
HPC5273A	<12	EIDE/AT	1,7 RLL	3.5 3H	128k	300k	4480	
HPC5273AK	<12	EIDE/ATA-2		3.5 3H	128k	300k	4480	Y
HPC5280A	<12	EIDE		3.5 3H	128k	300k	4480	Y
HPC5281A	<12	EIDE		3.5 3H	128k	300k	4480	Y
HPC5283A	<12	EIDE/AT	1,7 RLL	3.5 3H	128k	300k	4480	Y
HPC5421SK	8.7	SE SCSI	PRML	3.5HH	512k		7200	Y
HPC5421TK	8.7	SE SCSI-2W	PRML	3.5HH	512k		7200	Y
HPC5435A	<12	EIDE/AT	1,7 RLL	3.5 3H	64k	300k	4480	
HPC5435AK	<12	EIDE	1,7 RLL	3.5 3H	64k	300k	4480	
HPC5436AK	<12	EIDE/ATA-2	1,7 RLL	3.5 3H	128k	300k	4480	
HPD1296A	65	ST412/506	MFM	5.25HH		100k		Y
HPD1297A	40	ST412/506	MFM	5.25HH		100k		Y
HPD1660A	16	ESDI (15)	2,7 RLL	5.25HH	64k	150k		Y
HPD1661A	16	ESDI (15)	2,7 RLL	5.25HH	64k	150k		Y
HPD1674A	40	ST412/506	MFM	5.25FH		100k		Y
HPD1675A	40	ESDI (15)	2,7 RLL	5.25FH		100k		Y
HPD1676A	40	ESDI (15)	2,7 RLL	5.25FH		100k		Y
HPD1697A	17	IDE AT	1,7 RLL	3.5 3H				Y

Drive Model	Format Size MB	Head	Cyl	Sect/ Trac	Translate H/C/S	RWC/ WPC	Land Zone
HPD2076B	1050					—/—	
HPD2077A	2100					—/—	
HPD2389A	540					—/—	
HPD3340A	2100					—/—	
HPD3341A	4200					—/—	

HITACHI AMERICA

Drive Model	Format Size MB	Head	Cyl	Sect/ Trac	Translate H/C/S	RWC/ WPC	Land Zone
DK211A-51	510	6				Na/Na	Auto
DK211A-54	540	16	1047	63	16/1047/63	Na/Na	Auto
DK211A-68	680				16/1384/60	—/—	
DK211C-51	510	6				Na/Na	Auto
DK212A-10	1080	8				—/—	
DK212A-81	810	8				—/—	
DK213A-13	1350	10	2605			—/—	
DK213A-18	1800				16/3491/63	—/—	
DK221A-34	340	4				Na/Na	Auto
DK222A-27	270	2				—/—	
DK222A-54	540	4				—/—	
DK223A-11	1080				16/2095/63	—/—	
DK223A-81	810	6	2605			—/—	
DK224A-14	1440				16/2792/63	—/—	
DK225A-21	2160				16/4188/63	—/—	
DK226A-32	3240				16/6282/63	—/—	
DK227A-41	4090					—/—	
DK227A-50	5020					—/—	
DK237A-32	3240					—/—	
DK301-1	10	4	306	17		—/—	
DK301-2	15	6	306	17		—/—	
DK312C-20	209	9	1076	38		—/—	
DK312C-25	251	11	1076	38		—/—	
DK314C-41	419	14		17		—/—	
DK315C-10	1000	11				Na/Na	Auto
DK315C-11	1100	15				Na/Na	Auto
DK315C-14	1400	15				Na/Na	Auto
DK318H-91	9100	20				—/—	
DK319H-18	18200					—/—	
DK325C-57	573	6	2458	75		Na/Na	Auto
DK326C-10	1050	8				Na/Na	Auto
DK326C-10WD	1050	7				Na/Na	Auto
DK326C-6	601	4				Na/Na	Auto
DK326C-6WD	601	4				Na/Na	Auto
DK328C-10	1050	3				—/—	
DK328C-21	2100	5				—/—	
DK328C-43	4300	10				—/—	
DK328H-43	4370	10				—/—	
DK329H-91	9100					—/—	
DK503-2	10					—/—	
DK505-2	21	4	615	17		—/—	
DK511-3	29	5	699	17		—/300	699
DK511-5	41	7	699	17		—/300	699
DK511-8	67	10	823	17		—/400	822
DK512-12	94	7	823			Na/Na	
DK512-17	134	10	823			Na/Na	
DK512-8	67	5	823			Na/Na	Auto
DK512C-12	94	7	823			—/—	
DK512C-17	134	10	819	35		—/—	
DK512C-8	67	5	823			—/—	
DK512S-17	143					—/—	
DK514-38	330	14	903	51		Na/Na	
DK514C-38	322	14	898	50		—/—	
DK514S-38	332					—/—	
DK515-12	1229	15		69		Na/Na	Auto
DK515-78	673	14	1361	69		—/—	

Drive Model	Seek Time	Interface	Encode	Form Factor	cache kb	mtbf	RPM	Obsolete?
HPD2076B	10.5	SCSI-2Fast			256k	500k	5400	Y
HPD2077A	10.5	SCSI-2Fast			256k	500k	5400	Y
HPD2389A	14	IDE AT		3.5		300k	3600	Y
HPD3340A	8.4	SCSI-2			512k	1000k	5400	Y
HPD3341A	8.4	SCSI-2			512k	1000k	5400	Y

HITACHI AMERICA

Drive Model	Seek Time	Interface	Encode	Form Factor	cache kb	mtbf	RPM	Obsolete?
DK211A-51	12.6	IDE AT		2.5 4H	64k	300k	4464	Y
DK211A-54	12	ATA			64k		4464	Y
DK211A-68	12	IDE AT		2.5 4H	64k		4464	Y
DK211C-51	12.6	SCSI-2Fast		2.5 4H	512k	300k		Y
DK212A-10	12	EIDE/ATA-2	PRML8,9	2.5 4H	64k	300k	4464	Y
DK212A-81	12	EIDE/ATA-2	PRML8,9	2.5 4H	64k	300k	4464	Y
DK213A-13	12	ATA-2	PRML8,9	2.5 4H	128k	300k	4464	Y
DK213A-18	12	ATA-2		2.5 4H	128k		4464	Y
DK221A-34	12.6	IDE AT		2.5 4H	64k	300k	4464	Y
DK222A-27	12	EIDE/ATA-2	PRML8,9	2.5 4H	64k	300k	4464	Y
DK222A-54	12	EIDE/ATA-2	PRML8,9	2.5 4H	64k	300k	4464	Y
DK223A-11	12	ATA-2		2.5 4H	128k		4464	Y
DK223A-81	12	ATA-2	PRML8,9	2.5 4H	128k	300k	4464	Y
DK224A-14	12	ATA-2		2.5 4H	128k		4464	Y
DK225A-21	12	ATA-2 Fast		2.5 4H	128k		4464	
DK226A-32	12	ATA-3		2.5 4H	128k		4000	
DK227A-41		ATA-3	EPRML 16,17	2.5 4H	512k			
DK227A-50		ATA-3		2.5 4H	512k			
DK237A-32		ATA-3	EPRML 16,17	2.5	512k			
DK301-1	85	ST412/506	MFM	3.5HH				Y
DK301-2	85	ST412/506	MFM	3.5HH				Y
DK312C-20	17	SCSI	2,7 RLL	3.5HH		40k		Y
DK312C-25	17	SCSI	2,7 RLL	3.5HH		40k		Y
DK314C-41	17	SCSI	2,7 RLL	3.5HH	64k	150k		Y
DK315C-10	11.8	SCSI-2Fast		3.5HH	256k	400k		Y
DK315C-11	11	SCSI-2		3.5HH	256k	150k		Y
DK315C-14	11	SCSI-2Fast		3.5HH	256k	400k		Y
DK318H-91	9	Ultra SCSI		3.5HH	512k		7200	
DK319H-18	7.5	Ultra SCSI	PRML	3.5	512k		7200	
DK325C-57	12	SCSI-2	1,7 RLL	5.25HH		200k	4500	Y
DK326C-10	9.8	SCSI-2Fast		3.5 3H	448k	400k	6300	Y
DK326C-10WD	9.8	SCSI-2FstW		3.5 3H	448k	400k	6300	Y
DK326C-6	<10	SCSI-2Fast		3.5 3H	448k	400k		Y
DK326C-6WD	<10	SCSI-2FstW		3.5 3H	448k	400k		Y
DK328C-10	9.8	SE SCSI-2F		3.5 3H	512k	800k	5400	Y
DK328C-21	9.8	SE SCSI-2D		3.5 3H	512k	800k	5400	Y
DK328C-43	9.8	SE SCSI-2F		3.5 3H	512k	800k	5400	Y
DK328H-43	9	Ultra SCSI		3.5 3H	512k		7200	
DK329H-91	7.5	Ultra SCSI	PRML	3.5	512k		7200	
DK503-2			MFM	5.25HH				Y
DK505-2	85	ST412/506	MFM	5.25HH				Y
DK511-3	30	ST412/506	MFM	5.25FH				Y
DK511-5	26	ST412/506	MFM	5.25FH				Y
DK511-8	23	ST412/506	MFM	5.25FH				Y
DK512-12	23	ESDI	2,7 RLL	5.25FH		20k	3482	Y
DK512-17	23	ESDI	2,7 RLL	5.25FH		20k	3482	Y
DK512-8	23	ESDI	2,7 RLL	5.25FH		20k	3482	Y
DK512C-12	23	SCSI	2,7 RLL	5.25FH				Y
DK512C-17	23	SCSI	2,7 RLL	5.25FH				Y
DK512C-8	23	SCSI	2,7 RLL	5.25FH				Y
DK512S-17		SMD-E		5.25FH				Y
DK514-38	16	ESDI	2,7 RLL	5.25HH		30k	3600	Y
DK514C-38	16	SCSI	2,7 RLL	5.25FH		30k		Y
DK514S-38		SMD-E		5.25FH				Y
DK515-12	14	ESDI	2,7 RLL	5.25FH		150k		Y
DK515-78	16	ESDI	2,7 RLL	5.25FH		150k		Y

Drive Model	Format Size MB	Head	Cyl	Sect/ Trac	Translate H/C/S	RWC/ WPC	Land Zone
DK515C-78	670	14	1356	69		—/—	
DK515C-78D	673	14	1361	69		Na/Na	Auto
DK515S-78	673	14				—/—	
DK516-12	1230					—/—	
DK516-15	1320	15				Na/Na	Auto
DK516C-16	1340	15				—/—	
DK517C-26	2000	14				Na/Na	Auto
DK517C-37	2900	21				Na/Na	Auto
DK521-5	51	6	823	17		—/None	822
DK522-10	91	6	823	36		Na/Na	
DK522C-10	87	6	819	35		—/—	
DK524C-20	169	6	1105	51		—/—	

HYOSUNG

Drive Model	Format Size MB	Head	Cyl	Sect/ Trac	Translate H/C/S	RWC/ WPC	Land Zone
HC8085	71	8	1024	17		Na/Na	Auto
HC8128	109	8	1024	26		Na/Na	Auto
HC8170E	150	8	1024	36		Na/Na	Auto

IBM CORP. (STORAGE SYS DIV)

Drive Model	Format Size MB	Head	Cyl	Sect/ Trac	Translate H/C/S	RWC/ WPC	Land Zone
0661-371	326	14	949	48		Na/Na	Auto
0661-371	325	14	949	48		—/—	
0661-437	467					—/—	
0661-467	412	14	1199	48		Na/Na	Auto
0661-467	406	14	1199	48		—/—	
0661-467R	400	14	1199	48		—/—	
0662-A10	1052	6				—/—	
0662-S12	1062	6				—/—	
0662-S1D	1052					Na/Na	Auto
0662-SW1	1062	6				—/—	
0662-SWD	1062	6				—/—	
0663-E12	1044	14				—/—	
0663-E15	1206	16				—/—	
0663-E15R	1206	15	2463	66		—/—	
0663-H11	868	13	2051	66		—/—	
0663-H12	1004	15	2051	66		Na/Na	Auto
0663-L08	623	9	2051	66		Na/Na	Auto
0663-L11	868	13	2051	66		Na/Na	Auto
0663-L12R	1004	15	2051	66		Na/Na	Auto
0663-W2H	2412	15				—/—	
0664-CSH	4027	38	2328	211		—/—	
0664-DSH	4027	32				—/—	
0664-ESH	4027	38	2328	211		—/—	
0664-FSH	4027	32				—/—	
0664-M1H	2013	16				—/—	
0664-N1H	2013	16				—/—	
0664-P1S	1741	15	2304			—/—	
0665-38	31	5	733	17		Na/Na	Auto
0665-53	44	7	733	17		Na/Na	Auto
0667-61	52	5	582	35		Na/Na	Auto
0667-85	73	7	582	35		Na/Na	Auto
0669-133	133					—/—	
0671-315/S	315					—/—	
0671-S11	234	11	1224	34		Na/Na	Auto
0671-S15	319	15	1224	34		Na/Na	Auto
0681-1000	865	20	1458	58		Na/Na	Auto
0681-500	476	11	1458	58		Na/Na	Auto
115MB	118	7	915	36		—/—	Auto
120MB	120	8	920	32		—/—	
120MB	120	8	920	32		—/—	
1430	21	4	615	17		320/128	307
1431	31	5	733	17		733/733	
1470	31	5	733	17		733/733	
1471	31	5	733	17		733/733	

Drive Model	Seek Time	Interface	Encode	Form Factor	cache kb	mtbf	RPM	Obsolete?
DK515C-78	16	SCSI	2,7 RLL	5.25FH		150k		Y
DK515C-78D	16	SCSI	2,7 RLL	5.25FH		150k		Y
DK515S-78	16	E-SMD		5.25FH		150k		Y
DK516-12	14	ESDI		5.25FH		100k		Y
DK516-15	14	ESDI		5.25FH		150k		Y
DK516C-16	14	SCSI	2,7 RLL	5.25FH		150k		Y
DK517C-26	12	SCSI-2		5.25FH		150k		Y
DK517C-37	12	SCSI-2Fast		5.25FH	512k	400k		Y
DK521-5	25	ST412/506	MFM	5.25HH				Y
DK522-10	25	ESDI	2,7 RLL	5.25HH		30k		Y
DK522C-10	25	SCSI	2,7 RLL	5.25FH		30k		Y
DK524C-20	25	SCSI-2	2,7 RLL	5.25HH		40k	3600	Y

HYOSUNG

Drive Model	Seek Time	Interface	Encode	Form Factor	cache kb	mtbf	RPM	Obsolete?
HC8085	25	ST412/506		5.25FH		28k		Y
HC8128	25	ST412/506		5.25FH		28k		Y
HC8170E	25	ESDI		5.25FH		28k		Y

IBM CORP. (STORAGE SYS DIV)

Drive Model	Seek Time	Interface	Encode	Form Factor	cache kb	mtbf	RPM	Obsolete?
0661-371	12.5	SCSI-2		3.5HH	64k	300k		Y
0661-371	12	SCSI-2	RLL		64k	30k	4316	Y
0661-437		SCSI		3.5HH				Y
0661-467	11.5	SCSI-2		3.5HH	128k	300k		Y
0661-467	11	SCSI-2			128k	30k	4316	Y
0661-467R	11	SCSI-2		3.5FH	128k	50k	4316	Y
0662-A10	10	IDE AT		3.5 3H	512k	500k	5400	Y
0662-S12	10	SCSI-2Fast		3.5 3H	512k	800k	5400	Y
0662-S1D	10	SCSI-2FstD		3.5 3H	512k	800k	5400	Y
0662-SW1	10	SCSI-2FstW		3.5 3H	512k	800k	5400	Y
0662-SWD	10	SCSI-2FstW		3.5 3H	512k	800k	5400	Y
0663-E12	11	SCSI-2Fast		3.5HH	256k	50k	4317	Y
0663-E15	11	SCSI-2Fast		3.5HH	256k	50k	4317	Y
0663-E15R	9	SCSI-2		3.5FH	256k	75k	4316	Y
0663-H11	11	SCSI-2	RLL	3.5HH	256k	400k	4316	Y
0663-H12	11	SCSI-2	RLL	3.5HH	256k	400k	4316	Y
0663-L08	9.8	SCSI-2		3.5HH		400k		Y
0663-L11	11	SCSI-2	RLL	3.5HH	256k	400k	4316	Y
0663-L12R	11	SCSI-2	RLL	3.5FH	256k	75k	4316	Y
0663-W2H	9	SCSI-2Fast		5.25FH	256k	300k	4317	Y
0664-CSH	11	SCSI-2Fast		5.25FH		375k	5400	Y
0664-DSH	<11	SCSI-2Fast		5.25FH		375k	5400	Y
0664-ESH	11	SCSI-2Fast		5.25FH		375k	5400	Y
0664-FSH	<11	SCSI-2Fast		5.25FH		375k	5400	Y
0664-M1H	11	SCSI-2Fast		3.5HH	512k	750k	5400	Y
0664-N1H	11	SCSI-2FstW		3.5HH	512k	750k	5400	Y
0664-P1S	11	IPI-2		3.5HH		750k	5400	Y
0665-38	40	ST412/506	MFM	5.25FH				Y
0665-53	40	ST412/506	MFM	5.25FH				Y
0667-61	30	ESDI	RLL	5.25FH				Y
0667-85	30	ESDI		5.25FH				Y
0669-133		ESDI		5.25FH				Y
0671-315/S		ESDI		5.25FH				Y
0671-S11	21.5	SCSI		5.25FH				Y
0671-S15	21.5	SCSI		5.25FH				Y
0681-1000	13	SCSI	RLL	5.25FH		150k		Y
0681-500	13	SCSI	RLL	5.25FH		150k		Y
115MB	28	ESDI		5.25				Y
120MB	23	ST412/506	MFM	5.25FH				Y
120MB	23	ST412/506	MFM	5.25FH				Y
1430	80	ST412/506	MFM	5.25FH				Y
1431	40	ST412/506	MFM	5.25FH				Y
1470	40	ST412/506	MFM	5.25FH				Y
1471	40	ST412/506	MFM	5.25FH				Y

Drive Model	Format Size MB	Head	Cyl	Sect/ Trac	Translate H/C/S	RWC/ WPC	Land Zone
20MB	21	4	612	17		—/306	663
20MB PS/2	21	4	612	17		—/—	128
245MB	245					—/—	
30MB	31	4	615	25		—/300	663
314MB	319	15	1225	34		—/—	Auto
340MB	340					—/—	
44MB	44	7	733	17		—/300	733
527MB	527					—/—	
540MB	540	7	2466			—/—	
60MB	60	6	762	26		—/—	
70MB	75	7	583	36		—/—	Auto
DADA 25400	5400	5	9280			—/—	
DADA 26480	6480	6	9280			—/—	
DALA 3540	540	2	4892		16/1049/63	—/—	
DALA 3540	528				16/1049/63	—/—	
DALS 3540	541	2				—/—	
DAQA 32160	2160	4	6911			—/—	
DAQA 33240	3240	6	6911			—/—	
DBOA 2360	360	2	3478		16/700/63	—/—	
DBOA 2528	528				16/1024/63	—/—	
DBOA 2540	540	3	3478		16/1050/63	—/—	
DBOA 2720	722	4	3478		16/1400/63	—/—	
DCAA 32880	2880	4	8210			—/—	
DCAA 33610	3610	5	8210			—/—	
DCAA 34330	4330	6	8210			—/—	
DCAS 32160	2160	3	8120			—/—	
DCAS 34330	4330	6	8120			—/—	
DCHC 34550	4550	9				—/—	
DCHC 38700	8700	18				—/—	
DCHC 39100	9100	18				—/—	
DCHS 34550	4550	9				—/—	
DCHS 38700	8700	18				—/—	
DCHS 39100	9100	18				—/—	
DCMS 310800	10800	20				—/—	
DCRA 22160	2160				16/4200/63	—/—	
DDLA 21215	1215	3	5120			—/—	
DDLA 21620	1620	4	5120			—/—	
DDRS 34560	4500	5	8420			—/—	
DDRS 34560	4500	5	8420			—/—	
DDRS 39130	9100	10	8420			—/—	
DDRS 39130	9100	10	8420			—/—	
Deskstar 14GXP 10.1G	10100	7	13085			—/—	
Deskstar 14GXP 12.9G	12900	8	13085			—/—	
Deskstar 14GXP 14.4G	14400	10	13085			—/—	
Deskstar 16GP 10.1G	10100	6	13085			—/—	
Deskstar 16GP 12.9G	12900	8	13085			—/—	
Deskstar 16GP 16.8G	16800	10	13085			—/—	
Deskstar 16GP 3.2G	3200	2	13085			—/—	
Deskstar 16GP 4.3G	4300	3	13085			—/—	
Deskstar 16GP 6.4G	6400	4	13085			—/—	
Deskstar 16GP 8.4G	8400	5	13085			—/—	
Deskstar 1700AT	1700	2				—/—	
Deskstar 3	2160					—/—	
Deskstar 4 3.61G	3610	5	8210			—/—	
Deskstar 4 4.33G	4330	6	8210			—/—	
Deskstar 5 EIDE	4200	6				—/—	
Deskstar 5 EIDE	6400					—/—	
Deskstar 540AT	540					—/—	
Deskstar 8	8400					—/—	
Deskstar XP 1.	1080	2				—/—	
DFHC 31080	1126	4				—/—	
DFHC 32160	2255	8				—/—	
DFHC 32160	2255	8				—/—	

Drive Model	Seek Time	Interface	Encode	Form Factor	cache kb	mtbf	RPM	Obsolete?
20MB		ST412/506	MFM	5.25FH				Y
20MB PS/2	80	ST412/506	MFM	3.5HH				Y
245MB	15	IDE AT		4H				Y
30MB		ST412/506	MFM	5.25FH				Y
314MB	23	ESDI		5.25FH				Y
340MB	14	IDE AT		4H				Y
44MB		ST412/506	MFM	5.25FH				Y
527MB	9	IDE AT		3.5 3H				Y
540MB	9	SCSI-2		3.5 3H	256k	300k	6300	Y
60MB	27	ST412/506	MFM	5.25FH				Y
70MB	30	ESDI		5.25FH				Y
DADA 25400	12	ATA-4		4H	512k	300k	4200	
DADA 26480	12	ATA-4		4H	512k	300k	4200	
DALA 3540	12	ATA-2		3.5 3H	128k	350k	4500	Y
DALA 3540	12	ATA-2		3.5 3H	128k	350k	4500	Y
DALS 3540	12	SCSI-2Fast		3.5 4H	64k	350k	4500	Y
DAQA 32160	9.5	ATA-3		3.5 3H	128k		5400	
DAQA 33240	9.5	ATA-3		3.5 3H	128k		5400	Y
DBOA 2360	13	ATA-2		2.5 4H	32k	300k	4000	Y
DBOA 2528	13	ATA-2		2.5 4H	64k	300k	4000	Y
DBOA 2540	13	ATA-2		2.5 4H	32k	300k	4000	Y
DBOA 2720	13	ATA-2		2.5 4H	64k	300k	4000	Y
DCAA 32880	9.5	ATA-3	PRML	3.5 3H	128k	40k	5400	
DCAA 33610	9.5	ATA-3	PRML	3.5 3H	128k	40k	5400	
DCAA 34330	9.5	ATA-3	PRML	3.5 3H	128k	40k	5400	
DCAS 32160	8.5	SCSI-3Ultra	PRML	3.5 3H	512k		5400	
DCAS 34330	8.5	SCA-2	PRML	3.5 3H	512k		5400	
DCHC 34550	7.5	SSA	PRML	3.5 3H	512k		7200	
DCHC 38700	9	SSA		3.5FH	512k	1000k	7200	
DCHC 39100	7.5	SSA	PRML	3.5HH	512k		7200	
DCHS 34550	7.5	SCSI-2FstW	PRML	3.5 3H	512k		7200	
DCHS 38700	9	IPI-2		3.5FH	512k	1000k	7200	
DCHS 39100	7.5	SCSI-3Fst20	PRML	3.5HH	512k		7200	
DCMS 310800	9	SCSI-2FstW		3.5FH	512k	1000k	5400	
DCRA 22160	12	IDE AT		2.5 4H	96k		4200	Y
DDLA 21215	13	ATA-3	PRML	2.5 4H	128k		4000	Y
DDLA 21620	13	ATA-3	PRML	2.5 4H	128k		4000	Y
DDRS 34560	7.5	SCA-2	Prml16,17	3.5 3H	512k		7200	
DDRS 34560	7.5	SCA-2FstW	Prml16,17	3.5 3H	512k		7200	
DDRS 39130	7.5	Ultra SCSI3	Prml16,17	3.5 3H	512k		7200	
DDRS 39130	7.5	Ultra2 SCSI	Prml16,17	3.5 3H	512k		7200	
Deskstar14GXP 10.1G	9.5	ATA-4	PRML	3H	512k		7200	
Deskstar 14GXP 12.9G	9.5	ATA-4	PRML	3H	512k		7200	
Deskstar 14GXP 14.4G	9.5	ATA-4	PRML	3H	512k		7200	
Deskstar 16GP 10.1G	9.5	ATA-4	PRML	3H	512k		5400	
Deskstar 16GP 12.9G	9.5	ATA-4	PRML	3H	512k		5400	
Deskstar 16GP 16.8G	9.5	ATA-4	PRML	3H	512k		5400	
Deskstar 16GP 3.2G	9.5	ATA-4	PRML	3H	512k		5400	Y
Deskstar 16GP 4.3G	9.5	ATA-4	PRML	3H	512k		5400	Y
Deskstar 16GP 6.4G	9.5	ATA-4	PRML	3H	512k		5400	
Deskstar 16GP 8.4G	9.5	ATA-4	PRML	3H	512k		5400	
Deskstar 1700AT	12	ATA-2		3.5 3H	128k	350k	4500	Y
Deskstar 3	9.5	EIDE		3.5 3H	128k		5400	Y
Deskstar 4 3.61G	9.5	ATA-3	PRML	3.5 3H	128k	40k	5400	Y
Deskstar 4 4.33G	9.5	ATA-3	PRML	3.5 3H	128k	40k	5400	Y
Deskstar 5 EIDE	9.5	ATA-3 Fast		3.5 3H	128k		5400	Y
Deskstar 5 EIDE	9.5	EIDE		3.5 3H			7200	
Deskstar 540AT	12	ATA-2		3.5 3H	128k	350k	4500	Y
Deskstar 8	9.5	IDE		3.5 3H	512k			
Deskstar XP 1.	10.5	ATA-2		3.5 3H	512k	500k	5400	Y
DFHC 31080	9	SSA		3.5 3H	512k	1000k	7200	Y
DFHC 32160	9	SSA		3.5 3H	512k	1000k	7200	Y
DFHC 32160	9	SSA		3.5 3H	512k	1000k	7200	Y

Drive Model	Format Size MB	Head	Cyl	Sect/ Trac	Translate H/C/S	RWC/ WPC	Land Zone
DFHC 34320	4512	16				—/—	
DFHC 34320	4512	16				—/—	
DFHC C4x	4510	16				—/—	
DFHS 31080 S1F	1126	4				—/—	
DFHS 32160	2255	8				—/—	
DFHS 32160 S2D	2255					—/—	
DFHS 32160 S2F	2250					—/—	
DFHS 32160 S2W	2250					—/—	
DFHS 34320	4512	16				—/—	
DFHS 34320 S4D	4512					—/—	
DFHS 34320 S4F	4510					—/—	
DFHS 34320 S4W	4510					—/—	
DFMS 31080	1320	4				—/—	
DFMS 32160	2325	8				—/—	
DFMS 32600	2657	8				—/—	
DFMS 34320	4320	13				—/—	
DFMS 351AV	5106	16				—/—	
DFMS 35250	5318	16				—/—	
DFMS 35250 S5D	5318					—/—	
DFMS 35250 S5F	5318					—/—	
DFMS 35250 S5W	5318					—/—	
DGHL 39110	9100	10				—/—	
DGHS 31820	18200	20				—/—	
DGHS 39110	9100	10				—/—	
DGHS 39110	9100	10				—/—	
DGHU 39110	9100	10				—/—	
DGVS 39110	9100	12				—/—	
DHAA 2270	270	2	2788		16/524/63	—/—	
DHAA 2344	344	3	2788			—/—	
DHAA 2405	405	3	2788		16/785/63	—/—	
DHAA 2540	540	4	2788		16/1047/63	—/—	
DHAS 2270	270	2	2788			—/—	
DHAS 2344	344	3	2788			—/—	
DHAS 2405	405	3	2788			—/—	
DHAS 2540	540	4	2788			—/—	
DHEA 34330	4330	5	8209			—/—	
DHEA 34860	4860	6	8209			—/—	
DHEA 36480	6480	8	8209			—/—	
DHEA 38451	8450	8	9784			—/—	
DKLA 22160	2160	2	9280			—/—	
DKLA 23240	3240	3	9280			—/—	
DKLA 24320	4320	4	9280			—/—	
DLGA 22690	2690	7	5120			—/—	
DLGA 23080	3080	7	5120			—/—	
DMCA 21080	1080	3	4975			—/—	
DMCA 21440	1440	4	4975			—/—	
DORS 32160	2160	3				—/—	
DPEA 30540	540				16/1050/63	—/—	
DPEA 30810	812				16/1574/63	—/—	
DPEA 31080	1083				16/2100/63	—/—	
DPES 30540	540	4	4896			—/—	
DPES 30810	810	4	4896			—/—	
DPES 31080	1080	4	4896			—/—	
DPLA 24480	4480	7	6976			—/—	
DPLA 25120	5120	8	6976			—/—	
DPRA 20810	810	16	1572	63		—/—	
DPRA 21215	1215	16	2358	63		—/—	
DPRS 20810	810					—/—	
DPRS 21215	1215	16	2358	63		—/—	
DSAA 3270	270					—/—	
DSAA 3360	364					—/—	
DSAA 3540	548	3	3875			—/—	
DSAA 3720	720	3	3875			—/—	

Drive Model	Seek Time	Interface	Encode	Form Factor	cache kb	mtbf	RPM	Obsolete?
DFHC 34320	9	SSA		3.5HH	512k	1000k	7200	
DFHC 34320	9.5	SSA		3.5HH	512k	1000k	7200	
DFHC C4x	8	SSA		3.5 3H			7200	
DFHS 31080 S1F	9	SCSI-2 F/FW		3.5 3H	512k	1000k	7200	Y
DFHS 32160	9	SCSI-2 F/FW		3.5 3H	512k	1000k	7200	Y
DFHS 32160 S2D	7.5	SCSI-2Diff		3.5 3H	512k	1000k	7200	Y
DFHS 32160 S2F	7.5	SCSI-2Fast		3.5 3H	512k	1000k	7200	Y
DFHS 32160 S2W	7.5	SCSI-2FstW		3.5 3H	512k	1000k	7200	Y
DFHS 34320	9.5	SCSI-2 F/FW		3.5HH	512k	1000k	7200	Y
DFHS 34320 S4D	8	SCSI-2Diff		3.5HH	512k	1000k	7200	Y
DFHS 34320 S4F	8	SCSI-2Fast		3.5HH	512k	1000k	7200	Y
DFHS 34320 S4W	8	SCSI-2FstW		3.5HH	512k	1000k	7200	Y
DFMS 31080	7	SCSI-2Fast		3.5 3H	512k	1000k	5400	Y
DFMS 32160	9	SCSI-2Fast		3.5 3H	512k	1000k	5400	Y
DFMS 32600	9	SCSI-2Fast		3.5 3H	512k	1000k	5400	Y
DFMS 34320	9.5	SCSI-2Fast		3.5HH	512k	1000k	5400	Y
DFMS 351AV	9.5	SCSI-2 F/FW		3.5HH	512k	1000k	5400	
DFMS 35250	9.5	SCSI-2Fast		3.5HH	512k	1000k	5400	
DFMS 35250 S5D	8	SCSI-2Diff		3.5HH	512k	1000k	5400	
DFMS 35250 S5F	8	SCSI-2Diff		3.5HH	512k	1000k	5400	
DFMS 35250 S5W	8	SCSI-2FstW		3.5HH	512k	1000k	5400	
DGHL 39110	6.5	FC-AL	RLL 16,17	3.5 4H	1000k		7200	
DGHS 31820	6.5	SCA-2FstWD	RLL 16,17	3.5 3H	1000k		7200	
DGHS 39110	6.5	SSA	RLL 16,17	3.5 4H	1000k		7200	
DGHS 39110	6.5	SCA-2FstWD	RLL 16,17	3.5 4H	1000k		7200	
DGHU 39110	6.5	SCA-2FstWD	RLL 16,17	3.5 4H	1000k		7200	
DGVS 39110	6.3	ULSCSI40Fst	PRML		1mb		1000	
DHAA 2270	14	IDE AT		2.5 4H	32k	300k	3800	Y
DHAA 2344	14	IDE AT		2.5 4H	32k	300k	3800	Y
DHAA 2405	14	IDE AT		2.5 4H	32k	300k	3800	Y
DHAA 2540	14	IDE AT		2.5 4H	32k	300k	3800	Y
DHAS 2270	14	SCSI-2Fast		2.5 4H	32k	300k	3800	Y
DHAS 2344	14	SCSI-2Fast		2.5 4H	32k	300k	3800	Y
DHAS 2405	14	SCSI-2Fast		2.5 4H	32k	300k	3800	Y
DHAS 2540	14	SCSI-2Fast		2.5 4H	32k	300k	3800	Y
DHEA 34330	9.5	ATA-3	PRML	3.5 3H	512k	40k	5400	Y
DHEA 34860	9.5	ATA-3	PRML	3.5 3H	512k	40k	5400	Y
DHEA 36480	9.5	ATA-3	PRML	3.5 3H	512k	40k	5400	
DHEA 38451	9.5	ATA-3	PRML	3.5 3H	512k	40k	5400	
DKLA 22160	12	ATA-4			512k		4200	Y
DKLA 23240	12	ATA-4			512k		4200	
DKLA 24320	12	ATA-4			512k		4200	
DLGA 22690	12	ATA-3	PRML	2.5 4H	128k		4900	Y
DLGA 23080	12	ATA-3	PRML	2.5 4H	128k		4900	
DMCA 21080	13	ATA-3	PRML	2.5 4H	128k		4000	Y
DMCA 21440	13	ATA-3	PRML	2.5 4H	128k		4000	Y
DORS 32160	8.5	SCSI-3Ultra		3.5 3H	512k	800k	5400	Y
DPEA 30540	10.5	ATA-2		3.5 3H	448k	350k	5400	Y
DPEA 30810	10.5	IDE AT		3.5 3H	448k	350k	5400	Y
DPEA 31080	10.5	ATA-2		3.5 3H	448k	350k	5400	Y
DPES 30540	10.5	SCSI-2Fast		3.5 3H	512k	1000k	5400	Y
DPES 30810	10.5	SCSI-2Fast		3.5 3H	512k	1000k	5400	Y
DPES 31080	10.5	SCSI-2Fast		3.5 3H	512k	1000k	5400	Y
DPLA 24480	12	ATA-3	PRML	2.5 4H	512k		4900	
DPLA 25120	12	ATA-3	PRML	2.5 4H	512k		4900	
DPRA 20810	12	ATA-2		2.5 4H	64k	300k	4900	Y
DPRA 21215	12	ATA-2		2.5 4H	64k	300k	4900	Y
DPRS 20810	12	SCSI-2		2.5 4H	64k	300k	4900	Y
DPRS 21215	12	SCSI-2		2.5 4H	64k	300k	4900	Y
DSAA 3270	12	IDE AT		3.5 3H	96k	300k	4500	Y
DSAA 3360	12	IDE AT		3.5 3H	96k	300k	4500	Y
DSAA 3540	12	ATA-2		3.5 3H	128k	300k	4500	Y
DSAA 3720	12	ATA-2		3.5 3H	128k	300k	4500	Y

Drive Model	Format Size MB	Head	Cyl	Sect/ Trac	Translate H/C/S	RWC/ WPC	Land Zone
DSAS 3270	270					—/—	
DSAS 3360	364					—/—	
DSAS 3540	548	4	3875			—/—	
DSAS 3720	720	4	3875			—/—	
DSOA 20540	540				16/1050/63	—/—	
DSOA 20810	810				16/1575/63	—/—	
DSOA 21080	1080				16/2100/63	—/—	
DTCA 23240	3240	5	6976			—/—	
DTCA 24090	4090	6	6976			—/—	
DTNA 21800	1800	5	4928			—/—	
DTNA 22160	2160	6	4928			—/—	
DTTA 350320	3200	2	13085			—/—	
DTTA 350430	4300	3	13085			—/—	
DTTA 350640	6400	4	13085			—/—	
DTTA 350840	8400	5	13085			—/—	
DTTA 351010	10100	6	13085			—/—	
DTTA 351290	12900	8	13085			—/—	
DTTA 351680	16800	10	13085			—/—	
DTTA 371010	10100	7	13085			—/—	
DTTA 371290	12900	8	13085			—/—	
DTTA 371440	14400	10	13085			—/—	
DVAA 2810	810	6	2788		16/1571/63	—/—	
DVAA 2810	810				16/1571/63	—/—	
DVAS 2810	810	6	2788			—/—	
DYKA 22160	2160	3	8128			—/—	
DYKA 23240	3240	4	8128			—/—	
DYLA 26480	6480	8	8320			—/—	
DYLA 28100	8100	10	8320			—/—	
H1172-S2	172	2	2264			—/—	
H2172-A2	172	2	2264			Auto/Auto	NA
H2172-S2	172	2	2264			Auto/Auto	NA
H2258-A3	258	3	2264			—/—	
H2258-S3	258	3	2264			—/—	
H2344-A4	344	4	2264			—/—	
H2344-S4	344	4	2264			—/—	
H3133-A2	133	2	2420		15/1023/17	—/—	
H3171-A2	171	2	2420		10/984/34	—/—	
H3256-A3	256	3	2420		16/872/36	—/—	
H3342-A4	342	4	2420		16/872/48	—/—	
Travelstar 3GN 2.16G	2160	3	8128			—/—	
Travelstar 3GN 3.24G	3240	4	8128			—/—	
Travelstar 4GN 2.16G	2160	2	9280			—/—	
Travelstar 4GN 3.24G	3240	3	9280			—/—	
Travelstar 4GN 4.32G	4320	4	9280			—/—	
Travelstar 4GT 3.2G	3240	5	6976			—/—	
Travelstar 4GT 4.0G	4090	6	6976			—/—	
Travelstar 5GS 4.4G	4480	7	6976			—/—	
Travelstar 5GS 5.1G	5120	8	6976			—/—	
Travelstar 6GT 5.4G	5400	5	9280			—/—	
Travelstar 6GT 6.48G	6480	6	9280			—/—	
Travelstar 8GS 6.48G	6480	8	8320			—/—	
Travelstar 8GS 8.10G	8100	10	8320			—/—	
Ultrastar 18XP	18200	20				—/—	
Ultrastar 18XP	18200	20				—/—	
Ultrastar 18XP	18200	20				—/—	
Ultrastar 18XP	18200	20				—/—	
Ultrastar 2ES 2.1G	2160	3	8120			—/—	
Ultrastar 2ES 4.3G	4330	6	8120			—/—	
Ultrastar 2XP 4.51G	4512	9				—/—	
Ultrastar 2XP 9.1G	9100	18				—/—	
Ultrastar 9ES 4.5G	4500	5	8420			—/—	
Ultrastar 9ES 9.0G	9100	10	8420			—/—	
Ultrastar 9LP	9100	10				—/—	

Drive Model	Seek Time	Interface	Encode	Form Factor	cache kb	mtbf	RPM	Obsolete? ⇓
DSAS 3270	12	SCSI-2Fast		3.5 3H	96k	300k	4500	Y
DSAS 3360	12	SCSI-2Fast		3.5 3H	96k	300k	4500	Y
DSAS 3540	12	SCSI-2Fast		3.5 3H	128k	300k	4500	Y
DSAS 3720	12	SCSI-2Fast		3.5 3H	128k	300k	4500	Y
DSOA 20540	13	IDE AT		2.5 4H	128k		4000	Y
DSOA 20810	13	IDE AT		2.5 4H	128k		4000	Y
DSOA 21080	13	IDE AT		2.5 4H	128k		4000	Y
DTCA 23240	13	ATA-3	PRML	2.5 4H	512k		4000	
DTCA 24090	13	ATA-3	PRML	2.5 4H	512k		4000	
DTNA 21800	13	ATA-3	PRML	2.5 4H	128k		4000	Y
DTNA 22160	13	ATA-3	PRML	2.5 4H	128k		4000	
DTTA 350320	9.5	ATA-4	PRML	3H	512k		5400	Y
DTTA 350430	9.5	ATA-4	PRML	3H	512k		5400	Y
DTTA 350640	9.5	ATA-4	PRML	3H	512k		5400	
DTTA 350840	9.5	ATA-4	PRML	3H	512k		5400	
DTTA 351010	9.5	ATA-4	PRML	3H	512k		5400	
DTTA 351290	9.5	ATA-4	PRML	3H	512k		5400	
DTTA 351680	9.5	ATA-4	PRML	3H	512k		5400	
DTTA 371010	9.5	ATA-4	PRML	3H	512k		7200	
DTTA 371290	9.5	ATA-4	PRML	3H	512k		7200	
DTTA 371440	9.5	ATA-4	PRML	3H	512k		7200	
DVAA 2810	14	IDE AT		2.5 4H	32k	300k	3800	Y
DVAA 2810	14	IDE AT	1,7 RLL	2.5 4H	32k		3800	Y
DVAS 2810	14	SCSI-2Fast		2.5 4H	32k	300k	3800	Y
DYKA 22160	13	ATA-3	PRML	2.5 4H	128k	300k	4200	Y
DYKA 23240	13	ATA-3	PRML	2.5 4H	128k	300k	4200	
DYLA 26480	12	ATA-3	PRML	2.5 4H	512k	300k	4900	
DYLA 28100	12	ATA-3	PRML	2.5 4H	512k	300k	4900	
H1172-S2	14	SCSI		2.5 4H	32k	300k	3800	Y
H2172-A2	14	IDE AT		2.5 4H	32k	300k	3800	Y
H2172-S2	14	SCSI-2		2.5 3H	32k	300k	3800	Y
H2258-A3	14	IDE AT		2.5 4H	32k	300k	3800	Y
H2258-S3	14	SCSI		2.5 4H	32k	300k	3800	Y
H2344-A4	14	IDE AT		2.5 4H	32k	300k	3800	Y
H2344-S4	14	SCSI		2.5 4H	32k	300k	3800	Y
H3133-A2	14	IDE AT		3.5 3H	96k	250k	3600	Y
H3171-A2	14	IDE AT		3.5 3H	96k	250k	3600	Y
H3256-A3	14	IDE AT		3.5 3H	96k	250k	3600	Y
H3342-A4	14	IDE AT		3.5 3H	96k	250k	3600	Y
Travelstar 3GN 2.16G	13	ATA-3	PRML	2.5 4H	128k	300k	4200	Y
Travelstar 3GN 3.24G	13	ATA-3	PRML	2.5 4H	128k	300k	4200	
Travelstar 4GN 2.16G	12	ATA-4			512k		4200	Y
Travelstar 4GN 3.24G	12	ATA-4			512k		4200	
Travelstar 4GN 4.32G	12	ATA-4			512k		4200	
Travelstar 4GT 3.2G	13	ATA-3	PRML	2.5 4H	512k		4000	
Travelstar 4GT 4.0G	13	ATA-3	PRML	2.5 4H	512k		4000	
Travelstar 5GS 4.4G	12	ATA-3	PRML	2.5 4H	512k		4900	
Travelstar 5GS 5.1G	12	ATA-3	PRML	2.5 4H	512k		4900	
Travelstar 6GT 5.4G	12	ATA-4		4H	512k	300k	4200	
Travelstar 6GT 6.48G	12	ATA-4		4H	512k	300k	4200	
Travelstar 8GS 6.48G	12	ATA-3	PRML	2.5 4H	512k	300k	4900	
Travelstar 8GS 8.10G	12	ATA-3	PRML	2.5 4H	512k	300k	4900	
Ultrastar 18XP	6.5	SCA-2FstWD	RLL 16,17	3.5 3H	1000k		7200	
Ultrastar 18XP	6.5	SCA-2FstWD	RLL 16,17	3.5 3H	1000k		7200	
Ultrastar 18XP	6.5	SSA	RLL 16,17	3.5 3H	1000k		7200	
Ultrastar 18XP	6.5	FC-AL	RLL 16,17	3.5 3H	1000k		7200	
Ultrastar 2ES 2.1G	8.5	SCSI-3Ultra	PRML	3.5 3H	512k		5400	Y
Ultrastar 2ES 4.3G	8.5	SCSI-3Ultra	PRML	3.5 3H	512k		5400	
Ultrastar 2XP 4.51G	7.5	Ultra Wide		3.5 3H	512k		7200	
Ultrastar 2XP 9.1G	8.5	Ultra Wide		3.5 3H	512k		7200	
Ultrastar 9ES 4.5G	7.5	SCA-2FstW	Prml16,17	3.5 3H	512k		7200	
Ultrastar 9ES 9.0G	7.5	Ultra SCSI3	Prml16,17	3.5 3H	512k		7200	
Ultrastar 9LP	6.5	SCA-2FstWD	RLL 16,17	3.5 4H	1000k		7200	

Drive Model	Format Size MB	Head	Cyl	Sect/ Trac	Translate H/C/S	RWC/ WPC	Land Zone
Ultrastar 9LP	9100	10				—-/—-	
Ultrastar 9LP	9100	10				—-/—-	
Ultrastar 9LP	9100	10				—-/—-	
Ultrastar 9ZX	9100	12				—-/—-	
Ultrastar Ultra 2.16S	2160	3				—-/—-	
Ultrastar XP 2.25G	2250	4				—-/—-	
Ultrastar XP 4.51G	4510	8				—-/—-	
WD-12	10	4	306	17		296/296	
WD-2120	126	4	1248	50		—-/—-	
WD-240	42	2	1120	38		Na/Na	Auto
WD-240	43	2	1122	38		Na/Na	Auto
WD-240	42	2	1120	38		—-/—-	
WD-25	20	8	306	17		296/296	
WD-25A	20					—-/—-	
WD-25R	20					—-/—-	
WD-280	85	4	1120	38		—-/—-	
WD-3158	120	8	920	32		Na/Na	Auto
WD-3158(PS2/70)	120					—-/—-	
WD-3160	163	8	1021	39		Na/Na	Auto
WD-325	21	4	615	17		—-/—-	
WD-325K	20					—-/—-	
WD-325N(PS2/50)	21					—-/—-	
WD-325Q(PS2/30)	21					—-/—-	
WD-336P(PS2/30E)	31					—-/—-	
WD-336R(PS2/50Z)	31					—-/—-	
WD-380	81	4	1021	39		Na/Na	Auto
WD-380S(PS2/70)	81					—-/—-	
WD-387(PS2/70)	60	4	928	32		Na/Na	Auto
WD-387T(PS2/70)	60					—-/—-	
WD-L320(PS2/30E)	20					—-/—-	
WD-L330P(PS2/30E)	30					—-/—-	
WD-L330R(PS2/70)	30					—-/—-	
WD-L40	41	2	1038	39		Na/Na	Auto
WD-L40S(PS2/70)	41	2	1038	39		Na/Na	Auto
WDA-2120R	126	4	1243	50		—-/—-	
WDA-240	43	2	1122	38		Na/Na	Auto
WDA-260	63	2	1248	50		—-/—-	
WDA-280	87	4	1122	38		Na/Na	Auto
WDA-3160	81	4	1021	39		Na/Na	Auto
WDA-380	81	4	1021	39		Na/Na	Auto
WDA-L160	171	4	1923	44	8/966/44	—-/—-	
WDA-L40	41	2	1040	39		Na/Na	Auto
WDA-L42	42	2	1067	39		Na/Na	Auto
WDA-L80	85	2	1923	44		—-/—-	
WDS-240	43	2	1120	38		Na/Na	Auto
WDS-260	63	2	1248	50		—-/—-	
WDS-280	85	4	1120	38		—-/—-	
WDS-3100	104	2	1990	44		Na/Na	Auto
WDS-3160	163	8	1021	39		Na/Na	Auto
WDS-3168	160					—-/—-	
WDS-3200	209	4	1990	44		—-/—-	
WDS-380	81	4	1021	39		Na/Na	Auto
WDS-387	80					—-/—-	
WDS-L160	171	4	1923	44		—-/—-	
WDS-L40	41	2	1038	39		Na/Na	Auto
WDS-L42	42	2	1066	39		Na/Na	Auto
WDS-L80	85	2	1923	44		—-/—-	

IBM

06H3370	2250					—-/—-	
06H3372	2250					—-/—-	
06H5709	4510					—-/—-	
06H5710	5318					—-/—-	

Drive Model	Seek Time	Interface	Encode	Form Factor	cache kb	mtbf	RPM	Obsolete?
Ultrastar 9LP	6.5	SCA-2FstWD	RLL 16,17	3.5 4H	1000k		7200	
Ultrastar 9LP	6.5	SSA	RLL 16,17	3.5 4H	1000k		7200	
Ultrastar 9LP	6.5	FC-AL	RLL 16,17	3.5 4H	1000k		7200	
Ultrastar 9ZX	6.3	ULSCSI40FST	PRML		1mb		1000	
Ultrastar Ultra 2.16S	8.5	UltraSCSIW		3.5 3H	512k	800k	5400	Y
Ultrastar XP 2.25G	7.5	SCSI-2Fast		3.5 3H	512k	1000k	7200	Y
Ultrastar XP 4.51G	8	SCSI-2Fast		3.5HH	512k	1000k	7200	
WD-12		ST412/506	MFM	5.25FH		150k	3600	Y
WD-2120	16	IDE AT	RLL	2.5 4H		150k	3600	Y
WD-240	19	MCA		2.5 4H		150k	3600	Y
WD-240	19	MCA		2.5 4H		150k	3600	Y
WD-240	19	MCA		2.5 4H		150k	3600	Y
WD-25		ST412/506	MFM	5.25FH		150k	3600	Y
WD-25A		ST412/506	MFM	5.25FH		150k	3600	Y
WD-25R		ST412/506	MFM	5.25FH				Y
WD-280	17	MCA		2.5 4H		150k	3600	Y
WD-3158	23	MCA		3.5FH		45k		Y
WD-3158(PS2/70)		MCA		3.5HH				Y
WD-3160	16	MCA		3.5HH		110k		Y
WD-325	88	MCA		3.5HH				Y
WD-325K		ST412/506	MFM	3.5HH				Y
WD-325N(PS2/50)		MCA		3.5HH				Y
WD-325Q(PS2/30)		MCA		3.5HH				Y
WD-336P(PS2/30E)		MCA		3.5HH				Y
WD-336R(PS2/50Z)		MCA		3.5HH				Y
WD-380	16	MCA		3.5HH		110k		Y
WD-380S(PS2/70)		MCA		3.5HH				Y
WD-387(PS2/70)	23	MCA		3.5HH		45k		Y
WD-387T(PS2/70)		MCA		3.5HH				Y
WD-L320(PS2/30E)		MCA		3.5HH				Y
WD-L330P(PS2/30E)		MCA		3.5HH				Y
WD-L330R(PS2/70)		MCA		3.5HH				Y
WD-L40	17	MCA		3.5HH		90k		Y
WD-L40S(PS2/70)	17	MCA		3.5HH		90k		Y
WDA-2120R	16	IDE AT		2.5 3H		25k	3600	Y
WDA-240	19	IDE AT		2.5 4H		150k	3600	Y
WDA-260	16	IDE AT		2.5 4H		150k	3600	Y
WDA-280	19	IDE AT		2.5 4H		150k	3600	Y
WDA-3160	16	IDE AT		3.5HH		110k		Y
WDA-380	16	IDE AT		3.5HH		110k		Y
WDA-L160	16	IDE AT		3.5 4H		150k	3600	Y
WDA-L40	17	IDE AT	2,7 RLL	3.5 3H		90k		Y
WDA-L42	17	IDE AT	2,7 RLL	3.5 3H		90k		Y
WDA-L80	16	SCSI-2		3.5 4H		150k	3600	Y
WDS-240	19	SCSI		2.5 4H		150k	3600	Y
WDS-260	16	SCSI-2		2.5 4H		150k	3600	Y
WDS-280	17	SCSI		2.5 4H		150k	3600	Y
WDS-3100	12	SCSI-2		3.5 4H	32k	150k	4320	Y
WDS-3160	16	SCSI-2		3.5HH		110k		Y
WDS-3168		SCSI		3.5HH				Y
WDS-3200	12	SCSI-2		3.5 4H	32k	150k	4320	Y
WDS-380	16	SCSI-2		3.5HH		110k		Y
WDS-387		SCSI		3.5HH				Y
WDS-L160	16	SCSI-2		3.5 4H		150k	3600	Y
WDS-L40	17	SCSI-2		3.5FH		90k		Y
WDS-L42	17	SCSI-2		3.5 3H		80k		Y
WDS-L80	16	SCSI-2		3.5 4H		150k	3600	Y

IBM

Drive Model	Seek Time	Interface	Encode	Form Factor	cache kb	mtbf	RPM	Obsolete?
06H3370	7.5	SCSI-2Fast		3.5 3H	512k	1000k	7200	Y
06H3372	7.5	SCSI-2FstW		3.5 3H	512k	1000k	7200	Y
06H5709	8	SCSI-2FstW		3.5HH	512k		7200	
06H5710	8	SCSI-2FstW		3.5HH	512k	1000k	5400	

Drive Model	Format Size MB	Head	Cyl	Sect/ Trac	Translate H/C/S	RWC/ WPC	Land Zone
06H6111	1080	2				—/—	
06H6740	2255					—/—	
06H6741	4510					—/—	
06H6742	4512					—/—	
06H6749	5318					—/—	
06H6750	5318					—/—	
06H7141	540					—/—	
06H7142	540					—/—	
06H8558	540					—/—	
06H8724	1700	2				—/—	
06H8891	1080					—/—	
07H0386	125	3				—/—	
07H0387	2250	5				—/—	
07H0834	4510	10				—/—	
07H1124	2160	3				—/—	
07H1128	2160	3				—/—	
07H2689	9100	18				—/—	
32G3796	2000					—/—	
32G4194	245					—/—	
32G4195	340					—/—	
32G4196	527					—/—	
32G4198	1000					—/—	
32G4199	105					—/—	
32G4336	2000					—/—	
32G4338	2880					—/—	
3513364	364					—/—	
3513527	527					—/—	
70G7164	1000					—/—	
70G7424	170	2	2233			—/—	
70G8480	170	2	2111			—/—	
70G8481	340	4	2111			—/—	
70G8486	527					—/—	
70G8487	270					—/—	
70G8488	364					—/—	
70G8491	540	7	2466			—/—	
70G8492	1052	6				—/—	
70G8493	2014	16				—/—	
70G8494	2014	16				—/—	
70G8495	40					—/—	
70G8499	1440					—/—	
70G8500	1440					—/—	
70G8511	728	4	3875			—/—	
70G8512	1000	5				—/—	
70G8847	270					—/—	
70G8848	364					—/—	
70G8849	527					—/—	
70G8850	728					—/—	
70G9743	1000					—/—	
71G0666	1000	5				—/—	
71G6550	170	2	2111			—/—	
76H2687	4512	9				—/—	
76H2689	9100	18				—/—	
76H7246	4200	6				—/—	
82G5926	270					—/—	
82G5927	364					—/—	
82G5928	540					—/—	
82G5929	1000	5				—/—	
82G5930	270					—/—	
82G5931	364					—/—	
82G5932	540					—/—	
82G5933	728					—/—	
82G6106	527					—/—	
92F0428	1052	6				—/—	

Drive Model	Seek Time	Interface	Encode	Form Factor	cache kb	mtbf	RPM	Obsolete? ⇓
06H6111	10.5	ATA-2		3.5 3H	512k	500k	5400	Y
06H6740	7.5	SCSI-2Diff		3H		1000k	7200	Y
06H6741	8	SCSI-2Fast		3.5HH	512k		7200	
06H6742	8	SCSI-2Diff		3.5HH		1000k	7200	
06H6749	8	SCSI-2Diff		3.5HH		1000k	5400	
06H6750	8	SCSI-2Diff		3.5HH		1000k	5400	
06H7141	12	ATA-2		3.5 3H	128k	350k	4500	Y
06H7142	12	ATA-2		3.5 3H	128k	350k	4500	Y
06H8558	12	SCSI-2Fast		3.5 3H	128k	300k	4500	Y
06H8724	12	ATA-2		3.5 3H	128k	350k	4500	Y
06H8891	10.5	SCSI-2Fast		3.5 3H	512k	500k	5400	Y
07H0386	8.5	SCSI-2FstW		3.5 3H		800k	7200	Y
07H0387	8.5	SCSI-2FstW		3.5 3H		800k	7200	Y
07H0834	8.5	SCSI-2FstW		3.5HH		800k	7200	
07H1124	8.5	SCSI-2Fast		3.5 3H	512k	800k	5400	Y
07H1128	8.5	Ultra SCSIW		3.5 3H	512k	800k	5400	Y
07H2689	8.5	Ultra Wide		3.5 3H	512k		7200	
32G3796	9.5	SCSI-2FstW		3.5HH	512k	750k	5400	Y
32G4194	15	IDE AT		3.5 3H				Y
32G4195	14	IDE AT		3.5 3H				Y
32G4196	9	IDE AT		3.5 3H				Y
32G4198	8.6	SCSI-2Fast		3.5 3H	512k	800k	5400	Y
32G4199	15	PCMCIA		FH				Y
32G4336	9.5	SCSI-2Fast		3.5HH	512k	750k	5400	Y
32G4338	94	AT BUS		3.5 3H				Y
3513364	12	PCMCIA		HH		200k		Y
3513527	12	PCMCIA		HH		200k		Y
70G7164	8.6	SCSI-2Fast		3.5 3H	512k	800k	5400	Y
70G7424	14	IDE AT		3.5 3H	96k	250k	3322	Y
70G8480	13	SCSI-2Fast		3.5 3H	64k	250k	4011	Y
70G8481	13	SCSI-2		3.5 3H	64k	250k	4011	Y
70G8486	12	IDE AT		3.5 3H	96k	300k	4500	Y
70G8487	12	IDE AT		3.5 3H	96k	300k	4500	Y
70G8488	12	IDE AT		3.5 3H	96k	300k	4500	Y
70G8491	8.5	SCSI-2Fast		3.5 3H	256k	300k	6300	Y
70G8492	8.6	SCSI-2Fast		3.5 3H	512k	800k	5400	Y
70G8493	9.5	SCSI-2Fast		3.5HH	512k	750k	5400	Y
70G8494	9.5	SCSI-2FstW		3.5HH	512k	750k	5400	Y
70G8495	18	PCMCIA		FH				Y
70G8499	94	IDE AT		3.5 3H				Y
70G8500	94	IDE AT		5.25HH				Y
70G8511	12	IDE AT		3.5 3H	96k	300k	4500	Y
70G8512	8.5	IDE AT		3.5 3H	512k	800k	5400	Y
70G8847	12	IDE AT		3.5 3H	96k	300k	4500	Y
70G8848	12	IDE AT		3.5 3H	96k	300k	4500	Y
70G8849	12	IDE AT		3.5 3H	96k	300k	4500	Y
70G8850	12	IDE AT		3.5 3H	96k	300k	4500	Y
70G9743	8	SCSI-2FstW		3.5 3H	512k	800k	5400	Y
71G0666	8.5	IDE AT		3.5 3H	512k	800k	5400	Y
71G6550	13	SCSI-2Fast		3.5 3H	64k	250k	4011	Y
76H2687	7.5	Ultra Wide		3.5 3H	512k		7200	
76H2689	8.5	Ultra Wide		3.5 3H	512k		7200	
76H7246	9.5	ATA-3 Fast		3.5 3H	128k		5400	
82G5926	12	IDE AT		3.5 3H	96k	300k	4500	Y
82G5927	12	IDE AT		3.5 3H	96k	300k	4500	Y
82G5928	12	ATA-2		3.5 3H	128k	350k	4500	Y
82G5929	8.5	IDE AT		3.5 3H	512k	800k	5400	Y
82G5930	12	SCSI-2Fast		3.5 3H	96k	300k	4500	Y
82G5931	12	SCSI-2Fast		3.5 3H	96k	300k	4500	Y
82G5932	12	SCSI-2Fast		3.5 3H	96k	300k	4500	Y
82G5933	12	SCSI-2Fast		3.5 3H	96k	300k	4500	Y
82G6106	12	IDE AT		3.5 3H	96k	300k	4500	Y
92F0428	8.6	SCSI-2Fast		3.5 3H	512k	800k	5400	Y

Drive Model	Format Size MB	Head	Cyl	Sect/ Trac	Translate H/C/S	RWC/ WPC	Land Zone
92F0440	2014	16				—/—	
94G2413	1052	6				—/—	
94G2439	270					—/—	
94G2440	364					—/—	
94G2441	540					—/—	
94G2442	728					—/—	
94G2644	270					—/—	
94G2645	364					—/—	
94G2646	540					—/—	
94G2647	728					—/—	
94G2649	1120					—/—	
94G2650	2250					—/—	
94G2651	4510					—/—	
94G3052	1120					—/—	
94G3054	2250					—/—	
94G3055	2250					—/—	
94G3056	2255					—/—	
94G3057	4510					—/—	
94G3059	5318					—/—	
94G3183	1080	2				—/—	
94G3184	1080	2				—/—	
94G3186	1080	2				—/—	
94G3187	1080	2				—/—	
94G3192	2250					—/—	
94G3193	2250					—/—	
94G3195	4510					—/—	
94G3196	4510					—/—	
94G3197	5318					—/—	
94G3198	4510					—/—	
94G3199	2255					—/—	
94G3200	4512					—/—	
94G3201	5318					—/—	
94G3203	2255					—/—	
94G3204	4512					—/—	
94G3205	5318					—/—	
94G3787	5318					—/—	
94G3794	5318					—/—	
94G4196	527					—/—	

IMI

Drive Model	Format Size MB	Head	Cyl	Sect/ Trac	Translate H/C/S	RWC/ WPC	Land Zone
5006	5	2	306	17		307/214	
5007	5	2	306	17		—/—	
5012	10	4	306	17		307/214	
5018	15	6	306	17		307/214	
5021H	15			17		—/—	
7720	20			17		—/—	
7740	40			17		—/—	

INTEGRAL PERIPHERALS

Drive Model	Format Size MB	Head	Cyl	Sect/ Trac	Translate H/C/S	RWC/ WPC	Land Zone
105 (Viper)	105	4				—/—	
170 (Viper)	171	4				—/—	
1820 (Mustang)	21	2	615			Na/Na	Auto
1842 (Stingray)	42	3				Na/Na	Auto
1862	64	3		V		Na/Na	Auto
2100	1000	6			16/1900/63	Na/Na	Auto
260 (Viper)	262	4				—/—	
340 (Viper)	341	4				—/—	

IOMEGA CORPORATION

Drive Model	Format Size MB	Head	Cyl	Sect/ Trac	Translate H/C/S	RWC/ WPC	Land Zone
Jaz 1GB Ext PC/Mc	1070					—/—	
Jaz 1GB Int PC	1070					—/—	
Jaz 2GB Ext PCMac	2000					—/—	

Drive Model	Seek Time	Interface	Encode	Form Factor	cache kb	mtbf	RPM	Obsolete?
92F0440	9.5	SCSI-2Fast		3.5HH	512k	750k	5400	Y
94G2413	8.6	SCSI-2Fast		3.5 3H	512k	800k	5400	Y
94G2439	12	SCSI-2Fast		3.5 3H	96k	300k	4500	Y
94G2440	12	SCSI-2Fast		3.5 3H	96k	300k	4500	Y
94G2441	12	SCSI-2Fast		3.5 3H	96k	300k	4500	Y
94G2442	12	SCSI-2Fast		3.5 3H	96k	300k	4500	Y
94G2644	12	SCSI-2Fast		3.5 3H	96k	300k	4500	Y
94G2645	12	SCSI-2Fast		3.5 3H	96k	300k	4500	Y
94G2646	12	SCSI-2Fast		3.5 3H	96k	300k	4500	Y
94G2647	12	SCSI-2Fast		3.5 3H	96k	300k	4500	Y
94G2649	6	SCSI-2FstW		3.5 3H	512k	1000k	7200	Y
94G2650	7	SCSI-2FstW		3.5 3H	512k	1000k	7200	Y
94G2651	8	SCSI-2FstW		3.5HH	512k	1000k	7200	
94G3052	6.9	SCSI-2FstW		3.5 3H	512k	1000k	7200	Y
94G3054	7.5	SCSI-2Fast		3.5 3H	512k	1000k	7200	Y
94G3055	7.5	SCSI-2FstW		3.5 3H	512k	1000k	7200	Y
94G3056	7.5	SCSI-2FstW		3.5 3H	512k	1000k	7200	Y
94G3057	8	SCSI-2FstW		3.5HH	512k	1000k	7200	
94G3059	8	SCSI-2FstW		3.5HH	512k	1000k	5400	
94G3183	10.5	ATA-2		3.5 3H	512k	500k	5400	Y
94G3184	10.5	SCSI-2Fast		3.5 3H	512k	500k	5400	Y
94G3186	10.5	ATA-2		3.5 3H	512k	500k	5400	Y
94G3187	10.5	SCSI-2Fast		3.5 3H	512k	500k	5400	Y
94G3192	7.5	SCSI-2Fast		3.5 3H	512k	1000k	7200	Y
94G3193	7.5	SCSI-2FstW		3.5 3H	512k	1000k	7200	Y
94G3195	8	SCSI-2FstW		3.5HH	512k	1000k	7200	
94G3196	8	SCSI-2Fast		3.5HH	512k	1000k	7200	
94G3197	8	SCSI-2FstW		3.5HH	512k	1000k	5400	
94G3198	8	SCSI-2Fast		3.5HH	512k	1000k	7200	
94G3199	7.5	SCSI-2Diff		3H	512k	1000k	7200	Y
94G3200	8	SCSI-2Diff		3.5HH	512k	1000k	7200	
94G3201	8	SCSI-2Diff		3.5HH	512k	1000k	5400	
94G3203	7.5	SCSI-2Diff		3H	512k	1000k	7200	Y
94G3204	8	SCSI-2Diff		3.5HH	512k	1000k	7200	
94G3205	8	SCSI-2Diff		3.5HH	512k	1000k	5400	
94G3787	8	SCSI-2Fast		3.5HH	512k	1000k	5400	
94G3794	8	SCSI-2Fast		3.5HH	512k	1000k	5400	
94G4196	8	IDE AT		3.5 3H	512k	1000k	5400	Y

IMI

Drive Model	Seek Time	Interface	Encode	Form Factor	cache kb	mtbf	RPM	Obsolete?
5006	27	ST412/506	MFM					Y
5007	85	ST412/506	MFM	5.25FH				Y
5012	27	ST412/506	MFM					Y
5018	27	ST412/506	MFM					Y
5021H	85	ST412/506	MFM	5.25FH				Y
7720	85	ST412/506	MFM					Y
7740	85	ST412/506	MFM					Y

INTEGRAL PERIPHERALS

Drive Model	Seek Time	Interface	Encode	Form Factor	cache kb	mtbf	RPM	Obsolete?
105 (Viper)	15	PCMCIA-ATA	1,7 RLL	1.8 IN	32k	250k	4500	Y
170 (Viper)	12	PCMCIA-ATA	1,7 RLL	1.8 IN	32k	250k	4500	Y
1820 (Mustang)	18	IDE AT	1,7 RLL	1.8		100k		Y
1842 (Stingray)	18	IDE AT	1,7 RLL	1.8		100k		Y
1862	18	IDE AT	1,7 RLL			100k		Y
2100	12	ATA-2	1,7 RLL	2.5 4H	128k	250k	4200	Y
260 (Viper)	12	PCMCIA-ATA	1,7 PRML	1.8 IN	32k	250k	4500	Y
340 (Viper)	12	PCMCIA-ATA	1,7 PRML	1.8 IN	32k	250k	4500	Y

IOMEGA CORPORATION

Drive Model	Seek Time	Interface	Encode	Form Factor	cache kb	mtbf	RPM	Obsolete?
Jaz 1GB Ext PC/Mc	10-12	SCSI-2		3.5FH	256k	250k		
Jaz 1GB Int PC	10-12	SCSI-2		3.5FH	256k	250k	5394	
Jaz 2GB Ext PCMac	15.5	Ultra SCSI		3.5FH	512k			

Drive Model	Format Size MB	Head	Cyl	Sect/ Trac	Translate H/C/S	RWC/ WPC	Land Zone
Zip 100	100					—/—	
Zip 100	100					—/—	

JCT (SEE MAXCARD)

Drive Model	Format Size MB	Head	Cyl	Sect/ Trac	Translate H/C/S	RWC/ WPC	Land Zone
100	5			17		—/—	
1000	5			17		—/—	
1005	7			17		—/—	
1010	14			17		—/—	
105	5	2	306	17		—/—	
110	14			17		—/—	
120	20			17		—/—	

JTS CORPORATION

Drive Model	Format Size MB	Head	Cyl	Sect/ Trac	Translate H/C/S	RWC/ WPC	Land Zone
C1700-2AF	1700	4	3312			—/—	
C2000-2AF	2000	4	3882			—/—	
C2500-3AF	2500	6	4970			—/—	
C3000-3AF	3000	6	5824			—/—	
Champ Family	1000	4	5050			—/—	
Champ Family	1300	4	5050			—/—	
Champ Family	1700	6	5050			—/—	
Champ Family	2000	6	5050			—/—	
N1080-2AR	1080	4	4032			—/—	
N1440-3AR	1440	6	4032			—/—	
N1620-3AR	1620	6	4032			—/—	
N2160-3AR	2160	6	4435			—/—	

JVC COMPANIES OF AMERICA

Drive Model	Format Size MB	Head	Cyl	Sect/ Trac	Translate H/C/S	RWC/ WPC	Land Zone
JD-3842HA	21	2	436	48		—/—	
JD-3848HA	43	4	436	48		—/—	
JD-E2042M	42	2	973	43		Na/Na	Auto
JD-E2085M	85	4	973	43		Na/Na	Auto
JD-E2825P(A)	21	2	581	36		—/—	Auto
JD-E2825P(S)	21	2	581	36		—/—	Auto
JD-E2825P(X)	21	2	581	36		—/—	Auto
JD-E2850P(A)	42	3	791	35		—/—	Auto
JD-E2850P(S)	42	3	791	35		—/—	Auto
JD-E2850P(X)	42	3	791	35		—/—	Auto
JD-E3824TA	21	2	436	48		—/—	
JD-E3848HA	42	4	436	48		—/—	
JD-E3848P(A)	42	2	862	48		—/—	Auto
JD-E3848P(S)	42	2	862	48		—/—	Auto
JD-E3848P(X)	42	2	862	48		—/—	Auto
JD-E3896P(A)	84	4	862	48		—/—	Auto
JD-E3896P(S)	84	4	862	48		—/—	Auto
JD-E3896P(X)	84	4	862	48		—/—	Auto
JD-E3896V(A)	84	4	862	48		Na/Na	Auto
JD-E3896V(S)	84	4	862	48		Na/Na	Auto
JD-E3896V(X)	84	4	862	48		Na/Na	Auto
JD-F2042M	42	2	973	43		Na/Na	Auto

KALOK CORPORATION

Drive Model	Format Size MB	Head	Cyl	Sect/ Trac	Translate H/C/S	RWC/ WPC	Land Zone
KL1000	105	6	978	35		—/—	Auto
KL3100	105	6	820	48/35	6/979/35	Na/Na	Auto
KL3120	121	6	820	55/40	6/981/40	Na/Na	Auto
KL320	21	4	615	17		616/300	
KL330	33	4	615	26		617/617	
KL332	40	4	615			—/—	
KL340	43	6	820	17		—/—	
KL341	43	4	676	31		—/—	Auto
KL342	42	4	676	31		—/—	
KL343	43	4	676	31		645/645	Auto
KL360	66	6	820	26		—/—	

Drive Model	Seek Time	Interface	Encode	Form Factor	cache kb	mtbf	RPM	Obsolete?
Zip 100	20-29	IDE		3.5 3H	16k	100k	2941	
Zip 100	20-29	SCSI			16k	100k	2941	

JCT (SEE MAXCARD)

Drive Model	Seek Time	Interface	Encode	Form Factor	cache kb	mtbf	RPM	Obsolete?
100	110	ST412/506	MFM	5.25HH				Y
1000	110	Commodore	MFM	5.25HH				Y
1005	110	Commodore	MFM	5.25HH				Y
1010	130	Commodore	MFM	5.25HH				Y
105	110	ST412/506	MFM	5.25HH				Y
110	130	ST412/506	MFM	5.25HH				Y
120	100	ST412/506	MFM	5.25HH				Y

JTS CORPORATION

Drive Model	Seek Time	Interface	Encode	Form Factor	cache kb	mtbf	RPM	Obsolete?
C1700-2AF	<12	EIDE/ATA-3		3.5 3H	256k	500k	5400	Y
C2000-2AF	<12	EIDE/ATA-3		3.5 3H	256k	500k	5400	Y
C2500-3AF	<12	EIDE/ATA-3		3.5 3H	256k	500k	5400	
C3000-3AF	<12	EIDE/ATA-3		3.5 3H	256k	500k	5400	
Champ Family	<14	EIDE	1,7 RLL	3.5 4H	128k	500k	4500	Y
Champ Family	<14	EIDE	1,7 RLL	3.5 4H	128k	500k	4500	Y
Champ Family	<14	EIDE	1,7 RLL	3.5 4H	128k	500k	4500	Y
Champ Family	<14	EIDE	1,7 RLL	3.5 4H	128k	500k	4500	
N1080-2AR	<14	EIDE/ATA-3		3.0	128k	500k	4103	Y
N1440-3AR	<14	EIDE/ATA-3		3.0	128k	500k	4103	Y
N1620-3AR	<14	EIDE/ATA-3		3.0	128k	500k	4103	Y
N2160-3AR	<14	EIDE/ATA-3		3.0	128k	500k	4103	

JVC COMPANIES OF AMERICA

Drive Model	Seek Time	Interface	Encode	Form Factor	cache kb	mtbf	RPM	Obsolete?
JD-3842HA	28		2,7 RLL	3.5 3H		20k		Y
JD-3848HA	29		2,7 RLL	3.5 3H		20k		Y
JD-E2042M	16	IDE AT	1,7 RLL	2.5 4H	32k	130k	3118	Y
JD-E2085M	16	IDE AT	1,7 RLL	2.5 4H	32k	130k	3118	Y
JD-E2825P(A)	25	IDE AT	2,7 RLL	3.5 4H		30k	3109	Y
JD-E2825P(S)	25	SCSI	2,7 RLL	3.5 4H		30k	3109	Y
JD-E2825P(X)	25	IDE XT	2,7 RLL	3.5 4H		30k	3109	Y
JD-E2850P(A)	25	IDE AT	2,7 RLL	3.5 4H	32k	40k	3109	Y
JD-E2850P(S)	25	SCSI	2,7 RLL	3.5 4H	32k	40k	3109	Y
JD-E2850P(X)	25	IDE XT	2,7 RLL	3.5 4H	32k	40k	3109	Y
JD-E3824TA	28		2,7 RLL	3.5 3H		20k		Y
JD-E3848HA	29		2,7 RLL	3.5 3H		20k		Y
JD-E3848P(A)	25	IDE AT	2,7 RLL	3.5 4H		30k	2332	Y
JD-E3848P(S)	25	SCSI	2,7 RLL	3.5 4H		30k	2332	Y
JD-E3848P(X)	25	IDE XT	2,7 RLL	3.5 4H		30k	2332	Y
JD-E3896P(A)	25	IDE AT	2,7 RLL	3.5 4H		30k	3109	Y
JD-E3896P(S)	25	SCSI	2,7 RLL	3.5 4H		30k	3109	Y
JD-E3896P(X)	25	IDE XT	2,7 RLL	3.5 4H		30k	3109	Y
JD-E3896V(A)	25	IDE AT	2,7 RLL	3.5 3H		30k		Y
JD-E3896V(S)	25	SCSI	2,7 RLL	3.5 3H		30k		Y
JD-E3896V(X)	25	IDE XT	2,7 RLL	3.5 3H		30k		Y
JD-F2042M	16	IDE AT	1,7 RLL	2.5 4H	32k	130k	3118	Y

KALOK CORPORATION

Drive Model	Seek Time	Interface	Encode	Form Factor	cache kb	mtbf	RPM	Obsolete?
KL1000	25	IDE AT	2,7 RLL	3.5HH	32k	50k	3662	Y
KL3100	19	IDE AT	2,7 RLL	3.5HH		100k	3662	Y
KL3120	19	IDE AT	2,7 RLL	3.5HH		100k	3663	Y
KL320	40	ST412/506	MFM	3.5HH		43.5	3600	Y
KL330	40	ST412/506	2,7 RLL	3.5HH		43.5	3600	Y
KL332	48	MCA	2,7 RLL	3.5HH				Y
KL340	25	ST412/506	MFM	3.5HH		50		Y
KL341	33	SCSI	2,7 RLL	3.5HH	8k	40k	3375	Y
KL342	30	MCA	2,7 RLL	3.5HH		40k		Y
KL343	28	IDE AT	2,7 RLL	3.5HH	8k	100k	3375	Y
KL360	25	ST412/506	2,7 RLL	3.5HH		50k		Y

Drive Model	Format Size MB	Head	Cyl	Sect/ Trac	Translate H/C/S	RWC/ WPC	Land Zone
KL381	85	6	820			—/—	
KL383	84	6	815	34	6/815/33	Na/Na	
P3250	251	4	2048	80	16/961/32	Na/Na	Auto
P3360	362	4	791	56	16/791/56	Na/Na	Auto
P3540	540	4	1024	63		Na/Na	Auto
P5-125(A)	125	2	2048			Na/Na	Auto
P5-125(S)	125	2	2048			Na/Na	Auto
P5-250(A)	251	4	2048			Na/Na	Auto
P5-250(S)	251	4	2048			Na/Na	Auto

KYOCERA ELECTRONICS, INC.

Drive Model	Format Size MB	Head	Cyl	Sect/ Trac	Translate H/C/S	RWC/ WPC	Land Zone
KC20A	21	4	615	17		—/—	
KC20B	21	4	615	17		—/—	
KC30A	33	4	615	26		—/—	
KC30B	33	4	615	26		—/—	
KC40GA	40	2	1075	17	4/577	33/—	Auto
KC80C	87	8	787	28		Na/Na	
KC80GA	78	4	1069	36	8/577/33	Na/Na	Auto

LANSTOR

Drive Model	Format Size MB	Head	Cyl	Sect/ Trac	Translate H/C/S	RWC/ WPC	Land Zone
LAN-115		15	918	17		—/None	1023
LAN-140		8	1024	34		—/None	1023
LAN-180		8	1024	26		—/None	1023
LAN-64		8	1024	17		—/None	1023

LAPINE

Drive Model	Format Size MB	Head	Cyl	Sect/ Trac	Translate H/C/S	RWC/ WPC	Land Zone
LT10	10	2	615	17		616/—	
LT100 (not verified)	10					—/—	
LT20	20	4	615	17		616/—	
LT200	20	4	614	17		615/—	
LT2000	20	4	614	17		615/—	
LT300	32	4	614	17		615/—	
LT3065	10	4	306	17		306/128	
LT3512	10	4	306	17		306/128	
LT3522	10	4	306	17		307/—	
LT3532	32	4	614	26		—/615	
LT4000 (not verified)	40					—/—	
TITAN 20	21	4	615	17		—/—	615
TITAN 30	21	4	615			—/—	

MAXTOR CORPORATION

Drive Model	Format Size MB	Head	Cyl	Sect/ Trac	Translate H/C/S	RWC/ WPC	Land Zone
250837	837	5		66-132		—/—	
250840	840	5		43-67		—/—	
25084A	80	2		43-67	16/569/18	Na/Na	569
251005	1005	6		66-132		—/—	
251010	1010	6		66-132		—/—	
25128A	128	4	1092	NA	14/1024/17	Na/Na	Auto
251340	1340	8		66-132		—/—	
251350	1350	8		66-132	16/2616/63	—/—	
25252A	252	6			16/569/54	—/—	
25252S	251	6		67		Na/Na	Auto
2585A	85	4	1092	NA	10/981/17	Na/Na	Auto
2585S (not made)	85	4	1092	V		Na/Na	Auto
3053	44	5	1024	17		1024/512	Auto
3085	68	7	1170	17		1170/512	Auto
3130E	112	5	1250	36		1251/512	Auto
3130S	112	5	1255	35		1256/512	Auto
3180E	150	7	1250	35		1251/512	Auto
3180S	153	7	1255	36		1256/512	Auto
3380	338	15	1224	NA		Na/Na	Auto
7040A	41	2	1155	36	5/981/17	Na/Na	Auto
7040S	42	2	1155	36		Na/Na	Auto

Drive Model	Seek Time	Interface	Encode	Form Factor	cache kb	mtbf	RPM	Obsolete?
KL381	25	SCSI	2,7 RLL	3.5HH		50k		Y
KL383	25	IDE AT	2,7 RLL	3.5HH		50k		Y
P3250	16.5	IDE AT	1,7 RLL	3.5 4H	128k	250k	3600	Y
P3360	16.5	IDE AT	1,7 RLL	3.5 4H	128k	250k	3600	Y
P3540	11/16.5	IDE AT	1,7 RLL	3.5 4H	128k	250k	4200	Y
P5-125(A)	17	IDE AT	1,7 RLL			100k		Y
P5-125(S)	17	SCSI-2	1,7 RLL			100k		Y
P5-250(A)	17	IDE AT	1,7 RLL			100k		Y
P5-250(S)	17	SCSI-2	1,7 RLL			100k		Y

KYOCERA ELECTRONICS, INC.

Drive Model	Seek Time	Interface	Encode	Form Factor	cache kb	mtbf	RPM	Obsolete?
KC20A	65	ST412/506	MFM	3.5HH		40k		Y
KC20B	62	ST412/506	MFM	3.5HH		40k		Y
KC30A	65	ST412/506	2,7 RLL	3.5HH		40k		Y
KC30B	62	ST412/506	2,7 RLL	3.5HH		40k		Y
KC40GA	28	IDE AT	2,7 RLL	3.5HH		40k		Y
KC80C	28	SCSI	2,7 RLL	3.5HH		28k		Y
KC80GA	23	IDE AT	2,7 RLL	3.5HH		28k		Y

LANSTOR

Drive Model	Seek Time	Interface	Encode	Form Factor	cache kb	mtbf	RPM	Obsolete?
LAN-115								Y
LAN-140								Y
LAN-180								Y
LAN-64								Y

LAPINE

Drive Model	Seek Time	Interface	Encode	Form Factor	cache kb	mtbf	RPM	Obsolete?
LT10	27	ST412/506	MFM	3.5HH				Y
LT100 (not verified)	85	ST412/506		3.5HH				Y
LT20		ST412/506	MFM	3.5HH				Y
LT200	65	ST412/506	MFM	3.5HH				Y
LT2000		ST412/506	MFM	3.5HH				Y
LT300		ST412/506	2,7 RLL	3.5HH				Y
LT3065	65	ST412/506	2,7 RLL	3.5HH				Y
LT3512	65	ST412/506	2,7 RLL	3.5HH				Y
LT3522	27	ST412/506	MFM	3.5HH				Y
LT3532	65	ST412/506	2,7 RLL	3.5HH				Y
LT4000 (not verified)	27	SCSI		3.5HH				Y
TITAN 20		ST412/506	MFM	3.5HH				Y
TITAN 30			RLL?	3.5HH				Y

MAXTOR CORPORATION

Drive Model	Seek Time	Interface	Encode	Form Factor	cache kb	mtbf	RPM	Obsolete?
250837	14	IDE AT	1,7 RLL	2.5 4H	64k	300k	4464	Y
250840	12	IDE AT	1,7 RLL	2.5 4H	128k	350k	4247	Y
25084A	12	IDE AT	1,7 RLL	2.5 4H	128k	350k	4247	Y
251005	14	IDE AT	1,7 RLL	2.5 4H	64k	300k	4464	Y
251010	14	IDE AT	1,7 RLL	2.5 4H	64k	300k	4464	Y
25128A	14	IDE AT	1,7 RLL	2.5 4H		250k	3600	Y
251340	14	IDE AT	1,7 RLL	2.5 4H	64k	300k	4464	Y
251350	13	IDE AT	1,7 RLL	2.5 4H	64k	300k	4464	Y
25252A	12	IDE AT	1,7 RLL	2.5 4H	64k	350k	4247	Y
25252S	12	SCSI	1,7 RLL	2.5 4H	128k	350k	4247	Y
2585A	14	IDE AT	1,7 RLL	2.5 4H		250k	3600	Y
2585S (not made)	15	SCSI	1,7 RLL	2.5 4H		150k	3600	Y
3053	25	ST412/506	MFM	5.25HH		30k	3600	Y
3085	22	ST412/506	MFM	5.25HH		40k	3600	Y
3130E	17	ESDI	2,7 RLL	5.25HH		35k	3600	Y
3130S	17	SCSI	2,7 RLL	5.25HH		35k	3600	Y
3180E	17	ESDI	2,7 RLL	5.25HH		35k	3600	Y
3180S	17	SCSI	2,7 RLL	5.25HH		35k	3600	Y
3380	27	SCSI	RLL	5.25FH		20k	3600	Y
7040A	17	IDE AT	1,7 RLL	3.5 3H	32k	150k	3703	Y
7040S	17	SCSI	1,7 RLL	3.5 3H	32k	150k	3600	Y

Drive Model	Format Size MB	Head	Cyl	Sect/ Trac	Translate H/C/S	RWC/ WPC	Land Zone
7060A	65	2	1498	NA	16/467/17	Na/Na	Auto
7060S	60	2	1498	42		Na/Na	Auto
7080A	85	4	1166	36	10/981/17	Na/Na	Auto
7080S	85	4	1166	36		Na/Na	Auto
71000A	1002	3			16/1946/63	—/—	
71050A	1000	5		77-124	16/2045/63	—/—	
71050S	1000	5		77-124		—/—	
71084A	1084	4	4136	91-155	16/2105/63	Na/Na	Auto
71084AP	1084	4	4136	91-155	16/2105/63	Na/Na	Auto
7120A	125	4	1498	NA	16/936/17	Na/Na	Auto
7120S	125	4	1498	42		Na/Na	Auto
71260A	1200	6		77-124	16/2448/63	—/—	
71260AP	1260	5	4136	91-155	16/2632/63	Na/Na	Auto
71260S	1200					—/—	
7131A	125	2	2096		8/1002/32	Na/Na	Auto
71336A	1336	4	4721		16/2595/63	—/—	
71336AP	1336	4	4721		16/2595/63	—/—	
71350A	1350	4			16/2624/63	—/—	
71350AP	1350	4			16/2624/63	—/—	
7135AV	135	1		72-123	13/966/21	—/—	
71626A	1626	6	4136	91-155	16/3158/63	Na/Na	Auto
71626AP	1626	6	4136	91-155	16/3158/63	Na/Na	Auto
71670A	1670	5	4721		16/3224/63	—/—	
71670AP	1670	5	4721		16/3224/63	—/—	
71687AP	1687	5			16/3280/63	—/—	
7170A	171	4	1281	48-72	10/984/34	—/—	
7171A	172	2		V	15/866/26	—/—	
72004A	2004	6	4721		16/3893/63	—/—	
72004AP	2004	6	4721		16/3893/63	—/—	
72025AP	2025	6			16/3936/63	—/—	
7213A	213	4	1690	42	16/683/38	Na/Na	Auto
7213S	213	4	1690	42		Na/Na	Auto
7245A	234	4		48-72	16/967/31	—/—	
7245S	245	4		48-72		—/—	
72577AP	2577	8			16/4996/63	—/—	
72700AP	2700	8			16/5248/63	—/—	
7270AV	270	2		72-123	11/959/50	—/—	
7273A	273	3		V	16/1012/33	—/—	
7290A	290	4		60-96		Na/Na	Auto
7290S	290	4				—/—	
7345A	345	4			15/790/57	—/—	
7345S	345	4				—/—	
7405A	4051				16/989/50	—/—	
7405AV	405	3		72-123	16/989/50	—/—	
7420AV	420	3		72-123	16/1046/63	—/—	
7425AV	425	2	3721	76-144	16/1000/52	Na/Na	Auto
7540AV	540	4		72-123	16/1046/63	—/—	
7541A	541	2	4136	91-155	16/1052/63	Na/Na	Auto
7541AP	541	2	4136	91-155	16/1052/63	Na/Na	Auto
7546A	547	4		V	16/1024/63	—/—	
7546S	547	4		V		—/—	
7668A	668	2	4721		16/1297/63	—/—	
7668AP	668	2	4721		16/1297/63	—/—	
7850AV	850	4	3721	76-144	16/1648/63	Na/Na	Auto
8051A	41	4	745	26	5/981/17	Na/Na	981
8051S	40	4	793	28		Na/Na	Auto
80875A2	875	2			16/1700/63	—/—	
81080A3	1080	3			16/2100/63	—/—	
81081A2	1081	2			16/2100/63	—/—	
81275A3	1275	3			16/2480/63	—/—	
81280A2	1280	2			16/2481/63	—/—	
81312A3	1312	3			16/2548/63	—/—	
81620A3	1550	3			16/3150/63	—/—	

Drive Model	Seek Time	Interface	Encode	Form Factor	cache kb	mtbf	RPM	Obsolete?⇓
7060A	15	IDE AT	1,7 RLL	3.5 3H		150k	3600	Y
7060S	15	SCSI	1,7 RLL	3.5 3H		150k	3600	Y
7080A	17	IDE AT	1,7 RLL	3.5 3H	32k	150k	3703	Y
7080S	17	SCSI	1,7 RLL	3.5 3H	32k	150k	3600	Y
71000A	12	IDE AT	1,7 RLL	3.5 3H		300k		Y
71050A	12	EIDE	1,7 RLL	3.5 3H	256k	300k	4500	Y
71050S	12	SCSI	1,7 RLL	3.5 3H	256k	300k	4500	Y
71084A	12	IDE AT	1,7 RLL	3.5 3H	64k	300k	4480	Y
71084AP	12	IDE AT	1,7 RLL	3.5 3H	128k	300k	4480	Y
7120A	15	IDE AT	1,7 RLL	3.5 3H	64k	150k	3600	Y
7120S	15	SCSI	1,7 RLL	3.5 3H	64k	150k	3600	Y
71260A	12	EIDE	1,7 RLL	3.5 3H	256k	300k	4500	Y
71260AP	12	IDE AT	1,7 RLL	3.5 3H	128k	300k	4480	Y
71260S	14	ATA-2	1,7 RLL	3.5 3H	256k	300k	4500	Y
7131A	14	IDE AT	1,7 RLL	3.5 3H	64k	300k	3551	Y
71336A	<12	IDE AT	1,7 RLL	3.5 3H	64k	300k	4480	Y
71336AP	<12	IDE AT	1,7 RLL	3.5 3H	128k	300k	4480	Y
71350A	<12	IDE AT	1,7 RLL	3.5 3H	64k	300k	4480	Y
71350AP	<12	IDE AT	1,7 RLL	3.5 3H	128k	300k	4480	Y
7135AV	12	IDE AT	1,7 RLL	3.5 3H	32k	300k	3551	Y
71626A	12	IDE AT	1,7 RLL	3.5 3H	64k	300k	4480	Y
71626AP	12	IDE AT	1,7 RLL	3.5 3H	128k	300k	4480	Y
71670A	<12	IDE AT	1,7 RLL	3.5 3H	64k	300k	4480	Y
71670AP	<12	IDE AT	1,7 RLL	3.5 3H	128k	300k	4480	Y
71687AP	<12	IDE AT	1,7 RLL	3.5 3H	128k	300k	4480	Y
7170A	15	IDE AT	1,7 RLL	3.5 3H	64k	150k	3551	Y
7171A	14	IDE AT	1,7 RLL	3.5 3H	64k	300k	3551	Y
72004A	<12	IDE AT	1,7 RLL	3.5 3H	64k	300k	4480	Y
72004AP	<12	IDE AT	1,7 RLL	3.5 3H	128k	300k	4480	Y
72025AP	<12	IDE AT	1,7 RLL	3.5 3H	128k	300k	4480	Y
7213A	15	IDE AT	1,7 RLL	3.5 3H	64k	150k	3551	Y
7213S	15	SCSI	1,7 RLL	3.5 3H	64k	150k	3551	Y
7245A	15	IDE AT	1,7 RLL	3.5 3H	64k	300k	3551	Y
7245S	15	SCSI	1,7 RLL	3.5 3H	64k	250k	3551	Y
72577AP	<12	IDE AT	1,7 RLL	3.5 3H	128k	300k	4480	Y
72700AP	<12	IDE AT	1,7 RLL	3.5 3H	128k	300k	4480	Y
7270AV	12	IDE AT	1,7 RLL	3.5 3H	32k	300k	3551	Y
7273A	12	IDE AT	1,7 RLL	3.5 3H	256k	300k	4500	Y
7290A	14	IDE AT	1,7 RLL	3.5 3H	64k	300k	3551	Y
7290S	14	SCSI	1,7 RLL	3.5 3H	64k	300k	3551	Y
7345A	14	IDE AT	1,7 RLL	3.5 3H	64k	300k	3551	Y
7345S	14	SCSI	1,7 RLL	3.5 3H	64k	300k	3551	Y
7405A	12	IDE AT		3.5 3H	32k	300k	3551	Y
7405AV	12	IDE AT	1,7 RLL	3.5 3H	32k	300k	3551	Y
7420AV	12	IDE AT	1,7 RLL	3.5 3H	32k	300k	3551	Y
7425AV	12	IDE AT	1,7 RLL	3.5 3H	64k	300k	3551	Y
7540AV	12	IDE AT	1,7 RLL	3.5 3H	32k	300k	3551	Y
7541A	12	IDE AT	1,7 RLL	3.5 3H	64k	300k	4480	Y
7541AP	12	IDE AT	1,7 RLL	3.5 3H	128k	300k	4480	Y
7546A	12	IDE AT	1,7 RLL	3.5 3H	256k	300k	4500	Y
7546S	12	SCSI	1,7 RLL	3.5 3H	256k	300k	4500	Y
7668A	<12	IDE AT	1,7 RLL	3.5 3H	64k	300k	4480	Y
7668AP	<12	IDE AT	1,7 RLL	3.5 3H	128k	300k	4480	Y
7850AV	12	IDE AT	1,7 RLL	3.5 3H	64k	300k	3551	Y
8051A	28	IDE AT	2,7 RLL	3.5HH	32k	150k	3484	Y
8051S	28	SCSI	2,7 RLL	3.5HH		30k	3600	Y
80875A2	12	EIDE	1,7 RLL	3.5 3H	128k	400k	4480	Y
81080A3	12	ATA-3	1,7 RLL	3.5 3H	128k	400k	4480	Y
81081A2	11	ATA-3	RLL 8,9	3.5 3H	256k	400k	4480	Y
81275A3	12	IDE AT	1,7 RLL	3.5 3H	128k	400k	4480	Y
81280A2	10	ATA-3	RLL 8,9	3.5 3H	256k	500k	5400	Y
81312A3	12	EIDE	1,7 RLL	3.5 3H	128k	400k	4480	Y
81620A3	11	ATA-3	RLL 8,9	3.5 3H	256k	400k	4480	Y

Drive Model	Format Size MB	Head	Cyl	Sect/ Trac	Translate H/C/S	RWC/ WPC	Land Zone
81630A4	1630	4			16/3168/63	—/—	
81750A2	1750	2			15/3618/63	—/—	
81750A4	1750	4			16/3400/63	—/—	
81750D2	1750	2			15/3618/63	—/—	
82100A4	2100	4			16/4092/63	—/—	
82160A4	2160	4			16/4185/63	—/—	
82187A5	2187	5			16/4248/63	—/—	
82400A4	2400	4			16/4708/63	—/—	
82559A4	2559	4			16/4960/63	—/—	
82560A3	2560	3			15/5292/63	—/—	
82560A4	2560	4			16/4962/63	—/—	
82560D3	2560	3			15/5292/63	—/—	
82577A6	2577	6			16/5000/63	—/—	
82580A5	2580	5			16/5004/63	—/—	
82625A6	2625	6			16/5100/63	—/—	
83062A7	3062	7			16/5948/63	—/—	
83200A5	3200	5			16/6296/63	—/—	
83200A6	3200	6			15/6296/15	—/—	
83200A8	3200	8			16/6218/63	—/—	
83201A6	3201	6			16/6218/63	—/—	
83202A6	3202	6			15/6296/63	—/—	
83209A5	3209	5			16/6218/63	—/—	
83240A4	3240	4			15/6696/63	—/—	
83240D4	3240	4			15/6696/63	—/—	
83500A4	3500	4			15/7237/63	—/—	
83500A8	3500	8			16/6800/63	—/—	
83500D4	3500	4			15/7237/63	—/—	
83840A6	3840	6			16/7441/63	—/—	
84000A6	4000	6			16/7763/63	—/—	
84004A8	4004	8			16/7758/63	—/—	
84200A8	4200	8			16/8184/63	—/—	
8425S	21	4	612	17		616/128	664
84320A5	4320	5			15/8928/63	—/—	
84320A8	4320	8			16/8400/63	—/—	
84320D4	4320					—/—	
84320D5	4320	5			15/8928/63	—/—	
85120A8	5120	8			16/9924/63	—/—	
85121A8	5121	8			15/10585/63	—/—	
85210D6	5210	6			15/10856/63	—/—	
85250A6	5250	6			15/10856/63	—/—	
85250D6	5210	6			15/10856/63	—/—	
86480A8	6480	8			15/13392/63	—/—	
86480D6	6480					—/—	
86480D8	6480	8			15/13392/63	—/—	
87000A8	7000	8			15/14475/63	—/—	
87000D8	7000	8			15/14475/63	—/—	
90250D2	2559				16/4960/63	—/—	
90256D2	2560	2			16/4960/63	—/—	
90288D2	2880	2			16/5583/63	—/—	
90320D2	3240				15/6697/63	—/—	
90340D2	3400				16/6588/63	—/—	
90430D3	4311				15/8912/63	—/—	
90432D3	4320	3			16/8374/63	—/—	
90500D4	5003				16/9695/63	—/—	
90510D3	5100				16/9882/63	—/—	
90510D4	5122	5			16/9925/63	—/—	
90576D4	5760	4			16/11166/63	—/—	
90625D5	6254				16/12119/63	—/—	
90640D4	6481				15/13395/63	—/—	
90640E4	6448				15/13328/63	—/—	
90648D5	6480	5			16/12555/63	—/—	
90680D4	6800				16/13176/63	—/—	
90720D5	7200	5			16/13957/63	—/—	

Drive Model	Seek Time	Interface	Encode	Form Factor	cache kb	mtbf	RPM	Obsolete?
81630A4	12	IDE AT	1,7 RLL	3.5 3H	128k	400k	4480	Y
81750A2	10	ATA-4	RLL 8,9	3.5 3H	256k	500k	5200	Y
81750A4	12	EIDE	1,7 RLL	3.5 3H	128k	400k	4480	Y
81750D2	10	ATA-4	RLL 8,9	3.5 3H	256k	500k	5200	Y
82100A4	11	ATA-3	RLL 8,9	3.5 3H	256k	400k	4480	Y
82160A4	10	ATA-3	RLL 8,9	3.5 3H	256k	500k	5400	Y
82187A5	12	EIDE	1,7 RLL	3.5 3H	128k	400k	4480	Y
82400A4	10	ATA-3	RLL 8,9	3.5 3H	256k	500k	5400	Y
82559A4	10	ATA-3	RLL 8,9	3.5 3H	256k	500k	5400	Y
82560A3	10	ATA-4	RLL 8,9	3.5 3H	256k	500k	5200	Y
82560A4	10	ATA-3	RLL 8,9	3.5 3H	256k	500k	5400	Y
82560D3	10	ATA-4	RLL 8,9	3.5 3H	256k	500k	5200	Y
82577A6	12	IDE AT	1,7 RLL	3.5 3H	128k	400k	4480	Y
82580A5	11	ATA-3	RLL 8,9	3.5 3H	256k	400k	4480	Y
82625A6	12	EIDE	1,7 RLL	3.5 3H	128k	400k	4480	Y
83062A7	12	EIDE	1,7 RLL	3.5 3H	128k	400k	4480	Y
83200A5	10	ATA-3	RLL 8,9	3.5 3H	256k	500k	5400	Y
83200A6	10	ATA-3	RLL 8,9	3.5 3H	256k	500k	5400	Y
83200A8	12	EIDE	1,7 RLL	3.5 3H	128k	400k	4480	Y
83201A6	11	ATA-3	RLL 8,9	3.5 3H	256k	400k	4480	Y
83202A6	10	ATA-3	RLL 8,9	3.5 3H	256k	500k	5400	Y
83209A5	10	ATA-3	RLL 8,9	3.5 3H	256k	500k	5400	Y
83240A4	10	ATA-4	RLL 8,9	3.5 3H	256k	500k	5200	Y
83240D4	10	ATA-4	RLL 8,9	3.5 3H	256k	500k	5200	Y
83500A4	10	ATA-4	RLL 8,9	3.5 3H	256k	500k	5200	Y
83500A8	12	EIDE	1,7 RLL	3.5 3H	128k	400k	4480	Y
83500D4	10	ATA-4	RLL 8,9	3.5 3H	256k	500k	5200	Y
83840A6	10	ATA-3	RLL 8,9	3.5 3H	256k	500k	5400	Y
84000A6	10	ATA-3	RLL 8,9	3.5 3H	256k	500k	5400	Y
84004A8	10	ATA-3	RLL 8,9	3.5 3H	256k	500k	5400	Y
84200A8	10	ATA-3	RLL 8,9	3.5 3H	256k	500k	5400	Y
8425S	68	SCSI	MFM	3.5HH		20k	3600	Y
84320A5	10	ATA-4	RLL 8,9	3.5 3H	256k	500k	5200	Y
84320A8	11	ATA-3	RLL 8,9	3.5 3H	256k	400k	4480	Y
84320D4	10	EIDE	RLL 8,9	3.5 3H	256k	500k	5200	Y
84320D5	10	ATA-4	RLL 8,9	3.5 3H	256k	500k	5200	Y
85120A8	10	ATA-3	RLL 8,9	3.5 3H	256k	500k	5400	Y
85121A8	10	ATA-3	RLL 8,9	3.5 3H	256k	500k	5400	Y
85210D6	10	ATA-4	RLL 8,9	3.5 3H	256k	500k	5200	Y
85250A6	10	ATA-4	RLL 8,9	3.5 3H	256k	500k	5200	Y
85250D6	10	ATA-4	RLL 8,9	3.5 3H	256k	500k	5200	Y
86480A8	10	ATA-4	RLL 8,9	3.5 3H	256k	500k	5200	Y
86480D6	10	EIDE	RLL 8,9	3.5 3H	256k	500k	5200	Y
86480D8	10	ATA-4	RLL 8,9	3.5 3H	256k	500k	5200	Y
87000A8	10	ATA-4	RLL 8,9	3.5 3H	256k	500k	5200	Y
87000D8	10	ATA-4	RLL 8,9	3.5 3H	256k	500k	5200	Y
90250D2	9	EIDE		3.5	256k		5400	
90256D2	9	ATA-4	PRML	3.5 3H	252k	500k	5400	Y
90288D2	9	ATA-4	PRML	3.5 3H	252k	500k	5400	Y
90320D2	9	EIDE		3.5	256k		5400	
90340D2	9	EIDE		3.5	256k		5400	
90430D3	9	EIDE		3.5	256k		5400	
90432D3	9	ATA-4	PRML	3.5 3H	252k	500k	5400	Y
90500D4	9	EIDE		3.5			7200	
90510D3	9	EIDE		3.5	256k		5400	
90510D4	9	ATA-4	PRML	3.5 3H	252k	500k	5400	Y
90576D4	9	ATA-4	PRML	3.5 3H	252k	500k	5400	Y
90625D5	9	EIDE		3.5			7200	
90640D4	9	EIDE		3.5	256k		5400	
90640E4	9	EIDE		3.5	256k		5400	
90648D5	9	ATA-4	PRML	3.5 3H	252k	500k	5400	Y
90680D4	9	EIDE		3.5	256k		5400	
90720D5	9	ATA-4	PRML	3.5 3H	252k	500k	5400	Y

Drive Model	Format Size MB	Head	Cyl	Sect/ Trac	Translate H/C/S	RWC/ WPC	Land Zone
90750D6	7505				16/14542/63	—/—	
90840D5	8438				16/16383/63	—/—	
90840D6	8400	6			16/16276/63	—/—	
90840D7	8400				16/16277/63	—/—	
90840E5	8438				16/16383/63	—/—	
90845D6	8455	6			16/16383/63	—/—	
90875D7	8756				16/16966/63	—/—	
90910D8	9106				16/17645/63	—/—	
91000D8	10007				16/19390/63	—/—	
91008D7	10080	7			16/19540/63	—/—	
91010D6	10110				16/19590/63	—/—	
91010E6	10005				16/19386/63	—/—	
91020D6	10200				16/19765/63	—/—	
91152D8	11520	8			16/22332/63	—/—	
91190D7	11900				16/23059/63	—/—	
91202D8	12020				16/23291/63	—/—	
91202E8	12020				16/23291/63	—/—	
91350D8	13520				16/26197/63	—/—	
91350E8	1352016/135203					—/—	
91360D8	13600				16/26353/63	—/—	
9380E	338	15	1224	36		NA/512	Auto
9380S	336	15	1218	36		NA/512	Auto
9780E	676	15	1661	53		NA/512	Auto
9780S	676	15	1661	53		166/512	Auto
EXT4175	149	7	1224	34		Na/Na	Auto
EXT4280	234	11	1224	36		Na/Na	Auto
EXT4380	319	15	1224	34		Na/Na	Auto
LXT100A	90					—/—	
LXT100S	96	8	733	32		Na/Na	Auto
LXT200A	191	7	1320	NA	15/816/32	Na/Na	Auto
LXT200S	207	7	1320	33,53		Na/Na	Auto
LXT213A	203	7	1320	NA	16/683/38	Na/Na	Auto
LXT213S	213	7	1320	34-56		Na/Na	Auto
LXT340A	340	7	1560	47-72	16/654/63	Na/Na	Auto
LXT340S	340	7	1560	47-72		Na/Na	Auto
LXT437A (not made)	437	9	1560	V	16/842/63	Na/Na	Auto
LXT437S (not made)	437	9	1560	V		Na/Na	Auto
LXT50S	48	4	733	32		Na/Na	Auto
LXT535A	535	11	1024	63	16/1024/36	Na/Na	Auto
LXT535S	535	11	1560	47-72		Na/Na	Auto
MOBILEMAX 105MB	105	4	1254	28-50		—/—	
MOBILEMAX 131MB	131	4	1254	28-50		—/—	
MOBILEMAX 171MB	171	4	1254	28-50		—/—	
MOBILEMAX 262MB	262					—/—	
MX9217SDN	2170	9		100-14		—/—	
MX9217SDW	2170	9		100-14		—/—	
MX9217SSN	2170	9		100-14		—/—	
MX9217SSW	2170	9		100-14		—/—	
MXT1240S	1240	15	2512	NA		Na/Na	Auto
MXT540AL	547	7	2466		16/1024/63	Na/Na	Auto
MXT540SL	547	7	2466	NA		Na/Na	Auto
P0-12S PANTHER	1045	15	163261-103			Na/Na	Auto
P1-08E (not made)	696	9	1778	85		Na/Na	Auto
P1-12E (not made)	1051	15	1778	77		Na/Na	Auto
P1-13E (not made)	1160	15	1778			Na/Na	Auto
P1-16E (not made)	1331	19	1778			Na/Na	Auto
P1-17E (not made)	1470	19	1778	85		Na/Na	Auto
P1-17S PANTHER	1503	19	177870-101			Na/Na	Auto
RXT-800HD	786					—/—	
RXT-800HS	786					—/—	
RXT-800S	786					—/—	
XT1050	38	5	902	17		Na/Na	Auto
XT1065	52	7	918	17		Na/Na	Auto

Drive Model	Seek Time	Interface	Encode	Form Factor	cache kb	mtbf	RPM	Obsolete?
90750D6	9	EIDE		3.5			7200	
90840D5	9	EIDE		3.5	256k		5400	
90840D6	9	ATA-4	PRML	3.5 3H	252k	500k	5400	Y
90840D7	9	EIDE		3.5			7200	
90840E5	9	EIDE		3.5	256k		5400	
90845D6	9	ATA-4	PRML	3.5 3H	252k	500k	5400	Y
90875D7	9	EIDE		3.5			7200	
90910D8	9	EIDE		3.5			7200	
91000D8	9	EIDE		3.5			7200	
91008D7	9	ATA-4	PRML	3.5 3H	252k	500k	5400	Y
91010D6	9	EIDE		3.5	256k		5400	
91010E6	9	EIDE		3.5	256k		5400	
91020D6	9	EIDE		3.5	256k		5400	
91152D8	9	ATA-4	PRML	3.5 3H	252k	500k	5400	Y
91190D7	9	EIDE		3.5	256k		5400	
91202D8	9	EIDE		3.5	256k		5400	
91202E8	9	EIDE		3.5	256k		5400	
91350D8	9	EIDE		3.5	256k		5400	
91350E8	9	EIDE		3.5	256k		5400	
91360D8	9	EIDE		3.5	256k		5400	
9380E	16	ESDI	2,7 RLL	5.25FH		50k	3600	Y
9380S	16	SCSI	2,7 RLL	5.25FH		50k	3600	Y
9780E	17	ESDI	1,7 RLL	5.25FH		50k	3600	Y
9780S	17	SCSI	1,7 RLL	5.25FH		30k	3600	Y
EXT4175	27	ESDI	RLL	5.25FH		20k	3600	Y
EXT4280	27	ESDI	RLL	5.25FH		20k	3600	Y
EXT4380	27	ESDI	RLL	5.25FH		20k	3600	Y
LXT100A		IDE AT	1,7 RLL	3.5HH		150k	3600	Y
LXT100S	27	SCSI	2,7 RLL	3.5HH		150k	3600	Y
LXT200A	15	IDE AT	1,7 RLL	3.5HH		150k	3600	Y
LXT200S	15	SCSI	1,7 RLL	3.5HH		150k	3600	Y
LXT213A	15	IDE AT	1,7 RLL	3.5HH	32k	150k	3600	Y
LXT213S	15	SCSI-2	1,7 RLL	3.5HH	32k	150k	3600	Y
LXT340A	15	IDE AT	2,7 RLL	3.5HH	128k	150k	3600	Y
LXT340S	15	SCSI	2,7 RLL	3.5HH	128k	150k	3600	Y
LXT437A (not made)	12	IDE AT	2,7 RLL	3.5HH		150k	3600	Y
LXT437S (not made)	13	SCSI	2,7 RLL	3.5HH		150k	3600	Y
LXT50S	27	SCSI	2,7 RLL	3.5HH		40k	3600	Y
LXT535A	12	IDE AT	2,7 RLL	3.5HH	128k	150k	3600	Y
LXT535S	13	SCSI	2,7 RLL	3.5HH	128k	150k	3600	Y
MOBILEMAX 105MB	19	PCMCIA-ATA	1,7 RLL	1.8 4H	31k	300k	4464	Y
MOBILEMAX 131MB	19	PCMCIA-ATA	1,7 RLL	2.8 4H	31k	300k	4464	Y
MOBILEMAX 171MB	19	PCMCIA-ATA	1,7 RLL	2.8 4H	31k	300k	4464	Y
MOBILEMAX 262MB		PCMCIA-ATA	1,7 RLL	1.8 4H				Y
MX9217SDN	10.5	SE SCSI-2D	1,7 RLL	3.5 3H	512k	800k	5400	Y
MX9217SDW	10.5	SE SCSI-2DW	1,7 RLL	3.5 3H	512k	800k	5400	Y
MX9217SSN	10.5	SE SCSI-2	1,7 RLL	3.5 3H	512k	800k	5400	Y
MX9217SSW	10.5	SE SCSI-2W	1,7 RLL	3.5 3H	512k	800k	5400	Y
MXT1240S	9	SCSI-2Fast	1,7 RLL	3.5HH		300k	6300	Y
MXT540AL	9	IDE AT	1,7 RLL	3.5 3H		300k	6300	Y
MXT540SL	9	SCSI-2Fast	1,7 RLL	3.5 3H		300k	6300	Y
P0-12S PANTHER	13	SCSI-2	RLL	5.25FH	256k	150k	3600	Y
P1-08E (not made)	12	ESDI	RLL	5.25FH		100k	3600	Y
P1-12E (not made)	13	ESDI	RLL	5.25FH		100k	3600	Y
P1-13E (not made)	13	ESDI	RLL	5.25FH		100k	3600	Y
P1-16E (not made)	13	ESDI	RLL	5.25FH		100k	3600	Y
P1-17E (not made)	13	ESDI	RLL	5.25FH		100k	3600	Y
P1-17S PANTHER	13	SCSI-2	RLL	5.25FH	256k	150k	3600	Y
RXT-800HD		SCSI		5.25HH				Y
RXT-800HS		SCSI		5.25HH				Y
RXT-800S		SCSI		5.25HH				Y
XT1050	30	ST412/506	MFM	5.25FH		20k	3600	Y
XT1065	30	ST412/506	MFM	5.25FH		20k	3600	Y

Drive Model	Format Size MB	Head	Cyl	Sect/ Trac	Translate H/C/S	RWC/ WPC	Land Zone
XT1085	71	8	1024	17		Na/Na	Auto
XT1105	84	11	918	17		Na/Na	Auto
XT1120R	105	8	1024	25		Na/Na	Auto
XT1140	119	15	918	17		Na/Na	Auto
XT1240R	196	15	1024	25		Na/Na	Auto
XT2085	72	7	1224	17		Na/Na	Auto
XT2140	113	11	1224	17		Na/Na	Auto
XT2190	159	15	1224	17		Na/Na	Auto
XT3170	146	9	1224	26		Na/Na	Auto
XT3280	244	15	1224	26		Na/Na	Auto
XT3380	319	15	1224	34		Na/Na	Auto
XT4170E	157	7	1224	35/36	16	Na/Na	Auto
XT4170S	157	7	1224	35-36		Na/Na	Auto
XT4175	234	11	1224	34		Na/Na	Auto
XT4230E	203	9	1224	35/36		Na/Na	Auto
XT4280SF	338	15	1224	36		Na/Na	Auto
XT4380E	338	15	1224	36		Na/Na	Auto
XT4380S	338	15	1224	NA		Na/Na	Auto
XT81000E	889	15	1632	71		Na/Na	Auto
XT8380E	361	8	1632	53-54		Na/Na	Auto
XT8380EH	360	8	1632	54		Na/Na	Auto
XT8380S	361	8	1632	NA		Na/Na	Auto
XT8380SH	360	8	1632	NA		Na/Na	Auto
XT8610E	541	12	1632	53-54		Na/Na	Auto
XT8702S	616	15	1490	NA		Na/Na	Auto
XT8760E	676	15	1632	53-54		Na/Na	Auto
XT8760EH	676	15	1632	54		Na/Na	Auto
XT8760S	670	15	1632	NA		Na/Na	Auto
XT8760SH	670	15	1632	NA		Na/Na	Auto
XT8800E	694	15	1274	54		Na/Na	Auto

MEGA DRIVE SYSTEMS

Drive Model	Format Size MB	Head	Cyl	Sect/ Trac	Translate H/C/S	RWC/ WPC	Land Zone
M1-105	105	4	1219			—/—	
M1-120	122	2	1818			—/—	
M1-240	245	4	1818			—/—	
M1-52	52	2	1219			—/—	
MH-1G	1050	13	1974			—/—	
MH-340	338	9	1100			—/—	
MH-425	426	9	1520			—/—	
MH-535	525	9	1476			—/—	
P105	103	6	1019	33		Na/Na	Auto
P120	120	5	1123			Na/Na	Auto
P170	168	7	1123			Na/Na	Auto
P210	210	7	1156			Na/Na	Auto
P320	320	15	886			Na/Na	Auto
P42	42	3	834	33		Na/Na	Auto
P425	426	9	1512			Na/Na	Auto
P84	84	6	834	33		Na/Na	Auto

MEMOREX

Drive Model	Format Size MB	Head	Cyl	Sect/ Trac	Translate H/C/S	RWC/ WPC	Land Zone
321	5	2	320	17		321/128	
322	10	4	320	17		321/128	
323	15	6	320	17		321/128	
324	20	8	320	17		321/128	
450	10	2	612	17		321/350	
512	25	3	961	17		321/480	
513	41	5	961	17		321/480	
514	58	7	961	17		961?/480	

MICROPOLIS CORP

Drive Model	Format Size MB	Head	Cyl	Sect/ Trac	Translate H/C/S	RWC/ WPC	Land Zone
1050 AV LT	1000	9	2360	V		—/—	
1302	20	3	830	17		831/831	Auto

Drive Model	Seek Time	Interface	Encode	Form Factor	cache kb	mtbf	RPM	Obsolete?
XT1085	28	ST412/506	MFM	5.25FH		150k	3600	Y
XT1105	27	ST412/506	MFM	5.25FH		20k	3600	Y
XT1120R	27	ST412/506	2,7 RLL	5.25FH		150k	3600	Y
XT1140	27	ST412/506	MFM	5.25FH		150k	3600	Y
XT1240R	27	ST412/506	2,7 RLL	5.25FH		150k	3600	Y
XT2085	30	ST412/506	MFM	5.25FH		30k	3600	Y
XT2140	30	ST412/506	MFM	5.25FH		30k	3600	Y
XT2190	29	ST412/506	MFM	5.25FH		150k	3600	Y
XT3170	30	SCSI	RLL	3.5FH		20k	3600	Y
XT3280	30	SCSI	RLL	5.25FH		20k	3600	Y
XT3380	27	SCSI	RLL	5.25FH		20k	3600	Y
XT4170E	14	ESDI	1,7 RLL	5.25FH		150k	3600	Y
XT4170S	14	SCSI	1,7 RLL	5.25FH		150k	3600	Y
XT4175	27	ESDI	RLL	5.25FH		20k	3600	Y
XT4230E	16	ESDI	1,7 RLL	5.25FH		150k	3600	Y
XT4280SF	16	SCSI	1,7 RLL	5.25FH		150k	3600	Y
XT4380E	16	ESDI	1,7 RLL	5.25FH		150k	3600	Y
XT4380S	16	SCSI	1,7 RLL	5.25FH		150k	3600	Y
XT81000E	16	ESDI	1,7 RLL	5.25FH		150k	3600	Y
XT8380E	16	ESDI	1,7 RLL	5.25FH		150k	3600	Y
XT8380EH	13	ESDI	1,7 RLL	5.25FH		150k	3600	Y
XT8380S	14	SCSI	1,7 RLL	5.25FH		150k	3600	Y
XT8380SH	14	SCSI	1,7 RLL	5.25FH	256k	150k	3600	Y
XT8610E	16	ESDI	1,7 RLL	5.25FH		150k	3600	Y
XT8702S	17	SCSI	1,7 RLL	5.25FH		150k	3600	Y
XT8760E	16	ESDI	1,7 RLL	5.25FH		150k	3600	Y
XT8760EH	14	ESDI	1,7 RLL	5.25FH		150k	3600	Y
XT8760S	16	SCSI	1,7 RLL	5.25FH		150k	3600	Y
XT8760SH	14	SCSI	1,7 RLL	5.25FH	256k	150k	3600	Y
XT8800E	14	ESDI	1,7 RLL	5.25FH		150k	3600	Y

MEGA DRIVE SYSTEMS

Drive Model	Seek Time	Interface	Encode	Form Factor	cache kb	mtbf	RPM	Obsolete?
M1-105	17	SCSI	2,7 RLL	3.5HH	64k	60k	3662	Y
M1-120	16	SCSI	1,7 RLL	3.5HH	256k	250k	4306	Y
M1-240	16	SCSI	1,7 RLL	3.5HH	256k	250k	4306	Y
M1-52	17	SCSI	2,7 RLL	3.5HH	64k	60k	3662	Y
MH-1G	10	SCSI	1,7 RLL	3.5HH	256k	300k	5400	
MH-340	13	SCSI	1,7 RLL	3.5HH	64k	150k	4412	Y
MH-425	14	SCSI	1,7 RLL	3.5HH	64k	150k	4412	Y
MH-535	14	SCSI	1,7 RLL	3.5HH	256k	150k	4412	Y
P105	19	SCSI	2,7 RLL	3.5HH		50k		Y
P120	14	SCSI	1,7 RLL	3.5HH		50k		Y
P170	14	SCSI	1,7 RLL	3.5HH		50k		Y
P210	14	SCSI	1,7 RLL	3.5HH		50k		Y
P320	12.5	SCSI	1,7 RLL	3.5HH		150k		Y
P42	19	SCSI	2,7 RLL	3.5HH		50k		Y
P425	12	SCSI	1,7 RLL	3.5HH		75k		Y
P84	19	SCSI	2,7 RLL	3.5HH		50k		Y

MEMOREX

Drive Model	Seek Time	Interface	Encode	Form Factor	cache kb	mtbf	RPM	Obsolete?
321		ST412/506	MFM	5.25FH				Y
322		ST412/506	MFM	5.25FH				Y
323		ST412/506	MFM					Y
324		ST412/506	MFM					Y
450		ST412/506	MFM					Y
512		ST412/506	MFM					Y
513		ST412/506	MFM					Y
514		ST412/506	MFM					Y

MICROPOLIS CORP

Drive Model	Seek Time	Interface	Encode	Form Factor	cache kb	mtbf	RPM	Obsolete?
1050 AV LT	10	SCSI-2Fast	MZR	3.5FH	512k	300k	5400	Y
1302	30	ST412/506	MFM	5.25FH		20k	3600	Y

Drive Model	Format Size MB	Head	Cyl	Sect/ Trac	Translate H/C/S	RWC/ WPC	Land Zone
1303	35	5	830	17		831/831	Auto
1304	40	6	830	17		831/831	Auto
1323	35	4	1024	17		1025/1025	Auto
1323A	44	5	1024	17		1025/1025	Auto
1324	53	6	1024	17		1025/1025	Auto
1324A	62	7	1024	17		1025/1025	Auto
1325	71	8	1024	17		1025/1025	Auto
1325CT	71		1024	17		1025/1025	Auto
1333	35	4	1024	17		1025/1025	Auto
1333A	44	5	1024	17		1025/1025	Auto
1334	53	6	1024	17		1025/1025	Auto
1334A	62	7	1024	17		1025/1025	Auto
1335	71	8	1024	17		1025/1025	Auto
1352	32	2	1024	36		—/—	
1352A	41	3	1024	36		Na/Na	
1353	75	4	1024	36		Na/Na	Auto
1353A	94	5	1024	36		Na/Na	Auto
1354	113	6	1024	36		Na/Na	Auto
1354A	131	7	1024	36		Na/Na	Auto
1355	150	8	1024	36		Na/Na	Auto
1372A	52		1024	36		—/—	
1373	72	4	1024	36		1017/1017	Auto
1373A	91	5	1024	36		1017/1017	Auto
1374	109	6	1024	36		1017/1017	Auto
1374-6	135	6	1245	36		—/—	
1374A	127	7	1024	36		1017/1017	Auto
1375	145	8	1024	36		1017/1017	Auto
1516-10S	678	10	1840	72		Na/Na	
1517-13	922	13	1925	72		Na/Na	
1517-14	981	14	1925	71		—/—	
1517-15	1051	15	1925	71		—/—	
1518	1346					—/—	
1518-14	993	14	1925	72		Na/Na	
1518-15	1341	15	2104	83		Na/Na	Auto
1528	1342	15	2100	84		—/—	
1528-15	1342	15	2100	84		Na/Na	Auto
1528-15D	1300					—/—	
1538	871	15	1669	68		Na/Na	Auto
1538-15	910	15	1669	71		Na/Na	Auto
1548-15	1748	15	2112	V		Na/Na	Auto
1554-07	157	7	1224	36		Na/Na	Auto
1555-08	180	8	1224	36		Na/Na	Auto
1555-09	203	9	1224	36		Na/Na	Auto
1556-10	225	10	1224	36		Na/Na	Auto
1556-11	248	11	1224	36		Na/Na	Auto
1557-12	270	12	1224	36		Na/Na	Auto
1557-13	293	13	1224	36		Na/Na	Auto
1557-14	315	14	1224	36		1225/1225	
1557-15	338	15	1224	36		1225/1225	
1558	338		1224	36		—/—	
1558-13	293	14	1224	36		Na/Na	Auto
1558-14	315	14	1224	36		Na/Na	Auto
1558-15	338	15	1224	36		Na/Na	Auto
1560-8S	389	8	1632	54		—/—	
1564-07	315	7	1224	54		Na/Na	Auto
1565-08	360	8	1224	54		Na/Na	Auto
1565-09	406	9	1224	54		Na/Na	Auto
1566-10	451	10	1224	54		Na/Na	Auto
1566-11	496	11	1224	54		Na/Na	Auto
1567-12	541	12	1632	54		Na/Na	Auto
1567-13	586	13	1224	54		Na/Na	Auto
1567-14	631	14	1632	54		—/—	
1568	676		1632	54		—/—	

Drive Model	Seek Time	Interface	Encode	Form Factor	cache kb	mtbf	RPM	Obsolete? ⇓
1303	30	ST412/506	MFM	5.25FH		20k	3600	Y
1304	30	ST412/506	MFM	5.25FH		20k	3600	Y
1323	28	SP412/506	MFM	5.25FH		35k	3600	Y
1323A	28	ST412/506	MFM	5.25FH		35k	3600	Y
1324	28	ST412/506	MFM	5.25FH		35k	3600	Y
1324A	28	ST412/506	MFM	5.25FH		35k	3600	Y
1325	28	ST412/506	MFM	5.25FH		35k	3600	Y
1325CT	28	ST412/506	MFM	5.25FH		35k	3600	Y
1333	28	ST412/506	MFM	5.25 FD		25k	3600	Y
1333A	28	ST412/506	MFM	5.25FH		25k	3600	Y
1334	28	ST412/506	MFM	5.25FH		25k	3600	Y
1334A	28	ST412/506	MFM	5.25FH		25k	3600	Y
1335	28	ST412/506	MFM	5.25FH		25k	3600	Y
1352	23	ESDI	2,7 RLL	5.25FH				Y
1352A	23	ESDI	2,7 RLL	5.25FH				Y
1353	23	ESDI	2,7 RLL	5.25FH		150k	3600	Y
1353A	23	ESDI	2,7 RLL	5.25FH		150k	3600	Y
1354	23	ESDI	2,7 RLL	5.25FH		150k	3600	Y
1354A	23	ESDI	2,7 RLL	5.25FH		150k	3600	Y
1355	23	ESDI	2,7 RLL	5.25FH		150k	3600	Y
1372A		SCSI	2,7 RLL	5.25FH				Y
1373	23	SCSI	2,7 RLL	5.25FH		30k	3600	Y
1373A	23	SCSI	2,7 RLL	5.25FH		30k	3600	Y
1374	23	SCSI	2,7 RLL	5.25FH		30k	3600	Y
1374-6	16	SCSI	2,7 RLL	5.25HH		40k	3600	Y
1374A	23	SCSI	2,7 RLL	5.25FH		30k	3600	Y
1375	23	SCSI	2,7 RLL	5.25FH		30k	3600	Y
1516-10S	14	ESDI	2,7 RLL	5.25FH		150k		Y
1517-13	14	ESDI	2,7 RLL	5.25FH		150k		Y
1517-14	14	ESDI	2,7 RLL	5.25FH		150k		Y
1517-15	14	ESDI	2,7 RLL	5.25FH		150k		Y
1518	14.5	ESDI	1,7 RLL	5.25FH		150k		Y
1518-14	14	ESDI	2,7 RLL	5.25FH		150k		Y
1518-15	14	ESDI	2,7 RLL	5.25FH		150k	3600	Y
1528	14.5	SCSI-2		5.25FH	256k	150k	3600	Y
1528-15	14	SCSI-2		5.25FH		150k	3600	Y
1528-15D		SCSI-2Diff		5.25FH			3600	Y
1538		ESDI	1,7 RLL	5.25FH		150k	3600	Y
1538-15	15	ESDI	2,7 RLL	5.25FH		150k	3600	Y
1548-15	14	SCSI-2		5.25FH	256k	150k	3600	Y
1554-07	18	ESDI	2,7 RLL	5.25FH		150k	3600	Y
1555-08	18	ESDI	2,7 RLL	5.25FH		150k	3600	Y
1555-09	18	ESDI	2,7 RLL	5.25FH		150k	3600	Y
1556-10	18	ESDI	2,7 RLL	5.25FH		150k	3600	Y
1556-11	18	ESDI	2,7 RLL	5.25FH		150k	3600	Y
1557-12	18	ESDI	2,7 RLL	5.25FH		150k	3600	Y
1557-13	18	ESDI	2,7 RLL	5.25FH		150k	3600	Y
1557-14	18	ESDI	2,7 RLL	5.25FH		150k	3600	Y
1557-15	18	ESDI	2,7 RLL	5.25FH		150k	3600	Y
1558	19	ESDI	2,7 RLL	5.25FH		150k	3600	Y
1558-13	18	ESDI	2,7 RLL	5.25FH		150k	3600	Y
1558-14	18	ESDI	2,7 RLL	5.25FH		150k	3600	Y
1558-15	18	ESDI	2,7 RLL	5.25FH		150k	3600	Y
1560-8S	16	ESDI	2,7 RLL	5.25FH		150k		Y
1564-07	18	ESDI	2,7 RLL	5.25FH		150k	3600	Y
1565-08	18	ESDI	2,7 RLL	5.25FH		150k	3600	Y
1565-09	18	ESDI	2,7 RLL	5.25FH		150k	3600	Y
1566-10	18	ESDI	2,7 RLL	5.25FH		150k	3600	Y
1566-11	18	ESDI	2,7 RLL	5.25FH		150k	3600	Y
1567-12	18	ESDI	2,7 RLL	5.25FH		150k	3600	Y
1567-13	18	ESDI	2,7 RLL	5.25FH		150k	3600	Y
1567-14	16	ESDI	2,7 RLL	5.25FH		150k	3600	Y
1568	16	ESDI	2,7 RLL	5.25FH		150k	3600	Y

Drive Model	Format Size MB	Head	Cyl	Sect/ Trac	Translate H/C/S	RWC/ WPC	Land Zone
1568-13	586		1632	54		—/—	
1568-14	631	14	1632	54		Na/Na	Auto
1568-15	676	15	1632	54		Na/Na	Auto
1574-07	155	7	1224	36		Na/Na	Auto
1575-08	177	8	1224	36		Na/Na	Auto
1575-09	199	9	1224	36		Na/Na	Auto
1576-10	221	10	1224	36		1224/1224	Auto
1576-11	243	11	1224	36		1224/1224	Auto
1577-12	265	12	1224	36		1224/1224	Auto
1577-13	287	13	1224	36		1224/1224	Auto
1578	331		1224	36		1224/1224	Auto
1578-14	310	14	1224	36		1224/1224	
1578-15	332	15	1224	36		1224/1224	Auto
1585-8S	344	8	1628	54		—/—	
1586-11	490	11	1628	54		1632/1632	Auto
1587-12	540	12	1628	54		1632/1632	Auto
1587-13	579	13	1628	54		1632/1632	Auto
1587-13	585	13	1628	54		Na/Na	Auto
1588	668					—/—	
1588-14	624	14	1628	54		1632/1632	Auto
1588-15	667	15	1632	54		1632/1632	Auto
1588T-15	676	15	1632	54		Na/Na	Auto
1596-10S	668	10	1834	72		1835/1835	
1597-13	909	13	1919	72		1835/1835	
1598	1034					—/—	
1598-14	979	14	1919	72		1920/1920	
1598-15	1034	15	1928	71		1920/1920	Auto
1624	667	7	2089	V		Auto/Auto	NA
1624-7	667	7	2112			Na/Na	Auto
1653-4	92	4	1249	36		Na/Na	Auto
1653-5	115	5	1249	36		Na/Na	Auto
1653-6	138	6	1249			—/—	
1654	161		1249	36		—/—	
1654-6	138	6	1249	36		Na/Na	Auto
1654-7	161	7	1249	36		Na/Na	Auto
1663-4	197	4	1780	54		Na/Na	Auto
1663-5	246	5	1780	54		Na/Na	Auto
1664	345		1780	54		—/—	
1664-6	295	6	1780	54		Na/Na	Auto
1664-7	344	7	1780	54		Na/Na	Auto
1670-4	90	4	1245	36		—/—	
1670-5	90		1245	36		—/—	
1670-6	112		1245	36		—/—	
1670-7	135		1245	36		—/—	
1673-4	90	4	1249	36		1250/1250	Auto
1673-5	112	5	1249	36		1250/1250	Auto
1674	158		1249	36		—/—	
1674-6	135	6	1249	36		1250/1250	Auto
1674-7	157	7	1249	36		1250/1250	Auto
1683-4	193	4	1776	54		1777/1777	Auto
1683-5	242	5	1776	54		1777/1777	Auto
1684	340		1776	54		—/—	
1684-6	291	6	1776	54		1777/1777	Auto
1684-7	339	7	1780	54		1777/1777	Auto
1743-5	112	5	1140	28		Na/Na	
1744-6	135	6	1140	28		Na/Na	
1744-7	157	7	1140	28		Na/Na	
1745-8	180	8	1140	28		Na/Na	
1745-9	202	9	1140	28		Na/Na	
1760 AV LT	1700	15	2360	V		—/—	
1773-5	112	5	1140	28		1141/1141	
1774-6	135	6	1140	28		1141/1141	
1774-7	157	7	1140	28		1141/1141	

Drive Model	Seek Time	Interface	Encode	Form Factor	cache kb	mtbf	RPM	Obsolete? ⇓
1568-13	16	ESDI	2,7 RLL	5.25FH		150k	3600	Y
1568-14	16	ESDI	2,7 RLL	5.25FH		150k	3600	Y
1568-15	16	ESDI	2,7 RLL	5.25FH		150k	3600	Y
1574-07	16	SCSI	2,7 RLL	5.25FH		150k	3600	Y
1575-08	16	SCSI	2,7 RLL	5.25FH		150k	3600	Y
1575-09	16	SCSI	2,7 RLL	5.25FH		150k	3600	Y
1576-10	16	SCSI	2,7 RLL	5.25FH		150k	3600	Y
1576-11	16	SCSI	2,7 RLL	5.25FH		150k	3600	Y
1577-12	16	SCSI	2,7 RLL	5.25FH		150k	3600	Y
1577-13	16	SCSI	2,7 RLL	5.25FH		150k	3600	Y
1578	16	SCSI		5.25FH	64k	150k		Y
1578-14	16	SCSI	2,7 RLL	5.25FH		150k	3600	Y
1578-15	16	SCSI	2,7 RLL	5.25FH		150k	3600	Y
1585-8S	16	SCSI	2,7 RLL	5.25FH		150k		Y
1586-11	16	SCSI	2,7 RLL	5.25FH		150k		Y
1587-12	16	SCSI	2,7 RLL	5.25FH		150k		Y
1587-13	16	SCSI	2,7 RLL	5.25FH		150k		Y
1587-13	16	SCSI	2,7 RLL	5.25FH		150k		Y
1588	16	SCSI	2,7 RLL	5.25FH	256k	150k		Y
1588-14	16	SCSI	2,7 RLL	5.25FH		150k		Y
1588-15	16	SCSI	2,7 RLL	5.25FH		150k	3600	Y
1588T-15	16	SCSI	2,7 RLL	5.25FH		150k		Y
1596-10S	14	SCSI	2,7 RLL	5.25FH		150k		Y
1597-13	14	SCSI	2,7 RLL	5.25FH		150k		Y
1598	14.5	SCSI-2	2,7 RLL	5.25FH	256k	150k		Y
1598-14	14	SCSI	2,7 RLL	5.25FH		150k		Y
1598-15	14	SCSI-2	2,7 RLL	5.25FH		150k	3600	Y
1624	15	SCSI-2		5.25HH		150k		Y
1624-7	15	SCSI-2Fast		5.25HH		150k	3600	Y
1653-4	16	ESDI	2,7 RLL	5.25HH		150k	3600	Y
1653-5	16	ESDI	2,7 RLL	5.25HH		150k	3600	Y
1653-6		ESDI	2,7 RLL	5.25HH		150k		Y
1654	16	ESDI	2,7 RLL	5.25HH		150k		Y
1654-6	16	ESDI	2,7 RLL	5.25HH		150k	3600	Y
1654-7	16	ESDI	2,7 RLL	5.25HH		150k	3600	Y
1663-4	14	ESDI	2,7 RLL	5.25HH		150k		Y
1663-5	14	ESDI	2,7 RLL	5.25HH		150k		Y
1664	15	ESDI	2,7 RLL	5.25HH		150k		Y
1664-6	14	ESDI	2,7 RLL	5.25HH		150k		Y
1664-7	14	ESDI	2,7 RLL	5.25HH		150k	3600	Y
1670-4	16	SCSI		5.25HH		150k		Y
1670-5		SCSI		5.25HH		150k		Y
1670-6		SCSI		5.25HH		150k		Y
1670-7		SCSI		5.25HH		150k		Y
1673-4	16	SCSI	2,7 RLL	5.25HH		150k	3600	Y
1673-5	16	SCSI	2,7 RLL	5.25HH		150k	3600	Y
1674	16	SCSI	2,7 RLL	5.25HH		150k		Y
1674-6	16	SCSI	2,7 RLL	5.25HH		150k	3600	Y
1674-7	16	SCSI	2,7 RLL	5.25HH		150k	3600	Y
1683-4	14	SCSI	2,7 RLL	5.25HH		150k		Y
1683-5	14	SCSI	2,7 RLL	5.25HH		150k		Y
1684	15	SCSI	2,7 RLL	5.25HH		150k		Y
1684-6	14	SCSI	2,7 RLL	5.25HH		150k		Y
1684-7	14	SCSI	2,7 RLL	5.25HH		150k	3600	Y
1743-5	15	IDE AT	2,7 RLL	3.5HH				Y
1744-6	15	IDE AT	2,7 RLL	3.5HH				Y
1744-7	15	IDE AT	2,7 RLL	3.5HH				Y
1745-8	15	IDE AT	2,7 RLL	3.5HH				Y
1745-9	15	IDE AT	2,7 RLL	3.5HH				Y
1760 AV LT	10	SCSI-2Fast	MZR	3.5FH	512k	300k	5400	Y
1773-5	15	SCSI	2,7 RLL	3.5HH				Y
1774-6	15	SCSI	2,7 RLL	3.5HH				Y
1774-7	15	SCSI	2,7 RLL	3.5HH				Y

Drive Model	Format Size MB	Head	Cyl	Sect/ Trac	Translate H/C/S	RWC/ WPC	Land Zone
1775-8	180	8	1140	28		1141/1141	
1775-9	202	9	1140	28		1141/1141	
1908-15	1381	15	2112	V		Na/Na	Auto
1924-21	2100	21	2267	V		Na/Na	Auto
1924D	2100		2267	V		—/—	
1926	2158	15				Na/Na	Auto
1926-15	2158	15	2772	V		—/—	
1936	3022	21	2759	V		Na/Na	Auto
1936-21	3022	21	2772	V		Na/Na	Auto
1936AV	3022	21	2759	V		Na/Na	Auto
1936D	3022	21	2759	V		Na/Na	Auto
1991	9091	27	4446	V		—/—	
1991AV	9090	27	4477	V		—/—	
1991W	9090	27	4477	V		—/—	
1991WAV	9090	27	4477	V		—/—	
2100	512	15	2759			Na/Na	Auto
2105(A)	560	8	1745	V	16/1084/63	Na/Na	Auto
2105(S)	560	8	1745	V		Na/Na	Auto
2105-15	560	15	1747	V		Na/Na	Auto
2105A-15	560	15	1747	V		Na/Na	Auto
2108(A)	666	10	1745	V		Na/Na	Auto
2108(S)	666	10	1745	V		Na/Na	Auto
2112(A)	1050	15	1745	V	16/2034/63	Na/Na	Auto
2112(D)	1050	15	1744	V		Na/Na	Auto
2112(S)	1050	15	1745	V		Na/Na	Auto
2112-15	1050	15	1747	V		Na/Na	Auto
2112-DW	1050	15	1745	V		Na/Na	Auto
2112A-15	1050	15	1747	V		Na/Na	Auto
2121(A)				V		Na/Na	Auto
2121(S)				V		Na/Na	Auto
2205	585	5	2360			Na/Na	Auto
2205A	542	5				Na/Na	Auto
2207	701	9	2360	V		Na/Na	Auto
2210	1056	9	2360	V		Na/Na	Auto
2210 AV	1000	9	2360	V		—/—	
2210A	976	9	2360			Na/Na	Auto
2210AV	1056	9	2360			Na/Na	Auto
2210WD	1056	9	2360	V		—/—	
2217	1765	15	2360	V		Na/Na	Auto
2217 AV	1700	15	2360	V		—/—	
2217A	1626	15				Na/Na	Auto
2217AV	1765	15				Na/Na	Auto
2217WD	1765		2360	V		—/—	
3020	512	21	2759			Na/Na	Auto
3221	2050		3956	V		—/—	
3221AV	2050		3956	V		—/—	
3243	4294	19	4124	V		—/—	
3243AV	4290	19	4081			Na/Na	Auto
3243S	4294	19	3957	V		—/—	
3243W	4294	19	3956	V		—/—	
3243WAV	4294	19	3957	V		Na/Na	Auto
3243WD	4294	19	3956	V		—/—	
3243WDAV	4294	19	3956	V		—/—	
3387NS	8700		4811	V		—/—	
3387SS	8700		4811	V		—/—	
3387WS	8700		4811	V		—/—	
3391AV	9103		4811	V		—/—	
3391NS	9103		4811	V		—/—	
3391SS	9103		4811	V		—/—	
3391WAV	9103		4811	V		—/—	
3391WD	9103		4811	V		—/—	
3391WS	9103		4811	V		—/—	
3418NS	18250		7308	V		—/—	

Drive Model	Seek Time	Interface	Encode	Form Factor	cache kb	mtbf	RPM	Obsolete? ⇓
1775-8	15	SCSI	2,7 RLL	3.5HH				Y
1775-9	15	SCSI	2,7 RLL	3.5HH				Y
1908-15	11	SCSI-2Fast		5.25FH		150k	5400	Y
1924-21	12	SCSI-2Fast		5.25FH		250k	5400	Y
1924D	12	SCSI-2Fast		5.25FH		250k	5400	Y
1926	13	SCSI-2Fast		5.25FH	512k	250k	5400	Y
1926-15	13	SCSI-2		5.25FH		300k		Y
1936	12	SCSI-2Fast		5.25FH	256k	250k	5400	Y
1936-21	11.5	SCSI-2	2,7 RLL	5.25FH		300k		Y
1936AV	13	SCSI-2Fast	MZR	5.25FH	256k	250k	5400	Y
1936D	12	SCSI-2Fast		5.25FH	256k	250k	5400	Y
1991	12	SCSI-2Fast		5.25FH	512k	650k	5400	Y
1991AV	12	SCSI-2Fast	MZR	5.25FH	512k	650k	5400	Y
1991W	12	SCSI-2FstW	MZR	5.25FH	512k	650k	5400	Y
1991WAV	12	SCSI-2FstW	MZR	5.25FH	512k	650k	5400	Y
2100	13	SCSI-2Fast		5.25FH	512k	250k	5400	Y
2105(A)	10	IDE AT	RLL	3.5HH		300k		Y
2105(S)	10	SCSI-2	RLL	3.5HH		300k		Y
2105-15	10	SCSI-2Fast		3.5FH		300k	5400	Y
2105A-15	10	IDE AT		3.5FH		300k	5400	Y
2108(A)	10	IDE AT	RLL	3.5HH		300k		Y
2108(S)	10	SCSI-2	RLL	3.5HH		300k		Y
2112(A)	10	IDE AT	RLL	3.5FH		300k		Y
2112(D)	10	SCSI-2Diff		3.5HH		300k		Y
2112(S)	10	SCSI-2	RLL	5.25FH		300k		Y
2112-15	10	SCSI-2Fast	RLL	3.5FH		300k	5400	Y
2112-DW	10	SCSI-2FstW	RLL	3.5FH		300k	5400	Y
2112A-15	10	IDE AT	RLL	3.5FH		300k	5400	Y
2121(A)	10	IDE AT	RLL	3.5FH		300k		Y
2121(S)	10	SCSI-2	RLL	5.25FH		300k		Y
2205	10	SCSI-2Fast		3.5FH		300k	5400	Y
2205A	10	IDE AT		3.5FH	512k	300k	5400	Y
2207	10	SCSI-2Fast		3.5FH	512k	300k	5400	Y
2210	10	SCSI-2Fast		3.5FH	512k	300k	5400	Y
2210 AV		SCSI-2Fast	MZR	3.5HH	512k	300k	5400	Y
2210A	10	IDE AT		3.5FH	512k	300k	5400	Y
2210AV	10	SCSI-2Fast		3.5FH	512k	300k	5400	Y
2210WD	10	SCSI-2FstW		3.5FH		300k	5400	Y
2217	10	SCSI-2Fast		3.5FH		300k	5400	Y
2217 AV		SCSI-2Fast	MZR	3.5HH	512k	300k	5400	Y
2217A	10	IDE AT		3.5FH	512k	300k	5400	Y
2217AV	10	SCSI-2Fast		3.5FH	512k	300k	5400	Y
2217WD	10	SCSI-2FstW		3.5FH		300k	5400	Y
3020	13	SCSI-2Fast		5.25FH	512k	250k	5400	Y
3221	9	SCSI-2Fast		3.5FH		650k	7200	Y
3221AV	9	SCSI-2Fast		3.5FH		650k	7200	Y
3243	8.5	SCSI-2Fast		3.5FH	512k	650k	7200	Y
3243AV	9	SCSI-2Fast	MZR	3.5HH	512k	650k	7200	Y
3243S	9	SSA-SCSI		3.5FH	512k	650k	7200	Y
3243W	9	SCSI-2FstW	MZR	3.5HH	512k	650k	7200	Y
3243WAV	9	SCSI-2FstW	MZR	3.5HH	512k	650k	7200	Y
3243WD	9	SCSI-2Diff		3.5FH	512k	650k	7200	Y
3243WDAV	9	SCSI-2Diff		3.5FH	512k	650k	7200	Y
3387NS	8	Ultra SCSI3		3.5FH		650k	7200	Y
3387SS	8	Ultra SCSI3		3.5FH		650k	7200	Y
3387WS	8	Ultra SCSI3		3.5FH		650k	7200	Y
3391AV	8	Ultra SCSI3		3.5HH	2mb	650k	7200	Y
3391NS	8	Ultra SCSI3		3.5HH		650k	7200	Y
3391SS	8	SCSI-3Wide		3.5HH	2mb	650k	7200	Y
3391WAV	8	SCSI-3Wide		3.5HH	2mb	650k	7200	Y
3391WD	8	SCSI-3Wide		3.5HH		650k	7200	Y
3391WS	8	Ultra SCSI3		3.5HH		650k	7200	Y
3418NS	8	Ultra SCSI3		3.5 3H		1000k	7200	Y

Drive Model	Format Size MB	Head	Cyl	Sect/ Trac	Translate H/C/S	RWC/ WPC	Land Zone
3418SS	18250		7308	V		—/—	
3418WS	18250		7308	V		—/—	
3420AV	20270		7308	V		—/—	
3420WAV	20270		7308	V		—/—	
4110	1052	9				Na/Na	Auto
4110	1052		2415	V		—/—	
4110A	1057			V	16/1024/63	—/—	
4210	1000			V		—/—	
4221	2050		4150	V		—/—	
4221AV	2050	9	4050			Na/Na	Auto
4221W	2050	9	4150	V		Na/Na	Auto
4221WAV	2050	9	4150	V		Na/Na	Auto
4221WD	2050	9	4050	V		Na/Na	Auto
4221WDAV	2050	9	4150	V		—/—	
4341NS	4130		4811	V		—/—	
4341SS	4130		4811	V		—/—	
4341WS	4130		4811	V		—/—	
4345AV	4550		4811	V		—/—	
4345NS	4550		4811	V		—/—	
4345SD	4550		4811	V		—/—	
4345SS	4550		4811	V		—/—	
4345WAV	4550		4811	V		—/—	
4345WD	4550		4811	V		—/—	
4345WS	4550		4811	V		—/—	
4421	2147		4050	V		—/—	
4421AV	2050		4050	V		—/—	
4525A	2500	4	6807		16/4884/63	—/—	
4540A	4000	6	6807		16/7847/63	—/—	
4550A	5000	8	6807		16/9768/63	—/—	
4691AV	9100		7308	V		—/—	
4691NS	9100		7308	V		—/—	
4691SS	9100		7308	V		—/—	
4691WAV	9100		7308	V		—/—	
4691WS	9100		7308	V		—/—	
4721NS	2100		6565	V		—/—	
4743NS	4300		6565	V		—/—	
4743SS	4300		6565	V		—/—	
4743WS	4300		6565	V		—/—	
MICROSCIENCE INTERNATIONAL COR							
4050	44	5	1024	17		1025/1025	
4060	67	5	1024	26		—/—	
4070	62	7	1024	17		—/—	
4090	93	7	1024	26		—/—	
5040	45	3	855	35		Na/Na	Auto
5070	76	5	855	35		Na/Na	Auto
5070-20	86	5	960	35		Na/Na	Auto
5100	110	7	855	36		Na/Na	
5100-20	120	7	960	35		Na/Na	Auto
5160	159	7	1271	35		Na/Na	Auto
6100	110	7	855	36		Na/Na	Auto
7040	47	3	855	36		Na/Na	
7070-20	86	5	960	35		NA/960	960
7100	100	7	855	36		Na/Na	
7100-20	120	7	960	35		NA/960	960
7100-21	121	5	1077	44		NA/992	992
7200	200	7	1277	44		—/—	
7400	304	8	1904			Na/Na	Auto
8040	42	2	1024	40		Na/Na	Auto
8040MLC 48-000	42	2	1024	40		Na/Na	Auto
8080	85	2	1768	47		Na/Na	Auto
8200	152	4	1904			Na/Na	Auto
FH21200	1062	15	1921	72		Na/Na	Auto

Drive Model	Seek Time	Interface	Encode	Form Factor	cache kb	mtbf	RPM	Obsolete? ⇓
3418SS	8	SCSI-3Wide		3.5 3H		1000k	7200	Y
3418WS	8	SCSI-3Wide		3.5 3H		1000k	7200	Y
3420AV	8	Ultra SCSI3		3.5 3H	2mb	1000k	7200	Y
3420WAV	8	SCSI-3Wide		3.5 3H	2mb	1000k	7200	Y
4110	8.5	SCSI-2Fast		3.5 3H	512k	500k	5400	Y
4110	8.5	SCSI-2Fast		3.5 3H		500k	5400	Y
4110A	8.5	IDE AT		3.5 3H	512k	500k	5400	Y
4210		SCSI-2		3.5 3H			7200	Y
4221	9	SCSI-2Fast		3.5 3H	512k	650k	7200	Y
4221AV	9	SCSI-2Fast	MZR	3.5 3H	512k	650k	7200	Y
4221W	9	SCSI-2FstW	MZR	3.5 3H	512k	650k	7200	Y
4221WAV	9	SCSI-2FstW	MZR	3.5 3H	512k	650k	7200	Y
4221WD	9	SCSI-2Diff	MZR	3.5 3H	512k	650k	7200	Y
4221WDAV	9	SCSI-2Diff	MZR	3.5 3H	512k	650k	7200	Y
4341NS	8	Ultra SCSI3		3.5 3H		650k	7200	Y
4341SS	8	SCSI-3Wide		3.5 3H		650k	7200	Y
4341WS	8	SCSI-3Wide		3.5 3H		650k	7200	Y
4345AV	8	Ultra SCSI3		3.5 3H	2mb	650k	7200	Y
4345NS	8	Ultra SCSI3		3.5 3H		650k	7200	Y
4345SD	8	SCSI-3Wide		3.5 3H		650k	7200	Y
4345SS	8	SCSI-3Wide		3.5 3H		650k	7200	Y
4345WAV	8	SCSI-3Wide		3.5 3H	2mb	650k	7200	Y
4345WD	8	SCSI-3Wide		3.5 3H		650k	7200	Y
4345WS	8	SCSI-3Wide		3.5 3H		650k	7200	Y
4421	9	SCSI-2Fast		3.5 3H		650k	5400	Y
4421AV	9	SCSI-2Fast		3.5 3H		650k	5400	Y
4525A	10.5	EIDE	PRML	3.5 3H		400k	5200	Y
4540A	10.5	EIDE	PRML	3.5 3H		400k	5200	Y
4550A	10.5	EIDE	PRML	3.5 3H		400k	5200	Y
4691AV	7.9	Ultra SCSI3		3.5 3H	2mb	1000k	7200	Y
4691NS	7.9	Ultra SCSI3		3.5 3H	2mb	1000k	7200	Y
4691SS	7.9	SCSI-3Wide		3.5 3H	2mb	1000k	7200	Y
4691WAV	7.9	SCSI-3Wide		3.5 3H	2mb	1000k	7200	Y
4691WS	7.9	SCSI-3Wide		3.5 3H	2mb	1000k	7200	Y
4721NS	10	Ultra SCSI3		3.5 3H		400k	5400	Y
4743NS	10	Ultra SCSI3		3.5 3H		400k	5400	Y
4743SS	10	SCSI-3Wide		3.5 3H		400k	5400	Y
4743WS	10	SCSI-3Wide		3.5 3H		400k	5400	Y

MICROSCIENCE INTERNATIONAL COR

Drive Model	Seek Time	Interface	Encode	Form Factor	cache kb	mtbf	RPM	Obsolete?
4050	18	ST412/506	MFM	3.5HH		36k		Y
4060	18	ST412/506	2,7 RLL	3.5HH		36k		Y
4070	18	ST412/506	MFM	3.5HH		36k		Y
4090	18	ST412/506	2,7 RLL	3.5HH		36k		Y
5040	18	ESDI	2,7 RLL	3.5HH		36k		Y
5070	18	ESDI	2,7 RLL	3.5HH		36k		Y
5070-20	18	ESDI	2,7 RLL	3.5HH		36k		Y
5100	18	ESDI	2,7 RLL	3.5HH		36k		Y
5100-20	18	ESDI	2,7 RLL	3.5HH		60k		Y
5160	18	ESDI	2,7 RLL	3.5HH		36k		Y
6100	18	SCSI	2,7 RLL	3.5HH		36k		Y
7040	18	IDE AT	2,7 RLL	3.5HH				Y
7070-20	18	IDE AT	2,7 RLL	3.5HH		36k		Y
7100	18	IDE AT	2,7 RLL	3.5HH		36k		Y
7100-20	18	IDE AT	2,7 RLL	3.5HH		60k	3600	Y
7100-21	18	IDE AT	2,7 RLL	3.5HH		60k		Y
7200	18	IDE AT	2,7 RLL	3.5HH				Y
7400	15	IDE AT	2,7 RLL	3.5HH		100k		Y
8040	25	IDE AT	2,7 RLL	3.5 3H		20k		Y
8040MLC 48-000	25	IDE AT	2,7 RLL	3.5 3H		300k		Y
8080	17	IDE AT	2,7 RLL	3.5 3H		100k		Y
8200	16	IDE AT	2,7 RLL	3.5 3H		100k		Y
FH21200	14	ESDI	2,7 RLL	5.25FH		100k	3600	Y

Drive Model	Format Size MB	Head	Cyl	Sect/ Trac	Translate H/C/S	RWC/ WPC	Land Zone
FH21600	1418	15	2147	86		Na/Na	Auto
FH2414	366	8	1658	54		Na/Na	Auto
FH2777	687	15	1658	54		Na/Na	Auto
FH31200	1062	15	1921	72		Na/Na	Auto
FH31600	1418	15	2147	86		Na/Na	Auto
FH3414	366	8	1658	54		Na/Na	Auto
FH3777	687	15	1658	54		Na/Na	Auto
HH1050	44	5	1024	17		1025/1025	1023
HH1060	65	5	1024	26		1025/1025	
HH1075	62	7	1024	17		1025/1025	
HH1080	65	5	1024	26		—/—	
HH1090	80	7	1314	17		1315/1315	
HH1095	95	7	1024	26		1025/1025	
HH1120	122	7	1314	26		1315/1315	
HH2012	10	4	306	17		—/—	
HH2120	128	7	1024	35		Na/Na	
HH2160	160	7	1276	35		Na/Na	
HH312	10	4	306	17		307/307	
HH3120	121	5	1314	36		—/—	
HH315	10	4	306	17		307/307	
HH3160	170	7	1314	36		—/—	
HH325	21	4	612	17		613/613	615
HH330	32	4	612	26		613/613	
HH612	10	4	306	17		307/307	
HH625	21	4	612	17		613/613	
HH712	10	2	612	17		613/613	
HH712A	10	2	612	17		—/—	
HH725	21	4	612	17		613/613	615
HH738	32	4	612	26		613/613	
HH825	21	4	615	17		616/616	
HH830	33	4	615	26		616/616	

MINISCRIBE CORPORATION

Drive Model	Format Size MB	Head	Cyl	Sect/ Trac	Translate H/C/S	RWC/ WPC	Land Zone
1006	5	2	306	17		307/128	336
1012	10	4	306	17		307/128	336
2006	5	2	306	17		307/128	336
2012	10	4	306	17		307/128	336
3006	5	2	306	17		307/128	306
3012	10	2	612	19		613/128	656
3053	44	5	1024	17		1024/512	Auto
3085	68	7	1170	17		1170/512	Auto
3085E	72	3	1270	36		Na/Na	Auto
3085S	72	3	1255	125		Na/Na	Auto
3130E	112	5	1250	36		1251/512	Auto
3130S	112	5	1255	35		1256/512	Auto
3180E	150	7	1250	35		1251/512	Auto
3180S	153	7	1255	36		1256/512	Auto
3180SM	161	7	1250	36		Na/Na	Auto
3212	10	2	612	17		613/128	656
3212 PLUS	11	2	615	17		613/128	Auto
3412	10	4	306	17		307/128	336
3425	20	4	615	17		616/128	656
3425 PLUS	20	4	615	17		616/128	656
3425S	21	4	612	17		615/128	656
3438	32	4	615	26		616/128	656
3438 PLUS	32	4	615	26		616/128	656
3650	40	6	809	17		819/128	852
3650F	42	6	809	17		810/128	852
3650R	64	6	809	26		809/128	852
3675	63	6	809	26		810/128	852
4010	8	2	480	17		481/128	520
4020	16	4	480	17		481/128	520
5330	25	6	480	17		481/128	

Drive Model	Seek Time	Interface	Encode	Form Factor	cache kb	mtbf	RPM	Obsolete?
FH21600	14	ESDI	2,7 RLL	5.25FH		100k	3600	Y
FH2414	14	ESDI	2,7 RLL	5.25FH		100k		Y
FH2777	14	ESDI	2,7 RLL	5.25FH		50k	3600	Y
FH31200	14	SCSI	2,7 RLL	5.25FH		100k	3600	Y
FH31600	14	SCSI	2,7 RLL	5.25FH		100k	3600	Y
FH3414	14	SCSI	2,7 RLL	5.25FH		100k		Y
FH3777	14	SCSI	2,7 RLL	5.25FH		100k	3600	Y
HH1050	28	ST412/506	MFM	5.25HH		140k		Y
HH1060	28	ST412/506	2,7 RLL	5.25HH		140k		Y
HH1075	28	ST412/506	MFM	5.25HH				Y
HH1080	28	ST412/506	2,7 RLL	5.25HH		50k		Y
HH1090	28	ST412/506	MFM	5.25HH		40k		Y
HH1095	28	ST412/506	2,7 RLL	5.25HH				Y
HH1120	28	ST412/506	2,7 RLL	5.25HH		40k		Y
HH2012		ST412/506	MFM	5.25HH				Y
HH2120	28	ESDI (10)	2,7 RLL	5.25HH		40k		Y
HH2160	28	ESDI (10)	2,7 RLL	5.25HH		40k		Y
HH312	65	ST412/506	MFM	5.25HH				Y
HH3120	28	SCSI	2,7 RLL	5.25HH		40k		Y
HH315	65	ST412/506	MFM	5.25HH				Y
HH3160	28	SCSI	2,7 RLL	5.25HH		40k		Y
HH325	80	ST412/506	MFM	5.25HH				Y
HH330	105	ST412/506	2,7 RLL	5.25HH				Y
HH612	85	ST412/506	MFM	5.25HH				Y
HH625	65	ST412/506	MFM	5.25HH				Y
HH712	105	ST412/506	MFM	5.25HH				Y
HH712A	75	ST412/506	MFM	5.25HH				Y
HH725	105	ST412/506	MFM	5.25HH				Y
HH738	105	ST412/506	2,7 RLL	5.25HH				Y
HH825	65	ST412/506	MFM	5.25HH				Y
HH830	65	ST412/506	2,7 RLL	5.25HH				Y

MINISCRIBE CORPORATION

Drive Model	Seek Time	Interface	Encode	Form Factor	cache kb	mtbf	RPM	Obsolete?
1006	179	ST412/506	MFM	5.25FH		8k		Y
1012	179	ST412/506	MFM	5.25FH		8k		Y
2006	93	ST412/506	MFM	5.25FH		10k		Y
2012	85	ST412/506	MFM	5.25FH		10k		Y
3006		ST412/506	MFM	5.25HH				Y
3012	155	ST412/506	MFM	5.25HH		10k		Y
3053	25	ST412/506	MFM	5.25HH		30k	3600	Y
3085	22	ST412/506	MFM	5.25HH		40k	3600	Y
3085E	17	ESDI	2,7 RLL	5.25HH				Y
3085S	17	SCSI	2,7 RLL	5.25HH				Y
3130E	17	ESDI	2,7 RLL	5.25HH		35k	3600	Y
3130S	17	SCSI	2,7 RLL	5.25HH		35k	3600	Y
3180E	17	ESDI	2,7 RLL	5.25HH		35k	3600	Y
3180S	17	SCSI	2,7 RLL	5.25HH		35k	3600	Y
3180SM	17	SCSI(Mac)	RLL	5.25HH		35k		Y
3212	85	ST412/506	MFM	5.25HH		20k	3600	Y
3212 PLUS	53	ST412/506	MFM	5.25HH		20k	3600	Y
3412	60	ST412/506	MFM	5.25HH		11k		Y
3425	85	ST412/506	MFM	5.25HH		20k	3600	Y
3425 PLUS	53	ST412/506	MFM	5.25HH		20k	3600	Y
3425S	68	SCSI	MFM	5.25HH		20k		Y
3438	85	ST412/506	2,7 RLL	5.25HH		20k	3600	Y
3438 PLUS	53	ST412/506	2,7 RLL	5.25HH		20k	3600	Y
3650	61	ST412/506	MFM	5.25HH		25k	3600	Y
3650F	46	ST412/506	MFM	5.25HH		25k	3600	Y
3650R	61	ST412/506	2,7 RLL	5.25HH		25k	3600	Y
3675	61	ST412/506	2,7 RLL	5.25HH		25k		Y
4010	133	ST412/506	MFM	5.25FH		10k		Y
4020	133	ST412/506	MFM	5.25FH		10k		Y
5330	27	ST412/506	MFM	5.25FH				Y

Drive Model	Format Size MB	Head	Cyl	Sect/ Trac	Translate H/C/S	RWC/ WPC	Land Zone
5338	32	6	612	17		613/306	
5440	32	8	480	17		481/128	
5451	43	8	612	17		613/306	
6032	26	3	1024	17		1024/512	Auto
6053	44	5	1024	17		1024/512	Auto
6074	62	7	1024	17		1025/512	
6079	68	5	1024	26		1024/512	Auto
6085	71	8	1024	17		1024/512	Auto
6128	109	8	1024	26		1024/512	Auto
6170E	130	8	1024	34		Na/Na	Auto
6212	10	2	612	17		613/128	
7040A	40	2	1159	36	5/981/17	981/512	Auto
7040S	40	2	1156	36		Na/Na	Auto
7060A	65	2	1516	42	7/1024/17	Na/Na	Auto
7060S	65	2	1516	42		Na/Na	Auto
7080A	81	4	1159	36	10/981/17	981/512	Auto
7080S	81	4	1156	36		Na/Na	Auto
7120A	131	2	1516	85	14/1024/17	Na/Na	Auto
7120S	131	2	1516	85		Na/Na	Auto
7426	21	4	612	17		613/613	
8051A	41	4	745	26	4/745/28	746/128	Auto
8051S	43	4	745	26		746/128	Auto
80SC-MFM	21	4	615	17		—/—	
80SC-RLL	33	4	615	26		—/—	
8212	10	2	615	17		616/128	664
8225	20	2	771	26		772/128	810
8225A	21	2	747	28	4/615/17	Na/Na	Auto
8225AT	20	2	747	28		748/128	820
8225S	21	2	804	26		805/128	820
8225XT	20	2	805	26		806/128	820
8412	10	4	306	17		307/128	336
8425	21	4	615	17		616/128	664
8425F	20	4	615	17		616/128	664
8425S	21	4	612	17		616/128	664
8425XT	20	4	615	17		616/128	664
8434F	32	4	615	26		616/128	
8438	31	4	615	26		616/128	664
8438 PLUS	31	4	615	26		615/128	664
8438F	32	4	615	26		616/128	664
8438XT	31	4	615	26		Na/Na	664
8450	39	4	771	26		772/128	810
8450AT	42	4	745	28		746/128	820
8450S	42	4	804	26		805/128	820
8450XT	42	4	805	26		806/128	820
9000E	338	15	1224	36		Na/Na	Auto
9000S	347	15	1220	36		Na/Na	Auto
9230	203	9	1224	34		0/512	0
9230E	203	9	1224	36		Na/Na	Auto
9230S	203	9	1224	36		Na/Na	Auto
9380E	338	15	1224	36		NA/512	Auto
9380S	336	15	1218	36		NA/512	Auto
9380SM	319	15	1218			Na/Na	Auto
9424E	360	8	1661			Na/Na	Auto
9424S	355	8	1661			Na/Na	Auto
9780E	676	15	1661	53		NA/512	Auto
9780S	676	15	1661	53		166/512	Auto

MITSUBISHI ELECTRONICS

Drive Model	Format Size MB	Head	Cyl	Sect/ Trac	Translate H/C/S	RWC/ WPC	Land Zone
M2860-1	21			17		—/—	
M2860-2	50			17		—/—	
M2860-3	85			17		—/—	
MR335	69	7	743	26		—/—	
MR521	10	2	612	17		—/—	

Drive Model	Seek Time	Interface	Encode	Form Factor	cache kb	mtbf	RPM	Obsolete?
5338	27	ST412/506	MFM	5.25FH				Y
5440	27	ST412/506	MFM	5.25FH				Y
5451	27	ST412/506	MFM	5.25FH				Y
6032	28	ST412/506	MFM	5.25FH		25k	3600	Y
6053	28	ST412/506	MFM	5.25FH		25k	3600	Y
6074	28	ST412/506	MFM	5.25FH				Y
6079	28	ST412/506	2,7 RLL	5.25FH		25k	3600	Y
6085	28	ST412/506	MFM	5.25FH		25k	3600	Y
6128	28	ST412/506	2,7 RLL	5.25FH		25k	3600	Y
6170E	28	ESDI	RLL	5.25FH				Y
6212	27	ST412/506	MFM	5.25FH				Y
7040A	19	IDE AT	1,7 RLL	3.5 3H	32k	40k	3703	Y
7040S	19	SCSI	RLL	3.5 3H		40k		Y
7060A	15	IDE AT	1,7 RLL	3.5 3H		150k		Y
7060S	15	SCSI	1,7 RLL	3.5 3H		150k		Y
7080A	19	IDE AT	1,7 RLL	3.5 3H	32k	40k	3703	Y
7080S	19	SCSI	1,7 RLL	3.5 3H		150k		Y
7120A	15	IDE AT	1,7 RLL	3.5 3H		150k		Y
7120S	15	SCSI	1,7 RLL	3.5 3H		150k		Y
7426	27	ST412/506	MFM	3.5HH				Y
8051A	28	IDE AT	2,7 RLL	3.5HH	32k	150k	3484	Y
8051S	28	SCSI	2,7 RLL	3.5HH	32k	150k	3484	Y
80SC-MFM	68	ST412/506	MFM	3.5HH		20k	3600	Y
80SC-RLL	68	ST412/506	2,7 RLL	3.5HH		20k	3600	Y
8212	68	ST412/506	MFM	3.5HH		20k	3600	Y
8225	68	ST412/506	2,7 RLL	3.5HH		30k	3600	Y
8225A		IDE	2,7 RLL	3.5HH		30k	3600	Y
8225AT	40	IDE AT	2,7 RLL	3.5HH		30k	3600	Y
8225S	68	SCSI	2,7 RLL	3.5HH		30k	3600	Y
8225XT	68	IDE XT	2,7 RLL	3.5HH		30k	3600	Y
8412	50	ST412/506	MFM	3.5HH		20k	3600	Y
8425	68	ST506/412	MFM	3.5HH		20k	3600	Y
8425F	40	ST412/506	MFM	3.5HH		20k	3600	Y
8425S	68	SCSI	MFM	3.5HH		20k	3600	Y
8425XT	68	IDE XT	MFM	3.5HH		20k	3600	Y
8434F	40	ST412/506	RLL	3.5HH		20k	3600	Y
8438	68	ST412/506	RLL	3.5HH		20k	3600	Y
8438 PLUS	55	ST412/506	2,7 RLL	5.25HH		20k	3600	Y
8438F	40	ST412/506	2,7 RLL	5.25HH		20k	3600	Y
8438XT	68	IDE XT	RLL	3.5HH		20k	3600	Y
8450	45	ST412/506	2,7 RLL	3.5HH		20k	3600	Y
8450AT	40	IDE AT	2,7 RLL	3.5HH		30k	3600	Y
8450S	45	SCSI	2,7 RLL	3.5HH		20k	3600	Y
8450XT	68	IDE XT	2,7 RLL	3.5HH		20k	3600	Y
9000E	16	ESDI		5.25FH		30k		Y
9000S	16	SCSI		5.25FH		30k		Y
9230	16	ESDI	RLL	5.25FH				Y
9230E	16	ESDI	RLL	5.25FH				Y
9230S	16	SCSI	RLL	5.25FH				Y
9380E	16	ESDI	2,7 RLL	5.25FH		50k	3600	Y
9380S	16	SCSI	2,7 RLL	5.25FH		50k	3600	Y
9380SM	16	SCSI(Mac)	RLL	5.25FH		50k		Y
9424E	17	ESDI	2,7 RLL	5.25FH				Y
9424S	17	SCSI	2,7 RLL	5.25FH				Y
9780E	17	ESDI	1,7 RLL	5.25FH		50k	3600	Y
9780S	17	SCSI	1,7 RLL	5.25FH		30k	3600	Y

MITSUBISHI ELECTRONICS

Drive Model	Seek Time	Interface	Encode	Form Factor	cache kb	mtbf	RPM	Obsolete?
M2860-1		ST412/506	MFM					Y
M2860-2		ST412/506	MFM					Y
M2860-3		ST412/506	MFM					Y
MR335	20	ST412/506	MFM	3.5HH		30k		Y
MR521	85	ST412/506	MFM	5.25HH				Y

Drive Model	Format Size MB	Head	Cyl	Sect/ Trac	Translate H/C/S	RWC/ WPC	Land Zone
MR522	20	4	612	17		—/300	612
MR5310E	65	5	977	26		Na/Na	Auto
MR533	24	3	971	17		—/None	971
MR535	42	5	977	17		300/300	Auto
MR535-U00	42	5	977	17		300/300	
MR535R	65	5	977	26		Na/Na	Auto
MR535S	85	5	977	34		Na/Na	Auto
MR537S	65	5	977	26		Na/Na	Auto

MITSUMI ELECTRONICS CORP.

Drive Model	Format Size MB	Head	Cyl	Sect/ Trac	Translate H/C/S	RWC/ WPC	Land Zone
HD2509AA	92	4		>52		—/—	
HD2513AA	130	4		>52		—/—	

MMI

Drive Model	Format Size MB	Head	Cyl	Sect/ Trac	Translate H/C/S	RWC/ WPC	Land Zone
M106	5	2	306	17		—/128	
M112	10	4	306	17		—/128	
M125	20	8	306	17		—/128	
M212	10	4	306	17		—/128	
M225	20	8	306	17		—/128	
M306	5	2	306	17		—/128	
M312	10	4	306	17		—/128	
M325	20	8	306	17		—/128	
M350	42	8	612	17		—/288	

NCL AMERICA

Drive Model	Format Size MB	Head	Cyl	Sect/ Trac	Translate H/C/S	RWC/ WPC	Land Zone
SEE BRAND TECHNOLOGIES							—/—

NCR CORP

Drive Model	Format Size MB	Head	Cyl	Sect/ Trac	Translate H/C/S	RWC/ WPC	Land Zone
6091-5101	323	9				Na/Na	Auto
6091-5301	675	15				Na/Na	Auto
H6801-STD1-03-17	53	7	872	17		—/650	
H6801-STD1-07-17	45	3	868	34		Na/Na	Auto
H6801-STD1-10-17	104	8	776	33		Na/Na	Auto
H6801-STD1-12-17	42	2	1047	40		Na/Na	Auto
H6801-STD1-46-46	21	4	615	17		616/128	664
H6801-STD1-47-46	71	8	1024	17		1025/128	Auto
H6801-STD1-47-46	121	7	969	35		1025/128	Auto

NEC TECHNOLOGIES INC

Drive Model	Format Size MB	Head	Cyl	Sect/ Trac	Translate H/C/S	RWC/ WPC	Land Zone
D1711	42	2				—/—	
D1731	85	4				—/—	
D3126	21	4	615	17		616/256	
D3126H	21					—/—	
D3142	42	8	642	17		—/—	
D3146H	40	8	615	17		—/—	
D3661	118	7	915	36		Na/Na	Auto
D3713	345	16	670	63		—/—	
D3717	540	4	2924			—/—	
D3724	426	2			16/827/63	—/—	
D3725-351	730	4	3493		16/1416/63	—/—	
D3725-351	730	4			16/1416/63	—/—	
D3725-540	540	2			16/1416/63	Na/Na	Auto
D3727	1083	6	3493		16/2100/63	Na/Na	Auto
D3735	45	2	1084	41	4/542/41	—/—	Auto
D3741	40					—/—	
D3743	540	2			16/1048/63	—/—	
D3745-301	1080	4			16/2096/63	Na/Na	Auto
D3745-351	1080	4			16/2096/63	—/—	
D3747	1620	6	3678			—/—	
D3755	105	4	1250	41	8/625/41	—/—	Auto
D3756	105					—/—	

Drive Model	Seek Time	Interface	Encode	Form Factor	cache kb	mtbf	RPM	Obsolete? ⇓
MR522	85	ST412/506	MFM	5.25HH				Y
MR5310E	28	ESDI	2,7 RLL	5.25HH		30k		Y
MR533		ST412/506	MFM	5.25HH				Y
MR535	28	ST412/506	MFM	5.25HH		30k	3600	Y
MR535-U00	28	ST412/506	MFM	5.25HH		30k		Y
MR535R	28	ST412/506	2,7 RLL	5.25HH		30k	3600	Y
MR535S	28	SCSI	2,7 RLL	5.25HH		30k		Y
MR537S	28	SCSI	2,7 RLL	5.25HH		30k		Y

MITSUMI ELECTRONICS CORP.

Drive Model	Seek Time	Interface	Encode	Form Factor	cache kb	mtbf	RPM	Obsolete?
HD2509AA	16	IDE AT	1,7 RLL	2.5 4H	32k	150k	3600	Y
HD2513AA	16	IDE AT	1,7 RLL	2.5 4H	32k	150k	3600	Y

MMI

Drive Model	Seek Time	Interface	Encode	Form Factor	cache kb	mtbf	RPM	Obsolete?
M106	75	ST412/506	MFM	3.5HH				Y
M112	75	ST412/506	MFM	3.5HH				Y
M125	75	ST412/506	MFM	3.5HH				Y
M212	75	ST412/506	MFM	5.25HH				Y
M225	75	ST412/506	MFM	5.25HH				Y
M306	75	ST412/506	MFM	5.25HH				Y
M312	75	ST412/506	MFM	5.25HH				Y
M325	75	ST412/506	MFM	5.25HH				Y
M350	75	ST412/506	MFM	5.25HH				Y

NCL AMERICA

SEE BRAND TECHNOLOGIES

NCR CORP

Drive Model	Seek Time	Interface	Encode	Form Factor	cache kb	mtbf	RPM	Obsolete?
6091-5101	27	SCSI	2,7 RLL	5.25				Y
6091-5301	25	SCSI	2,7 RLL	5.25				Y
H6801-STD1-03-17	28	ST412/506	MFM	3.5HH		20k		Y
H6801-STD1-07-17	18	IDE AT	2,7 RLL	3.5HH		20k		Y
H6801-STD1-10-17	25	IDE AT	2,7 RLL	3.5HH		150k		Y
H6801-STD1-12-17	25	IDE AT	2,7 RLL	3.5 3H		150k		Y
H6801-STD1-46-46	68	ST412/506	MFM	3.5HH		20k		Y
H6801-STD1-47-46	28	ST412/506	MFM	5.25FH		40k		Y
H6801-STD1-47-46	16	ESDI (10)	2,7 RLL	5.25FH		100k		Y

NEC TECHNOLOGIES INC

Drive Model	Seek Time	Interface	Encode	Form Factor	cache kb	mtbf	RPM	Obsolete?
D1711	19	IDE/PCMCIA	1,7 RLL	4H	32k	100k	5400	Y
D1731	19	IDE/PCMCIA	1,7 RLL	4H	32k	100k	5400	Y
D3126	85	ST412/506	MFM	3.5HH				Y
D3126H		ST412/506	MFM	3.5HH				Y
D3142	28	ST412/506	MFM	3.5HH		30k		Y
D3146H	35	ST412/506	MFM	3.5HH				Y
D3661	20	ESDI (10)	2,7 RLL	3.5HH		30k		Y
D3713	12	IDE			64k			Y
D3717	12	IDE AT	1,7 RLL	3.5 3H	96k	250k	4500	Y
D3724	14	IDE	PRML 8,9	3.5 3H	256k	300k	4090	Y
D3725-351	11	IDE AT	1,7 RLL	3.5 3H	128k	300k	4090	Y
D3725-351	11	IDE AT	1,7 RLL	3.5 3H	128k	300k	4090	Y
D3725-540	11	IDE AT	1,7 RLL	3.5 3H	128k	300k	4090	Y
D3727	11	IDE AT	1,7 RLL	3.5 3H	128k	300k	4090	Y
D3735	25	IDE AT	1,7 RLL	3.5 3H		50k	3456	Y
D3741		IDE AT		3.5HH				Y
D3743	11	IDE	PRML 8,9	3.5 3H	128k	300k	4500	Y
D3745-301	11	IDE	PRML 8,9	3.5 3H	128k	300k	4500	Y
D3745-351	11	IDE	PRML 8,9	3.5 3H	128k	300k	4500	Y
D3747	11	IDE AT	PRML	3.5 3H	128k	300k	4500	Y
D3755	25	IDE AT	1,7 RLL	3.5 3H		50k	3456	Y
D3756		IDE AT		3.5HH				Y

Drive Model	Format Size MB	Head	Cyl	Sect/ Trac	Translate H/C/S	RWC/ WPC	Land Zone
D3761	114	7	915	35	7/915/35	—/—	
D3781	425	9	1464	63	9/1464/63	—/—	Auto
D3817	540	4				—/—	
D3825	730	4				—/—	
D3825	1083	6				—/—	
D3827	1083	6				—/—	
D3835	45	2	1084	41		—/—	Auto
D3841	45	8	440	25		—/—	
D3843	540	2				—/—	
D3845	1080	4				—/—	
D3847	1620	6				—/—	
D3855	105	4	1250	41		—/—	Auto
D3856	105					—/—	
D3861	114	7	915	35		—/—	
D3881	425	9	1464	63		—/—	Auto
D3896	2160	9				—/—	
D5114	5	2	306	17		—/—	
D5124	10	4	309	17		310/310	664
D5126	20	4	612	17		613/None	664
D5126H	21	4	612	17		613/None	664
D5146	40	8	615	17		616/None	664
D5146H	42	8	615	17		616/None	664
D5392	1322	16	615	17		—/—	
D5452	71	10	823	17		—/—	
D5652	143	10	823	34		Na/Na	
D5655	140	7	1224	35		Na/Na	1230
D5662	300	15	1224	35		Na/Na	
D5682	664	15	1633	53		Na/Na	Auto
D5862	301	15	1224	53		Na/Na	
D5882	664	15	1633	53		—/—	Auto
D5892	1404	19	1678	86		—/—	
DSE1700A	1706	4			16/3306/63	—/—	
DSE2010A	2010	6			16/3900/63	—/—	
DSE2100A	2100	5			16/4092/63	—/—	
DSE2550A	2550	6			16/4960/63	—/—	

NEI

Drive Model	Format Size MB	Head	Cyl	Sect/ Trac	Translate H/C/S	RWC/ WPC	Land Zone
RD3127	10	2	612	17		—/—	
RD3255	20	4	612	17		—/—	
RD4127	10	4	306	17		—/—	
RD4255	20	8	306	17		—/—	

NEWBURY DATA

Drive Model	Format Size MB	Head	Cyl	Sect/ Trac	Translate H/C/S	RWC/ WPC	Land Zone
NDR1065	55	7	918	17		—/—	
NDR1085	71	8	1024	17		—/None	1023
NDR1105	87	11	918	17		—/None	1023
NDR1140	120	15	918	17		—/None	1023
NDR2085	74	7	1224	17		1224/1224	
NDR2140	117	11	1224	17		1224/1224	
NDR2190	160	15	1224	17		—/None	1223
NDR3170S	146	9	1224	26		Na/Na	Auto
NDR320	21	4	615	17		—/None	615
NDR3280S	244	15	1224	26		—/—	
NDR3380S	319	15	1224	34		Na/Na	Auto
NDR340	42	8	615	17		—/None	615
NDR4175	179	7	1224	36		Na/Na	
NDR4380	338	15	1224	36		Na/Na	
NDR4380S	319	15	1224	34		—/—	
PENNY 340	42	8	615	17		615/615	

Drive Model	Seek Time	Interface	Encode	Form Factor	cache kb	mtbf	RPM	Obsolete? ⇓
D3761	20	IDE AT	2,7 RLL	3.5HH		30k		Y
D3781	15	IDE AT	1,7 RLL	3.5HH	64k	50k	3600	Y
D3817	12	SCSI-2	1,7 RLL	3.5 3H	96k	250k	4500	Y
D3825	11	SCSI-2	1,7 RLL	3.5 3H	64k	300k	4090	Y
D3825	11	SCSI-2	1,7 RLL	3.5 3H	32k	300k	4090	Y
D3827	11	SCSI-2	1,7 RLL	3.5 3H	64k	300k	4090	Y
D3835	25	SCSI	1,7 RLL	3.5 3H		50k	3456	Y
D3841	28	SCSI	1,7 RLL	3.5HH		30k		Y
D3843	11	SCSI-2	PRML 8,9	3.5 3H	64k	300k	4500	Y
D3845	11	SCSI-2	PRML	3.5 3H	64k	300k	4500	Y
D3847	11	SCSI-2	PRML8,9	3.5 3H	64k	300k	4500	Y
D3855	25	SCSI	1,7 RLL	3.5 3H		50k	3456	Y
D3856		SCSI		3.5HH				Y
D3861	20	SCSI	2,7 RLL	3.5HH		30		Y
D3881	15	SCSI	1,7 RLL	3.5HH	64k	50k	3600	Y
D3896	9	SCSI-2	1,7 RLL	3.5 3H	1024k	800k	7200	Y
D5114		ST412/506	MFM	5.25HH				Y
D5124	80	ST412/506	MFM	5.25HH				Y
D5126	80	ST412/506	MFM	5.25HH				Y
D5126H	40	ST412/506	MFM	5.25HH				Y
D5146	40	ST412/506	MFM	5.25HH				Y
D5146H	40	ST412/506	MFM	5.25HH				Y
D5392	14	IPI-2		5.25FH		100k		Y
D5452		ST412/506	MFM	5.25FH				Y
D5652	23	ESDI	2,7 RLL	5.25HH				Y
D5655	18	ESDI	2,7 RLL	5.25HH		30k		Y
D5662	18	ESDI	2,7 RLL	5.25FH		30k		Y
D5682	16	ESDI	RLL 1,7	5.25FH		50k	3600	Y
D5862	18	SCSI		5.25FH		30k		Y
D5882	16	SCSI	1,7 RLL	5.25FH		50k	3600	Y
D5892	14	SCSI	1,7 RLL	5.25FH		100k		Y
DSE1700A	11	IDE	PRML 8,9	3.5 3H	128k	300k	5200	
DSE2010A	11	IDE	PRML 8,9	3.5 3H	128k	300k	5200	Y
DSE2100A	11	IDE	PRML 8,9	3.5 3H	128k	300k	5200	
DSE2550A	11	IDE	PRML 8,9	3.5 3H	128k	300k	5200	

NEI

Drive Model	Seek Time	Interface	Encode	Form Factor	cache kb	mtbf	RPM	Obsolete?
RD3127		ST412/506	MFM	5.25				Y
RD3255		ST412/506	MFM	5.25				Y
RD4127		ST412/506	MFM	5.25				Y
RD4255		ST412/506	MFM	5.25				Y

NEWBURY DATA

Drive Model	Seek Time	Interface	Encode	Form Factor	cache kb	mtbf	RPM	Obsolete?
NDR1065	25	ST412/506	MFM	5.25FH				Y
NDR1085	26	ST412/506	MFM	5.25FH				Y
NDR1105	25	ST412/506	MFM	5.25FH				Y
NDR1140	25	ST412/506	MFM	5.25FH				Y
NDR2085		ST412/506	MFM	5.25FH				Y
NDR2140		ST412/506	MFM	5.25FH				Y
NDR2190	28	ST412/506	MFM	5.25FH				Y
NDR3170S	28	SCSI	2,7 RLL	5.25FH				Y
NDR320		ST412/506	MFM	5.25FH				Y
NDR3280S	28	SCSI	2,7 RLL	5.25FH				Y
NDR3380S	28	SCSI	2,7 RLL	5.25FH		50k		Y
NDR340	40	ST412/506	MFM	3.5HH				Y
NDR4175	28	ESDI	2,7 RLL	5.25FH				Y
NDR4380	28	ESDI	2,7 RLL	5.25FH				Y
NDR4380S	28	SCSI	RLL	5.25FH				Y
PENNY 340		ST412/506	MFM	5.25HH				Y

Drive Model	Format Size MB	Head	Cyl	Sect/ Trac	Translate H/C/S	RWC/ WPC	Land Zone
NPL							
4064	5			17		—/—	
4127	10			17		—/—	
4191S	15			17		—/—	
4255	20			17		—/—	
4362	30			17		—/—	
NP02-13	11	4	320	17		NA/0	320
NP02-26A/26S	22	4	640	17		NA/0	640
NP02-52A	44	8	640	17		NA/640	640
NP03-20	16	6	306	17		NA/0	306
NP04-13T	10	6		17		—/—	
NP04-55	45	7	754	17		NA/0	754
NP04-85	72	11	754	17		NA/0	754
OKIDATA							
OD526	31	4	640	26		651/651	
OD540	51	6	640	26		651/651	
OLIVETTI							
HD662/11	10	2	612	17		—/—	
HD662/12	20	4	612	17		—/—	
XM3220	21	4	612	17		NA/128	656
XM5210	10	2	612	17		—/—	
XM5220/2	20	4	612	17		—/—	
XM5221	21	4	615	17		NA/256	700
XM5340	42	6	820	17		256/256	819
XM5360	42	6	820	17		128/128	819
XM563-12	10					—/—	
OPTIMA TECHNOLOGY CORP							
Concorde 1050	990	15				Na/Na	Auto
Concorde 1350	1342					Na/Na	Auto
Concorde 23	22130					—/—	
Concorde 23W	22130					—/—	
Concorde 635	640	14				Na/Na	Auto
Concorde 9000	8669					Na/Na	Auto
Concorde 9000W	8669					Na/Na	Auto
Diskovery 1000	1001					Na/Na	Auto
Diskovery 1000	2040					Na/Na	Auto
Diskovery 130	137					Na/Na	Auto
Diskovery 1800DHW	1763					Na/Na	Auto
Diskovery 200	200					Na/Na	Auto
Diskovery 2100W	2040					Na/Na	Auto
Diskovery 325	321					Na/Na	Auto
Diskovery 40	45					Na/Na	Auto
Diskovery 4100	4095					Na/Na	Auto
Diskovery 4100W	4095					Na/Na	Auto
Diskovery 420	416	8				Na/Na	Auto
Diskovery 500	520					Na/Na	Auto
Diskovery 9000	8683					—/—	
Diskovery 9000W	8683					—/—	
MINIPAK 100	104	4				Na/Na	Auto
MINIPAK 1000	1001					Na/Na	Auto
MINIPAK 200	209	8				Na/Na	Auto
MINIPAK 2100	2040					Na/Na	Auto
MINIPAK 2100	2040					Na/Na	Auto
MINIPAK 300	320					Na/Na	Auto
MINIPAK 40	45					Na/Na	Auto
MINIPAK 4100	4095					Na/Na	Auto
MINIPAK 500	520					Na/Na	Auto

Drive Model	Seek Time	Interface	Encode	Form Factor	cache kb	mtbf	RPM	Obsolete? ⇓
NPL								
4064		ST412/506	MFM	5.25FH				Y
4127		ST412/506	MFM	5.25FH				Y
4191S		ST412/506	MFM	5.25FH				Y
4255		ST412/506	MFM	5.25FH				Y
4362		ST412/506	MFM	5.25FH				Y
NP02-13	95	ST412/506	MFM	5.25FH				Y
NP02-26A/26S	40	ST412/506	MFM	5.25HH				Y
NP02-52A	40	ST412/506	MFM	5.25HH				Y
NP03-20	85	ST412/506	MFM	3.5FH				Y
NP04-13T	85	ST412/506	MFM	5.25FH				Y
NP04-55	35	ST412/506	MFM	5.25FH				Y
NP04-85	35	ST412/506	MFM	3.5HH				Y
OKIDATA								
OD526	85	ST412/506	2,7 RLL	5.25HH				Y
OD540	85	ST412/506	2,7 RLL	5.25HH				Y
OLIVETTI								
HD662/11	27	ST412/506	MFM	5.25HH				Y
HD662/12	27	ST412/506	MFM	5.25HH				Y
XM3220	85	ST412/506	MFM	3.5HH				Y
XM5210	65	ST412/506	MFM	5.25HH				Y
XM5220/2	85	ST412/506	MFM	5.25FH				Y
XM5221	40	ST412/506	MFM	5.25HH				Y
XM5340	40	ST412/506	MFM	5.25HH				Y
XM5360	40	ST412/506		5.25HH				Y
XM563-12		ST412/506	MFM	5.25FH				Y
OPTIMA TECHNOLOGY CORP								
Concorde 1050	15	SCSI	2,7 RLL	5.25		150k		Y
Concorde 1350	14	SCSI	2,7 RLL	5.25		150k		Y
Concorde 23	13.2	Ultra SCSI		5.25FH		800k		
Concorde 23W	13.2	Ultra Wide		5.25FH		800k		
Concorde 635	16	SCSI	2,7 RLL	5.25		150k		Y
Concorde 9000	11	SCSI-2Fast	2,7 RLL	5.25FH		500k	5400	
Concorde 9000W	11	SCSI-2FstW	2,7 RLL	5.25FH		500k	5400	
Diskovery 1000	9	SCSI-2Fast	2,7 RLL	3.5 4H		800k	5400	Y
Diskovery 1000	8	SCSI-2Fast	2,7 RLL	3.5 4H		500k	5400	Y
Diskovery 130	20	SCSI	2,7 RLL	5.25		50k		Y
Diskovery 1800DHW	8	SCSI-2FstW	2,7 RLL	3.5HH		500k		Y
Diskovery 200	15	SCSI	2,7 RLL	5.25		150k		Y
Diskovery 2100W	8	SCSI-2FstW	2,7 RLL	3.5 4H		500k	7200	Y
Diskovery 325	14	SCSI	2,7 RLL	5.25		150k		Y
Diskovery 40	25	SCSI	2,7 RLL	5.25		50k		Y
Diskovery 4100	8	SCSI-2Fast	2,7 RLL	3.5HH		800k	7200	
Diskovery 4100W	8	SCSI-2FstW	2,7 RLL	3.5HH		800k	7200	
Diskovery 420	16	SCSI	2,7 RLL	5.25		100k		Y
Diskovery 500	12	SCSI-2Fast	2,7 RLL	3.5 4H		300k	5411	Y
Diskovery 9000	9	Ultra SCSI		3.5HH		1000k		
Diskovery 9000W	9	Ultra Wide		3.5HH		1000k		
MINIPAK 100	25	SCSI	2,7 RLL	3.5HH		30k		Y
MINIPAK 1000	9	SCSI-2Fast	2,7 RLL	3.5 4H		800k	5400	Y
MINIPAK 200	20	SCSI	2,7 RLL	3.5HH		40k		Y
MINIPAK 2100	8	SCSI-2Fast	2,7 RLL	3.5 4H		500k	7200	Y
MINIPAK 2100	8	SCSI-2FstW	2,7 RLL	3.5 4H		500k	7200	Y
MINIPAK 300	13	SCSI	2,7 RLL	3.5HH		150k		Y
MINIPAK 40	25	SCSI	2,7 RLL	3.5HH		30k		Y
MINIPAK 4100	8	SCSI-2Fast	2,7 RLL	3.5HH		800k	7200	
MINIPAK 500	12	SCSI-2Fast	2,7 RLL	3.5 4H		300k	5411	Y

Drive Model	Format Size MB	Head	Cyl	Sect/ Trac	Translate H/C/S	RWC/ WPC	Land Zone
ORCA TECHNOLOGY CORP							
320A	370	9				Na/Na	Auto
320S	370	9				Na/Na	Auto
400A	470	9				Na/Na	Auto
400S	470	9				Na/Na	Auto
760E	760	15	1564			Na/Na	Auto
760S	760	15	1564			Na/Na	Auto
OTARI							
SEE DISCTRON						—/—	
PACIFIC MAGTRON							
MT3050	50	2	1062	46		—/—	
MT3100	100	4	1062	46		—/—	
MT4115E	115	4	1597			—/—	
MT4115S	115	4	1597			—/—	
MT4140E	140	5	1597			—/—	
MT4140S	140	5	1597			—/—	
MT4170E	170	6	1597			—/—	
MT4170S	170	6	1597			—/—	
MT5760E	676	15	1632	54		Na/Na	Auto
MT5760S	673	15	1632	54		Na/Na	Auto
MT6120S	1050	15	1927	71		Na/Na	Auto
PANASONIC							
JU116	20	4	615	17		616/616	
JU128	42	7	733	17		734/734	
PLUS DEVELOPMENT							
HARDCARD 20	21	4	615	17		Na/Na	Auto
HARDCARD 40	42	8	612	17		Na/Na	Auto
HARDCARD II-40	40	5	925	17		Na/Na	Auto
HARDCARD II-80	80	10	925	17		Na/Na	Auto
HARDCARD II-XL105	105	15	806	17		—/—	
HARDCARD II-XL50	52	10	601	17		—/—	
IMPULSE 105AT/LP	105	16	755	17	16/755/17	—/—	Auto
IMPULSE 105S	105	6	1019			—/—	Auto
IMPULSE 105S/LP	105	4	1056			—/—	Auto
IMPULSE 120AT	120	5	1123	42	9/814/32	—/—	Auto
IMPULSE 120S	120	5	1123	42		—/—	Auto
IMPULSE 170AT	169	7	1123	42	10/966/34	—/—	Auto
IMPULSE 170S	169	7	1123	42		—/—	Auto
IMPULSE 210AT	174	7	1156	42	13/873/36	—/—	Auto
IMPULSE 210S	174	7	1156	42		—/—	Auto
IMPULSE 330AT	331					—/—	Auto
IMPULSE 330S	331					—/—	Auto
IMPULSE 40AT	41	5	965	17	5/968/17	Na/Na	Auto
IMPULSE 40S	42	3	834			—/—	Auto
IMPULSE 425AT	425					—/—	Auto
IMPULSE 425S	425					—/—	Auto
IMPULSE 52AT/LP	52	8	751	17	8/751/17	—/—	Auto
IMPULSE 52S/LP	52	2				—/—	Auto
IMPULSE 80AT	83	10	965	17	6/611/17	Na/Na	Auto
IMPULSE 80AT/LP	85	16	616	17	6/611/17	—/—	Auto
IMPULSE 80S	84	6	918			—/—	Auto
IMPULSE 80S/LP	85	4				—/—	Auto
PRAIRIETEK CORP							
PRAIRIE 120	21	2	615	34		—/—	
PRAIRIE 140	42	4	615	34		Na/Na	Auto
PRAIRIE 220A	20	4	612	16		—/—	

Drive Model	Seek Time	Interface	Encode	Form Factor	cache kb	mtbf	RPM	Obsolete?
ORCA TECHNOLOGY CORP								
320A	12	IDE AT	2,7 RLL	3.5HH		100k		Y
320S	12	SCSI	2,7 RLL	3.5HH		100k		Y
400A	12	IDE AT	2,7 RLL	3.5HH		100k		Y
400S	12	SCSI	2,7 RLL	3.5HH		100k		Y
760E	14	ESDI	2,7 RLL	5.25		50k		Y
760S	14	SCSI	2,7 RLL	5.25		50k		Y
OTARI								
SEE DISCTRON								
PACIFIC MAGTRON								
MT3050	20	IDE AT	2,7 RLL	5.25HH		60k		Y
MT3100	20	IDE AT	2,7 RLL	5.25HH		60k		Y
MT4115E	16	ESDI	2,7 RLL	5.25HH		100k		Y
MT4115S	16	SCSI	2,7 RLL	5.25HH		100k		Y
MT4140E	16	ESDI	2,7 RLL	5.25HH		100k		Y
MT4140S	16	SCSI	2,7 RLL	5.25HH		100k		Y
MT4170E	16	ESDI	2,7 RLL	5.25HH		100k		Y
MT4170S	16	SCSI	2,7 RLL	5.25HH		100k		Y
MT5760E	14	ESDI (15)	1,7 RLL	5.25FH		150k		Y
MT5760S	14	SCSI	1,7 RLL	5.25FH		150k		Y
MT6120S	14	SCSI	1,7 RLL	5.25FH		150k		Y
PANASONIC								
JU116	85	ST412/506	MFM	3.5HH		5		Y
JU128	35	ST412/506	MFM	3.5HH		5		Y
PLUS DEVELOPMENT								
HARDCARD 20	40	IDE AT	2,7 RLL	3.5 3H		60k		Y
HARDCARD 40	40	IDE AT	2,7 RLL	3.5 3H		60k		Y
HARDCARD II-40	25	IDE AT	2,7 RLL	3.5 3H				Y
HARDCARD II-80	25	IDE AT	2,7 RLL	3.5 3H				Y
HARDCARD II-XL105	17	IDE AT	2,7 RLL	CARD 3H				Y
HARDCARD II-XL50	17	IDE AT	2,7 RLL	CARD 3H				Y
IMPULSE 105AT/LP	17	IDE AT	2,7 RLL	3.5 3H		60k		Y
IMPULSE 105S	19	SCSI-2	2,7 RLL	3.5HH		50k		Y
IMPULSE 105S/LP	17	SCSI-2	2,7 RLL	3.5 3H		60k		Y
IMPULSE 120AT	15	IDE AT	1,7 RLL	3.5HH		50k	3605	Y
IMPULSE 120S	15	SCSI-2	1,7 RLL	3.5HH		50k	3605	Y
IMPULSE 170AT	15	IDE AT	1,7 RLL	3.5HH		50k	3605	Y
IMPULSE 170S	15	SCSI-2	1,7 RLL	3.5HH		50k	3605	Y
IMPULSE 210AT	15	IDE AT	1,7 RLL	3.5HH		50k	3605	Y
IMPULSE 210S	15	SCSI-2	1,7 RLL	3.5HH		50k	3605	Y
IMPULSE 330AT	14	IDE AT	1,7 RLL	3.5HH		75k		Y
IMPULSE 330S	14	SCSI-2	1,7 RLL	3.5HH		75k		Y
IMPULSE 40AT	19	IDE AT	2,7 RLL	3.5HH		50k	3660	Y
IMPULSE 40S	19	SCSI-2	2,7 RLL	3.5HH		50k	3660	Y
IMPULSE 425AT	14	SCSI-2	1,7 RLL	3.5HH		75k		Y
IMPULSE 425S	14	SCSI-2	1,7 RLL	3.5HH		75k		Y
IMPULSE 52AT/LP	17	IDE AT	2,7 RLL	3.5 3H		60k	3660	Y
IMPULSE 52S/LP	17	SCSI-2	2,7 RLL	3.5 3H		60k		Y
IMPULSE 80AT	19	IDE AT	2,7 RLL	3.5HH		50k	3660	Y
IMPULSE 80AT/LP	17	IDE AT	2,7 RLL	3.5 3H		60k	3660	Y
IMPULSE 80S	19	SCSI-2	2,7 RLL	3.5HH		50k	3660	Y
IMPULSE 80S/LP	17	SCSI-2	2,7 RLL	3.5 3H		60k		Y
PRAIRIETEK CORP								
PRAIRIE 120	23	IDE AT	2,7 RLL	2.5 4H		20k		Y
PRAIRIE 140	23	IDE AT	2,7 RLL	2.5 4H		20k		Y
PRAIRIE 220A	28	IDE AT	2,7 RLL	2.5 3H		20k		Y

Drive Model	Format Size MB	Head	Cyl	Sect/ Trac	Translate H/C/S	RWC/ WPC	Land Zone
PRAIRIE 220S	20	4	612	16		—/—	
PRAIRIE 240	42	4	615	34		—/—	
PRAIRIE 242A	42	4	615	34		Na/Na	Auto
PRAIRIE 242S	42	4	615	34		Na/Na	Auto
PRAIRIE 282A	82	4		34		Na/Na	Auto
PRAIRIE 282S	82	4		34		Na/Na	Auto

PRIAM CORPORATION

Drive Model	Format Size MB	Head	Cyl	Sect/ Trac	Translate H/C/S	RWC/ WPC	Land Zone
3504	32	4	820	26		—/—	
502	46	7	755	17		756/756	
504	46	7	755	17		756/756	
514	117	11	1224	17		—/—	
519	160	15	1224	17		1225/1225	
519	244	11	1224	26		—/—	
617	153	7	1225			Na/Na	
628	241	11	1225			Na/Na	
638	329	15	1225			Na/Na	
717	153	7	1225			1226/1226	
728	241	11	1225			1226/1226	
738	329	15	1225			1226/1226	
ID/ED040	42	5	987	17		—/—	
ID/ED045	50	5	1166	17		—/—	
ID/ED060	62	7	1018	17		—/—	
ID/ED062	71	7	1166	17		—/—	
ID/ED075	74	5	1166	25		—/—	
ID/ED100	122	7	1314	26		—/—	
ID/ED1000	1046	15	1919	71		—/—	Auto
ID/ED120	121	7	1024	33		Na/Na	Auto
ID/ED130	159	15	1224	17		—/—	
ID/ED150	160	7	1276	35		Na/Na	Auto
ID/ED160	158	7	1225	36		Na/Na	Auto
ID/ED230	235	15	1224	25		—/—	
ID/ED240	243	15	1220	26		—/—	
ID/ED250	248	11	1225	36		Na/Na	Auto
ID/ED660	675	15	1628	54		—/—	Auto
ID100	103	7	1166	25		—/—	
ID1000	1034	15	1919	71		Na/Na	Auto
ID120	119	7	1024	33		Na/Na	
ID130	132	15	1224	17		—/—	
ID150	158	7	1276	35		Na/Na	
ID160	158	7	1218	36		—/—	Auto
ID160H	156	7	1225	36		Na/Na	Auto
ID20	25	3	987	17		—/—	
ID230	233	15	1224	25		—/—	
ID250	246	11	1225	36		Na/Na	
ID330	339	15	1218	36		—/—	
ID330D	337	15	1225	36		Na/Na	
ID330E	337	15	1218	36		—/—	
ID330E-PS/2	330	15	1195	36		—/—	
ID330S	338	15	1225	36		Na/Na	Auto
ID340H	340	7	1218	36		—/—	Auto
ID40	42	5	987	17		—/—	
ID40AT	40	5	1018	17		—/—	
ID45	44	5	1018	17		—/—	
ID45H	44	5	1024	17		—/—	
ID60	59	7	1018	17		—/—	
ID60AT	59	7	1018	17		—/—	
ID62	62	7	1166	17		—/—	
ID660	660	15	1632	54		Na/Na	Auto
ID75	73	5	1166	25		—/—	
V130	39	3	987	26		988/988	987
V150	42	5	987	17		988/988	987
V160	50	5	1166	17		1167/1167	

Drive Model	Seek Time	Interface	Encode	Form Factor	cache kb	mtbf	RPM	Obsolete?
PRAIRIE 220S	28	SCSI	2,7 RLL	2.5 3H		20k		Y
PRAIRIE 240	28	IDE AT	2,7 RLL	2.5 3H		20k		Y
PRAIRIE 242A	23	IDE XT-AT	2,7 RLL			20k		Y
PRAIRIE 242S	23	SCSI	2,7 RLL			20k		Y
PRAIRIE 282A	28	IDE AT	2,7 RLL			20k		Y
PRAIRIE 282S	23	SCSI	2,7 RLL			20k		Y
PRIAM CORPORATION								
3504	27	ST412/506	2,7 RLL	3.5HH				Y
502	22	ST412/506	MFM	5.25FH				Y
504	22	ST412/506	MFM	5.25FH				Y
514	22	ST412/506	MFM	5.25FH				Y
519	22	ST412/506	MFM	5.25FH		40k		Y
519	22	ST412/506	2,7 RLL	5.25FH		40		Y
617	20	ESDI	2,7 RLL	5.25FH		40k		Y
628	20	ESDI	2,7 RLL	5.25FH		40k		Y
638	20	SCSI	2,7 RLL	5.25FH		40k		Y
717	20	SCSI	2,7 RLL	5.25FH		40k		Y
728	20	SCSI	2,7 RLL	5.25FH		40k		Y
738	20	SCSI	2,7 RLL	5.25FH		40k		Y
ID/ED040	23	ST412/506	MFM	5.25FH		40k		Y
ID/ED045	23	ST412/506	MFM	5.25FH		40k		Y
ID/ED060	30	ST412/506	MFM	5.25FH		40k		Y
ID/ED062	23	ST412/506	MFM	5.25FH		40k		Y
ID/ED075	23	ST412/506		5.25FH		40k		Y
ID/ED100	15	ST412/506	2,7 RLL	5.25HH		40k		Y
ID/ED1000	14	SCSI		5.25FH		150k		Y
ID/ED120	28	ESDI	2,7 RLL	5.25HH				Y
ID/ED130	13	ST412/506	MFM	5.25FH		40k		Y
ID/ED150	28	ESDI	2,7 RLL	5.25HH				Y
ID/ED160	18	ESDI	2,7 RLL	5.25FH				Y
ID/ED230	11	ST412/506		5.25FH		40k		Y
ID/ED240	28	ST412/506	2,7 RLL	5.25FH				Y
ID/ED250	18	ESDI		5.25FH				Y
ID/ED660	16	SCSI		5.25FH		150k		Y
ID100	15	ST412/506	2,7 RLL	5.25FH		40k		Y
ID1000	14	ESDI		5.25FH		150k		Y
ID120	28	ESDI	2,7 RLL	5.25FH				Y
ID130	13	ST412/506	MFM	5.25FH		40k		Y
ID150	28	ESDI	2,7 RLL	5.25FH				Y
ID160	28	SCSI		5.25FH		150k		Y
ID160H	28	ESDI	2,7 RLL	5.25FH		150k		Y
ID20	23	ST412/506	MFM	5.25FH		40k		Y
ID230	11	ST412/506	2,7 RLL	5.25FH		40k		Y
ID250	18	ESDI	2,7 RLL	5.25FH				Y
ID330	18	SCSI	2,7 RLL	5.25FH				Y
ID330D	18	ESDI	2,7 RLL	5.25FH				Y
ID330E	18	ESDI	2,7 RLL	5.25FH				Y
ID330E-PS/2	18	PS/2	2,7 RLL	5.25FH				Y
ID330S	18	SCSI	2,7 RLL	5.25FH				Y
ID340H	14	ESDI	2,7 RLL	5.25FH		150k		Y
ID40	23	ST412/506	MFM	5.25FH		40k		Y
ID40AT	23	ST412/506	MFM	5.25FH		150k		Y
ID45	23	ST412/506	MFM	5.25FH		150k		Y
ID45H	25	ST412/506	MFM	5.25HH		40k		Y
ID60	30	ST412/506	MFM	5.25FH		40k		Y
ID60AT	23	ST412/506	MFM	5.25FH		150k		Y
ID62	23	ST412/506	MFM	5.25FH		40k		Y
ID660	16	ESDI	2,7 RLL	5.25FH		150k		Y
ID75	23	ST412/506	2,7 RLL	5.25FH		40k		Y
V130		ST412/506	2,7 RLL	5.25FH				Y
V150		ST412/506	MFM	5.25FH				Y
V160		ST412/506	MFM	5.25FH				Y

Drive Model	Format Size MB	Head	Cyl	Sect/ Trac	Translate H/C/S	RWC/ WPC	Land Zone
V170	60	7	987	17		988/988	987
V170R	91	7	987	26		988/988	987
V185	72	7	1166	17		1167/1167	1165
V519	159	15	1224	17		—/None	1223
PROCOM TECHNOLOGY							
ATOM-AT1300	1350					—/—	
ATOM-AT1302	1350					—/—	
ATOM-AT2001	2160					—/—	
ATOM-AT3000	3050					—/—	
ATOM-AT340	340					Na/Na	Auto
ATOM-AT4000	4090				16/7944/63	—/—	
ATOM-AT500	528					—/—	
ATOM-AT5000	5020				16/10380/63	—/—	
ATOM-AT6000	6480				15/13424/63	—/—	
ATOM-AT800	811					—/—	
ATOM-AT8000	8100				16/15880/63	—/—	
BRAVOPAQ120	124	14	1024	17		—/—	Auto
BRAVOPAQ40	42	5	977	17		—/—	Auto
HIPER 145	150	8	1024	36		—/—	
HIPER 155	160	9	966	36		—/—	
HIPER 20	21	4	615	17		—/—	
HIPER 30	33	4	615	26		—/—	
HIPER 330	337	15	1224	36		—/—	
HIPER 380	388	16	755	63		—/—	
HIPER 48	48	6	615	26		—/—	
HIPER/II 155	157	64	150	32		—/—	
HIPER/II 380	383	64	365	32		—/—	
HIPER/II 65	65	9	925	17		—/—	
MD100	104	64	102	32		—/—	
MD1003 (external)	1080	4	4826	116		—/—	
MD120	122					—/—	
MD20	21	64	21	32		—/—	
MD200	209	32	200	32		—/—	
MD2003 (external)	2160	5	2149	148		—/—	
MD2103 (external)	2147	11	3711	86-125		—/—	
MD2103W (external)	2100					—/—	
MD240	245					—/—	
MD30	30	64	30	32		—/—	
MD320	337	64	317	32		—/—	
MD420	442	64	415	32		—/—	
MD4303 (external)	4350					—/—	
MD4303W (external)	4294					—/—	
MD45	45	64	45	32		—/—	
MD544 (external)	541	2	4901	108		—/—	
MD80	83	64	80	32		—/—	
MD9103	9100					—/—	
MD9103W	9100					—/—	
MTD1000	1037	64	989	32		—/—	
MTD1350	1420					—/—	
MTD1900	1900					—/—	
MTD2000	2040					—/—	
MTD2800	2800					—/—	
MTD320-10	337	64	317	32		—/—	
MTD585	601	64	573	32		—/—	
MTD650	676	64	650	32		—/—	
MTD9000 (external)	9090					—/—	
PAT100	110	14	535	29		—/—	Auto
PAT40	42	4	805	26		—/—	Auto
PH.D20	21	4	615	17		—/—	
PH.D2520	21	4	615	17		—/—	
PH.D2545	45	7	733	17		—/—	
PH.D30	33	4	615	26		—/—	

Drive Model	Seek Time	Interface	Encode	Form Factor	cache kb	mtbf	RPM	Obsolete?
V170	28	ST412/506	MFM	5.25FH				Y
V170R	28	ST412/506	MFM	5.25FH				Y
V185	28	ST412/506	MFM	5.25FH				Y
V519	20		MFM	5.25FH				Y

PROCOM TECHNOLOGY

Drive Model	Seek Time	Interface	Encode	Form Factor	cache kb	mtbf	RPM	Obsolete?
ATOM-AT1300	13	ATA-2		2.5 4H	128k	300k	4200	Y
ATOM-AT1302	13	ATA-2		2.5 4H	128k	300k	4200	Y
ATOM-AT2001	13	ATA-2		2.5 4H	128k	300k	4200	
ATOM-AT3000	13	ATA-2		2.5 4H	128k	300k	4852	
ATOM-AT340	16	IDE		2.5 4H	120k	300k		Y
ATOM-AT4000	12	ATA-3			512k	300k	4000	
ATOM-AT500	13	IDE		2.5 4H	128k	300k		Y
ATOM-AT5000	12	ATA-3			512k	300k	4000	
ATOM-AT6000	13	ATA-3			512k	300k	4200	
ATOM-AT800	13	IDE		2.5 4H	128k	300k		Y
ATOM-AT8000	12	ATA-3			512k	300k	4900	
BRAVOPAQ120	19	IDE AT	RLL	3.5HH		150k		Y
BRAVOPAQ40	25	IDE AT	RLL	3.5HH		150k		Y
HIPER 145	23	ESDI		5.25FH		30		Y
HIPER 155	16.5	SCSI	RLL	5.25FH		100k		Y
HIPER 20	40	ST412/506	MFM			150k		Y
HIPER 30	28	ST412/506	RLL			150k		Y
HIPER 330	18	ESDI		5.25FH		30k		Y
HIPER 380	16	SCSI	RLL	5.25FH		100k		Y
HIPER 48	28	ST412/506	RLL			150k		Y
HIPER/II 155	16.5	ESDI	RLL	5.25FH		100k		Y
HIPER/II 380	16	ESDI	RLL	5.25FH		100k		Y
HIPER/II 65	28	ST412/506	MFM	5.25FH		40k		Y
MD100	18	SCSI	RLL			70k		Y
MD1003 (external)	12.5	SCSI-2Fast		3.5	128k	300k	5376	Y
MD120	17	SCSI	RLL					Y
MD20	28	SCSI	RLL			150k		Y
MD200	18	SCSI	RLL			70k		Y
MD2003 (external)	8.5	SCSI-2Fast		3.5	448k	800k	5400	Y
MD2103 (external)	9	SCSI-2Fast		3.5	512k	800k	7200	Y
MD2103W (external)	8	SCSI-2FstW		3.5			7200	Y
MD240	17	SCSI	RLL	3.5				Y
MD30	28	SCSI	RLL			150k		Y
MD320	12	SCSI	RLL			100k		Y
MD420	14	SCSI	RLL	3.5 3H		100k		Y
MD4303 (external)	8	SCSI-2Fast		3.5HH	512k	1000k	7200	
MD4303W (external)	8	SCSI-2FstW		3.5	1024k	800k	7200	
MD45	28	SCSI	RLL			150k		Y
MD544 (external)	12	SCSI-2Fast		3.5	64k	300k	5400	Y
MD80	24	SCSI	RLL			150k		Y
MD9103	9	Ultra SCSI			2024k	1000k	7200	
MD9103W	9	WIDE SCSI			2024k	1000k	7200	
MTD1000	15	SCSI	RLL ZBR			100k		Y
MTD1350	15	SCSI	RLL ZBR			100k		Y
MTD1900	12.9	SCSI	RLL ZBR			100k		Y
MTD2000	11	SCSI-2Fast	RLL ZBR				5400	
MTD2800	11	SCSI-2Fast	RLL ZBR				5400	
MTD320-10	10.7	SCSI	RLL ZBR			100k		Y
MTD585	16.5	SCSI	RLL ZBR			100k		Y
MTD650	15.5	SCSI	RLL ZBR			100k		Y
MTD9000 (external)	11	SCSI-2Fast			1024k	500k		
PAT100	15	IDE AT	RLL	3.5HH		150		Y
PAT40	25	IDE AT	RLL	5.25HH		150k		Y
PH.D20	40	ST412/506	MFM	3.5HH		150		Y
PH.D2520	40	ST412/506	MFM	3.5HH		30k		Y
PH.D2545	25	ST412/506	MFM	3.5HH		30k		Y
PH.D30	28	ST412/506	RLL	3.5HH		150		Y

Drive Model	Format Size MB	Head	Cyl	Sect/ Trac	Translate H/C/S	RWC/ WPC	Land Zone
PH.D30-CE	33	4	615	26		—/—	
PH.D3020	21	4	615	17		—/—	
PH.D45	45	7	773	17		—/—	
PH.D48	49	6	615	26		—/—	
PH.D5045	45	7	773	17		—/—	
PIRA 100	101	8	776	33		—/—	
PIRA 120	124	14	1024	17		—/—	Auto
PIRA 200	210	12	954	36		—/—	Auto
PIRA 40	42	5	977	17		—/—	Auto
PIRA 50-120	210	14	1024	36		—/—	Auto
PIRA 50-200	210	12	954	36		—/—	Auto
PIRA 50-270	270					—/—	
PIRA 50-340	340					—/—	
PIRA 50-420	420					—/—	
PIRA 55-270	270					—/—	
PIRA 55-340	340					—/—	
PIRA 55-420	420					—/—	
PR-IDE1000	1080					—/—	
PR-IDE1200	1200					—/—	
PR-IDE1600	1629					—/—	
PR-IDE16800	16800					—/—	
PR-IDE2000	2113					—/—	
PR-IDE210	210					—/—	
PR-IDE270	270					—/—	
PR-IDE3000	3200				16/6296/63	—/—	
PR-IDE340	340					—/—	
PR-IDE420	420					—/—	
PR-IDE4300U	4300				16/8400/63	—/—	
PR-IDE500	510					—/—	
PR-IDE6400U	6400				16/12592/63	—/—	
PR-IDE800	800					—/—	
PR-IDE8400	8400				16/16383/63	—/—	
PROPAQ/N100	101	8	776	33		—/—	Auto
PROPAQ/N120-19	124	14	1024	17		—/—	Auto
PROPAQ/N185-15	189	12	1023	33		—/—	Auto
PROPAQ/N40	40	4	805	26		—/—	Auto
PROPAQ/N40N	40	6	560	26		—/—	Auto
PROPAQ/S100	101	8	776	33		—/—	Auto
PROPAQ/S120-19	124	14	1024	17		—/—	Auto
PROPAQ/S185-15	189	12	1023	33		—/—	Auto
PROPAQ/S40	40	4	805	26		—/—	Auto
PROPAQ/S40N	40	6	560	26		—/—	Auto
PROPAQ100	101	8	776	33		—/—	Auto
PROPAQ120-19	124	14	1024	17		—/—	Auto
PROPAQ185-15	189	12	1023	33		—/—	Auto
PROPAQ185-15	189	5				Na/Na	Auto
PROPAQ40	40	4	805	26		—/—	Auto
PROPAQ40N	40	6	560	26		—/—	Auto
PROTON 523	520					—/—	
SI100	104	64	102	32		—/—	
SI1000	1037	64		32		—/—	
SI1000/S5	1037	8				Na/Na	Auto
SI1003	1080	4	4826	116		—/—	
SI1003/C	1080	4	4826	116		—/—	
SI200	209	64	200	32		—/—	
SI200/PS3	209	4				Na/Na	Auto
SI2003	2160	5	2149	148		—/—	
SI2003/C	2160	5	2149	148		—/—	
SI2103	2147	11	3711	86-125		—/—	
SI2103/C	2147	11	3711	86-125		—/—	
SI2103W/C	2100					—/—	
SI320-10	337	64	317	32		—/—	
SI320H	331	64	339	32		—/—	

Drive Model	Seek Time	Interface	Encode	Form Factor	cache kb	mtbf	RPM	Obsolete?
PH.D30-CE	28	ST412/506	RLL	3.5HH		150		Y
PH.D3020	40	ST412/506	MFM	3.5HH		30k		Y
PH.D45	25	ST412/506	MFM	3.5HH		150		Y
PH.D48	28	ST412/506	RLL	3.5HH		150		Y
PH.D5045	25	ST412/506	MFM	3.5HH		150k		Y
PIRA 100	25	IDE AT		3.5HH		20k		Y
PIRA 120	18	IDE AT	RLL	3.5HH		150		Y
PIRA 200	15	IDE AT	RLL	3.5HH		150k		Y
PIRA 40	28	IDE AT	RLL	3.5HH		150k		Y
PIRA 50-120	19	IDE AT	RLL	3.5HH		150k		Y
PIRA 50-200	15	IDE AT	RLL	3.5HH		150k		Y
PIRA 50-270	14	IDE AT	RLL	3.5HH		150k		Y
PIRA 50-340	15	IDE AT	RLL	3.5HH		150k		Y
PIRA 50-420	14	IDE AT	RLL	3.5HH		150k		Y
PIRA 55-270	14	IDE				150k		Y
PIRA 55-340	15	IDE				150k		Y
PIRA 55-420	14	IDE				150k		Y
PR-IDE1000	14	ATA-2			128k	350k	3800	Y
PR-IDE1200	10	IDE		3H				Y
PR-IDE1600	12	ATA-2			128k	300k	4480	Y
PR-IDE16800	8.5	Ultra ATA			464k	300k	5400	
PR-IDE2000	10.5	ATA-2			256k	500k	5400	Y
PR-IDE210	14	IDE		3H				Y
PR-IDE270	14	IDE		3H				Y
PR-IDE3000	9.5	Ultra ATA			256k	300k	5400	
PR-IDE340	12	IDE		3H				Y
PR-IDE420	14	IDE		3H				Y
PR-IDE4300U	8.5	Ultra ATA			476k	300k	5400	
PR-IDE500	12	IDE		3H				Y
PR-IDE6400U	8.5	Ultra ATA			476k	300k	5400	
PR-IDE800	12	IDE		3H				Y
PR-IDE8400	8.5	Ultra ATA			476k	300k	5400	
PROPAQ/N100	25	IDE AT	RLL	3.5HH		100k		Y
PROPAQ/N120-19	19	IDE AT	RLL	3.5HH		150k		Y
PROPAQ/N185-15	15	IDE AT	RLL	3.5HH		150k		Y
PROPAQ/N40	25	IDE AT	RLL	3.5HH		100k		Y
PROPAQ/N40N	25	IDE AT	RLL	3.5HH		150k		Y
PROPAQ/S100	25	IDE AT	RLL	3.5HH		20k		Y
PROPAQ/S120-19	19	IDE AT	RLL	3.5HH		150k		Y
PROPAQ/S185-15	15	IDE AT	RLL	3.5HH		150k		Y
PROPAQ/S40	25	IDE AT	RLL	3.5HH		100k		Y
PROPAQ/S40N	25	IDE AT	RLL	3.5HH		150k		Y
PROPAQ100	25	IDE AT	RLL	3.5HH		100k		Y
PROPAQ120-19	19	IDE AT	RLL	3.5HH		150k		Y
PROPAQ185-15	15	IDE AT	RLL	3.5HH		150k		Y
PROPAQ185-15		IDE AT	RLL	3.5HH		70k		Y
PROPAQ40	25	IDE AT	RLL	3.5HH		100k		Y
PROPAQ40N	25	IDE AT	RLL	3.5HH		150k		Y
PROTON 523	12	PCMCIA-ATA						
SI100	18	SCSI	RLL			70k		Y
SI1000	15	SCSI	RLL	5.25FH		100k		
SI1000/S5	15	SCSI		5.25		40k		
SI1003	12	SCSI-2Fast		3.5 3H	128k	300k	5376	
SI1003/C	12	SCSI-2Fast		3.5 3H	128k	300k	5376	
SI200	18	SCSI	RLL			70k		Y
SI200/PS3	18	SCSI	2,7 RLL	3.5HH		70k		Y
SI2003	8.5	SCSI-2Fast		3.5 3H		800k	5400	
SI2003/C	8.5	SCSI-2Fast		3.5 3H		800k	5400	
SI2103	9	SCSI-2Fast		3.5 3H	512k	800k	7200	
SI2103/C	9	SCSI-2Fast		3.5 3H	512k	800k	7200	
SI2103W/C	8	SCSI-2FstW		3.5 3H			7200	
SI320-10	10.7	SCSI	RLL	5.25FH		100k		Y
SI320H	14	SCSI	RLL	5.25HH		100k		Y

Drive Model	Format Size MB	Head	Cyl	Sect/ Trac	Translate H/C/S	RWC/ WPC	Land Zone
SI420H	435	64	415	32		—-/—-	
SI4303	4350	10	5288	165		—-/—-	
SI4303/C	4350	10	5288	165		—-/—-	
SI4303W/C	4300					—-/—-	
SI45	48	64	45	32		—-/—-	
SI544	541	2	4901	108		—-/—-	
SI544/C	544	2	4901	108		—-/—-	
SI585	601	64	415	32		—-/—-	
SI585/PS5	601	8				Na/Na	Auto
SI585/S5	601	8				Na/Na	Auto
SI650	662	64	632	32		—-/—-	
SI80	83	64	80	32		—-/—-	
SI9000/S5	9090					—-/—-	
SI9103	9100	20	5273153-23			—-/—-	
SI9103W	9100	20	5273153-23			—-/—-	
SI9103W/C	9100	20	5273153-23			—-/—-	

PTI (PERIPHERAL TECHNOLOGY)

Drive Model	Format Size MB	Head	Cyl	Sect/ Trac	Translate H/C/S	RWC/ WPC	Land Zone
PL100 TURBO	105	4				Na/Na	Auto
PL200 TURBO	210	7				Na/Na	Auto
PL32 TURBO	320	14				Na/Na	Auto
PT225	21	4	615	17		—-/—-	
PT234	28	4	820	17		—-/—-	
PT238A	32	4	615	26		Na/Na	
PT238R	32	4	615	26		—-/—-	
PT238S	32	4	615	26		—-/—-	
PT251A	51	4	820	26		—-/—-	
PT251R	44	4	820	26		—-/—-	
PT251S	44	4	820	26		—-/—-	
PT338	32	6	615	17		—-/—-	
PT351	42	6	820	17		—-/—-	
PT357A	49	6	615	26		—-/—-	
PT357R	49	6	615	26		—-/—-	
PT357S	49	6	615	26		—-/—-	
PT376A	65	6	820	26		Na/Na	
PT376R	65	6	820	26		—-/—-	
PT376S	65	6	820	26		—-/—-	
PT4102A	87	8	820	26		—-/—-	
PT4102R	87	8	820	26		—-/—-	
PT4102S	87	8	820	26		—-/—-	
PT468	57	8	820	17		—-/—-	

QUANTUM CORPORATION

Drive Model	Format Size MB	Head	Cyl	Sect/ Trac	Translate H/C/S	RWC/ WPC	Land Zone
ATLAS II 2.2S	2275	5		V		—-/—-	
ATLAS II 4.5S	4550	10		V		—-/—-	
ATLAS II 9.1S	9100	20		V		—-/—-	
ATLAS III 18.2S	18200	20				—-/—-	
ATLAS III 18.2S	18200	20				—-/—-	
ATLAS III 18.2S	18200	20				—-/—-	
ATLAS III 4.5S	4550	5				—-/—-	
ATLAS III 4.5S	4550	5				—-/—-	
ATLAS III 4.5S	4550	5				—-/—-	
ATLAS III 9.1S	9100	10				—-/—-	
ATLAS III 9.1S	9100	10				—-/—-	
ATLAS III 9.1S	9100	10				—-/—-	
ATLAS XP31070S	1075	5		80-134		—-/—-	
ATLAS XP32150S	2150	10		80-134		—-/—-	
ATLAS XP34300S	4350	20		80-134		—-/—-	
BIGFOOT 1275	1275	2		144-23	16/2492/63	—-/—-	
BIGFOOT 2.1	2110	4		149-27		—-/—-	
BIGFOOT 2550	2550	4		144-23	16/4994/63	—-/—-	
BIGFOOT CY 2.1	2111	2			16/4092/63	—-/—-	
BIGFOOT CY 4.3	4335	4			15/8960/63	—-/—-	

Drive Model	Seek Time	Interface	Encode	Form Factor	cache kb	mtbf	RPM	Obsolete?
SI420H	16	SCSI	RLL	5.25FH		100k		Y
SI4303	8	SCSI-2Fast		3.5HH	512k	1000k	7200	
SI4303/C	8	Ultra SCSI			512k	1000k	7200	
SI4303W/C	8	SCSI-2FstW		3.5HH	1024k	800k	7200	
SI45	28	SCSI	RLL			150k		Y
SI544	12	SCSI-2Fast		3.5 3H	64k	300k	5400	
SI544/C	12	SCSI-2Fast		3.5 3H	64k	300k	5400	
SI585	16.5	SCSI	RLL	5.25FH		100k		
SI585/PS5	17	SCSI		5.25		100k		
SI585/S5	17	SCSI		5.25		100k		
SI650	15.5	SCSI	RLL	5.25FH		100k		
SI80	24	SCSI	RLL			150k		Y
SI9000/S5	11	SCSI-2Fast		5.25FH	1024k	500k		
SI9103	9	Ultra SCSI			2024k	1000k	7200	
SI9103W	9	SCSI WIDE			2024k	1000k	7200	
SI9103W/C	9	WIDE SCSI			2024k	1000k	7200	

PTI (PERIPHERAL TECHNOLOGY)

Drive Model	Seek Time	Interface	Encode	Form Factor	cache kb	mtbf	RPM	Obsolete?
PL100 TURBO	19	SCSI	2,7 RLL	3.5HH		60k		Y
PL200 TURBO	19	SCSI	2,7 RLL	3.5HH		50k		Y
PL32 TURBO	12	SCSI	2,7 RLL	3.5HH		100k		Y
PT225	35	ST412/506	MFM	3.5HH				Y
PT234	35	ST412/506	MFM	3.5HH				Y
PT238A	35	IDE AT	2,7 RLL	3.5HH				Y
PT238R	35	ST412/506	2,7 RLL	3.5HH				Y
PT238S	35	SCSI	2,7 RLL	3.5HH				Y
PT251A	35	IDE AT	2,7 RLL	3.5HH		25k		Y
PT251R	35	ST412/506	2,7 RLL	3.5HH		25		Y
PT251S	35	SCSI	2,7 RLL	3.5HH		25k		Y
PT338	35	ST412/506	MFM	3.5HH		25k		Y
PT351	35	ST412/506	MFM	3.5HH				Y
PT357A	35	IDE AT	2,7 RLL	3.5HH		25k		Y
PT357R	35	ST412/506	2,7 RLL	3.5HH				Y
PT357S	35	SCSI	2,7 RLL	3.5HH		25k		Y
PT376A	35	IDE AT	2,7 RLL	3.5HH		25k		Y
PT376R	35	ST412/506	2,7 RLL	3.5HH		25k		Y
PT376S	35	SCSI	2,7 RLL	3.5HH		25k		Y
PT4102A	35	IDE AT	2,7 RLL	3.5HH		25k		Y
PT4102R	35	ST412/506	2,7 RLL	3.5HH		25k		Y
PT4102S	35	SCSI		3.5HH		25k		Y
PT468	35	ST412/506	MFM	3.5HH		25k		Y

QUANTUM CORPORATION

Drive Model	Seek Time	Interface	Encode	Form Factor	cache kb	mtbf	RPM	Obsolete?
ATLAS II 2.2S	8	SCSI-3	1,7 RLL	3.5 3H	512k	1000k	7200	Y
ATLAS II 4.5S	8	SCSI-3	1,7 RLL	3.5 3H	512k	1000k	7200	Y
ATLAS II 9.1S	8	SCSI-3	1,7 RLL	3.5HH	1024k	1000k	7200	Y
ATLAS III 18.2S	7.5	ULTRA 2 LVD	Prml16,17	3.5HH	1024k		7200	
ATLAS III 18.2S	7.5	FC	Prml16,17	3.5HH	1024k		7200	
ATLAS III 18.2S	7.5	UL SE SCSI3	Prml16,17	3.5HH	1024k		7200	
ATLAS III 4.5S	7.5	UL SE SCSI3	Prml16,17	3.5 3H	1024k		7200	
ATLAS III 4.5S	7.5	FC	Prml16,17	3.5 3H	1024k		7200	
ATLAS III 4.5S	7.5	Ultra 2 LVD	Prml16,17	3.5 3H	1024k		7200	
ATLAS III 9.1S	7.5	FC	Prml16,17	3.5 3H	1024k		7200	
ATLAS III 9.1S	7.5	UL SE SCSI3	Prml16,17	3.5 3H	1024k		7200	
ATLAS III 9.1S	7.5	ULTRA 2 LVD	Prml16,17	3.5 3H	1024k		7200	
ATLAS XP31070S	8	SCSI-2Fast	1,7 RLL	3.5 3H	1024k	800k	7200	Y
ATLAS XP32150S	8	SCSI-2Fast	1,7 RLL	3.5 3H	1024k	800k	7200	Y
ATLAS XP34300S	8	SCSI-2Fast	1,7 RLL	3.5HH	1024k	800k	7200	Y
BIGFOOT 1275	15.5	ATA-2 Fast	Prml16,17	5.25 4H	128k	300k	3600	Y
BIGFOOT 2.1	15.5	ATA-2 Fast	Prml16,17	5.25 4H	128k	300k	3600	Y
BIGFOOT 2550	15.5	ATA-2 Fast	Prml16,17	5.25 4H	128k	300k	3600	
BIGFOOT CY 2.1	<12	ATA-2 Fast		5.25 4H	128k	300k	3600	Y
BIGFOOT CY 4.3	<14	ATA-2 Fast		5.25 3H	128k	300k	3600	

Drive Model	Format Size MB	Head	Cyl	Sect/ Trac	Translate H/C/S	RWC/ WPC	Land Zone
BIGFOOT CY 6.4	6510	6			15/13456/63	—/—	
CAPELLA VP31110S	1108	4		97-149		—/—	
CAPELLA VP32210S	2216	8		97-149		—/—	
DAYTONA 127AT	127	2		54-92	9/677/41	Na/Na	Auto
DAYTONA 127S	127	2		54-92		Na/Na	Auto
DAYTONA 170AT	256	3		54-92	10/538/62	Na/Na	Auto
DAYTONA 170S	170	3		54-92		Na/Na	Auto
DAYTONA 256AT	256	4		54-92	11/723/63	Na/Na	Auto
DAYTONA 256S	256	4		54-92		Na/Na	Auto
DAYTONA 341AT	341	6		54-92	15/1011/44	Na/Na	Auto
DAYTONA 341S	341	6		54-92		Na/Na	Auto
DAYTONA 514AT	514	8		54-92	16/996/63	Na/Na	Auto
DAYTONA 514S	514	8		54-92		Na/Na	Auto
DSP3053LS	535	4		59-119		—/—	
DSP3107LS	1070	8		59-119		—/—	
DSP3133LS	1337	10		59-119		—/—	
DSP3210S	2148	16		59-119		—/—	
ELS127AT	127	3	1536	V	16/919/17	Na/Na	Auto
ELS127S	127	3	1536	V		Na/Na	Auto
ELS170AT	170	4	1536	V	15/1011/22	Na/Na	Auto
ELS170S	170	4	1536	V		Na/Na	Auto
ELS42AT	42	1	1536	V	5/968/17	Na/Na	Auto
ELS42S	42	1	1536	V		Na/Na	Auto
ELS85AT	85	2	1536	V	10/977/17	Na/Na	Auto
ELS85S	85	2	1536	V		Na/Na	Auto
Empire 1080S	1080	8				Na/Na	Auto
Empire 1400S	1400	8		72-137		—/—	
Empire 2100S	2100	12		72-137		—/—	
Empire 540S	540	4				Na/Na	Auto
Empire II VP32181S	2180	5				—/—	
Empire II VP34360S	4360	10				—/—	
Empire II VP39100S	9100	20	311586-126			—/—	
EUROPA 1080AT	1080	8		66-110	15/2362/60	—/—	
EUROPA 540AT	540	4		66-110	15/1179/60	—/—	
EUROPA 810AT	810	6		66-110	15/1771/63	—/—	
Fireball 1080AT	1089	4		88-177	16/2112/63	—/—	
Fireball 1080S	1093	4		88-177		—/—	
Fireball 1280AT	1280	4		95-177	16/2484/63	—/—	
Fireball 1280S	1280	4		95-177		—/—	
Fireball 540AT	544	2		88-177	16/1056/63	—/—	
Fireball 540S	545	2		88-177		—/—	
Fireball 640AT	640	2		95-177	16/1244/63	—/—	
Fireball 640S	640	2		95-177		—/—	
Fireball SE 2.1AT 1-98	2111	2			16/4092/63	—/—	
Fireball SE 2.1S 1-98	2111	2				—/—	
Fireball SE 3.2AT 1-98	3228	3			16/6256/63	—/—	
Fireball SE 3.2S 1-98	3228	3				—/—	
Fireball SE 4.3AT 1-98	4310	4			9/14848/63	—/—	
Fireball SE 4.3S 1-98	4310	4				—/—	
Fireball SE 6.4AT 1-98	6448	6			15/13328/63	—/—	
Fireball SE 6.4S 1-98	6448	6				—/—	
Fireball SE 8.4AT 1-98	8455	8			16/16383/63	—/—	
Fireball SE 8.4S 1-98	8455	8				—/—	
Fireball ST 1.6	1614	2			16/3128/63	—/—	
Fireball ST 2.1	2111	3				—/—	
Fireball ST 3.2AT	3228	4			16/6256/63	—/—	
Fireball ST 3.2S	3228	4				—/—	
Fireball ST 4.3AT	4310	6			9/14848/63	—/—	
Fireball ST 4.3S	4310	6				—/—	
Fireball ST 6.4AT	6448	8			15/13328/63	—/—	
Fireball ST 6.4S	6448	8				—/—	
Fireball TM 1.0	1089	2		104-23	16/2112/63	—/—	
Fireball TM 1.2AT	1281	2		104-23	16/2484/63	—/—	

Drive Model	Seek Time	Interface	Encode	Form Factor	cache kb	mtbf	RPM	Obsolete?
BIGFOOT CY 6.4	<14	ATA-2 Fast		5.25 3H	128k	300k	3600	
CAPELLA VP31110S	9	SCSI-2Fast	1,7 RLL	3.5 3H	1024k	800k	5400	Y
CAPELLA VP32210S	9	SCSI-2Fast	1,7 RLL	3.5 3H	1024k	800k	5400	Y
DAYTONA 127AT	17	IDE AT	1,7 RLL	2.5 4H	96k	350k	4500	Y
DAYTONA 127S	17	SCSI-2	1,7 RLL	2.5 4H	96k	350k	4500	Y
DAYTONA 170AT	17	IDE AT	1,7 RLL	2.5 4H	96k	350k	4500	Y
DAYTONA 170S	17	SCSI-2	1,7 RLL	2.5 4H	96k	350k	4500	Y
DAYTONA 256AT	17	IDE AT	1,7 RLL	2.5 4H	96k	350k	4500	Y
DAYTONA 256S	17	SCSI-2	1,7 RLL	2.5 4H	96k	350k	4500	Y
DAYTONA 341AT	17	IDE AT	1,7 RLL	2.5 4H	96k	350k	4500	Y
DAYTONA 341S	17	SCSI-2	1,7 RLL	2.5 4H	96k	350k	4500	Y
DAYTONA 514AT	17	IDE AT	1,7 RLL	2.5 4H	96k	350k	4500	Y
DAYTONA 514S	17	SCSI-2	1,7 RLL	2.5 4H	96k	350k	4500	Y
DSP3053LS	9.5	SCSI-2Fast	1,7 RLL	3.5 3H	512k	500k	5400	Y
DSP3107LS	9.5	SCSI-2Fast	1,7 RLL	3.5 3H	512k	500k	5400	Y
DSP3133LS	9.5	SCSI-2Fast	1,7 RLL	3.5 3H	512k	500k	5400	Y
DSP3210S	9.5	SCSI-2Fast	1,7 RLL	3.5HH	1024k	500k	5400	Y
ELS127AT	17	IDE AT	1,7 RLL	3.5 3H	32k	250k	3663	Y
ELS127S	17	SCSI	1,7 RLL	3.5 3H	32k	250k	3663	Y
ELS170AT	17	IDE AT	1,7 RLL	3.5 3H	32k	250k	3663	Y
ELS170S	17	SCSI	1,7 RLL	3.5 3H	32k	250k	3663	Y
ELS42AT	19	IDE AT	2,7 RLL	3.5 3H		250k		Y
ELS42S	19	SCSI	2,7 RLL	3.5 3H		250k		Y
ELS85AT	17	IDE XT	2,7 RLL	3.5 3H		250k		Y
ELS85S	17	SCSI	2,7 RLL	3.5 3H		250k		Y
Empire 1080S	9.5	SCSI-3		3.5 3H	512k	500k	5400	Y
Empire 1400S	11	SCSI-3Fast	Prml0,4,4	3.5 3H	512k	500k	5400	Y
Empire 2100S	11	SCSI-3Fast	Prml0,4,4	3.5HH	512k	500k	5400	Y
Empire 540S	9.5	SCSI-3		3.5 3H	512k	500k	5400	Y
Empire II VP32181S	9	SCSI-3	PRML	3.5 3H	512k	1000k	5400	Y
Empire II VP34360S	9	SCSI-3	PRML	3.5 3H	512k	1000k	5400	Y
Empire II VP39100S	9	SCSI-3	PRML	3.5HH	512k	1000k	5400	Y
EUROPA 1080AT	14	ATA-2 Fast	PRML	2.5 4H	128k	350k	3800	Y
EUROPA 540AT	14	ATA-2 Fast	PRML	2.5 4H	128k	350k	3800	Y
EUROPA 810AT	14	ATA-2 Fast	PRML	2.5 4H	128k	350k	3800	Y
Fireball 1080AT	12	ATA-2 Fast	Prml16,17	3.5 3H	128k	500k	5400	Y
Fireball 1080S	12	SCSI-3	Prml	3.5 3H	128k	500k	5400	Y
Fireball 1280AT	12	ATA-2 Fast	Prml16,17	3.5 3H	128k	400k	5400	Y
Fireball 1280S	12	SCSI-3	Prml16,17	3.5 3H	128k	400k	5400	Y
Fireball 540AT	12	ATA-2	PRML	3.5 3H	128k	500k	5400	Y
Fireball 540S	12	SCSI-3	PRML	3.5 3H	128k	500k	5400	Y
Fireball 640AT	12	ATA-2 Fast	Prml16,17	3.5 3H	128k		5400	Y
Fireball 640S	12	SCSI-3	Prml16,17	3.5 3H	128k		5400	Y
Fireball SE 2.1AT 1-98	9.5	ULTRA ATA		3.5 3H	128k		5400	Y
Fireball SE 2.1S 1-98	9.5	Ultra SCSI3		3.5 3H	128k		5400	Y
Fireball SE 3.2AT 1-98	9.5	ULTRA ATA		3.5 3H	128k		5400	Y
Fireball SE 3.2S 1-98	9.5	Ultra SCSI3		3.5 3H	128k		5400	Y
Fireball SE 4.3AT 1-98	9.5	ULTRA ATA		3.5 3H	128k		5400	Y
Fireball SE 4.3S 1-98	9.5	Ultra SCSI3		3.5 3H	128k		5400	Y
Fireball SE 6.4AT 1-98	9.5	ULTRA ATA		3.5 3H	128k		5400	Y
Fireball SE 6.4S 1-98	9.5	Ultra SCSI3		3.5 3H	128k		5400	Y
Fireball SE 8.4AT 1-98	9.5	ULTRA ATA		3.5 3H	128k		5400	Y
Fireball SE 8.4S 1-98	9.5	Ultra SCSI3		3.5 3H	128k		5400	Y
Fireball ST 1.6	<10	ULTRA ATA		3.5 3H	128k	400k	5400	Y
Fireball ST 2.1	<10	Ultra SCSI3		3.5 3H	128k	400k	5400	Y
Fireball ST 3.2AT	<10	ULTRA ATA		3.5 3H	128k	400k	5400	
Fireball ST 3.2S	<10	Ultra SCSI3		3.5 3H	128k	400k	5400	
Fireball ST 4.3AT	<10	ULTRA ATA		3.5 3H	128k	400k	5400	
Fireball ST 4.3S	<10	Ultra SCSI3		3.5 3H	128k	400k	5400	
Fireball ST 6.4AT	<10	ULTRA ATA		3.5 3H	128k	400k	5400	
Fireball ST 6.4S	<10	Ultra SCSI3		3.5 3H	128k	400k	5400	
Fireball TM 1.0	12	ATA-2 Fast	Prml16,17	3.5 3H	128k	400k	4500	Y
Fireball TM 1.2AT	12	ATA-2 Fast	Prml16,17	3.5 3H	128k	400k	4500	Y

Drive Model	Format Size MB	Head	Cyl	Sect/ Trac	Translate H/C/S	RWC/ WPC	Land Zone
Fireball TM 1.2S	1281	2		104-23		—/—	
Fireball TM 2.1AT	2111	4		104-23	16/4092/63	—/—	
Fireball TM 2.1S	2111	4		104-23		—/—	
Fireball TM 2.5AT	2564	4		104-23	16/4969/63	—/—	
Fireball TM 3.2AT	3216	5		104-23	16/6232/63	—/—	
Fireball TM 3.2S	3216	5		104-23		—/—	
Fireball TM 3.8AT	3860	6		104-23	16/7480/63	—/—	
GODRIVE 120AT	127	4	1097	V	13/731/26	Na/Na	Auto
GODRIVE 120S	127	4	1097	V		Na/Na	Auto
GODRIVE 40AT	43	2	957		6/820/17	—/—	Auto
GODRIVE 40S	43	2	957			—/—	Auto
GODRIVE 60AT	63	2	1097	V	9/526/26	Na/Na	Auto
GODRIVE 60S	63	2				Na/Na	Auto
GODRIVE 80AT	84	2		NA	10/991/17	Na/Na	Auto
GODRIVE 80S	84	2				Na/Na	Auto
GODRIVE GLS127AT	127	3			9/677/41	Na/Na	Auto
GODRIVE GLS127S	127	3				Na/Na	Auto
GODRIVE GLS170AT	170	4			10/538/62	Na/Na	Auto
GODRIVE GLS170S	170	4				Na/Na	Auto
GODRIVE GLS256AT	256	6			11/723/63	Na/Na	Auto
GODRIVE GLS256S	256	6				Na/Na	Auto
GODRIVE GLS85AT	85	2			10/722/23	Na/Na	Auto
GODRIVE GLS85S	85	2				Na/Na	Auto
GODRIVE GRS160AT	169	4			10/966/34	Na/Na	Auto
GODRIVE GRS160S	169	4				Na/Na	Auto
GODRIVE GRS80AT	84	2		45-73	5/966/34	Na/Na	Auto
GODRIVE GRS80S	84	2				Na/Na	Auto
GrandPrix XP32140S	2140	10				—/—	
GrandPrix XP32151S	2150	10		118		—/—	
GrandPrix XP34280S	4280	20				—/—	
GrandPrix XP34301S	4300	20		118		—/—	
HARDCARD EZ 42	42	5	977	17		Na/Na	Auto
Lightning 365AT	366	2		61-128	12/976/61	Na/Na	Auto
Lightning 365S	365	2		64-128		Na/Na	Auto
Lightning 540AT	541	4		61-128	16/1120/59	Na/Na	Auto
Lightning 540S	541	3		64-128		Na/Na	Auto
Lightning 730AT	731	4		61-128	16/1416/63	Na/Na	Auto
Lightning 730S	732	4		54-128		Na/Na	Auto
MAVERICK 270AT	271	2		58-118	14/944/40	Na/Na	Auto
MAVERICK 270S	271	2		58-118		Na/Na	Auto
MAVERICK 540AT	541	4		58-118	16/1049/63	Na/Na	Auto
MAVERICK 540S	542	4		58-118		Na/Na	Auto
PIONEER SG 1.0	1082	2			16/2097/63	—/—	
PIONEER SG 2.1	2111	4			16/4092/63	—/—	
PRODRIVE 100E	103					Na/Na	
PRODRIVE 1050S	1050	12	2442	NA		Na/Na	Auto
PRODRIVE 105AT	104	4	1219	17	16/755/17	Na/Na	Auto
PRODRIVE 105S	105	6	1019			—/—	Auto
PRODRIVE 120AT	120	5	1123		9/814/32	Na/Na	Auto
PRODRIVE 120S	120	5	1123			—/—	Auto
PRODRIVE 1225S	1225	14	2444	NA		Na/Na	Auto
PRODRIVE 145E	145					Na/Na	
PRODRIVE 160AT	168	4	839			Na/Na	Auto
PRODRIVE 160S	168	4	839			Na/Na	Auto
PRODRIVE 170AT	168	7	1123		10/968/34	Na/Na	Auto
PRODRIVE 170S	168	7	1123			—/—	Auto
PRODRIVE 1800S	1800	14				Na/Na	Auto
PRODRIVE 210AT	209	7	1156		13/873/36	Na/Na	Auto
PRODRIVE 210S	210	7	1156			—/—	Auto
PRODRIVE 330AT	331	7	1156			—/—	Auto
PRODRIVE 330S	331	7	1156			—/—	Auto
PRODRIVE 40AT	42	3	834		5/965/17	Na/Na	Auto
PRODRIVE 40S	42	3	834			—/—	Auto

Drive Model	Seek Time	Interface	Encode	Form Factor	cache kb	mtbf	RPM	Obsolete?
Fireball TM 1.2S	12	Ultra SCSI3	Prml16,17	3.5 3H	128k	400k	4500	Y
Fireball TM 2.1AT	10.5	ULTRA ATA	Prml16,17	3.5 3H	128k	400k	4500	Y
Fireball TM 2.1S	10.5	Ultra SCSI3	Prml16,17	3.5 3H	128k	400k	4500	Y
Fireball TM 2.5AT	10.5	ULTRA ATA	Prml16,17	3.5 3H	128k	400k	4500	
Fireball TM 3.2AT	10.5	ULTRA ATA	Prml16,17	3.5 3H	128k	400k	4500	
Fireball TM 3.2S	10.5	Ultra SCSI3	Prml16,17	3.5 3H	128k	400k	4500	
Fireball TM 3.8AT	10.5	ULTRA ATA	Prml16,17	3.5 3H	128k	400k	4500	
GODRIVE 120AT	17	IDE AT	1,7 RLL	2.5 3H	32k	150k		Y
GODRIVE 120S	17	SCSI	1,7 RLL	2.5 3H	32k	150k		Y
GODRIVE 40AT	19	IDE AT	1,7 RLL	2.5 4H	32k	80k		Y
GODRIVE 40S	19	SCSI	1,7 RLL	2.5 4H	32k	80k		Y
GODRIVE 60AT	19	IDE AT	1,7 RLL	2.5 3H		150k		Y
GODRIVE 60S	17	SCSI	1,7 RLL	2.5 4H		150k		Y
GODRIVE 80AT	19	IDE AT	1,7 RLL	2.5 4H		80k		Y
GODRIVE 80S	19	SCSI	1,7 RLL	2.5 4H		80k		Y
GODRIVE GLS127AT	17	IDE AT		2.5	128k	350k		Y
GODRIVE GLS127S	17	SCSI-2		2.5	128k	350k		Y
GODRIVE GLS170AT	17	IDE AT		2.5	128k	350k		Y
GODRIVE GLS170S	17	SCSI-2		2.5	128k	350k		Y
GODRIVE GLS256AT	17	IDE AT		2.5	128k	350k		Y
GODRIVE GLS256S	17	SCSI-2		2.5	128k	350k		Y
GODRIVE GLS85AT	17	IDE AT		2.5	128k	350k		Y
GODRIVE GLS85S	17	SCSI-2		2.5	128k	350k		Y
GODRIVE GRS160AT	17	IDE AT		2.5	32k	150k		Y
GODRIVE GRS160S	17	SCSI		2.5	32k	150k		Y
GODRIVE GRS80AT	17	IDE AT	1,7 RLL	2.5 4H	32k	150k	3600	Y
GODRIVE GRS80S	17	SCSI		2.5	32k	150k		Y
GrandPrix XP32140S	9	SCSI-3Fast		3.5HH	512k	800k	7200	Y
GrandPrix XP32151S	10	SCSI-3	Prml0,4,4	3.5HH	512k	800k	7200	Y
GrandPrix XP34280S	9	SCSI-3Fast		3.5HH	512k	800k	7200	Y
GrandPrix XP34301S	10	SCSI-3	Prml0,4,4	3.5HH	512k	800k	7200	Y
HARDCARD EZ 42		IDE AT						Y
Lightning 365AT	11	IDE AT	1,7 RLL	3.5 3H	128k	300k	4500	Y
Lightning 365S	11	SCSI-2	1,7 RLL	3.5 3H	128k	300k	4500	Y
Lightning 540AT	11.5	IDE AT	1,7 RLL	3.5 3H	128k	300k	4500	Y
Lightning 540S	11.5	SCSI-2	1,7 RLL	3.5 3H	128k	300k	4500	Y
Lightning 730AT	11.5	IDE AT	1,7 RLL	3.5 3H	128k	300k	4500	Y
Lightning 730S	11.5	SCSI-2	1,7 RLL	3.5 3H	128k	300k	4500	Y
MAVERICK 270AT	14	IDE AT	1,7 RLL	3.5 3H	128k	300k	3600	Y
MAVERICK 270S	14	SCSI-2	1,7 RLL	3.5 3H	128k	300k	3600	Y
MAVERICK 540AT	14	IDE AT	1,7 RLL	3.5 3H	128k	300k	3600	Y
MAVERICK 540S	14	SCSI-2	1,7 RLL	3.5 3H	128k	300k	3600	Y
PIONEER SG 1.0	12	ATA-2 Fast		3.5 3H	64k	300k	4500	Y
PIONEER SG 2.1	12	ATA-2 Fast		3.5 3H	64k	300k	4500	Y
PRODRIVE 100E	19	ESDI	2,7 RLL	3.5HH				Y
PRODRIVE 1050S	10	SCSI		3.5HH	512k	350k	4500	Y
PRODRIVE 105AT	17	IDE AT	2,7 RLL	3.5HH		60k		Y
PRODRIVE 105S	19	SCSI	2,7 RLL	3.5HH	64k	50k		Y
PRODRIVE 120AT	15	IDE AT	1,7 RLL	3.5HH	64k	50k	3605	Y
PRODRIVE 120S	15	SCSI	1,7 RLL	3.5HH	64k	50k		Y
PRODRIVE 1225S	10	SCSI		3.5HH	512k	350k	4500	Y
PRODRIVE 145E	19	ESDI		3.5HH				Y
PRODRIVE 160AT	19	IDE AT	1,7 RLL	3.5 4H		80k		Y
PRODRIVE 160S	19	SCSI	1,7 RLL	3.5 4H		80k		Y
PRODRIVE 170AT	15	IDE AT	1,7 RLL	3.5HH	56k	50k	3605	Y
PRODRIVE 170S	15	SCSI	1,7 RLL	3.5HH	64k	50k		Y
PRODRIVE 1800S	10	SCSI		3.5HH	512k	350k	4500	Y
PRODRIVE 210AT	15	IDE AT	1,7 RLL	3.5HH	56k	50k	3605	Y
PRODRIVE 210S	15	SCSI	1,7 RLL	3.5HH	64k	50k	3606	Y
PRODRIVE 330AT	14	IDE AT	1,7 RLL	3.5HH	64k	150k	3606	Y
PRODRIVE 330S	14	SCSI	1,7 RLL	3.5HH	64k	150k		Y
PRODRIVE 40AT	19	IDE AT	2,7 RLL	3.5HH	64k	50k		Y
PRODRIVE 40S	19	SCSI	2,7 RLL	3.5HH	64k	50k		Y

Drive Model	Format Size MB	Head	Cyl	Sect/ Trac	Translate H/C/S	RWC/ WPC	Land Zone
PRODRIVE 425AT	426	9	1520	V	16/1021/51	Na/Na	Auto
PRODRIVE 425S	426	9				—/—	Auto
PRODRIVE 525S	525	6	2446	NA		Na/Na	Auto
PRODRIVE 700S	700	8	2443	NA		Na/Na	Auto
PRODRIVE 80AT	84	6	834	35	10/965/17	Na/Na	Auto
PRODRIVE 80S	84	6	834	35		—/—	Auto
PRODRIVE LPS105AT	105	4	1219		16/755/17	Na/Na	Auto
PRODRIVE LPS105S	105	4	1219			—/—	Auto
PRODRIVE LPS120AT	122	2			5/901/53	Na/Na	Auto
PRODRIVE LPS120S	122	2	1818			—/—	
PRODRIVE LPS127AT	128	2		65-91	16/919/17	Na/Na	Auto
PRODRIVE LPS127S	127	2				Na/Na	Auto
PRODRIVE LPS170AT	171	2		52-91	15/1011/22	Na/Na	Auto
PRODRIVE LPS170S	170	2				Na/Na	Auto
PRODRIVE LPS210AT	211	2		55-104	15/723/38	Na/Na	Auto
PRODRIVE LPS240AT	245	4			13/723/51	Na/Na	Auto
PRODRIVE LPS240S	245	4	1818	V	13/723/51	Na/Na	Auto
PRODRIVE LPS270AT	270	2		V	14/944/40	Na/Na	Auto
PRODRIVE LPS270S	270	2				Na/Na	Auto
PRODRIVE LPS340AT	342	4			15/1011/44	Na/Na	Auto
PRODRIVE LPS340S	342	4				Na/Na	Auto
PRODRIVE LPS420AT	420	4		55-104	16/1010/51	Na/Na	Auto
PRODRIVE LPS525AT	525	6			16/1017/63	Na/Na	Auto
PRODRIVE LPS525S	525	6				Na/Na	Auto
PRODRIVE LPS52AT	52	2	1219		8/751/17	Na/Na	Auto
PRODRIVE LPS52S	52	2	1219			—/—	Auto
PRODRIVE LPS540AT	541	4		V	16/1120/59	Na/Na	Auto
PRODRIVE LPS540S	541	4				Na/Na	Auto
PRODRIVE LPS80AT	85				16/616/17	Na/Na	Auto
PRODRIVE LPS80S	86	4				Na/Na	Auto
SATURN VP31080S	1080	5				—/—	
SATURN VP32170S	2170	10				—/—	
SIROCCO 1700AT	1700	4		90-180	16/3309/63	Na/Na	Auto
SIROCCO 1700S	1700	4		90-180		Na/Na	Auto
SIROCCO 2550AT	2550	6		90-180	16/4969/63	Na/Na	Auto
SIROCCO 2550S	2550	6		90-180		Na/Na	Auto
TRAILBLAZER 420AT	422	2		76-141	16/1010/51	—/—	
TRAILBLAZER 420S	425	2		76-141		—/—	
TRAILBLAZER 635AT	636	3		76-141	16/1234/63	—/—	
TRAILBLAZER 635S	636	3		76-141		—/—	
TRAILBLAZER 850AT	850	4		76-141	16/1647/63	—/—	
TRAILBLAZER 850S	852	4		76-141		—/—	
VIKING 2.1S	2180	4				—/—	
VIKING 4.3S	4360	8				—/—	
VIKING II 4.5 1-98	4550	5				—/—	
VIKING II 4.5 1-98	4550	5				—/—	
VIKING II 9.1 1-98	9100	10				—/—	
VIKING II 9.1 1-98	9100	10				—/—	

RICOH

Drive Model	Format Size MB	Head	Cyl	Sect/ Trac	Translate H/C/S	RWC/ WPC	Land Zone
RH5130	10	2	612	17		613/400	
RH5260	10	2	615	17		—/—	
RH5261	10	2	612	17		—/—	
RH5500	100	2	1285	76		Na/Na	Auto
RS9150AR	100	2	1285	76		Na/Na	Auto

RMS

Drive Model	Format Size MB	Head	Cyl	Sect/ Trac	Translate H/C/S	RWC/ WPC	Land Zone
RMS503	2.5	2	153	17		77/77	
RMS506	5	4	153	17		77/77	
RMS509	8	6	153	17		77/77	
RMS512	10	8	153	17		77/77	

Drive Model	Seek Time	Interface	Encode	Form Factor	cache kb	mtbf	RPM	Obsolete?
PRODRIVE 425AT	14	IDE AT	1,7 RLL	3.5HH	56k	150k	3606	Y
PRODRIVE 425S	14	SCSI	1,7 RLL	3.5HH	64k	150k	3606	Y
PRODRIVE 525S		SCSI		3.5HH				Y
PRODRIVE 700S	10	SCSI		3.5HH	512k	350k	4500	Y
PRODRIVE 80AT	19	IDE AT	2,7 RLL	3.5HH	64k	50k		Y
PRODRIVE 80S	19	SCSI	2,7 RLL	3.5HH	64k	50k		Y
PRODRIVE LPS105AT	17	IDE AT	2,7 RLL	3.5 3H	64k	60k		Y
PRODRIVE LPS105S	17	SCSI	2,7 RLL	3.5 3H	64k	60k		Y
PRODRIVE LPS120AT	16	IDE AT	1,7 RLL	3.5 3H	256k	250k		Y
PRODRIVE LPS120S	16	SCSI	1,7 RLL	3.5 3H	256k	250k	4306	Y
PRODRIVE LPS127AT	14	IDE AT	1,7 RLL	3.5 3H	128k	300k	3600	Y
PRODRIVE LPS127S	14	SCSI-2	1,7 RLL	3.5 3H	128k	300k	3600	Y
PRODRIVE LPS170AT	14	IDE AT	1,7 RLL	3.5 3H	128k	300k	3600	Y
PRODRIVE LPS170S	14	SCSI-2	1,7 RLL	3.5 3H	128k	300k	3600	Y
PRODRIVE LPS210AT	15	IDE AT	1,7 RLL	3.5 3H	128k	300k	3600	Y
PRODRIVE LPS240AT	16	IDE AT	1,7 RLL	3.5 3H	256k	250k	4306	Y
PRODRIVE LPS240S	17	SCSI	1,7 RLL	3.5 3H	256k	250k	4306	Y
PRODRIVE LPS270AT	14	IDE AT	1,7 RLL	3.5 3H	128k	300k	3600	Y
PRODRIVE LPS270S	12	SCSI-2	1,7 RLL	3.5 3H	128k	300k	4500	Y
PRODRIVE LPS340AT	12	IDE AT		3.5 3H	128k	300k	3600	Y
PRODRIVE LPS340S	12	SCSI-2		3.5 3H	128k	300k	3600	Y
PRODRIVE LPS420AT	13	IDE AT	1,7 RLL	3.5 3H	128k	300k	3600	Y
PRODRIVE LPS525AT	10	IDE AT		3.5 3H	512k	350k	4500	Y
PRODRIVE LPS525S	10	SCSI		3.5 3H	512k	350k	4500	Y
PRODRIVE LPS52AT	17	IDE AT	2,7 RLL	3.5 3H	64k	60k		Y
PRODRIVE LPS52S	17	SCSI	2,7 RLL	3.5 3H	64k	60k		Y
PRODRIVE LPS540AT	14	IDE AT	1,7 RLL	3.5 3H	128k	300k	3600	Y
PRODRIVE LPS540S	12	SCSI-2	1,7 RLL	3.5 3H	128k	300k	4500	Y
PRODRIVE LPS80AT		IDE AT		3.5 3H				Y
PRODRIVE LPS80S	19	SCSI	2,7 RLL	3.5 3H		60k		Y
SATURN VP31080S	8.5	SCSI-2	1,7 RLL	3.5 3H	512k		5400	Y
SATURN VP32170S	8.5	SCSI-3Fast	1,7 RLL	3.5 3H	512k		5400	Y
SIROCCO 1700AT	11	ATA-2	Prml16,17	3.5 3H	128k	400k	4500	Y
SIROCCO 1700S	11	SCSI-2	Prml16,17	3.5 3H	128k	400k	4500	Y
SIROCCO 2550AT	11	ATA-2	Prml16,17	3.5 3H	128k		4500	
SIROCCO 2550S	11	SCSI-3	Prml16,17	3.5 3H	128k		4500	
TRAILBLAZER 420AT	14	ATA-2 Fast	1,7 RLL	3.5 3H	128k	300k	4500	Y
TRAILBLAZER 420S	14	SCSI-2Fast	1,7 RLL	3.5 3H	128k	300k	4500	Y
TRAILBLAZER 635AT	14	ATA-2 Fast	1,7 RLL	3.5 3H	128k	300k	4500	Y
TRAILBLAZER 635S	14	SCSI-3	1,7 RLL	3.5 3H	128k	300k	4500	Y
TRAILBLAZER 850AT	14	ATA-2 Fast	1,7 RLL	3.5 3H	128k	300k	4500	Y
TRAILBLAZER 850S	14	SCSI-2Fast	1,7 RLL	3.5 3H	128k	300k	4500	Y
VIKING 2.1S	8.5	Ultra SCSI3		3.5 3H	512k	800k	7200	Y
VIKING 4.3S	8.5	Ultra SCSI3		3.5 3H	512k	800k	7200	Y
VIKING II 4.5 1-98	8	ULTRA 2 LVD	Prml16,17	3.5 3H	512k		7200	
VIKING II 4.5 1-98	8	UL SE SCSI3	Prml16,17	3.5 3H	512k		7200	
VIKING II 9.1 1-98	8	UL SE SCSI3	Prml16,17	3.5 3H	512k		7200	
VIKING II 9.1 1-98	8	ULTRA 2 LVD	Prml16,17	3.5 3H	512k		7200	

RICOH

Drive Model	Seek Time	Interface	Encode	Form Factor	cache kb	mtbf	RPM	Obsolete?
RH5130	85	ST412/506	MFM					Y
RH5260	85	ST412/506	MFM					Y
RH5261	85	SCSI	MFM					Y
RH5500	25	SCSI	2,7 RLL	5.25HH		20k		Y
RS9150AR	25	SCSI	2,7 RLL	5.25HH		20k		Y

RMS

Drive Model	Seek Time	Interface	Encode	Form Factor	cache kb	mtbf	RPM	Obsolete?
RMS503		ST412/506	MFM	5.25				Y
RMS506		ST412/506	MFM	5.25				Y
RMS509		ST412/506	MFM	5.25FH				Y
RMS512		ST412/506	MFM	5.25				Y

Drive Model	Format Size MB	Head	Cyl	Sect/ Trac	Translate H/C/S	RWC/ WPC	Land Zone
RODIME SYSTEMS, INC							
COBRA 1000E (Mac)	1000					—/—	Auto
COBRA 110AT	110	4				—/—	Auto
COBRA 210AT	210	5				—/—	Auto
COBRA 330E (Mac)	330					—/—	Auto
COBRA 40AT	40	2	1170	36	4/585/36	—/—	Auto
COBRA 650E (Mac)	650					—/—	Auto
COBRA 80AT	80	4	1159	36	8/579/36	—/—	Auto
RO101	6	2	192	17		96/192	
RO102	12	4	192	17		96/192	
RO103	18	6	192	17		96/192	
RO104	24	8	192	17		96/192	
RO200	11	4	320	17		—/132	
RO201	5	2	321	17		132/300	
RO201E	11	2	640	17		264/300	
RO202	10	4	321	17		132/300	
RO202E	21	4	640	17		264/300	640
RO203	15	6	321	17		132/300	321
RO203E	32	6	640	17		264/300	640
RO204	21	8	320	17		132/300	321
RO204E	43	8	640	17		264/300	640
RO251	5	2	306	17		307/307	
RO252	11	4	306	17		64/128	
RO3045	37	5	872	17		873/—	
RO3055	45	6	872	17		873/—	
RO3055A	49					—/—	
RO3055T	45	3	1053	26		Na/Na	Auto
RO3057S	45	5	680			—/—	
RO3058A	45	3	868	17	3/868/34	—/—	
RO3058T	45	3	868	17		—/—	
RO3059A	46	2	1216	17		—/—	
RO3059T	46	2	1216	34		—/—	
RO3060R	50	2	1216	17		—/—	
RO3065	53	7	872	17		—/650	
RO3070S	71					—/—	
RO3075R	59	6	750			—/650	
RO3085R	69	7	750			—/650	
RO3085S	69	7	750			—/650	
RO3088A	75	5	868	34	5/868/34	—/—	
RO3088T	75	5	868	34		—/—	
RO3089A	70	3	1216	34		—/—	
RO3089T	70	3	1216	34		—/—	
RO3090T	75	5	1053	28		Na/Na	Auto
RO3095A	80	3	1216	34	5/923/34	—/—	
RO3099A	80	4	1030		15/614/17	Na/Na	Auto
RO3099AP	80	4	1030		15/614/17	Na/Na	Auto
RO3128A	105	7	868	34		—/—	
RO3128T	105	7	868	17		—/—	
RO3129A	105	5	1090			—/—	
RO3129T	105	5	1090	17		—/—	
RO3130S	105	7	1047	30		—/—	
RO3130T	105	7	1053	28		Na/Na	Auto
RO3135A	112	7	923	34	7/923/34	—/—	
RO3139A	112	5	1168	17	15/861/17	—/—	
RO3139AP	112	5	1168		15/861/17	Na/Na	Auto
RO3139S	112	5	1148			Na/Na	Auto
RO3139TP	112	5	1148			Na/Na	Auto
RO3258TS	210					—/—	
RO3259A	210				15/976/28	—/—	
RO3259AP	212	9	1235		15/990/28	Na/Na	Auto
RO3259T	210					—/—	
RO3259TP	210	9	1148	V		Na/Na	Auto
RO3259TS	210	9	1216			Na/Na	Auto

Drive Model	Seek Time	Interface	Encode	Form Factor	cache kb	mtbf	RPM	Obsolete?
RODIME SYSTEMS, INC								
COBRA 1000E (Mac)	15	SCSI			45k	100k	3600	Y
COBRA 110AT	19	IDE AT	2,7 RLL	3.5HH		40k		Y
COBRA 210AT		IDE AT	2,7 RLL	3.5HH		40k		Y
COBRA 330E (Mac)	14.5	SCSI			45k	50k	3600	Y
COBRA 40AT	19	IDE AT	2,7 RLL	3.5HH		40k		Y
COBRA 650E (Mac)	16.5	SCSI	2,7 RLL		45k	50k	3600	Y
COBRA 80AT	20	IDE AT	2,7 RLL	3.5HH		40k		Y
RO101		ST412/506	MFM	5.25FH				Y
RO102		ST412/506	MFM	5.25FH				Y
RO103	55	ST412/506	MFM	5.25FH				Y
RO104		ST412/506	MFM	5.25FH				Y
RO200		ST412/506	MFM	5.25FH				Y
RO201	85	ST412/506	MFM	5.25FH				Y
RO201E	55	ST412/506	MFM	5.25FH				Y
RO202	85	ST412/506	MFM	5.25HH				Y
RO202E	55	ST412/506	MFM	5.25FH				Y
RO203	85	ST412/506	MFM	5.25HH				Y
RO203E	55	ST412/506	MFM	5.25FH				Y
RO204	85	ST412/506	MFM	5.25FH				Y
RO204E	55	ST412/506	MFM	5.25FH				Y
RO251	85	ST412/506	MFM	5.25HH				Y
RO252	85	ST412/506	MFM	5.25HH				Y
RO3045	28	ST412/506	MFM	3.5HH				Y
RO3055	28	ST412/506	MFM	3.5HH				Y
RO3055A		IDE AT	2,7 RLL	3.5HH				Y
RO3055T		SCSI	RLL	3.5HH				Y
RO3057S	28	SCSI	2,7 RLL	3.5HH				Y
RO3058A	18	IDE AT	2,7 RLL	3.5HH		20k		Y
RO3058T	18	SCSI	2,7 RLL	3.5HH		20k		Y
RO3059A	18	IDE AT	2,7 RLL	3.5HH		20k		Y
RO3059T	18	SCSI	2,7 RLL	3.5HH		20k		Y
RO3060R	28	ST412/506	2,7 RLL	3.5HH		20k		Y
RO3065	28	ST412/506	MFM	3.5HH		20k		Y
RO3070S	28	SCSI	2,7 RLL	3.5HH				Y
RO3075R	28	ST412/506	2,7 RLL	3.5HH		20k		Y
RO3085R	28	ST412/506	2,7 RLL	3.5HH		20k		Y
RO3085S	28	SCSI	2,7 RLL	3.5HH				Y
RO3088A	18	IDE AT	2,7 RLL	3.5HH		20k		Y
RO3088T	18	SCSI	2,7 RLL	3.5HH		20k		Y
RO3089A	18	IDE AT	2,7 RLL	3.5HH		20k		Y
RO3089T	18	SCSI	2,7 RLL	3.5HH		20k		Y
RO3090T		SCSI	2,7 RLL	3.5HH				Y
RO3095A	18	IDE AT	2,7 RLL	3.5HH		20k		Y
RO3099A	19	IDE AT	2,7 RLL	3.5HH				Y
RO3099AP	19	IDE AT	2,7 RLL	3.5HH				Y
RO3128A	18	IDE AT	2,7 RLL	3.5HH		20k		Y
RO3128T	18	SCSI	2,7 RLL	3.5HH		20k		Y
RO3129A	18	IDE AT	2,7 RLL	3.5HH		20k		Y
RO3129T	18	SCSI	2,7 RLL	3.5HH		20k		Y
RO3130S	22	SCSI	2,7 RLL	5.25HH		20k		Y
RO3130T	22	SCSI	2,7 RLL	5.25HH		20k		Y
RO3135A	19	IDE AT	2,7 RLL	3.5HH		20k		Y
RO3139A	18	IDE AT	2,7 RLL	3.5HH		20k		Y
RO3139AP	18	IDE AT	2,7 RLL	3.5HH		20k		Y
RO3139S	18	SCSI		3.5HH				Y
RO3139TP		SCSI	RLL ZBR	3.5HH				Y
RO3258TS		SCSI		3.5HH				Y
RO3259A	18	IDE AT	2,7 RLL	3.5HH				Y
RO3259AP		IDE AT		3.5HH				Y
RO3259T	18	SCSI	2,7 RLL	3.5HH				Y
RO3259TP		SCSI	2,7 RLL	3.5HH				Y
RO3259TS	18	SCSI	2,7 RLL	3.5HH				Y

Drive Model	Format Size MB	Head	Cyl	Sect/ Trac	Translate H/C/S	RWC/ WPC	Land Zone
RO351	5	2	306	17		307/307	
RO352	11	4	306	17		64/128	
RO365	21	4	612	17		613/613	
RO5040S	38	3		17		—/—	
RO5065	63	5		17		—/—	
RO5075E	65	3	1224	35		Na/Na	Auto
RO5075S	76					—/—	
RO5078S	62	3	1224	33		Na/Na	Auto
RO5090	89	7	1224	17		—/—	
RO5095R	81	5	1224	26		Na/Na	Auto
RO5125-1F2	106	5	1219	34		Na/Na	Auto
RO5125E	106	5	1224	34		—/—	
RO5125S	106	5	1219	34		Na/Na	Auto
RO5128S	103	5	1224	33		Na/Na	Auto
RO5130R	114	7	1224	26		—/—	
RO5178S	144	7	1219			—/—	
RO5180-1F2	148	7	1219	34		Na/Na	Auto
RO5180E	149	7	1224	34		—/—	
RO5180S	144	7	1219	34		—/—	
RO652	20	4	306	33		Na/Na	Auto
RO652A	20					—/—	
RO652B	20	4	306	33		—/—	
RO752	20	4	306	33		Na/Na	Auto
RO752A	25					—/—	
SAMSUNG							
ACB20811A (Rel. 10-96)	810					—/—	
ACE21021A (Rel. 10-96)	1020					—/—	
PLS30540A	547	3				—/—	
PLS30730A	731	4				—/—	
PLS30850A	731	4				—/—	
PLS30850S	731	4				—/—	
PLS30854A	850	4	386872-132		16/1647/63	—/—	
PLS30854S	850	4	3868	VAR		—/—	
PLS31084A	1080	5	384072-144		16/2093/63	—/—	
PLS31084S	1080	5	384072-144		16/2093/63	—/—	
PLS31100A	1100	6				—/—	
PLS31100S	1100	6				—/—	
PLS31274A	1273	5	384472-132			—/—	
PLS31274S	1273	5	384472-132			—/—	
SHD2040N	44	4	820	26		—/544	819
SHD2041	47	4	820	28		Na/Na	Auto
SHD30280A	280	2	276872-120			Na/Na	Auto
SHD30420A	421	3	276872-120			—/—	
SHD30560A	561	4	276872-120		16/1086/63	—/—	
SHD30560A	528	4			16/1024/63	—/—	
SHD3061A	60	2	1478	40	7/993/17	Na/Na	Auto
SHD3062A	121	4	1479	40	15/927/17	Na/Na	Auto
SHD3101A	105				8/766/33	Na/Na	Auto
SHD3101B	105	4	1282	40		Na/Na	Auto
SHD3121A	125	2	1956	79		—/—	
SHD3122A	251	4	1956	79		—/—	
SHD3171A	178	2				—/—	
SHD3172A	356	4	2223	96		—/—	
SHD3202S	212	7	1376	43		Na/Na	Auto
SHD3210S	212	7	1376	43		Na/Na	Auto
SHD3211A	213	2	2570	55-95		—/—	
SHD3212A	426					Na/Na	Auto
SHD3272A	545	4				—/—	
SHD3272S	545	4				—/—	
STG31271A	1280	4			16/2483/63	—/—	
STG31601A	1610				16/3104/63	—/—	
TBR-31084A	1080				16/2092/63	—/—	

Drive Model	Seek Time	Interface	Encode	Form Factor	cache kb	mtbf	RPM	Obsolete?
RO351	85	ST412/506	MFM	3.5HH				Y
RO352	85	ST412/506	MFM	3.5HH				Y
RO365		ST412/506	MFM	3.5HH				Y
RO5040S	28	SCSI	MFM	5.25HH				Y
RO5065	28	ST412/506	MFM	5.25HH				Y
RO5075E	28	ESDI		5.25HH				Y
RO5075S	28	SCSI		5.25HH				Y
RO5078S		SCSI		5.25HH				Y
RO5090	28	ST412/506	MFM	5.25HH				Y
RO5095R		ST412/506	2,7 RLL	5.25HH				Y
RO5125-1F2	18	SCSI	2,7 RLL	5.25HH		20k		Y
RO5125E	18	ESDI	2,7 RLL	5.25HH		25k		Y
RO5125S	28	SCSI	2,7 RLL	5.25HH		20k		Y
RO5128S		SCSI		5.25HH				Y
RO5130R	28	ST412/506	2,7 RLL	5.25FH		20k		Y
RO5178S	19	SCSI	2,7 RLL	5.25HH				Y
RO5180-1F2	19	SCSI	2,7 RLL	5.25HH		20k		Y
RO5180E	18	ESDI	2,7 RLL	5.25HH		25k		Y
RO5180S	28	SCSI	2,7 RLL	5.25HH				Y
RO652	85	SCSI	2,7 RLL	3.5HH				Y
RO652A	85	SCSI		3.5HH				Y
RO652B	85	SCSI	2,7 RLL	3.5HH				Y
RO752	85	SCSI		5.25HH				Y
RO752A	85	SCSI		5.25HH				Y
SAMSUNG								
ACB20811A	12	ATA-2 Fast		2.5				Y
ACE21021A	12	ATA-2 Fast		2.5				Y
PLS30540A	11	IDE AT		3.5 3H	256k	300k	4500	Y
PLS30730A	11	IDE AT		3.5 3H	256k	300k	4500	Y
PLS30850A	11	IDE AT		3.5 3H	256k	300k	4500	Y
PLS30850S	11	SCSI		3.5 3H	256k	300k	4500	Y
PLS30854A	11	EIDE	1,7 RLL	3.5 3H	256k	300k	4500	Y
PLS30854S	11	SCSI-2Fast	1,7 RLL	3.5 3H	256k	300k	4500	Y
PLS31084A	11	ATA-2	1,7 RLL	3.5 3H	256k	300k	4500	Y
PLS31084S	11	SCSI-2	1,7 RLL	3.5 3H	256k	300k	4500	Y
PLS31100A	11	IDE AT		3.5 3H	256k	300k	4500	Y
PLS31100S	11	SCSI		3.5 3H	256k	300k	4500	Y
PLS31274A	11	ATA-2	1,7 RLL	3.5 3H	256k	300k	4500	Y
PLS31274S	11	SCSI-2	1,7 RLL	3.5 3H	256k	300k	4500	Y
SHD2040N	39	ST412/506	2,7 RLL	3.5HH		30k	3568	Y
SHD2041	29	IDE AT	2,7 RLL	3.5HH		30k	3525	Y
SHD30280A	12	ATA	1,7 RLL	3.5HH	64k	250k	3600	Y
SHD30420A	12	IDE AT	1,7 RLL	3.5 3H	128k	250k	3600	Y
SHD30560A	12	IDE AT	1,7 RLL	3.5 3H	128k	250k	3600	Y
SHD30560A	12	IDE AT	1,7 RLL	3.5 3H	128k	250k	3600	Y
SHD3061A	16	IDE AT	1,7 RLL	3.5 3H		200k		Y
SHD3062A	16	IDE AT	1,7 RLL	3.5 3H		200k		Y
SHD3101A	19	IDE AT	1,7 RLL	3.5 3H	32k	40k	3600	Y
SHD3101B	19	IDE AT	1,7 RLL	3.5 3H	32k	40k	3600	Y
SHD3121A	16	IDE AT	1,7 RLL	3.5 3H	64k	250k	3600	Y
SHD3122A	16	IDE AT	1,7 RLL	3.5 3H	64k	250k	3600	Y
SHD3171A	13	IDE AT	1,7 RLL	3.5 3H	64k	250k	3600	Y
SHD3172A	13	IDE AT	1,7 RLL	3.5 3H	64k	250k	3600	Y
SHD3202S	16	SCSI	1,7 RLL	3.5HH		50k		Y
SHD3210S	16	SCSI	1,7 RLL	3.5HH		50k		Y
SHD3211A	13	IDE AT	1,7 RLL	3.5 3H	64k	250k	3600	Y
SHD3212A	13	ATA		3.5HH	128k			Y
SHD3272A	12	IDE AT	1,7 RLL		256k		4510	Y
SHD3272S	12	SCSI-2Fast	1,7 RLL		256k		4510	Y
STG31271A	12	ATA-2 Fast		3.5 3H	128k	300k	4500	Y
STG31601A	12	ATA-2 Fast		3.5	128k		4500	Y
TBR-31084A	11	IDE AT	1,7 RLL					Y

Drive Model	Format Size MB	Head	Cyl	Sect/ Trac	Translate H/C/S	RWC/ WPC	Land Zone
TBR31080A	1080	4			16/2092/63	—/—	
TBR31080S	1080	4			16/2092/63	—/—	
TBR31081A	1080	4	4308			—/—	
TBR31081S	1080	4	4308			—/—	
VG33402A	3400	4				—/—	
VG34202A	4250	5				—/—	
VG35102A	5100	6				—/—	
WN310820A	1080	2	6022			—/—	
WN312016A	1207	3	5389			—/—	
WN312021A	1207	2	6077			—/—	
WN31273A	1270	2	6333			—/—	
WN316025A	1620	3	5891			—/—	
WN32101S	2160	6				—/—	
WN321620A	2160	4	6022			—/—	
WN32162U	2160	4	5909			—/—	
WN32543A	2540	4	6331			—/—	
WN33203A	3175	5	6331			—/—	
WN34003A	4000	6				—/—	
WN34003U	4000	6				—/—	
WNR31601A	1610	4	5589			—/—	
WNR32100A	2104	6				—/—	
WNR32101A	2060	5	5589			—/—	
WNR32501A	2415	6				—/—	

SEAGATE TECHNOLOGIES

Drive Model	Format Size MB	Head	Cyl	Sect/ Trac	Translate H/C/S	RWC/ WPC	Land Zone
Barracuda 9LP	9100	10				—/—	
Elite12G	1050	17				—/—	Auto
Sabre1123	964	19				—/—	Auto
Sabre1150	990	19				—/—	Auto
Sabre1230	1050	15	1635			—/—	Auto
Sabre2270	1948	19				—/—	Auto
Sabre2500	2145	19				—/—	Auto
Sabre368	368	10	1635			—/—	Auto
Sabre500	500	10	1217			—/—	Auto
Sabre736	741	15	1217			—/—	Auto
Sabre850	851	15	1635			—/—	Auto
ST1057A	53	3	1024	17	6/1024/17	Na/Na	Auto
ST1057N	49	3	940	34		—/—	Auto
ST1090A	79	5	1072	29	16/335/29	Na/Na	Auto
ST1090N	79	5	1068	29		Na/Na	Auto
ST1096N	84	7	906	26		Na/Na	Auto
ST1100	83	9	1072	17		1073/1073	Auto
ST1102A	89	5	1024	17	10/1024/17	Na/Na	Auto
ST1102N	84	5	965	34		—/—	Auto
ST1106R	91	7	977	26		Na/Na	Auto
ST1111A	98	5	1072	36	10/536/36	Na/Na	Auto
ST1111E	98	5	1072	36		Na/Na	Auto
ST1111N	98	5	1068	36		Na/Na	Auto
ST11200N	1054	15	1872	73		Na/Na	Auto
ST11200ND	1050	15	1877			—/—	
ST11201N (not made)	1054	15	1872	73		—/—	Auto
ST11201ND	1050	15	1877			—/—	
ST1126A	111	7	1072	29	16/469/29	Na/Na	Auto
ST1126N	107	7	1068	29		Na/Na	Auto
ST1133A	117	5	1272	36	10/636/36	Na/Na	Auto
ST1133NS	113	5	1268	36		Na/Na	Auto
ST1144A	131	7	1024	32	15/1001/17	Na/Na	Auto
ST1144N	126	7		32		—/—	Auto
ST1150R	128	9	1072	26		NA/300	Auto
ST1156A	138	7	1072	36	14/536/36	Na/Na	Auto
ST1156E	138	7	1072	36		Na/Na	Auto
ST1156N	138	7	1068	36		Na/Na	Auto
ST1156NS	138	7	1068	36		—/—	Auto

Drive Model	Seek Time	Interface	Encode	Form Factor	cache kb	mtbf	RPM	Obsolete?
TBR31080A	9	ATA-2 Fast		3.5 3H	256k	500k	5400	Y
TBR31080S	9	SCSI		3.5 3H	256k	500k	5400	Y
TBR31081A	9	ATA-2 Fast	1,7 RLL		256k	500k	5400	Y
TBR31081S	9	SCSI-2Fast	1,7 RLL		256k	500k	5400	Y
VG33402A	11				128k	500k	5400	
VG34202A	11				128k	500k	5400	
VG35102A	11				128k	500k	5400	
WN310820A	<10	ATA-2 Fast	RLL 8,9	3.5 3H	128k	500k	4500	Y
WN312016A	11	ATA-2 Fast	RLL 8,9	3.5 3H	128k	500k	5400	Y
WN312021A	10	ATA-2 Fast	RLL 8,9	3.5 3H	128k	500k	4500	Y
WN31273A	11	ATA-2 Fast	RLL 8,9	3.5 3H	128k	500k	4500	Y
WN316025A	10	ATA-2 Fast	RLL 8,9	3.5 3H	128k	500k	4500	Y
WN32101S	11	SCSI-2Fast		3.5 3H	128k	500k	5400	
WN321620A	<10	ATA-2 Fast	RLL 8,9	3.5 3H	128k	500k	5400	
WN32162U	9.5		RLL 8,9	3.5 3H	512k	500k	5400	
WN32543A	10	ATA-2 Fast	RLL 8,9	3.5 3H	128k	500k	5400	
WN33203A	10	ATA-2 Fast	RLL 8,9	3.5 3H	128k	500k	5400	
WN34003A	10	ATA-2 Fast	RLL 8,9	3.5 3H	128k	500k	5400	
WN34003U	10	Ultra SCSI	RLL 8,9	3.5 3H	128k	500k	5400	
WNR31601A	11	ATA-2 Fast	RLL 8,9	3.5 3H	128k	500k	5400	Y
WNR32100A	11			3.5 3H	128k	500k	5400	
WNR32101A	11	ATA-2 Fast	RLL 8,9	3.5 3H	128k	500k	5400	Y
WNR32501A	11	ATA-2 Fast	PRML	3.5 3H	128k	500k	5400	

SEAGATE TECHNOLOGIES

Drive Model	Seek Time	Interface	Encode	Form Factor	cache kb	mtbf	RPM	Obsolete?
Barracuda 9LP	7.1	Ultra2 SCSI	PRML		1mb		7200	
Elite12G	12	SMD	RLL	5.25FH		100k		Y
Sabre1123	15	SMD	RLL	8.0 FH		100k		Y
Sabre1150	15	IPI-2	RLL	8.0 FH		100k		Y
Sabre1230	15	SMD/SCSI	RLL	8.0 FH		100k		Y
Sabre2270	12	SMD	RLL	8.0 FH		100k		Y
Sabre2500	12	SMD/SCSI	RLL	8.0 FH		100k		Y
Sabre368	182	SMD/SCSI	RLL	8.0 FH		100k		Y
Sabre500	18	SMD/SCSI	RLL	8.0 FH		100k		Y
Sabre736	15	SMD/SCSI	RLL	8.0 FH		50k		Y
Sabre850	15	SMD/SCSI	RLL	8.0 FH		50k		Y
ST1057A	19	IDE AT	RLL ZBR	3.5HH	8/32k	50k	3528	Y
ST1057N	19	SCSI-2	2,7 RLL	3.5HH	8/32k	50k	3528	Y
ST1090A	15	IDE AT	2,7 RLL	3.5HH		70k	3600	Y
ST1090N	15	SCSI	RLL	3.5HH		70k	3600	Y
ST1096N	20	SCSI	2,7 RLL	3.5HH	8k	150k	3600	Y
ST1100	15	ST412/506	MFM	3.5HH		150k	3600	Y
ST1102A	19	IDE AT	RLL ZBR	3.5HH	8k	150k	3528	Y
ST1102N	19	SCSI-2	RLL ZBR	3.5HH	8/32k	50k	3528	Y
ST1106R	24	ST412/506	RLL	3.5HH		50k	3600	Y
ST1111A	15	IDE AT	2,7 RLL	3.5HH		70k	3600	Y
ST1111E	15	ESDI (10)	2,7 RLL	3.5HH		150k	3600	Y
ST1111N	15	SCSI	RLL	3.5HH		70k	3600	Y
ST11200N	11	SCSI-2Fast	RLL ZBR	3.5HH	256k	200k	5411	Y
ST11200ND	12	SCSI-2Fast	1,7 RLL	3.5HH	256k	200k	5400	Y
ST11201N (not made)	10	SCSI-2FstW	Zbr1,7RLL	3.5HH	256k	200k	5411	Y
ST11201ND	12	SCSI-2FstW	1,7 RLL	3.5HH	256k	200k	5400	Y
ST1126A	15	IDE AT	2,7 RLL	3.5HH	32k	150k	3600	Y
ST1126N	15	SCSI	RLL	3.5HH	64k	150k	3600	Y
ST1133A	15	IDE AT	2,7 RLL	3.5HH	64k	150k	3600	Y
ST1133NS	15	SCSI	RLL	3.5HH		70k	3600	Y
ST1144A	19	IDE AT	RLL ZBR	3.5HH	32k	150k	3528	Y
ST1144N	19	SCSI-2	RLL ZBR	3.5HH	8/32k	50k	3528	Y
ST1150R	15	ST412/506	RLL	3.5HH		150k	3600	Y
ST1156A	15	IDE AT	2,7 RLL	3.5HH		70k	3600	Y
ST1156E	15	ESDI	RLL	3.5HH		70k	3600	Y
ST1156N	15	SCSI	RLL	3.5HH		70k	3600	Y
ST1156NS	15	SCSI-2	2,7 RLL	3.5HH		70k	3600	Y

Drive Model	Format Size MB	Head	Cyl	Sect/ Trac	Translate H/C/S	RWC/ WPC	Land Zone
ST1162A	143	9	1072	29	16/603/29	Na/Na	Auto
ST1162N	138	9	1068	29		Na/Na	Auto
ST11700N	1430	13	2626			—/—	
ST11700ND	1430	13	2626			—/—	
ST11701N	1430	13	2626			—/—	
ST11701ND	1430	13	2626			—/—	
ST11750N	1437		2756			—/—	
ST11750ND	1437		2756			—/—	
ST11751N	1437		2756			—/—	
ST11751ND	1437		2756			—/—	
ST118273FC	18200	20				—/—	
ST118273LC	18200	20				—/—	
ST118273LW	18200	20				—/—	
ST118273N	18200	20				—/—	
ST118273W	18200	20				—/—	
ST118273WC	18200	20				—/—	
ST118273WD	18200	20				—/—	
ST1186A	164	7	1272	36	12/742/36	Na/Na	Auto
ST1186NS	159	7	1268	36		Na/Na	Auto
ST11900N	1700	15	2621	83		Na/Na	Auto
ST11900NC	1700	15	2621	83		Na/Na	Auto
ST11900ND	1700	15	2621	83		Na/Na	Auto
ST11900W	1700	15	2621	83		Na/Na	Auto
ST11900WC	1700	15	2621	83		Na/Na	Auto
ST11900WD	1700	15	2621	83		Na/Na	Auto
ST11950N	1690	15	2706	81		Na/Na	Auto
ST11950ND	1690					Na/Na	Auto
ST11950W	1690	15	2706	81		Na/Na	Auto
ST11950WD	1690					Na/Na	Auto
ST1201A	177	9	1072	36	9/804/48	Na/Na	Auto
ST1201E	177	9	1072	36		Na/Na	Auto
ST1201N	172	9	1068	36		Na/Na	Auto
ST1201NS	177	9	1068	36		—/—	Auto
ST1239A	211	9	1272	36	14/817/36	Na/Na	Auto
ST1239NS	204	9	1268	36		Na/Na	Auto
ST124	21	4	615	17		616/616	670
ST12400N	2148	19	2621	83		Na/Na	Auto
ST12400NC	2148	19	2621	83		Na/Na	Auto
ST12400ND	2100	19	2626			—/—	
ST12400ND	2148	19	2621	83		Na/Na	Auto
ST12400W	2148	19	2621	84		Na/Na	Auto
ST12400WC	2148	19	2621	84		Na/Na	Auto
ST12400WD	2148	19	2621	84		Na/Na	Auto
ST12401N	2100	19	2626			Na/Na	Auto
ST12401ND	2100	19	2626			—/—	
ST12450W	1849	18	2710	149		Na/Na	Auto
ST12450WD	1781					Na/Na	Auto
ST125-0	21	4	615	17		Na/Na	Auto
ST125-1	21	4	615	17		Na/Na	Auto
ST12550N	2139	19	2707	81		Na/Na	Auto
ST12550ND	2139		2756			Na/Na	Auto
ST12550W	2139	19	2707	81		Na/Na	Auto
ST12550WD	2139					Na/Na	Auto
ST12551N	2100		2756			—/—	
ST12551ND	2100		2756			—/—	
ST125A-0	21	4	404	26	4/615/17	Na/Na	Auto
ST125A-1	21	4	404	26	4/615/17	Na/Na	Auto
ST125N-0	21	4	407	26		None/NA	NA
ST125N-1	21	4	407	26		Na/Na	Auto
ST125R	21.5	4	404	26		—/—	
ST1274A	230	4	407	26	4/407/26	—/—	
ST137R	33	4	615	26		—/—	Auto
ST138-0	32	6	615	17		Na/Na	Auto

Drive Model	Seek Time	Interface	Encode	Form Factor	cache kb	mtbf	RPM	Obsolete? ⇓
ST1162A	15	IDE AT	2,7 RLL	3.5HH	32k	150k	3600	Y
ST1162N	15	SCSI	2,7 RLL	3.5HH	64k	70k	3600	Y
ST11700N	9	SCSI-2Fast	1,7 RLL	3.5HH	256k	500k	5400	Y
ST11700ND	10	SCSI-2Fast	1,7 RLL	3.5HH	256k	500k	5400	Y
ST11701N	9	SCSI-2FstW	1,7 RLL	3.5HH	256k	500k	5400	Y
ST11701ND	10	SCSI-2FstW	1,7 RLL	3.5HH	256k	500k	5400	Y
ST11750N	8	SCSI-2Fast	1,7 RLL	3.5HH	1024k	500k	7200	Y
ST11750ND	9	SCSI-2Fast	1,7 RLL	3.5HH	1024k	500k	7200	Y
ST11751N	8	SCSI-2Fast	1,7 RLL	3.5HH	1024k	500k	7200	Y
ST11751ND	9	SCSI-2Fast	1,7 RLL	3.5HH	1024k	500k	7200	Y
ST118273FC	7.1	FiberChanel	16/17Epr4	3.5HH	1024k		7200	
ST118273LC	7.1	Ultra2 SCSI	16/17Epr4	3.5HH	1024k		7200	
ST118273LW	7.1	Ultra2 SCSI	16/17Epr4	3.5HH	1024k		7200	
ST118273N	7.1	Ultra SCSI	16/17Epr4	3.5HH	512k		7200	
ST118273W	7.1	Ultra SCSI	16/17Epr4	3.5HH	512k		7200	
ST118273WC	7.1	SCA-2	16/17Epr4	3.5HH	512k		7200	
ST118273WD	7.1	Ultra2 SCSI	16/17Epr4	3.5HH	1024k		7200	
ST1186A	15	IDE AT	2,7 RLL	3.5HH	32k	150k	3600	Y
ST1186NS	15	SCSI	2,7 RLL	3.5HH	64k	150k	3600	Y
ST11900N	10	SCSI-2Fast	1,7 RLL	3.5HH		500k	5411	Y
ST11900NC	10	SCSI-2Fast	1,7 RLL	3.5HH		500k	5411	Y
ST11900ND	10	SCSI-2Fast	1,7 RLL	3.5HH		500k	5411	Y
ST11900W	10	SCSI-2FstW	RLL ZBR	3.5HH		500k	5411	Y
ST11900WC	10	SCSI-2FstW	RLL ZBR	3.5HH		500k	5411	Y
ST11900WD	10	SCSI-2FstW	RLL ZBR	3.5HH		500k	5411	Y
ST11950N	9	SCSI-2Fast	RLL ZBR	3.5HH	1024k	500k	7200	Y
ST11950ND	9	SCSI-2Fast		3.5HH	1024k	500k	7200	Y
ST11950W	9	SCSI-2FstW	RLL ZBR	3.5HH	1024k	500k	7200	Y
ST11950WD	9	SCSI-2FstW		3.5HH	1024k	500k	7200	Y
ST1201A	15	IDE AT	2,7 RLL	3.5HH	32k	150k	3600	Y
ST1201E	15	ESDI (10)	2,7 RLL	3.5HH		150k	3600	Y
ST1201N	15	SCSI	2,7 RLL	3.5HH	64k	150k	3600	Y
ST1201NS	15	SCSI-2	2,7 RLL	3.5HH		70k		Y
ST1239A	15	IDE AT	2,7 RLL	3.5HH	32k	150k	3600	Y
ST1239NS	15	SCSI-2	2,7 RLL	3.5HH	64k	150k	3600	Y
ST124	40	ST412/506	MFM	3.5HH		150k	3600	Y
ST12400N	9	SCSI-2Fast	RLL ZBR	3.5HH	256k	500k	5411	Y
ST12400NC	9	SCSI-2Fast	RLL ZBR	3.5HH	256k	500k	5411	Y
ST12400ND	10	SCSI-2Fast	1,7 RLL	3.5HH	256k	500k	5400	Y
ST12400ND	9	SCSI-2Fast	RLL ZBR	3.5HH	256k	500k	5411	Y
ST12400W	10.5	SCSI-2FstW	RLL ZBR	3.5HH	256k	500k	5411	Y
ST12400WC	10.5	SCSI-2FstW	RLL ZBR	3.5HH	256k	500k	5411	Y
ST12400WD	10.5	SCSI-2FstW	RLL ZBR	3.5HH	256k	500k	5411	Y
ST12401N	9	SCSI-2FstW	1,7 RLL	3.5HH	256k	500k	5411	Y
ST12401ND	10	SCSI-2Fast	1,7 RLL	3.5HH	256k	500k	5400	Y
ST12450W	9	SCSI-2FstW	1,7RLLZbr	3.5HH	1024k	500k	7200	Y
ST12450WD	9	SCSI-2FstW		3.5HH	1024	500k	7200	Y
ST125-0	40	ST412/506	MFM	3.5HH		150k	3600	Y
ST125-1	28	ST412/506	MFM	3.5HH		150k	3600	Y
ST12550N	8	SCSI-2Fast	1,7 RLL	3.5HH	1024k	500k	7200	Y
ST12550ND	9	SCSI-2Fast	1,7 RLL	3.5HH	1024k	500k	7200	Y
ST12550W	9	SCSI-2FstW	1,7RLLZbr	3.5HH	1024k	500k	7200	Y
ST12550WD	9	SCSI-2FstW		3.5HH	1024k	500k	7200	Y
ST12551N	8	SCSI-2Fast	1,7 RLL	3.5HH	1024k	500k	7200	Y
ST12551ND	9	SCSI-2Fast	1,7 RLL	3.5HH	1024k	500k	7200	Y
ST125A-0	40	IDE AT	RLL	3.5HH	2k	150k	3600	Y
ST125A-1	28	IDE AT	RLL	3.5HH	2k	150k	3600	Y
ST125N-0	40	SCSI	RLL	3.5HH	2k	150k	3600	Y
ST125N-1	28	SCSI	RLL	3.5HH	2k	150k	3600	Y
ST125R		ST412/506	2,7 RLL	3.5HH		150k		Y
ST1274A	18	IDE AT	2,7 RLL	3.5HH		70k		Y
ST137R	40	ST412/506	2,7 RLL	3.5HH		70k		Y
ST138-0	40	ST412/506	MFM	3.5HH	2k	150k	3600	Y

Drive Model	Format Size MB	Head	Cyl	Sect/ Trac	Translate H/C/S	RWC/ WPC	Land Zone
ST138-1	32	6	615	17		Na/Na	Auto
ST138A-0	32	4	604	26	6/615/17	Na/Na	Auto
ST138A-1	32	4	604	26	6/615/17	Na/Na	Auto
ST138N-0	32	4	615	26		Na/Na	Auto
ST138N-1	32	4	615	26		Na/Na	Auto
ST138R-0	32	4	615	26		Na/Na	Auto
ST138R-1	32	4	615	26		Na/Na	Auto
ST1400A	331	7	1475	NA	12/1018/53	Na/Na	Auto
ST1400N	331	7	1476	62		Na/Na	Auto
ST1401A	340	9	1132		15/726/61	Na/Na	Auto
ST1401N	338	9	1100	66		Na/Na	Auto
ST14207N Cayman	4294	20	4016	104		Na/Na	Auto
ST14207W Cayman	4294	20	4016	104		Na/Na	Auto
ST14209N Cayman	4295	20	3999	104		Na/Na	Auto
ST14209W Cayman	4295	20	3999	104		Na/Na	Auto
ST1480A	426	9	1474	NA	15/895/62	Na/Na	Auto
ST1480N	426	9	1476	62		Na/Na	Auto
ST1480NV	426	9	1478	V		Na/Na	Auto
ST1481N	426	9	1476	62		Na/Na	Auto
ST151	42	5	977	17		Na/Na	Auto
ST15150DC	4294	21	3711			Na/Na	Auto
ST15150FC	4294	21	3711			Na/Na	Auto
ST15150N	4294	21	3711	81		Na/Na	Auto
ST15150ND	4294	21	3711			Na/Na	Auto
ST15150W	4294	21	3711			Na/Na	Auto
ST15150WC	4294	21	3711			Na/Na	Auto
ST15150WD	4294	21	3711			Na/Na	Auto
ST15230DC	4294	19	3892			Na/Na	Auto
ST15230N	4294	19	3892			Na/Na	Auto
ST15230ND	4294	19	3892			Na/Na	Auto
ST15230W	4294	19	3892			Na/Na	Auto
ST15230WC	4294	19	3892			Na/Na	Auto
ST15230WD	4294	19	3892			Na/Na	Auto
ST15320N	4294					—/—	
ST157A-0	45	6	560	26	7/733/17	Na/Na	Auto
ST157A-1	45	6	560	26	7/733/17	Na/Na	Auto
ST157N-0	49	6	615	26		Na/Na	Auto
ST157N-1	49	6	615	26		Na/Na	Auto
ST157R-0	49	6	615	26		Na/Na	Auto
ST157R-1	49	6	615	26		Na/Na	Auto
ST1581N	525	9	1476	77		Na/Na	Auto
ST177N	60	5	921	26		Na/Na	Auto
ST1830N	702	13	1325			—/—	Auto
ST18771DC	8700	20	5333			Na/Na	Auto
ST18771FC	8700	20	5333			Na/Na	Auto
ST18771N	8700	20	5333			Na/Na	Auto
ST18771ND	8700	20	5333			Na/Na	Auto
ST18771W	8700	20	5333			Na/Na	Auto
ST18771WC	8700	20	5333			Na/Na	Auto
ST18771WD	8700	20	5333			Na/Na	Auto
ST19101DC	9100	16	6526			—/—	
ST19101FC	9100	16	6526			—/—	
ST19101N	9100	16	6526			—/—	
ST19101W	9100	16	6526			—/—	
ST19101WC	9100	16	6526			—/—	
ST19101WD	9100	16	6526			—/—	
ST19171DC	9100	20	5274			—/—	
ST19171FC	9100	20	5274			—/—	
ST19171N	9100	20	5274			—/—	
ST19171W	9100	20	5274			—/—	
ST19171WC	9100	20	5274			—/—	
ST19171WD	9100	20	5274			—/—	
ST1950N	803	13	1575			—/—	Auto

Drive Model	Seek Time	Interface	Encode	Form Factor	cache kb	mtbf	RPM	Obsolete?
ST138-1	28	ST412/506	MFM	3.5HH	2k	70k	3600	Y
ST138A-0	40	IDE AT	2,7 RLL	3.5HH	2k	150k	3600	Y
ST138A-1	28	IDE AT	2,7 RLL	3.5HH	2k	150k	3600	Y
ST138N-0	40	SCSI	2,7 RLL	3.5HH	2k	150k	3600	Y
ST138N-1	28	SCSI	2,7 RLL	3.5HH	2k	150k	3600	Y
ST138R-0	40	ST412/506	2,7 RLL	3.5HH	2k	150k	3600	Y
ST138R-1	28	ST412/506	2,7 RLL	3.5HH	2k	150k	3600	Y
ST1400A	14	IDE AT	1,7 RLL	3.5HH	64k	150k	4412	Y
ST1400N	14	SCSI-2	1,7RLLZbr	3.5HH	64k	150k	4412	Y
ST1401A	12	IDE AT	1,7 RLL	3.5HH	64k	150k	4412	Y
ST1401N	12	SCSI-2	1,7RLLZbr	3.5HH	64k	150k	4412	Y
ST14207N Cayman	9	SCSI-2Fast	1,7 RLL	3.5HH	512k	1000k	7200	
ST14207W Cayman	9	SCSI-2FstW	1,7 RLL	3.5HH	512k	1000k	7200	
ST14209N Cayman	9	Ultra SCSI	1,7 RLL	3.5HH	512k	1000k	7200	
ST14209W Cayman	9	Ultra SCSIW	1,7 RLL	3.5HH	512k	1000k	7200	
ST1480A	14	IDE AT	ZBR	3.5HH	64k	150k	4412	Y
ST1480N	14	SCSI-2	ZBR	3.5HH	64k	150k	4412	Y
ST1480NV	14	SCSI-2	1,7 RLL	3.5HH	64k	150k	4412	Y
ST1481N	14	SCSI-2Fast	1,7RLLZbr	3.5HH	64k	150k	4412	Y
ST151	24	ST412/506	MFM	3.5HH		150k	3600	Y
ST15150DC	9	SCSI-2Diff	1,7 RLL	3.5HH	1024k	800k	7200	
ST15150FC	9	FC	1,7 RLL	3.5HH	1024k	800k	7200	
ST15150N	9	SCSI-2Fast	1,7 RLL	3.5HH	1024k	800k	7200	
ST15150ND	9	SCSI-2Diff	1,7 RLL	3.5HH	1024k	800k	7200	
ST15150W	9	SCSI-2FstW	1,7 RLL	3.5HH	1024k	800k	7200	
ST15150WC	9	SCSI-2FstW	1,7 RLL	3.5HH	1024k	800k	7200	
ST15150WD	9	SCSI-2Diff	1,7 RLL	3.5HH	1024k	800k	7200	
ST15230DC	10	SCSI-2FstW	1,7RLLZbr	3.5HH	512k	800k	5411	
ST15230N	9	SCSI-2Fast	1,7 RLL	3.5HH	512k	800k	5411	
ST15230ND	9	SCSI-2Fast	1,7 RLL	3.5HH	512k	800k	5411	
ST15230W	10	SCSI-2FstW	1,7RLLZbr	3.5HH	512k	800k	5411	
ST15230WC	10	SCSI-2FstW	1,7RLLZbr	3.5HH	512k	800k	5411	
ST15230WD	10	SCSI-2FstW	1,7RLLZbr	3.5HH	512k	800k	5411	
ST15320N	10	ASA-2		3.5HH			5400	
ST157A-0	40	IDE AT	2,7 RLL	3.5HH	2k	150k	3600	Y
ST157A-1	28	IDE AT	2,7 RLL	3.5HH	2k	150k	3600	Y
ST157N-0	40	SCSI	2,7 RLL	3.5HH	2k	150k	3600	Y
ST157N-1	28	SCSI	2,7 RLL	3.5HH	2k	150k	3600	Y
ST157R-0	40	ST412/506	2,7 RLL	3.5HH	2k	150k	3600	Y
ST157R-1	28	ST412/506	2,7 RLL	3.5HH	2k	150k	3600	Y
ST1581N	14	SCSI-2Fast	RLL ZBR	3.5HH	64k	150k	4412	Y
ST177N	24	SCSI	RLL	3.5HH	8k	150k	3600	Y
ST1830N		SCSI-2Fast	1,7RLLZbr	3.5HH	256k	200k	4535	Y
ST18771DC	9	Ultra SCSI	Prml0,6,6	3.5HH	512k	1000k	7200	
ST18771FC	9	FC	Prml0,6,6	3.5HH	512k	1000k	7200	
ST18771N	9	Ultra SCSI	Prml0,6,6	3.5HH	512k	1000k	7200	
ST18771ND	9	Ultra SCSI	Prml0,6,6	3.5HH	512k	1000k	7200	
ST18771W	9	Ultra SCSI	Prml0,6,6	3.5HH	512k	1000k	7200	
ST18771WC	9	Ultra SCSI	Prml0,6,6	3.5HH	512k	1000k	7200	
ST18771WD	9	Ultra SCSI	Prml0,6,6	3.5HH	512k	1000k	7200	
ST19101DC	8	Ultra SCSI	Prml0,4,4	3.5HH	512k	1000k	1003	
ST19101FC	8	FC	Prml0,4,4	3.5HH	1024k	1000k	1003	
ST19101N	8	Ultra SCSI	Prml0,4,4	3.5HH	512k	1000k	1003	
ST19101W	8	Ultra SCSI	Prml0,4,4	3.5HH	512k	1000k	1003	
ST19101WC	8	Ultra SCSI	Prml0,4,4	3.5HH	512k	1000k	1003	
ST19101WD	8	Ultra SCSI	Prml0,4,4	3.5HH	512k	1000k	1003	
ST19171DC	9	FC-AL	Prml0,4,4	3.5HH	512k	1000k	7200	
ST19171FC	9	FC-AL	Prml0,4,4	3.5HH	512k	1000k	7200	
ST19171N	9	Ultra SCSI	Prml0,4,4	3.5HH	512k	1000k	7200	
ST19171W	9	FC-AL	Prml0,4,4	3.5HH	512k	1000k	7200	
ST19171WC	9	FC-AL	Prml0,4,4	3.5HH	512k	1000k	7200	
ST19171WD	9	FC-AL	Prml0,4,4	3.5HH	512k	1000k	7200	
ST1950N		SCSI-2Fast	1,7RLLZbr	3.5HH	256k	200k	4535	Y

Drive Model	Format Size MB	Head	Cyl	Sect/ Trac	Translate H/C/S	RWC/ WPC	Land Zone
ST1980N	860	13	1730	74		Na/Na	Auto
ST1980NC	860	13	1730			Na/Na	Auto
ST1980ND	860	13	1730			—/—	
ST206	5	2	306	17		307/128	
ST2106E	89	5	1024	34		Na/Na	Auto
ST2106N	91	5	1022	36		Na/Na	Auto
ST2106NM	94	5	1022	35		Na/Na	Auto
ST212	10	4	306	17		307/128	319
ST2125N	107	3	1544	45		Na/Na	Auto
ST2125NM	107	3	1544	45		Na/Na	Auto
ST2125NV	107	3	1544	45		Na/Na	Auto
ST213	10	2	615	17		616/300	670
ST2182E	160	4	1453	54		Na/Na	Auto
ST2209N	179	5	1544	45		Na/Na	Auto
ST224N	21	2				—/—	
ST225	21	4	615	17		None/300-614	670
ST225N	21	4	615	17		Na/Na	670
ST225R	21	2	667	31		Na/Na	670
ST2274A	241	5	1747	54	16/536/55	Na/Na	Auto
ST2383A	338	7	1747	54	16/737/56	Na/Na	Auto
ST2383E	338	7	1747	54		Na/Na	Auto
ST2383N	332	7	1261	74		Na/Na	Auto
ST2383ND	332	7	1261	NA		Na/Na	Auto
ST2383NM	332	7	1261	NA		Na/Na	Auto
ST238R	32	4	615	26		Na/Na	670
ST2502N	435	7	1755	NA		Na/Na	Auto
ST2502ND	435	7	1765	NA		Na/Na	Auto
ST2502NM	435	7	1765	NA		Na/Na	Auto
ST2502NV	435	7	1765	NA		Na/Na	Auto
ST250N	42	4	667			Na/Na	Auto
ST250R	42	4	667	31		Na/Na	Auto
ST251-0	42	6	820	17		Na/Na	Auto
ST251-1	42	6	820	17		Na/Na	Auto
ST251N-0	43	4	820	26		Na/Na	Auto
ST251N-1	43	4	820	26		Na/Na	Auto
ST251R	43	4	820	26		Na/Na	Auto
ST252	42	6	820	17		Na/Na	Auto
ST253	43	5	989	17		NA/128	Auto
ST274A	65	5	948	26	8/940/17	Na/Na	Auto
ST277N-0	65	6	820	26		Na/Na	Auto
ST277N-1	65	6	628	34		Na/Na	Auto
ST277R-0	65	6	820	26		Na/Na	Auto
ST277R-1	65	6	820	26		Na/Na	Auto
ST278R	65	6	820	26		Na/Na	Auto
ST279R	65	5	989	26		NA/128	Auto
ST280A	71	5	1032	26	10/516/27	Na/Na	Auto
ST296N	85	6	820	34		Na/Na	Auto
ST3025A	21	1	615	17	2/808/26	Na/Na	Auto
ST3025N	21	1	1616	26		Na/Na	Auto
ST3051A	43	6	820	17	6/820/17	Na/Na	Auto
ST3057A	53	*	1024	17		Na/Na	Auto
ST3057N	49	3	940	34		Na/Na	Auto
ST3096A	90	10	1024	17	8/836/26	Na/Na	Auto
ST3096N	84	3	1024	35		Na/Na	Auto
ST31010A	1082	2			16/2098/63	Na/Na	Auto
ST31012A	1082	2			16/2098/63	Na/Na	Auto
ST310240A	10200	4				—/—	
ST31051N	1060	4	4176			Na/Na	Auto
ST31051W	1060	4	4176			Na/Na	Auto
ST31051WC	1060	4	4176			Na/Na	Auto
ST31055N	1060	4	4176			—/—	
ST31055W	1060	4	4176			—/—	
ST31055WC	1060	4	4176			—/—	

Drive Model	Seek Time	Interface	Encode	Form Factor	cache kb	mtbf	RPM	Obsolete?
ST1980N	10	SCSI-2Fast	1,7RLLZbr	3.5HH	256k	200k	5411	Y
ST1980NC	11	SCSI-2Fast		3.5HH	256k	200k	5400	Y
ST1980ND	11	SCSI-2Fast	1,7 RLL	3.5HH	256k	200k	5400	Y
ST206		ST412/506	MFM	5.25FH				Y
ST2106E	18	ESDI (10)	2,7 RLL	5.25HH		100k	3600	Y
ST2106N	18	SCSI	2,7 RLL	5.25HH	32k	100k	3600	Y
ST2106NM	18	SCSI	2,7 RLL	5.25HH	32k	100k		Y
ST212	65	ST412/506	MFM	5.25FH		11k	3600	Y
ST2125N	18	SCSI	2,7RLLZbr	5.25HH	32k	100k	3600	Y
ST2125NM	18	SCSI	2,7RLLZbr	5.25HH	32k	100k	3600	Y
ST2125NV	18	SCSI	2,7RLLZbr	5.25HH	32k	100k	3600	Y
ST213	65	ST412/506	MFM	5.25FH		20k	3600	Y
ST2182E	16	ESDI (15)	2,7 RLL	5.25HH		100k	3600	Y
ST2209N	18	SCSI	2,7RLLZbr	5.25HH	32k	100k	3600	Y
ST224N	70	SCSI	2,7 RLL	5.25HH		100k		Y
ST225	65	ST412/506	MFM	5.25HH		100k	3600	Y
ST225N	65	SCSI	MFM	5.25HH		100k	3600	Y
ST225R	70	ST412/506	2,7 RLL	5.25HH		100k	3000	Y
ST2274A	16	IDE AT	2,7 RLL	5.25HH	32k	100k	3600	Y
ST2383A	16	IDE AT	2,7 RLL	5.25HH	32k	100k	3600	Y
ST2383E	16	ESDI	2,7 RLL	5.25HH		100k	3600	Y
ST2383N	14	SCSI	2,7RLLZbr	5.25HH	64k	100k	3600	Y
ST2383ND	14	SCSI	RLL ZBR	5.25HH	64k	100k	3600	Y
ST2383NM	14	SCSI	RLL ZBR	5.25HH	64k	100k	3600	Y
ST238R	65	ST412/506	RLL	5.25HH		100k	3600	Y
ST2502N	16	SCSI	2,7RLLZbr	5.25HH	64k	100k	3600	Y
ST2502ND	16	SCSI	RLL ZBR	5.25HH	64k	100k		Y
ST2502NM	16	SCSI	RLL ZBR	5.25HH	64k	100k		Y
ST2502NV	16	SCSI	RLL ZBR	5.25HH	64k	100k		Y
ST250N	70	SCSI	2,7 RLL	5.25HH		100k		Y
ST250R	70	ST412/506	2,7 RLL	5.25HH		100k	3600	Y
ST251-0	40	ST412/506	MFM	5.25HH		100k	3600	Y
ST251-1	28	ST412/506	MFM	5.25HH		100k	3600	Y
ST251N-0	40	SCSI	RLL	5.25HH		70k	3600	Y
ST251N-1	28	SCSI	RLL	5.25HH		70k	3600	Y
ST251R	40	ST412/506	2,7 RLL	5.25HH		100k		Y
ST252	40	ST412/506	MFM	5.25HH		100k	3600	Y
ST253	28	ST412/506	MFM	5.25HH		40k	3600	Y
ST274A	29	IDE AT	RLL	5.25HH		40k	3600	Y
ST277N-0	40	SCSI	RLL	5.25HH	2k	70k	3600	Y
ST277N-1	28	SCSI	RLL	5.25HH	2k	70k	3600	Y
ST277R-0	40	ST412/506	2,7 RLL	5.25HH		70k	3600	Y
ST277R-1	28	ST412/506	2,7 RLL	5.25HH		70k	3600	Y
ST278R	40	ST412/506	2,7 RLL	5.25HH		70k	3600	Y
ST279R	28	ST412/506	RLL	5.25HH		40k	3600	Y
ST280A	29	IDE AT	RLL	5.25HH		40k	3600	Y
ST296N	28	SCSI	2,7 RLL	5.25HH	8k	70k	3600	Y
ST3025A	19	IDE AT	2,7 RLL	3.5 3H	8/32k	50k	3600	Y
ST3025N	19	SCSI-2	2,7 RLL	3.5 3H	8/32k	50k	3600	Y
ST3051A	16	IDE AT	2,7 RLL	3.5 3H	32k	150k	3211	Y
ST3057A	19	IDE AT	2,7 RLL	3.5 3H	8/32k	50k	3600	Y
ST3057N	19	SCSI-2	2,7 RLL	3.5 3H	8/32k	50k	3600	Y
ST3096A	14	IDE AT	2,7 RLL	3.5 3H	32k	150k	3211	Y
ST3096N	20	SCSI-2	2,7 RLL	3.5 3H	8/32k	50k	3528	Y
ST31010A	12.5	ATA-2 Fast	1,7 RLL	3.5 3H	128k	300k	4500	Y
ST31012A	12.5	Ultra ATA-3	1,7 RLL	3.5 3H	128k	300k	4500	Y
ST310240A	11	Ultra ATA	16/17Epr4	3.5 3H	128k	300k	5400	
ST31051N	10.5	SCSI-3Fast	RLL 0,4,4	3.5 3H	256k	800k	5411	Y
ST31051W	10.5	SCSI-3Fast	RLL 0,4,4	3.5 3H	512k	800k	5411	Y
ST31051WC	10.5	SCSI-3Fast	RLL 0,4,4	3.5 3H	512k	800k	5411	Y
ST31055N	9	Ultra SCSI	RLL 0,4,4	3.5 3H	256k	800k	5411	Y
ST31055W	9	Ultra SCSI	RLL 0,4,4	3.5 3H	512k	800k	5411	Y
ST31055WC	9	Ultra SCSI	RLL 0,4,4	3.5 3H	512k	800k	5411	Y

Drive Model	Format Size MB	Head	Cyl	Sect/ Trac	Translate H/C/S	RWC/ WPC	Land Zone
ST31060A	1065	6	3640		16/2064/63	Na/Na	Auto
ST31060N	1062	8	2757	94		Na/Na	Auto
ST31060W	1062	8	2757	94		Na/Na	Auto
ST31080N	1080	6	3658	96		Na/Na	Auto
ST31081A	1081	4	3924		16/2097/63	Na/Na	Auto
ST31082A	1082				4/2097/63	—/—	
ST31200N	1052	9	2700	84		Na/Na	Auto
ST31200NC	1052					Na/Na	Auto
ST31200ND	1052	9	2626			Na/Na	Auto
ST31200W	1052	9	2700	84		Na/Na	Auto
ST31200WC	1052	9	2700	84		Na/Na	Auto
ST31200WD	1052	9	2700	84		Na/Na	Auto
ST3120A	107	12	1024	NA	12/1024/17	Na/Na	Auto
ST31220A	1083	6	3876		16/2099/63	Na/Na	Auto
ST31230DC	1050	5	3892			Na/Na	Auto
ST31230N	1050	5	3892			Na/Na	Auto
ST31230NC	1050	5	3898			Na/Na	Auto
ST31230ND	1050	5	3892			Na/Na	Auto
ST31230W	1050	5	3892			Na/Na	Auto
ST31230WC	1050	5	3898			Na/Na	Auto
ST31230WD	1050	5	3898			Na/Na	Auto
ST31231N	1060	5	3992			Na/Na	Auto
ST3123A	106	2			12/1024/17	Na/Na	Auto
ST31250N	1021	5	3711	107		Na/Na	Auto
ST31250ND	1021	5	3711			Na/Na	Auto
ST31250W	1021	5	3711			Na/Na	Auto
ST31250WC	1021	5	3711			Na/Na	Auto
ST31250WD	1021	5	3711			Na/Na	Auto
ST31270A	1283	6	3876		16/2485/63	Na/Na	Auto
ST31274A	1279	6	3659		16/2479/63	Na/Na	Auto
ST31275A	1275	6	3640		16/2477/63	Na/Na	Auto
ST31276A	1281	4	4893		16/2482/63	Na/Na	Auto
ST31277A	1281	4			16/2482/63	Na/Na	Auto
ST3144A	130	15	1001	17	15/1001/17	Na/Na	Auto
ST3145A	130	2				Na/Na	Auto
ST31621A	1621	6	3924		16/3146/63	Na/Na	Auto
ST31640A	1625		4834		16/3150/63	Na/Na	Auto
ST31720A	1700	2			16/3306/63	—/—	
ST31721A	1704	4			16/3303/63	Na/Na	Auto
ST31722A	1704	4			16/3303/63	Na/Na	Auto
ST31930N	1700	7	3898			Na/Na	Auto
ST31930ND	1700	7	3898			Na/Na	Auto
ST3195A	170	4			10/981/34	Na/Na	Auto
ST32105N Cayman	2147	10	3948	106		Na/Na	Auto
ST32105W Cayman	2147	10	3948	106		Na/Na	Auto
ST32107N Cayman	2147	10	3999	104		Na/Na	Auto
ST32107W Cayman	2147	10	3999	104		Na/Na	Auto
ST32110A	2100	1			16/4092/63	—/—	
ST3211A	213	2	2388		16/685/38	Na/Na	Auto
ST32120A	2111	4			16/4092/63	Na/Na	Auto
ST32122A	2100	4			16/4092/63	—/—	
ST32132A	2113	6			6/4095/63	—/—	
ST32140A	2113	8	4834		16/4200/63	Na/Na	Auto
ST32151N	2148	8	4176			Na/Na	Auto
ST32151W	2148	8	4176			Na/Na	Auto
ST32151WC	2148	8	4176			Na/Na	Auto
ST32155N	2148	8	4176			—/—	
ST32155W	2148	8	4176			—/—	
ST32155WC	2148	8	4176			—/—	
ST32161A	2147	8	4474		16/4095/63	Na/Na	Auto
ST32171DC	2150	6	5178			Na/Na	Auto
ST32171FC	2150	6	5178			Na/Na	Auto
ST32171N	2150	6	5178			Na/Na	Auto

Drive Model	Seek Time	Interface	Encode	Form Factor	cache kb	mtbf	RPM	Obsolete? ⇓
ST31060A	14	ATA	RLL ZBR	3.5 3H	64k	300k	3600	Y
ST31060N	9	SCSI-2Fast	1,7 RLL	3.5 3H	512k	500k	5400	Y
ST31060W	9	SCSI-2FstW	1,7 RLL	3.5 3H	512k	500k	5400	Y
ST31080N	11	SCSI-2Fast	1,7 RLL	3.5 3H	256k	500k	5400	Y
ST31081A	14	ATA	1,7 RLL	3.5 3H	64k	300k	3600	Y
ST31082A	12.5	ATA-3	1,7 RLL	3.5 3H	64k	300k	4500	Y
ST31200N	10	SCSI-2Fast	1,7RLLZbr	3.5 3H	256k	500k	5411	Y
ST31200NC	10.5	SCSI-2Fast		3.5 3H	256k	500k	5400	Y
ST31200ND	10	SCSI-2Fast	1,7 RLL	3.5 3H	256k	500k	5400	Y
ST31200W	10.5	SCSI-2FstW	1,7RLLZbr	3.5 3H	256k	500k	5411	Y
ST31200WC	10.5	SCSI-2FstW	1,7RLLZbr	3.5 3H	256k	500k	5411	Y
ST31200WD	10.5	SCSI-2FstW	1,7RLLZbr	3.5 3H	256k	500k	5411	Y
ST3120A	15	IDE AT	RLL ZBR	3.5 3H	32k	150k	3211	Y
ST31220A	12	ATA-2 Fast	1,7 RLL	3.5 3H	256k	300k	4500	Y
ST31230DC	10.5	SCSI-2Diff	1,7 RLL	3.5 3H	512k	800k	5411	Y
ST31230N	10.5	SCSI-2Fast	1,7 RLL	3.5 3H	512k	800k	5411	Y
ST31230NC	10.5	SCSI-2Fast	1,7RLLZbr	3.5 3H		800k	5411	Y
ST31230ND	10.5	SCSI-2Diff	1,7 RLL	3.5 3H	512k	800k	5411	Y
ST31230W	10.5	SCSI-2FstW	1,7 RLL	3.5 3H	512k	800k	5411	Y
ST31230WC	10.5	SCSI-2FstW	1,7 RLL	3.5 3H	512k	800k	5411	Y
ST31230WD	10.5	SCSI-2Diff	1,7 RLL	3.5 3H	512k	800k	5411	Y
ST31231N	10	SCSI-2Fast	RLL ZBR	3.5 3H	256k	800k	5411	Y
ST3123A	16	IDE AT	1,7RLLZbr	3.5 3H	32k	250k	3811	Y
ST31250N	9	SCSI-2Fast	1,7 RLL	3.5 3H	512k	800k	7200	Y
ST31250ND	9	SCSI-2Diff	1,7 RLL	3.5 3H	512k	800k	7200	Y
ST31250W	9	SCSI-2FstW	1,7 RLL	3.5 3H	512k	800k	7200	Y
ST31250WC	9	SCSI-2Diff	1,7 RLL	3.5 3H	512k	800k	7200	Y
ST31250WD	9	SCSI-2Diff	1,7 RLL	3.5 3H	512k	800k	7200	Y
ST31270A	12	ATA	RLL ZBR	3.5 3H	256k	300k	4500	Y
ST31274A	12	ATA	RLL ZBR	3.5 3H	256k	300k	4500	Y
ST31275A	14	ATA	RLL ZBR	3.5 3H	64k	300k	3600	Y
ST31276A	12	ATA	RLL ZBR	3.5 3H	64k	300k	4500	Y
ST31277A	12.5	ATA-2 Fast	1,7 RLL	3.5 3H	128k	300k	4500	Y
ST3144A	16	IDE AT	2,7 RLL	3.5 3H	32k	150k	3211	Y
ST3145A	16	IDE AT	1,7 RLL	3.5 3H		250k	3811	Y
ST31621A	14	ATA	RLL ZBR	3.5 3H	64k	300k	3600	Y
ST31640A	10	ATA-2 FAST	1,7 RLL	3.5 3H	256k	300k	5400	Y
ST31720A	12	ATA-2	1,7 RLL	3.5 3H	128k	500k	4500	
ST31721A	12.5	ATA-2 Fast	1,7 RLL	3.5 3H	128k	300k	4500	
ST31722A	12.5	Ultra ATA-3	1,7 RLL	3.5 3H	128k	300k	4500	
ST31930N	10.5	SCSI-2Fast	1,7RLLZbr	3.5 3H		800k	5411	Y
ST31930ND	10.5	SCSI-2Fast	1,7RLLZbr	3.5 3H		800k	5411	Y
ST3195A	16	IDE AT	1,7RLLZbr	3.5 3H	64k	250k	3811	Y
ST32105N Cayman	9.5	SCSI-2Fast	1,7 RLL	3.5 3H	512k	1000k	5400	
ST32105W Cayman	9.5	SCSI-2FstW	1,7 RLL	3.5 3H	512k	1000k	5400	
ST32107N Cayman	8.5	SCSI-2Fast	1,7 RLL	3.5 3H	512k	1000k	7200	
ST32107W Cayman	8.5	SCSI-2FstW	1,7 RLL	3.5 3H	512k	1000k	7200	
ST32110A	11	Ultra ATA	Prml16,17	3.5 3H	128k	300k	5400	
ST3211A	14	ATA	RLL ZBR	3.5 3H	32k	250k	3600	Y
ST32120A	12.5	ATA-2 Fast	1,7 RLL	3.5 3H	128k	300k	4500	
ST32122A	12	Ultra ATA	1,7RLL2/3	3.5 3H	128k	300k	4500	
ST32132A	12.5	ATA-3	Prml0,12,8	3.5 3H	128k	300k	4500	
ST32140A	10	ATA-2 Fast	1,7 RLL	3.5 3H	128k	500k	5400	
ST32151N	10.5	SCSI-3Fast	RLL 0,4,4	3.5 3H	256k	800k	5411	
ST32151W	10.5	SCSI-3Fast	RLL 0,4,4	3.5 3H	512k	800k	5411	
ST32151WC	10.5	SCSI-3Fast	RLL 0,4,4	3.5 3H	512k	800k	5411	
ST32155N	9	Ultra SCSI	RLL 0,4,4	3.5 3H	256k	800k	5411	
ST32155W	9	Ultra SCSI	RLL 0,4,4	3.5 3H	512k	800k	5411	
ST32155WC	9	Ultra SCSI	RLL 0,4,4	3.5 3H	512k	800k	5411	
ST32161A	10.5	ATA	RLL ZBR	3.5 3H	128k	500k	5400	
ST32171DC	9	Ultra SCSI	PRML	3.5 3H	512k	1000k	7200	
ST32171FC	9	FC-AL	PRML	3.5 3H	512k	1000k	7200	
ST32171N	9	Ultra SCSI	PRML	3.5 3H	512k	1000k	7200	

Drive Model	Format Size MB	Head	Cyl	Sect/ Trac	Translate H/C/S	RWC/ WPC	Land Zone
ST32171ND	2150	6	5178			Na/Na	Auto
ST32171W	2150	6	5178			Na/Na	Auto
ST32171WC	2150	6	5178			Na/Na	Auto
ST32171WD	2150	6	5178			Na/Na	Auto
ST32271DC	2260	6	5178			—/—	
ST32271N	2260	6	5178			—/—	
ST32271W	2260	6	5178			—/—	
ST32271WC	2260	6	5178			—/—	
ST32271WD	2260	6	5178			—/—	
ST32272DC	2260	4	6311			—/—	
ST32272N	2260	4	6311			—/—	
ST32272W	2260	4	6311			—/—	
ST32272WC	2260	4	6311			—/—	
ST32272WD	2260	4	6311			—/—	
ST3240A	211	2				Na/Na	Auto
ST32430DC	2147	9	3892			Na/Na	Auto
ST32430N	2147	9	3892			Na/Na	Auto
ST32430NC	2147	9	3898			Na/Na	Auto
ST32430ND	2147	9	3898			Na/Na	Auto
ST32430W	2147	9	3892			Na/Na	Auto
ST32430WC	2147	9	3892			Na/Na	Auto
ST32430WD	2147	9	3892			Na/Na	Auto
ST3243A	214	4	1024	34	12/1024/34	Na/Na	Auto
ST3250A	213	2			12/1024/34	Na/Na	Auto
ST32520A	2500	2			15/4888/63	—/—	
ST32530A	2558	6			16/4958/63	Na/Na	Auto
ST32531A	2557	6			6/4956/63	—/—	
ST32532A	2557	6			16/4956/63	Na/Na	Auto
ST32550DC	2147	11	3711	V		Na/Na	Auto
ST32550N	2147	11	3711	V		Na/Na	Auto
ST32550ND	2147	11	3711	V		Na/Na	Auto
ST32550W	2147	11	3510	108		Na/Na	Auto
ST32550W	2147	11	3711	V		Na/Na	Auto
ST32550WC	2147	11	3711	V		Na/Na	Auto
ST32550WD	2147	11	3711	V		Na/Na	Auto
ST325A,X	21	2	615	17	4/615/17	Na/Na	Auto
ST325N	21	2	654	32		Na/Na	Auto
ST325X	21	2	615	17		Na/Na	
ST3270A	271	2	2595		14/600/63	Na/Na	Auto
ST3271A	265	2	2805		10/977/53	Na/Na	Auto
ST3283A	245				14/978/35	Na/Na	Auto
ST3283N	248	5	1691	57		Na/Na	Auto
ST3285N	248	3	1691			Na/Na	Auto
ST3290A	260				15/1001/34	Na/Na	Auto
ST3291A	272	4			14/761/50	Na/Na	Auto
ST3295A	273	2			14/761/50	Na/Na	Auto
ST33220A	3227	4			16/6253/63	Na/Na	Auto
ST33221A	3200	2			16/6253/63	—/—	
ST33232A	3200	6			16/6253/63	—/—	
ST33240A	3227	8			8/6253/63	—/—	
ST3385A	340	5	767	62	14/767/62	Na/Na	Auto
ST3390A	341				14/768/62	Na/Na	Auto
ST3390N	344	3	2676	83		Na/Na	Auto
ST3391A	341	4			14/768/62	Na/Na	Auto
ST3420A	427	4	2388		16/826/63	Na/Na	Auto
ST34217N	4294	10	6028			—/—	
ST34217W	4294	10	6028			—/—	
ST34217WC	4294	10	6028			—/—	
ST34217WD	4294	10	6028			—/—	
ST3425A	425	2	3687		16/839/62	Na/Na	Auto
ST34321A	4300	2			15/8894/63	—/—	
ST34340A	4303	8			8/8894/63	—/—	
ST34342A	4300	8			15/8894/63	—/—	

Drive Model	Seek Time	Interface	Encode	Form Factor	cache kb	mtbf	RPM	Obsolete?
ST32171ND	9	Ultra SCSI	PRML	3.5 3H	512k	1000k	7200	
ST32171W	9	Ultra SCSI	PRML	3.5 3H	512k	1000k	7200	
ST32171WC	9	Ultra SCSI	PRML	3.5 3H	512k	1000k	7200	
ST32171WD	9	Ultra SCSI	PRML	3.5 3H	512k	1000k	7200	
ST32271DC	9	Ultra SCSI	PRML	3.5 3H	512k	1000k	7200	
ST32271N	9	Ultra SCSI	PRML	3.5 3H	512k	1000k	7200	
ST32271W	9	Ultra SCSI	PRML	3.5 3H	512k	1000k	7200	
ST32271WC	9	Ultra SCSI	PRML	3.5 3H	512k	1000k	7200	
ST32271WD	9	Ultra SCSI	PRML	3.5 3H	512k	1000k	7200	
ST32272DC	9	Ultra SCSI	PRML	3.5 3H	512k	1000k	7200	
ST32272N	9	Ultra SCSI	PRML	3.5 3H	512k	1000k	7200	
ST32272W	9	Ultra SCSI	PRML	3.5 3H	512k	1000k	7200	
ST32272WC	9	Ultra SCSI	PRML	3.5 3H	512k	1000k	7200	
ST32272WD	9	Ultra SCSI	PRML	3.5 3H	512k	1000k	7200	
ST3240A	8	IDE AT	RLL ZBR	3.5 3H	120k	300k	3811	Y
ST32430DC	10.5	SCSI-2Fast	1,7 RLL	3.5 3H	512k	800k	5411	
ST32430N	10.5	SCSI-2Fast	1,7 RLL	3.5 3H	512k	800k	5411	
ST32430NC	10.5	SCSI-2Fast	1,7RLLZbr	3.5 3H		800k	5411	Y
ST32430ND	10.5	SCSI-2Diff	1,7 RLL	3.5 3H	512k	800k	5411	
ST32430W	10.5	SCSI-2FstW	1,7RLLZbr	3.5 3H	512k	800k	5411	
ST32430WC	10.5	SCSI-2FstW	1,7 RLL	3.5 3H	512k	800k	5411	
ST32430WD	10.5	SCSI-2Diff	1,7 RLL	3.5 3H	512k	800k	5411	
ST3243A	16	IDE AT	1,7RLLZbr	3.5 3H	32k	250k	3811	Y
ST3250A	15	IDE AT	1,7RLLZbr	3.5 3H	120k	300k	3811	Y
ST32520A	13	Ultra ATA	Eprml1617	3.5 3H	256k	300k	5400	
ST32530A	10.5	ATA	ZBR Prml	3.5 3H	128k	500k	5376	
ST32531A	12	ATA-3	1,7 RLL	3.5 3H	128k	300k	5400	
ST32532A	12.5	Ultra ATA-3	1,7 RLL	3.5 3H	128k	300k	4500	
ST32550DC	8	SCSI-2Diff	1,7 RLL	3.5 3H	512k	800k	7200	
ST32550N	8	SCSI-2Fast	1,7 RLL	3.5 3H	512k	800k	7200	
ST32550ND	8	SCSI-2Diff	1,7 RLL	3.5 3H	512k	800k	7200	
ST32550W	8	SCSI-2FstW	RLL ZBR	3.5 3H	512k	800k	7200	Y
ST32550W	8	SCSI-2FstW	1,7 RLL	3.5 3H	512k	800k	7200	
ST32550WC	8	SCSI-2FstW	1,7 RLL	3.5 3H	512k	800k	7200	
ST32550WD	8	SCSI-2Diff	1,7 RLL	3.5 3H	512k	800k	7200	
ST325A,X	28	IDE AT	2,7RLLZbr	3.5HH	8/32k	150k	3048	Y
ST325N	28	SCSI	2,7 RLL	3.5HH	2k/8k	50k	3600	Y
ST325X	45	IDE XT	2,7 RLL	3.5HH	8/32k	150k	3600	Y
ST3270A	15	ATA	RLL ZBR	3.5 3H	32k	250k	3400	Y
ST3271A	10.5	ATA	RLL ZBR	3.5 3H	256k	300k	4500	Y
ST3283A	12	IDE AT	RLL ZBR	3.5 3H	128k	200k	4500	Y
ST3283N	12	SCSI-2Fast	RLL ZBR	3.5 3H	128k	250k	4500	Y
ST3285N	12	SCSI-2Fast	1,7RLLZbr	3.5 3H	128k	250k	4500	Y
ST3290A	16	IDE AT	1,7 RLL	3.5 3H		250k	3811	Y
ST3291A	13	IDE AT	1,7RLLZbr	3.5 3H	120k	300k	3811	Y
ST3295A	14	IDE AT	1,7 RLL	3.5 3H	120k	300k	3811	Y
ST33220A	11	Ultra ATA-3	Prml16,17	3.5 3H	128k	300k	4490	
ST33221A	11	Ultra ATA	Eprml1617	3.5 3H	128k	33k	5400	
ST33232A	12	Ultra ATA	1,7RLL2/3	3.5 3H	128k	300k	4500	
ST33240A	12	ATA-3	1,7 RLL	3.5 3H	128k	300k	4500	
ST3385A	12	IDE AT	1,7RLLZbr	3.5 3H	256k	250k	4500	Y
ST3390A	12	IDE AT	1,7 RLL	3.5 3H		250k	4500	Y
ST3390N	12	SCSI-2Fast	1,7RLLZbr	3.5 3H	256k	250k	4500	Y
ST3391A	14	IDE AT	1,7RLLZbr	3.5 3H	120k	300k	3811	Y
ST3420A	14	ATA	RLL ZBR	3.5 3H	32k	250k	3600	Y
ST34217N	9	Ultra SCSI	8,9RLL	3.5 3H	512k	1000k	7200	
ST34217W	9	Ultra SCSI	8,9RLL	3.5 3H	512k	1000k	7200	
ST34217WC	9	Ultra SCSI	8,9RLL	3.5 3H	512k	1000k	7200	
ST34217WD	9	Ultra SCSI	8,9RLL	3.5 3H	512k	1000k	7200	
ST3425A	14	ATA	RLL ZBR	3.5 3H	64k	300k	3600	Y
ST34321A	11	Ultra ATA	Eprml1617	3.5 3H	128k	300k	5400	
ST34340A	12	ATA-3	1,7 RLL	3.5 3H	128k	300k	4500	
ST34342A	12	Ultra ATA	1,7RLL2/3	3.5 3H	128k	300k	4500	

Drive Model	Format Size MB	Head	Cyl	Sect/ Trac	Translate H/C/S	RWC/ WPC	Land Zone
ST34371DC	4350	10	5288			Na/Na	Auto
ST34371FC	4350	10	5288			Na/Na	Auto
ST34371N	4350	10	5288			Na/Na	Auto
ST34371ND	4350	10	5288			Na/Na	Auto
ST34371W	4350	10	5288			Na/Na	Auto
ST34371WC	4350	10	5288			Na/Na	Auto
ST34371WD	4350	10	5288			Na/Na	Auto
ST34501DC	4550	8	6526			—/—	
ST34501FC	4550	8	6526			—/—	
ST34501FC	4550	8	6526			—/—	
ST34501N	4550	8	6526			—/—	
ST34501W	4550	8	6526			—/—	
ST34501WC	4550	8	6526			—/—	
ST34501WD	4550	8	6526			—/—	
ST34502FC	4550	6				—/—	
ST34502LC	4550	6				—/—	
ST34502LW	4550	6				—/—	
ST34520A	4550	4			15/9408/63	—/—	
ST34520N	4550	4				—/—	
ST34520W	4550	4				—/—	
ST34520WC	4550	4				—/—	
ST34555N	4550	8	6311	176		Na/Na	Auto
ST34555W	4550	8	6311	176		Na/Na	Auto
ST34571DC	4550					—/—	
ST34571FC	4550					—/—	
ST34571N	4550					—/—	
ST34571W	4550					—/—	
ST34571WD	4550					—/—	
ST34572DC	4550	8	6311			—/—	
ST34572N	4550	8	6311			—/—	
ST34572W	4550	8	6311			—/—	
ST34572WC	4550	8	6311			—/—	
ST34572WD	4550	8	6311			—/—	
ST34573FC	4550	5				—/—	
ST34573LC	4550	5				—/—	
ST34573LW	4550	5				—/—	
ST34573N	4550	5				—/—	
ST34573W	4550	5				—/—	
ST34573WC	4550	5				—/—	
ST34573WD	4550	5				—/—	
ST3491A	428	4			15/899/62	Na/Na	Auto
ST3500A	426	7	1547		15/895/62	Na/Na	Auto
ST3500N	426	7	1547	V		Na/Na	Auto
ST35040A	5008	8	6536		15/10352635	Na/Na	Auto
ST35130A	5121	6			15/10585/63	Na/Na	Auto
ST351A,X	43	2	820	17	6/820/17	Na/Na	Auto
ST352A,X	42	2		17	5/980/17	Na/Na	Auto
ST3541A	541	2	3925		16/1048/63	Na/Na	Auto
ST3543A	542	4	2574		16/1050/63	Na/Na	Auto
ST3544A	540	4	2805		16/1048/63	Na/Na	Auto
ST3550A	452	5	1018	62	14/1018/62	Na/Na	Auto
ST3550N	456	5	2126	83		Na/Na	Auto
ST3600A	528	7	1872		16/1024/63	Na/Na	Auto
ST3600N	525	7	1872	79		Na/Na	Auto
ST3600ND	525	7	1872			Na/Na	Auto
ST3610N	535	7	1872	79		Na/Na	Auto
ST3610NC	535					Na/Na	Auto
ST3610ND	535	7	1872			Na/Na	Auto
ST3620N	545	5	2700	78		Na/Na	Auto
ST3620NC	545	5	2700	78		Na/Na	Auto
ST3620ND	545	5	2700	78		Na/Na	Auto
ST3620W	546	5	2700	78		Na/Na	Auto
ST3630A	631	4			16/1223/63	NA/NA	Auto

Drive Model	Seek Time	Interface	Encode	Form Factor	cache kb	mtbf	RPM	Obsolete? ⇓
ST34371DC	9	Ultra SCSI	RLL 0,4,4	3.5 3H	512k	1000k	7200	
ST34371FC	9	FC-AL	RLL 0,4,4	3.5 3H	512k	1000k	7200	
ST34371N	9	Ultra SCSI	RLL 0,4,4	3.5 3H	512k	1000k	7200	
ST34371ND	9	Ultra SCSI	RLL 0,4,4	3.5 3H	512k	1000k	7200	
ST34371W	9	Ultra SCSI	RLL 0,4,4	3.5 3H	512k	1000k	7200	
ST34371WC	9	Ultra SCSI	RLL 0,4,4	3.5 3H	512k	1000k	7200	
ST34371WD	9	Ultra SCSI	RLL 0,4,4	3.5 3H	512k	1000k	7200	
ST34501DC	7.5	Ultra SCSI	Prml0,4,4	3.5 3H	512k	1000k	1003	
ST34501FC	7.5	FC	Prml0,4,4	3.5 3H	1024k	1000k	1003	
ST34501FC	7.5	FiberChanel	Prml0,4,4	3.5 3H	512k	1000k	1003	
ST34501N	7.5	Ultra SCSI	Prml0,4,4	3.5 3H	512k	1000k	1003	
ST34501W	7.5	Ultra SCSI	Prml0,4,4	3.5 3H	512k	1000k	1003	
ST34501WC	7.5	Ultra SCSI	Prml0,4,4	3.5 3H	512k	1000k	1003	
ST34501WD	7.5	Ultra SCSI	Prml0,4,4	3.5 3H	512k	1000k	1003	
ST34502FC	5.4	FiberChanel	8/9 PR4	3.5 3H	1024	1000k	1002	
ST34502LC	5.4	Ultra2 SCSI	8/9 PR4	3.5 3H	1024	1000k	1002	
ST34502LW	5.4	Ultra2 SCSI	8/9 PR4	3.5 3H	1024	1000k	1002	
ST34520A	9.5	Ultra ATA	Eprml1617	3.5 3H	512	400k	7200	
ST34520N	9.5	Ultra SCSI3	Eprml1617	3.5 3H	512	800k	7200	
ST34520W	9.5	Ultra SCSI3	Eprml1617	3.5 3H	512	800k	7200	
ST34520WC	9.5	SCA-2	Eprml1617	3.5 3H	512	800k	7200	
ST34555N	9.1	Ultra SCSI	ZBR Prml	3.5 3H	512k	800k	7200	
ST34555W	9.1	Ultra SCSIW	ZBR Prml	3.5 3H	512k	800k	7200	
ST34571DC	8.8	SCSI-2Diff		3.5 3H			7200	
ST34571FC	8.8	FC-AL		3.5 3H			7200	
ST34571N	8.8	Ultra SCSI		3.5 3H			7200	
ST34571W	8.8	Ultra SCSI		3.5 3H			7200	
ST34571WD	8.8	SCSI-2Diff		3.5 3H			7200	
ST34572DC	9	Ultra SCSI	PRML	3.5 3H	512k	1000k	7200	
ST34572N	9	Ultra SCSI	PRML	3.5 3H	512k	1000k	7200	
ST34572W	9	Ultra SCSI	PRML	3.5 3H	512k	1000k	7200	
ST34572WC	9	Ultra SCSI	PRML	3.5 3H	512k	1000k	7200	
ST34572WD	9	Ultra SCSI	PRML	3.5 3H	512k	1000k	7200	
ST34573FC	7.1	FiberChanel	16/17Epr4	3.5 3H	1024		7200	
ST34573LC	7.1	Ultra2 SCSI	16/17Epr4	3.5 3H	1024		7200	
ST34573LW	7.1	Ultra2 SCSI	16/17Epr4	3.5 3H	1024		7200	
ST34573N	7.1	Ultra SCSI	16/17Epr4	3.5 3H	512k		7200	
ST34573W	7.1	Ultra SCSI	16/17Epr4	3.5 3H	512k		7200	
ST34573WC	7.1	Ultra SCSI	16/17Epr4	3.5 3H	512k		7200	
ST34573WD	7.1	Ultra2 SCSI	16/17Epr4	3.5 3H	1024		7200	
ST3491A	14	ATA Fast	1,7RLLZbr	3.5 3H	120k	300k	3811	Y
ST3500A	10	AT BUS	RLL ZBR	3.5 3H	256k	200k	4535	Y
ST3500N	11	SCSI-2Fast	1,7RLLZbr	3.5 3H	240k	200k	4535	Y
ST35040A	10	ATA-2 Fast	PRML8,9	3.5 3H	512k	300k	5397	
ST35130A	11	Ultra ATA-3	Prml1617	3.5 3H	128k	300k	4490	
ST351A,X	28	IDE AT	2,7 RLL	3.5 3H	32k	150k	3048	Y
ST352A,X	28	AT/XT	2,7RLLZbr	3.5 3H		150k	3048	Y
ST3541A	14	ATA	RLL ZBR	3.5 3H	64k	300k	3600	Y
ST3543A	15	ATA	RLL ZBR	3.5 3H	64k	250k	3600	Y
ST3544A	12	ATA	RLL ZBR	3.5 3H	256k	300k	4500	Y
ST3550A	12	IDE AT	1,7RLLZbr	3.5 3H	256k	250k	4500	Y
ST3550N	12	SCSI-2Fast	1,7RLLZbr	3.5 3H	256k	250k	4500	Y
ST3600A	11	IDE AT	1,7 RLL	3.5 3H	256k	200k	4535	Y
ST3600N	12	SCSI-2Fast	1,7RLLZbr	3.5 3H	256k	200k	4467	Y
ST3600ND	12	SCSI-2Fast	1,7 RLL	3.5 3H	256k	200k	5400	Y
ST3610N	12	SCSI-2Fast	1,7RLLZbr	3.5 3H	256k	200k	5400	Y
ST3610NC	12	SCSI-2Fast		3.5 3H		200k	5400	Y
ST3610ND	12	SCSI-2Fast	1,7 RLL	3.5 3H	256k	200k	5400	Y
ST3620N	10.5	SCSI-2Fast	1,7RLLZbr	3.5 3H	256k	500k	5411	Y
ST3620NC	10.5	SCSI-2Fast	1,7RLLZbr	3.5 3H	256k	500k	5411	Y
ST3620ND	10.5	SCSI-2Fast	1,7RLLZbr	3.5 3H	256k	500k	5411	Y
ST3620W	10	SCSI-2FstW	RLL ZBR	3.5 3H	256k	500k	5411	Y
ST3630A	14	ATA	RLL ZBR	3.5 3H	120k	300k	3811	Y

Drive Model	Format Size MB	Head	Cyl	Sect/ Trac	Translate H/C/S	RWC/ WPC	Land Zone
ST3635A	635	3	3640		16/1238/63	NA/NA	Auto
ST3636A	640	2	4893		16/1241/63	NA/NA	Auto
ST36450A	6448	10	6536		15/13328/63	NA/NA	Auto
ST36451A	6448	10	6536		15/13328/63	NA/NA	Auto
ST36530A	6500	6			15/13456/63	—/—	
ST36530N	6550	6				—/—	
ST36530W	6550	6				—/—	
ST36530WC	6550	6				—/—	
ST36531A	6505	6			15/13446/63	NA/NA	Auto
ST36540A	6505	8			15/13446/63	NA/NA	Auto
ST3655A	528	5			16/1024/63	NA/NA	Auto
ST3655N	545	5	2393	89		NA/NA	Auto
ST3660A	545				16/1057/63	NA/NA	Auto
ST3780A	722	4	3876		16/1399/63	NA/NA	Auto
ST3850A	850	4			16/1648/63	NA/NA	Auto
ST3851A	850	4	3640		16/1654/63	NA/NA	Auto
ST3852A	850	1			16/1653/63	—/—	
ST3853A	852	4	3659		16/1652/63	NA/NA	Auto
ST38641A	8606	8			16/16383/63	NA/NA	Auto
ST39102FC	9100	12				—/—	
ST39102LC	9100	12				—/—	
ST39102LW	9100	12				—/—	
ST39140A	9100	8			16/16383/63	—/—	
ST39140N	9100	8				—/—	
ST39140W	9100	8				—/—	
ST39140WC	9100	8				—/—	
ST39173FC	9190	10				—/—	
ST39173LC	9190	10				—/—	
ST39173LW	9190	10				—/—	
ST39173N	9190	10				—/—	
ST39173W	9190	10				—/—	
ST39173WC	9190	10				—/—	
ST39173WD	9190	10				—/—	
ST4026	20	4	615	17		NA/NA	Auto
ST4038	31	5	733	17		NA/300	Auto
ST4038N	30	5	733			NA/NA	977
ST4051	40	5	977	17		NA/NA	Auto
ST4053	44	5	1024	17		NA/NA	Auto
ST406	5	2	306	17		NA/128	319
ST4077N	67	5	1024	26		1025/1025	
ST4077R	65	5	1024	26		1025/1025	
ST4085	71	8	1024	17		NA/NA	Auto
ST4086	72	9	925	17		NA/NA	Auto
ST4096	80	9	1024	17		NA/NA	Auto
ST4096N	83	4				—/—	Auto
ST4097	80	9	1024	17		NA/NA	Auto
ST410800N	9090	27	4925	133		NA/NA	Auto
ST410800ND	9090	27	4925	133		NA/NA	Auto
ST410800W	9090	27	4925	133		NA/NA	Auto
ST410800WD	9090	27	4925	133		NA/NA	Auto
ST41097J	1097	17	2101			NA/NA	Auto
ST412	10	4	306	17		307/128	319
ST41200N	1037	15	1931	71		NA/NA	Auto
ST41200ND	1037	15	1931	NA		NA/NA	Auto
ST41200NM	1037	15	1931	NA		NA/NA	Auto
ST41200NV	1037	15	1931	NA		NA/NA	Auto
ST41201J	1200U	17	2101			NA/NA	Auto
ST41201K	1200U	17	2101	NA		NA/NA	Auto
ST4135R	115	9	960	26		NA/128	Auto
ST4144N	122	9	1024	26		NA/NA	1023
ST4144R	122	9	1024	26		NA/NA	Auto
ST41520N	1370	17	2101	NA		NA/NA	Auto
ST41520ND	1370	17	2101	NA		NA/NA	Auto

Drive Model	Seek Time	Interface	Encode	Form Factor	cache kb	mtbf	RPM	Obsolete?
ST3635A	14	ATA	RLL ZBR	3.5 3H	64k	300k	3600	Y
ST3636A	12.5	ATA	RLL ZBR	3.5 3H	64k	300k	4500	Y
ST36450A	10	ATA-2 Fast	PRML8,9	3.5 3H	512k	300k	5397	
ST36451A	10	Ultra ATA-3	PRML8,9	3.5 3H	512k	300k	5397	
ST36530A	9.5	Ultra ATA	Eprml1617	3.5 3H	512k	400k	7200	
ST36530N	9.5	Ultra SCSI3	Eprml1617	3.5 3H	512k	800k	7200	
ST36530W	9.5	Ultra SCSI3	Eprml1617	3.5 3H	512k	800k	7200	
ST36530WC	9.5	SCA-2	Eprml1617	3.5 3H	512k	800k	7200	
ST36531A	10.5	Ultra ATA-3	Prml16,17	3.5 3H	128k	300k	5400	
ST36540A	11	Ultra ATA-3	Prml16,17	3.5 3H	128k	300k	4490	
ST3655A	12	IDE AT	1,7 RLL	3.5 3H	256k	250k	4500	Y
ST3655N	12	SCSI-2Fast	1,7RLLZbr	3.5 3H		250k	4500	Y
ST3660A	14	ATA Fast	1,7 RLL	3.5 3H	120k	300k	3811	Y
ST3780A	14	IDE AT	RLL ZBR	3.5 3H	256k	300k	4500	Y
ST3850A	14	ATA	1,7 RLL	3.5 3H	120k	300k	3811	Y
ST3851A	14	ATA	RLL ZBR	3.5 3H	64k	300k	3600	Y
ST3852A	12	ATA-2	1,7 RLL	3.5 3H	128k	500k	4500	Y
ST3853A	12	ATA	RLL ZBR	3.5 3H	256k	300k	4500	Y
ST38641A	10.5	Ultra ATA-3	Prml16,17	3.5 3H	128k	300k	5400	
ST39102FC	5.4	FiberChanel	8/9 PR4	3.5 3H	1024	1000k	1002	
ST39102LC	5.4	Ultra2 SCSI	8/9 PR4	3.5 3H	1024	1000k	1002	
ST39102LW	5.4	Ultra2 SCSI	8/9 PR4	3.5 3H	1024	1000k	1002	
ST39140A	9.5	Ultra ATA	Eprml1617	3.5 3H	512k	400k	7200	
ST39140N	9.5	Ultra SCSI3	Eprml1617	3.5 3H	512k	800k	7200	
ST39140W	9.5	Ultra SCSI3	Eprml1617	3.5 3H	512k	800k	7200	
ST39140WC	9.5	SCA-2	Eprml1617	3.5 3H	512k	800k	7200	
ST39173FC	7.1	FiberChanel	16/17Epr4	3.5 3H	1024k		7200	
ST39173LC	7.1	Ultra2 SCSI	16/17Epr4	3.5 3H	1024k		7200	
ST39173LW	7.1	Ultra2 SCSI	16/17Epr4	3.5 3H	1024k		7200	
ST39173N	7.1	Ultra SCSI	16/17Epr4	3.5 3H	512k		7200	
ST39173W	7.1	Ultra SCSI	16/17Epr4	3.5 3H	512k		7200	
ST39173WC	7.1	SCA-2	16/17Epr4	3.5 3H	512k		7200	
ST39173WD	7.1	Ultra2 SCSI	16/17Epr4	3.5 3H	1024k		7200	
ST4026	40	ST412/506	MFM	5.25FH		15k	3600	Y
ST4038	40	ST412/506	MFM	5.25FH		25k	3600	Y
ST4038N		SCSI		5.25FH				Y
ST4051	40	ST412/506	MFM	5.25FH		15k	3600	Y
ST4053	28	ST412/506	MFM	5.25FH		40k	3600	Y
ST406	85	ST412/506	MFM	5.25FH		11k	3600	Y
ST4077N	28	SCSI	2,7 RLL	5.25FH				Y
ST4077R	28	ST412/506	2,7 RLL	5.25FH				Y
ST4085	28	ST412/506	MFM	5.25FH		40k	3600	Y
ST4086	28	ST412/506	MFM	5.25FH		40k	3600	Y
ST4096	28	ST412/506	MFM	5.25FH		40k	3600	Y
ST4096N	17	SCSI		5.25FH				Y
ST4097	28	ST412/506	MFM	5.25FH		40k	3600	Y
ST410800N	12	SCSI-2Fast	1,7 RLL	5.25FH	1024k	500k	5400	Y
ST410800ND	12	SCSI-2Fast	1,7 RLL	5.25FH	1024k	500k	5400	
ST410800W	12	SCSI-2FstW	1,7 RLL	5.25FH	1024k	500k	5400	
ST410800WD	12	SCSI-2FstW	1,7 RLL	5.25FH	1024k	500k	5400	
ST41097J	11	SMD-O/E	2,7 RLL	5.25FH		150k	5400	Y
ST412	85	ST412/506	MFM	5.25FH		110k		Y
ST41200N	15	SCSI-2	1,7RLLZbr	5.25FH	256k	150k	3600	Y
ST41200ND	15	SCSI-2	RLL ZBR	5.25FH	256k	150k		Y
ST41200NM	15	SCSI-2	RLL ZBR	5.25FH	256k	150k		Y
ST41200NV	15	SCSI-2	RLL ZBR	5.25FH	256k	150k		Y
ST41201J	11	SMD-O/E	2,7 RLL	5.25FH		150k	5400	Y
ST41201K	11	IPI-2	2,7 RLL	5.25FH		150k	5400	Y
ST4135R	28	ST412/506	RLL	5.25FH		40k	3600	Y
ST4144N	28	SCSI	2,7 RLL	5.25FH				Y
ST4144R	28	ST412/506	2,7 RLL	5.25FH		40k	3600	Y
ST41520N	11	SCSI-2	2,7RLLZbr	5.25FH	48k	150k	5400	Y
ST41520ND	11	SCSI-2	ZBR	5.25FH	48k	150k	5400	Y

Drive Model	Format Size MB	Head	Cyl	Sect/ Trac	Translate H/C/S	RWC/ WPC	Land Zone
ST41600N	1370	17	2101	NA		NA/NA	Auto
ST41600ND	1370	17	2101	NA		NA/NA	Auto
ST41601N	1370	17	2101	V		NA/NA	Auto
ST41601ND	1370	17	2101	V		NA/NA	Auto
ST41650N	1415	15	2107	87		NA/NA	Auto
ST41650ND	1415	15	2107	NA		NA/NA	Auto
ST41651N	1415	15	2107	87		NA/NA	Auto
ST41651ND	1415	15	2107	NA		NA/NA	Auto
ST41800K	1986U	18	2627	NA		NA/NA	Auto
ST4182E	151	9	969	34		NA/NA	Auto
ST4182N	155	9	967	36		NA/NA	Auto
ST4182NM	155	9	967	36		NA/NA	Auto
ST419	15	6	306	32		307/128	319
ST4192E	169	8	1147	36		NA/NA	
ST4192N	168	8	1147	36		1148/1148	
ST42000N,ND	1792	16	2627	83		NA/NA	Auto
ST42100N	1900	15	2573	96		NA/NA	Auto
ST423451N	23.2	28	6876	237		—/—	
ST423451W	23.2	28	6876	237		—/—	
ST423451WD	23.2	28	6876	237		—/—	
ST42400N,ND	2129	19	2627	83		NA/NA	Auto
ST425	20	8	306	17		307/128	
ST43200K	3386u	20	2738			NA/NA	Auto
ST43200N	3338			NA		NA/NA	Auto
ST43400N	2912	21	2738	99		NA/NA	Auto
ST43400ND	2912	21	2738	99		NA/NA	Auto
ST43401N	2912	21	2738			NA/NA	Auto
ST43401ND	2912	21	2738			NA/NA	Auto
ST43402N	2912	21	2738	99		NA/NA	Auto
ST43402ND	2912	21	2738	99		NA/NA	Auto
ST4350N	300	9	1412	46		NA/NA	Auto
ST4350NM	307	9	1412	NA		NA/NA	Auto
ST4376N	330	9	1549	45		NA/NA	Auto
ST4376NM	330	9	1549	NA		NA/NA	Auto
ST4376NV	330	9	1549	NA		NA/NA	Auto
ST4383E	319	13	1412	34		NA/NA	Auto
ST4384E	319	15	1224	34		NA/NA	Auto
ST4385N	330	15	791	55		NA/NA	Auto
ST4385NM	330	15	791	NA		NA/NA	Auto
ST4385NV	330	15	791	NA		NA/NA	Auto
ST4442E	368	15	1412	34		NA/NA	Auto
ST4702N	601	15	1546	50		NA/NA	Auto
ST4702NM	601	15	1546	NA		NA/NA	Auto
ST4766E	664	15	1632	53		NA/NA	Auto
ST4766N	676	15	1632	54		NA/NA	Auto
ST4766NM	663		1632	54		NA/NA	Auto
ST4766NV	663		1632	54		NA/NA	Auto
ST4767E	676	15	1399	63		NA/NA	Auto
ST4767N	665	15	1356	64		NA/NA	Auto
ST4767ND	665	15	1356	64		NA/NA	Auto
ST4767NM	665	15	1356	64		NA/NA	Auto
ST4767NV	665	15	1356	64		NA/NA	Auto
ST4769E	631	15	1552	53		NA/NA	Auto
ST506	5	4	153	17		128/128	157
ST51080A	1080	4	4771		16/2114/63	NA/NA	Auto
ST51080N	1000					—/—	
ST51270A	1282	4	5414		16/2485/63	NA/NA	Auto
ST52160A	2113	4			16/4095/63	—/—	
ST52520A	2560	4			16/4970/63	—/—	
ST5540A	541	2	4834		16/1050/63	NA/NA	Auto
ST5660A	545	4	3420		16/1057/63	NA/NA	Auto
ST5660N	545	4	3420	77		NA/NA	Auto
ST5850A	855	4	4085		16/1656/63	NA/NA	Auto

Drive Model	Seek Time	Interface	Encode	Form Factor	cache kb	mtbf	RPM	Obsolete?
ST41600N	11	SCSI-2	2,7RLLZbr	5.25FH	48k	150k	5400	Y
ST41600ND	11	SCSI-2	ZBR	5.25FH	48k	150k	5400	Y
ST41601N	11	SCSI-2Fast	2,7RLLZbr	5.25FH	256k	150k	5400	Y
ST41601ND	11	SCSI-2Fast	2,7 RLL	5.25FH	256k	150k	5400	Y
ST41650N	15	SCSI-2	1,7RLLZbr	5.25FH	256k	150k	3600	Y
ST41650ND	15	SCSI-2Diff	RLL ZBR	5.25FH	256k	150k		Y
ST41651N	15	SCSI-2Fast	1,7RLLZbr	5.25FH	256k	150k	3600	Y
ST41651ND	15	SCSI-2Diff	1,7 RLL	5.25FH	256k	150k		Y
ST41800K	11	IPI-2	2,7 RLL	5.25FH		150k	5400	Y
ST4182E	16	ESDI	RLL	5.25FH		100k	3600	Y
ST4182N	16	SCSI	2,7 RLL	5.25FH	32k	100k	3600	Y
ST4182NM	16	SCSI	2,7 RLL	5.25FH	32k	100k	3600	Y
ST419	85	ST412/506	MFM	5.25FH		11k		Y
ST4192E	17	ESDI	2,7 RLL	5.25FH		20k		Y
ST4192N	17	SCSI	2,7 RLL	5.25FH		20k		Y
ST42000N,ND	11	SCSI-2Fast	2,7RLLZbr	5.25FH		150k	5400	Y
ST42100N	13	SCSI-2FstW	1,7RLLZbr	5.25FH	256k	150k	3600	Y
ST423451N	13	Ultra SCSI	Prml0,4,4	5.25FH	2048k	500k	5400	
ST423451W	13	Ultra SCSI	Prml0,4,4	5.25FH	2048k	500k	5400	
ST423451WD	13	Ultra SCSI	Prml0,4,4	5.25FH	2048k	500k	5400	
ST42400N,ND	11	SCSI-2Fast	2,7RLLZbr	5.25FH	512k	150k	5400	Y
ST425		ST412/506	MFM	5.25FH				Y
ST43200K	11	IPI-2	1,7 RLL	5.25FH	512k	200k	5400	Y
ST43200N	11	IPI-2	RLL ZBR	5.25FH		300k		Y
ST43400N	11	SCSI-2Fast	1,7 RLL	5.25FH	512k	200k	5400	Y
ST43400ND	11	SCSI-2Fast	1,7 RLL	5.25FH	512k	200k	5400	Y
ST43401N	11	SCSI-2FstW	1,7 RLL	5.25FH	512k	200k	5400	Y
ST43401ND	11	SCSI-2FstW	1,7 RLL	5.25FH	512k	200k	5400	Y
ST43402N	11	SCSI-2 2POR	1,7RLLZbr	5.25FH	2048k	200k	5400	Y
ST43402ND	11	SCSI-2 2POR	1,7RLLZbr	5.25FH	394k	200k	5400	Y
ST4350N	16	SCSI	2,7RLLZbr	5.25FH	32k	100k	3600	Y
ST4350NM	16	SCSI	RLL ZBR	5.25FH	32k	100k		Y
ST4376N	17	SCSI	2,7RLLZbr	5.25FH	32k	100k	3600	Y
ST4376NM	17	SCSI	RLL ZBR	5.25FH	32k	100k		Y
ST4376NV	17	SCSI	RLL ZBR	5.25FH	32k	100k		Y
ST4383E	18	ESDI	2,7 RLL	5.25FH		100k	3600	Y
ST4384E	14	ESDI	2,7 RLL	5.25FH		100k	3600	Y
ST4385N	10	SCSI	2,7RLLZbr	5.25FH	32k	100k	3600	Y
ST4385NM	10	SCSI	RLL ZBR	5.25FH	32k	100k		Y
ST4385NV	10	SCSI	RLL ZBR	5.25FH	32k	100k		Y
ST4442E	16	ESDI	RLL	5.25FH		100k	3600	Y
ST4702N	16	SCSI	2,7RLLZbr	5.25FH	32k	100k	3600	Y
ST4702NM	16	SCSI	RLL ZBR	5.25FH	32k	100k		Y
ST4766E	16	ESDI (15)	RLL	5.25FH		150k	3600	Y
ST4766N	15	SCSI	RLL	5.25FH	32k	150k	3600	Y
ST4766NM	15	SCSI	2,7 RLL	5.25FH	32k	150k		Y
ST4766NV	15	SCSI	2,7 RLL	5.25FH	32k	150k		Y
ST4767E	11	ESDI (24)	1,7 RLL	5.25FH		150k	4800	Y
ST4767N	11	SCSI-2	1,7RLLZbr	5.25FH	256k	150k	4800	Y
ST4767ND	11	SCSI-2	RLL ZBR	5.25FH	256k	150k	4800	Y
ST4767NM	11	SCSI-2	RLL ZBR	5.25FH	256k	150k	4800	Y
ST4767NV	11	SCSI-2	RLL ZBR	5.25FH	256k	150k	4800	Y
ST4769E	14	ESDI	1,7 RLL	5.25FH		150k	4800	Y
ST506	85	ST412/506	MFM	3.5 4H		11k		Y
ST51080A	10	ATA-2 Fast	1,7 RLL	3.5 4H	256k	300k	5400	Y
ST51080N		SCSI		2.5 4H				Y
ST51270A	10.5	ATA	RLL ZBR	3.5 4H	128k	300k	5376	Y
ST52160A	11	ATA-2 Fast	PRML	3.5 3H	128k	500k	5400	
ST52520A	11	ATA-2 Fast	PRML	3.5 4H	128k	500k	5400	
ST5540A	10.5	ATA	RLL ZBR	3.5 4H	128k	300k	5376	Y
ST5660A	12	IDE AT	1,7 RLL	3.5 4H		300k	4500	Y
ST5660N	12	SCSI-2Fast	1,7RLLZbr	3.5 4H		300k	4500	Y
ST5850A	11	ATA-2 Fast	1,7 RLL	3.5 4H	256k	300k	5400	Y

Drive Model	Format Size MB	Head	Cyl	Sect/ Trac	Translate H/C/S	RWC/ WPC	Land Zone
ST5851A	854	4	4834		16/1656/63	NA/NA	Auto
ST6165J	165	10	823			—/—	
ST6315J	315	19	823			—/—	
ST6344J	344	24	711			—/—	Auto
ST6515J	516	24	711			—/—	Auto
ST6515K	516u	24	711			—/—	
ST6516J	516	24	711			—/—	
ST683J	83	5	823			—/—	
ST7050P	42	2				NA/NA	Auto
ST706	5	2	306	17		307/128	
ST81123J	1123U	15	1635			—/—	Auto
ST81154K	1154U	14	1635			—/—	
ST81236J	1236	15	1635			—/—	Auto
ST81236K	1236	15	1635			—/—	
ST81236N	1056	15	1635	NA		NA/NA	Auto
ST82030J	2030U	19	2120			—/—	
ST82030K	2030U	19	2120			—/—	
ST82038J	2038U	19	2611			—/—	Auto
ST82105K	2105U	16	2611			—/—	
ST82272J	2272U	19	2611			—/—	Auto
ST82368K	2368U	18	2611			—/—	
ST82500J	2500	19	2611			—/—	Auto
ST82500K	2500 (U)	19	2611			—/—	
ST82500N	2140	19	2611	NA		NA/NA	Auto
ST83050K	3050U	18	2655	NA		NA/NA	Auto
ST83050N	3050U	18	2655	NA		NA/NA	Auto
ST83073J	3073u	19	2655			NA/NA	
ST83220K	3220U	19	2655	NA		NA/NA	Auto
ST8368J	368U	10	1217			—/—	
ST8368N	316	10	1217	NA		NA/NA	Auto
ST8500J	500U	10	1217			—/—	
ST8500N	427	10	1217	NA		NA/NA	Auto
ST8741J	741U	15	1635			—/—	
ST8741N	637	15	1635	NA		NA/NA	Auto
ST8851J	851	15	1381			—/—	Auto
ST8851K	851	15	1381			—/—	
ST8851N	727	15	1381			NA/NA	Auto
ST8885N	727			NA		NA/NA	Auto
ST9025A	21	4	1024		4/615/17	NA/NA	Auto
ST9051A	43	4	654	32	6/820/17	NA/NA	Auto
ST9052A	42	16	1024	63	5/980/17	—/—	
ST9077A	64	4	802	39	11/669/17	NA/NA	Auto
ST9080A	64	2		38	4/823/38	NA/NA	Auto
ST9096A	85	4		34	10/980/17	NA/NA	Auto
ST9100A	85					NA/NA	Auto
ST9100AG	85	2		63	14/748/16	NA/NA	Auto
ST91080A	1083	6			16/2100/63	NA/NA	Auto
ST91350AG	1350				16/2616/63	—/—	
ST9140AG	127	4			15/980/17	NA/NA	Auto
ST91420A	1442	4			16/2794/63	NA/NA	Auto
ST91430A	1449	6			16/2808/63	NA/NA	Auto
ST9144A	128	6			15/980/17	NA/NA	Auto
ST9145A	128	4	1463		15/980/17	NA/NA	Auto
ST9145AG	127	4	1463		15/980/17	NA/NA	Auto
ST9150AG	131	2			13/419/47	NA/NA	Auto
ST91685AG	1680				8/3256/63	—/—	
ST9190AG	171	4			16/873/24	NA/NA	Auto
ST92130A	2163	6			16/4191/63	NA/NA	Auto
ST92255AG	2250				10/4360/63	—/—	
ST9235AG	209	6	985	32	13/985/32	NA/NA	Auto
ST9235N	209	13	985	NA		NA/NA	Auto
ST9240AG	210	4			8/988/52	NA/NA	Auto
ST9295AG	261	16	1024	63		—/—	

Drive Model	Seek Time	Interface	Encode	Form Factor	cache kb	mtbf	RPM	Obsolete?
ST5851A	10.5	ATA	RLL ZBR	3.5 4H	128k	300k	5376	Y
ST6165J	30	SMD	2,7 RLL	8		10k	3600	Y
ST6315J	20	SMD-E	MFM	9		30k	3600	Y
ST6344J	18	SMD-O/E	MFM	9		30k	3600	Y
ST6515J	18	SMD	2,7 RLL	9		30k	3600	Y
ST6515K	18	IPI-2	2,7 RLL	9		30k	3600	Y
ST6516J	18	SMD-E	2,7 RLL	9		30k	3600	Y
ST683J	30	SMD	2,7 RLL	8		8k	3600	Y
ST7050P	18	PCMCIA/ATA	1,7 RLL	1.8 4H	32k	300k	3545	Y
ST706		ST412/506	MFM	5.25FH				Y
ST81123J	15	SMD-E	2,7 RLL	8		150k	3600	Y
ST81154K	15	IPI-2	2,7 RLL	8		150k	3600	Y
ST81236J	15	SMD-E	2,7 RLL	8		150k	3600	Y
ST81236K	15	IPI-2	2,7 RLL	8		150k	3600	Y
ST81236N	15	SCSI	2,7 RLL	8		150k	3600	Y
ST82030J	11	SMD-O/E	2,7 RLL	8		150k	3600	Y
ST82030K	11	IPI-2	2,7 RLL	8		150k	3600	Y
ST82038J	12	SMD-E	2,7 RLL	8		150k	3600	Y
ST82105K	12	IPI-2	2,7 RLL	8		80k	3600	Y
ST82272J	12	SMD-E	2,7 RLL	8		150k		Y
ST82368K	12	IPI-2	2,7 RLL	8		80k	3600	Y
ST82500J	12	SMD-E	2,7 RLL	8		150k	3600	Y
ST82500K	12	IPI-2	2,7 RLL	8		150k	3600	Y
ST82500N	12	SCSI	2,7 RLL	8		150k	3600	Y
ST83050K	12	IPI-2	1,7 RLL	8		150k	4365	Y
ST83050N	12	IPI-2	1,7 RLL	8		150k	4365	Y
ST83073J	12	SMD-O/E	1,7 RLL	8 FH		150k	4235	Y
ST83220K	12	IPI-2	1,7 RLL	8		150k	4365	Y
ST8368J	18	SMD-E	2,7 RLL	8 FH		35k	3600	Y
ST8368N	18	SCSI	2,7 RLL	8		30k	3600	Y
ST8500J	18	SMD-E	2,7 RLL	8		30k	3600	Y
ST8500N	18	SCSI	2,7 RLL	8		30k	3600	Y
ST8741J	15	SMD-E	2,7 RLL	8		50k	3600	Y
ST8741N	15	SCSI	2,7 RLL	8		50k	3600	Y
ST8851J	15	SMD-E	2,7 RLL	8		100k	3600	Y
ST8851K	15	IPI-2	2,7 RLL	8		100k	3600	Y
ST8851N	12	SCSI	2,7 RLL	8		100k	3600	Y
ST8885N	15	SCSI		8		150k		Y
ST9025A	<20	IDE AT	2,7 RLL	2.5 4H		150k	3631	Y
ST9051A	<20	IDE AT	2,7 RLL	2.5 4H	32k	150k	3631	Y
ST9052A	16	IDE AT	2,7 RLL	2.5 4H	32k	150k	3450	Y
ST9077A	19	IDE AT	2,7 RLL	2.5 4H	32k	150k	3546	Y
ST9080A	16	IDE AT	2,7RLLZbr	2.5 4H	32k	150k	3449	Y
ST9096A	16	IDE AT	2,7RLLZbr	2.5 4H	64k	150k	3450	Y
ST9100A	16	IDE AT		2.5 4H	120k	300k		Y
ST9100AG	16	IDE AT	1,7RLLZbr	2.5 4H	120k	300k	3545	Y
ST91080A	12	ATA-3	PRML8,9	2.5 4H	103k	300k	4508	Y
ST91350AG	12	ATA-2 Fast	RLL ZBR	2.5 4H	103k	300k	4508	Y
ST9140AG	16	IDE AT	1,7RLLZbr	2.5 4H	120k	300k	3545	Y
ST91420A	12	ATA-3	Prml16,17	2.5 4H	103k	350k	4508	Y
ST91430A	12	ATA-3	Prml16,17	2.5 4H	103k	300k	4508	Y
ST9144A	16	IDE AT	2,7RLLZbr	2.5 4H	64k	150k	3450	Y
ST9145A	16	AT BUS	RLL ZBR	2.5 4H	32k	150k	3449	Y
ST9145AG	16	IDE AT	2,7RLLZbr	2.5 4H	32k	150k	3449	Y
ST9150AG	16	IDE AT	1,7RLLZbr	2.5 4H	120k	300k	3980	Y
ST91685AG	12	ATA-2 Fast	RLL ZBR	2.5 4H	103k	300k	4508	Y
ST9190AG	16	IDE AT	1,7RLLZbr	2.5 4H	120k	300k	3545	Y
ST92130A	12	ATA-3	Prml16,17	2.5 4H	103k	350k	4508	
ST92255AG	12	ATA-2 Fast		2.5 4H	103k	300k	4508	
ST9235AG	16	IDE AT	RLL ZBR	2.5 4H	64k	150k	3449	Y
ST9235N	16	SCSI	2,7RLLZbr	2.5 4H	64k	150k	3449	Y
ST9240AG	16	ATA Fast	1,7RLLZbr	2.5 4H	120k	300k	3980	Y
ST9295AG	16	IDE AT	2,7 RLL	2.5 4H	120k	300k	3450	Y

Drive Model	Format Size MB	Head	Cyl	Sect/ Trac	Translate H/C/S	RWC/ WPC	Land Zone
ST9295N (never made)	250	NA	NA	NA		—/—	
ST9300AG	262	4			15/569/60	NA/NA	Auto
ST9342A	345	6	1598		16/667/63	NA/NA	Auto
ST9352A	350	4	2225		12/905/63	NA/NA	Auto
ST9385AG	341	6			16/934/51	NA/NA	Auto
ST9420AG	420				16/988/32	—/—	
ST9422A	421	4	2393		16/816/63	NA/NA	Auto
ST9546A	540	6			16/1047/63	NA/NA	Auto
ST9550AG	455	6			16/942/59	NA/NA	Auto
ST9655AG	524	6			14/1016/63	NA/NA	Auto
ST9810AG	811	4			16/1572/63	NA/NA	Auto
ST9816AG	810				16/1571/63	NA/NA	Auto
ST9840AG	840	4			16/1628/63	NA/NA	Auto

SEQUEL, INC

Drive Model	Format Size MB	Head	Cyl	Sect/ Trac	Translate H/C/S	RWC/ WPC	Land Zone
5300	3000	21		V		—/—	
5350	3572	25		V		—/—	
5400	4000	26		V		—/—	
EXT4175	149	7	1224	34		NA/NA	Auto
EXT4280	234	11	1224	36		NA/NA	Auto
EXT4380	319	15	1224	34		NA/NA	Auto
XT1050	38	5	902	17		NA/NA	Auto
XT1065	52	7	918	17		NA/NA	Auto
XT1085	71	8	1024	17		NA/NA	Auto
XT1105	84	11	918	17		NA/NA	Auto
XT1120R	105	8	1024	25		NA/NA	Auto
XT1140	119	15	918	17		NA/NA	Auto
XT1240R	196	15	1024	25		NA/NA	Auto
XT2085	72	7	1224	17		NA/NA	Auto
XT2140	113	11	1224	17		NA/NA	Auto
XT2190	159	15	1224	17		NA/NA	Auto
XT3170	146	9	1224	26		—/—	
XT3280	244	15	1224	26		—/—	
XT3380	319	15	1224	34		—/—	
XT4170E	157	7	1224	35/36	16	NA/NA	Auto
XT4170S	157	7	1224	35-36		NA/NA	Auto
XT4380E	338	15	1224	36		NA/NA	Auto
XT4380S	338	15	1224	36		NA/NA	Auto
XT8380E	361	8	1632	53-54		NA/NA	Auto
XT8380S	361	8	1632	54		NA/NA	Auto
XT8760E	676	15	1632	53-54		NA/NA	Auto
XT8760EH	676	15	1632	54		NA/NA	Auto
XT8760S	670	15	1632	NA		NA/NA	Auto
XT8760SH	670	15	1632	NA		NA/NA	Auto
XT8800E	694	15	1274	54		NA/NA	Auto

SHUGART

Drive Model	Format Size MB	Head	Cyl	Sect/ Trac	Translate H/C/S	RWC/ WPC	Land Zone
1002	5			17		—/—	
1004	10			17		—/—	
1006	30					—/—	
4004	14			17		—/—	
4008	29			17		—/—	
4100	56			17		—/—	
604	5	4	160	17		128/128	Auto
606	7	6	160	17		128/128	Auto
612	10	4	306	17		307/128	Auto
706	6	2	320	17		321/128	Auto
712	10	4	320	17		321/128	Auto
725	20					—/—	

Drive Model	Seek Time	Interface	Encode	Form Factor	cache kb	mtbf	RPM	Obsolete?
ST9295N (never made)	16	SCSI	2,7 RLL	2.5 4H	64k	150k	3450	Y
ST9300AG	16	ATA Fast	1,7RLLZbr	2.5 4H	120k	300k	3980	Y
ST9342A	13	ATA	RLL ZBR	2.5 4H	32k	150k	4000	Y
ST9352A	12	ATA	RLL ZBR	2.5 4H	32k	300k	3750	Y
ST9385AG	16	ATA Fast	1,7RLLZbr	2.5 4H	120k	300k	3980	Y
ST9420AG	16	ATA-2 Fast		2.5 4H	120k	300k	4500	Y
ST9422A	12	ATA	RLL ZBR	2.5 4H	64k	300k	3600	Y
ST9546A	16	ATA	RLL ZBR	2.5 4H	120k	300k	4500	Y
ST9550AG	16	IDE AT	1,7RLLZbr	2.5 4H	120k	300k	3980	Y
ST9655AG	16	ATA Fast	1,7RLLZbr	2.5 4H	120k	300k	3980	Y
ST9810AG	14	ATA	PRML8,9	2.5 4H	120k	300k	3968	Y
ST9816AG	16	ATA-2 Fast		2.5 4H	120k	300k	4500	Y
ST9840AG	14	ATA	PRML8,9	3.5 4H	107k	300k	4500	Y
SEQUEL, INC								
5300	12	SCSI-2FstW	1,7 RLL	5.25FH	512	300k	5400	Y
5350	12	SCSI-2FstW	1,7 RLL	5.25FH	512k	300k	5400	Y
5400	12	SCSI-2FstW	1,7 RLL	5.25FH	1024k	300k	5400	Y
EXT4175	27	ESDI	RLL	5.25FH		20k	3600	Y
EXT4280	27	ESDI	RLL	5.25FH		20k	3600	Y
EXT4380	27	ESDI	RLL	5.25FH		20k	3600	Y
XT1050	30	ST412/506	MFM	5.25FH		20k	3600	Y
XT1065	30	ST412/506	MFM	5.25FH		20k	3600	Y
XT1085	28	ST412/506	MFM	5.25FH		150k	3600	Y
XT1105	27	ST412/506	MFM	5.25FH		20k	3600	Y
XT1120R	27	ST412/506	2,7 RLL	5.25FH		150k	3600	Y
XT1140	27	ST412/506	MFM	5.25FH		150k	3600	Y
XT1240R	27	ST412/506	2,7 RLL	5.25FH		150k	3600	Y
XT2085	30	ST412/506	MFM	5.25FH		30k	3600	Y
XT2140	30	ST412/506	MFM	5.25FH		30k	3600	Y
XT2190	29	ST412/506	MFM	5.25FH		150k	3600	Y
XT3170	30	SCSI	RLL	5.25FH		20k	3600	Y
XT3280	30	SCSI		5.25FH		20k	3600	Y
XT3380	27	SCSI		5.25FH		20k	3600	Y
XT4170E	14	ESDI	1,7 RLL	5.25FH		150k	3600	Y
XT4170S	14	SCSI	1,7 RLL	5.25FH		150k	3600	Y
XT4380E	16	ESDI	1,7 RLL	5.25FH		150k	3600	Y
XT4380S	16	SCSI	1,7 RLL	5.25FH		150k	3600	Y
XT8380E	16	ESDI	1,7 RLL	5.25FH		150k	3600	Y
XT8380S	14	SCSI	1,7 RLL	5.25FH		150k	3600	Y
XT8760E	16	ESDI	1,7 RLL	5.25FH		150k	3600	Y
XT8760EH	14	ESDI	1,7 RLL	5.25FH		150k	3600	Y
XT8760S	16	SCSI	1,7 RLL	5.25FH		150k	3600	Y
XT8760SH	14	SCSI	1,7 RLL	5.25FH	256k	150k	3600	Y
XT8800E	14	ESDI	1,7 RLL	5.25FH		150k	3600	Y
SHUGART								
1002		ST412/506	MFM	8.0 FH				Y
1004		ST412/506	MFM	8.0 FH				Y
1006		ST412/506	MFM	8.0				Y
4004		ST412/506	MFM					Y
4008		ST412/506	MFM	14.0				Y
4100		ST412/506	MFM					Y
604	27	ST412/506	MFM	5.25FH				Y
606	27	ST412/506	MFM	5.25FH				Y
612	27	ST412/506	MFM	5.25FH				Y
706	27	ST412/506	MFM	5.25FH				Y
712	27	ST412/506	MFM	5.25FH				Y
725		ST412/506	MFM	5.25HH				Y

Drive Model	Format Size MB	Head	Cyl	Sect/ Trac	Translate H/C/S	RWC/ WPC	Land Zone
SIEMENS							
1200	174	8	1216	35		NA/NA	Auto
1300	261	12	1216	35		NA/NA	Auto
2200	174	8	1216			NA/NA	Auto
2300	261	12	1216	35		NA/NA	Auto
4410	322	11	1100	52		NA/NA	Auto
4420	334	11	1100	54		NA/NA	Auto
5710	655	15				NA/NA	Auto
5720	655	15				NA/NA	Auto
5810	777	16				NA/NA	Auto
5820	777	16				NA/NA	Auto
6200	1200					NA/NA	Auto
7520	655	15				NA/NA	Auto
SONY							
2020A	20					—/—	
2040A	40					—/—	
3080L	80					—/—	
SRD2040Z	42	4	624			—/—	
SRD3040C	42.9					—/—	
SRD3040Z	42.9					—/—	
SRD3080C	85.8					—/—	
SRD3080Z	85.8					—/—	
STORAGE DIMENSIONS							
AT100	109	8	1024	26		—/None	1023
AT1000S	1000	15				NA/NA	Auto
AT100S	105	3				NA/NA	Auto
AT120	119	15	918	17		NA/NA	Auto
AT133	133	15	1024	17		—/None	1023
AT140	142	8	1024	34		—/None	1023
AT155E	158	9	1224	36		—/—	
AT155S	156	9	1224	36		—/—	
AT160	160	15	1224	17		—/None	1023
AT200	204	15	1024	26		—/None	1023
AT200S	204	7				—/—	
AT320S	320	15	1224	36		—/—	
AT335E	338	15	1224	36		—/—	
AT40	44	5	1024	17		—/None	1023
AT650E	651	15	1632	54		—/—	
AT650S	651	15	1632	54		—/—	
AT70	70			17		—/—	
CDASM-1051F	1000					—/—	
CDASM-2105F	2100					—/—	
CDASM-4005F	4300					—/—	
DMH-A02W	2100					—/—	
DMH-A04W	4300					—/—	
DMH-B02W	2100					—/—	
DMH-B04W	4300					—/—	
DMH-B09W	9100					—/—	
LAN1050F	1050					—/—	
LAN2101F	2101					—/—	
LAN2105F	2105					—/—	
LAN4005	4300					—/—	
LAN9000F	9000					—/—	
MAC-195	195	7				NA/NA	Auto
PS155E	156	9	1224	36		—/—	
PS155S	156	9	1224	36		—/—	
PS320S	320	15	1224	36		—/—	
PS335E	338	15	1224	36		—/—	
PS650S	651	15	1632	16		—/—	
XT100	109	8	1024	26		—/None	1023

Drive Model	Seek Time	Interface	Encode	Form Factor	cache kb	mtbf	RPM	Obsolete?
SIEMENS								
1200	25	ESDI	2,7 RLL	5.25FH				Y
1300	25	ESDI	2,7 RLL	5.25FH				Y
2200	25	ESDI	2,7 RLL	5.25FH				Y
2300	25	ESDI	2,7 RLL	5.25FH				Y
4410	18	ESDI	2,7 RLL	5.25FH		30k		Y
4420	16	SCSI	2,7 RLL	5.25FH		40k		Y
5710	16	ESDI	2,7 RLL	5.25FH				Y
5720	16	SCSI	2,7 RLL	5.25FH				Y
5810	18	ESDI	2,7 RLL	5.25FH				Y
5820	18	SCSI	2,7 RLL	5.25FH				Y
6200	14	SCSI	2,7 RLL	5.25FH				Y
7520	16	SCSI	2,7 RLL	5.25FH				Y
SONY								
2020A		SCSI		3.5HH				Y
2040A		SCSI		3.5HH				Y
3080L		SCSI		3.5 3H				Y
SRD2040Z	29	SCSI		3.5HH		25k	3600	Y
SRD3040C	18	IDE AT		3.5 3H	64k	50k	2975	Y
SRD3040Z	18	SCSI		3.5 3H	8k	50k	2975	Y
SRD3080C	18	IDE AT		3.5 3H	64k	50k	2975	Y
SRD3080Z	18	SCSI		3.5 3H	8k	50k	2975	Y
STORAGE DIMENSIONS								
AT100		ST412/506	2,7 RLL					Y
AT1000S		SCSI				100k		Y
AT100S	19	SCSI	2,7 RLL	3.5HH		150k		Y
AT120	26	ST412/506	MFM	5.25FH		40k		Y
AT133		ST412/506	MFM					Y
AT140								Y
AT155E	14	ESDI	2,7 RLL	5.25FH		40k		Y
AT155S	17	SCSI	2,7 RLL	5.25FH		40k		Y
AT160	28	ST412/506	MFM	5.25FH		40k		Y
AT200		ST412/506	2,7 RLL	3.5HH				Y
AT200S	16	SCSI	2,7 RLL	3.5HH		150k		Y
AT320S	17	SCSI	2,7 RLL	5.25FH		40k		Y
AT335E	16	ESDI	2,7 RLL	5.25FH		40k		Y
AT40		ST412/506	MFM					Y
AT650E	16	ESDI	2,7 RLL	5.25FH		40k		Y
AT650S	16	SCSI	2,7 RLL	5.25FH		40k		Y
AT70	27	ST412/506	MFM	5.25FH		40k		Y
CDASM-1051F	9.5	SCSI-2Fast		3.5		800k	5400	
CDASM-2105F	8.5	SCSI-2Fast		3.5		800k	7200	
CDASM-4005F	8.5	SCSI-2Fast		3.5		800k	7200	
DMH-A02W	9.5	SCSI-2Fast		3.5		800k	5400	
DMH-A04W	9.5	SCSI-2Fast		3.5		800k	5400	
DMH-B02W	8.5	SCSI-2FstW		3.5		800k	7200	
DMH-B04W	8.5	SCSI-2FstW		3.5		800k	7200	
DMH-B09W	9	SCSI-2FstW		3.5		1000k	7200	
LAN1050F	9.5	SCSI-2Fast		3.5		500k	5400	
LAN2101F	9	SCSI-2Fast		3.5		500k	5400	
LAN2105F	8.5	SCSI-2Fast		3.5		500k	7200	
LAN4005	8.5	SCSI-2Fast		3.5		800k	7200	
LAN9000F	11.5	SCSI-2Fast		5.25FH		500k	5400	
MAC-195	15	SCSI	2,7 RLL	3.5HH		150k		Y
PS155E	14	ESDI	2,7 RLL	5.25FH		70k		Y
PS155S	14	SCSI	2,7 RLL	5.25FH		70k		Y
PS320S	16	SCSI	2,7 RLL	5.25FH		150k		Y
PS335E	15	ESDI	2,7 RLL	5.25FH		70k		Y
PS650S	15	SCSI	2,7 RLL	5.25FH		100k		Y
XT100		ST412/506	2,7 RLL					Y

Drive Model	Format Size MB	Head	Cyl	Sect/ Trac	Translate H/C/S	RWC/ WPC	Land Zone
XT120	119	15	918	17		—/None	
XT200	204	15	1024	26		—/None	1023
XT40	44	5	1024	17		—/None	1023
XT70	71	8	1024	17		—/None	1023

SYQUEST TECHNOLOGY

Drive Model	Format Size MB	Head	Cyl	Sect/ Trac	Translate H/C/S	RWC/ WPC	Land Zone
EZ135 (removable)	135					—/—	
EZ135 (removable)	135					—/—	
EZ230 (removable)	230					—/—	
EZ230 (removable)	230	1	4092			—/—	
SQ105 (removable)	105					—/—	
SQ200	200					—/—	
SQ225F	20			17		—/—	
SQ270 (removable)	270					—/—	
SQ306F	5			17		—/—	
SQ306R	5	2	306	17		—/—	
SQ306RD	5	2	306	17		307/307	
SQ3105 (removable)	105	2			16/420/32	—/—	
SQ312	10	2	615	17		—/—	
SQ312RD	10	2	615	17		616/616	
SQ319	10	2	612	17		—/—	
SQ325	21	4	612	17		612/612	
SQ325F	20	4	615	17		616/616	
SQ3270 (removable)	256	2			16/1024/32	—/—	
SQ338F	30	6	615	17		616/616	
SQ340AF	38	6	640	17		616/616	
SQ5110C (removable)	89					—/—	
SQ5200C (removable)	200					—/—	
SQ555 (removable)	44					—/—	
SQ88	88					—/—	
SYJET 1.3 (removeable)	1300					—/—	
SYJET 1.5 (removeable)	1500					—/—	
SYJET 1.5 (removeable)	1500					—/—	
SYJET 650 (removeable)	650					—/—	

TANDON COMPUTER CORPORATION

Drive Model	Format Size MB	Head	Cyl	Sect/ Trac	Translate H/C/S	RWC/ WPC	Land Zone
TM2085	74	9	1004	17		1005/1005	
TM2128	115	9	1004	26		1005/1005	
TM2170	154	9	1344	26		1345/1345	
TM244	41	4	782	26		783/783	
TM246	62	6	782	26		783/783	
TM251	5	2	306	17		—/—	
TM252	10	4	306	17		307/307	
TM261	10	2	615	17		616/616	
TM262	21	4	615	17		616/616	Auto
TM262R	20	2	782	26		783/783	
TM264	41	4	782	26		783/783	
TM3085	71	8	1024	17		1024/1024	
TM3085R	105	8	1024	26		1024/1024	
TM344	41	4	782	26		783/783	
TM346	62	6	782	26		783/783	
TM361	10	2	615	17		616/616	
TM362	20	4	615	17		616/616	615
TM362R	20	2	782	26		783/783	
TM364	41	4	782	26		783/783	
TM501	5	2	306	17		128/153	
TM502	10	4	306	17		128/153	
TM503	15	6	306	17		128/153	
TM601	3					—/—	
TM602S	5	4	153	17		128/128	
TM602SE	12		153	17		—/—	
TM603S	10	6	153	17		128/128	

Drive Model	Seek Time	Interface	Encode	Form Factor	cache kb	mtbf	RPM	Obsolete?
XT120		ST412/506	MFM					Y
XT200		ST412/506	2,7 RLL					Y
XT40		ST412/506	MFM					Y
XT70		ST412/506	MFM					Y

SYQUEST TECHNOLOGY

Drive Model	Seek Time	Interface	Encode	Form Factor	cache kb	mtbf	RPM	Obsolete?
EZ135 (removable)	13	SCSI-2	1,7 RLL	3.5 3H	64k	200k	3600	Y
EZ135 (removable)	13	ATA-2	1,7 RLL	3.5 3H	64k	200k	3600	Y
EZ230 (removable)	13.5	SCSI		3.5HH		200k		
EZ230 (removable)	13.5	EIDE		3.5 3H	32k	200k	3600	
SQ105 (removable)	14.5	IDE AT		3.5 3H	64k	100k	3600	Y
SQ200	18			5.25HH	64k	200k		Y
SQ225F	99	ST412/506	MFM	5.25HH				Y
SQ270 (removable)	13.5	IDE AT		3.5 3H	128k	100k	3600	Y
SQ306F	99	ST412/506	MFM	5.25HH				Y
SQ306R	99	ST412/506	MFM	5.25HH				Y
SQ306RD	99	ST412/506	MFM	5.25HH				Y
SQ3105 (removable)	14.5	ATA-2	1,7 RLL	3.5 3H	64k	100k	3600	Y
SQ312	80	ST412/506	MFM	5.25HH				Y
SQ312RD	80	ST412/506	MFM	5.25HH				Y
SQ319	80	ST412/506	RLL	5.25HH				Y
SQ325	80	ST412/506	MFM	5.25HH				Y
SQ325F	99	ST412/506	MFM	5.25HH				Y
SQ3270 (removable)	13.5	ATA-2	1,7 RLL	3.5 3H	128k	100k	3600	Y
SQ338F	80	ST412/506	MFM	5.25HH				Y
SQ340AF	80	ST412/506	MFM	5.25HH				Y
SQ5110C(removable)	20	SCSI-2		5.25HH	64k	100k	3220	Y
SQ5200C(removable)	18	SCSI-2		5.25HH	64k	100k	3220	Y
SQ555 (removable)	20	SCSI-2		5.25HH	64k	100k	3220	Y
SQ88	20			5.25HH	32k	100k		Y
SYJET1.3(removeable)	<11	SCSI		3.5HH	256k	250k	5400	Y
SYJET 1.5(removeable)	12	SCSI-2		3.5HH	512k	250k		
SYJET 1.5(removeable)	12	EIDE		3.5HH	512k	250k		
SYJET 650(removeable)	<11	SCSI		3.5HH	256k	250k	5400	Y

TANDON COMPUTER CORPORATION

Drive Model	Seek Time	Interface	Encode	Form Factor	cache kb	mtbf	RPM	Obsolete?
TM2085	25	SCSI	MFM	5.25FH				Y
TM2128	25	SCSI	2,7 RLL	5.25FH				Y
TM2170	25	SCSI	2,7 RLL	5.25FH				Y
TM244	37	ST412/506	2,7 RLL	3.5HH				Y
TM246	37	ST412/506	2,7 RLL	3.5HH				Y
TM251		ST412/506	MFM	5.25				Y
TM252	85	ST412/506	MFM	5.25HH				Y
TM261		ST412/506	MFM	5.25				Y
TM262	65	ST412/506	MFM	3.5HH				Y
TM262R	85	ST412/506	2,7 RLL	3.5HH				Y
TM264	85	ST412/506	2,7 RLL	3.5HH				Y
TM3085	35	ST412/506	MFM	5.25				Y
TM3085R	35	ST412/506	2,7 RLL	5.25				Y
TM344	35	ST412/506	2,7 RLL	3.5HH				Y
TM346	35	ST412/506	2,7 RLL	3.5HH				Y
TM361	27	ST412/506	MFM	5.25				Y
TM362	85	ST412/506	MFM	5.25				Y
TM362R	85	ST412/506	2,7 RLL	3.5HH				Y
TM364	85	ST412/506	2,7 RLL	3.5HH				Y
TM501	85	ST412/506	MFM	5.25FH				Y
TM502	85	ST412/506	MFM	5.25FH				Y
TM503	85	ST412/506	MFM	5.25FH				Y
TM601		ST412/506	MFM	5.25FH				Y
TM602S	85	ST412/506	MFM	5.25FH				Y
TM602SE		ST412/506	MFM	5.25FH				Y
TM603S		ST412/506	MFM	5.25FH				Y

Drive Model	Format Size MB	Head	Cyl	Sect/ Trac	Translate H/C/S	RWC/ WPC	Land Zone
TM603SE	12	6	230	17		128/128	
TM702	20	4	615	26		616/616	Auto
TM702AT	21	4	615	17		616/616	615
TM703	30	5	733	17		734/734	695
TM703AT	31	5	733	17		733/733	733
TM703C	25	17	733			—/—	
TM705	41	5	962	17		—/None	962
TM755	42	5	981	17		982/982	981
TANDY CORP							
25-1045	28					—/—	Auto
25-1046	43	4	782	27		NA/NA	Auto
25-4130	100	4	1219			NA/NA	Auto
TEAC AMERICA, INC.							
SD150	10	4	306	17		—/—	
SD240	43	2	1000	42		NA/NA	Auto
SD260	63	2	1226	50		NA/NA	Auto
SD3105A	105	4	1282	40	8/641/40	NA/NA	Auto
SD3105S	105	4	1282	40		NA/NA	Auto
SD3210A	215	4	1695	62	8/847/62	NA/NA	Auto
SD3210S	215	4	1695	62		NA/NA	Auto
SD3240	245	4	1930		8/965/62	—/—	
SD3250N (removable)	252					NA/NA	Auto
SD3360N (removable)	363					NA/NA	Auto
SD340A	43	2	1050	40	4/525/40	NA/NA	Auto
SD340HA	43	2	1050	40		NA/NA	Auto
SD340HS	43	2	1050	40		—/—	Auto
SD340S	43	2	1050	40		NA/NA	Auto
SD3540N (removable)	540					NA/NA	Auto
SD380	86	4	1025	40	8/965/62	NA/300	1025
SD380HA	86	4	1050	40		NA/NA	Auto
SD380HS	86	4	1050	40		—/—	Auto
SD380S	86	4	1050	40		—/—	Auto
SD510	10	4	306	17		128/128	
SD520	20	4	615	17		128/128	
SD540	40	8	615	17		—/—	
TEXAS INSTRUMENTS							
525-122	20					—/—	
DB260	212	8				NA/NA	Auto
DB380	333	15				64/64	
TI5	5	4	153	17		64/64	
TOSHIBA AMERICA, INC.							
HDD2616	2160					—/—	
HDD2619	3008				16/6409/63	—/—	
HDD2712	1350					—/—	
HDD2714	1440	4			16/2800/63	—/—	
HDD2716	2160				16/4200/63	—/—	
HDD2718	2160	4			16/4200/63	—/—	
HDD2912	3250	6			16/5850/63	—/—	
MK-2526FB	528					—/—	
MK-2528FB	704					—/—	
MK-2728FB	1080					—/—	
MK1001MAV	1080				16/2100/63	—/—	
MK1002MAV	108516/2100/3					—/—	
MK1034FC	107	4	1345		8/664/39	—/—	
MK1122FC	43	2	977		5/988/17	—/—	
MK130	53	7	733			—/—	
MK1301MAV	1350					—/—	

Drive Model	Seek Time	Interface	Encode	Form Factor	cache kb	mtbf	RPM	Obsolete?
TM603SE		ST412/506	MFM	5.25FH				Y
TM702	27	ST412/506	MFM	5.25FH				Y
TM702AT	27	ST412/506	MFM	5.25FH				Y
TM703	40	ST412/506	MFM	5.25FH				Y
TM703AT	40	ST412/506	MFM	5.25FH				Y
TM703C		ST412/506	MFM	5.25FH				Y
TM705		ST412/506	MFM	5.25FH				Y
TM755	27	ST412/506	MFM	5.25FH				Y

TANDY CORP

Drive Model	Seek Time	Interface	Encode	Form Factor	cache kb	mtbf	RPM	Obsolete?
25-1045	28	IDE XT		3.5HH				Y
25-1046	28	IDE XT	2,7 RLL	3.5HH		40k		Y
25-4130	17	IDE XT	2,7 RLL	3.5HH				Y

TEAC AMERICA, INC.

Drive Model	Seek Time	Interface	Encode	Form Factor	cache kb	mtbf	RPM	Obsolete?
SD150		ST412/506	MFM	5.25				Y
SD240	19	IDE AT	1,7 RLL	2.5	32k	100k	3600	Y
SD260	19	IDE AT	1,7 RLL	2.5	32k	100k	3600	Y
SD3105A	19	IDE AT	2,7 RLL	3.5 3H	64k	30k	3600	Y
SD3105S	19	SCSI	2,7 RLL	3.5 3H	64k	30k	3600	Y
SD3210A	17	IDE AT	1,7 RLL	3.5 3H	65k	100k	3600	Y
SD3210S	17	SCSI	1,7 RLL	3.5 3H	63k	100k	3600	Y
SD3240	17	IDE AT	1,7 RLL	3.5 3H	64k	100k	3600	Y
SD3250N (removable)	17	IDE		5.25HH		250k	3600	Y
SD3360N (removable)	17	IDE		5.25HH		250k	3600	Y
SD340A	23	IDE AT	2,7 RLL	3.5 3H	64k	30k	2358	Y
SD340HA	19	IDE AT	2,7 RLL	3.5 3H		30k	2358	Y
SD340HS	19	SCSI	2,7 RLL	3.5 3H		30k	2358	Y
SD340S	23	SCSI	2,7 RLL	3.5 3H	28k	30k	2358	Y
SD3540N (removable)	11	IDE		5.25HH		250k	4201	Y
SD380	22	IDE AT	2,7 RLL	3.5 3H		30k	2358	Y
SD380HA	19	IDE AT	2,7 RLL	3.5 3H		30k	2358	Y
SD380HS	19	SCSI	2,7 RLL	3.5 3H		30k	2358	Y
SD380S	22	SCSI	2,7 RLL	3.5 3H		30k	2358	Y
SD510	27	ST412/506	MFM	5.25FH				Y
SD520	27	ST412/506	MFM	5.25FH				Y
SD540	40			5.25HH		20k	3600	Y

TEXAS INSTRUMENTS

Drive Model	Seek Time	Interface	Encode	Form Factor	cache kb	mtbf	RPM	Obsolete?
525-122		ST412/506	MFM	5.25FH				Y
DB260	16	SCSI		3.5HH				Y
DB380	16	SCSI	MFM	5.25FH				Y
TI5	27	ST412/506	MFM	5.25FH				Y

TOSHIBA AMERICA, INC.

Drive Model	Seek Time	Interface	Encode	Form Factor	cache kb	mtbf	RPM	Obsolete?
HDD2616	13	ATA-2		2.5 4H	128k	300k		
HDD2619	13	ATA-2		2.5 4H	128k		4852	
HDD2712	13	ATA-2		2.5 4H	128k			
HDD2714	13	ATA-2		3.5 3H	128k		4200	
HDD2716	13	ATA-3		2.5 4H	128k	300k	4200	
HDD2718	13	ATA-4		2.5 4H	128k	300k	4200	
HDD2912	13	ATA-4		2.5 4H	512k	300k	4200	
MK-2526FB	12	SCSI-2		2.5	128k			Y
MK-2528FB	12	SCSI-2		2.5	128k			Y
MK-2728FB	12	SCSI-2		2.5	128k			
MK1001MAV	13	ATA-2		2.5 4H	128k		4200	
MK1002MAV	13	ATA-2		2.5 4H	128k		4200	
MK1034FC	16	IDE AT	2,7 RLL	3.5 3H	64k	40k	3414	Y
MK1122FC	23	IDE AT	2,7 RLL	2.5 4H	32k	40k	3600	Y
MK130	25	ST412/506	MFM	3.5HH		30k		Y
MK1301MAV	13	ATA-2 Fast	PRML	2.5 4H	128k	300k	4200	

Drive Model	Format Size MB	Head	Cyl	Sect/ Trac	Translate H/C/S	RWC/ WPC	Land Zone
MK132FA	18					—/—	
MK133FA	30					—/—	
MK134FA	44	7	733	17		—/—	
MK134FA(R)	65	7	733	26		—/—	
MK1401MAV	1440					—/—	
MK1403MAN	1440	4			16/2800/63	—/—	
MK1422FCV	86	2	988		10/988/17	—/—	
MK1522FCV	126	2	812		8/812/38	NA/NA	Auto
MK153FA	74	5	830	35		NA/NA	
MK153FA-I	74	5	830	35		NA/NA	Auto
MK153FB	76	5	830	35		—/—	
MK154FA	104	7	830	35		NA/NA	
MK154FA-I	104	7	830	35		NA/NA	Auto
MK154FB	106	7	830	35		—/—	
MK156FA	148	10	830	35		NA/NA	
MK156FB	152	10	830	35		—/—	
MK158FA	173u	10	830			—/—	
MK1624FCV	213	4			16/684/38	NA/NA	Auto
MK1722FCV	131	2			8/842/38	—/—	
MK1724FCV	262	4	841		16/842/38	NA/NA	Auto
MK1824FBV	352	4	2050			—/—	
MK1824FCV	353	4		63	16/682/63	NA/NA	Auto
MK182FB	83	5	823			—/—	
MK184FB	116	7	823			—/—	
MK186FB	166	10	823			—/—	
MK1924FBV	543	4	2920			—/—	
MK1924FCV	543	4			16/1053/63	NA/NA	Auto
MK1926FBV	815	6	2920			—/—	
MK1926FCV	815	6			16/1579/63	NA/NA	Auto
MK2024FC	86	4	977	43	10/988/17	NA/NA	Auto
MK2101MAN	2160				16/4200/63	—/—	
MK2103MAV	2160				16/4200/63	—/—	
MK2104MAV	2160	4			16/4200/63	—/—	
MK2105MAT	2160	4			16/2100/63	—/—	
MK2124FC	130	4	934	55	16/934/17	NA/NA	Auto
MK2224FB	213	4	1560	83		NA/NA	Auto
MK2224FC	213	4	684		16/684/38	NA/NA	Auto
MK2326FB	340	6	1830	74		NA/NA	Auto
MK2326FC	340	6			14/969/49	NA/NA	Auto
MK2326FCH	340					—/—	
MK232FB	45	3	845	35		—/—	Auto
MK232FBS	45	3	845	35		—/—	
MK232FC	45	3	845	35		NA/NA	
MK233FB	75	5	845	35		—/—	Auto
MK234FB	106	7	845	35		—/—	Auto
MK234FBS	106	7	845	35		—/—	
MK234FC	106	7	845	35	7/845/35	—/—	Auto
MK234FCH	106	7	845	35	7/845/35	—/—	
MK2428FB	524	8	1920	83		NA/NA	Auto
MK2428FC	524	8		63	16/1016/63	NA/NA	Auto
MK250FA	382	10	1224	35		NA/NA	
MK250FB	382	10	1224	35		NA/NA	
MK2526FC	528	6		63	16/1023/63	NA/NA	Auto
MK2528FC	704	8			16/1365/63	NA/NA	Auto
MK253FA	162					—/—	
MK253FB	158					—/—	
MK254FA	227					—/—	
MK254FB	221					—/—	
MK256FA	325					—/—	
MK256FB	316					—/—	
MK256FB	315					—/—	
MK2616	2160	10				—/—	
MK2628FC	811	8			16/1571/63	NA/NA	Auto

Drive Model	Seek Time	Interface	Encode	Form Factor	cache kb	mtbf	RPM	Obsolete? ⇓
MK132FA		ST412/506	MFM	3.5HH				Y
MK133FA		ST412/506	MFM	3.5HH				Y
MK134FA	25	ST412/506	MFM	3.5HH		30k	3600	Y
MK134FA(R)	23	ST412/506	2,7 RLL	3.5HH				Y
MK1401MAV	13	ATA-2		2.5 4H	128k		4200	
MK1403MAN	13	ATA-2		3.5 3H	128k		4200	
MK1422FCV	15	IDE AT		2.5 4H	32k	150k	3600	Y
MK1522FCV	15	IDE AT		2.5 4H	128k	150k	3600	Y
MK153FA	23	ESDI	2,7 RLL	5.25FH		30k	3600	Y
MK153FA-I	23	ESDI	2,7 RLL	5.25FH		30k		Y
MK153FB	23	SCSI	2,7 RLL	5.25FH	32k	30k	3600	Y
MK154FA	23	ESDI	2,7 RLL	5.25FH		30k	3600	Y
MK154FA-I	23	ESDI	2,7 RLL	5.25FH		30k		Y
MK154FB	23	SCSI	2,7 RLL	5.25FH	32k	30k	3600	Y
MK156FA	23	ESDI	2,7 RLL	5.25FH		30k		Y
MK156FB	23	SCSI	2,7 RLL	5.25FH	32k	30k	3600	Y
MK158FA	23	ESDI	2,7 RLL	5.25FH		30k	3600	Y
MK1624FCV	13	IDE AT		2.5 4H	128k	150k	4000	Y
MK1722FCV	13	IDE AT		2.5 4H	128k		4000	Y
MK1724FCV	12	IDE AT		2.5 4H	128k	150k	4000	Y
MK1824FBV	13	SCSI-2	1,7 RLL	2.5 4H	128k	300k	4200	Y
MK1824FCV	13	ATA-2		2.5 4H	128k	300k	4200	Y
MK182FB	18	SMD/CMD	2,7 RLL	8.00 FH		20k	3600	Y
MK184FB	18	SMD/CMD	2,7 RLL	8.00 FH		20k	3600	Y
MK186FB	18	SMD/CMD	2,7 RLL	8.00 FH		20k	3600	Y
MK1924FBV	13	SCSI-2	8,9RLL	2.5 4H	128k		4200	Y
MK1924FCV	13	ATA-2		2.5 4H	128k	300k	4200	Y
MK1926FBV	13	SCSI-2	8,9RLL	2.5 4H	128k		4200	Y
MK1926FCV	13	ATA-2		2.5 4H	128k	300k	4200	Y
MK2024FC	19	IDE AT	2,7 RLL	2.5 4H	32k	80k	3600	Y
MK2101MAN	13	ATA-2	PRML	2.5 4H	128k	300k	4200	
MK2103MAV	13	ATA-3		2.5 4H	128k	300k	4200	
MK2104MAV	13	ATA-4		2.5 4H	128k	300k	4200	
MK2105MAT	13	ATA-4		2.5 4H	128k	300k	4200	
MK2124FC	17	IDE AT	2,7 RLL	2.5 4H	32k	150k	3600	Y
MK2224FB	12	SCSI-2Fast		2.5 4H	128k	150k	4000	Y
MK2224FC	12	IDE AT		2.5 4H	128k	150k	4000	Y
MK2326FB	12	SCSI-2Fast		2.5 4H	128k	150k	4200	Y
MK2326FC	12	IDE AT		2.5 4H	128k	150k	4200	Y
MK2326FCH		IDE AT		2.5 4H				Y
MK232FB	25	SCSI		3.5HH		30k	3600	Y
MK232FBS	19	SCSI	2,7 RLL	3.5HH		30k		Y
MK232FC	25	IDE AT	2,7 RLL	3.5HH		30k		Y
MK233FB	25	SCSI	2,7 RLL	3.5HH		30	3600	Y
MK234FB	25	SCSI	2,7 RLL	3.5HH		30k	3600	Y
MK234FBS	19	SCSI	2,7 RLL	3.5HH		30k		Y
MK234FC	25	IDE AT	2,7 RLL	3.5HH		30k	3600	Y
MK234FCH	25	IDE AT	2,7 RLL	3.5HH		30k		Y
MK2428FB	12	SCSI-2Fast		2.5 4H	512k	150k	4000	Y
MK2428FC	12	IDE AT		2.5 4H	512k	150k	4000	Y
MK250FA	18	ESDI	2,7 RLL	5.25FH		30k		Y
MK250FB	18	SCSI	2,7 RLL	5.25FH		30k		Y
MK2526FC	13	IDE AT		2.5 4H	128k		4200	Y
MK2528FC	13	IDE AT		2.5 4H	128k		4200	Y
MK253FA		ESDI		5.25FH				Y
MK253FB		SCSI		5.25FH				Y
MK254FA		ESDI		5.25FH				Y
MK254FB		SCSI		5.25FH				Y
MK256FA		ESDI		5.25FH				Y
MK256FB		SCSI		5.25FH				Y
MK256FB		SCSI		5.25FH				Y
MK2616	13	ATA-2		2.5 4H	128k		4200	
MK2628FC	13	ATA-2		2.5 4H	128k	300k	4200	Y

Drive Model	Format Size MB	Head	Cyl	Sect/ Trac	Translate H/C/S	RWC/ WPC	Land Zone
MK2712	1350					—/—	
MK2720FC	1350	10			16/2633/63	NA/NA	Auto
MK2728FC	1080	8			16/1579/63	—/—	
MK286FC	374	11	823			—/—	
MK288FC	510	15	823			—/—	
MK3003MAN	3008				16/6409/63	—/—	
MK3205MAV	3250	6			16/5850/63	—/—	
MK3303	3300					—/—	
MK355FA	405	9	1661	53		—/—	
MK355FB	405	9	1661	53		—/—	
MK356FA	495					—/—	
MK358FA	675	15	1661	53		—/—	
MK358FB	675	15	1661	53		—/—	
MK388FA	720	15	1162			—/—	
MK438FB	900	11	1980			NA/NA	Auto
MK537FB	1064	13	1980	NA		NA/NA	Auto
MK538FB	1230	15	1980	NA		NA/NA	Auto
MK53FA	36	5	830	17		—/512	830
MK53FA(M)	36	5	830	17		830/512	830
MK53FA(R)	43	5	830	26		831/831	
MK53FB	36	5	830	17		830/512	
MK53FB(M)	36	5	830	17		830/512	
MK53FB(R)	64	5	830	26		831/831	
MK53FB-I	36	5	830	17		830/512	
MK54FA(M)	60	7	830	17		831/512	830
MK54FA(R)	90	7	830	26		831/831	
MK54FB(M)	60	7	830	17		830/512	
MK54FB(R)	90	7	830	26		831/831	
MK54FB-I	50	7	830	17		830/512	830
MK556FA	152	10	830			NA/NA	
MK56FA(M)	86	10	830	17		831/831	
MK56FA(R)	129	10	830	26		—/512	830
MK56FB(M)	86	10	830	17		830/512	
MK56FB(R)	129	10	830	26		831/831	
MK56FB-I	72	10	830	17		830/512	830
MK72PC	72	10	830	17		—/—	
MK72PCR	109	10	830	26		—/—	
MKM0351E	36	5	830	17		830/512	830
MKM0351J	36	5	830	17		830/512	830
MKM0352E	50	7	830	17		—/512	830
MKM0352J	50	7	830	17		—/512	830
MKM0353E	72	10	830	17		830/512	830
MKM0353J	72	10	830	17		830/512	830
MKM0363A	74	5	830	35		NA/NA	Auto
MKM0363J	74	5	830	35		NA/NA	Auto
MKM0364A	104	7	830	35		NA/NA	Auto
MKM0364J	104	7	830	35		NA/NA	Auto
MKM0381E	36	5	830	17		830/512	
MKM0381J	36	5	830	17		830/512	830
MKM0382E	50	7	830	17		—/512	830
MKM0382J	50	7	830	17		—/512	830
MKM0383E	72	10	830	17		830/512	830
MKM0383J	72	10	830	17		830/512	830

TULIN

Drive Model	Format Size MB	Head	Cyl	Sect/ Trac	Translate H/C/S	RWC/ WPC	Land Zone
TL213	10	2	640	17		656/656	640
TL226	22	4	640	17		656/656	656
TL238	22	4	640	17		—/None	640
TL240	33	6	640	17		656/656	656
TL258	32	6	640	17		—/None	640
TL326	22	4	640	17		641/641	640
TL340	33	6	640	17		641/641	640

Drive Model	Seek Time	Interface	Encode	Form Factor	cache kb	mtbf	RPM	Obsolete?
MK2712	13	ATA-2		2.5 4H	128k		4200	
MK2720FC	13	ATA-2		2.5 4H	128k	300k	4200	
MK2728FC	13	ATA-2		2.5 4H	128k	300k	4200	
MK286FC	18	HSMD	2,7 RLL	8.00 FH		35k	3600	Y
MK288FC	18	HSMD	2,7 RLL	8.00 FH		35k	3600	Y
MK3003MAN	13	ATA-2		2.5 4H	128k		4852	
MK3205MAV	13	ATA-4		2.5 4H	512k	300k	4200	
MK3303	13	ATA-3		2.5 4H		300k	4852	
MK355FA	16	ESDI	1,7 RLL	5.25FH	64k	30k	3600	Y
MK355FB	16	SCSI	2,7 RLL	5.25FH	64k	30k	3600	Y
MK356FA		SCSI	RLL	5.25FH				Y
MK358FA	16	ESDI	1,7 RLL	5.25FH	64k	30k	3600	Y
MK358FB	16	SCSI-2	2,7 RLL	5.25FH	64k	30k		Y
MK388FA	18	HSMD	2,7 RLL	8.00 FH		35k	3600	Y
MK438FB	12	SCSI-2	1,7 RLL	3.5HH	512k	200		
MK537FB	12	SCSI-2	1,7 RLL	3.5HH	512k	200k		
MK538FB	12	SCSI-2	1,7 RLL	3.5HH	512k	200k		
MK53FA	30	ST412/506	MFM	5.25FH		20k		Y
MK53FA(M)	25	ST412/506	MFM	5.25FH		20k		Y
MK53FA(R)	30	ST412/506	2,7 RLL	5.25FH		20k		Y
MK53FB	25	ST412/506	MFM	5.25FH		20k		Y
MK53FB(M)	25	ST412/506	MFM	5.25FH		20k		Y
MK53FB(R)	25	ST412/506	2,7 RLL	5.25FH		20k		Y
MK53FB-I	25	ST412/506	MFM	5.25FH		20k		Y
MK54FA(M)	30	ST412/506	MFM	5.25FH		20k		Y
MK54FA(R)	25	ST412/506	2,7 RLL	5.25FH		20k		Y
MK54FB(M)	25	ST412/506	MFM	5.25FH		20k		Y
MK54FB(R)	25	ST412/506	2,7 RLL	5.25FH		20k		Y
MK54FB-I	25	ST412/506	MFM	5.25FH		20k		Y
MK556FA	23	ESDI		5.25FH		30k		Y
MK56FA(M)	30	ST412/506	MFM	5.25FH		20k		Y
MK56FA(R)	30	ST412/506	2,7 RLL	5.25FH		20k		Y
MK56FB(M)	25	ST412/506	MFM	5.25FH		20k		Y
MK56FB(R)	25	ST412/506	2,7 RLL	5.25FH		20k		Y
MK56FB-I	25	ST412/506	MFM	5.25FH		20k		Y
MK72PC	25	ST412/506	MFM	3.5HH				Y
MK72PCR	25	ST412/506	2,7 RLL	3.5HH				Y
MKM0351E	25	ST412/506	MFM	5.25FH		20k		Y
MKM0351J	25	ST412/506	MFM	5.25FH		20k		Y
MKM0352E	30	ST412/506	MFM	5.25FH		20k		Y
MKM0352J	30	ST412/506	MFM	5.25FH		20k		Y
MKM0353E	25	ST412/506	MFM	5.25FH		20k		Y
MKM0353J	25	ST412/506	MFM	5.25FH		20k		Y
MKM0363A	23	ESDI	2,7 RLL	5.25FH		30k		Y
MKM0363J	23	SCSI	2,7 RLL	5.25FH		30k		Y
MKM0364A	23	ESDI	2,7 RLL	5.25FH		30k		Y
MKM0364J	23	ESDI	2,7 RLL	5.25FH		30k		Y
MKM0381E	25	ST412/506	MFM	5.25FH		20k		Y
MKM0381J	25	ST412/506	MFM	5.25FH		20k		Y
MKM0382E	30	ST412/506	MFM	5.25FH		20k		Y
MKM0382J	30	ST412/506	MFM	5.25FH		20k		Y
MKM0383E	25	ST412/506	MFM	5.25FH		20k		Y
MKM0383J	25	ST412/506	MFM	5.25FH		20k		Y

TULIN

Drive Model	Seek Time	Interface	Encode	Form Factor	cache kb	mtbf	RPM	Obsolete?
TL213	27	ST412/506	MFM	5.25HH				Y
TL226	85	ST412/506	MFM	5.25HH				Y
TL238		ST412/506	MFM	5.25HH				Y
TL240	85	ST412/506	MFM	5.25HH				Y
TL258		ST412/506	MFM	5.25HH				Y
TL326	40	ST412/506	MFM	5.25HH				Y
TL340	40	ST412/506	MFM	5.25HH				Y

Drive Model	Format Size MB	Head	Cyl	Sect/ Trac	Translate H/C/S	RWC/ WPC	Land Zone
VERTEX (SEE PRIAM)							
						—/—	
WESTERN DIGITAL							
PhD1000	1083					—/—	
PhD1400	1400					—/—	
PhD2100	2168					—/—	
PIRANHA 105A	1104	4				NA/NA	Auto
PIRANHA 105S	1104	4				NA/NA	Auto
PIRANHA 210A	210					—/—	
PIRANHA 210S	210					—/—	
WD140	40					—/—	
WD2120	125					—/—	
WD262	20	4	615	17		616/616	616
WD280	80					—/—	
WD344R	40	4	782	26		783/783	783
WD362	20	4	615	17		616/616	616
WD382R	20	2	782	26		783/783	782
WD383R	30	4	615	26		616/616	616
WD384R	40	4	782	26		783/783	783
WD544R	40	4	782	26		783/783	783
WD562-5	21	4	615	17		—/—	
WD582R	20	2	782	26		783/783	783
WD583R	30	4	615	26		616/616	616
WD584R	40	4	782	26		783/783	783
WD93018-A	21					—/—	
WD93020-XE1	20	4	615	17		NA/NA	616
WD93023-A	21					—/—	
WD93024-A	21	2	782	27	4/615/17	NA/NA	783
WD93024-X	21	2	782	27		NA/NA	783
WD93028-A	21	2	782	27		NA/NA	783
WD93028-AD	21	2	782	27	4/615/17	NA/NA	783
WD93028-X	21	2	782	27		NA/NA	783
WD93034-X	32	3	782	27		NA/NA	783
WD93038-X	32	3	782	27		NA/NA	783
WD93044-A	43	4	782	27	5/977/17	NA/NA	783
WD93044-X	43	4	782	27		NA/NA	862
WD93048-A	40	4	782	27		NA/NA	783
WD93048-AD	43	4	782	27		NA/NA	783
WD93048-X	43	4	782	27		NA/NA	783
WD95024-A	21	2	782	27	4/615/17	NA/NA	783
WD95024-X	21	2	782	27		783/783	783
WD95028-A	20	2	782	27		NA/NA	783
WD95028-AD	21	2	782	27		783/783	783
WD95028-X	20	2	782	27		NA/NA	783
WD95034-X	32	3	782	27		783/783	783
WD95038-X	30	3	782	27		NA/NA	783
WD95044-A	43	4	782	27	4/782/27	783/783	783
WD95044-X	43	4	782	27	4/782/27	783/783	783
WD95048-A	40	4	782	27	4/782/27	NA/NA	783
WD95048-AD	43	4	782	27	4/782/27	NA/NA	783
WD95048-X	40	4	782	27	4/782/27	NA/NA	783
WDAB130 (Tidbit)	31	5	733	17	4/916/17	734/734	Auto
WDAB140	42	2	1390		5/980/17	—/—	
WDAB260 (Tidbit)	62	4	1020	17		NA/NA	Auto
WDAC11000	1056	16	2046	63		—/—	
WDAC11200	1282				16/2484/63	—/—	
WDAC1170 (Caviar)	170	2	2233	56-96	6/1010/55	NA/NA	Auto
WDAC1210 (Caviar)	212	2	2720	55-99	12/989/35	NA/NA	Auto
WDAC1270 (Caviar)	270	2			12/917/48	NA/NA	Auto
WDAC1365 (Caviar)	364	2			16/708/63	—/—	
WDAC140 (Caviar)	42	2	1082	39	5/980/17	NA/NA	Auto

VERTEX (SEE PRIAM)

WESTERN DIGITAL

Drive Model	Seek Time	Interface	Encode	Form Factor	cache kb	mtbf	RPM	Obsolete?
PhD1000	14	PCMIDE		3.0 4H	128k	300k	4536	Y
PhD1400	14	PCMIDE		3.0 4H	256k	350k	4000	Y
PhD2100	14	PCMIDE		3.0 4H	256k	350k	4000	
PIRANHA 105A	15	IDE AT	2,7 RLL	3.5HH		50k		Y
PIRANHA 105S	15	SCSI	2,7 RLL	3.5HH		50k		Y
PIRANHA 210A		IDE AT		3.5HH				Y
PIRANHA 210S		SCSI		3.5HH				Y
WD140		IDE AT		3.5 3H				Y
WD2120		IDE AT		3.5 3H				Y
WD262	80	ST412/506	MFM	3.5HH				Y
WD280		IDE AT		3.5 3H				Y
WD344R	40	ST412/506	2,7 RLL	3.5HH				Y
WD362	80	ST412/506	MFM	3.5HH				Y
WD382R	85	ST412/506	2,7 RLL	3.5HH				Y
WD383R	85	ST412/506	2,7 RLL	3.5HH				Y
WD384R	85	ST412/506	2,7 RLL	3.5HH				Y
WD544R	40	ST412/506	2,7 RLL	3.5HH				Y
WD562-5	80	ST412/506	MFM	3.5HH		40k		Y
WD582R	85	ST412/506	2,7 RLL	3.5HH				Y
WD583R	85	ST412/506	2,7 RLL	3.5HH				Y
WD584R	85	ST412/506	2,7 RLL	3.5HH				Y
WD93018-A		IDE AT		3.5HH				Y
WD93020-XE1	85	IDE XT	2,7 RLL	3.5HH				Y
WD93023-A		IDE AT		3.5HH				Y
WD93024-A	28	IDE AT	2,7 RLL	3.5HH		40k		Y
WD93024-X	39	IDE XT	2,7 RLL	3.5HH	1k	50k		Y
WD93028-A	70	IDE AT	2,7 RLL	3.5HH		40k		Y
WD93028-AD	69	IDE AT	2,7 RLL	3.5HH		40k		Y
WD93028-X	70	IDE XT	2,7 RLL	3.5HH		40k		Y
WD93034-X	39	IDE XT	2,7 RLL	3.5HH	1k	50k		Y
WD93038-X	70	IDE XT	2,7 RLL	3.5HH		40k		Y
WD93044-A	28	IDE AT	2,7 RLL	3.5HH	640k	40k		Y
WD93044-X	39	IDE XT	2,7 RLL	3.5HH	1k	50k		Y
WD93048-A	69	IDE AT	2,7 RLL	3.5HH		40k		Y
WD93048-AD	69	IDE AT	2,7 RLL	3.5HH		40k		Y
WD93048-X	70	IDE XT	2,7 RLL	3.5HH		40k		Y
WD95024-A	28	IDE AT	2,7 RLL	5.25HH		40k		Y
WD95024-X	39	IDE XT	2,7 RLL	3.5HH	1k	50k		Y
WD95028-A	70	IDE AT	2,7 RLL	5.25HH		40k		Y
WD95028-AD	69	IDE AT	2,7 RLL	5.25HH		40k		Y
WD95028-X	70	IDE XT	2,7 RLL	5.25HH		40k		Y
WD95034-X	39	IDE XT	2,7 RLL	3.5HH	1k	50k		Y
WD95038-X	70	IDE XT	2,7 RLL	5.25HH		40k		Y
WD95044-A	28	IDE AT	2,7 RLL	5.25HH		40k		Y
WD95044-X	39	IDE XT	2,7 RLL	3.5HH	1k	50k		Y
WD95048-A	70	IDE AT	2,7 RLL	5.25HH		40k		Y
WD95048-AD	69	IDE AT	2,7 RLL	5.25HH		40k		Y
WD95048-X	70	IDE XT	2,7 RLL	5.25HH		40k		Y
WDAB130 (Tidbit)	19	IDE AT-XT	2,7 RLL	2.50 4H	32k			Y
WDAB140	16	IDE	2,7 RLL	2.5 4H				Y
WDAB260 (Tidbit)	19	IDE XT-AT	2,7 RLL	2.5 4H		50k		Y
WDAC11000	<12							Y
WDAC11200	11	EIDE		3.5 3H	256k	350k	5200	Y
WDAC1170 (Caviar)	13	IDE AT	1,7 RLL	3.5 3H	32k	250k	3322	Y
WDAC1210 (Caviar)	13	IDE AT	1,7 RLL	3.5 3H	64k	250k	3314	Y
WDAC1270 (Caviar)	11	IDE AT		3.5 3H	64k	250k	4500	Y
WDAC1365 (Caviar)	10	IDE AT		3.5 3H	64k	300k	4500	Y
WDAC140 (Caviar)	18	IDE AT	2,7 RLL	3.5 3H	32k	50k		Y

Drive Model	Format Size MB	Head	Cyl	Sect/ Trac	Translate H/C/S	RWC/ WPC	Land Zone
WDAC1425 (Caviar)	427	2			16/827/63	—/—	
WDAC160 (Caviar)	62	7	1024	17	7/1024/17	1023/1023	Auto
WDAC21000 (Caviar)	1083	4			16/2100/63	—/—	
WDAC2120 (Caviar)	125	8	872	35	8/872/35	872/872	Auto
WDAC21200 (Caviar)	1282	4			16/2484/63	—/—	
WDAC21600 (Caviar)	1625	4			16/3148/63	NA/NA	Auto
WDAC2170 (Caviar)	171	4	1584	48-56	6/1010/55	NA/NA	Auto
WDAC21700	1707	16	3308	63		—/—	
WDAC2200 (Caviar)	213	4	1971	48-56	12/989/35	NA/NA	Auto
WDAC22000	2000				16/3876/63	—/—	
WDAC22100	2112				16/4092/63	—/—	
WDAC2250 (Caviar)	256	3	2233	56-96	9/1010/55	NA/NA	Auto
WDAC22500	2559				16/4960/63	—/—	
WDAC2340 (Caviar)	341	4	2233	56-96	12/1010/55	NA/NA	Auto
WDAC2420 (Caviar)	425	4	2720	55-99	15/989/56	NA/NA	Auto
WDAC2540 (Caviar)	540	3			16/1048/63	NA/NA	Auto
WDAC2635 (Caviar)	640	3			16/1240/63	—/—	
WDAC2700 (Caviar)	730	4			16/1416/63	—/—	
WDAC280 (Caviar)	85	10	980	17	10/980/17	NA/NA	981
WDAC2850 (Caviar)	854	4			16/1654/63	—/—	
WDAC31000 (Caviar)	1084	6			16/2100/63	—/—	
WDAC310100 (Caviar)	10141	6			16/16383/63	—/—	
WDAC31200 (Caviar)	1282	6			16/2484/63	—/—	
WDAC31600 (Caviar)	1625	6			16/3148/63	—/—	
WDAC3210 (Caviar)	1250					—/—	
WDAC32100 (Caviar)	2112	5			16/4092/63	NA/NA	Auto
WDAC32500 (Caviar)	2560	6			16/4960/63	NA/NA	Auto
WDAC33100 (Caviar)	3166					—/—	
WDAC34000	4001				16/7752/63	—/—	
WDAC34300	4304				15/8896/63	—/—	
WDAC35100	516315/106723					—/—	
WDAC36400	6449				15/13328/63	—/—	
WDAH260 (Tidbit)	62	4	1024	17	7/1024/17	NA/NA	Auto
WDAH280	86	4	1390	V	10/980/17	NA/NA	Auto
WDAL1100	100					—/—	
WDAL2120	120	15	1001	17	8/872/35	NA/NA	Auto
WDAL2170	170					—/—	
WDAL2200	200					—/—	
WDAL2540	541	4			16/1048/63	—/—	
WDAP2120 (Piranha)	125	8	872	35		NA/NA	Auto
WDAP4200 (Pirahna)	212	8	1280	41	12/987/35	NA/NA	Auto
WDCU140	42	2	1050	30-50	5/980/17	NA/NA	Auto
WDE2170-0003	2170					—/—	
WDE2170-0007	2170					—/—	
WDE2170-0008	2170					—/—	
WDE2170-0023	2170					—/—	
WDE4360-0003	4360					—/—	
WDE4360-0007	4360					—/—	
WDE4360-0008	4360					—/—	
WDE4360-0023	4360					—/—	
WDE4550	4550	6				—/—	
WDE9100-0003 (50-pin)	9105	12				—/—	
WDE9100-0007 (68-pin)	9105	12				—/—	
WDE9100-0008 (80-pin)	9105	12				—/—	
WDE9100-0016 (68-pin)	9105	12				—/—	
WDE9100-0017 (80-pin)	9105	12				—/—	
WDMI130-44 (44 PIN)	31	2	920	33		NA/NA	Auto
WDMI130-72 (72 PIN)	30	2	928	32		NA/NA	Auto
WDMI4120-72 (72 PIN)	125	8	925	33		NA/NA	Auto
WDSC8320 (Condor)	320	14	949	48		NA/NA	Auto
WDSC8400 (Condor)	400	15	1199	48		NA/NA	Auto
WDSP2100 (Piranha)	104	4	1265	41		NA/NA	Auto
WDSP4200 (Piranha)	209	8	1265	41		NA/NA	Auto

Drive Model	Seek Time	Interface	Encode	Form Factor	cache kb	mtbf	RPM	Obsolete? ⇓
WDAC1425 (Caviar)	10	IDE AT		3.5 3H	64k	300k	4500	Y
WDAC160 (Caviar)	17	IDE AT	2,7 RLL	3.5 3H			3605	Y
WDAC21000 (Caviar)	<11	EIDE		3.5 3H	128k	300k	5200	Y
WDAC2120 (Caviar)	15	IDE AT	2,7 RLL	3.5 3H	32k	100k	3600	Y
WDAC21200 (Caviar)	<11	EIDE		3.5 3H	128k	300k	5200	Y
WDAC21600 (Caviar)	12	EIDE		3.5 3H	128k	300k	5200	Y
WDAC2170 (Caviar)	14	IDE AT	2,7 RLL	3.5 3H	32k	100k	3652	Y
WDAC21700	<12							Y
WDAC2200 (Caviar)	14	IDE AT	2,7 RLL	3.5 3H	64k	100k	3652	Y
WDAC22000	11	EIDE		3.5 3H	256k	350k	5200	Y
WDAC22100	12	EIDE		3H	128k	300k	5200	
WDAC2250 (Caviar)	13	IDE AT	1,7 RLL	3.5 3H	64k	250k	3322	Y
WDAC22500	11	EIDE		3.5 3H	256k	350k	5200	
WDAC2340 (Caviar)	13	IDE AT	1,7 RLL	3.5 3H	128k	250k	3322	Y
WDAC2420 (Caviar)	13	IDE AT	1,7 RLL	3.5 3H	128k	250k	3314	Y
WDAC2540 (Caviar)	11	IDE AT		3.5 3H	64k	300k	4500	Y
WDAC2635 (Caviar)	10	IDE AT		3.5 3H	64k	300k	4500	Y
WDAC2700 (Caviar)	10	IDE AT		3.5 3H	64k	300k	4500	Y
WDAC280 (Caviar)	17	IDE AT	2,7 RLL	3.5 3H	32k	100k	3595	Y
WDAC2850 (Caviar)	10	EIDE		3.5 3H	64k	300k	4500	Y
WDAC31000 (Caviar)	10	IDE AT		3.5 3H	128k	250k	4500	Y
WDAC310100 (Caviar)	9.5	EIDE	Prml16,17	3.5 3H	512k	400k	5400	Y
WDAC31200 (Caviar)	10	IDE AT		3.5 3H	64k	250k	4500	Y
WDAC31600 (Caviar)	<11	EIDE		3.5 3H	128k	300k	5200	Y
WDAC3210 (Caviar)	13	IDE		3H	128k		4500	Y
WDAC32100 (Caviar)	<12	EIDE		3.5 3H	128k	300k	5200	
WDAC32500 (Caviar)	<12	EIDE		3.5 3H	128k	300k	5200	
WDAC33100 (Caviar)	<12	EIDE		3.5 3H	128k	300k	5200	
WDAC34000	11.5	EIDE		3H	256k	350k	5200	
WDAC34300	11	EIDE		3.5 3H	256k	350k	5400	
WDAC35100	11	EIDE		3.5 3H	256k	350k	5400	
WDAC36400	9.5	EIDE		3.5 3H	256k	350k	5400	
WDAH260 (Tidbit)	19	IDE XT-AT	2,7 RLL	2.5 4H		50k	3383	Y
WDAH280	19	IDE XT-AT	2,7 RLL	2.5 4H		50k		Y
WDAL1100	17	IDE AT		2.5 4H	32k	100k		Y
WDAL2120	<16	IDE AT	2,7 RLL	2.5 3H		100k		Y
WDAL2170	16	IDE AT		2.5 4H	32k	100k		Y
WDAL2200	17	IDE AT		2.5 4H	32k	100k		Y
WDAL2540	13	EIDE		2.5 4H	128k	300k	4500	Y
WDAP2120 (Piranha)	15	IDE AT	2,7 RLL	3.5 3H		100k	3605	Y
WDAP4200 (Pirahna)	14	IDE AT	2,7 RLL	3.5HH	64k	50k		Y
WDCU140	19	PCMCIA-ATA	1,7 RLL	1.8 4H	32k	255k	4503	Y
WDE2170-0003	8	UltraFast	PRML	3.5	512k	1000k	7200	
WDE2170-0007	8	UltraFastW	PRML	3.5	512k	1000k	7200	
WDE2170-0008	8	SCA-2	PRML	3.5	512k	1000k	7200	
WDE2170-0023	8	Ultra FstDf	PRML	3.5	512k	1000k	7200	
WDE4360-0003	8	UltraFast	PRML	3.5	512k	1000k	7200	
WDE4360-0007	8	UltraFastW	PRML	3.5	512k	1000k	7200	
WDE4360-0008	8	SCA-2	PRML	3.5	512k	1000k	7200	
WDE4360-0023	8	Ultra FstDf	PRML	3.5	512k	1000k	7200	
WDE4550	7.8	Ultra2 SCSI	PRML	3.5 3H	512k	1000k	7200	
WDE9100-0003(50-pin)	7.9	ULSCSIFST	PRML	3.5 3H	512k	1000k	7200	
WDE9100-0007 (68-pin)	7.9	ULSCSIFstWD	PRML	3.5 3H	512k	1000k	7200	
WDE9100-0008 (80-pin)	7.9	SCA-2	PRML	3.5 3H	512k	1000k	7200	
WDE9100-0016 (68-pin)	7.9	ULSCSIFstWD	PRML	3.5 3H	1000k	1000k	7200	
WDE9100-0017 (80-pin)	7.9	SCA-2	PRML	3.5 3H	1000k	1000k	7200	
WDMI130-44 (44 PIN)	19	MCA	RLL	3.5 3H		45k		Y
WDMI130-72 (72 PIN)	19	MCA	RLL	3.5 3H		45k		Y
WDMI4120-72 (72 PIN)	23	MCA	2,7 RLL	3.5 3H		45k		Y
WDSC8320 (Condor)	12	SCSI-2	1,7 RLL	3.5HH	64k	150k	4316	Y
WDSC8400 (Condor)	16	SCSI-2	1,7 RLL	3.5HH	128k	150k	4316	Y
WDSP2100 (Piranha)	14	SCSI-2	2,7 RLL	3.5HH	64k	50k		Y
WDSP4200 (Piranha)	14	SCSI-2	2,7 RLL	3.5HH	64k	50k		Y

Drive Model	Format Size MB	Head	Cyl	Sect/ Trac	Translate H/C/S	RWC/ WPC	Land Zone
WDTM262R (Tandon)	20	2	782	26		783/783	784
WDTM364 (Tandon)	41	4	782	26		783/783	784

XEBEC

Drive Model	Format Size MB	Head	Cyl	Sect/ Trac	Translate H/C/S	RWC/ WPC	Land Zone
OWL I	25	4				—/—	
OWL II	38	4				—/—	
OWL III	52	4				—/—	
XE3100	105	6	979	35		—/—	

Y-E DATA AMERICA, INC

Drive Model	Format Size MB	Head	Cyl	Sect/ Trac	Translate H/C/S	RWC/ WPC	Land Zone
YD3042	43	4	788	28		789/789	Auto
YD3081B	45	2	1057	42		NA/NA	Auto
YD3082	87	8	788	28		789/789	Auto
YD3082B	90	4	1057	42		NA/NA	Auto
YD3083B	136	6	1057	42		NA/NA	Auto
YD3084B	181	8	1057	42		NA/NA	Auto
YD3161B	45	2	1057	42		NA/NA	Auto
YD3162B	90	4	1057	42		NA/NA	Auto
YD3181B	45	2	1057	42		NA/NA	Auto
YD3182B	90	4	1057	42		NA/NA	Auto
YD3530	32	5	731	17		732/732	Auto
YD3540	42	7	733	32		732/732	Auto
YD3541	45	8	731	15		732/732	Auto

ZENTEC

Drive Model	Format Size MB	Head	Cyl	Sect/ Trac	Translate H/C/S	RWC/ WPC	Land Zone
DRACO	518	6	2142	V		—/—	
ZH3100(A)	86					NA/NA	Auto
ZH3100(S)	86					NA/NA	Auto
ZH3140(A)	121					NA/NA	Auto
ZH3140(S)	121					NA/NA	Auto
ZM3180	170					—/—	
ZM3272	260	4	2076	55		—/—	
ZM3360	340					—/—	
ZM3540	518					—/—	
ZQ2140	126	4	1410	44		—/—	

Drive Model	Seek Time	Interface	Encode	Form Factor	cache kb	mtbf	RPM	Obsolete? ⇓
WDTM262R (Tandon)	85	ST412/506	2,7 RLL	3.5HH				Y
WDTM364 (Tandon)	85	ST412/506	2,7 RLL	3.5HH				Y

XEBEC

Drive Model	Seek Time	Interface	Encode	Form Factor	cache kb	mtbf	RPM	Obsolete?
OWL I	55	SCSI	MFM	5.25HH				Y
OWL II	40	SCSI	MFM	5.25HH				Y
OWL III	38	SCSI	MFM	5.25HH				Y
XE3100		IDE AT						Y

Y-E DATA AMERICA, INC

Drive Model	Seek Time	Interface	Encode	Form Factor	cache kb	mtbf	RPM	Obsolete?
YD3042	28	SCSI	2,7 RLL	3.5HH		40k		Y
YD3081B	28	SCSI	2,7 RLL	3.5HH		30k		Y
YD3082	28	SCSI	2,7 RLL	3.5HH		40k		Y
YD3082B	28	SCSI	2,7 RLL	3.5HH		30k		Y
YD3083B	28	SCSI	2,7 RLL	3.5HH		30k		Y
YD3084B	28	SCSI	2,7 RLL	3.5HH		30k		Y
YD3161B	19	IDE AT	2,7 RLL	3.5 3H		40k		Y
YD3162B	19	IDE AT	2,7 RLL	3.5 3H		40k		Y
YD3181B	19	SCSI	2,7 RLL	3.5 3H		40k		Y
YD3182B	19	SCSI	2,7 RLL	3.5 3H		40k		Y
YD3530	26	ST412/506	MFM	3.5HH				Y
YD3540	29	ST412/506	MFM	3.5HH		20k	3600	Y
YD3541	29	SCSI	2,7 RLL	3.5HH		20k	3600	Y

ZENTEC

Drive Model	Seek Time	Interface	Encode	Form Factor	cache kb	mtbf	RPM	Obsolete?
DRACO	12	SCSI-2Fast	1,7 RLL	3.5 3H	512k	150k	4200	Y
ZH3100(A)	20	IDE AT		3.5HH		50k		Y
ZH3100(S)	20	SCSI		3.5HH		50k		Y
ZH3140(A)	20	IDE AT		3.5HH		50k		Y
ZH3140(S)	20	SCSI		3.5HH		50k		Y
ZM3180	12	IDE AT		3.5 3H		150k		Y
ZM3272	13	IDE AT	1,7 RLL	3.5 3H	64k	150k	3600	Y
ZM3360	12	IDE AT		3.5 3H		150k		Y
ZM3540	12	IDE AT		3.5 3H		150k		Y
ZQ2140	18	IDE AT	1,7 RLL	2.5 4H	32k	150k	3600	Y

Hard Drive Source Notes

Information contained in the hard drive chapter was derived from numerous sources, including the manufacturers of the drives. When compiling tables this large, the chance for typing and resource error is great. The authors and publisher would greatly appreciate being notified of any inaccurate or missing information. Some of the older drives (especially those from companies who have gone out of business) are very difficult to obtain accurate and verifiable specifications for. If you have access to old specification sheets, etc please send us a copy so that we may add the information to future editions.

The following are important resources:

ONTRACK Computer Systems Disk Manager Series
Eden Prairie, Minnesota, 1985 to 1990
The Hard Disk Technical Guide by Douglas T. Anderson
PCS Publications, Clearwater, FL, 1990, 1991
The Micro House Encyclopedia of Hard Drives edited
by Douglas T. Anderson, Boulder, CO, 1990 to 1995
Numerous public domain and BBS hard drive listings.
SpeedStor Hard Disk Preparation/Diagnostics
Storage Dimensions, 1985, 1988
Numerous manufacturer specification sheets
Reseller's Resource - Hard Drives, Volume 2, No 1
Technology Publishing, Inc, Livonia, MI January 1990
Buyer's Guide-Hard Drives 40MB to 400MB
Computer Shopper, March 1990
THEREF by F. Robert Falbo, Rome, New York, 1991
Western Digital BBS Listing, 6-6-91

Chapter 8

CD-ROM Drive Specifications

CD-ROM Drive Manufacturers

The following table is a general summary of companies that have manufactured and/or are still manufacturing CD-ROM drives. If you have information concerning the status of any of these companies, such as "XYZ Company went bankrupt in August, 1990" or "XYZ Company was bought by Q Company," please let us know so we can keep this section current. If a phone number is listed in the Status column, the company is in business.

Manufacturer	Phone
Acer, Inc.	408-432-6200
ACS Innovation, Inc.	408-566-0900
Addonics Technology	510-438-6530
Aiwa America Inc.	210-512-3600
Alps Electric, Inc.	408-432-6000
	Stopped manufacturing CD-ROM's in 1995
Apple Computer	408-996-1010
BTC (Behavor Tech Computer Corp.)	510-657-3956
CD Technology	408-752-8500
Chinon America	See Tech Media?
Creative Labs, Inc.	408-428-6600
Delta Microsystems	Unknown
Denon America	Unknown
Diamond Multimedia	408-325-7000
Digital Equipment Corp	508-841-3111
	Supported by Compaq Computer
Fidelity International Tech	732-417-2230
Funai Electric Co., Ltd.	81 720 70 4303 (Japan)
Goldstar Electronics Intern'l	201-816-2000
Hewlett-Packard Company	301-670-4300
Hi-Val	714-953-3000
Hitachi America, Ltd.	800-448-2244
IBM	800-772-2227
JVC	714-816-6500
Laser Magnetic Storage	719-593-7900
Lite-On Inc.	408-946-4873
Mashusta	201-348-7000
Micro Design International	407-677-8333
Micro Solutions	815-756-3411

Manufacturer	Status
Mitsumi Electronics Corp	408-970-0700
Nakamichi America Corp	310-538-8150
NEC Technologies	516-753-7000
NewCom, Inc.	818-597-3200
Nomai, USA	408-542-5900
Ocean Information Systems, Inc.	626-339-8888
Optical Access Intern'l	800-433-5133
Optics Storage	65-3823100 (Japan)
Panasonic Communications	800-742-8086
Panasonic Computer Peripherals	201-348-7000
Peripheral Land	Unknown
Philips Electronics	408-570-5600
Pinnacle Micro, Inc.	714-789-3000
Pioneer Communications	201-327-6400
Pioneer New Media Tech, Inc.	310-952-2111
Plasmon Data, Inc.	612-946-4100
Plextor	408-980-1838
Procom Technology	714-852-1000
Ricoh Corp	408-432-8800
Samsung Electronics America	201-229-7000
Sanyo Electric	81-64 432 949 (Japan)
Severn Companies	301-794-9680
Smart and Friendly	818-772-8001
Sony Corp	201-930-1000
Sun Microsystems	213-726-0303
Teac America, Inc.	213-726-0303
Tech Media	800-379-9069
Texel	Unknown
Todd Enterprises	516-777-8633
Toshiba America	714-583-3000
Wearnes Systems Tech, inc.	408-432-1888
Yamaha Systems Tech, Inc.	408-467-2300
Young Minds Inc.	909-335-1350

Total number of CDROM Drives = 562

CD-ROM Drive Syntax and Notation

The following are descriptions of the information contained in the CD-ROM drive tables. Telephone and BBS numbers for drive manufacturers are listed in the Phone Book chapter of this book. (page)

1. Drive Type .... CD-ROM=Standard CD-ROM
   CD-R=Write once CD-ROM
   CD-RW=Re-writeable CD-ROM
   Changer=multi disk CD-ROM
2. Speed .... Drive speed, for example "32X". 1X is defined as the speed of audio CDs, so 32X is 32 times faster than audio speed. In actuality, a drive rated at 32X is 32X maximum and the average is approximately 8X.
3. Interface.................. Type of communication interface between the computer and drive
4. Buffer .... Size of memory buffer in drive. More provides better performance.
5. Discs .... Number of discs the drive holds
6. Loader .... Type of disc carrier. Magazine, Tray, Caddy (including brand of caddy, if available)
7. Plug & Play .... If indicated, drive is Plug & Play. Note: Many of the listed drives are Plug & Play, but could not be verified so they are not listed as Plug & Play.
8. Internal/External ..... Internal=mounted inside computer
   External=mounted in an external drive case.

Sequoia needs your help! If you have specifications on new or obsolete CD-ROM drives, please send them to us so that we can include them in future editions of this book.

Manufacturer Model	Drive Type Discs, Loader	Speed Plug&Play	Interface Internal/External	Buffer

Acer, Inc.

Model	Drive Type / Discs, Loader	Speed / Plug&Play	Interface / Internal/External	Buffer
CD-612A	CD-ROM 1 disc,Tray	12X	IDE Internal	128kb
CD-620A	CD-ROM 1 disc,Tray	20X	EIDE/ATAPI Internal	128kb
CD-624A	CD-ROM 1 disc	24X	EIDE/ATAPI Internal	128kb
CD-912E	CD-ROM 1 disc	12X	EIDE/ATAPI Internal	128kb
CD-916 E	CD-ROM 1 disc,Tray	16X	ATAPI IDE Internal	
CD-916E	CD-ROM 1 disc,Tray	16X	Atapi IDE Internal	
CD-920E	CD-ROM 1 disc,Tray	20X	ATAPI IDE Internal	128kb
CD-924E	CD-ROM 1 disc,Tray	24X	ATAPI IDE Internal	128kb
CRW620	CD-R 1 disc	2Xw/6Xr Plug & Play	SCSI-2 Internal	1Mb

ACS Innovation, Inc.

Model	Drive Type / Discs, Loader	Speed / Plug&Play	Interface / Internal/External	Buffer
CDR 7501-INT	CD-R disc,Tray	4X	Adaptec SCSI Internal	1Mb
COMPRO CDR-7502B	CD-R 1 disc	4Xw/8Xr	SCSI	1Mb
COMPRO CR-585B	CD-ROM 1 disc,Tray	24X	IDE/ATAPI Internal	128kb
COMPRO LMD-584	CD-ROM 1 disc,Tray	12X	IDE/ATAPI Internal	128kb

Addonics Technology

Model	Drive Type / Discs, Loader	Speed / Plug&Play	Interface / Internal/External	Buffer
PCD4X4	CD-ROM	4X		
PCD6X	CD-ROM	6X		
PCDS6X	CD-ROM 1 disc	6X	Portable	128kb

Aiwa America Inc.

Model	Drive Type / Discs, Loader	Speed / Plug&Play	Interface / Internal/External	Buffer
ACD-630	CD-ROM 3 disc	4X		256kb

Alps Electric (USA), Inc

Model	Drive Type / Discs, Loader	Speed / Plug&Play	Interface / Internal/External	Buffer
Alps 4X Internal	CD-ROM-changer 4 disc,Magazine	4X	EIDE Internal	128kb
CD544C	CD-ROM-changer 4 disc,Magazine	4X	EIDE	128kb

Altec Lansing Technologies Inc

Model	Drive Type / Discs, Loader	Speed / Plug&Play	Interface / Internal/External	Buffer
AMC2800	CD-ROM 1 disc	8X	PC Card External	

Apple

Model	Drive Type / Discs, Loader	Speed / Plug&Play	Interface / Internal/External	Buffer
AppleCD SC	CD-ROM 1 disc,Sony Caddy	1X	SCSI External	64kb

Manufacturer Model	Drive Type Discs, Loader	Speed Plug&Play	Interface Internal/External	Buffer

BTC (Behavior Tech Computer Corp)

Model	Drive Type / Discs, Loader	Speed / Plug&Play	Interface / Internal/External	Buffer
BCE 621E	CD-RW 1 disc, Tray	2Xw/6Xr Plug & Play	EIDE Internal	
BCD 739	CD-ROM 1 disc, Tray	8X Plug & Play	EIDE Internal	128kb
BCD 8X	CD-ROM 1 disc, Tray	8X Plug & Play	EIDE Internal	256kb
BCD 10X	CD-ROM 1 disc, Tray	10X Plug & Play	EIDE Internal	256kb
BCD 16X	CD-ROM 1 disc, Tray	16X Plug & Play	EIDE Internal	256kb
BCD 20X	CD-ROM 1 disc, Tray	20X Plug & Play	EIDE Internal	256kb
BCD 24X	CD-ROM 1 disc, Tray	24X Plug & Play	EIDE Internal	256kb
BCD 36X	CD-ROM 1 disc, Tray	36X Plug & Play	EIDE Internal	256kb
BCD 40X	CD-ROM 1 disc, Tray	40X Plug & Play	EIDE Internal	256kb

CD Technology

Model	Drive Type / Discs, Loader	Speed / Plug&Play	Interface / Internal/External	Buffer
CD-T3201MAC	CD-ROM 1 disc,Sony Caddy	2X	SCSI External	64kb
CD-T3201MCA	CD-ROM 1 disc,Sony Caddy	2X	SCSI External	64kb
CD-T3201PC	CD-ROM 1 disc,Sony Caddy	2X	SCSI External	64kb
Porta-Drive	CD-ROM 1 disc	4X	SCSI External	256kb

Chinon America

Model	Drive Type / Discs, Loader	Speed / Plug&Play	Interface / Internal/External	Buffer
CDA-535	CD-ROM 1 disc,Caddy	2X	SCSI	320kb
CDS-430	CD-ROM 1 disc,Tray	1X	Non-SCSI Parallel External	32kb
CDS-435	CD-ROM 1 disc,Tray	2X	Internal	
CDS-525	CD-ROM 1 disc	2X	SCSI Internal	64kb
CDS-535	CD-ROM 1 disc,Tray	2X	SCSI-2 Internal	256kb
CDS-545	CD-ROM 1 disc	4X	EIDE Internal	128kb
CDX-431	CD-ROM 1 disc,Sony Caddy	1X	SCSI External	32kb
CDX-535	CD-ROM 1 disc,Caddy	2X	SCSI Internal & External	320kb

Creative Labs, Inc.

Model	Drive Type / Discs, Loader	Speed / Plug&Play	Interface / Internal/External	Buffer
Blaster CD 4X	CD-ROM 1 disc	4X	IDE	256kb
Blaster CD 6X	CD-ROM 1 disc	6X Plug & Play	IDE Internal	256kb
Blaster CD 8X	CD-ROM 1 disc,Tray	8X Plug & Play	IDE Internal	
CD 200	CD-ROM disc,Tray	2X	Creative/MKE Internal	64kb

Manufacturer Model	Drive Type Discs, Loader	Speed Plug&Play	Interface Internal/External	Buffer
CD 220E	CD-ROM disc,Tray	2X	Creative/MKE Internal	64kb
CD 420E	CD-ROM disc,Tray	4X	IDE/ATAPI Internal	256kb
CD 420P	CD-ROM	4X	Creative PCMCIA External	256kb
CDR 7730	CD-ROM disc,Tray	4X	IDE/ATAPI Internal	128kb
CR 521	CD-ROM 1 disc,Caddy	1X	Creative/Panasonic Internal	64kb
CR 523	CD-ROM 1 disc,Caddy	1X	Creative/Panasonic Internal	64kb
CR 531	CD-ROM 1 disc,Caddy	1X	Creative/Panasonic Internal	64kb
PC-DVD Encore Dxr2	DVD-ROM 1 disc,Tray	20X	IDE Internal	256kb
PC-DVD X2	DVD-ROM 1 disc,Tray	20X	IDE Internal	256kb
Vibra 24X	CD-ROM 1 disc,Tray	24X	IDE Internal	

Delta Microsystems

Model	Drive Type / Discs, Loader	Speed / Plug&Play	Interface / Internal/External	Buffer
SS-600C	CD-ROM 1 disc,Sony Caddy	1X	SCSI External	8kb

Denon America

Model	Drive Type / Discs, Loader	Speed / Plug&Play	Interface / Internal/External	Buffer
DRD-253M	CD-ROM 1 disc,Sony Caddy	1X	SCSI External	32kb
DRD-253S	CD-ROM 1 disc,Sony Caddy	1X	SCSI External	32kb

Diamond Multimedia Systems, Inc

Model	Drive Type / Discs, Loader	Speed / Plug&Play	Interface / Internal/External	Buffer
8000	CD-ROM 1 disc	8X Plug & Play	EIDE	256kb

Digital Equipment Company

Model	Drive Type / Discs, Loader	Speed / Plug&Play	Interface / Internal/External	Buffer
DS-RRD46-VA	CD-ROM 1 disc,Tray	12X	SCSI Internal	
DS-RRD46-VU	CD-ROM 1 disc,Tray	12X	SCSI Internal	
Qbus RRD40 Slave	CD-ROM 1 disc,Philips Ca	1X	Qbus External	2kb
Qbus RRD40	CD-ROM 1/drive disc,Phillips Caddy	1X	Qbus 4 Drives/Con External	2kb
SCSI RRD40	CD-ROM 1/drive disc,Phillips Ca	1X	SCSI 4 Drives/Controller External	2kb
SCSI RRD40 Slave	CD-ROM 1 disc,Philips Ca	1X	SCSI External	2kb

DynaTek Automation Sys, Inc

Model	Drive Type / Discs, Loader	Speed / Plug&Play	Interface / Internal/External	Buffer
CDM400	CD-R 1 disc	4Xw/4Xr	SCSI Internal	512kb

Fidelity Intl Technologies

Model	Drive Type / Discs, Loader	Speed / Plug&Play	Interface / Internal/External	Buffer
TCD4X-P	CD-ROM disc,Tray	4X Plug & Play	Parallel SCSI-2 External	

Manufacturer Model	Drive Type Discs, Loader	Speed Plug&Play	Interface Internal/External	Buffer
TCD6X-P	CD-ROM 1 disc,Tray	6X Plug & Play	Parallel SCSI-2 External	

Funai Electric Co., Ltd.

Model	Drive Type	Speed	Interface	Buffer
E2420	CD-ROM 1 disc,Tray	2X	ISA Internal	
E2450	CD-ROM 1 disc,Tray	2X	ISA Internal	
E2550UA	CD-ROM 1 disc,Tray	2X	Creative/MKE Internal	64kb
E2650	CD-ROM 1 disc,Tray	6X	IDE/ATAPI Internal	
E2750UA	CD-ROM 1 disc,Tray	4X	IDE/ATAPI Internal	128kb
E2800UA	CD-ROM 1 disc,Tray	2X	Creative/MKE Internal	64kb
E2850	CD-ROM 1 disc,Tray	8X	IDE/ATAPI Internal	
E2960	CD-ROM 1 disc,Tray	16X	IDE/ATAPI Internal	

Goldstar Electronics Int'l, Inc.

Model	Drive Type	Speed	Interface	Buffer
GCD-R320B	CD-ROM 1 disc,Tray	2X	SCSI Internal	64kb
GCD-R540B	CD-ROM 1 disc,Tray	4X	IDE/ATAPI Internal	128kb
GCD-R56B	CD-ROM 1 disc,Tray	6X Plug & Play	IDE Internal	
GCD-R580B	CD-ROM 1 disc,Tray	8X	EIDE Internal	256kb

Hewlett-Packard Company

Model	Drive Type	Speed	Interface	Buffer
CD-2896A	CD-ROM 1 disc,Tray	4X	IDE/ATAPI Internal	256kb
CD-4020i	CD-R 1 disc,Tray	2Xw/4Xr	SCSI-2 Internal	1Mb
CD-6020ep	CD-R 1 disc,Tray	2Xw/6Xr	parallel SCSI-2 External	1Mb
CD-6020es	CD-R 1 disc,Tray	2Xw/6Xr	SCSI-2 External	1Mb
CD-6020i	CD-R 1 disc,Tray	2Xw/6Xr	SCSI-2 Internal & External	1Mb
CD-7200e	CD-RW 1 disc,Tray	2Xw/6Xr	Parallel External	1Mb
CD-7200i	CD-R 1 disc,Tray	2Xw/6Xr	EIDE/ATAPI Internal	1Mb
HP-6100/600/A	CD-ROM 1 disc,Philips Ca	1X	HP-1B (IEEE-488) External	12kb
HP-7110e	CD-RW 1 disc,Tray	2X/2X/6X	Parallel Port External	1Mb
HP-7110i	CD-RW 1 disc,Tray	2X/2X/6X	IDE Internal	1Mb

Hi-Val

Model	Drive Type	Speed	Interface	Buffer
16X	CD-ROM-changer 5 disc,Direct load	16X Plug & Play	EIDE External	
16X HotPort	CD-ROM	16X	IDE	

Manufacturer Model	Drive Type	Speed	Interface	Buffer	Discs, Loader	Plug&Play	Internal/External
					1 disc,Tray		External
24X	CD-ROM	24X	IDE/ATAPI	128kb	1 disc		
24X (Portable)	CD-ROM	24X	PCMCIA Type II		1 disc,Tray	Plug & Play	External
24X MAX	CD-ROM	24X	EIDE		1 disc,Slot	Plug & Play	External
2X6 EPP	CD-R	2Xw/6Xr	EPP to IDE	1Mb	1 disc,Tray	Plug & Play	External
2X6 HotPort	CD-R	2Xw/6Xr	EIDE/ATAPI	1Mb	1 disc,Tray		External
2X8	CD-R	2Xw/6Xr	EIDE/ATAPI		1 disc	Plug & Play	Internal
2X8 EPP	CD-R	2Xw/8Xr	EPP to EIDE	1Mb	1 disc,Tray	Plug & Play	External
2X8 HotPort	CD-R	2Xw/8Xr	IDE	1Mb	1 disc,Tray		External
32X MAX	CD-ROM	32X			1 disc,Tray	Plug & Play	Internal
HV6200	CD-RW	2Xw/6Xr	SCSI -2	1Mb	1 disc,Tray		Internal

Hitachi America, Ltd.

Manufacturer Model	Drive Type	Speed	Interface	Buffer	Discs, Loader	Plug&Play	Internal/External
CD-MAC	CD-ROM	1X	SCSI	64kb	1 disc,Sony Caddy		External
CDR-1503BZ	CD-ROM	1X	Non-SCSI Parallel	32kb	1 disc,Tray		External
CDR-1503S	CD-ROM	1X	Non-SCSI Parallel	32kb	1 disc,Tray		External
CDR-1503S	CD-ROM						External
CDR-1503S-MC	CD-ROM	1X	Non-SCSI Parallel	8kb	1 disc,Tray		External
CDR-1503S-PC	CD-ROM	1X	Non-SCSI Parallel	32kb	1 disc,Tray		External
CDR-1520S	CD-ROM						External
CDR-1600	CD-ROM	1X	Non-SCSI Parallel	32kb	1 disc,Sony Caddy		External
CDR-1600SY	CD-ROM	1X	Non-SCSI Parallel	32kb	1 disc,Sony Caddy		External
CDR-1600SZ	CD-ROM	1X	Non-SCSI Parallel	32kb	1 disc,Sony Caddy		External
CDR-1650	CD-ROM	1X	SCSI	64kb	1 disc,Sony Caddy		External
CDR-1700S	CD-ROM		SCSI		1 disc		Internal
CDR-1750S	CD-ROM		SCSI		1 disc		Internal
CDR-1900S	CD-ROM	2X	Hitachi Bus	128kb	1 disc,Caddy		External
CDR-1950S	CD-ROM	2X	SCSI-2	256kb	1 disc,Caddy		
CDR-2500S	CD-ROM						External
CDR-3500	CD-ROM						Internal
CDR-3600	CD-ROM						

Manufacturer Model	Drive Type	Speed	Interface	Buffer	Discs, Loader	Plug&Play	Internal/External
							Internal
CDR-3650	CD-ROM						
CDR-3700	CD-ROM						Internal
CDR-3750	CD-ROM						
CDR-6700	CD-ROM	2X	Hitachi Bus	128kb	1 disc,Caddy		Internal
CDR-6750	CD-ROM	2X	SCSI-2	256kb	1 disc,Caddy		
CDR-7730	CD-ROM	4X	IDE/ATAPI	256kb	1 disc,Tray		Internal
CDR-7830	CD-ROM		IDE				
CDR-7930	CD-ROM	8X	EIDE/ATAPI		1 disc,Tray		Internal
CDR-8130	CD-ROM	16X	EIDE/ATAPI	128kb	1 disc,Tray		Internal
CDR-8330	CD-ROM	24X	EIDE/ATAPI	128kb	1 disc,Tray		Internal
CDR-8430	CD-ROM	32X	EIDE/ATAPI		1 disc		Internal
GD-1000	DVD-ROM	8X	EIDE/ATAPI	256kb	1 disc,Tray		Internal

IBM

Model	Drive Type	Speed	Interface	Buffer	Discs, Loader	Plug&Play	Internal/External
32X	CD-ROM	32X	IDE	128kb	1 disc,Tray		Internal
4X CD	CD-ROM	4X	SCSI-2	256kb	1 disc,Tray		External
4X PCMCIA	CD-ROM	4X	PCMCIA Type II	128kb		Plug & Play	External
4X PCMCIA Stereo	CD-ROM	4X	PCMCIA Type II	128kb		Plug & Play	External
Ext CD-ROM Drive	CD-ROM	1X	SCSI	64kb	1 disc,Sony Caddy		External
IDE 16XMAX	CD-ROM	16X	IDE/ATAPI	128kb	1 disc,Tray		Internal
PS/2 Ext CD-ROM Drive	CD-ROM	1X	SCSI	64kb	1 disc,Sony Caddy		External
SCSI 8X	CD-ROM	8X	SCSI	256kb	1 disc,Tray		Internal
SCSI 8X	CD-ROM	8X	SCSI	256kb	1 disc,Tray		External
TP 365X/XD 8X Int IDE	CD-ROM	8X	IDE/ATAPI	128kb	1 disc,Tray		Internal
TP 380/385 8X IDE	CD-ROM	8X	IDE	256kb	1 disc,Tray		Internal
TP 600 24X-10X Int IDE	CD-ROM	24X	IDE	256kb	1 disc,Tray		Internal
TP 770 24X-10X Int	CD-ROM	24X	IDE	128kb	1 disc,Tray		Internal

JVC Company of America

Model	Drive Type	Speed	Interface	Buffer	Discs, Loader	Plug&Play	Internal/External
Personal RomMaker	CD-R	2X		1Mb	1 disc,Caddy		
XR-R100	CD-ROM	1X	SCSI	24k			

Manufacturer Model	Drive Type	Speed	Interface	Buffer	Discs, Loader	Plug&Play	Internal/External
					1 disc,JVC Caddy		External

Laser Magnetic Storage

Model	Drive Type	Speed	Interface	Buffer	Discs, Loader	Plug&Play	Internal/External
CM 121	CD-ROM	1X	Serial	2kb	1 disc,Philips Ca		External
CM 131	CD-ROM	1X	SCSI	2kb	1 disc,Philips Ca		External
CM 206	CD-ROM		AT		1 disc,Tray		Internal
CM 215	CD-ROM		SCSI		1 disc,Tray		Internal
CM 221	CD-ROM	1X	Serial	2kb	1 disc,Philips Ca		External
CM 231	CD-ROM	1X	SCSI	64kb	1 disc,Philips Ca		External

Lite-On Technology

Model	Drive Type	Speed	Interface	Buffer	Discs, Loader	Plug&Play	Internal/External
LTN-2221	CD-ROM	12X	ATAPI/EIDE	256kb	1 disc		Internal
LTN-26216X	CD-ROM	16X	ATAPI/EIDE	256kb	1 disc		Internal

Micro Design International

Model	Drive Type	Speed	Interface	Buffer	Discs, Loader	Plug&Play	Internal/External
600C DX4	CD-ROM	4X	SCSI-2	256kb	1 disc		Internal
600C LaserBank	CD-ROM	1X	SCSI	None	1 disc,Sony Caddy		External
SCSI Express	CD-ROM-changer	8X	SCSI-2	256kb	1 disc,Slot		External

Micro Solutions

Model	Drive Type	Speed	Interface	Buffer	Discs, Loader	Plug&Play	Internal/External
24X Backpack Bantam	CD-ROM	24X	Parallel Port	128kb	1 disc,Tray	Plug & Play	
32X Backpack	CD-ROM	32X	Paralledl Port	256kb	1 disc,Tray	Plug & Play	
8X Bantam Backpack	CD-ROM	8X	Parallel Port	256kb	1 disc,Tray	Plug & Play	
Backpack pd/cd	CD-ROM	4X	Parallel Port	128kb	1 disc,Tray	Plug & Play	External

Mitsumi Electronics Corp

Model	Drive Type	Speed	Interface	Buffer	Discs, Loader	Plug&Play	Internal/External
CR-2600TE	CD-R	2Xw/6Xr	EIDE ATAPI	1Mb	1 disc,Tray		Internal
FX120	CD-ROM	12X	ATAPI IDE	256kb	1 disc		Internal
FX1200	CD-ROM	12X	IDE/ATAPI	256kb	1 disc,Tray		Internal
FX1600	CD-ROM	16X	IDE/ATAPI	256kb	1 disc,Tray		Internal
FX240	CD-ROM	12X/24X	EIDE		disc,Caddyless		
FX2400	CD-ROM	24X	IDE/ATAPI	256kb	1 disc,Tray		Internal
FX400	CD-ROM	4X	IDE/ATAPI	128kb	1 disc,Tray		Internal
FX600S	CD-ROM	6X	EIDE	128kb	1 disc,Tray		Internal

Manufacturer Model	Drive Type Discs, Loader	Speed Plug&Play	Interface Internal/External	Buffer
FX800	CD-ROM	8X	IDE/ATAPI	256kb
	1 disc,Tray		Internal	
FXN01DE	CD-ROM	2X	ATAPI/IDE	
	1 disc,Tray			

Nakamichi America Corp

Model	Drive Type	Speed	Interface	Buffer
MJ-4.4	CD-ROM-changer	4X	EIDE	128kb
	1 disc		External	
MJ-4.8s	CD-ROM-changer	8X	SCSI-2	256kb
	4 disc,Slot	Plug & Play	Internal	

NEC Technologies

Model	Drive Type	Speed	Interface	Buffer
CDR-1350A	CD-ROM	6X	EIDE	128kb
	2 disc		Internal	
CDR-1400A	CD-ROM		IDE	
	1 disc			
CDR-1450	CD-ROM	8X	EIDE	128kb
	2 disc		Internal	
CDR-1460A	CD-ROM	8X	SCSI	256kb
	disc,Integrated Drawer		Plug & Play	Internal
CDR-1610A	CD-ROM		SCSI-2	256kb
	1 disc			
CDR-1810A	CD-ROM		SCSI-2	256kb
CDR-210	CD-ROM	2X	SCSI	64kb
	1 disc,Tray		Internal	
CDR-211	CD-ROM	2X		
CDR-250	CD-ROM	2X	IDE	256kb
	1 disc,Tray		Internal	
CDR-250DX	CD-ROM	2X	IDE	256kb
	1 disc,Tray		Internal	
CDR-251	CD-ROM		IDE	
CDR-260	CD-ROM	2X	EIDE	
	1 disc			
CDR-271	CD-ROM	4X	IDE/ATAPI	256kb
	disc,Tray			
CDR-272	CD-ROM			
CDR-300 (MAC)	CD-ROM	2X	SCSI	256kb
	1 disc,Tray		External	
CDR-3460A	CD-ROM	8X	SCSI	256kb
	1 disc,Integrated Drawer		Plug & Play	External
CDR-401	CD-ROM	3.3X	SCSI-2	256kb
	1 disc,Top Loading			
CDR-501	CD-ROM	4X	SCSI-2	256kb
	1 disc,Cartridge		Internal	
CDR-502	CD-ROM	6X	SCSI-2	256kb
	1 disc,Cartridge	Plug & Play	Internal	
CDR-510	CD-ROM	3X	SCSI	256kb
	1 disc,Tray		Internal	
CDR-511	CD-ROM	4X	SCSI-2	256kb
	1 disc,Caddy		Internal	
CDR-601	CD-ROM	4X	SCSI-2	256kb
	1 disc,Cartridge		External	
CDR-602	CD-ROM	6X	SCSI-2	256kb
	1 disc,Cartridge	Plug & Play	Internal & External	

Manufacturer Model	Drive Type Discs, Loader	Speed Plug&Play	Interface Internal/External	Buffer
CDR-900	CD-ROM	4X	SCSI-2/SCSI1 Switch Selectable	256kb
	1 disc,Cartridge		External	
CDR-C251	CD-ROM-changer	4X	EIDE	128kb
	4 disc,Direct Feed		Internal	
CDR-C302	CD-ROM-changer	4X	SCSI-2	128kb
	2 disc,Drawer		External	
CDR3 Intersect	CD-ROM	1X	SCSI	2kb
	1 disc,Top Load		External	
CDR7 Intersect	CD-ROM	1X	SCSI	64kb
	1 disc,Sony Caddy		External	
MultiSpin 2Xc	CD-ROM-changer	2X	SCSI-2	64kb
	7 disc,Drawer		External	
MultiSpin 3Xi	CD-ROM	3X	SCSI-2	
	1 disc		Internal & External	
MultiSpin 3Xp Plus	CD-ROM	3.3X	SCSI-2	256kb
	1 disc,Tray		Internal	
MultiSpin 4X4	CD-ROM	4X	EIDE	128kb
	1 disc,Slot		Internal	
MultiSpin 4Xc	CD-ROM-changer	4X	SCSI-2	128kb
	7 disc,Tray		External	
MultiSpin 4Xi	CD-ROM	4X	SCSI-2	256kb
	1 disc,Caddy		Internal	
MultiSpin 6V	CD-ROM	6X	EIDE	128kb
	1 disc,Tray		Internal	
MultiSpin 6Xe	CD-ROM	6X	SCSI-2	256kb
	1 disc,Cartridge	Plug & Play	Internal	
MultiSpin 6Xi	CD-ROM	6X	SCSI-2	256kb
	1 disc,Cartridge	Plug & Play	Internal	
MultiSpin 8V	CD-ROM	8X	ATAPI 2.5	128kb
	1 disc,Tray		Internal	
MultiSpin 8Xe	CD-ROM	8X	SCSI-2	256kb
	1 disc,Tray		External	
MultiSpin 8Xi	CD-ROM	8X	SCSI-2	256kb
	1 disc,Tray		Internal	

NewCom, Inc.

Manufacturer Model	Drive Type Discs, Loader	Speed Plug&Play	Interface Internal/External	Buffer
CDRW622 CopyCat	CD-RW	2Xw/6Xr	IDE/ATAPI	1Mb
	1 disc		Internal	
NewCom 10X	CD-ROM	10X	IDE/ATAPI	
	1 disc		Internal	
NewCom 12X	CD-ROM	12X	IDE/ATAPI	
	1 disc		Internal	
NewCom 20X	CD-ROM	20X	IDE/ATAPI	
	1 disc		Internal	
NewCom 24X	CD-ROM	24X	IDE/ATAPI	
	1 disc	Plug & Play	Internal	
NewCom 32X	CD-ROM	32X	EIDE/ATAPI	
	1 disc		Internal	
NewCom 6X	CD-ROM	6X	IDE/ATAPI	
	1 disc		Internal	
NewCom 8X	CD-ROM	8X	IDE/ATAPI	
	1 disc		Internal	

Manufacturer Model	Drive Type Discs, Loader	Speed Plug&Play	Interface Internal/External	Buffer

Nomai USA

Model	Drive Type / Discs, Loader	Speed / Plug&Play	Interface / Internal/External	Buffer
680R.W.	CD-RW	2Xw/6Xr	AdvanSystems, Ultra-SCSI	
	1 disc,Tray		Internal & External	

Ocean Information Systems,Inc

Model	Drive Type / Discs, Loader	Speed / Plug&Play	Interface / Internal/External	Buffer
CDR-688	CD-ROM	8X	EIDE/ATAPI	256kb
	1 disc,Tray		Internal	
CDR-810	CD-ROM	10X	EIDE/ATAPI	256kb
	1 disc,Tray		Internal	
CDR-812	CD-ROM	12X	EIDE/ATAPI	256kb
	1 disc,Tray		Internal	
CDR-820	CD-ROM	20X	EIDE/ATAPI	256kb
	1 disc,Tray		Internal	
CDR-824	CD-ROM	24X	EIDE/ATAPI	256kb
	1 disc,Tray		Internal	
RW260	CD-R	2Xw/6Xr	EIDE/ATAPI	1Mb
	1 disc,Tray	Plug & Play	Internal	

Optical Access International

Model	Drive Type / Discs, Loader	Speed / Plug&Play	Interface / Internal/External	Buffer
Access CD	CD-ROM	1X	SCSI	64kb
	1 disc,Philips Ca		External	

Optics Storage

Model	Drive Type / Discs, Loader	Speed / Plug&Play	Interface / Internal/External	Buffer
Mach 5111	DVD-ROM	24X	IDE/ATAPI	256kb
	1 disc	Plug & Play	Internal	
Maverick 8424	CD-ROM	12X	Fast SCSI-2	256kb
	1 disc,Tray	Plug & Play	External	
Maverick 8622	CD-ROM	12X	EIDE	256kb
	1 disc,Tray		Internal	
Maverick 8622	CD-ROM	12X	Fast SCSI-2	256kb
	1 disc,Tray	Plug & Play	Internal	
Maverick 8623	CD-ROM	12X	IDEATAPI	128kb
	1 disc,Tray		Internal	
Maverick 8831	CD-ROM	16X	IDE/ATAPI	128kb
	1 disc,Tray	Plug & Play	Internal	
Maverick 8841	CD-ROM	16X	IDE/ATAPI	128kb
	1 disc,Tray	Plug & Play	Internal	

Panasonic Communications

Model	Drive Type / Discs, Loader	Speed / Plug&Play	Interface / Internal/External	Buffer
CD-616P	CD-ROM	16X	EIDE/ATAPI	128kb
	1 disc,Tray		Internal	
KXL-783A	CD-ROM	8X	PCMCIA	128kb
	1 disc,Tray		Portable	
KXL-D720	CD-ROM		PCMCIA	128kb
	1 disc		Portable	
KXL-D721	CD-ROM	2X		
KXL-D740	CD-ROM	4X	PC Card Type II	128kb
	1 disc,Tray		Portable	
KXL-D742	CD-ROM	4X	PCMCIA	128kb
	1 disc,Tray		Portable	
KXL-D745	CD-ROM	4X	PCMCIA Type II SCSI-2	128kb
	1 disc,Tray		Portable	
LF-1000	PD/CD-ROM	4X	SCSI-2	256kb
	1 disc,Tray		External	

Manufacturer Model	Drive Type Discs, Loader	Speed Plug&Play	Interface Internal/External	Buffer
LF-1004	PD/CD-ROM 1 disc,Tray	4X	SCSI-2 Internal	256kb
LK-MC501B	CD-ROM		SCSI Internal	
LK-MC501S	CD-ROM		SCSI External	
LK-MC509S	CD-ROM 1 disc	2X	SCSI External	
LK-MC521B	CD-ROM		AT BUS Internal	
LK-MC521S	CD-ROM		AT BUS External	
LK-MC579BP	CD-ROM 1 disc,Tray	2X	EIDE/ATAPI Internal	
LK-MC604S	CD-ROM 1 disc,Tray	4.5X	SCSI External	256kb
LK-MC605BP	CD-ROM 1 disc,Disk	4X	SCSI Internal	
LK-MC605S	CD-ROM 1 disc,Tray	4X	SCSI External	
LK-MC682BP	CD-ROM 1 disc	32X	IDE Internal	256kb
LK-MC684BP	CD-ROM 1 disc,Tray	4.5X	EIDE/ATAPI Internal	256kb
LK-MV8581BP	DVD-ROM		ATAPI Internal	
LK-MW602BP	CD-R disc,1	2Xw/4Xr	SCSI-2	256kb
SQ-TC500N	CD-ROM-changer 5 disc,Tray	4X	EIDE/ATAPI Internal	256kb
SQ-TC512N	CD-ROM-changer 5 disc,Tray	12X	EIDE/ATAPI Internal	

Panasonic Computer Peripheral Co.

Model	Drive Type	Speed	Interface	Buffer
Big 5	CD-ROM-changer 5 disc,Tray	10X	ATAPI/IDE Internal	128kb

Peripheral Land

Model	Drive Type	Speed	Interface	Buffer
PLI CD-ROM	CD-ROM 1 disc,Sony Caddy	1X	SCSI External	64kb

Phillips Electronics

Model	Drive Type	Speed	Interface	Buffer
CDD 2000/12	CD-R 1 disc,Tray	2Xw/4Xr	SCSI Internal	1Mb
CDD 2000/20	CD-R 1 disc,Tray	2Xw/4Xr	SCSI External	1Mb
CDD 2600	CD-R 1 disc,Tray	2Xw/6Xr	SCSI-2 Internal	1Mb
CDD 300	CD-ROM 1 disc	2X	IDE/ATAPI Internal	256kb
CDD 3610	CD-R 1 disc,Tray	2Xw/6Xr	EIDE Internal	1Mb
CM 405	CD-ROM 1 disc	2X	SCSI-2 Internal	
CM 425	CD-ROM 1 disc	2X	SCSI-2 External	
DRD 5200	CD-ROM/DVD	24X/2XDVD	ATAPI/IDE	256kb

Manufacturer Model	Drive Type	Speed	Interface	Buffer	Discs, Loader	Plug&Play	Internal/External
					1 disc,Tray		
OmniWriter 12	CD-RW	2X/2X/6X	ATAPI	1Mb	1 disc,Tray		Internal
OmniWriter 20	CD-RW	2X/2X/6X	SCSI	1Mb	1 disc,Tray		External
OmniWriter 40	CD-RW	2X/2X/6X	SCSI	1Mb	1 disc,Tray		External
PCA 202 CD (Europe)	CD-ROM	20X	ATAPI/IDE	128kb	1 disc	Plug & Play	Internal
PCA 243 CD (Europe)	CD-ROM	24X	ATAPI/IDE	128kb	1 disc	Plug & Play	Internal
PCA-52CR	CD-ROM	5X	ATAPI/IDE	128kb	1 disc	Plug & Play	
PCA-62CR	CD-ROM	6X	IDE	256kb	1 disc		Internal
PCA-80SC	CD-ROM	8X	SCSI-2		1 disc,Tray	Plug & Play	Internal & External

Pinnacle Micro, Inc.

Model	Drive Type	Speed	Interface	Buffer	Discs, Loader	Plug&Play	Internal/External
10Xtreme	CD-ROM	10X	IDE		1 disc		Internal
RCD 4X12 Ext Mac Pro	CD-R	4Xw/12Xr	SCSI-2	256kb	1 disc		External
RCD 4X12 Ext PC Pro	CD-R	4Xw/12Xr	SCSI-2	256kb	1 disc		External
RCD 4X12 Int PC Pro	CD-R	4Xw/12Xr	SCSI-2	1Mb	1 disc		Internal
RCD 4X4 Int IPCA	CD-R	4Xw/4Xr	SCSI-2	256kb	1 disc,Tray		Internal
RCD 5020	CD-R	2X	SCSI-2	1Mb	1 disc,Tray		Internal & External
RCD 5040	CD-R	2Xw/4Xr	SCSI-2	512kb	1 disc,Caddy		Internal & External
RCD-1000	CD-R	2X	SCSI-1	1Mb	1 disc,Caddy		External
RCD-5040	CD-ROM	4X					Internal & External

Pioneer New Media Tech, Inc

Model	Drive Type	Speed	Interface	Buffer	Discs, Loader	Plug&Play	Internal/External
DR-U124X	CD-ROM	4.4X	SCSI-2	128kb	1 disc,Tray		Internal
DRM-1004X	CD-R jukebox	12X	SCSI-2		100 disc,Mail Slot		Internal
DRM-1804X	CD-ROM-changer	4X			18 disc,Caddy		External
DRM-5002R2W	CD-R jukebox	2.4Xw/2.4r			500 disc		Internal
DRM-5003R1W	CD-R jukebox				500 disc		Internal
DRM-5004X	CD-R jukebox	4.4Xr	SCSI-3		500 disc		
DRM-600	CD-ROM	1X	SCSI		6 disc,6 Disc Mag		External
DRM-602X	CD-ROM	2X	SCSI-2	256kb	6 disc,Magazine		
DRM-604X	CD-ROM	4X	SCSI-2	128kb	6 disc,6 Disc Mag		Internal
DRM-624X	CD-ROM-changer	4.4X	SCSI-2	128kb			

Manufacturer / Model	Drive Type / Discs, Loader	Speed / Plug&Play	Interface / Internal/External	Buffer
	6 disc,Magazine		Internal & External	
DRM-6324X	CD-ROM-changer	24X	SCSI 2	
	6 Disc Mag disc		Internal	
DW-S114X	CD-R	4X	SCSI	1Mb
	1 disc,Tray		Internal & External	
Super 10X	CD-ROM	10X	ATAPI	
	1 disc			

Plasmon Data, Inc.

Model	Drive Type / Discs, Loader	Speed / Plug&Play	Interface / Internal/External	Buffer
CDR-4220	CD-R	4X	SCSI	1Mb
	1 disc,Tray	Plug & Play	External	
CDR-4240(Afterburner)	CD-R	2Xw/4Xr	EIDE	1Mb
	1 disc,Tray	Plug & Play	External	
PD2000e	CD-ROM	4X	SCSI	
	1 disc		External	
RF4102	CD-R	2X		2Mb
	1 disc,Tray			

Plextor

Model	Drive Type / Discs, Loader	Speed / Plug&Play	Interface / Internal/External	Buffer
12/20PleX	CD-ROM	12X/20X	SCSI	
	1 disc,Caddyless		External	
12PleX	CD-ROM	12X	SCSI	512kb
	1 disc,Caddy or Tray		Internal & External	
6PleX	CD-ROM	6X	SCSI-2	256kb
	1 disc,Caddy		Internal & External	
8PleX	CD-ROM	8X	SCSI	256kb
	1 disc,Caddy	Plug & Play	Internal	
DM-3020	CD-ROM		SCSI-1	64kb
	1 disc,Caddy		Internal	
DM-3021	CD-ROM		SCSI-1	64kb
	1 disc,Caddy		Internal	
DM-5020	CD-ROM		SCSI-1	64kb
	1 disc,Caddy		External	
DM-5021	CD-ROM		SCSI-1	64kb
	1 disc,Caddy		External	
DM-5024	CD-ROM		SCSI-2	64kb
	1 disc,Caddy		External	
PX-12CS	CD-ROM	12X	Fast SCSI	512kb
	1 disc,Caddy	Plug & Play	Internal & External	
PX-12CSe	CD-ROM	12X	SCSI-2	512kb
	1 disc,Caddy	Plug & Play	External	
PX-12CSe/ISA	CD-ROM	12X	SCSI-2	512kb
	1 disc,Caddy	Plug & Play	External	
PX-12CSi	CD-ROM	12X	SCSI	512kb
	1 disc,Caddy		Internal	
PX-12CSi/ISA	CD-ROM	12X	SCSI-2	512kb
	1 disc,Caddy	Plug & Play	Internal	
PX-12TSe	CD-ROM	12X	SCSI-2	512kb
	1 disc,Tray		External	
PX-12TSe/MAC	CD-ROM	12X	SCSI-2	512kb
	1 disc,Tray		External	
PX-12TSe/PCI	CD-ROM	12X	SCSI-2; 32-bit EISA	512kb
	1 disc,Tray		External	
PX-12TSi	CD-ROM	12X	Fast SCSI	512kb
	1 disc,Tray	Plug & Play	Internal	
PX-12TSi/ISA	CD-ROM	12X	SCSI-2; 16-bit ISA	512kb
	1 disc,Tray		Internal	
PX-12TSi/PCI	CD-ROM	12X	SCSI-2; 32-bit EISA	512kb

Manufacturer Model	Drive Type Discs, Loader	Speed Plug&Play	Interface Internal/External	Buffer
	1 disc,Tray		Internal	
PX-20TSe	CD-ROM	20X	Fast SCSI	512kb
	1 disc,Tray	Plug & Play	External	
PX-20TSi	CD-ROM	20X	Fast SCSI	512kb
	1 disc,Tray	Plug & Play	Internal	
PX-20TSi/e	CD-ROM	12/20X	Fast SCSI	512kb
	1 disc,Tray	Plug & Play	Internal & External	
PX-32CSe	CD-ROM	32X	SCSI-2	512kb
	1 disc,Caddy	Plug & Play	External	
PX-32CSe/ISA	CD-ROM	32X	SCSI-2; 16-bit ISA	512kb
	1 disc,Caddy	Plug & Play	External	
PX-32CSe/MAC	CD-ROM	32X	SCSI-2	512kb
	1 disc,Caddy	Plug & Play	External	
PX-32CSe/PCI	CD-ROM	32X	SCSI-2; 32-bit PCI	512kb
	1 disc,Caddy	Plug & Play	External	
PX-32CSe/PCMCIA	CD-ROM	32X	Ultra SCSI; PCMCIA	512kb
	1 disc,Tray	Plug & Play	External	
PX-32CSi/e	CD-ROM	32X/14X	Ultra SCSI CAV	512kb
	1 disc,Caddy	Plug & Play	Internal & External	
PX-32CSi/ISA	CD-ROM	32X	Ultra SCSI-2; 16-bit ISA	512kb
	1 disc,Caddy	Plug & Play	Internal	
PX-32CSi/PCI	CD-ROM	32X	Ultra SCSI-2; 32-bit PCI	512kb
	1 disc,Caddy	Plug & Play	Internal	
PX-32TSe	CD-ROM	32X	SCSI-2	512kb
	1 disc,Tray	Plug & Play	External	
PX-32TSe/ISA	CD-ROM	32X	SCSI-2, 16-bit ISA	512kb
	1 disc,Tray	Plug & Play	External	
PX-32TSe/MAC	CD-ROM	32X	SCSI-2	512kb
	1 disc,Tray	Plug & Play	External	
PX-32TSi/e	CD-ROM	32X	Ultra SCSI CAV	512kb
	1 disc,Tray	Plug & Play	Internal & External	
PX-32TSi/ISA	CD-ROM	32X	Ultra-SCSI CAV; 16-bit ISA	512kb
	1 disc,Tray	Plug & Play	Internal	
PX-32TSi/PCI	CD-ROM	32X	Ultra-SCSI; 32-bit PCI	512kb
	1 disc,Tray	Plug & Play	Internal	
PX-43CE	CD-ROM	4.5X	SCSI-2	256kb
	1 disc,Caddy	Plug & Play	Internal	
PX-43CH	CD-ROM	4X	SCSI BUS	1Mb
	1 disc,Caddy		Internal	
PX-43CS	CD-ROM	4.5X	SCSI	256kb
	1 disc,Caddy		Internal	
PX-45CE	CD-ROM	4.5X	SCSI-2	256kb
	1 disc,Caddy	Plug & Play	External	
PX-45CH	CD-ROM	4X	SCSI-BUS	1Mb
	1 disc,Caddy		External	
PX-45CS	CD-ROM	4.5X	SCSI-2	256kb
	1 disc,Caddy		External	
PX-63CS	CD-ROM	6X	SCSI Fast Wide	256kb
	1 disc,Caddy		Internal	
PX-65CS	CD-ROM	6X	SCSI Fast Wide	256kb
	1 disc,Caddy		External	
PX-83CS	CD-ROM	8X	Fast SCSI-2	256kb
	1 disc,Caddy	Plug & Play	Internal	
PX-85CS	CD-ROM	8X	Fast SCSI	256kb
	1 disc,Caddy	Plug & Play	External	
PX-R24CS	CD-R	2Xw/4Xr	SCSI-2	512kb
	1 disc,Caddy	Plug & Play	Internal	

Manufacturer / Model	Drive Type / Discs, Loader	Speed / Plug&Play	Interface / Internal/External	Buffer
PX-R24CSe	CD-R 1 disc,Tray	2Xw/4Xr	SCSI-2 External	256kb
PX-R412CE	CD-RW 1 disc,Caddy/Tray	20X	SCSI-2 External	2Mb
PX-R412CE/ISA	CD-RW 1 disc,Caddy	12X	Fast SCSI-2; 16 bit ISA External	2Mb
PX-R412CE/PCI	CD-RW 1 disc,Caddy	12X	Fast SCSI-2; 32 bit PCI External	2Mb
PX-R412CE/PCMCIA	CD-RW 1 disc,Caddy	12X	Fast SCSI-2; PCMCIA External	2Mb
PX-R412CE/SW	CD-RW 1 disc,Caddy	12X	Fast SCSI-2 External	2Mb
PX-R412CI/E	CD-RW 1 disc,Caddy	4Xw/12Xr Plug & Play	Fast SCSI Internal	2Mb
PX-R412CI/ISA	CD-RW 1 disc,Caddy	4Xw/12Xr	SCSI-2; 16 bit Internal	2Mb
PX-R412CI/PSI	CD-RW 1 disc,Caddy	4Xw/12Xr	SCSI-2; 32 bit PCI Internal	2Mb
PX-R412CI/SW	CD-RW 1 disc,Caddy	4Xw/12Xr	SCSI-2 Internal	2Mb

Procom Technology

Model	Drive Type / Discs, Loader	Speed	Interface	Buffer
8X	CD-ROM 1 disc,Caddy	8X	SCSI Internal	
MCD 650 MAC	CD-ROM 1 disc,Procom Cad	1X	SCSI External	
MCD 650 XT/AT/MCA	CD-ROM 1 disc,Procom Cad	1X	SCSI External	None
MCD-4X	CD-ROM 1 disc	4X	SCSI-2 Internal	256kb

Ricoh Corporation

Model	Drive Type / Discs, Loader	Speed	Interface	Buffer
MP-6200A	CD-RW 1 disc	2Xw/6Xr	IDE/ATAPI Internal	1Mb
MP-6200S	CD-RW 1 disc	2Xw/6Xr	SCSI-2 External	1Mb
RO-1420C	CD-R 1 disc,Caddy	2Xw/4Xr	SCSI Internal	1Mb

Samsung Electronics America

Model	Drive Type / Discs, Loader	Speed	Interface	Buffer
SCR-1638	CD-ROM 1 disc,Tray	16X	EIDE/ATAPI Internal	128kb
SCR-2030	CD-ROM 1 disc,Tray	24X	EIDE/ATAPI Internal	512kb
SCR-2430	CD-ROM 1 disc,Tray	24X	EIDE/ATAPI Internal	512kb
SCR-2431	CD-ROM 1 disc,Tray	24X	EIDE/ATAPI Internal	512kb
SCR-2437	CD-ROM 1 disc,Tray	10X/24X	EIDE/ATAPI Internal	256kb
SCR-3230	CD-ROM 1 disc,Tray	12.8X/32X	EIDE/ATAPI Internal	512kb
SCR-630	CD-ROM 1 disc	4X	ATAPI/EIDE Internal	256kb
SCR-830	CD-ROM			

Manufacturer Model	Drive Type Discs, Loader	Speed Plug&Play	Interface Internal/External	Buffer

Sanyo Electric

Model	Drive Type / Discs, Loader	Speed	Interface / Internal/External	Buffer
C3G	CD-ROM 1 disc,Caddy	4X	EIDE	256kb
CRD-254SH	CD-ROM 1 disc,Drawer	4X	SCSI	256kb
ROM 3000 U	CD-ROM 1 disc,Sony Caddy	1X	Non-SCSI Parallel External	32kb
ROM 3001 U	CD-ROM 1 disc,Sony Caddy	1X	SCSI External	32kb

Severn Companies

Model	Drive Type / Discs, Loader	Speed	Interface / Internal/External	Buffer
SC-922 Gemini	CD-ROM 1 disc,Sony Caddy	1X	Non-SCSI Parallel External	None

Smart and Friendly

Model	Drive Type / Discs, Loader	Speed	Interface / Internal/External	Buffer
CD Rocket 8X	CD-R 1 disc,Caddy	8Xw/20Xr	SCSI-2 Internal	2Mb
CD SpeedWriter Deluxe	CD-R 1 disc,Tray	4Xw/12Xr	SCSI-2 External	1Mb
CD SpeedWriter	CD-R 1 disc,Tray	4Xw/12Xr	SCSI-2 Internal	1Mb
CD-R 1002	CD-R 1 disc,Caddy	2Xw/2Xr	SCSI-2 External	1Mb
CD-R 1004	CD-ROM 1 disc,Caddy		SCSI-2	512kb
CD-R 4000	CD-ROM 1 disc,Caddy	4X	SCSI-2 External	512kb
CD-R 4006	CD-R 1 disc,Caddy	4Xw/6Xr	SCSI-2 Internal & External	2Mb
CD-RW 226 Plus	CD-ROM 1 disc,Tray	2X	SCSI-2 External	1Mb
CD-RW 226 Plus	CD-ROM 1 disc,Tray	2X	SCSI-2 Internal	1Mb
CD-RW 426 Deluxe	CD-R 1 disc,Tray	4Xw/6Xr	EIDE Internal	2Mb
CD-RW 426 Deluxe	CD-R 1 disc,Tray	4Xw/6Xr	SCSI-2 External	2Mb
CDJ 4008	CD-ROM-changer 1 disc,Slot	8X	SCSI-2 External	256kb
CDJ7004	CD-ROM 7 disc,Tray	4X		
CR-R 2006 Plus	CD-R 1 disc	2Xw/6Xr	SCSI-2	1Mb

Sony Corporation

Model	Drive Type / Discs, Loader	Speed	Interface / Internal/External	Buffer
CDU-111	CD-ROM 1 disc,Tray	6X	IDE/ATAPI Internal	128kb
CDU-311-10	CD-ROM 1 disc,Tray	8X	IDE/ATAPI Internal	256kb
CDU-311-20	CD-ROM 1 disc,Tray	8X	IDE/ATAPI Internal	128kb
CDU-31A-02	CD-ROM 1 disc,Tray	2X	Sony BUS Internal	64kb
CDU-31A-03	CD-ROM 1 disc,Tray	2X	Sony BUS Internal	64kb
CDU-33A	CD-ROM 1 disc,Tray	2X	Sony BUS Internal	64kb

Manufacturer Model	Drive Type Discs, Loader	Speed Plug&Play	Interface Internal/External	Buffer
CDU-415	CD-ROM 1 disc,Tray	8X/12X	SCSI Internal	256kb
CDU-510	CD-ROM 1 disc,Caddy		Sony BUS Internal	
CDU-511	CD-ROM 1 disc,Tray	13.6X	IDE/ATAPI Internal	128kb
CDU-531	CD-ROM 1 disc,Caddy		Sony BUS Internal	8kb
CDU-535	CD-ROM 1 disc,Caddy		Sony BUS External	8kb
CDU-541	CD-ROM 1 disc,Caddy	2X	SCSI BUS Internal	64kb
CDU-55E	CD-ROM 1 disc,Tray	2X Plug & Play	ATAPI External	256kb
CDU-55S	CD-ROM 1 disc,Tray	2.4X	SCSI-2 Internal	256kb
CDU-561	CD-ROM 1 disc,Caddy	2X	SCSI-2 BUS Internal	256kb
CDU-571	CD-ROM 1 disc,Slot	13.6X	IDE/ATAPI Internal	128kb
CDU-6100	CD-ROM 1 disc,Sony Caddy	1X	Non-SCSI Parallel External	8kb
CDU-6101	CD-ROM 1 disc,Sony Caddy	1X	Non-SCSI Parallel External	8kb
CDU-611	CD-ROM 1 disc,Tray	24X	IDE/ATAPI Internal	256kb
CDU-6110	CD-ROM 1 disc,Sony Caddy	1X	SCSI External	8kb
CDU-6111	CD-ROM 1 disc,Sony Caddy	1X	SCSI External	8kb
CDU-625	CD-ROM 1 disc,Caddy	24X	SCSI Internal	
CDU-6511	CD-ROM 1 disc,Tray	2X	SCSI-2 External	256kb
CDU-6811	CD-ROM 1 disc,Caddy	2X	SCSI-2 External	256kb
CDU-7101	CD-ROM 1 disc,Sony Caddy	1X	Non-SCSI Parallel External	8kb
CDU-711	CD-ROM 1 disc,Caddy	32X	EIDE Internal	256kb
CDU-7201	CD-ROM 1 disc,Caddy		Sony BUS External	8kb
CDU-7205/N	CD-ROM 1 disc,Caddy		Sony BUS External	8kb
CDU-7211	CD-ROM 1 disc,Caddy		SCSI BUS External	64kb
CDU-7305	CD-ROM 1 disc		Sony BUS External	64kb
CDU-7305-3A	CD-ROM 1 disc		Sony BUS External	64kb
CDU-7511	CD-ROM 1 disc,Tray	2X	SCSI-2 External	256kb
CDU-75E	CD-ROM 1 disc,Tray	4X	IDE	256kb
CDU-76E	CD-ROM 1 disc,Tray	4X Plug & Play	ATAPI (Mode 2) Internal	256kb
CDU-76S	CD-ROM 1 disc,Tray	4X	SCSI-2 Internal	256kb

Manufacturer Model	Drive Type Discs, Loader	Speed Plug&Play	Interface Internal/External	Buffer
CDU-77E	CD-ROM 1 disc,Tray	4X	IDE/ATAPI Internal	128kb
CDU-7811	CD-ROM 1 disc,Caddy	2X	SCSI-2 BUS External	256kb
CDU-8003	CD-ROM 1 disc,Caddy	2X	SCSI Internal & External	256kb
CDU-920S	CD-R 1 disc,Caddy	2X	SCSI-2 Internal	1Mb
CDU-9211S	CD-R 1 disc,Caddy	2X	SCSI-2 External	1Mb
CDU-921S	CD-R 1 disc,Caddy	2X	SCSI-2 External	1Mb
CDU-924S	CD-R 1 disc,Caddy	2Xw/4Xr	Single-ended SCSI-2	1Mb
CDU-926S	CD-R 1 disc,Tray	2Xw/6Xr	Single-ended SCSI-2	512kb
CDU-928E	CD-ROM 1 disc,Tray	8X	IDE/ATAPI Internal	512kb
CDU-928E/C	CD-R 1 disc,Caddy	2Xw/8Xr	IDE/ATAPI Internal	512kb
CDU-928E/H	CD-R 1 disc,Caddy	2Xw/8Xr	IDE/ATAPI Internal	512kb
CDU-948S/C	CD-R 1 disc,Caddy	4Xw/8Xr	SCSI-2 Internal	
CDU-948S/CH	CD-R 1 disc,Caddy	4Xw/8Xr	SCSI-2 Internal	
CDW-900E	CD-R disc,1	2X	SCSI Internal	
CSD-760E	CD-ROM 1 disc,Tray	4X	ATAPI (Mode 2) Internal	256kb
CSD-760S	CD-ROM 1 disc,Tray	4X	SCSI-II Internal	256kb
CSD-7611M	CD-ROM 1 disc,Tray	4X	SCSI-II External	256kb
CSD-76SB	CD-ROM 1 disc,Tray	4X	SCSI-II Internal	256kb
CSD-880E	CD-ROM 1 disc,Tray	8X	IDE/ATAPI Internal	256kb
CSD-88EN	CD-ROM 1 disc,Tray	8X	IDE/ATAPI Internal	256kb
CSP-920S	CD-ROM 1 disc,Tray	2X	SCSI-2 Internal	1Mb
CSP-9211S	CD-R 1 disc,Caddy	2X	SCSI-2 External	1Mb
CSP-940S	CD-R 3 disc,Caddy	2X	SCSI-2 Internal	1Mb
CSP-9411S	CD-R 5 disc,Caddy	2X	SCSI Internal	1Mb
CSP-960H	CD-R 1 disc	2Xw/6Xr	Single-ended SCSI-2 Internal	512kb
CSP-960S	CD-R 1 disc	2Xw/6Xr	Single-ended SCSI-2 Internal	512kb
CSP-9611H	CD-R	2Xw/6Xr	Single-ended SCSI-2	512kb

Manufacturer / Model	Drive Type	Discs, Loader	Speed	Plug&Play	Interface	Internal/External	Buffer
		1 disc				External	
DDU100E/C	DVD-ROM	1 disc	1XDVD/8XCD	Plug & Play	IDE/ATAPI	Internal	512kb
DDU100E/K	DVD-ROM	1 disc	1XDVD/8XCD	Plug & Play	IDE/ATAPI	Internal	512kb
PRD-250WN	CD-ROM	1 disc,Slot	4X	Plug & Play	SCSI-2 to PC Card	External	128kb
PRD-650WN	CD-ROM	1 disc,Tray	6X	Plug & Play	SCSI-2	Portable	128kb

Sun Microsystems

Model	Drive Type	Discs, Loader	Speed	Plug&Play	Interface	Internal/External	Buffer
Sun CD	CD-ROM	1 disc,Sony Caddy	1X		SCSI	External	64kb

Teac America, Inc.

Model	Drive Type	Discs, Loader	Speed	Plug&Play	Interface	Internal/External	Buffer
CD-211PE	CD-ROM	1 disc,Clam Shell Tray	10X		PCMCIA Type II	Portable	128kb
CD-220E	CD-ROM	1 disc	20X		IDE/ATAPI	Internal	
CD-224E	CD-ROM	1 disc	24X		IDE/ATAPI	Internal	
CD-28L	CD-ROM	1 disc	8X	Plug & Play	IDE/ATAPI	Internal	128kb
CD-316E	CD-ROM	1 disc	16X		IDE/ATAPI	Internal	128kb
CD-36E	CD-ROM	1 disc,Drawer	8X		ATAPI	Internal	128kb
CD-38E	CD-ROM	1 disc,Drawer	8X	Plug & Play	IDE/ATAPI	Internal	128kb
CD-50	CD-ROM	1 disc,Caddy	2X		SCSI-2	Internal	64kb
CD-512E	CD-ROM	1 disc,Tray	12X	Plug & Play	IDE/ATAPI	Internal	128kb
CD-512S	CD-ROM	1 disc,Tray	12X		SCSI-2	Internal	512kb
CD-516E	CD-ROM	1 disc,Tray	16X	Plug & Play	IDE/ATAPI	Internal	128kb
CD-516S	CD-ROM	1 disc,Tray	16X	Plug & Play	SCSI-2	Internal	512kb
CD-524E	CD-ROM	1 disc,Tray	24X		ATAPI	Internal	128kb
CD-532E	CD-ROM	1 disc,Tray	32X		IDE/ATAPI	Internal	128kb
CD-532S	CD-ROM	1 disc,Tray	32X		Ultra SCSI-3	Internal	512kb
CD-55A	CD-ROM	1 disc,Tray	4X		Teac AT	Internal	64kb
CD-56E	CD-ROM	1 disc,Tray	6X	Plug & Play	IDE/ATAPI	Internal	256kb
CD-56S	CD-ROM	1 disc,Tray	6X	Plug & Play	SCSI-2	Internal	256kb
CD-58E	CD-ROM	1 disc,Tray	8X	Plug & Play	IDE/ATAPI	Internal	128kb
CD-58S	CD-ROM		8X		SCSI-2		128kb

Manufacturer	Drive Type	Speed	Interface	Buffer
Model	Discs, Loader	Plug&Play	Internal/External	
	1 disc,Tray		Internal	
CD-C68E	CD-ROM-changer	8X	IDE/ATAPI	128kb
	6 disc,Direct Loading	Plug & Play	Internal	
CD-R50S	CD-R	4Xw/4Xr	Fast SCSI-2	1Mb
	1 disc,Tray		Internal	
CD-R55S	CD-R	4Xw/12Xr	Fast SCSI-2	1Mb
	1 disc,Tray		Internal	
MC-1200	CD-ROM	8X	SCSI-2	256kb
	200 disc,4 Magazine		External	
MC-1600	CD-ROM	8X	SCSI-2	256kb
	600 disc,12 Magazines		External	

Texel

Model	Drive Type	Speed	Interface	Buffer
DM-3024	CD-ROM	2X	SCSI	64kb
	1 disc,Caddy		Internal	
DM-3028	CD-ROM	2X	SCSI	64kb
	1 disc,Caddy		Internal	
DM-5000 LM	CD-ROM	1X	SCSI	64kb
	1 disc,Sony Caddy		External	
DM-5000 LS	CD-ROM	1X	SCSI	64kb
	1 disc,Sony Caddy		External	
DM-5028	CD-ROM	2X	SCSI-2	64kb
	1 disc,Caddy		External	

Todd Enterprises

Model	Drive Type	Speed	Interface	Buffer
TCDR-3000	CD-ROM	1X	Non-SCSI Parallel	32kb
	1 disc,Sony Caddy		External	
TCDR-3004	CD-ROM	1X	Non-SCSI Parallel	32kb
	1 disc,Sony Caddy		External	
TCDR-6000	CD-ROM	1X	Non-SCSI Parallel	32kb
	1 disc,Sony Caddy		External	

Toshiba America

Model	Drive Type	Speed	Interface	Buffer
CD400T	CD-ROM	4X	IDE/ATAPI	256kb
	1 disc,Tray		Internal	
TXM-3401E	CD-ROM	2X	SCSI-2	256kb
	1 disc		External	
TXM-3401P	CD-ROM	2X	SCSI-2	256kb
	1 disc		Portable	
TXM-3501A4	CD-ROM	4X	SCSI	
	disc,Caddy			
TXM-3501E	CD-ROM	4X	SCSI-2	256kb
	1 disc,Caddy		External	
TXM-3601D	CD-ROM	4X	SCSI-2	
	1 disc		External	
TXM-3701D	CD-ROM	6.7X	SCSI-2	256kb
	1 disc,Tray		External	
TXM-3801F1	CD-ROM	15X	SCSI-2	256kb
	1 disc,Tray		External	
TXM-4101L	CD-ROM	2X		
	1 disc,Tray		External	
TXM-5201D	CD-ROM	3.4X	SCSI-2	256kb
	1 disc,Tray		External	
TXM-5701F1	CD-ROM	12X	SCSI-2	256kb
	1 disc	Plug & Play	External	
TXM-6201F1	CD-ROM	24X	SCSI	
	1 disc		External	

Manufacturer Model	Drive Type Discs, Loader	Speed Plug&Play	Interface Internal/External	Buffer
XM-1402B	CD-ROM	6X		
XM-1502B	CD-ROM 1 disc,Tray	11X	ATAPI Internal	128kb
XM-1602B	CD-ROM 1 disc,Tray	20X	ATAPI Internal	128kb
XM-1702B	CD-ROM 1 disc,Tray	24X	ATAPI Internal	128kb
XM-2402B	CD-ROM 1 disc,Tray	11X	ATAPI Internal	128kb
XM-3201A1-MAC	CD-ROM 1 disc,Sony Caddy	2X Plug & Play	SCSI External	64kb
XM-3201A1-PCF	CD-ROM 1 disc,Sony Caddy	2X	SCSI External	64kb
XM-3201A1-PS2	CD-ROM 1 disc,Sony Caddy	2X	SCSI External	64kb
XM-3401B	CD-ROM 1 disc,Caddy	2X	SCSI-2 Internal	256kb
XM-3501B	CD-ROM 1 disc,Caddy	4X	SCSI-2 Internal	256kb
XM-3501E	CD-ROM 1 disc	4X	SCSI External	
XM-3601D1	CD-ROM 1 disc,Tray	4.5X	SCSI External	256kb
XM-3701	CD-ROM 1 disc,Tray	6.7X	SCSI-2 External	256kb
XM-3701B	CD-ROM 1 disc,Tray	6.7X	SCSI-2 Internal	256kb
XM-3801B	CD-ROM 1 disc,Tray	15X Plug & Play	SCSI-2 Internal	256kb
XM-4100A	CD-ROM 1 disc,Tray	2X	SCSI-2 Portable	64kb
XM-4101B	CD-ROM 1 disc,Tray	2X	SCSI-2 Internal	64kb
XM-5071	CD-ROM 1 disc	12X	SCSI Internal	
XM-5100-A-MAC	CD-ROM 1 disc,Tray	2X	SCSI External	64kb
XM-5100-A-PCF	CD-ROM 1 disc,Tray	2X	SCSI External	64kb
XM-5100-A-PS2	CD-ROM 1 disc,Tray	2X	SCSI External	64kb
XM-5201B	CD-ROM 1 disc,Tray	3.4X	SCSI-2 Internal	256kb
XM-5302B	CD-ROM 1 disc,Tray	4X	ATAPI/IDE Internal	256kb
XM-5401	CD-ROM 1 disc,Tray	4X	SCSI-2 Internal & External	256kb
XM-5402	CD-ROM 1 disc,Tray	4X	IDE/ATAPI Internal	256kb
XM-5522	CD-ROM 1 disc,Tray	6X	ATAPI Internal	128kb
XM-5602	CD-ROM 1 disc,Tray	8X	ATAPI Internal	256kb
XM-5701B	CD-ROM 1 disc,Tray	12X Plug & Play	SCSI-2 Internal	256kb
XM-5702B	CD-ROM 1 disc,Tray	12X Plug & Play	ATAPI Internal	256kb

Manufacturer Model	Drive Type Discs, Loader	Speed Plug&Play	Interface Internal/External	Buffer
XM-6002B	CD-ROM	16X	EIDE/ATAPI	256kb
	1 disc		Internal	
XM-6102B	CD-ROM	12-24X	ATAPI	256kb
	1 disc		Internal	
XM-6201B	CD-ROM	32X	SCSI-2	256kb
	1 disc,Tray		Internal	
XM-6201F1	CD-ROM	32X	SCSI-2	256kb
	1 disc,Tray		External	
XM-6202B	CD-ROM	32X	EIDE/ATAPI	256kb
	1 disc,Tray		Internal	

Wearnes Peripherals Intl

Model	Drive Type / Discs, Loader	Speed	Interface / Internal/External	Buffer
CDD-1020	CD-ROM	10X	IDE	256kb
	1 disc,Tary		Internal & External	
CDD-120	CD-ROM	2X	IDE/ATAPI	256kb
	1 disc,Tray			
CDD-240	CD-ROM	4X	IDE	256kb
	1 disc		Internal	
CDD-320	CD-ROM	6X	IDE	256kb
	1 disc,Tray		Internal	
Multi-Taskin' CDD-1620	CD-ROM	16X	EIDE/ATAPI	128kb
	1 disc,Tray			

Workstatin Solutions

Model	Drive Type / Discs, Loader	Speed	Interface / Internal/External	Buffer
OFS-CD	CD-ROM	1X	SCSI	None
	1 disc,Sony Caddy		External	

Yamaha Systems Tech, Inc.

Model	Drive Type / Discs, Loader	Speed	Interface / Internal/External	Buffer
CDR100	CD-ROM	2X	SCSI-2	512kb
	1 disc,Caddy		Internal & External	
CDR100-Recordable	CD-R	4X	SCSI-2	512kb
	1 disc,Caddy		Internal & External	
CDR102	CD-ROM	2X	SCSI-2	512kb
	1 disc,Caddy		Internal & External	
CDR400	CD-R	4Xw/6Xr	SCSI-2	2Mb
	1 disc,Tray		Internal & External	
CDR400AT	CD-R	1Xw/6Xr	SCSI 2	2Mb
	1 disc,Tray		Internal & External	
CDR400TI	CD-R	4Xw/6Xr	SCSI-2	256kb
	1 disc,Tray		Internal	
CDR401	CD-R	2Xw/4Xr	EIDE/ATAPI	2Mb
	1 disc,Tray		Internal	
CRW4001	CD-RW	2Xw/6Xr	EIDE/ATAPI	2Mb
	1 disc,Tray		Internal	
CRW4260	CD-ROM	6X	SCSI-2	2Mb
	1 disc,Tray		External	
CRW4260TI	CD-ROM	6X	SCSI-2	256kb
	1 disc,Tray		Internal	

Young Minds

Model	Drive Type / Discs, Loader	Speed	Interface / Internal/External	Buffer
YMI Dual Drv for Sun	CD-ROM	1X	SCSI	64kb
	2 disc,Sony Caddy		External	
YMI Single Drv	CD-ROM	1X	SCSI	64kb
	1 disc,Sony Caddy		External	

Chapter 9

Floppy Drive Specifications

Many thanks to Bottom Line Industries, 9556 Cozycroft Ave, Chatsworth, California, 91311, (818) 700-1922, (800) 344-6044 for providing Sequoia with additional floppy drive information included in this chapter. If you need to have a floppy or hard drive rebuilt or would like to purchase a rebuilt floppy or hard drive, Bottom Line Industries is an excellent source!

Floppy Drive Manufacturers

The following table is a general summary of companies that have manufactured and/or are still manufacturing floppy drives. If you have information concerning the status of any of these companies, such as "XYZ Company went bankrupt in August, 1990" or "XYZ Company was bought by Q Company", please let us know so we can keep this section current. If a phone number is listed in the Status column, the company is in business.

Manufacturer	Status
Alps	800-449-2577
Aurora Tech	781-290-4800
Bachelor	Unknown
BASF	See Emtech Corp
Burroughs	Unknown
Calcomp	741-821-2000
Canon	714-438-3000
C.D.C	Unknown
Century Data	919-821-5696; Not a manufacturer
Chinon	See Tech Media?
Citizen	310-453-0614
Disc Tec	407-671-5500
Emtech Corp	800-343-4600
Epson	310-782-0770
Fuji	510-438-9700; Do not manufacture floppy or hard drives anymore.
Fujitsu	408-432-6333; Made in Japan
Hewlett Packard	301-670-4300
Hi-Tech (China)	886-2-773-3555
Hitachi	914-332-5800
IBM	914-765-1900
Iomega	801-778-1000
JVC	714-816-6500 Never manufactured floppy drives
MFE	Unknown
MPI	708-460-0555
Maple Tech	Unknown
Memorex	800-636-8352
Micropolis	Out of Business
Mitac	510-656-3333
Mitsubishi	408-730-5900 Corporate

Manufacturer	Status
Mitsumi	972-550-7300
NEC	516-753-7000
Newtronic	Unknown
Okidata	609-235-2600
Olivetti	Out of Business
Pacific Rim	Out of Business
Panasonic	800-854-4536
Persci	Unknown
Pertec	Unknown
Phillips	719-593-7900
Qume	See WYSE Tech
Remex	Unknown
Samsung	201-229-7000
Sanyo (Japan)	81-64-432-949
Seiko	408-922-5806; Never manufactured floppy drives
Shugart	520-294-0898
Siemans	Out of Business
Sierra, Inc	414-638-1851
Sony	201-930-1000
Tandon	Filed Chapter 11 bankruptcy 9-95; See Sierra, Inc for support
Teac Corp.	213-726-0303
Tec	Unknown
Tech Media	800-379-0077
Tecmate	Unknown
Texas Peripherals	Unknown
Toshiba	714-457-0777
Victor	800-628-2420
Weltec	302-737-1260
World Storage	Unknown
WYSE Tech	408-473-1200
Y-E Data	847-753-6650

GENERAL FLOPPY DRIVE SPECS

Formatted Capacity	Sides	Tracks	Sectors	ID Byte	Media Type*	Media Agent
5-1/4 inch diameter						
160 kb**	1	40	8	FE	SSDD	Ferrite
180 kb**	1	40	9	FC	SSDD	Ferrite
320 kb**	2	40	8	FF	DSDD	Ferrite
360 kb	2	40	9	FD	DSDD	Ferrite
1.2 Mb	2	80			DSQD	Ferrite
1.2 Mb	2	80	15	F9	DSHD	Cobalt
3-1/2 inch diameter						
720 kb	2	80	9	F9	DSDD	Cobalt
1.44 Mb	2	80	18	F0	DSHD	Cobalt
2.8 Mb	2	80	36	F0	DSEHD	Barium

* SS = Single Sided, DS=Double Sided
  DD = Double Density
  HD = High Density
  QD = Quad Density (now obsolete)
  EHD or ED = Extra High Density

** Obsolete drives

Maximum Entries in the Root Directory:
5-1/4 DD and 3.5 DD = 112 Entries
5-1/4 HD and 3.5 HD = 224 Entries
3.5 EHD = 240 Entries

All floppy drives currently produced rotate at 300 RPM, except for the 1.2Mb, 5-1/4 HD drives, which rotate at 360 RPM.

All floppy drives are formatted at 512 Bytes Per Sector.

Floppy disks have 2 FATs, 12 Bit Type

Sequoia needs your help! If you have specifications on new or obsolete floppy drives, please send them to us for future editions of this book.

FLOPPY DRIVE SPECS BY MODEL

Manufacturer	Model Number	Width (Inch)	Height (Inch)	Format Capacity	Media Density
Alps	413(PS2)	3.50	Half	720kb	DSDD
	713(PS2)	3.50	Half	1.44Mb	DSHD
	723	3.50	Third	1.44Mb	DSHD
	723(PS2)	3.50	Half	1.44Mb	DSHD
	2124	5.25	Half	180kb	SSDD
	2124A	5.25	Full	360kb	DSDD
	2624-BKI	5.25	Half	360kb	DSDD
	DF328N	3.50	Qtr	2.88Mb	DSHD
	DFC 222 B02A,01A	5.25	Half	360kb	DSDD
	DFC 222A05A	5.25	Half	360kb	DSDD
	DFC 642 B01B	5.25	Half	1.2Mb	DSDD
Aurora Tech	FD350(SCSI)	3.50	Half		
	FD525(SCSI)	5.25	Half		
Bachelor	FD-104	5.25	Half	360kb	DSDD
BASF	6106	5.25	Full	180kb	SSDD
	6128	5.25	Half	360kb	DSDD
	6138	5.25	Half	720kb	DSDD
Burroughs	B9489-1	8.00	Full	1.6Mb	DSDD
Calcomp	142	8.00	Full	800kb	SSDD
	143	8.00	Full	1.6Mb	DSDD
Canon	221	5.25	Half	720kb	DSDD
	530	5.25	Half	720kb	DSDD
	531	5.25	Half	360kb	DSDD
	3361	3.50	Qtr	1.44Mb	DSHD
	5201	5.25	Half	360kb	DSDD
	5501	5.25	Third	1.2Mb	DSDD
	5511	5.25/3.5	Half	1.2/1.44Mb	DUAL
C.D.C	9404	8.00	Full	800kb	SSDD
	9406-3	8.00	Full	800kb	SSDD
	9406-4	8.00	Full	1.6Mb	DSDD
	9408	5.25	Full	180kb	SSDD
	9409	5.25	Full	360kb	DSDD
	9409T	5.25	Full	720kb	DSQD
	9428	5.25	Half	360kb	DSDD
	9428-01	5.25	Half	180kb	SSDD
	9428-02	5.25	Half	360kb	DSDD
	9429	5.25	Half	720kb	DSQD
	9429-01	5.25	Half	360kb	SSQD
	BR8B1A	5.25	Full	360kb	DSDD
Century Data	140	8.00	Full	800kb	SSDD
Chinon	506-L	5.25	Half	1.2Mb	DSDD
	C354	3.50	Half	720kb	DSDD
	FP357	3.50	Half	1.4Mb	DSHD
	FR506	5.25	Full	1.2Mb	DSHD
	FX354	3.50	1.0	720kb	DSDD
	FZ357	3.50	1.0	1.4Mb	DSHD
	FZ358	3.50	1.0	1.4mb	DSHD

FLOPPY DRIVE SPECS BY MODEL

Manufacturer	Model Number	Width (Inch)	Height (Inch)	Format Capacity	Media Density
Chinon (cont.)					
	FZ502	5.25	Half	360kb	DSDD
	FZ506	5.25	Full	1.2Mb	DSHD
	C359	3.50	Half	1.4Mb	DSHD
	F,FZ,C502	5.25	Half	360kb	DSDD
	C506	5.25	Half	1.2Mb	DSHD
Citizen	OSDA-01D	3.50	Third	720kb	DSQD
	OSDA-14A	3.50	Third	1.44Mb	DSHD
	OSDA-39D	3.50	Third	1.44Mb	DSDD
	OSDA-51B	3.50	Third	1.44Mb	DSHD
	OSDA-52B	3.50	Third	1.44Mb	DSHD
	OSDA-53B	3.50	Third	1.44Mb	DSHD
	OSDA-77D	3.50	Third	720kb	DSQD
	OSDA-81F	3.50	Half	1.44Mb	DSHD
	OSDA-90E-U	3.50	Third	720kb	DSQD
	OPDB-22A	3.50	Half	720kb	DSQD
	OSDD-05B	3.50	Third	720kb	DSQD
	OSDD-57	3.50	Third	720kb	DSQD
	OSDD-57B	3.50	Third	720kb	DSQD
	U1DA-14A	3.50	Qtr	1.44Mb	DSHD
	V1DA-10A	3.50	Qtr	1.44Mb	DSHD
	V1DA-27A	3.50	Qtr	1.44Mb	DSHD
	V1DA-31B	3.50	Qtr	1.44Mb	DSHD
	V9DA-55A	3.50	Qtr	1.44Mb	DSHD
	V9DA-55B	3.50	Qtr	1.44Mb	DSHD
	V9DA-71B	3.50	Qtr	1.44Mb	DSHD
Compaq..........	LS-120	3.50		1.44/120Mb	
Digital	PBXRX-AA	3.50	1.0	1.44Mb	DSHD
	PBXRX-AB	3.50	1.0	1.44Mb	DSHD
	FR-PC7XR-AA	3.50		1.44Mb	DSHD
	FR-PC7XR-BA	5.25		1.2Mb	DSHD
Emtech Corp (See BASF)					
Epson	170-SMD	3.50	Half	400kb	SSDD
	180	3.50	Half	720kb	DSDD
	200P-053	3.50	Half	720kb	DSDD
	200P-055	3.50	Half	720kb	DSDD
	200P-073	3.50	Half	720kb	DSDD
	280	3.50	Half	720kb	DSDD
	300	3.50	Third	1.44Mb	DSHD
	340	3.50	Third	1.44Mb	DSHD
	400 W/FRAME	3.50	Third	1.44Mb	DSHD
	400P-4	3.50	Third	1.44Mb	DSHD
	500	5.25	Half	360kb	DSDD
	521	5.25	Half	360kb	DSDD
	521L	5.25	Half	360kb	DSDD
	621L	5.25	Half	360kb	DSDD
	700/800	5.25/3.5		1.2/1.44Mb	DUAL
	1000	3.50	Third	1.44Mb	DSHD
	1000P	3.50	Qtr	1.44Mb	DSHD

FLOPPY DRIVE SPECS BY MODEL

Manufacturer	Model Number	Width (Inch)	Height (Inch)	Format Capacity	Media Density
Epson (cont.)					
	DYO-211	3.50	Half	1.44Mb	DSHD
	DYO-212	3.50	Half	1.44Mb	DSHD
	SD-321	5.25	Third	360kb	DSDD
	SD-520	5.25	Half	360kb	DSDD
	SD-521	5.25	Half	360kb	DSDD
	SD-581	5.25	Half		
	SD-621L	5.25	Half	328kb	DSDD
	SD-680L	5.25	Half	1.02Mb	DSHD
	SMD-1040	3.50	0.7	1.44Mb	DSHD
	SMD-1060	3.50	0.7	2.8Mb	
	DSEHD				
	SMD-1340	3.50	1.0	1.44Mb	DSHD
	SMD-340	3.50	1.0	1.47Mb	DSHD
	SMD-349	3.50	Half	1.4Mb	DSHD
	SMD-380	3.50	1.0	656kb	DSDD
	SMD-389	3.50	Half	720kb	DSDD
	SMD-400P-4	3.50	Third	1.44Mb	DSHD
Fuji/Toshiba	FDD4206AOK	3.50	Half	720kb	DSDD
	FDD421GOK	3.50	1.0	720kb	DSDD
	FDD5452BOK	5.25	Half	360kb	DSDD
	FDD6471LOK	5.25	Half	360kb	DSDD
	FDD6474H1	5.25	Half	360kb	DSDD
Fujitsu	2551 A08	5.25	Half	360kb	DSDD
	2552K	5.25	Half	720kb	DSQD
	2553A,K	5.25	Half	1.2Mb	DSHD
	2553 K03B	5.25	Half	1.2Mb	DSQD
	2554	5.25	Half	720kb	DSQD
	M2537K	3.50	Third	1.44Mb	DSHD
	N02B-0112-B001	3.50	Half	720kb	DSDD
	N02B-0112-B201	3.50	Half	720kb	DSDD
Hewlett Packard					
	OmniB 800	3.50		1.44Mb	DSHD
	J455-3	5.25	Half	360kb	DSDD
	J475-1	5.25	Half	1.2Mb	DSQD
Hi-Tech	548-25	5.25	Half	180kb	SSDD
	548-50	5.25	Half	360kb	DSDD
	548-A	5.25	Half	360kb	DSDD
	596-10	5.25	Full	720kb	DSQD
Hitachi	HFD 305S	5.25	Half	360kb	SSDD
	FD532EIU	5.25	Half	2.4Mb	DSHD
	FDD412A	5.25	Half	1.2Mb	DSDD
IBM	0384-002	5.25	Full	360kb	DSDD
JVC	MDP-100	5.25	Half	720kb	DSQD
	SS01JG	5.25	Half	360kb	DSDD
Maple Tech	MT-502	5.25	Half	360kb	DSQD

FLOPPY DRIVE SPECS BY MODEL

Manufacturer	Model Number	Width (Inch)	Height (Inch)	Format Capacity	Media Density
Matsushita	EME-263TL	3.50	Qtr	1.44Mb	DSHD
	EME-278T	3.50	Qtr	1.44Mb	DSHD
	EME-278TA	3.50	Qtr	1.44Mb	DSHD
Memorex	651	8.00	Full	1.2Mb	DSDD
MFE	M700	8.00	Full	1.6Mb	DSDD
	M750	8.00	Full	1.6Mb	DSDD
Micropolis	1006-4N	5.25	Full	720kb	DSQD
	1015-2	5.25	Full	360kb	SSDD
	1015-4	5.25	Full	720kb	DSQD
	1015-6	5.25	Full	720kb	DSQD
	1016-2	5.25	Full	360kb	DSDD
	1115-4	5.25	Full	720kb	DSQD
	1115-5	5.25	Full	360kb	SSDD
	1115-6	5.25	Full	720kb	DSQD
	1117-6	5.25	Full	720kb	DSQD
Microsolutions	Backpack	3.50		1.44Mb	DSHD
Mitac	MC-490	5.25	Half	360kb	DSDD
Mitsubishi	2894	8.00	Full	1.6Mb	DSDD
	2894-63	8.00	Full	1.6Mb	DSDD
	2896	8.00	Half	1.6Mb	DSDD
	2896-63	8.00	Half	1.6Mb	DSDD
	353AF	3.50	Half	720kb	DSDD
	353B-12	3.50	Half	720kb	DSDD
	353B-82	3.50	Third	720kb	DSDD
	353B,C	3.50	Half	720kb	DSDD
	353C	3.50	Third	720kb	DSDD
	353-12	3.50	Third	720kb	DSDD
	355A,B,C	3.50	1.0	1.4Mb	DSHD
	355B-52	3.50	Half	1.44Mb	DSHD
	355B-82UF	3.50	Half	1.44Mb	DSHD
	355BA-82UF/W51/4	3.50	Half	1.44Mb	DSHD
	355BA-88UF/W51/4	3.50	Half	1.44Mb	DSHD
	355B-88UF	3.50	Half	1.44Mb	DSHD
	355C-12	3.50	Third	1.44Mb	DSHD
	355C-215	3.50	Third	1.44Mb	DSHD
	355C-222	3.50	Third	1.44Mb	DSHD
	355C-258MC	3.50	Third	1.44Mb	DSHD
	355C-352	3.50	Third	1.44Mb	DSHD
	355C-37/W51/4	3.50	Third	1.44Mb	DSHD
	355C-526	3.50	Third	1.44Mb	DSHD
	355C-58UF	3.50	Third	1.44Mb	DSHD
	355C599MA(PS2)	3.50	Half	1.4Mb	DSHD
	355C599MB(PS2)	3.50	Half	1.4Mb	DSHD
	355C599MR4(PS2)	3.50	Half	1.4Mb	DSHD
	355C599MQ4(PS2)	3.50	Half	1.4Mb	DSHD
	355C599MQ41(PS2)	3.50	Half	1.4Mb	DSHD
	355C-82UF/W51/4	3.50	Half	1.44Mb	DSHD
	355C-88UF/W51/4	3.50	Half	1.44Mb	DSHD
	355F258	3.50	Third	1.4Mb	DSHD

FLOPPY DRIVE SPECS BY MODE

Manufacturer	Model Number	Width (Inch)	Height (Inch)	Format Capacity	Media Density
Mitshbishi (cont.)					
	355F3250UG	3.50	Full	1.44Mb	DSHD
	355F3252UG	3.50	Full	1.44Mb	DSHD
	355F3258UG	3.50	Full	1.44Mb	DSHD
	355H-120MG	3.50	Half	1.44Mb	DSHD
	355H-212MG	3.50	Half	1.44Mb	DSHD
	355H-218MG	3.50	Half	1.44Mb	DSHD
	355H-240MG	3.50	Half	1.44Mb	DSHD
	355H-242MG	3.50	Half	1.44Mb	DSHD
	355H-248MG	3.50	Half	1.44Mb	DSHD
	355W99M1(PS2)	3.50	Half	1.4Mb	DSHD
	355W99M2(PS2)	3.50	Half	1.4Mb	DSHD
	355W99M3(PS2)	3.50	Half	1.4Mb	DSHD
	355W99WI(PS2)	3.50	Half	1.4Mb	DSHD
	356F-250UG	3.50	Full	2.8MB	DSHD
	356F-252UG	3.50	Full	2.8MB	DSHD
	356F-258UG	3.50	Full	2.8MB	DSHD
	4851	5.25	Half	360kb	DSDD
	4852	5.25	Full	720kb	DSQD
	4853	5.25	Half	720kb	DSQD
	4854	5.25	Half	1.2Mb	DSQD
	501A	5.25	Half	360kb	DSDD
	501B	5.25	Half	360kb	DSDD
	501C	5.25	Half	360kb	DSDD
	503	5.25	Half	720kb	DSQD
	504A	5.25	Half	1.2Mb	DSHD
	504B	5.25	Half	1.2Mb	DSQD
	504C	5.25	Half	1.2Mb	DSQD
	504S	5.25	Half	1.2Mb	DSQD
	LS-120	3.50	Full	1.44Mb	DSHD
Mitsumi		3.50		720kb	DSDD
	D352E	3.50	1.0	1.44Mb	DSHD
	D353F2	3.50	Half	1.44Mb	DSHD
	D353F2E	3.50	Half	1.44Mb	DSHD
	D353G	3.50	Half	1.44Mb	DSHD
	D353P3	3.50		1.44Mb	DSHD
	D353T3	3.50	1.0	1.44Mb	DSHD
	D353T5	3.50	Full	1.44Mb	DSHD
	D353T7	3.50		1.44Mb	DSHD
	D358F2	3.50	Half	1.2Mb	DSQD
	D358P3	3.50		1.2Mb	DSQD
	D358T3	3.50	1.0	1.2Mb	DSQD
	D359C	3.50	Qtr	1.44Mb	DSHD
	D359F2	3.50	Half	1.44Mb	DSHD
	D359F2E	3.50	Half	1.44Mb	DSHD
	D359G	3.50	Half	1.44Mb	DSHD
	D359P3	3.50		1.44Mb	DSHD
	D359T2	3.50	Third	1.44Mb	DSHD
	D359T3	3.50	Third	1.44Mb	DSHD
	D359T5	3.50	Third	1.44Mb	DSHD
	D359T7	3.50	Full	1.44Mb	DSHD
	D503	5.25	Half	360kb	DSDD
	D509V	5.25	Half	1.2Mb	DSQD

FLOPPY DRIVE SPECS BY MODEL

Manufacturer	Model Number	Width (Inch)	Height (Inch)	Format Capacity	Media Density
Mitsumi (cont.)					
	D509V3	5.25		1.2Mb	DSQD
	D509V5	5.25	Full	1.2Mb	DSQD
	D539W	5.25/3.5	Half	1.2/1.44Mb	DUAL
	DP119F2	3.50	Full	1.44Mb	DSHD
MPI	501	5.25	Half	180kb	SSDD
	502B	5.25	Half	360kb	DSDD
	51M	5.25	Full	180kb	SSDD
	51S	5.25	Full	180kb	SSDD
	52M	5.25	Full	360kb	DSDD
	52S	5.25	Full	360kb	DSDD
	91M	5.25	Full	360kb	SSQD
	92M-002	5.25	Full	720kb	DSQD
	B101M-S	5.25	Full	180kb	SSQD
	B102M-S	5.25	Full	360kb	DSQD
	B51S	5.25	Full	180kb	SSDD
	B52S	5.25	Full	360kb	DSDD
	B91S	5.25	Full	360kb	SSQD
	B92M	5.25	Full	720kb	DSQD
	B92S	5.25	Full	720kb	DSQD
NEC	1035	3.50	Half	720kb	DSDD
	1036A	3.50	Third	720kb	DSDD
	1037A	3.50	Third	720kb	DSDD
	1053	5.25	Half	360kb	DSDD
	1055	5.25	Half	720kb	DSQD
	1137H	3/50	Third	1.44Mb	DSHD
	1155C	5.25	Half	1.2Mb	DSQD
	1157C	5.25	Half	1.2Mb	DSQD
	1158C	5.25	Third	1.2Mb	DSHD
	1165A	8.00	Half	1.6Mb	DSDD
	1165FQ	8.00	Half	1.6Mb	DSDD
	5138A	3.50	Third	1.44Mb	DSHD
	FD1035	3.50	Half	720kb	DSDD
	FD1138H	3.50	.75	1.44Mb	DSHD
	FD1139H	3.50	0.6	1.44Mb	DSHD
	FD1148H	3.50	0.78	1.44Mb	DSHD
	FD1155C	5.25	1.6	1.2Mb	DSQD
	FD1157C	5.25	1.6	1.2Mb	DSQD
	FD1158C	5.25	1.6	1.2Mb	DSQD
	FD1165F	8.00	Half	1.6Mb	DSDD
	FD1165H	8.00	Half	1.6Mb	DSDD
	FD1165S	8.00	Half	1.6Mb	DSDD
	FD1177C	5.25	1.6	1.2Mb	DSHD
	FD1231H	3.50	1.0	1.44Mb	DSHD
	FD1238H	3.50	0.5	1.44Mb	DSHD
	FD1335H	3.50	1.0	1.44Mb	DSHD
	FD5839H	5.25/3.5	1.63	1.2/1.44Mb	DUAL
Newtronic	D357	3.50	Third	720kb	DSDD
Okidata	3305	5.25	Half	360kb	DSDD
	3305BU	5.25	Third	360kb	DSDD
	3305U	5.25	Half	360kb	DSDD

FLOPPY DRIVE SPECS BY MODEL

Manufacturer	Model Number	Width (Inch)	Height (Inch)	Format Capacity	Media Density
	3315B	5.25	Half	360kb	DSDD
Olivetti	4311	5.25	Half	360kb	DSDD
	4311-3	5.25	Half	360kb	DSDD
Pacific Rim	P35	3.50	1.0	1.44Mb	DSHD
	U1.2	5.25	Half	1.2Mb	DSHD
	U1.44	3.50		1.44Mb	DSHD
	U4	3.50	1.0	2.88Mb	DSEHD
	U720	3.50		720kb	DSDD
	U360	5.25	Half	360kb	DSDD
Panasonic	253	3.50	Third	720kb	DSDD
	257	3.50	Third	1.44Mb	DSHD
	257 W/FRAME	3.50	Third	1.44Mb	DSHD
	455 (-5=bk,-7=gr)	5.25	Half	360kb	DSDD
	465	5.25	Half	720kb	DSQD
	475	5.25	Half	1.2Mb	DSQD
	551	5.25	Half	360kb	DSDD
	595	5.25	Half	1.2Mb	DSQD
Persci	277(6N)	8.00	Full	1.2Mb	SSDD
	299	8.00	Full	2.0Mb	DSDD
Pertec	FD200	5.25	Full	180kb	SSDD
	FD250	5.25	Full	360kb	DSDD
	FD400	8.00	Full	800Kb	SSDD
	FD410	8.00	Full	800kb	SSDD
	FD500	8.00	Full	800kb	SSDD
	FD510	8.00	Full	800kb	DSDD
	FD511	8.00	Full	800kb	SSDD
	FD514-U2	8.00	Full	800kb	SSDD
	FD650	8.00	Full	1.6Mb	DSDD
Phillips	3121		Half	360kb	SSDD
	3132	5.25	Half	360kb	DSDD
	3133	5.25	Half	720kb	DSDD
	3134	5.25	Half	1.0Mb	DSDD
Qume	142	5.25	Half	360kb	DSDD
	242	8.00	Half	1.6Mb	DSDD
	542	5.25	Full	360kb	DSDD
	841	8.00	Full	800kb	DSDD
	842	8.00	Full	1.6Mb	DSDD
	DT/5	5.25	Full	360kb	DSDD
	DT/8	8.00	Full	1.6Mb	DSDD
Remex	RFD 2000	8.00	Full	800kb	SSDD
	RFD 4000	8.00	Full	1.6Mb	DSDD
	RFD 480	5.25	Half	360kb	DSDD
Richoh	5100	5.25	Half	720kb	DSQD
	RF8160	8.00	Half	1.6Mb	DSDD
Samsung	SFD500K	5.25	Half	360kb	DSDD
	SFD-560DT	5.25	Half	1.2Mb	DSHD
	SFD-321DT	3.50	Half	1.44Mb	DSHD

FLOPPY DRIVE SPECS BY MODEL

Manufacturer	Model Number	Width (Inch)	Height (Inch)	Format Capacity	Media Density
Sanyo	500C	5.25	Half	360kb	DSDD
	FDA5200	5.25	Half	360kb	DSDD
Seiko	8640	5.25	Full	640kb	DSDD
Shugart	SA200	5.25	Half	180kb	SSDD
	SA210	5.25	Half	360kb	DSDD
	SA215	5.25	Half	180kb	DSDD
	SA300	3.50	Half	360kb	SSDD
	SA390	5.25	Full	180kb	SSDD
	SA400	5.25	Full	180kb	SSDD
	SA400L	5.25	Full	180kb	SSDD
	SA410	5.25	Full	360kb	SSQD
	SA450	5.25	Full	360kb	DSDD
	SA455	5.25	Half	360kb	DSDD
	SA460	5.25	Full	720kb	DSQD
	SA465	5.25	Half	720kb	DSQD
	SA475	5.25	Half	1.2Mb	DSQD
	SA551	5.25	Half	360kb	DSDD
	SA561	5.25	Half	720kb	DSQD
	SA800-1	8.00	Full	800kb	SSDD
	SA800-1R	8.00	Full	800kb	SSDD
	SA800-2	8.00	Full	800k	SSDD
	SA800-2R	8.00	Full	800kb	SSDD
	SA800-4	8.00	Full	800kb	SSDD
	SA801	8.00	Full	800kb	SSDD
	SA801-R	8.00	Full	800kb	SSDD
	SA810	8.00	Half	800kb	SSDD
	SA850	8.00	Full	1.6Mb	DSDD
	SA850R	8.00	Full	1.6Mb	DSDD
	SA851	8.00	Full	1.6Mb	DSDD
	SA851R	8.00	Full	1.6Mb	DSDD
	SA860	8.00	Half	1.6Mb	DSDD
	SA860-1	8.00	Half	1.6Mb	DSDD
	SA900-1	8.00	Full	800kb	SSDD
	SA901	8.00	Full	800kb	SSSD
Siemans	FDD100-5	5.25	Full	180kb	SSDD
	FDD100-8	8.00	Full	800kb	SSDD
	FDD220-8	8.00	Full	800kb	SSDD
	FDD121-5	5.25	Full	360kb	SSDD
	FDD196-5	5.25	Full	360kb	SSDD
	FDD221-5	5.25	Full	360kb	DSDD
Sony	120-04	3.50	Third	1.44Mb	DSHD
	17W	3.50	Third	1.44Mb	DSHD
	17W-5PF	3.50	Third	1.44Mb	DSHD
	17W-10/W51/4	3.50	Third	1.44Mb	DSHD
	17W-34/W51/4	3.50	Third	1.44Mb	DSHD
	17W-42/W51/4	3.50	Third	1.44Mb	DSHD
	17W-55	3.50	Half	1.44Mb	DSHD
	17W-90	3.50	Third	1.44Mb	DSHD
	17W-WFP	3.50	Third	1.44Mb	DSHD
	40W-00(PS2)	3.50	Third	2.8Mb	DSHD
	40W-9E	3.50	Third	2.8Mb	DSHD

FLOPPY DRIVE SPECS BY MODEL

Manufacturer	Model Number	Width (Inch)	Height (Inch)	Format Capacity	Media Density
Sony (cont.)					
	40W-15	3.50	Half	2.8Mb	DSHD
	40W-KO	3.50	Half	2.8Mb	DSHD
	420-6	3.50	Third	1.44Mb	DSHD
	53	3.50	Third	1.44Mb	DSHD
	53W	3.50	Third	720kb	DSQD
	63W	3.50	Third	720kb	DSQD
	73W	3.50	3Qtr	1.44Mb	DSHD
	73W-34D/W51/4	3.50	3Qtr	1.44Mb	DSHD
	77W(PS2)	3.50	Third	1.44Mb	DSHD
	MFD51W	3.50	Third	800kb	DSQD
	MPR17W	3.50	1.0		
	MPF320	3.50	Half	1.44Mb	DSHD
	MPF40W	3.50	1.0	2.88Mb	DSHD
	MPF420	3.50	1.0	1.44Mb	DSHD
	MPF520	3.50	1.0	1.44Mb	DSHD
	MPF920	3.50	1.0	1.44Mb	DSHD
Tandon	TM100-1A	5.25	Full	180kb	SSDD
	TM100-2A	5.25	Full	360kb	DSDD
	TM100-3	5.25	Full	360kb	SSQD
	TM100-3M	5.25	Full	360kb	SSQD
	TM100-4	5.25	Full	720kb	DSQD
	TM100-4A	5.25	Full	720kb	DSQD
	TM101-2	5.25	Full	360kb	DSDD
	TM101-3	5.25	Full	360kb	SSQD
	TM101-4	5.25	Full	720kb	DSQD
	TM50-1	5.25	Half	180kb	SSDD
	TM50-2	5.25	Half	360kb	DSDD
	TM55-1	5.25	Half	180kb	SSDD
	TM55-2	5.25	Half	360kb	DSDD
	TM55-4	5.25	Half	720kb	DSQD
	TM65-1L	5.25	Half	180kb	SSDD
	TM65-2L	5.25	Half	360kb	DSDD
	TM65-4	5.25	Half	720kb	DSQD
	TM65-8	5.25	Half	1.2Mb	DSQD
	TM75-2	5.25	Half	360kb	DSDD
	TM75-8	5.25	Half	1.2Mb	DSQD
	TM848-1	8.00	Half	800kb	SSDD
	TM848-1E	8.00	Half	800kb	SSDD
	TM848-2	8.00	Half	1.6Mb	DSDD
	TM848-2E	8.00	Half	1.6Mb	DSDD
	TM965-2	5.25	Full	360kb	DSDD
Teac	35F	3.50	Half	720kb	DSDD
	35FN	3.50	Half	720kb	DSDD
	35HFN	3.50	Third	1.44Mb	DSHD
	50A	5.25	Full	180kb	SSDD
	53B	5.25	Half	360kb	DSDD
	54B	5.25	Half	360kb	DSDD
	55A	5.25	Half	180kb	SSDD
	55B	5.25	Half	360kb	DSDD
	55BR	5.25	Half	360kb	DSDD
	55BV	5.25	Half	360kb	DSDD

FLOPPY DRIVE SPECS BY MODEL

Manufacturer	Model Number	Width (Inch)	Height (Inch)	Format Capacity	Media Density
Teac (cont.)					
	55E	5.25	Half	360kb	DSDD
	55FR	5.25	Half	720kb	DSQD
	55FV	5.25	Half	720kb	DSDD
	55G	5.25	Half	1.2Mb	DSQD
	55GFR	5.25	Half	1.2Mb	DSQD
	55GR	5.25	Half	1.2Mb	DSHD
	55GS (SCSI)	5.25	Half	1.2Mb	DSHD
	55GV	5.25	Half	1.2Mb	DSQD
	55GVF	5.25	Half	1.2Mb	DSQD
	135FN	3.50	Third	720kb	DSDD
	135HF	3.50	Third	1.44Mb	DSHD
	135HFN	3.50	Third	720kb	DSDD
	155GF	5.25	1.0	1.2Mb	DSHD
	235F	3.50	Third	720kb	DSDD
	235GF	3.50	1.0	1.6Mb	DSDD
	235HF	3.50	Third	1.44Mb	DSHD
	235HG	3.50	Third	1.44Mb	DSHD
	235HS (SCSI)	3.50	1.0	1.44Mb	DSHD
	235J	3.50	1.0	2.88Mb	DSEHD
	235JS (SCSI)	3.50	1.0	2.88Mb	DSEHD
	334	3.50	1.0	1.44Mb	DSHD
	335F	3.50	0.75	720kb	DSDD
	335HF	3.50	0.75	1.44Mb	DSHD
	335HS (SCSI)	3.50	0.75	1.4Mb	DSHD
	335J	3.50	0.75	2.88Mb	DSEHD
	335JS (SCSI)	3.50	0.75	2.88Mb	DSEHD
	505	5.25/3.5	Half	1.2/1.44Mb	DSHD
	05HF-030	3.50	Third	1.44Mb	DSHD
	05HF-532U	3.50	Third	1.44Mb	DSHD
	FD-04	5.25	Half	1.44Mb	
	FD-05	5.25	Half	1.44Mb	
	HiFD	3.50	1.0	1.44Mb	DSHD
Tec	FB501	5.25	Half	180kb	SSDD
	FB503	5.25	Half	360kb	DSDD
	FB504	5.25	Half	720kb	DSQD
Tecmate	1103	5.25	Half	3.3Mb	DSDD
Texas Peripherals					
	10-5355-001	5.25	Full	180kb	SSDD
Toshiba	0202A	5.25	Full	720kb	DSQD
	0242A	5.25	Half	360kb	DSDD
	0401GR	5.25	Half	360kb	DSDD
	0801GR	5.25	Half	1.2Mb	DSDD
	0802GR	5.25	Half	1.2Mb	DSHD
	352TH	3.50	Third	720kb	DSQD
	3527H	3.50	Third	720kb	DSDD
	3527TH	3.50	Third	720kb	DSQD
	3561GR	3.50	Third	1.44Mb	DSHD
	3564	3.50	Third	1.44Mb	DSHD
	3567	3.50	Third	1.44Mb	DSHD
	4210	3.50	Third	720kb	DSDD

FLOPPY DRIVE SPECS BY MODEL

Manufacturer	Model Number	Width (Inch)	Height (Inch)	Format Capacity	Media Density
Toshiba (cont.)					
	4202-AOK	3.50	Third	720kb	DSDD
	4207-AOK	3.50	1.0	720kb	DSQD
	4207-AOK	3.50	Third	720kb	DSDD
	4261	3.50	Third	720kb	DSQD
	4449-AOZ(PS2)	3.50	Half	720kb	DSQD
	5401	5.25	Half	360kb	DSDD
	5406	5.25	Half	360kb	DSDD
	5426	5.25	Half	360kb	DSDD
	5451	5.25	Half	360kb	DSDD
	5454	5.25	Half	360kb	DSDD
	5471	5.25	Half	360kb	DSDD
	5472	5.25	Half	360kb	DSDD
	5474	5.25	Half	360kb	DSDD
	5629	5.25	Half	720kb	DSQD
	5861	5.25	Half	1.2Mb	DSHD
	5862	5.25	Half	1.2Mb	DSHD
	5863	5.25	Half	1.2Mb	DSHD
	5881	5.25	Half	1.2Mb	DSHD
	5882	5.25	Half	1.2Mb	DSQD
	6371	5.25	Half	360kb	DSDD
	6374	5.25	Half	360kb	DSDD
	6379R3B	5.25	Half	360kb	DSDD
	6471	5.25	Half	360kb	DSDD
	6474-T2P	5.25	Half	360kb	DSDD
	6782	5.25	Half	1.2Mb	DSHD
	6784	5.25	Half	1.2Mb	DSHD
	6881	5.25	Half	1.2Mb	DSHD
	6882	5.25	Half	1.2Mb	DSHD
	6890	5.25	Half	1.2Mb	DSHD
	M48D-12	5.25	Half	360kb	DSDD
	ND-04	5.25	Half	360kb	DSDD
	ND-08	5.25	Half	1.2Mb	DSHD
	ND-352T,S	3.50	1.0	720kb	DSDD
	ND-354A	3.50	1.0	720kb	DSDD
	ND-356	3.50	Third	1.44Mb	DSHD
	ND-3565-A	3.50	Third	1.44Mb	DSHD
	ND-3571	3.50	1.0	2.88Mb	DSEHD
	PD-211	3.50	1.0	2.88Mb	DSEHD
Victor	TM100-3	5.25	Full	360kb	SSQD
	TM100-4	5.25	Full	720kb	DSQD
Weltec	M16-A22	5.25	Half	1.0Mb	DSDD
	M16-P12	5.25	Half	720kb	DSDD
	M-16-R12	5.25	Half	1.0Mb	DSDD
	M16-R12/910	5.25	Half	720kb	DSDD
	M48D-1	5.25	Half	360kb	DSDD
	M48D-14	5.25	Half	360kb	DSDD
	N96-12	5.25	Half	720kb	DSDD
World Storage	FD100-5	5.25	Full	180kb	SSDD
	FD100-8	8.00	Full	800kb	SSDD
	FD200-5	5.25	Full	360kb	DSDD
	FD200-8	8.00	Full	1.6Mb	DSDD

FLOPPY DRIVE SPECS BY MODEL

Manufacturer	Model Number	Width (Inch)	Height (Inch)	Format Capacity	Media Density
YE-Data..........	YD180	8.00	Half	1.6Mb	DSDD
	YD280	5.25	Full	720kb	DSQD
	YD380	5.25	Half	1.2Mb	DSHD
YE-Data (cont.)					
	YD380B	5.25	Half	1.2Mb	DSHD
	YD380C	5.25	Half	1.2Mb	DSHD
	YD580	5.25	Half	360kb	DSDD
	YD580B	5.25	Half	360kb	DSDD
	YD701	3.50	Third	1.44Mb	DSHD
	YD701(PS2)	3.50	Third	1.44Mb	DSHD

Chapter 10

Glossary

Access Time: The amount of time it takes the computer to ask for data and the data is transferred to the computer. The access time is determined by latency, the seek time and command overhead.

Acoustic Coupler: Type of a modem where the telephone receiver is placed into the modem (cradle). The other method is a direct-connect modem where the telephone line connects into the modem and then into the computer.

Actuator (Access Mechanism): The motor whose purpose is to control read/write head movement by making the access arm move from the center to the edge of the platters.

Address Mark: A number assigned to each possible head position so the disk drive control circuit knows the location of the head assembly.

AGP (**A**ccelerated **G**raphics **P**ort): High performance video graphics card targeted at 3D graphics applications.

Analog to Digital Converter (**ADC**): A device that converts voltage into sets of signals that are representative of decimal or binary.

ARLL (**A**dvanced **R**un **L**ength **L**imited): Also known as RLL 3,9 and Rll 1,7.

Areal Density: Tells how much data can be put into an area. The range is 5,000 to 170,000/square millimeters. (*Formula* -Bits per inch x tracks per inch = bits per inch squared)

ASCII (**A**merican **S**tandard **C**ode for **I**nformation **I**nterchange): Set of standard codes used by the microcomputer industry.

ATA (**AT A**ttachment): An IDE type attachment designed in March, 1989. A 16 bit controller is on the drive.

ATA-2: An improved ATA product, which increased performance.

Asynchronous Data: Data sent in parallel mode without a clock pulse.

Atapi (**ATA P**ocket **I**nterface): Used for peripherals like CD-ROM drives and tape drives.

Auto Parking: When power is shut off, the motor forces the read/write head away from data on the disk and sends it to a special landing zone to be held.

Average Latency: Measurement of how much time it takes for a bit of data to rotate under the head (approximately 8.4 milliseconds for most drives).

Bank Switching: A capability given to a computer by installing more memory than it was originally made for and enables a computer to use more than one set of memory chips at different times, yet with the same address.

BCAI: **B**yte **C**ount **A**fter **I**ndex

Bezel: Cover for the front of the hard drive.

Board: This piece of hardware is the printed circuit board, also know as a card.

BPI (**B**its **p**er **I**nch): Measurement of bit density.

BFI (**B**ytes **f**rom **I**ndex):

Binary Number: The Base 2 mathematics system of counting— only two digits of 1 and 0 (or the mechanical or electrical status of on and off).

BIOS (**B**asic **I**nput/**O**utput **S**ystem): BIOS is the operating system for microcomputers that controls the input and output. It tells the drive controller to put the information being sent to a specific head, in a certain cylinder and into a certain sector. It also allows the computer to communicate with peripherals.

Bit: An information storage unit. (8 bits = 1 byte)

Bit Cell: The length of track used to store 1 bit.

Bit Density (**B**its **p**er **I**nch or **BPI**): Measurement indicating how many bits in 1 inch of track on a disk surface.

Block: A group of bytes that is equal to 1 physical sector of data.

BPS (**B**its **P**er Second): Measurement indicating speed of data processes.

Bubble Memory: Hardware (form of disk drive) that is expensive, durable and lightweight and is used in laptop computers for RAM. Also the memory is still present when power is turned off.

Bus: The main communication path in a computer consisting of all the parallel wires to which the memory, the CPU, and all input/output devices are connected.

Buffer: Part of the RAM that holds data temporarily; data that is going from one device to another.

Byte: The memory space required to store one character, usually 8 bits. [8 bits (storage unit) plus one parity = 1 byte]

Cache Memory (Caching): A temporary storage area between the computer's CPU and its main RAM. Its purpose is to increase the speed of data transfer by decreasing repetition.

Capacity: The number of hard disk megabytes available.

Carriage Assembly: Holds the roller bearings and read/write heads.

CD-ROM (**C**ompact **D**isk **R**ead **O**nly **M**emory): Circular plastic coated disks that store large capacity of data in binary code and uses laser technology (optical) to read the disk.

Chipset: The set of motherboard computer chips, excluding processor and memory. e.g. Intel BX chipset.

CHS (**C**ylinders, **H**eads, **S**ectors Per Track): The three parameters used to determine a hard drive capacity (physical = actual number; logical = translation).

Cluster (Allocation Unit): A small group of sectors that is allocated by the system when disk space is needed.

CMOS (**C**eramic **M**etal **O**xide **S**emiconductor): A type of chip which needs only small amounts of electric power.

Coaxial cable: Type of cable that is surrounded by insulation and a conductive shield and keeps the cable from conducting electrical noise.

Composite Video Signal: A signal that is transmitted on only one wire and is the kind of signal used in a TV monitor. An RGB (Red, Green, Blue) is the signal that is transmitted on three separate wires for each primary color and is a monitor type used by computers.

Controller: The link between the disk and the computers program. A printed circuit board that takes bits of data and converts them into bytes and words.

CPU (**C**entral **P**rocessing **U**nit): The primary microprocessor in the computer.

CRC (**C**yclic-**R**edundancy-**C**heck): A computer protection scheme that checks for corrupt data.

CRT (Cathode **R**ay **T**ube): The device that sprays electrons on a screen in certain patterns by use of a magnetic field making visualization of the TV and computer monitors possible.

Cylinder: Each platter on a hard disk has tracks on it. They are numbered (starting with "0") from the outside edge in. When you stack the platters on top of each other (like in a 4-platter hard disk), the tracks that line up with each other, and have the same number then they are called a cylinder.

Daisy Chain: Method to connect more than one drive to a controller.

Data: Storage information in computer.

Data Compression: A term used for data that is altered by a software program and is subsequently stored in a smaller space.

Dead Reckoning: A type of mechanism used in early PCs that makes the read/write heads go to a predetermined position.

Differential: A hard drive developed for an option to single-ended SCSI to increase bus length.

DIMM: **D**ual **I**nline **M**emory **M**odule, motherboard memory.

Disk: A piece of hardware that stores data using iron oxide which acts like a tiny bar magnet and codes the disk with binary language. Each disk stores data in concentric circles called tracks and each track is divided into a sector. A set of tracks is called a sector. There are different sizes of floppy disks (diskettes), various hard disks and another form is the optical disk (CD-ROM).

Disk Cache: A temporary storage area for data between the RAM and disk drive which allows for faster data retrieval.

Disk Drive: The hardware in the computer that reads and writes on disks.

DMA (**D**irect **M**emory **A**ccess): A method that increases speed of transfer of data between peripherals and RAM by data going from memory to peripherals without the use of the CPU.

DMI (**D**esktop **M**anagement **I**nterface): Software that allows hardware to communicate through a standard protocol, thereby creating a higher level of compatibility.

DOS (**D**isk **O**perating **S**ystem): A term used by computer manufacturers as a name for various operating systems (i.e., MS-DOS by Microsoft for the 16-bit microcomputers).

DRAM (**D**ynamic **R**andom **A**ccess **M**emory): The memory device (chips) in most computers that requires a refresh signal to be periodically sent to it.

Drive Type: A numerical assignment for a disk drive that tells a standard configuration of CHS.

Drum: Hardware device in the computer where data is stored on rotating metal cylinders. Also refers to part of a Laser printer

DTR (**D**ata **T**ransfer **R**ate) Measurement indicating how fast the controller reads a file once it finds it. Usually recorded as bits per second or bytes per second.

DVD (**D**igital **V**ideo **D**isc or **D**igital **V**ersatile **D**isc): A massive storage device for video and other processes that require removable data storage.

ECC (**E**rror **C**orrection **C**ode): Allows computer to detect and correct any pattern of errors in a sector.

EIDE (**E**nhanced **I**ntegrated **D**rive **E**lectronics): Interface for home machines, internal hard drives and CD-Rom drives based on ATA-2 and ATAPI standards.

EISA (**E**xtended **I**ndustry **S**tandard **A**rchitecture): Modification of IBM's ISA by Compaq Corporation.

Embedded Servo System: Servo data is embedded with data on every cylinder.

Encoding Schemes: The two types are MFM and RLL.

EPROM (**E**rasable **P**rogrammable **R**ead **O**nly **M**emory): A memory chip that uses ultraviolet light to erase a programmed disk and then be reprogrammed. A PROM is a memory chip that can only be written to once, cannot be reprogrammed.

ESDI (**E**nhanced **S**mall **D**evice **I**nterface): Controller that evolved from ST412/ST506 interface in the 1983 by Maxtor . The clock-data separator is on the drive in the ESDI where it is on the controller in the

ST 412/ST506. Therefore, the encoding and decoding is done on the drive, which allows for increased capacities and faster communication. It has a data transfer rate of 10-24 megabits/second.

Expanded Memory: A method of breaking the 640k memory barrier using bank switching technology.

Extended Memory: Memory at addresses greater than 1 megabyte on PC computers.

Extension: Part of a file name in an operating system that defines the type and operation or the file and is typically a period followed by three characters, e.g., (name of file).EXE, (name of file).COM or (name of file).BAS.

FAST: A Fast designation of the SCSI-2 (aka Fast SCSI) specification indicates a double speed transmission of 10 MB/s verses the SCSI-2 transfer rate of 5MB/s.

FAT (**F**ile **A**llocation **T**able): Tells what clusters are allocated to what files and their availability.

FCI (**F**lux **C**hanges **P**er **I**nch): -in RLL: 1 FCI = 1.5 BPI -in MFM: 1 FCI = 1BPI

Fibre Channel: A fast data transfer interface for storage networks or data centers.

File-Protect Ring: Device (when removed) insures that data cannot be written onto a magnetic tape.

Flux: Magnetic impulses on the disk surface.

FM Encoding (**F**requency **M**odulation **E**ncoding): Archaic way of encoding data which was replaced by MFM encoding.

Formatting: Records header data (sector number, head and cylinder address) on a disk so it is: ready to be read or written to.

Form Factor: The physical external size of a hard drive (desk top usually 3.5 inches or 5.25 inches; portable and lap tops usually 2.5 inches).

Full Duplex: The transmission of data in both directions simultaneously.

Full-Height Drive: A drive with the following physical parameters: 3.25 inch height, 8.00 inch depth, and 5.75 inch width.

G: Unit of force.

Gigabyte: A unit of storage equal to 1,073,741,824 bytes or approximately 1,000 megabytes.

Half Duplex: The transmission of data one direction at a time.

Half-Height Drive: A drive with the following physical parameters: 1.625 inch height, 4.0 or 8.0 inch depth, and 4.0 or 5.75 inch width.

Hard Card: Type of an internal hard disk that is built into a card and can be plugged into a computer through a slot.

Hard Disk: A rigid aluminum magnetic disk which is coated with iron oxide and is written to with codes using binary language. It is part of the internal hardware of a computer and has a large data storage capacity.

Hardware: Term for the tangible parts of a computer, i.e., monitor, wires, hard drive, modem, etc.

Head: The part of the hard disk that reads and writes data on the platters -usually 1 read/write head per each side of a platter (i.e., 2 platters = 4 heads). [Three types: composition, monolithic and thin film]

Head Parking: (see Auto Parking)

Head Crash: Catastrophic failure that occurs when the head lands on the disk and it is damaged. This rarely occurs, but when it does it is usually because particles get between the head and the disk or else its due to rough handling.

High Density (**HD**): The amount of disk storage capacity.

Host Adapter Card: A printed circuit board (card) added to a microcomputer in order to add a connector port.

IDE (**I**ntegrated **D**rive **E**lectronics): The most commonly used hard drive interface in which you do not need a controller card because the RLL or MFM controller is built in the drive and plugs directly into the motherboard. [Three types: XT IDE (8-bit); AT (ATA) IDE (16-bit); MCA IDE (16-bit) -basically an advanced RLL used on high capacity (>500Mb) drivesare connected.

IEEE 1394: (**I**nstitute of **E**lectrical and **E**lectronics **E**ngineers) (aka FireWire®, 1394-1995, iLink) Interface for audio-video devices and computer peripherals, either a 4-pin or 6-pin type.

Index Pulse: The start point for each disk track.

Index Time: A measurement of time it takes for a disk to take one revolution.

Interface: That which allows communication between computer devices.

Interleave: Makes a hard drive read data slower by renumbering the sectors and allowing the CPU to read six sectors in only one rotation. [PC/XT interleave = 3 or 4 -PC/AT interleave = 1 or 2 -XT interleave = 3]

I/O (**I**nput/**O**utput): The input is the data that is imported into the computer for it to process through an input device (i.e., keyboard). The output is the end result information the computer generates and is displayed on a terminal, printout, disks, etc.

IOCHRDY (**I**nput/**O**utput **C**hannel **R**eady): The CPU sends this signal to a peripheral to notify it that more resources are ready for transfer.

IPI (**I**ntelligent **P**eripheral **I**nterface): The 8 and 14 mainframe and minicomputers use this hard drive interface.

ISA (**I**ndustry **S**tandard **A**rchitecture): PC, XT and AT systems use these types of 8 and 16 bit expansion slots.

Isochronous Data Transfer: The type of data transfer that is sent at a fixed rate and is matched to a clock. Used with video-audio connections to devices such as a TV.

JCL (**J**ob **C**ontrol **L**anguage): The specific command language that tells the computer what to do.

Kilobyte: -1,024 bytes

LAN (**L**ocal **A**rea **N**etwork): Hardware and software package that allows computers to connect to each other and share data and peripherals.

Landing Zone: The part of the disk used as a safe area for head parking when the power is shut off.

Latency: The amount of time (in milliseconds) it takes the platter to make a half revolution.

LBA (**L**ogical **B**lock **A**ddressing): Method of accessing the disk via a logical sector number which points to a location on a drive instead of using CHS information.

LCD (**L**iquid **C**rystal **D**isplay): Type of hardware display or screen in some laptop computers or in some digital watches. The screen is very narrow compared to a monitor.

Low-Level Format: This step connects the drive and controller once physical installation is finished. [XT system - use DOS' debug utility -80 x 86 system - use 80 x 86 advanced diagnostics -IDS system] Do not low-level unless instructed by manufacturer because it is usually done at the factory.

Low Voltage Differential: A type of differential that uses lower voltage to conserve power.

LUN (**L**ogical **U**nit **N**umber): The units number on a daisy chain. It equals to SCSI ID number.

Mainframe computer: A large computer that supports up to 500 uses at one time, e.g. IBM 3081.

MCA (**M**icro **C**hannel **A**rchitecture): A type of BUS introduced in 1987 that is used in most PS/2 models.

Megabyte (MB): The unit used to express hard drive capacity.[equal to 1,048,576 bytes]

Megabytes/second (MB/s): 100 MS/second = 800 Megabits/second (Mb/s).

Memory –The part of computer where information is stored.

Memory Chips: Additional chips that can be installed into a computer to increase RAM.

Minicomputer: The middle sized computer that can support up to 100 users at one time, e.g. IBM System/3.

MFM Encoding (**M**odified **F**requency **M**odulation **E**ncoding): The data encoding method that decreases bit cell size by omitting each clock pulse associated with a bit cell (except those cells holding 0) so that data is stored evenly on a platter. The standard MFM has 17 SPT and 512 bytes per sector.

Modem (**Mod**ulator-**Dem**odulator): A hardware device that encodes data and allow an interface between different mediums, e.g. a telephone line to – (communication) modem (communication)- to computer.

Motherboard: The name of the computer's circuit board.

Mouse: Hardware input device that is connected to a computer by a wire and requires software and internal hardware to communicate. There are many models of this device and largely depends upon one's preference.

MTBF (**M**ean **T**ime **B**etween **F**ailures): Measurement of time (in power-on hours) a drive should last between hardware failures.

MTTR (**M**ean **T**ime **T**o **R**epair): The estimated time to repair or service a drive.

Narrow: Term used to indicate a SCSI device for a 8-bit data path.

NCP (**N**etwork **C**ontrol **P**rogram): Communication control software that runs the Front End Processor.

NDIS (**N**etwork **D**river **I**nterface **S**pecification: An interface that alows the loading of multiple protocol stacks at a server or workstation so there can be communication between a variety of protocols.

NRZ (**N**on-**R**eturn to **Z**ero): The data-encoding strategy where a pattern of pulses is converted into half pulses without any information loss.

OCR (**O**ptical **C**haracter **R**eader): A device that can recognize typewritten or handwritten data on a piece of paper.

OEM (**O**riginal **E**quipment **M**anufacturer): Term which refers to company that makes a product and sells it to a reseller.

Parallel Transmission: Sending sets of signals alongside of each other. E.g. parallel signals in a printer are transmitted on 8 wires at the same time.

Parity: Computer error recognition method where there is a 9th bit that accompanies every byte in a system RAM; therefore, if a bit is stored or read wrong the parity will be wrong so the PC realizes there is an error.

Partition: Divides the physical hard drive into logical volumes —maximum of 4 partitions per disk.

PCI (**P**eripheral **C**omponent **I**nterconnect): -Intel created this BUS to use with most Pentium and 486 systems.

PCMCIA (**PC** **M**emory **C**ard **I**nternational **A**ssociation): Association which sets standards for the interconnection between plug-in modules and electronic systems.

Peripheral: The equipment attached to the computer (i.e., modem, printer).

PIO (**P**rocessor **I**nput/**O**utput): Data is transferred between memory and peripherals by an input/output command and without communicating with the hard drive.

Plated Thin Film Disks: A magnetic disk memory media whose surface is coated with metallic alloy instead of oxide.

Platter: A part of the hard disk which is usually made of aluminum alloy (sometimes ceramic or glass) and has data recorded on it -usually between 2-8 platters per hard disk, but can be a maximum of 14. Each platter on a hard disk contains tracks. The platters are numbered (starting with 0) from the outside edge inward.

Port: A connection area between the CPU and another device that communicates together.

POST (**P**ower-**O**n **S**elf **T**est): The series of system tests that the BIOS runs at power-up where errors are recognized and reported with an error message or a beep sound stating everything checks out "okay."

Print Spooler: Program that stores output in a buffer for a printer and then sends it to the printer.

Pulse: Signals that go up and down.

Radial: Attaching multiple drives to a controller.

RAID (**R**edundant **A**rray of **I**nexpensive **D**isks): Method of combining two or more drives together to obtain more speed or better data protection. The four major types of RAID are striping, mirroring, parity using another disk for error protection, and parity using several disks for error protection. The hard drives are placed in a central cabinet as modules.

RAM (**R**andom **A**ccess **M**emory): A short term data storage area where information is stored prior to being stored on the disk so that it can be accessed at an instant by the microprocessor.

Raster: Name for the scan pattern that is seen on the CRT.

Recalibrate: A disk drive function to return heads to track zero.

RLL (**R**un-**L**ength **L**imited): -A type of hard-disk controller that increases storage by 50% over MFM by creating 25 or 26 sectors per track (instead of the usual 17 SPT) by decreasing level of data-checking information that is stored on disk -types of RLL: (1,7)RLL: 25% larger capacity than MFM encoding (2,7)RLL: 50% larger capacity than MFM encoding (3,9)RLL: doubles storage capacity over MFM encoding.

Rotational Speed: Meida spin rate which is usually 3,500 RPM for 5.25 inch or 3.5 inch Winchester drives.

Root Directory: The master directory of the hard or floppy disk that holds FAT and operating system boot files.

RWC (**R**educed **W**rite **C**urrent): The input signal that lessens the degree of write current at the drive head.

SA-400 Interface: Industry's standard floppy interface.

SCA (**S**ingle **C**onnector **A**ttachment): Way to connect drive modules to the cabinet in SCSI RAID systems.

SCSI (**S**mall **C**omputer **S**ystem **I**nterface): A standard bus developed in 1970s by Shugart for high-speed connections to peripherals. A SCSI allows up to seven peripherals to communicate in a daisy-chain with a PC, do most of the interfacing, and uses a wide ribbon cable and 50-pin connector.

SCSI-2: A bus that can carry 32 bits because it has a wider cable than the SCSI and this allows data to transfer twice as fast.

SCSI-3: A bus that can support wide and narrow cables at very high speeds.

SDRAM: **S**ynchronous **D**ynamic **R**andom **A**ccess **M**emory, faster memory than DRAM.

Sector: A chunk of stored data that is one section of track -DOS has 512 bytes of data per sector.

Sector Header: The address portion of a sector that holds three numbers: head, cylinder and physical sector numbers.

Sector Sparing: Method of decreasing the capacity of a drive by decreasing the number of sectors on each track by one and then putting defect information on it.

Seek: Radial movement of the heads to a specific cylinder or track address.

Seek Time: Time required for the heads to move between current location and where needed data is stored on the track.

Serial Transmission: Sending signals one after another over a single wire, like moving beads on a string.

SIMM: **S**ingle **I**nline **M**emory **M**odule, e.g. motherboard memory.

Single-Ended: There is a significant difference between the signal and the ground in this type of signal transmission.

Slack Space: Area at the end of a cluster that does not have data stored in it.

Slot: Term for an area or socket in a microcomputer that will accept a plug-in card.

Snow: A visible impairment on a monitor (screen) that is cause by small flickering white spots.

Spindle: Hub structure that disks are attached to that is controlled by signals and cause the hub structure to rotate the platters at a constant speed.

Spindle Motor: Motor that makes the platters spin at a usual rate of 3,600 RPM (some models at 4,800-7,200 RPM).

Spindle Motor Ground Strap: Older model hard drives have this strap attached to the circuit board which is used to press against the spindle motor to decrease static and sometimes causes a scraping or high pitched drive noise that can be eliminated by oiling.

ST-506/ST412 is a standard interface developed in 1980 by Seagate Technologies where the controller mechanism is on the controller card and is used with MFM drives smaller than 152 Mb or RLL drives less than 233Mb.

Step: An increase or decrease of the head positioning arm so that the head can move in or out one track.

Stiction: Name for the most common type of hard drive failure which occurs when the drive heads get stuck to the platter causing the drive to stick and not turn.

Step Pulse: Controller pulse that tells the stepper motor to start a step operation.

Stepper Motor Actuator: Moves the heads back and forth over the platters by rotating the motor a step at a time.

Supercomputer: The largest of computers that can handle many uses with rapid I/O, e.g. Intel iPSC parallel processor.

Synchronous Data: Data sent with a clock pulse.

Tape: A magnetic storage device that is in ribbon form.

Tape Drive:The hardware device that converts data stored on tapes into signals received by the computer.

Terminal: The hardware (such as the keyboard and monitor) that allows I/O with the computer.

Terminating Resistors, aka **Terminator**: Circuitry added at the end of the bus which absorbs reflected signals, improves data integrity and/or provides an electrical signal termination for the controller.

TPI (**T**racks **P**er **I**nch): Measures track density.

Track: The circular rings on the surface of the platter that hold data and are made by the head when the disk spins. The number of tracks per cylinder is equal to two times the number of platters.

Track Density (**TPI**): Number of many tracks on a disk surface.

Transfer Rate: How long it takes data to be sent to the motherboard from the drive.

Ultra: A SCSI designation of the first version of the SCSI-3. Subsequent Ultra2 SCSI, and Ultra3 are further refinements of the interface, not faster devices.

USB (**U**niversal **S**erial **B**us): An interface for keyboards and modems that is easily expandable.

VIB (**V**olume **I**nformation **B**lock):

Video Card: Is a plug-in circuit board that allows certain types of data to be displayed on a monitor.

Virtual Memory: Swap files are used to supplement RAM with hard drive space.

Virtual Split: Logical split of the disk drive.

Virus: Term for a computer program whose intent is to corrupt stored data by automatically copying itself.

Voice-Coil Actuator: Purpose is to control coil movement to and from a magnet resulting in sound from the speaker cone.

WAN (**W**ide **A**rea **N**etwork): Large global network.

Wide: Term used to indicate a SCSI device for a 16-bit data path and requires a 68-pin cable. Can transmit twice as much data as a narrow.

Winchester Drive: Non-removable hard drive originally used in 1960 IBM drives with 30Mb fixed and 30Mb removeable storage.

Worm Drive (**W**rite **O**nce **R**ead **M**emory): A removeable media optical drive that writes data to the drive once (cannot be changed), but can be read many times.

WP (**W**rite **P**recompensation): Helps with bit crowding when using high density data on a small cylinder.

ZBR (**Z**one **B**it **R**ecording): Formating done at the factory where the sector per track depends on the cylinder circumference.

Chapter 11

Printer Control Codes

Printer Control Codes

Since the PC boom started, there have been more than a thousand different printer makes and models released. With each new generation of printer, more and more bells and whistles have been introduced. All of a printers' functions can normally be accessed through a set of decimal or hex control codes and this chapter has been designed to provide the reader with some of the more standardized control code sets. "Standardized" simply means that the particular printer listed in this chapter has codes that are also used by other manufacturers, for example, the Panasonic 2124, 24 pin, dot matrix printer, can be configured to use either Epson LQ860 codes or IBM Proprinter X24E codes.

Please note that your particular printer may have additional, specialized codes which are unique to your printer and are not included in the standardized set. If in doubt, always refer to the printer manual that came with your printer.

Some control codes included in this chapter have been drastically simplified, particularly in the "Graphics" sections. Simplified sections are noted and you are told to refer to the manual that came with your printer for more details.

DIABLO 630 PRINTER CODES

Code	Hex	Decimal	Command
Page Format Control:			
ESC 9	1B 39	27 57	Set left margin at current position
ESC Ø	1B 3Ø	27 48	Set right margin at current position
ESC T	1B 54	27 84	Set top margin at current position
ESC L	1B 4C	27 76	Set bottom margin at current position.
ESC C	1B 43	27 67	Clear top and bottom margins
ESC FF #	1B ØC #	27 12 #	Set lines/page, # is 1 to 126 lines
Horizontal Movement and Spacing Control:			
CR	ØD	13	Carriage return
ESC M	1B 4D	27 77	Enable auto justify
ESC =	1B 3D	27 61	Enable auto center
ESC ?	1B 3F	27 63	Enable auto carriage return
ESC !	1B 21	27 33	Disable auto carriage return
ESC /	1B 2F	27 47	Enable auto backward printing
ESC \	1B 5C	27 92	Disable auto backward printing
ESC <	1B 3C	27 6Ø	Enable reverse printing
ESC >	1B 3E	27 62	Disable reverse printing
ESC 5	1B 35	27 53	Enable forward printing
ESC 6	1B 36	27 54	Enable backward printing
SP	2Ø	32	Space
BS	Ø8	Ø8	Backspace
ESC BS	1B Ø8	27 Ø8	Backspace 1/12Ø inch
HT	Ø9	Ø9	Horizontal tab
ESC HT #	1B Ø9 #	27 Ø9 #	Absolute horizontal tab, # is column 1 to 126
ESC DC1 #	1B 11 #	27 17 #	Spacing offset, # is 1 to 126 (1/12Ø" units), where #1 = offset 1 to # 63 = offset 63, # 64 = offset Ø, # 65 = offset –1 to # 126 = offset –62
ESC 1	1B 31	27 49	Set horizontal tab stop at current position
ESC 8	1B 38	27 56	Clear horizontal tab at current position
ESC 2	1B 32	27 5Ø	Clear all vertical and horizontal tab stops
ESC US #	1B 1F #	27 31 #	Set horizontal motion index, # is 1 to 126, where (#–1)/12Ø inch is the column spacing.
ESC S	1B 53	27 83	Return HMI control to spacing switch

DIABLO 630 PRINTER CODES

Code	Hex	Decimal	Command
Vertical Movement and Spacing Control:			
LF	ØA	1Ø	Line feed
ESC LF	1B ØA	27 1Ø	Reverse line feed
ESC U	1B 55	27 85	Half line feed
ESC D	1B 44	27 68	Reverse half line feed
FF	ØC	12	Form feed
VT	ØB	11	Vertical tab
ESC VT #	1B ØB #	27 11 #	Absolute vertical tab, # is line 1 to 126
ESC _	1B 2D	27 45	Set vertical tab stop current position
ESC 2	1B 32	27 5Ø	Clear all vertical and horizontal tab stops
ESC RS #	1B 1E #	27 3Ø #	Set vertical motion index, # is 1 to 126, where #1/48 inch is the line spacing.
Character Selection:			
ESC P	1B 5Ø	27 8Ø	Enable proportional print spacing
ESC Q	1B 51	27 81	Disable proportional print spacing
ESC SO DC2	1B ØE 12	27 14 18	Enable printwheel down–load mode
DC4	14	28	Exit printwheel down-load
SO	ØE	14	Enable ESC mode, supplementary characters
SI	ØF	15	Disable ESC mode, primary characters
ESC A	1B 41	27 65	Select red ribbon (secondary font)
ESC B	1B 42	27 66	Select black ribbon (primary font)
ESC X	1B 58	27 88	Cancel all WP modes except Proportional
ESC Y	1B 59	27 89	Printwheel Spoke Ø char.
ESC Z	1B 5A	27 9Ø	Printwheel Spoke 95 char.
Character Highlight Selection:			
ESC E	1B 45	27 69	Enable underscore print
ESC R	1B 52	27 82	Disable underscore print
ESC O	1B 4F	27 79	Enable bold printing
ESC W	1B 57	27 87	Enable shadow printing
ESC &	1B 26	27 38	Disable bold and shadow printing
Graphics:			
ESC 3	1B 33	27 51	Enable graphics mode
ESC 4	1B 34	27 52	Disable graphics mode
ESC G	1B 47	27 71	Enable HyPLOT mode
Miscellaneous:			
ESC CR P	1B ØD 5Ø	27 13 8Ø	Reset all modes to default
ESC SUB I	1B 1A 49	27 27 73	Reset all modes to default
ESC EM	1B 19	27 25	Enable auto sheet feeder

DIABLO 630 PRINTER CODES

Code	Hex	Decimal	Command
ESC SUB	1B 1A	27 26	Enable remote diagnostics
ESC N	1B 4E	27 78	Restore normal carriage settling time
ESC %	1B 25	27 37	Increase carriage settling time
ESC 7	1B 37	27 55	Enable print suppression
ESC SO M	1B ØE 4D	27 14 77	Enable program mode

EPSON FX–80 PRINTER CODES (9 PIN)

Code	Hex	Decimal	Command
Page Format Control:			
ESC l #	1B 6C #	27 1Ø8 #	Set Left Margin at Col #
ESC Q #	1B 51 #	27 81 #	Set Right Margin at Col #
ESC C #	1B 43 #	27 67 #	Set Form Length to # Lines (or n inches)
ESC C Ø #	1B 43 ØØ #	27 67 Ø #	Set Form Length to # inches
ESC N #	1B 4E #	27 78 #	Set Skip–over Perforation to # lines
ESC O	1B 4F	27 79	Turn Skip–over Perforation Off
Horizontal Movement and Spacing Control:			
CR	ØD	13	Carriage return
BS	Ø8	Ø8	Backspace
HT	Ø9	Ø9	Horizontal tab
ESC a Ø	1B 61 ØØ	27 97 Ø	Alignment Left Justified
ESC a 1	1B 61 Ø1	27 97 1	Alignment Auto Centering
ESC a 2	1B 61 Ø2	27 97 2	Alignment Right Justified
ESC a 3	1B 61 Ø3	27 97 3	Alignment Auto Justified
ESC D # Ø	1B 44 # Ø	27 68 # ØØ	Set Horizontal Tab(s), # can be 1 or a series of tabs
ESC D Ø	1B 44 Ø	27 68 ØØ	Release Horizontal Tab
ESC e Ø #	1B 44 Ø #	27 68 ØØ #	Set Horizontal Unit Tab(s), # is repeating Tab distance in columns.
ESC e ØØ	1B 44 ØØ	27 68 ØØ ØØ	Release Horiz Tab Unit
ESC f Ø #	1B 66 ØØ #	27 1Ø2 Ø #	Move print position # cols
ESC \ #1#2	1B 5C #1#2	27 92 #1#2	Move print position in increments of 1/12Ø inch
ESC $ #1#2	1B 24 #1#2	27 36 #1#2	Move print position in1/6Ø inch increments from left margin
ESC SP #	1B 2Ø #	27 32 #	Add space after each character in units of 1/24Ø inch where # is from 1 to 63
ESC <	1B 3C	27 6Ø	One Line Unidirectional Printing Mode On
ESC U	1B 55	27 85	Select Continuous Print Unidirectional Mode

EPSON FX–80 PRINTER CODES (9 PIN)

Code	Hex	Decimal	Command
Vertical Movement and Spacing Control:			
LF	ØA	1Ø	Line feed
ESC j #	1B 6A #	27 1Ø6 #	Reverse Line Feed of #/216 Inch
ESC J #	1B 4A #	27 74 #	Forward Line Feed of #/216 inches
ESC f 1 #	1B 66 Ø1 #	27 1Ø2 1 #	Forward Line Feed # lines
FF	ØC	12	Form feed
ESC Ø	1B 3Ø	27 48	Set Line Spacing to 1/8" (9 points or 8 lpi)
ESC 1	1B 31	27 49	Set Line Spacing to 7/72" (7 points)
ESC 2	1B 32	27 5Ø	Set Line Spacing to 1/6" (12 points, 6 lpi)
ESC 3 #	1B 33 #	27 51 #	Set Line Spacing to #/216"
Vertical Movement and Spacing Control: (Continued)			
ESC A #	1B 41 #	27 65 #	Set Line Spacing to # Points (#/72 inch)
VT	ØB	11	Vertical tab
ESC b#1#2#3 Ø	1B 62 #1#2#3 ØØ	27 98 #1#2#3 Ø	Set Vertical Tabs Format Units in Specific Channel, see the manual for details
ESC b #1 Ø	1B 62 #1 ØØ	27 98 #1 Ø	Release Vertical Tab Format Unit
ESC / #	1B 2F #	27 47 #	Select Vertical Tab Channel #
ESC B#1#2Ø	1B 42 #1#2 Ø	27 66 #1 #2 Ø	Set Vertical Tabs for Channel #1, #2 etc
ESC B Ø	1B 42 Ø	27 66 Ø	Release Vertical Tabs for Channels
ESC e 1 #	1B 65 Ø1 #	27 1Ø1 1 #	Set Vertical Tab Unit at # of equal space intervals
ESC e 1 1	1B 65 Ø1 Ø1	27 1Ø1 1 1	Release Vertical Tab Unit of equal space intervals
Character Selection:			
ESC I 1	1B 49 Ø1	27 73 1	Select Characters (Ø–31, 128–159) to Print
ESC I Ø	1B 49 ØØ	27 73 Ø	Disable Characters (Ø–31, 128–159) from Printing
ESC M	1B 4D	27 77	Enable Elite Pitch Mode
ESC P	1B 5Ø	27 8Ø	Enable Pica Pitch Mode
ESC o	1B 6F	27 111	Enable Elite Pitch Mode
ESC n	1B 6E	27 11Ø	Enable Pica Pitch Mode
ESC w #	1B 77 #	27 119 #	Direct Pitch Selection, #=Ø is 1Øcpi, #=1 is 12cpi, #=2 is 15cpi, #=3 is 17cpi, #=4 is proport.
ESC p 1	1B 7Ø Ø1	27 112 1	Select Proportional Spac
ESC p Ø	1B 7Ø ØØ	27 112 Ø	Release Proportional Spa
ESC W 1	1B 57 Ø1	27 87 1	Select Expanded Pitch
ESC W Ø	1B 57 ØØ	27 87 Ø	Release Expanded Pitch

EPSON FX–80 PRINTER CODES (9 PIN)

Code	Hex	Decimal	Command
Character Selection: (Continued)			
SO or ESC SO	ØE	14	Enable 1–line Expanded Print Mode
DC4	14	28	Disable one–line Expanded Print Mode
SI or ESC SI	ØF	15	Enable Compressed Print
DC2	12	18	Disable Compressed Print
ESC :	1B 3A	27 58	Duplicate Internal Font
ESC ! #	1B 21 #	27 33 #	Print Mode Selection, # determines mode, #=128 is underline, #=64 is italic, #=32 is double wide, #=16 is double strike, #=8 is bold, #=4 is compressed, #=2 is proport–ional, #=1 is Elite, #=Ø is Pica. Add numbers for multiples, eg, 129 is Underlined Elite
ESC %	1B 25	27 37	Select Character Set Bank
ESC &	1B 26	27 38	Define User Font
ESC 6	1B 36	27 54	Enable printing High Bit Symbols (Dec128–Dec159)
ESC 7	1B 37	27 55	Disable printing High Bit Symbols (Dec128–Dec159)
ESC 4	1B 34	27 52	Enable Italics printing
ESC 5	1B 35	27 53	Disable Italics printing
ESC R #	1B 52 #	27 82 #	Select International Char–acter Set, #=Ø is USA, 1 is France, 2 is Germany, 3 is England, 4 is Denmark A, 5 is Sweden, 6 is Italy, 7 is Spain, 8 is Japan, 9 is Norway, 1Ø is Denmark B
ESC S 1	1B 53 Ø1	27 83 1	Select Subscripting
ESC S Ø	1B 53 ØØ	27 83 Ø	Select Superscripting
ESC T	1B 54	27 84	Release Super or Subscripting
Character Highlight Selection:			
ESC – 1	1B 2D Ø1	27 45 1	Turn underline mode on
ESC – Ø	1B 2D ØØ	27 45 Ø	Turn underline mode off
ESC E	1B 45	27 69	Enable Bold Print Mode
ESC F	1B 46	27 7Ø	Disable Bold Print Mode
ESC G	1B 47	27 71	Enable Double–strike
ESC H	1B 48	27 72	Disable Double–strike
Graphics:			
For values for #1 and #2 below, see printer manuals			
ESC K#1#2	1B 4B #1#2	27 75 #1#2	Enable Single–density Graphics Mode, 6Ø dpi
ESC L#1#2	1B 4C #1#2	27 76 #1#2	Enable Double–density Graphics Mode, 12Ø dpi
ESC Y#1#2	1B 59 #1#2	27 89 #1#2	Enable Double–density, 12Ø dpi, High–speed Graphics Mode
ESC Z #1#2	1B 5A #1#2	27 9Ø #1#2	Enable Quadruple – density Graphics Mode, 24Ø dpi

EPSON FX–80 PRINTER CODES (9 PIN)

Code	Hex	Decimal	Command
Graphics: (Continued)			
ESC * #1#2#3	1B 2A #1#2#3	27 42 #1#2#3	Set Graphics Mode
ESC ^ #1#2#3	1B 5E #1 #2 #3	27 94 #1#2#3	9 pin Graphics Mode
ESC ? #1#2	1B 3F #1#2	27 63 #1#2	Bit Image Mode Reassignment
Miscellaneous:			
CAN	18	24	Cancel
DC1	11	17	Remote Printer Select
DC3	13	19	Remote Printer Deselect
DEL	7F	127	Delete
ESC @	1B 4Ø	27 64	Master Reset
ESC #	1B 23	27 35	Read Bit 7 of Received Word Normally
ESC =	1B 3D	27 61	Set Received Bit 7 to Ø
ESC >	1B 3E	27 62	Set Received Bit 7 to 1
ESC 8	1B 38	27 56	Out of Paper Sensor Off
ESC 9	1B 39	27 57	Out of Paper Sensor On
ESC i	1B 69	27 1Ø5	Enable Immediate Printing
ESC s	1B 73	27 115	Half Speed Printing
ESC s 1	1B 73 Ø1	27 115 1	Sets Half Speed Printing
ESC s Ø	1B 73 ØØ	27 115 Ø	Releases Half Speed Printing
ESC EM #	1B 19 #	27 25 #	Paper Cassette Selection, #=E is envelope, #=1 is Lower Cassette, #=2 is Upper Cassette, #=R is eject page

EPSON LQ860 PRINTER CODES (24 PIN)

Code	Hex	Decimal	Command
Page Format Control:			
ESC l #	1B 6C #	27 1Ø8 #	Set Left Margin at Col #
ESC Q #	1B 51 #	27 81 #	Set Right Margin at Col #
ESC C #	1B 43 #	27 67 #	Set Form Length to # Lines (or n inches)
ESC C Ø #	1B 43 ØØ #	27 67 Ø #	Set Form Length to # inches
ESC N #	1B 4E #	27 78 #	Set Skip–over Perforation to # lines
ESC O	1B 4F	27 79	Turn Skip–over Perforation Off
Horizontal Movement and Spacing Control:			
CR	ØD	13	Carriage return
BS	Ø8	Ø8	Backspace
HT	Ø9	Ø9	Horizontal tab
ESC a Ø	1B 61 ØØ	1B 61 Ø	Alignment Left Justified
ESC a 1	1B 61 Ø1	1B 61 1	Alignment Auto Centering

EPSON LQ860 PRINTER CODES (24 PIN)

Code	Hex	Decimal	Command
Horizontal Movement and Spacing Control (cont.):			
ESC a 2	1B 61 Ø2	1B 61 2	Alignment Right Justified
ESC a 3	1B 61 Ø3	1B 61 3	Alignment Auto Justified
ESC D #Ø	1B 44 # Ø	27 68 # ØØ	Set Horizontal Tab(s), # can be 1 or a series of tabs
ESC D Ø	1B 44 Ø	27 68 ØØ	Release Horizontal Tab
ESC e Ø #	1B 44 Ø #	27 68 ØØ #	Set Horizontal Unit Tab(s), # is repeating Tab distance in columns
ESC e ØØ	1B 44 ØØ	27 68 ØØ ØØ	Release Horiz Tab Unit
ESC f Ø #	1B 66 ØØ #	27 1Ø2 Ø #	Move print position # cols
ESC \ #1#2	1B 5C #1#2	27 92 #1#2	Move print position in increments of 1/12Ø inch
ESC $ #1#2	1B 24 #1#2	27 36 #1#2	Move print position in 1/6Ø inch increments from left margin
ESC SP #	1B 2Ø #	27 32 #	Add space after each character in units of 1/24Ø inch where # is from 1 to 63
ESC <	1B 3C	27 6Ø	One Line Unidirectional Printing Mode On
ESC U	1B 55	27 85	Select Continuous Print Unidirectional Mode
ESC U Ø	1B 55 ØØ	27 85 Ø	Releases unidirectional printing
ESC U 1	1B 55 Ø1	27 85 1	Sets unidirectional printing
Vertical Movement and Spacing Control:			
LF	ØA	1Ø	Line feed
ESC j #	1B 6A #	27 1Ø6 #	Reverse Line Feed of #/216 Inch
ESC J #	1B 4A #	27 74 #	Forward Line Feed of #/216 inches
ESC f 1 #	1B 66 Ø1 #	27 1Ø2 1 #	Forward Line Feed # lines
FF	ØC	12	Form feed
ESC Ø	1B 3Ø	27 48	Set Line Spacing to 1/8" (9 points or 8 lpi)
ESC 1	1B 31	27 49	Set Line Spacing to 7/72" (7 points)
ESC 2	1B 32	27 5Ø	Set Line Spacing to 1/6" (12 points, 6 lpi)
ESC 3 #	1B 33 #	27 51 #	Set Line Spacing to #/216"
ESC A #	1B 41 #	27 65 #	Set Line Spacing to # Points (#/72 inch)
ESC + #	1B 2B	27 43	Sets paper feed to #/360 inch
VT	ØB	11	Vertical tab
ESC b #1#2#3 Ø	1B 62 #1#2#3 ØØ	27 98 #1#2#3 Ø	Set Vertical Tabs Format Units in Specific Channel, see the manual for details
ESC b #1 Ø	1B 62 #1 ØØ	27 98 #1 Ø	Release Vertical Tab Format Unit

EPSON LQ860 PRINTER CODES (24 PIN)

Code	Hex	Decimal	Command
ESC / #	1B 2F #	27 47 #	Select Vertical Tab Channel #
ESC B #1#2Ø	1B 42 #1#2 Ø	27 66 #1 #2 Ø	Set Vertical Tabs for Channel #1, #2 etc
ESC B Ø	1B 42 Ø	27 66 Ø	Release Vertical Tabs for Channels
ESC e 1 #	1B 65 Ø1 #	27 1Ø1 1 #	Set Vertical Tab Unit at # of equal space intervals
ESC e 1 1	1B 65 Ø1 Ø1	27 1Ø1 1 1	Release Vertical Tab Unit of equal space intervals
Character Selection:			
ESC I 1	1B 49 Ø1	27 73 1	Select Characters (Ø–31, 128–159) to Print
ESC I Ø	1B 49 ØØ	27 73 Ø	Disable Characters (Ø–31, 128–159) from Printing
ESC M	1B 4D	27 77	Enable Elite Pitch Mode
ESC P	1B 5Ø	27 8Ø	Enable Pica Pitch Mode
ESC o	1B 6F	27 111	Enable Elite Pitch Mode
ESC n	1B 6E	27 11Ø	Enable Pica Pitch Mode
ESC w #	1B 77 #	27 119 #	Direct Pitch Selection, #=Ø is 1Øcpi, #=1 is 12cpi, #=2 is 15cpi, #=3 is 17cpi, #=4 is proport.
ESC p 1	1B 7Ø Ø1	27 112 1	Select Proportional Spacing
ESC p Ø	1B 7Ø ØØ	27 112 Ø	Release Proportional Spacing
ESC W 1	1B 57 Ø1	27 87 1	Select Expanded Pitch
ESC W Ø	1B 57 ØØ	27 87 Ø	Release Expanded Pitch
SO or ESC SO	ØE	14	Enable 1–line Expanded Print Mode
DC4	14	28	Disable one–line Expanded Print Mode
SI or ESC SI	ØF	15	Enable Compressed Print
DC2	12	18	Disable Compressed Print
ESC :	1B 3A	27 58	Duplicate Internal Font
ESC : Ø # Ø	1B 3A ØØ	27 58	Copies internal ROM CG font into download CG
ESC ! #	1B 21 #	27 33 #	Print Mode Selection, # determines mode, #=128 is underline, #=64 is italic, #=32 is double wide, #=16 is double strike, #=8 is bold, #=4 is compressed, #=2 is proportional, #=1 is Elite, #=Ø is Pica. Add numbers for multiples, eg, 129 is Underlined Elite
ESC %	1B 25	27 37	Select Character Set Bank
ESC % Ø	1B 25	27 37	Selects ROM CG
ESC % 1	1B 25	27 37	Selects download CG
ESC &	1B 26	27 38	Define User Font
ESC 6	1B 36	27 54	Enable printing High Bit

EPSON LQ860 PRINTER CODES (24 PIN)

Code	Hex	Decimal	Command
Character Selection: (Continued)			
			Symbols (Dec128–Dec159)
ESC 7	1B 37	27 55	Disable printing High Bit Symbols (Dec128–Dec159)
ESC 4	1B 34	27 52	Enable Italics printing
ESC 5	1B 35	27 53	Disable Italics printing
ESC R #	1B 52 #	27 82 #	Select International Character Set, #=Ø is USA, 1 is France, 2 is Germany,3 is England, 4 is Denmark A, 5 is Sweden, 6 is Italy, 7 is Spain, 8 is Japan, 9 is Norway, 1Ø is Denmark B
ESC S 1	1B 53 Ø1	27 83 1	Select Subscripting
ESC S Ø	1B 53 ØØ	27 83 Ø	Select Superscripting
ESC T	1B 54	27 84	Release Super or Subscripting
ESC t #	1B 74	27 116	Selects character set, #=Ø is Italic set, #=1 is Graphic set #=2 remaps downloaded characters from 0-127 to 128-255
ESC g	1B 67	27 1Ø3	Sets micron (15 cpi) printing
ESC x #	1B 78	27 12Ø	Selects print quality, #=Ø is Draft mode, #=1 is LQ mode, #2 is SLQ mode.
ESC k #	1B 6B	27 1Ø7	Selects print typeface (NOTE: these may vary between printers.) #=Ø is Roman #=1 is Sans Serif #=2 is Courier #=3 is Prestige #=4 is Script #=5 is OCR-B #=6 is Bold PS #=7 is Orator
Character Highlight Selection:			
ESC – 1	1B 2D Ø1	27 45 1	Turn underline mode on
ESC – Ø	1B 2D ØØ	27 45 Ø	Turn underline mode off
ESC E	1B 45	27 69	Enable Bold Print Mode
ESC F	1B 46	27 7Ø	Disable Bold Print Mode
ESC G	1B 47	27 71	Enable Double–strike
ESC H	1B 48	27 72	Disable Double–strike
ESC w 1	1B 77 Ø1	27 119 1	Sets Double-High Printing
ESC w Ø	1B 77 ØØ	27 119 Ø	Releases Double - High Printing
ESC q #	1B 71	27 113	Sets Outline & Shadow Printing

EPSON LQ860 PRINTER CODES (24 PIN)

Code	Hex	Decimal	Command
Graphics:			
For values for #1 and #2 below, see printer manuals			
ESC K#1#2	1B 4B #1#2	27 75 #1#2	Enable Single–density Graphics Mode, 6Ø dpi
ESC L#1#2	1B 4C #1#2	27 76 #1#2	Enable Double–density Graphics Mode, 12Ø dpi
ESC Y#1#2	1B 59 #1#2	27 89 #1#2	Enable Double–density, 12Ø dpi, High–speed Graphics Mode
ESC Z #1#2	1B 5A #1#2	27 9Ø #1#2	Enable Quadruple – density Graphics Mode, 24Ø dpi
ESC * #1#2#3	1B 2A #1#2#3	27 42 #1#2#3	Set Graphics Mode
ESC ∧ #1#2#3	1B 5E #1 #2 #3	27 94 #1#2#3	9 pin Graphics Mode
ESC ? #1#2	1B 3F #1#2	27 63 #1#2	Bit Image Mode Reassignment
Miscellaneous:			
CAN	18	24	Cancel
DC1	11	17	Remote Printer Select
DC3	13	19	Remote Printer Deselect
DEL	7F	127	Delete
ESC @	1B 4Ø	27 64	Master Reset
ESC "#"	1B 23	27 35	Set to receive Bit 8 as is.
ESC =	1B 3D	27 61	Set Received Bit 7 to Ø
ESC >	1B 3E	27 62	Set Received Bit 7 to 1
ESC 8	1B 38	27 56	Out of Paper Sensor Off
ESC 9	1B 39	27 57	Out of Paper Sensor On
ESC i	1B 69	27 1Ø5	Enable Immediate Printing
ESC s	1B 73	27 115	Half Speed Printing
ESC EM #	1B 19 #	27 25 #	Paper Cassette Selection, #=E is envelope, #=1 is Lower Cassette, #=2 is Upper Cassette, #=R is eject page
BEL	Ø7	7	Sounds the buzzer for approx. Ø.5 seconds
ESC r #	1B 72	27 114	Selects print color (Note: may vary between printers) #=Ø is Black #=1 is Red #=2 is Blue #=3 is Violet #=4 is Yellow #=5 is Orange #=6 is Green

NEC PINWRITER PRINTER CODES

NEC Pinwriters use most of the same codes as the Epson LQ1500, except for the following FS Codes:

Code	Hex	Decimal	Command
FS 3 #	1C 33 #	28 51 #	Line space Ø-255 #/360
FS C #	1C 43 #	28 67 #	Set Font Cartridge, #=Ø is resident font, #=1 is slot 1, #=2 is slot 2
FS E #	1C 45 #	28 69 #	Ø=Cancel horiz enlarge., 1=2X horiz enlargement, 2=3X horiz enlargement
FS F	1C 46	28 7Ø	Release Enhanced Print
FS I #	1C 49 #	28 73 #	Ø=Italic Set, 1=IBM Set
FS R	1C 52	28 82	Set Reverse Line Feed
FS S #	1C 53 #	28 83 #	Ø=Draft 12,1=high speed
FS V 1	1C 56 31	28 86 49	Set double vertical enlarge
FS V Ø	1C 56 3Ø	28 86 48	Release double vertical enlargement
FS Z #1 #2	1C 6Ø #1 #2	28 9Ø #1 #2	Set 360 dpi graphics
FS @	1C 4Ø	28 64	Initialize except user buffer

HP LASERJET PCL3 CODES

Code	Hex	Decimal	Command
Page Format Control:			
ESC & l Ø O	1B 26 6C 3Ø 4F	27 38 1Ø8 48 79	Portrait Orient.
ESC & l 1O	1B 26 6C 31 4F	27 38 1Ø8 49 79	Landscape Orient.
ESC & l #P	1B 26 6C # 5Ø	27 38 1Ø8 # 8Ø	Page length, # of lines
ESC & l #E	1B 26 6C # 45	27 38 1Ø8 # 69	Top Margin, # of lines
ESC & l #F	1B 26 6C # 46	27 38 1Ø8 # 7Ø	Text Length, # of lines
ESC & l 1L	1B 26 6C 31 4C	27 38 1Ø8 49 76	Skip Perforation, On
ESC & l ØL	1B 26 6C 3Ø 4C	27 38 1Ø8 48 76	Skip Perforation, Off
ESC & l #D	1B 26 6C # 44	27 38 1Ø8 # 68	Lines Per Inch, # of lines/inch
ESC & l #C	1B 26 6C # 43	27 38 1Ø8 # 67	Vertical Motion Index # of 1/48 inch
ESC &k#H	1B 26 6B # 48	27 38 1Ø7 # 72	Horizontal Motion Index, # of 1/12Ø inch
ESC &a#L	1B 26 61 # 4C	27 38 97 # 76	Left Margin, Left column #
ESC &a#M	1B 26 61 # 4D	27 38 97 # 77	Right Margin, Right column #
ESC 9	1B 39	27 57	Clear Margins
Horizontal Movement and Spacing Control:			
BS	Ø8	8	Backspace
CR	ØD	13	Carriage Return
ESC & k # G	1B 26 6B # 47	27 38 1Ø7 # 71	CR/LF/FF Line

HP LASERJET PCL3 CODES

Code	Hex	Decimal	Command
			Termination Action

Line Termination Action

#	CR	LF	FF
Ø	CR	LF	FF
1	CR+LF	LF	FF
2	CR	CR+LF	CR+FF
3	CR+LF	CR+LF	CR+FF

Code	Hex	Decimal	Command
ESC & s Ø C	1B 26 73 3Ø 43	27 38 115 48 67	Set Wrap Around
ESC & s 1 C	1B 26 73 31 43	27 38 115 49 67	Release Wrap Around
ESC & a # C	1B 26 61 # 43	27 38 97 # 67	Move Print Posi–tion to Column #
ESC & a # H	1B 26 61 # 48	27 38 97 # 72	Move Print Position Horizontal # of Decipoints
ESC *p # X	1B 2A 7Ø # 58	27 42 112 # 88	Move Print Position Horizontal # of Dots

Vertical Movement and Spacing Control:

Code	Hex	Decimal	Command
LF	ØA	1Ø	Line Feed
FF	ØC	12	Formfeed
ESC =	1B 3D	27 61	Half Line Feed
ESC & a # R	1B 26 61 # 52	27 38 97 # 82	Move Print Position to Row #
ESC & a # V	1B 26 61 # 56	27 38 97 # 86	Move Print Position Vertical # of Decipoints
ESC * p # Y	1B 2A 7Ø # 59	27 42 112 # 89	Move Print Position Vertical # of Dots

Font Selection:

Code	Hex	Decimal	Command
ESC (# X	1B 28 # 58	27 4Ø # 88	Symbol Set, Primary, # is Character ID
ESC) # X	1B 29 # 58	27 41 # 88	Symbol Set, Secondary, # is Character ID

Character ID's:

Roman-8bit = 8U
Kana-8bit = 8K,
Math-8bit = 8M
ANSI-8bit = 9U
USASCII = ØU
Line Draw = ØB
Math Symbols = ØA
US Legal = 1U
Roman Ext = ØE
ISO Denmark = ØD
ISO Italy = ØI
ISO United Kingdom = 1E
ISO France = ØF
ISO Germany = ØG
ISO Sweden = ØS
ISO Spain = 1S

Code	Hex	Decimal	Command
ESC (s Ø P	1B 28 73 3Ø 5Ø	27 4Ø 115 48 8Ø	Spacing, Primary Fixed
ESC (s 1 P	1B 28 73 31 5Ø	27 4Ø 115 49 8Ø	Spacing, Primary Proportional
ESC) s Ø P	1B 29 73 3Ø 5Ø	27 41 115 48 8Ø	Spacing, Secondary Fixed

HP LASERJET PCL3 CODES

Code	Hex	Decimal	Command
ESC) s 1 P	1B 29 73 31 5Ø	27 41 115 49 8Ø	Spacing, Secondary Proportional
ESC (s # H	1B 28 73 # 48	27 4Ø 115 # 72	Print Pitch, Primary, # is characters/inch
ESC) s # H	1B 29 73 # 48	27 41 115 # 72	Print Pitch, Secondary, # is characters/inch
ESC & k # S	1B 26 6B # 53	27 38 1Ø7 # 83	Print Pitch, Prim. & Secondary, #=Ø is 1Ø cpi, #=1 is 16.66 cpi
ESC (s # V	1B 28 73 # 56	27 4Ø 115 # 86	Print Point Size, Primary, # is points
ESC) s # V	1B 29 73 # 56	27 41 115 # 86	Print Point Size, Secondary, # is points
ESC (s Ø S	1B 28 73 3Ø 53	27 4Ø 115 48 83	Print Style, Primary, Upright
ESC (s1S	1B 28 73 31 53	27 4Ø 115 49 83	Print Style, Primary, Italic
ESC) sØS	1B 29 73 3Ø 53	27 41 115 48 83	Print Style, Secondary, Upright
ESC) s1S	1B 29 73 31 53	27 41 115 49 83	Print Style, Secondary, Italic
ESC (s # B	1B 28 73 # 42	27 4Ø 115 # 66	Stroke Weight, Primary, # is –7 to +7
ESC) s # B	1B 29 73 # 42	27 41 115 # 66	Stroke Weight, Secondary, # is –7 to +7 –1 to –7=light, Ø =Medium, 1 to 7 =Bold
ESC (s # T	1B 28 73 # 54	27 4Ø 115 # 84	Typeface, Primary # is typeface
ESC) s # T	1B 29 73 # 54	27 41 115 # 84	Typeface, Secondary # is typeface:

Typeface ID's:

Ø=Line printer	6=Gothic
1=Pica	7=Script
2=Elite	8=Prestige
3=Courier	9=Caslon
4=Swiss 721	1Ø=Orator
5=Dutch	23=Century 7Ø

Font Control:

SI	ØF	15	Shift In Primary
SO	ØE	14	Shift In Secondary
ESC (# X	1B 28 # 58	27 4Ø # 88	Define Font, Primary # is Font ID number
ESC) # X	1B 29 # 58	27 41 # 88	Define Font,Secondary # is Font ID numbr
ESC *c # F	1B 2A 63 # 46	27 42 99 # 7Ø	Font/Character Control, see printer manual

Code	Hex	Decimal	Command
Font Control (cont.):			
ESC (# @	1B 28 # 4Ø	27 4Ø # 64	Primary Font, Default see printer manual
ESC) # @	1B 29 # 4Ø	27 41 # 64	Secondary Font Default, see printer manual
ESC *c # D	1B 2A 63 # 44	27 42 99 # 68	Define Font ID, # is the ID
ESC) s # W	1B 29 73 # 57	27 41 115 # 87	Font Header, # is byte number of font attribute
ESC *c # E	1B 2A 63 # 45	27 42 99 # 69	Define Character Code to download # is Ø to 255
ESC (s # W	1B 28 73 # 57	27 4Ø 115 # 87	Produce Download Character see printer manual
Character Highlight Selection:			
ESC & d D	1B 26 64 44	27 38 1ØØ 68	Turn underline on
ESC & d @	1B 26 64 4Ø	27 38 1ØØ 64	Turn underline off
Graphics:			
ESC * t # R	1B 2A 74 # 52	27 42 116 # 82	Resolution, # is 75, 1ØØ, 15Ø, or 3ØØ Dots/inch
ESC * r # A	1B 2A 72 # 41	27 42 114 # 65	Graphics Start, #=Ø is start vertical from left end of print area, #=1 is start from present position.
ESC * b # W	1B 2A 62 # 57	27 42 98 # 87	Sending Graphics data, # is number of bytes of bit image data.
ESC * r B	1B 2A 72 42	27 42 114 66	End Raster Graphics Mode
ESC * c # A	1B 2A 63 # 41	27 42 99 # 65	Set Horizontal Rule Width to # dots (1 dot=1/3ØØ inch)
ESC * c # H	1B 2A 63 # 48	27 42 99 # 72	Set Horizontal Rule Width to # decipoints (1 deci-point=1/72Ø inch)
ESC*c # B	1B 2A 63 # 42	27 42 99 # 66	Set Vertical Rule Width to # dots (1 dot=1/3ØØ inch)
ESC *c # V	1B 2A 63 # 56	27 42 99 # 86	Set Vertical Rule Width to # deci-points (1 decipoint= 1/72Ø inch)
ESC *c # G	1B 2A 63 # 47	27 42 99 # 71	Set Gray Scale or Hatch Pattern ID #, see printer manual for a sample of each

Code	Hex	Decimal	Command
			pattern/hatch and its associated ID #
ESC *c # P	1B 2A 63 # 5Ø	27 42 99 # 8Ø	Set Print Pattern #
Macro's:			
ESC &f # Y	1B 26 66 # 59	27 38 1Ø2 # 89	Set Macro ID #
ESC &f Ø X	1B 26 66 3Ø 58	27 38 1Ø2 48 88	Start Macro
ESC &f 1 X	1B 26 66 31 58	27 38 1Ø2 49 88	End Macro
ESC &f 2 X	1B 26 66 32 58	27 38 1Ø2 5Ø 88	Jump to Macro
ESC &f 3 X	1B 26 66 33 58	27 38 1Ø2 51 88	Call Macro
ESC &f 4 X	1B 26 66 34 58	27 38 1Ø2 52 88	Set Overlay Macro
ESC &f 5 X	1B 26 66 35 58	27 38 1Ø2 53 88	Release Overlay Macro
ESC &f 6 X	1B 26 66 36 58	27 38 1Ø2 54 88	Release all Macro
ESC &f 7 X	1B 26 66 37 58	27 38 1Ø2 55 88	Release all temporary Macro
ESC &f 8 X	1B 26 66 38 58	27 38 1Ø2 56 88	Release current Macro
ESC &f 9 X	1B 26 66 39 58	27 38 1Ø2 57 88	Assign temporary attribute to Macro
ESC &f 1ØX	1B 26 66 31 3Ø 58	27 38 1Ø2 49 48 88	Assign permanent attribute to Macro
Miscellaneous:			
ESC Y	1B 59	27 89	Set Display Function of control codes
ESC Z	1B 5A	27 9Ø	Release Display Function of control codes
ESC & p # X	1B 26 7Ø # 58	27 38 112 # 88	Transparent Print Data (no ESC commands exist)
ESC & f Ø S	1B 26 66 3Ø 53	27 38 1Ø2 48 83	Push Printing Position. Puts present printing position on the top of the stack
ESC & f 1 S	1B 26 66 31 53	27 38 1Ø2 49 83	Pop Printing Position. Recall stored printing position and put on the top of the stack
ESC & l # X	1B 26 6C # 58	27 38 1Ø8 # 88	Set Number of Copies to #
ESC & l # H	1B 26 6C # 48	27 38 1Ø8 # 72	Paper Input Control.
	#=Ø is Feed out current page #=1 is Lower Cassette supplies paper #=3 is Envelope feeder supplies envelope #=4 is Upper Cassette supplies paper		
ESC E	1B 45	27 69	Reset Printer
ESC z	1B 7A	27 122	Start Printer Self Test

HP LASERJET PCL5 CODES

Code	Hex	Decimal	Command
Page Format Control:			
ESC & l ØO	1B 26 6C 3Ø 4F	27 38 1Ø8 48 79	Portrait Orient.
ESC & l 2O	1B 26 6C 32 4F	27 38 1Ø8 5Ø 79	Reverse Portrait
ESC & l 1O	1B 26 6C 31 4F	27 38 1Ø8 49 79	Landscape Orient.
ESC & l 3O	1B 26 6C 33 4F	27 38 1Ø8 51 79	Reverse Landscape
ESC & l #P	1B 26 6C # 5Ø	27 38 1Ø8 # 8Ø	Page length, # of lines
ESC & l #E	1B 26 6C # 45	27 38 1Ø8 # 69	Top Margin, # of lines
ESC & l #F	1B 26 6C # 46	27 38 1Ø8 # 7Ø	Text Length, # of lines
ESC & l 1L	1B 26 6C 31 4C	27 38 1Ø8 49 76	Skip Perforation, Set on
ESC & l ØL	1B 26 6C 3Ø 4C	27 38 1Ø8 48 76	Skip Perforation, Set off
ESC & l #D	1B 26 6C # 44	27 38 1Ø8 # 68	Lines Per Inch, # of lines/inch
ESC & l #C	1B 26 6C # 43	27 38 1Ø8 # 67	Vertical Motion Index, # of 1/48 inch
ESC &k #H	1B 26 6B # 48	27 38 1Ø7 # 72	Horizontal Motion Index, # of 1/12Ø inch
ESC &a #L	1B 26 61 # 4C	27 38 97 # 76	Left Margin, Left column #
ESC &a #M	1B 26 61 # 4D	27 38 97 # 77	Right Margin, Right column #
ESC &a #P	1B 26 61 #...#5Ø	27 38 97 #...# Ø8Ø	# Degrees of Rotation(counter-clockwise/90 degree increments only)
ESC 9	1B 39	27 57	Clear Margins
Horizontal Movement and Spacing Control:			
BS	Ø8	8	Backspace
CR	ØD	13	Carriage Return
ESC &k # G	1B 26 6B # 47	27 38 1Ø7 # 71	CR/LF/FF Line Termination Action

Line Termination Action

#	CR	LF	FF
Ø	CR	LF	FF
1	CR+LF	LF	FF
2	CR	CR+LF	CR+FF
3	CR+LF	CR+LF	CR+FF

Code	Hex	Decimal	Command
ESC &sØC	1B 26 73 3Ø 43	27 38 115 48 67	Set Wrap Around
ESC &s1C	1B 26 73 31 43	27 38 115 49 67	Release Wrap Around
ESC &a #C	1B 26 61 # 43	27 38 97 # 67	Move Print Posi–tion to Column #
ESC &a #H	1B 26 61 # 48	27 38 97 # 72	Move Print Position Horizontal # of Decipoints

HP LASERJET PCL5 CODES

Code	Hex	Decimal	Command
ESC *p #X	1B 2A 7Ø # 58	27 42 112 # 88	Move Print Position Horizontal # of Dots
ESC & l # U	1B 26 6C #...# 55	27 Ø38 1Ø8 #...# Ø85	Long-edge (left) Offset Registration

Vertical Movement and Spacing Control:

Code	Hex	Decimal	Command
LF	ØA	1Ø	Line Feed
FF	ØC	12	Formfeed
ESC =	1B 3D	27 61	Half Line Feed
ESC &a #R	1B 26 61 # 52	27 38 97 # 82	Move Print Position to Row #
ESC &a #V	1B 26 61 # 56	27 38 97 # 86	Move Print Position Vertical # of Decipoints
ESC *p #Y	1B 2A 7Ø # 59	27 42 112 # 89	Move Print Position Vertical # of Dots
ESC & l # Z	1B 26 6C #...# 5A	27 Ø38 1Ø8 #...# Ø9Ø	Short-edge (top) Offset Registration

Font Selection:

Code	Hex	Decimal	Command
ESC (#	1B 28 #	27 4Ø #	Symbol Set, Primary, # is Character ID
ESC) #	1B 29 #	27 41 #	Symbol Set, Secondary, # is Character ID

Character ID's:

Ø D = ISO 6Ø:Norwegian 1
1E=ISO 4:United Kingdom
1F=ISO 69:French
G=ISO 21:German
Ø1=ISO 15:Italian
6J=Microsoft Publishing
7J=DeskTop
1ØJ=PS Text
13J=Ventura International
14J=Ventura US
9L=Ventura ITC Zapf Dingbats
1ØL=PS ITC Zapf Dingbats
11L=ITC Zapf Dingbats(S100)
12L=ITC Zapf Dingbats(S200)
13L=ITC Zapf Dingbats(S300)

5M=PS Math
6M=Ventura Math
8M=Math-8
ØN=ECMA-94 Latin 1
ØS=ISO 11:Swedish
2S=ISO 17:Spanish
ØU=ISO 6:ASCII
1U=Legal
8U=Roman8
9U=Windows
1ØU=PC-8
11U=PC-8 D/N
12U=PC 850
15U=Pi Font

Code	Hex	Decimal	Command
ESC (s ØP	1B 28 73 3Ø 5Ø	27 4Ø 115 48 8Ø	Spacing, Primary Fixed
ESC (s 1P	1B 28 73 31 5Ø	27 4Ø 115 49 8Ø	Spacing, Primary Proportional
ESC) s ØP	1B 29 73 3Ø 5Ø	27 41 115 48 8Ø	Spacing,

Code	Hex	Decimal	Command
Font Selection: (Continued)			
ESC) s 1P	1B 29 73 31 5Ø	27 41 115 49 8Ø	Secondary Fixed Spacing, Secondary Proportional
ESC (s # H	1B 28 73 # 48	27 4Ø 115 # 72	Print Pitch, Primary, # is characters/inch
ESC) s # H	1B 29 73 # 48	27 41 115 # 72	Print Pitch, Secondary, # is characters/inch
ESC & k # S	1B 26 6B # 53	27 38 1Ø7 # 83	Print Pitch, Prim.
ESC & k ØS	1B 26 6B 31 53	27 38 1Ø7 49 83	1Ø.Ø CPI
ESC & k 1S	1B 26 6B 31 53	27 38 1Ø7 49 83	16.66 CPI
ESC & k 2S	1B 26 6B 32 53	27 38 1Ø7 50 83	Compressed (16.5 - 16.7 CPI)
ESC & k 4S	1B 26 6B 34 53	27 38 1Ø7 52 83	Elite (12.Ø CPI)
ESC (s # V	1B 28 73 # 56	27 4Ø 115 # 86	Print Point Size, Primary, # is points
ESC) s # V	1B 29 73 # 56	27 41 115 # 86	Print Point Size, Secondary, # is points
ESC (sØS	1B 28 73 3Ø 53	27 4Ø 115 48 83	Upright (Solid)
ESC (s1S	1B 28 73 31 53	27 4Ø 115 49 83	Italic
ESC (s4S	1B 28 73 34 53	27 4Ø 115 52 83	Condensed
ESC (s5S	1B 28 73 35 53	27 4Ø 115 53 83	Condensed Italic
ESC (s8S	1B 28 73 38 53	27 4Ø 115 56 83	Compressed (Extra Condensed)
ESC (s24S	1B 28 73 32 34 53	27 4Ø 115 5Ø 52 83	Expanded
ESC (s32S	1B 28 73 33 32 53	27 40 115 51 5Ø 83	Outline
ESC (s64S	1B 28 73 36 34 53	27 4Ø 115 54 52 83	Inline
ESC (s128S	1B 28 73 31 32 38 53	27 4Ø 115 49 5Ø 56 83	Shadowed
ESC (s16ØS	1B 28 73 31 36 30 53	27 4Ø 115 49 54 48 83	Outline Shadowed
ESC (s # B	1B 28 73 # 42	27 4Ø 115 # 66	Stroke Weight, Primary, # is –7 to +7

Stroke Weights

-7=Ultra Thin	1=Semi Bold
-6=Extra Thin	2=Demi Bold
-5=Thin	3=Bold
-4=Extra Light	4=Extra Bold
-3=Light	5=Black
-2=Demi Light	6=Extra Black
-1=Semi Light	7=Ultra Black
Ø=Medium (book or text)	

Code	Hex	Decimal	Command
ESC) s # B	1B 29 73 # 42	27 41 115 # 66	Stroke Weight, Secondary, # is –7 to +7, –1 to –7=light

Code	Hex	Decimal	Command
			Ø =Medium 1 to 7 =Bold
ESC (s # T	1B 28 73 # 54	27 4Ø 115 # 84	Typeface,Primary # is typeface (see below)
ESC) s # T	1B 29 73 # 54	27 41 115 # 84	Typeface,Secondary # is typeface:

Typeface ID's:

Ø=Line printer	7=Script
1=Pica	8=Prestige
2=Elite	9=Caslon
3=Courier	1Ø=Orator
4=Swiss 721	23=Century 7Ø
5=Dutch	4 14 8 = Universe
6=Gothic	4 1Ø 1 = CG Times

Font Control:

Code	Hex	Decimal	Command
SI	ØF	15	Shift In Primary
SO	ØE	14	Shift In Secondary
ESC (# X	1B 28 # 58	27 4Ø # 88	Define Font, Primary # is the Font ID number
ESC) # X	1B 29 # 58	27 41 # 88	Define Font, Secondary, # is the Font ID numbr
ESC *c # F	1B 2A 63 # 46	27 42 99 # 7Ø	Font/Character Control, see printer manual
ESC (# @	1B 28 # 4Ø	27 4Ø # 64	Primary Font Default, see printer manual
ESC) # @	1B 29 # 4Ø	27 41 # 64	Secondary Font Default, see printer manual
ESC *c # D	1B 2A 63 # 44	27 42 99 # 68	Define Font ID, # is the ID
ESC) s # W	1B 29 73 # 57	27 41 115 # 87	Font Header, # is byte number of font attribute
ESC *c # E	1B 2A 63 # 45	27 42 99 # 69	Define Character Code to download, # is Ø to 255
ESC (s # W	1B 28 73 # 57	27 4Ø 115 # 87	Produce Download Character, see printer manual
ESC*c # R	1B 2A 63 #...#52	27 4Ø 99 #...# 82	ID #
ESC (f # W	1B 2A 66 #...#46	27 4Ø 1Ø2 #...87	# of Bytes
ESC * c ØS	1B 24 63 30 53	27 4Ø 99 48 83	Delete all symbol sets
ESC *c1S	1B 2A 63 31 53	27 4Ø 99 49 83	Delete all temporary symbol sets
ESC *c 2S	1B 2A 63 32 53	27 4Ø 99 5Ø 83	Delete current soft symbol sets (last ID#)

Code	Hex	Decimal	Command
Font Control: (Continued)			
ESC *c 4S	1B 2A 63 34 53	27 4Ø 9Ø 52 83	Make current soft symbol set temporary
ESC *c 5S	1B 2A 63 35 53	27 4Ø 9Ø 53 83	Make current soft symbol set permanent
Character Highlight Selection:			
ESC & d D	1B 26 64 44	27 38 1ØØ 68	Turn underline on
ESC & d @	1B 26 64 4Ø	27 38 1ØØ 64	Turn underline off
Graphics:			
ESC *r #A	1B 2A 72 # 41	27 42 114 # 65	Graphics Start,
	#=Ø is start vertical from left end of print area #=1 is start from present position.		
ESC *c #A	1B 2A 63 # 41	27 42 99 # 65	Set Horizontal Rule Width to # dots (1 dot=1/3ØØ inch)
SC *c #B	1B 2A 63 # 42	27 42 99 # 66	Set Vertical Rule Width to # dots (1 dot=1/3ØØ inch)
ESC *c # H	1B 2A 63 # 48	27 42 99 # 72	Set Horizontal Rule Width to # decipoints (1 deci-point=1/72Ø inch)
ESC *c # V	1B 2A 63 # 56	27 42 99 # 86	Set Vertical Rule Width to # deci-points (1 decipoint= 1/72Ø inch)
ESC%Ø A	1B 25 3Ø 41	27 37 48 65	Use previous PCL cursor position
ESC%1A	1B 25 31 41	27 37 49 65	Use current HP-GL/2 pen position for cursor position
ESC%ØB	1B 25 30 42	27 37 48 66	Use previous HP-GL/2 pen position. Use current PCL cursor position
ESC*c #K	1B 2A 63 #...# 48	27 42 99 #...#75	Horizontal size in inches
ESC*c #L	1B 2A 63 #...# 4C	27 42 99 #...#76	Vertical size in inches
ESC*cØT	1B 2A 63 3Ø 54	27 42 99 84	Set anchor point to cursor position
ESC*c #X	1B 2A 63 #...# 58	27 42 99 #...# 88	Decipoints Horiz.
ESC*c #Y	1B 2A 63 #...#59	27 42 99 #...#89	Decipoints Vert.
ESC* t 75R	1B 2A 74 37 35 52	27 42 116 55 53 82	75 dots/inch
ESC* t 1Ø ØR	1B 2A 74 31 3Ø 3Ø 52	27 42 116 49 48 48 82	100 dots/inch
ESC* t 15 ØR	1B 2A 74 31 35 3Ø 52	27 42 116 49 53 48 82	150 dots/inch

HP LASERJET PCL5 CODES

Code	Hex	Decimal	Command
ESC* t 3 Ø ØR	1B 2A 74 33 30 3Ø 52	27 42 116 51 48 48 82	300 dots/inch
ESC* rØF	1B 2A 72 3Ø 46	27 42 114 48 7Ø	Follows orientation
ESC* r3F	1B 2A 72 33 46	27 42 114 51 7Ø	Follows physical page
ESC* b#Y	1B 2A 62 #...# 59	27 42 98 #...# 89	# of Raster Lines of vertical movement
ESC*bØM	1B 2A 62 30 4D	27 42 98 48 77	Unencoded
ESC*b1M	1B 2A 62 31 4D	27 42 98 49 77	Run-Length Encoded
ESC*b2M	1B 2A 62 32 4D	27 42 98 50 77	Tagged Image Format
ESC*b3M	1B 2A 62 33 4D	27 42 98 51 77	Delta Row
ESC*b5M	1B 2A 62 35 4D	27 42 98 53 77	Adaptive compression
ESC*b#W	1B 2A 62 #...# 57	27 42 98 #...# 87	# of Bytes
ESC*r B	1B 2A 72 42	27 42 114 66	End Raster Graphics
ESC*r # T	1B 2A 72 #...# 54	27 42 114 #...# 84	# Raster Rows
ESC*r # S	1B 2A 72 #...#53	27 42 114 #...# 83	# Pixels of the specified resolution
ESC*vØ T	1B 2A 76 30 54	27 42 118 48 84	Solid Black (default)
ESC*v1T	1B 2A 76 31 54	27 42 118 49 84	Solid White
ESC*v2T	1B 2A 76 32 54	27 42 118 5Ø 84	HP-defined shading pattern
ESC*v3T	1B 2A 76 33 54	27 42 118 51 84	HP-defined Cross-Hatched Pattern
ESC*y4T	1B 2A 76 34 54	27 42 118 52 84	User defined pattern
ESC*vØ N	1B 2A 76 3Ø 4E	27 42 118 48 78	Transparent Source
ESC*v1N	1B 2A 76 31 4E	27 42 118 49 78	Opaque Source
ESC*vØ O	1B 2A 76 3Ø 4F	27 42 118 48 79	Transparent Pattern
ESC*v1O	1B 2A 76 31 4F	27 42 118 49 79	Opaque Pattern
ESC*cØP	1B 2A 63 3Ø 5Ø	27 42 99 48 8Ø	Solid Black
ESC*c1P	1B 2A 63 31 5Ø	27 42 99 49 8Ø	Erase (solid white fill)
ESC*c2P	1B 2A 63 32 5Ø	27 42 99 50 8Ø	Shaded Fill
ESC*c3P	1B 2A 63 33 5Ø	27 42 99 51 8Ø	Cross-hatched Fill
ESC*c5P	1B 2A 63 35 5Ø	27 42 99 53 8Ø	Current Pattern
ESC*c#G	1B 2A 63 #...# 47	27 42 99 #...# 71	% Shading or Type of Pattern
ESC*c2G	1B 2A 63 32 47	27 42 99 5Ø 71	2% Gray
ESC*c1ØG	1B 2A 63 31 3Ø 47	27 42 99 49 48 71	1Ø% Gray
ESC*c15G	1B 2A 63 31 35 47	27 42 99 49 53 71	15% Gray
ESC*c3ØG	1B 2A 63 33 3Ø 47	27 42 99 51 48 71	3Ø% Gray
ESC*c45G	1B 2A 63 34 35 47	27 42 99 52 53 71	45% Gray
ESC*7ØG	1B 2A 63 37 3Ø 47	27 42 99 55 48 71	7Ø% Gray
ESC*c9ØG	1B 2A 63 39 3Ø 47	27 42 99 57 48 71	9Ø% Gray
ESC*c1ØØG	1B 2A 6 331 3Ø 3Ø 47	27 42 99 49 48 48 71	1ØØ% Gray
ESC*c1G	1B 2A 63 31 47	27 42 99 49 71	1 Horiz. Line
ESC*c2G	1B 2A 63 32 47	27 42 99 5Ø 71	2 Vert Lines
ESC*c3G	1B 2A 63 33 47	27 42 99 51 71	3 Diagonal Lines
ESC*c4G	1B 2A 63 34 47	27 42 99 52 71	4 Diagonal Lines
ESC*c5G	1B 2A 63 35 47	27 42 99 53 71	5 Square Grid
ESC*c6G	1B 2A 63 36 47	27 42 99 54 71	6 Diagonal Grid
ESC*c# W	1B 2A 63 31 51	27 42 99 #...# 87	# of Bytes
ESC*c# Ø Q	1B 2A 63 32 51	27 42 99 48 81	Delete all patterns
ESC*c#1Q	1B 2A 63 31 51	27 42 99 49 81	Delete all temporary patterns

Code	Hex	Decimal	Command
Graphics: (Continued)			
ESC*c#2Q	1B 2A 63 32 81	27 42 99 50 81	Delete current pat.
ESC*c#4Q	1B 2A 63 34 51	27 42 99 52 81	Make pattern temporary
ESC*c#5Q	1B 2A 63 34 51	27 42 99 53 81	Make pattern permanent
ESC*pØR	1B 2A 7Ø 3Ø 52	27 42 112 48 82	Rotate with orientation
ESC*p1 R	1B 2A 7Ø 31 52	27 42 112 49 82	Follow physical page
Macros:			
ESC &f#Y	1B 26 66 # 59	27 38 1Ø2 # 89	Set Macro ID #
ESC &fØX	1B 26 66 3Ø 58	27 38 1Ø2 48 88	Start Macro
ESC &f1X	1B 26 66 31 58	27 38 1Ø2 49 88	End Macro
ESC &f2X	1B 26 66 32 58	27 38 1Ø2 5Ø 88	Jump to Macro
ESC &f3X	1B 26 66 33 58	27 38 1Ø2 51 88	Call Macro
ESC &f4X	1B 26 66 34 58	27 38 1Ø2 52 88	Set Overlay Macro
ESC &f5X	1B 26 66 35 58	27 38 1Ø2 53 88	Release Overlay Macro
ESC &f6X	1B 26 66 36 58	27 38 1Ø2 54 88	Release all Macro
ESC &f7X	1B 26 66 37 58	27 38 1Ø2 55 88	Release all temporary Macro
ESC &f8X	1B 26 66 38 58	27 38 1Ø2 56 88	Release current Macro
ESC &f9X	1B 26 66 39 58	27 38 1Ø2 57 88	Assign temporary attribute to Macro
ESC &f1ØX	1B 26 66 31 3Ø 58	27 38 1Ø2 49 48 88	Assign permanent attribute to Macro
Miscellaneous:			
ESC Y	1B 59	27 89	Set Display Function of control codes
ESC Z	1B 5A	27 9Ø	Release Display Function of control codes
ESC & p # X	1B 26 7Ø # 58	27 38 112 # 88	Transparent Print Data (no ESC commands exist)
ESC &fØS	1B 26 66 3Ø 53	27 38 1Ø2 48 83	Push Printing Position. Puts present printing position on the top of the stack
ESC &f1S	1B 26 66 31 53	27 38 1Ø2 49 83	Pop Printing Position. Recall stored printing position and put on the top of the stack
ESC & l #X	1B 26 6C # 58	27 38 1Ø8 # 88	Set Number of Copies to #
ESC & l #H	1B 26 6C # 48	27 38 1Ø8 # 72	Paper Input Control.

#=Ø is feed out current page
#=1 is Lower Cassette supplies paper
#=3 is Envelope

HP LASERJET PCL5 CODES

Code	Hex	Decimal	Command
ESC & l ØH	1B 26 6C 30 48	27 Ø38 1Ø8 Ø48 Ø72	Eject Page
ESC & l 1H	1B 26 6C 31 48	27 Ø38 1Ø8 Ø49 Ø72	MP Tray
ESC & l 2H	1B 26 6C 32 48	27 Ø38 1Ø8 Ø5Ø Ø72	Manual Feed
ESC & l 3H	1B 26 6C 33 48	27 Ø38 1Ø3 Ø51 Ø72	Manual Envelope Feed
ESC & l 4H	1B 26 6C 34 48	27 Ø38 1Ø8 Ø52 Ø72	Lower Tray
ESC & l 6H	1B 26 6C 36 48	27 Ø38 1Ø8 Ø54 Ø72	Lower Cassette feeder supplies envelope, #=4 is Upper Cassette supplies paper
ESC & l 1G	1B 26 6C 31 47	27 Ø38 1Ø8 Ø49 Ø71	Upper Output Bin
ESC & l 1A	1B 26 6C 31 41	27 Ø38 1Ø8 Ø49 Ø65	Executive
ESC & l 2A	1B 26 6C 32 41	27 Ø38 1Ø8 Ø5Ø Ø65	Letter size
ESC & l 3A	1B 26 6C 33 41	27 Ø38 1Ø8 Ø51 Ø65	Legal size
ESC & l 26A	1B 26 6C 32 36 41	27 Ø38 1Ø8 Ø5Ø Ø54 Ø65	A4 size
ESC & l 8ØA	1B 26 6C 38 30 41	27 Ø38 1Ø8 Ø56 Ø48 Ø65	Monarch size
ESC & l 81A	1B 26 6C 38 31 41	27 Ø38 1Ø8 Ø56 Ø49 Ø65	COM 10 size
ESC & l 9ØA	1B 26 6C 39 3Ø 41	27 Ø38 1Ø8 Ø57 Ø48 Ø65	DL size
ESC & l 91A	1B 26 6C 39 31 41	27 Ø38 1Ø8 Ø57 Ø49 Ø65	C5 size
ESC E	1B 45	27 69	Reset Printer
ESC z	1B 7A	27 122	Start Printer Self Test

HP-GL GRAPHICS LANGUAGE CODES

HP-GL Command	Description [Parameters]	Syntax
ESC %#A	Enter PCL Mode	Ø-Retain previous PCL cursor position 1-Use current HP-GL/2 pen position
ESC E	Reset	None
AA	Arc Absolute	AA X,Y,arc angle (,chord tolerance)

[X,Y = coordinates, range -32768 to +32767]
[arc angle = coordinates, range -360 to 360 degrees]
[Chord Tolerance - Angle, range 0.1 to 180 degrees
Deviation, range -32768 to +32767]

AP . . . . Automatic Pen Operations AP n; or AP;
[n = coordinates, range 0 to 31]

HP-GL GRAPHICS LANGUAGE CODES

HP-GL Command	Description *[Parameters]*	Syntax
AR	Arc Relative	AR X,Y arc angle (,chord tolerance)
	[X,Y = coordinates, range -32768 to 32767] [arc angle = coordinates, range -360 to +360 degrees] [Chord Tolerance - Angle, range 0.1 to 180 degrees] Deviation, range -32768 to +32767]	
CA	Designate Alternate Character Set	CA set; or CA;
	[set = coordinates, range 0-9, 30-39, 61, 99, 100 & 101]	
CI	Circle	CI radius(,chord tolerance)
	[Radius = coordinates, range -32768 to 32767] [Chord Tolerance-angle, range 0.1 to 180 degrees Deviation, range -32768 to 32767]	
CM	Character Selction Mode	CM switch mode (,fallback mode); or CM;
	[Switch Mode = coordinates, range 0 to 3] [Fallback Mode = coordinates, range 0 or 1]	
CP	Character Plot	CP spaces,lines; or CP
	[spaces = coordinates, range -32768.9999 to +32767.9999] [lines = coordinates, range -32768.9999 to +32767.9999]	
CS	Designate Standard Character Set	CS set; or CS;
	[set = coordinates, range 0-9, 30-39, 61, 99, 100 & 101]	
CT	Chord Tolerance	CT n; or CT;
	[n = coordinates, range 0 to 1]	
DC	Digitize Clear	DC;
DF	Default	DF;
DI	Direction Absolute	DI run,rise; or DI;
	[run = coordinates, range -32768.9999 to +32767.9999] [rise = coordinates, range -32768.9999 to +32767.9999]	
DP	Digitize Point	DP;
DR	Direction Relative	DR run,rise; or DR;
	[run =coordinates, range -32768.9999 to +32767.9999] [rise = coordinates, range -32768.9999 to +32767.9999]	
DS	Designate Character Set Into Slot	DS slot,set; or DS;
	[slot = coordinates, range 0 to 1 (HP modes) 0 to 3 (ISO modes) set = coordinates, range 0-9, 30-39, 61, 99, 100 & 101]	
DT	Define Label Terminator	DT label terminator
	[label terminator = coordinates, range any character except NUL, ENQ, LF, ESC, and ; (decimal codes 0, 5, 10, 27, and 59, respectively]	
DV	Direction Vertical	DV n; or DV;
	[n = coordinates, range 0 or 1]	
EA	Edge Rectangle Absolute	EA X,Y;
	[X,Y coordinates, range -32768 to +32767]	
EP	Edge Polygon	EP;
ER	Edge Rectangle Relative	ER X,Y;

HP-GL Command	Description [Parameters]	Syntax
	[X,Y coordinates, range -32768 to +32767]	
ES	Extra Space	ES spaces(,lines); or ES;
	[spaces = coordinates, range -.05 to +1 char. plot cells]	
	[lines = coordinates, range -.05 to +2 char. plot cells]	
EW	Edge Wedge	EW radius,start angle,sweep angle, (,chord tolerance)
	[radius = coordinates, range -32768 to +32767]	
	[start angle = coordinates, range -360 to +360 degrees]	
	[sweep angle = coordinates, -360 to +360 degrees]	
	[chord tolerance-angle = coordinates range 0.1 to 180 deg.]	
	[deviation = coordinates, range -32768 to +32767]	
FI	Primary Font	Font ID
FP	Fill Polygon	FP;
FT	Fill Type	FT type(,spacing (,angle)); or FT;
	[type = coordinates, range 1-4]	
	[spacing = coordinates, range 0 to 32767]	
	[angle = coordinates, range 0 to 90 degrees]	
GM	Graphics Memory	GM (polygon buffer) (,reserved buffer) (,reserved buffer) (,reserved buffer) (,pen sort buffer); or GM;
	[polygon buffer = coordinates, range 0 to 31887 bytes]	
	[reserved = coordinates, range 0]	
	[reserved = coordinates, range 0]	
	[reserved = coordinates, range 0]	
	[pen sort buffer = coordinates, range 12 to 31889 bytes]	
IM	Input Mask	IM E-mask value (,S-mask value (,P-mask value)); or IM;
	[E-mask value = coordinates, range 0 to 255]	
	[S-mask value = coordinates, range 0 to 255]	
	[P-mask value = coordinates, range 0 to 255]	
IN	Initialize	IN;
IP	Input P1 and P2	IP P1x,P1y(,P2x,P2t); or IP
	[X,Y = coordinates, range -32678 to 32767 plotter units]	
IV	Invoke Character Slot	IV (slot, (left)); or IV;
	[slot = coordinates, range 0 to 1 (HP modes) 0 to 3 (ISO modes)]	
	[left = coordinates, range 0 to 1]	
IW	Input Window	IW X1,Y1,X2,Y2; or IW;
	[X1,Y1,X2,Y2 = coordinates, range -32768 to 32767]	
LB	Label	LB c...x CHR$(3)
	[c...c = coordinates, range any ASCII character]	
LO	Label Origin	LO position number;
	[position number = coordinates, range 1 to 9 or 11 to 19]	
LT	Line Type	LT pattern number (, pattern length); or LT;

HP-GL Command	Description *[Parameters]*	Syntax
	[pattern number = coordinates, range -6 to +6] [pattern length = coordinates, range 0 to 100 percentage]	
NR........	Not Ready	NR;
OA........	Output Actual Pen Status [X,Y = coordinates, range -32678 to +32767] [pen status = coordinates, range 0 (up) or 1 (down)]	OA; X,Y, pen status
OC........	Output Commanded Pen Status [X,Y = coordinates, range -32678 to +32767] [pen status = coordinates, range 0 (up) or 1 (down)]	OC; X,Y, pen status
OD........	Output Digitized Point and Pen Status [X,Y = coordinates, range -32678 to 32767] [pen status = coordinates, range 0 (up) or 1 (down)]	OD; X,Y, pen status
OE........	Output Error [error number = coordinates, range 0 to 7]	OE; error number
OF........	Output Factors [40,40 = coordinates, range none]	OF; 40,40
OH........	Output Hard-Clip Limits [YLL, YLL,YUR,YUR = coordinates, range -32678 to +32767]	OH; XLL,YLL,YUR,YUR
OI	Output Identification [model number = coordinates, range 7575A or 7576A]	OI; model number
OO	Output Options [none = coordinates, range 0 or 1]	OO; n,n,n,n,n,n,n,n
OP........	Output P1 and P2 [P1X, P1Y, P2X, P2Y = coordinates, range -32678 to +32767]	OP: P1X, P1Y, P2X, P2Y
OS........	Output Status [status number = coordinates, range 0 to 255]	OS; status number
OT........	Output Carousel Type [-1, 255 = coordinates, range none]	OT; -1, 255
OW	Output Window [YLL,YLL,XUR,YUR = coordinates, range -32678 to +32767]	OW; XLL, YLL, XUR, YUR
PA........	Plot Absolute [X,Y = coordinates, range -32768 to +32767]	PA X,Y (. . . ,X,Y) or PA;
PD........	Pen Down [X,Y = coordinates, range -32768 to +32767]	PD X,Y(,...); or PD;
PE........	Encoded Polyline [flag = coordinates, range ‘:’,’<‘,’>’,’=’, or ‘7’] [value = coordinates, range flag dependent] [X,Y = coordinates, range -32768 to 32767]	PE (flag)(value)X,Y... (flag)value)X,Y);
PM........	Polygon Mode [n = coordinates, range 0,1, and 2]	PM n; or PM;
PR........	Plot Relative	PR X,Y(,...); or PR;

HP-GL GRAPHICS LANGUAGE CODES

HP-GL Command	Description [Parameters]	Syntax
	[X,Y increments = coordinates, range -8388608.9999 to +8388607.9999]	
PT	Pen Thickness	PT pen thickness; or PT;
	[pen thickness = coordinates, range 0.1 to 5.0 millimetres]	
PU	Pen Up	PU X,Y(,...); or PU;
	[X,Y = coordinates, range -32768 to +32767]	
RA	Fill Rectangle Absolute	RA X,Y;
	[X,Y = coordinates, range -32768 to +32767]	
RO	Rotate Coord System	RO n; or RO;
	[n = coordinates, range 0 or 90 degrees]	
RR	Fill Relative Rectangle	RR X,Y
	[X,Y increments = coords, range -32768 to +32767]	
SA	Select Alt. Character Set	SA;
SC	Scale	SC Xmin,Xmax, Ymin,Ymax; or SC
	[Xmin, Xmax, Ymin, Ymax = coordinates, range -8388608 to +8388607]	
SG	Select Pen Group	SG pen number;
	[pen number = coordinates, range 0 to 8]	
SI	Absolute Character Size	SI width, height; or SI;
	[width = coordinates, range -110 to +110] [height = coordinates, range -100 to +110]	
SL	Slant Character	SL tangent; or SL;
	[tangent = coordinates, range -3.5 to +3.5]	
SM	Symbol Mode	SM character(character); or SM;
	[character = coordinates, range most printing characters (decimal codes 33-58 and 60-126)]	
SP	Select Pen	SP pen number; or SP;
	[pen number = coordinates, range 0 to 8]	
SR	Relative Character Size	SR width, height; or SR;
	[width = coordinates, range -100 to 100 percent of P2X - P1X] [height = coordinates, range -100 to 100 percent of P2X - P1X]	
SS	Select Std Character Set	SS;
TL	Tick Length	TL positive tick(,negative tick); or TL;
UC	User-defined Character	UC (pen control,)X-increment, Y-increment(,...)(,pen control)(,...); or UC;
VS	Velocity Select	VS pen velocity(,pen number); or VS;
	[pen velocity = coordinates, range 1 to 80] [pen number = coordinates, range 1 to 8]	

HP-GL GRAPHICS LANGUAGE CODES

HP-GL Command	Description *[Parameters]*	Syntax
WG	Wedge Fill [radius = coordinates, range -32768 to +32767] [start angle = coordinates, range -360 to +360 degrees] [sweep angle = coordinates, range -360 to +360 degrees] [chord tolerance-angle = coordinates, range 0.1 to 180 degrees] [chord deviation = coordinates, range -32768 to +32767]	WG radius, start angle, sweep angle(,chord tolerance);
XT	X-Tick	XT;
YT	Y-Tick	YT;

IBM PROPRINTER PRINTER CODES

Code	Hex	Decimal	Command
Page Format Control:			
ESC C Ø #	1B 43 ØØ #	27 67 Ø #	Page Length, # is in Inch
ESC C #	1B 43 #	27 67 #	Page Length, # is in Lines
ESC X #1#2	1B 58 #1#2	27 88 #1#2	Left/Right Margins Set, #1 is left inches, #2 is right inches
ESC N #	1B 4E #	27 78 #	Skip Perforation Set, # is Top + Bottom
ESC O	1B 4F	27 79	Skip Perforation Release
ESC 4	1B 34	27 52	Top of Page Set
Horizontal Movement and Spacing Control:			
BS	Ø8	8	Backspace
CR	ØD	13	Carriage Return
ESC D # Ø	1B 44 # ØØ	27 68 # Ø	Horizontal Tab Set, # is the column, can use more than one #
ESC D Ø	1B 44 ØØ	27 68 Ø	Horizontal Tab Release
HT	Ø9	9	Horizontal Tab, moves to next preset tab
ESC R	1B 52	27 82	Reset all Tabs
Vertical Movement and Spacing Control:			
ESC Ø	1B 3Ø	27 48	Set Line Spacing to 1/8 inch (9 points or 8 lpi)
ESC 1	1B 31	27 49	Set Line Spacing to 7/72 inch (7 points)

IBM PROPRINTER PRINTER CODES

Code	Hex	Decimal	Command
ESC 2	1B 32	27 5Ø	Execute a Line Feed, must follow ESC A # command
ESC 3 #	1B 33 #	27 51 #	Set Line Spacing to #/216 inch
ESC A #	1B 41 #	27 65 #	Set Line Spacing to # Points (#/72 inch)
LF	ØA	1Ø	Line feed
ESC 5 1	1B 35 Ø1	27 53 1	Set Auto Line Feed
ESC 5 Ø	1B 35 ØØ	27 53 Ø	Release Auto Line Feed
ESC j #	1B 6A #	27 1Ø6 #	Reverse Line Feed of #/216 Inches
ESC J #	1B 4A #	27 74 #	Forward Line Feed of #/216 Inches
FF	ØC	12	Form feed
ESC B # Ø	1B 42 # ØØ	27 66 # Ø	Vertical Tab Set, # is the line, can use more than one #
ESC B Ø	1B 42 ØØ	27 66 Ø	Vertical Tab Release
VT	ØB	11	Vertical Tab, moves to next preset tab
ESC R	1B 52	27 82	Reset all Tabs
Character Selection:			
DC2	12	18	Pica Pitch (12 pt,1Ø cpi)
ESC :	1B 3A	27 58	Elite Pitch (1Ø pt, 12 cpi)
SI	ØF	15	Compressed Print
ESC SI	1B ØF	27 15	Compressed Print
SO	ØE	14	Set Double Width for a single line
ESC SO	1B ØE	27 14	Set Double Width for a single line
DC4	14	2Ø	Release Double Width for a single line
ESC WØ	1B 57 ØØ	27 87 Ø	Release Double Wide Line
ESC W1	1B 57 Ø1	27 87 1	Set Double Width Line
ESC SØ	1B 53 ØØ	27 83 Ø	Set Superscript Mode On
ESC S1	1B 53 Ø1	27 83 1	Set Subscript Mode On
ESC T	1B 54	27 84	Release Superscript and Subscript
ESC 7	1B 37	27 55	Set IBM Character Set 1
ESC 6	1B 36	27 54	Set IBM Character Set 2
ESC ^	1B 5E	27 94	Select 1 Character from the All Character Chart
ESC \ #1 #2	1B 5C	27 92	Select Print Continuously from All Character Chart for a total of (#2 X 256) + #1

IBM PROPRINTER PRINTER CODES

Code	Hex	Decimal	Command
Character Highlight Selection:			
ESC – 1	1B 2D Ø1	27 45 1	Turn Underline Mode On
ESC – Ø	1B 2D ØØ	27 45 Ø	Turn Underline Mode Off
ESC _ 1	1B 5F Ø1	27 95 1	Enable Overline Mode
ESC _ Ø	1B 5F ØØ	27 95 Ø	Disable Overline Mode
ESC E	1B 45	27 69	Enable Bold Print Mode
ESC F	1B 46	27 7Ø	Disable Bold Print Mode
ESC G	1B 47	27 71	Enable Double–strike
ESC H	1B 48	27 72	Disable Double–strike
Graphics:			
For values of #1 and #2 below, see printer manuals			
ESC K#1#2	1B 4B #1#2	27 75 #1#2	Enable Single–density Graphics Mode, 6Ø dpi
ESC L#1#2	1B 4C #1#2	27 76 #1#2	Enable Double–density Graphics Mode, 12Ø dpi
ESC Y#1#2	1B 59 #1#2	27 89 #1#2	Enable Double–density, 12Ø dpi, High–speed Graphics Mode
ESC Z#1#2	1B 5A #1#2	27 9Ø #1#2	Enable Quad–density Graphics Mode, 24Ø dpi
Miscellaneous:			
CAN	18	24	Cancel
DC1	11	17	Remote Printer Select
ESC Q3	1B 51 Ø3	27 8 3	Remote Printer Deselect
ESC EM #	1B 19 #	27 25 #	Paper Cassette Selection, #=E is envelope #=1 is Lower Cassette #=2 is Upper Cassette #=R is eject page
NUL	ØØ	Ø	Null
BEL	Ø7	7	Sound Beeper

Chapter 12

PC Industry Directory

Anyone who has worked on a directory before knows that they can be a nightmare to keep current. This directory is no exception. If you find errors, changes, additions, etc, please fax them to us at 303-972-0158 or Toll Free (800) 873-7126 or email tglover@sequoiapub.com .

The following information, if it was available, has been included for each company in this directory

Company Name....................................Main Phone
Secondary Entry **(such as product)..........Phone**
Street Address
City, State Zip
Tech Phone: **Toll Free Phone:**
Fax: **Fax on Demand:**
World Wide Web Address:
email Address:
Compuserve Address:
Microsoft Net Address:
America Online Address:
BBS:
Other Phone Numbers:

01 Communique ....**(905) 795-2888**
1450 Meyerside Dr, Ste 500
Mississauga, ON L5T 2N5
Web Page: http://www.o1com.com
email: help@o1com.com

1776 Inc ....**(310) 215-1776**
8632 S Sepulveda Blvd, Suite 203
Los Angeles, CA 90045
Tech:(310) 215-1776
Fax:(310) 216-1107
Web Page: http://www.1776soft.com
email: info@1776soft.com

1st Class Software ....**(905) 302-9988**
30 Gooderham Drive
Georgetown, ON L7G 5R6 Canada
Fax:(905) 702-8648 Fax on Demand:(905) 302-9988
Web Page: http://www.1st-class-software.com
email: support@1st-class-software.com

1st Tech Corp ....**(512) 258-3570**
12201 Technology Blvd, Suite 130
Austin, TX 78727-6118
Tech:(512) 258-3570 Toll Free:(800) 533-1744
Fax:(512) 258-3689

20/20 Software ....**(503) 520-0504**
8196 SW Hall Blvd, Suite 200
Beaverton, OR 97008
Tech:(503) 520-0504 Toll Free:(800) 735-2020
Fax:(503) 520-9118
Web Page: http://www.twenty.com
email: support@twenty.com
Compuserve: GO TWENTY or 74774,222

3Com Corp ....**(801) 320-7000**
605 N. 5600 West, PO Box 16020
Salt Lake City, UT 84116
Tech:(800) 876-3266 Toll Free:(800) 527-8677
Fax on Demand:(800) 527-8677
Tech Fax:(801) 320-6020
Other Address: Microsoft keyword Megahertz
Web Page: http://www.megahertz.com
email: techsuport@mhz.com
Compuserve: go megahertz
America Online: keyword Megahertz
BBS:(801) 320-8840

3Com Corp .... **(408) 764-5000**
5400 Bayfront Plaza, P O Box 58145
Santa Clara, CA 95052-8145
Tech:(800) 876-3266 Toll Free:(800) 876-3266
Fax:(408) 764-5001 Fax on Demand:(800) 638-3266
Web Page: http://www.3com.com
email: ftp.3com.com
BBS:(408) 980-8204

3D Planet .... **(818) 222-7800**
23586 Calabasas Rd, Ste. 100
Calabasas, CA 91302
Fax:(818) 222-7092
Web Page: http://www.3dplanet.com
email: sales@3dplanet.com

3D Visions (see Visual Numerics)

3Dfx Interactive .... **(408) 935-4400**
4435 Fortran Dr
San Jose, CA 95134
Web Page: http://www.3dfx.com

3Dlabs Inc .... **(408) 436-3455**
181 Metro Drive, Suite 520
San Jose, CA 95110
Tech:(408) 436-3455
Fax:(408) 436-3458
Web Page: http://www.3dlabs.com
email: webmaster@3dlabs.com

3DO Company, The .... **(650) 261-3000**
600 Galveston Drive
Redwood City, CA 94063-4746
Tech:(650) 261-3454 Toll Free:(800) 336-3506
Fax:(650) 261-3419
Other Address: http://www.3do.com
Web Page: http://www.nwcomputing.com
email: customer-service@3do.com
Compuserve: GAMEDPUB
America Online: NEWWORLD

3DTV Corp .... **(541) 988-9634**
1863 Pioneer Pkwy East, # 303
Springfield, OR 97477
Fax:(541) 988-9627
FAX:(415) 680-1678
Web Page: http://www.3dmagic.com
email: mstarks@ibm.net

3M Data Storage Products Div ....(612) 736-1866
3M Center Building, Building 224-5N-38
Saint Paul, MN 55144
Tech:(800) 328-9438 Toll Free:(800) 328-6276
Fax:(800) 437-6264
Web Page: http://www.mmm.com/market/omc/index.ht
email: office@mmm.com

4Q Technologies ....(626) 333-6688
14425 Don Julian Road
City Of Industry, CA 91746
Fax:(626) 333-6637
Web Page: http://www.4qtech.com
email: techsupport@4qtech.com

7th Level Inc...(972) 498-8100
1110 E Collins Blvd, Suite 122
Richardson, TX 75081
Tech:(972) 498-8060
Fax:(972) 437-2717
Web Page: http://www.7thlevel.com
email: support@7thlevel.com
Compuserve: 74774,14

8 X 8 ....(408) 727-1885
2445 Mission College Blvd
Santa Clara, CA 95054
Tech:(408) 727-1676 Toll Free:(888) 843-9898
Fax:(408) 980-0432
Web Page: http://www.8x8.com
email: support@8x8.com
BBS:(408) 727-0952

@Home Network.................................(650) 569-5000
425 Broadway St
Redwood City, CA 94063
Fax:(650) 569-5100
Web Page: http://www.home.net

A & G Graphics Interface Inc...................(617) 492-0120
37 Medford St.
Somerville, MA 02143
Tech:(617) 498-0170
Fax:(617) 492-2133
Fax:(617) 491-6971
Web Page: http://www.customvoice.com
email: customvc@world.std.com
Compuserve: GO SPEECH

A-Prompt Corp .... **(800) 523-9511**
1541 Alta Drive, Suite 203
Whitehall, PA 18052
Toll Free:(800) 523-9511
Fax:(610) 770-0536

A.R.S. .... **(800) 443-5894**
PO Box 6701
Charlottesville, VA 22906
Toll Free:(800) 443-5894
Fax:(804) 973-2004
Web Page: http://www.luckycat.com/

A4 Technology Inc .... **(909) 468-0071**
20256 Paseo Robles
Walnut, CA 91789
Tech:(909) 468-0071
Fax:(909) 468-2231
Web Page: http://www.a4tech.com
email: rma@a4tech.com

Abacus Accounting Systems Inc .... **(403) 424-8100**
10303 Jasper Ave, Suite 1800
Edmonton, AB T5J 3N6 Canada
Toll Free:(800) 992-0616
Fax:(403) 423-4848
Canada:(800) 665-6657
Web Page: http://www.abacus-group.com
email: tech@abacus-group.com

Abacus Concepts (see StatView Software)

Abacus Software Inc .... **(616) 698-0330**
5370 52nd Street SE
Grand Rapids, MI 49512
Tech:(616) 698-0330 Toll Free:(800) 451-4319
Fax:(616) 698-0325
Web Page: http://www.abacuspub.com
email: tech@abacuspub.com
Compuserve: GO ABACUS

Abaton (see Everex Systems)

Abbeon Cal .... **(805) 966-0810**
123 Gray Ave
Santa Barbara, CA 93101
Toll Free:(800) 922-0977
Fax:(805) 966-7659
Web Page: http://www.wwwa.com/abbeon
email: abbeon@ix.netcom.com
America Online: abbeon@aol.com

Ability Systems Corp ....(215) 657-4338
1422 Arnold Ave
Roslyn, PA 19001
Fax:(215) 657-7815
Web Page: http://www.gus.com/emp/asc/asc.html

ABL Electronics Corp ....(410) 584-2700
350 Clubhouse Dr
Hunt Valley, MD 21031
Fax:(410) 584-2790
Web Page: http://www.ablcables.com

ABL Electronics Corp ....(630) 595-8088
1060 Entry Dr
Bensenville, IL 60106
Fax:(630) 595-6222
Web Page: http://www.ablcables.com

ABL Electronics Corp ....(510) 887-1802
23595 Cabot Blvd, Suite 116
Hayward, CA 94545
Fax:(510) 887-1856
Web Page: http://www.ablcables.com

ABL Electronics Corp ....(602) 897-6702
4625 S. Lakeshore Dr
Tempe, AZ 85282-7169
Fax:(602) 897-6703
Web Page: http://www.ablcables.com

Able Soft (Out of Business)

Abra Cadabra Software (see Best! Software)

Absoft Corp..(248) 853-0050
2781 Bond Street
Rochester Hills, MI 48309
Tech:(248) 853-0095
Fax:(248) 853-0108
Web Page: http://www.absoft.com
email: support@absoft.com

Absolute Battery Co..................................(908) 534-1560
50 Tannery Road, Suite 2
Somerville, NJ 08876
Tech:(908) 534-1560 Toll Free:(800) 829-8296
Fax:(908) 534-1792

Abstract Technologies Inc ....(512) 441-4040
4032 South Lamar Blvd, Suite 500-142
Austin, TX 78704-7900
Tech:(512) 441-4040

Fax:(512) 416-0310
Web Page: http://www.abstract.co.nz
email: info@abstract.com

ACCdotCom Inc ··································(800) 242-0739
10078 Tyler Court, Suite D
Ijamsville, MD 21754
Tech:(888) 444-7854 Toll Free:(800) 242-0739
Fax:(410) 635-9801
Tech Fax:(301) 831-8289
Web Page: http://www.accsystems.com
email: info@accsystems.com

Accent Software International Ltd ··········(719) 576-2610
2864 S. Circle Dr, Suite 340
Colorado Springs, CO 80906
Tech:(800) 535-5216 Toll Free:(800) 535-5216
Fax:(719) 576-2604
Web Page: http://www.accentsoft.com
email: info@accentsoft.com
Compuserve: GO ACCENT or 74774.264

Access Micro Products (see All American Semi

Access Software Inc. ·····························(801) 359-2900
4750 Wiley Post Way, Bldg 1 Suite 200
Salt Lake City, UT 84116-2837
Tech:(800) 793-8324 Toll Free:(800) 800-4880
Fax:(801) 596-9128
Web Page: http://www.accesssoftware.com
email: info@accesssoftware.com
Compuserve: GO GAMEPUB
America Online: linkspro1@aol.com

AccessData Corp··································(801) 377-5410
2500 N. University, Suite 200
Provo, UT 84604-3864
Tech:(800) 489-5199 Toll Free:(800) 489-5199
Fax:(801) 377-5426
Web Page: http://www.accessdata.com
email: info@accessdata.com

Accolade Inc ···(408) 985-1700
5300 Stevens Creek Blvd
San Jose, CA 95129-1032
Tech:(408) 296-8400 Toll Free:(800) 245-7744
Fax:(408) 246-0231
Web Page: http://www.accolade.com
email: techelp@accolade.com
America Online: accolade@aol.com

AccountMate Software Corp ····················**(415) 381-1011**
20 Sunnyside Ave, Suite C
Mill Valley, CA 94941-1928
Tech:(415) 381-1793 Toll Free:(800) 877-8896
Fax:(415) 381-6902
Other Address: http://www.accountmate.com (2nd Web)
Web Page: http://www.sourcemate.com
email: info@accountmate.com

Accton Technology Corp ·························**(408) 452-8900**
1962 Zanker Road
San Jose, CA 95112-4216
Toll Free:(800) 926-9288
Fax:(408) 452-8988 Fax on Demand:(408) 452-8811
Web Page: http://www.accton.com
email: don@accton.com

Acculogic Inc (see ACC Tech. Group)····**(619) 530-8170**
9265 Dowdy Drive, Suite 219
San Diego, CA 92126
Toll Free:(800) 234-7811
Fax:(619) 586-1540
Web Page: http://www.acculogic.com
email: sales@acculogic.com

Accurate Research Inc·····························**(408) 523-4788**
762 Palomar Ave
Sunnyvale, CA 94086
Tech:(408) 523-4788 Toll Free:(800) 799-8802
Fax:(408) 523-4789
Other Address: http://www.supertutor.com
Web Page: http://www.drcdrom.com
email: tech@supertutor.com

Accurite Technologies Inc························**(510) 668-4900**
48460 Lakeview Blvd
Fremont, CA 94538-6532
Fax:(510) 668-4905
Web Page: http://www.accurite.com
email: tech@accurite.com
Compuserve: 72360,2620

Accutek Inc. (see IEC Electronics Corp.)

ACE Contact Manager·····························**(619) 673-5313**
16835 W. Bernardo Drive, Suite 209
San Diego, CA 92127
Tech:(619) 673-5313 Toll Free:(800) 833-8892
Fax:(619) 673-7399
Web Page: http://www.sfsace.com

Acecad Inc .... **(408) 655-1900**
791 Foam Street
Monterey, CA 93940
Tech:(408) 655-9911 Toll Free:(800) 676-4223
Fax:(408) 655-1919
Web Page: http://www.acecad.com
email: support@acecad.com
BBS:(408) 655-1988

Acer America Corp .... **(408) 432-6200**
2641 Orchard Pkwy
San Jose, CA 95134
Tech:(800) 445-6495 Toll Free:(800) 733-2237
Fax:(408) 922-2933 Fax on Demand:(800) 554-2494
Extra Support:(900) 555-2237
Web Page: http://www.acer.com
email: tsup@smtplink.acer.com
Compuserve: GO ACER
America Online: GO ACER
BBS:(408) 428-0140

Acer Sertek Inc .... **(408) 733-3174**
116 S. Wolfe Road
Sunnyvale, CA 94086
Tech:(408) 733-3174
Fax:(408) 733-2569
Web Page: http://www.ussertek.com

Aces Research Inc .... **(510) 683-8855**
46750 Fremont Blvd, # 107
Fremont, CA 94538
Tech:(510) 661-2093
Fax:(510) 683-8875
Web Page: http://www.acesxprt.com
email: info@acesxprt.com

Aci Us Inc .... **(408) 252-4444**
20883 Stevens Creek Blvd
Cupertino, CA 95014
Tech:(800) 881-3466 Toll Free:(800) 881-3466
Fax:(408) 252-4829 Fax on Demand:(408) 252-7215
Other Address: AppleLink
Web Page: http://www.acius.com
Compuserve: GO ACIUS

ACL Staticide .... **(847) 981-9212**
1960 E. Devon Ave
Elk Grove Village, IL 60007
Tech:(847) 981-9212 Toll Free:(800) 782-8420
Fax:(847) 981-9278

Web Page: http://www.aclstaticide.com
email: info@aclstaticide.com

Acme Electric Corp ································(716) 968-2400
9962 Route 446
Cuba, NY 14727
Toll Free:(800) 325-5848
Fax:(716) 968-3948

Acorn Computers (see Text 100 Corp.)

ACS Innovation, Inc. ································(408) 566-0900
3171 Jay Street
Santa Clara, CA 95054-3308 USA
Fax:(408) 566-0909
Web Page: http://www.acscompro.com

ACT Networks Inc ································(805) 388-2474
188 Camino Ruiz
Camarillo, CA 93012
Toll Free:(800) 367-2281
Fax:(805) 388-3504
Customer Service Fax:(805) 389-3180
Web Page: http://www.acti.com
email: info@acti.com

Action Image Systems Technology Inc ··(908) 232-2166
1007 Sunny Slope Drive
Mountainside, NJ 07092-2145
Tech:(908) 232-2166
Fax:(908) 232-1621
email: aist@ix.net.com.com

Action Technologies Inc ································(510) 521-6190
1301 Marina Village Pkwy, Suite 100
Alameda, CA 94501-1028
Tech:(510) 521-6190 Toll Free:(800) 967-5356
Fax:(510) 769-0596
Web Page: http://www.actiontech.com
email: techsupport@actiontech.com

Active Voice Corp ································(206) 441-4700
2901 Third Ave
Seattle, WA 98121
Fax:(206) 441-4784
Other Address: http://www.activevoice.com
Web Page: http://www.avoice.com
email: support@activevoice.com

Activision ································(310) 255-2000
3100 Ocean Park Blvd.

Santa Monica, CA 90405
Tech:(310) 255-2050 Toll Free:(800) 477-3650
Fax:(310) 255-2151 Fax on Demand:(310) 255-2153
Web Page: http://www.activision.com
email: support@activision.com
Compuserve: GO ACTIVISION; 76004,2122
Microsoft: Activision
America Online: KEYWORD ACTIVISION
BBS:(310) 255-2146

Adaptec Inc ....(408) 945-8600
691 S. Milpitas Blvd
Milpitas, CA 95035
Tech:(408) 934-7274 Toll Free:(800) 959-7274
Fax:(408) 262-2533 Fax on Demand:(303) 684-3400
Tech Fax:(408) 957-6776
Web Page: http://www.adaptec.com
email: support@adaptec.com
Compuserve: GO ADAPTEC
BBS:(408) 945-7727

Adaptiv Software Corp..............................(714) 960-2211
125 Pacifica, Suite 250
Irvine, CA 92718-9601
Tech:(714) 960-4949 Toll Free:(800) 598-1222
Fax:(714) 789-7320
Web Page: http://www.adaptiv.com
email: techsupport@adaptiv.com

Adaptive Solutions Inc..............................(503) 690-1236
1400 N.W. Compton Drive, Suite 340
Beaverton, OR 97006-1992
Toll Free:(800) 482-6277
Fax:(503) 690-1249
Web Page: http://www.asi.com
email: info@asi.com

Adax Inc ....(510) 548-7047
614 Bancroft Way
Berkeley, CA 94710
Fax:(510) 548-5526
Web Page: http://www.adax.com
email: sales@adax.com

ADC Fibermux corp (see ADC Kentrox)

ADC Kentrox ....(503) 643-1681
14375 NW Science Park Drive
Portland, OR 97229
Tech:(800) 733-5511 Toll Free:(800) 733-5511

Fax:(503) 641-3341
Web Page: http://www.kentrox.com
email: info@kentrox.com

Addison Wesley Longman ....(781) 944-3700
1 Jacob Way
Redding, MA 01867-3999
Toll Free:(800) 447-2226
Fax:(781) 944-9338
Fax:(800) 333-3328
Web Page: http://www.awl.com

Addonics Technologies Corp ....(510) 438-6530
48434 Milmont Drive
Fremont, CA 94538
Tech:(888) 584-8324 Toll Free:(800) 787-8580
Fax:(510) 353-2020
Web Page: http://www.addonics.com
email: techinfo@addonics.com

Addtron Technology Co. Ltd. ....(888) 233-8766
4425 Cushing Parkway
Fremont, CA 94538
Toll Free:(888) 233-8766
Fax:(510) 668-0699
Web Page: http://www.addtron.com
email: sales@addtron.com

Adept Computer Solutions Inc ....(619) 597-1776
10951 Sorrento Valley Road, Suite 1G
San Diego, CA 92121
Tech:(619) 597-1776 Toll Free:(800) 578-6277
Fax:(619) 597-1774
Web Page: http://www.streetwizard.com
email: support@streetwizard.com

ADI Systems Inc ....(408) 944-0100
2115 Ringwood Court
San Jose, CA 95131
Tech:(408) 944-0100 Toll Free:(800) 228-0530
Fax:(408) 944-0300
Web Page: http://www.adiusa.com
email: techsupport@adiusa.com

Adobe Systems Inc ....(408) 536-6000
345 Park Ave
San Jose, CA 95110-2704
Tech:(800) 833-6687 Toll Free:(800) 628-2320
Fax:(408) 537-6000 Fax on Demand:(206) 628-5737
Tech Support:(800) 872-3623

Web Page: http://www.adobe.com
Compuserve: GO ADOBE

Acrobat .... (206) 675-6304
After Effects .... (206) 675-6310
ATM .... (206) 675-6306
Dimensions .... (206) 675-6316
File Utilities, Mastersoft (discontinued) .... (312) 527-4357
Font Chameleon (discontinued)
FontFiddler (discontinued)
FontHopper (discontinued)
FontMinder (discontinued)
FontMonger (discontinued)
Former Ares Font Products (discontinued)
FrameMaker .... (206) 675-6312
FrameMaker + SGML .... (206) 675-6315
FrameViewer .... (206) 675-6353
Gallery Effects .... (206) 675-6358
Illustrator .... (206) 675-6307
Image Club Products .... (403) 262-8008
ImageReady .... (206) 675-6370
InfoPublisher .... (206) 675-6360
Intellidraw (discontinued)
PageMaker .... (206) 675-6301
PageMill .... (206) 675-6313
Persuasion .... (206) 675-6308
PhotoDelux .... (206) 675-6309
PhotoDelux Business .... (206) 675-6371
Photoshop .... (206) 675-6303
Photostyler (discontinued)
PostScript Drivers .... (206) 675-6314
Premiere .... (206) 675-6305
Printer Drivers .... (206) 675-6314
Streamline .... (206) 675-6317
Type Manager/Type Products .... (206) 675-6306
Type Twister .... (206) 675-6367

ADS Technologies .... (562) 926-1928
13909 Bettencourt St
Cerritos, CA 90703
Tech:(800) 888-5244 Toll Free:(800) 888-5244
Fax:(562) 926-0518
Web Page: http://www.addtron.com

Adtran .... (205) 971-8000
901 Explorer Blvd, PO Box 140000
Huntsville, AL 35814-4000
Tech:(205) 971-8716 Toll Free:(800) 827-0807
Fax:(205) 971-7941

tech support:(800) 726-8663
Web Page: http://www.adtran.com
email: info@adtran.com

Advanced Digital Info. Corp. (ADIC) ·······(425) 881-8004
10201 Willows Road, PO Box 97057
Redmond, WA 98073-9757
Tech:(425) 883-4357 Toll Free:(800) 336-1233
Fax:(425) 881-2296
Web Page: http://www.adic.com
email: support@adic.com
BBS:(425) 883-3211

Advanced Digital Systems (ADS) ···········(562) 926-1928
13909 Bettencourt St
Cerritos, CA 90703
Tech:(562) 926-4338 Toll Free:(800) 888-5244
Fax:(562) 926-0518
Web Page: http://www.adstech.com
email: adstech@adstech.com
Compuserve: 72662,521

Advanced Graphics Software Inc ···········(760) 634-8360
132 N. El Camino Real, Suite 338
Encinitas, CA 92024
Toll Free:(800) 795-4754
Fax:(760) 634-8363
Fax:(760) 931-9313
Web Page: http://www.slidewrite.com
email: support@SlideWrite.com

Advanced Gravis Computer Tech Ltd ····(604) 431-5020
3750 N. Fraser Way, # 101
Burnaby, BC V5J5E-9 canada
Tech:(610) 266-9505 Toll Free:(800) 535-4242
Fax:(604) 431-5155
Tech Fax:(610) 231-1022
Web Page: http://www.gravis.com
email: pcstick@gravis.com

Advanced Logic Research (ALR Inc) ·····(714) 581-6770
9401 Jeronimo
Irvine, CA 92618
Tech:(800) 257-1230 Toll Free:(800) 444-4257
Fax:(714) 581-9240
Tech Fax:(714) 458-0532
Web Page: http://www.alr.com
email: tech@alr.com
BBS:(714) 458-6834

Advanced Matrix Technology Inc ··········(805) 388-5799
747 Calle Plano
Camarillo, CA 93012-8598
Tech:(805) 388-5799
Fax:(805) 389-3657
Web Page: http://www.amtprinters.com
email: sales@amtprinters.com

Advanced Media ··································(714) 957-1616
695 Town Center Drive, Suite 250
Costa Mesa, CA 92626
Tech:(714) 957-1616 Toll Free:(800) 292-4264
Fax:(714) 957-5977
Web Page: http://www.advancedmedia.com
email: info@advancedmedia.com

Advanced Media ··································(516) 244-1616
80 Orville Drive
Bohemia, NY 11716
Fax:(516) 244-1415
Web Page: http://www.advancedmedia.com
email: info@advancedmedia.com

Advanced Micro Devices ······················(408) 732-2400
One AMD Place, PO Box 3453
Sunnyvale, CA 94088-3453
Tech:(408) 749-3060 Toll Free:(800) 538-8450
Fax on Demand:(800) 222-9323
Web Page: http://www.amd.com

Advanced Micro Devices ······················(408) 435-0202
1623 Buckeye Drive
Milpitas, CA 95035
Tech:(408) 749-3060 Toll Free:(800) 222-9323
Fax:(408) 732-2400
Other Address: http://www.amd.com
Web Page: http://www.nexgen.com

Advanced Network Solutions ················(425) 644-6082
5527 Preston-Fall City Road
Fall City, WA 98024
Toll Free:(800) 837-4180
Fax:(425) 222-7622
Web Page: http://www.halcyon.com/routers/advance
email: routers@compuslan.com

Advanced RISC Machines Inc ················(408) 399-5199
985 University Ave, Suite 5
Los Gatos, CA 95030
Fax:(408) 399-8854

Web Page: http://www.arm.com
email: info@arm.com

Advanced Software (see Prairie Group)

Advanced Storage Concepts Inc ....(409) 744-2129
2720 Terminal Drive
Galveston, TX 77554
Tech:(409) 744-2129
Fax:(409) 744-2181
Web Page: http://www.c-com.net/~asci
email: support@advstor.com

AdvanSys Inc ....(408) 383-9400
1150 Ringwood Court
San Jose, CA 95131
Tech:(800) 525-7440 Toll Free:(800) 525-7443
Fax:(408) 383-9612 Fax on Demand:(408) 383-9753
Tech Support:(408) 467-2930
Web Page: http://www.advansys.com
email: support@advansys.com
BBS:(408) 383-9540

Advantage Memory ....(714) 453-8111
25A Technology Drive, Building 2
Irvine, CA 92618
Tech:(800) 245-5299 Toll Free:(800) 245-5299
Fax:(714) 453-8158
Tech Fax:(714) 453-1357
Web Page: http://www.advantagememory.com
email: sales@advantagememory.com

Advisor Publications Inc ....(619) 278-5600
5675 Ruffin Road, PO Box 429002
San Diego, CA 92123
Toll Free:(800) 336-6060
Fax:(619) 278-0300
Web Page: http://www.advisor.com
email: order@advisor.com
Compuserve: 70007,1614 or GO DBA

AEC Management (see AEC Software)

AEC Software...(703) 450-1980
22611-113 Markey Ct
Sterling, VA 20166
Tech:(703) 450-2318 Toll Free:(800) 346-9413
Fax:(703) 450-9786
Other Address: AppleLink
Web Page: http://www.aecsoft.com
email: support@aecsoft.com

Compuserve: 72541.60@compuserve.com

Aeronics Inc ··(512) 258-2303
12741 Research Blvd, Suite 500
Austin, TX 78759
Tech:(512) 258-2303
Fax:(512) 258-4392
Web Page: http://www.aeronics.com

AgData ··(530) 846-6203
891 Hazel Street
Gridley, CA 95948
Toll Free:(888) 327-6257
Web Page: http:/www.manznet.com/agdata/agdata.ht
email: agdata@manznet.com

Agfa Compugraphics (Bayer Corp) ········(201) 440-2500
100 Challanger Road
Ridgefield Park, NJ 07660
Tech:(800) 879-2432 Toll Free:(800) 424-8973
Fax:(201) 440-5733 Fax on Demand:(800) 879-2432
Fax:(201) 342-4742
Web Page: http://www.agfa.com

Agile Networks Inc ·····························(978) 287-9000
300 Baker Ave
Concord, MA 01742
Toll Free:(800) 286-9526
Fax:(978) 287-9050
Web Page: http://www.agile.com
email: info@agile.com

Ahead Systems Inc ······························(510) 623-0900
44244 Fremont Blvd
Fremont, CA 94538
Fax:(510) 623-0960
email: ahead@ix.netcom.com

Ahern Communications Corp ················(617) 471-1100
60 Washington Court
Quincy, MA 02169
Tech:(617) 471-1100 Toll Free:(800) 451-5067
Fax:(617) 328-9070
Web Page: http://aherncorp.com
email: info@aherncorp.com

Aim Tech (see Asymetrix Learning Systems)

Aims Lab ···(510) 661-2525
46740 Lakeview Blvd
Fremont, CA 94538

Fax:(510) 252-1572
Web Page: http://www.aimslab.com
email: support@aimslab.com

Air Media ··(212) 843-0000
11 E. 26th St, 16th Floor
New York, NY 10010-1402
Tech:(212) 843-0000 Toll Free:(800) 238-4738
Fax:(212) 545-7992
Web Page: http://www.airmedia.com
email: support@airmedia.com

AITech International ·······························(510) 226-8960
47941 Fremont Blvd
Fremont, CA 94538
Tech:(510) 226-9246
Fax:(510) 226-8996
Web Page: http://www.aitech.com
email: Tech_Support@aitech.com

Aiwa America Inc. ···································(201) 512-3600
800 Corporate Drive
Mahwah, NJ 07430 USA
Fax:(201) 460-2218
Web Page: http://www.aiwa.com

Akia Corp ··(512) 339-4804
11406 Metric Blvd, Ste 190
Austin, TX 78758
Toll Free:(800) 244-5019
Fax:(512) 339-4886
Web Page: http://www.akia.com
email: support@akia.com

Aladdin Knowledge Systems Inc. ···········(212) 564-5678
350 Fifth Ave, Suite 6614
New York, NY 10118
Toll Free:(800) 223-4277
Fax:(212) 564-3377
Web Page: http://www.aks.com
email: sales@us.aks.com

Aladdin Soft Secur. **(see Aladdin Knowl Sys)**

Aladdin Systems Inc ································(408) 761-6200
165 Westridge Drive
Watsonville, CA 95076-4167
Tech:(408) 761-6200
Fax:(408) 761-6206
Web Page: http://www.aladdinsys.com
email: support@aladdinsys.com

Compuserve: GO ALADDIN or 75300,1666
America Online: keyword: Aladdin

Alberta Printed Circuits LTD ····················(403) 250-3406
1112-40th Avenue N.E., Suite 3
Calgary, AL T2E 5T8 Canada
Web Page: http://www.apcircuits.com
email: staff@apcircuits.com
BBS:(403) 291-9342

Aldus Corp (see Adobe Systems Inc)

Alexander LAN Inc ··································(603) 880-8800
100 Perimeter Road
Nashua, NH 03063-1301
Fax:(603) 880-8881
Web Page: http://www.alexander.com
email: support@alexander.com
Compuserve: GO ALEXANDER 102212,3061

Algorithm Inc ···(770) 232-4949
7230 McGinnis Ferry Road, Suite 200
Atlanta, GA 30174-1245
Fax:(770) 232-4951
Web Page: http://www.algorithm.com
email: postmaster@algorith.com

Alias Wavefront ·······································(416) 362-9181
210 King St E
Toronto, ON M5C 1P1
Tech:(800) 465-0868
Fax:(416) 369-6138
Fax:(416) 369-6140
Web Page: http://www.aw.sgi.com
email: info@aw.sgi.com

Alki Software Corp ·································(206) 286-2600
300 Queen Anne Ave N., Suite 410
Seattle, WA 98109
Tech:(206) 286-2780 Toll Free:(800) 669-9673
Fax:(206) 286-2785
Web Page: http://www.alki.com
email: sales@alki.com

All American Semi··································(408) 441-1300
230 Devcon Drive
San Jose, CA 95112

All Components Inc································(972) 233-0203
14990 Landmark Blvd, Suite 300
Addison, TX 75240

Tech:(972) 233-0203 Toll Free:(800) 779-0234
Fax:(972) 851-1998
Web Page: http://www.allcomponents.com
email: support@allcomponents.com

Allaire Corp ....(617) 761-2000
One Alewife Center
Cambridge, MA 02140
Tech:(617) 761-2100 Toll Free:(888) 939-2545
Fax:(617) 761-2001
Web Page: http://www.allaire.com
email: info@allaire.com

Allegiant (Out of Business)

Allegro New Media (see SBT Accounting Sys)

Allegro Systems Ltd (Out of Business)

Alliance Research (see ORA Electronics)

Allied Telesyn International (ATI) ....(408) 730-0950
950 Kifer Road
Sunnyvale, CA 94086
Tech:(800) 428-4835 Toll Free:(800) 424-4284
Fax:(408) 736-0100
Web Page: http://www.alliedtelesyn.com
email: TSI@alliedtelesyn.com
Compuserve: GO ALLIED
BBS:(425) 483-7979

AllMicro (see Fore Front Direct)

Allsop Computer Accessories (see Allsop Inc)

Allsop Inc. ....(360) 734-9090
4201 Meridian, PO Box 23
Bellingham, WA 98226
Tech:(360) 734-9090 Toll Free:(800) 426-4303
Fax:(360) 734-9858
Web Page: http://www.allsop.com
email: acs@allsop.com

Alltech Electronics ....(714) 543-5011
1300 E. Edinger, Suite E
Santa Ana, CA 92705
Fax:(714) 543-0553
Web Page: http://www.alltechelectronics.com

Almo Corp ....(215) 698-4000
9815 Roosevelt Blvd
Philadelphia, PA 19114-1082
Tech:(800) 600-2566 Toll Free:(800) 878-5758

Fax:(215) 698-4080
Fax:(215) 698-4037
Web Page: http://www.almo.com

Almo Distributing (see Almo Corp)

ALOS Micrographics Corporation ....(914) 457-4400
118 Bracken Road
Montgomery, NY 12549-2604
Tech:(914) 457-4400 Toll Free:(800) 431-7105
Fax:(914) 457-9083
Web Page: http://www.alosmc.com
email: alosmc@alosmc.com
Compuserve: 75052 2760

Alpha Microsystems ....(714) 641-6373
2722 S. Fairview St
Santa Ana, CA 92704
Tech:(714) 957-8500
Other:(714) 641-6370
Web Page: http://www.alphaconnect.com

Alpha Software Corp ....(781) 229-2924
168 Middlesex Turnpike
Burlington, MA 01803-4483
Tech:(800) 229-3460 Toll Free:(800) 451-1018
Fax:(781) 272-4876 Fax on Demand:(800) 368-8872
Tech Fax:(617) 272-8222
Web Page: http://www.alphasoftware.com
email: customerservice@alphasoftware.com
Compuserve: 75300,3656 or GO ALPHA
America Online: ALPHATECH

AlphaBlox Corp ....(650) 526-1700
800 Maude Ave.
Mountain View, CA 94043
Tech:(650) 526-1700 Toll Free:(888) 256-9669
Fax:(650) 526-1701
Web Page: http://www.alphablox.com
email: info@alphablox.com

Alpharel Inc ....(619) 625-3000
9339 Carroll Park Drive
San Diego, CA 92121
Tech:(800) 633-6784 Toll Free:(800) 992-6784
Fax:(619) 546-7671
Web Page: http://www.alpharel.com
email: info@alpharel.com

Alps Electric USA ....(408) 432-6000
3553 North First Street

San Jose, CA 95134
Tech:(800) 449-2577 Toll Free:(800) 825-2577
Fax:(408) 432-6035
customer service:(800) 950-2577
Other Address: http://www.alpsusa.com
Web Page: http://www.alps.com
email: alpsusa@ccmail.alpsusa.com
Compuserve: GO ALPS
BBS:(408) 432-6424

ALR Inc (see Advanced Logic Research)

Alsoft Inc. ....(281) 353-4090
PO Box 927
Spring, TX 77383-0927
Tech:(281) 353-1510 Toll Free:(800) 257-6381
Fax:(281) 353-9868
Web Page: http://www.alsoft.com
email: Tech.Support@Alsoft.com

Alta Technology Corp ....(801) 562-1010
9500 South 500 West, Suite 212
Sandy, UT 84070-6655
Tech:(801) 562-1010
Fax:(801) 254-2020
Web Page: http://www.altatech.com
email: support@altatech.com

AltaVista Software ....(800) 344-4825
Tech:(800) 336-7890 Toll Free:(800) 336-7890
Web Page: http://www.altavista.software.digital.

Altec Lansing ....(717) 296-4434
PO Box 277
Milford, PA 18337-0277
Tech:(800) 258-3288 Toll Free:(800) 258-3288
Fax:(717) 296-1222
Web Page: http://www.altecmm.com
email: support2@altecmm.com
BBS:(717) 296-1360

Altex Computers & Electronics ....(210) 655-8882
11342 IH 35 North
San Antonio, TX 78233
Toll Free:(800) 531-5369
Fax:(210) 637-3250
Web Page: http://www.altex.com
email: altex2@altex.com

Altex Electronics Corp ....(512) 814-8882
2650 South Padre Island Drive

Corpus Christi, TX 78415
Toll Free:(800) 531-5369
Fax:(512) 814-8812
Web Page: http://www.altex.com
email: altex2@txdirect.net

Austin .... (512) 832-9131
10705 Metric
Austin, TX 78758
Toll Free:(800) 531-5369
Fax:(512) 832-9131
Web Page: http://www.altex.com
email: altex2@txdirect.net

Dallas .... (972) 267-8882
3215 Belmeade
Carrollton, TX 75006
Toll Free:(800) 531-5369
Fax:(972) 267-0770
Fax:(972) 386-9182
Web Page: http://www.altex.com
email: altex2@txdirect.net

Mail Order .... (210) 637-3200
11342 IH35 North
San Antonio, TX 78233-5792
Toll Free:(800) 531-5369
Fax:(210) 637-3264
Web Page: http://www.altex.com
email: altex2@txdirect.net

San Antonio .... (210) 340-3557
10731 Gulfdale
San Antonio, TX 78216
Toll Free:(800) 531-5369
Fax:(210) 340-2409
Web Page: http://www.altex.com
email: altex2@txdirect.net

Always Technology Corp (out of business)

Amber Wave Systems (see U.S. Robotics)

Amdahl Corp .... **(408) 746-6000**
1250 E. Arques Ave
Sunnyvale, CA 94088-3470
Toll Free:(800) 538-8460
Fax:(408) 773-0833
Fax:(408) 738-1051
Web Page: http://www.amdahl.com
email: www@amdahl.com

Amdek Corp (see Wyse Technology)

America Online ....(703) 448-8700
8619 Westwood Center Drive
Vienna, VA 22182
Tech:(800) 827-3338 Toll Free:(800) 827-6364
Fax:(703) 883-1509
Web Page: http://www.aol.com

American Bible Society ....(212) 408-1200
1865 Broadway
New York, NY 10023
Tech:(212) 408-1200 Toll Free:(800) 322-4253
Fax:(212) 408-1512
Web Page: http://www.americanbible.org

American Business Info ....(402) 593-4500
5711 S. 86th Circle
Omaha, NE 68127
Tech:(402) 593-4651 Toll Free:(800) 321-0869
Fax:(402) 331-6184
Web Page: http://www.abii.com
email: internet@abii.com

American Business System ....(978) 250-9600
315 Littleton Rd, PO Box 460
Chelmsford, MA 01824
Tech:(978) 250-9600 Toll Free:(800) 356-4034
Fax:(978) 250-8027
Web Page: http://www.abs-software.com
email: abssales@cybercom.net
BBS:(978) 250-6999

American Covers Inc ....(801) 553-0600
102 W 12200 South
Draper, UT 84020
Tech:(800) 228-8987 Toll Free:(800) 228-8987
Fax:(801) 553-1212
Web Page: http://www.americancovers.com

American Cybernetics ....(602) 968-1945
1830 W. University Drive, Suite 112
Tempe, AZ 85281
Tech:(602) 968-1945 Toll Free:(800) 899-0100
Fax:(602) 966-1654
Web Page: http://www.amcyber.com
email: tech@multiedit.com
Compuserve: 71333,10 or GO CYBERNET

American Ink Jet Corp ....(978) 670-9200
13 Alexander Road
Billerica, MA 01821

Tech:(978) 670-9200 Toll Free:(800) 332-6538
Fax:(978) 670-5637
Web Page: http://www.amjet.com
email: support@amjet.com

American Megatrends Inc ....(770) 246-8600
6145-F Northbelt Pkwy
Norcross, GA 30071-2976
Tech:(770) 246-8645 Toll Free:(800) 828-9264
Fax:(770) 246-8791 Fax on Demand:(770) 246-8787
Other Address: http://www.ami.com
Web Page: http://www.megatrends.com
email: support@ami.com
BBS:(770) 246-8780

American MPC Research Inc. ....(562) 801-0108
9816 Alburtis Ave
Santa Fe Springs, CA 90670
Tech:(562) 801-0108
Fax:(562) 801-0138
Web Page: http://www.americanmpc.com
email: ampc@vividnet.com

American Ntl Standards Institute ....(212) 642-4900
11 West 42nd St
New York, NY 10036
Fax:(212) 398-0023
Web Page: http://www.ansi.org/

American Power Conversion Corp ....(401) 789-5735
132 Fairgrounds Road
West Kingston, RI 02892
Tech:(800) 800-4272 Toll Free:(800) 541-8896
Fax:(401) 789-3710
Fax:(401) 789-3180
Web Page: http://www.apcc.com
email: retail@apcc.com
Compuserve: GO APCSUPPORT

American Small Bus Computer (see DesignCad)

AmeriQuest Technologies ....(215) 658-8900
2465 Maryland Road
Willow Grove, PA 19090
Tech:(888) 274-9111 Toll Free:(888) 274-9111
Fax:(215) 658-8979
Web Page: http://www.ameriquest.com

Amicus Networks Inc ....(512) 418-8828
9390 Research Blvd, Suite 2-110
Austin, TX 78759

Fax:(512) 418-8829
Web Page: http://www.amicus.com
email: info@amicus.com

AMP Inc ....(717) 986-7777
PO Box 3608
Harrisburg, PA 17105-3608
Tech:(800) 526-0721 Toll Free:(800) 522-6752
Fax:(717) 986-7575 Fax on Demand:(800) 522-6752
FaxBack:(717) 986-3500
Web Page: http://www.amp.com
email: product.info@amp.com

Ampex Corp ....(650) 367-2011
500 Broadway
Redwood City, CA 94063-3199
Tech:(800) 227-8402 Toll Free:(800) 752-7590
Web Page: http://www.ampex.com
email: info@ampex.com

Amptron International Inc ....(626) 912-5789
1239 Hatcher Ave
City Of Industry, CA 91748
Web Page: http://www.amptron.com

AMS Tech ....(626) 814-8851
12881 Ramona Blvd
Irwindale, CA 91706
Tech:(800) 886-3536 Toll Free:(800) 980-8889
Fax:(626) 814-0782 Fax on Demand:(800) 868-8148
Web Page: http://www.amstech.com
email: TechSupport@amstech.com

ANA Tech ....(303) 973-6722
10499 Bradford Road
Littleton, CO 80127
Fax:(303) 973-7092
Web Page: http://www.anatech.scanners.com
email: custsvc@anatech.ingr.com

Analog Devices Inc ....(781) 329-4700
One Technology Way, PO Box 9106
Norwood, MA 02062-9106
Tech:(800) 426-2564 Toll Free:(800) 262-5643
Fax:(781) 461-3091
Web Page: http://www.analog.com

Anaserve Inc ....(949) 250-7262
1300 Bristol North, Suite 220
Newport Beach, CA 92660
Tech:(949) 250-7263 Toll Free:(800) 711-6030

Fax:(949) 250-7265
Web Page: http://www.anaserve.com
email: sales@anawave.com

Anawave Software Inc (see Anaserve Inc)

Andataco ....(619) 453-9191
10140 Mesa Rim Road
San Diego, CA 92121
Tech:(619) 453-9809 Toll Free:(800) 334-9191
Fax:(619) 453-9294
Web Page: http://www.andataco.com
email: cs@anataco.com

Anderson Investor's Software Inc ....(314) 918-0990
7530 Watson Road
Saint Louis, MO 63119
Toll Free:(800) 286-4106
Fax:(314) 918-0980
Web Page: http://www.invest-soft.com
email: comments@invest-soft.com
Compuserve: 74111,3702
Microsoft: INVESTORS-SOFTWARE
America Online: KEYWORD ANDINVEST

Andrea Electronics Corp ....(718) 729-8500
11-40 45th Rd
Long Island City, NY 11101
Toll Free:(800) 442-7787
Fax:(718) 729-9174
Web Page: http://www.andreaelectronics.com

Andromeda Research ....(513) 831-9708
PO Box 222
Milford, OH 45150
Fax:(513) 831-7562

Andyne Computing Ltd (see Hummingbird)

ANGOSS Software International ....(416) 593-1122
34 St. Patrick Street, Suite 200
Toronto, ON M5T 1V1 Canada
Tech:(416) 593-1122
Fax:(416) 593-5077
Web Page: http://www.angoss.com
email: support@angoss.com
Compuserve: 71333,3661

AniCom Inc (see Animated Communications)

Animated Communications ....(919) 967-2890
PO Box 16907

Chapel Hill, NC 27516
Toll Free:(800) 949-4559
Fax:(919) 933-9503
Web Page: http://www.3dchor.com
email: info@3dchor.com
Compuserve: GO ANICOM

Annabooks .... **(619) 674-6155**
11838 Bernardo Plaza Court, Suite #102
San Diego, CA 92128-2414
Toll Free:(800) 462-1042
Fax:(619) 673-1432
Other Address: http://www.annasoft.com
Web Page: http://www.annabooks.com
email: info@annasoft.com
Compuserve: 73204,3405

Ansoft Corp .... **(412) 261-3200**
Four Station Square, Suite 660
Pittsburgh, PA 15219-1119
Tech:(412) 261-3200 Toll Free:(800) 323-9504
Fax:(412) 471-9427
Web Page: http://www.ansoft.com
email: info@ansoft.com

Antec Inc .... **(510) 770-1200**
2859 Bayview Dr
Fremont, CA 94538-6520
Tech:(800) 222-6832 Toll Free:(888) 542-6832
Fax:(510) 770-1288
Web Page: http://www.antec-inc.com
email: antec@antec-inc.com

Anthem Technology Systems .... **(408) 453-1200**
1680 McCandles Drive
Milpitas, CA 95035
Toll Free:(800) 359-3580
Fax:(408) 441-4504
email: http://www.anthem.com

Anvil Cases .... **(626) 968-4100**
15650 Salt Lake Ave
City Of Industry, CA 91745-1114
Toll Free:(800) 359-2684
Fax:(626) 968-1703
Web Page: http://www.anvilcase.com
email: info@anvilcase.com

AOL (see America Online)

Aonix ....(415) 543-0900
595 Market Street, 12th Floor
San Francisco, CA 94105
Tech:(415) 543-0900 Toll Free:(800) 972-6649
Fax:(415) 543-0145
Web Page: http://www.thomsoft.com
email: info@aonix.com

Apcom Systems Inc. ....(408) 739-8676
457 East Evelyn Ave, Suite H
Sunnyvale, CA 94086
Toll Free:(888) 422-0867
Fax:(408) 739-7169
Web Page: http://www.apcom.com
email: sales@apcom.com

Apertus Technologies Inc (see Carleton Corp)

Apex Data Inc (see Smart Modular Tech.)

APEX Software Corp ....(412) 681-4343
4516 Henry Street, Suite 202
Pittsburgh, PA 15213
Tech:(412) 681-4738 Toll Free:(800) 858-2739
Fax:(412) 681-4384
Web Page: http://www.apexsc.com
email: support@apexsc.com
Compuserve: GO APEX; 74774,1311

Apogee Software Inc ....(408) 369-9001
1999 S. Bascom Ave, Suite 325
Campbell, CA 95008-2207
Toll Free:(800) 854-6705
Fax:(408) 369-9018
Web Page: http://www.apogee.com
email: info@apogee.com

Apple Computer ....(408) 996-1010
20525 Mariani Avenue
Cupertino, CA 95014 USA
Fax:(408) 974-3770
Web Page: http://www.apple.com

Application Techniques Inc ....(978) 433-5201
10 Lomar Park Drive
Pepperell, MA 01463-1416
Tech:(978) 433-8464 Toll Free:(800) 433-5201
Fax:(978) 433-8466
Web Page: http://www.screencapture.com
email: support@screencapture.com

Applied Computer Systems (see Apcom Systems)

Applied Microsystems Corp ··················· **(425) 882-2000**
5020 148th Ave NE, PO Box 9702
Redmond, WA 98073-9702
Tech:(800) 275-4262 Toll Free:(800) 426-3925
Fax:(425) 883-3049
Web Page: http://www.amc.com
email: info@amc.com

Applix Inc ·· **(508) 870-0300**
112 Turnpike Rd
Westboro, MA 01581
Tech:(800) 827-7549 Toll Free:(800) 827-7549
Fax:(508) 366-2278
Web Page: http://www.applix.com
email: applixinfo@applix.com

Appro International Inc ··························· **(408) 941-8100**
446 S. Abbott Ave
Milpitas, CA 95035
Toll Free:(800) 927-5464
Fax:(408) 941-8111
Fax:(408) 941-8112
Web Page: http://www.appro.com
email: info@appro.com

Approach Software (see Lotus Develop)

APS Technologies ································· **(816) 483-1600**
6131 Deramus
Kansas City, MO 64120
Tech:(816) 483-6200 Toll Free:(800) 235-8935
Other:(800) 395-5871
Web Page: http://www.apstech.com
email: support@apstech.com

Apsylog Inc ·· **(510) 275-0200**
3000 Executive Pkwy, Suite 440
San Ramon, CA 94588
Tech:(510) 275-0200 Toll Free:(800) 277-9564
Fax:(510) 275-0225
Web Page: http://www.apsylog.com
email: inquiry@apsylog.com
Compuserve: 71563,3614

AR Industries (see Road Warrior Int.)

Arabesque Software (see NetManage)

Arc Soft Inc ·· **(510) 440-9901**
46601 Fremont Blvd

Fremont, CA 94538
Toll Free:(800) 762-8657
Fax:(510) 440-1270
Web Page: http://www.arcsoft.com
email: feedback@arcsoft.com

Arcada Software (see Seagate Softw..) ··(407) 333-7500
37 Skyline Drive, Suite 1101
Lake Mary, FL 32746
Tech:(800) 468-2587 Toll Free:(800) 327-2232
Fax:(407) 333-7730
Web Page: http://www.smg.seagatesoftware.com
email: support@arcada.com
Compuserve: GO ARCADA
BBS:(407) 444-9979

Archive Software (see Seagate)

Arco Computer Products Inc ··················(954) 925-2688
2750 N. 29th Ave, Suite 316
Hollywood, FL 33020
Tech:(954) 925-2688 Toll Free:(800) 458-1666
Fax:(954) 925-2889
Web Page: http://www.arcoide.com
email: arco@arcoide.com
BBS:(954) 925-2791

Ardent Software Inc ·································(508) 366-3888
50 Washington St
Westborough, MA 01581-1021
Tech:(800) 729-3553 Toll Free:(800) 486-9636
Fax:(508) 366-3669
Tech Fax:(508) 389-8750
Web Page: http://www.vmark.com
email: info@ardentsoftware.com

Areal Technology Inc (Out Of Business)

Argent Software···(408) 996-0938
1098 November Drive
Cupertino, CA 95014
Fax:(408) 343-1191
Web Page: http://www.argent.com
email: argent@argent.com

Arista Enterprises ··································(516) 435-0200
125 Commerce Drive
Hauppauge, NY 11788
Tech:(800) 274-7824 Toll Free:(800) 274-7824
Fax:(516) 435-4545

Aristo Computers Inc .... (503) 626-6333
6700 SW 105th Ave, Suite 300
Beaverton, OR 97008
Toll Free:(800) 327-4786
Fax:(503) 626-6492
Web Page: http://www.aristocom.com
email: sales@aristocom.com
BBS:(503) 520-0168

Aristosoft Inc. (see Software Made Simple)

Arlington .... (847) 362-1001
1001 Technology Way
Libertyville, IL 60048
Fax:(847) 362-3773
Web Page: http://www.arli.com
email: customer@arli.com

Arnet Corp (see Digi International)

Arrow Electronics Inc .... (516) 391-1300
25 Hub Drive
Melville, NY 11747
Fax:(516) 391-1640
Web Page: http://www.arrow.com

Arrowfield International Inc .... (714) 669-0101
2812-A Walnut Ave
Tustin, CA 92780
Fax:(714) 669-0526
Web Page: http://www.arrowfieldinc.com
email: arowfld@ix.netcom.com

Ars Nova Software .... (425) 889-0927
PO Box 637
Kirkland, WA 98083-0637
Tech:(425) 889-0927 Toll Free:(800) 445-4866
Fax:(425) 889-8699 Fax on Demand:(425) 828-2132
Web Page: http://www.ars-nova.com
email: info@ars-nova.com

Artecon Inc .... (760) 931-5500
6305 El Camino Real, PO Box 9000
Carlsbad, CA 92009-1606
Tech:(760) 931-5500 Toll Free:(800) 872-2783
Fax:(760) 931-5527
Tech Support:(800) 833-2783
Web Page: http://www.artecon.com
email: support@artecon.com

Artesyn Technologies .... **(608) 831-5500**
8310 Excelsior Dr
Madison, WI 53717
Toll Free:(800) 356-9602
Fax:(608) 831-4249
Fax:(608) 831-8844
Web Page: http://www.artesyn.com
email: cpsupport@artesyn.com

Articulate Systems .... **(781) 935-5656**
600 W. Cummings Park, Suite 4500
Woburn, MA 01801
Tech:(781) 935-2220 Toll Free:(800) 443-7077
Fax:(781) 935-0490
Other Address: AppleLink: Voice
Web Page: http://www.artsys.com
email: support@artsys.com

Artisoft Inc. .... **(520) 670-7100**
2202 N. Forbes Blvd
Tucson, AZ 85745
Tech:(520) 670-4287 Toll Free:(800) 846-9726
Fax:(520) 670-7101 Fax on Demand:(520) 884-1397
Automated Support:(520) 670-7000
Web Page: http://www.artisoft.com
email: sales@artisoft.com
Compuserve: GO ARTISOFT
BBS:(520) 884-8648

Artist Graphics Inc .... **(612) 631-7800**
900 Long Lake Rd
St Paul, MN 55112
Tech:(612) 631-7888 Toll Free:(800) 627-8478
Fax:(612) 631-7802
Web Page: http://www.artgraphics.com
email: sales@artgraphics.com
BBS:(612) 631-7664

Asante Technologies .... **(408) 435-8388**
821 Fox Lane
San Jose, CA 95131-1601
Tech:(800) 622-7464 Toll Free:(800) 662-9686
Fax:(408) 432-7511 Fax on Demand:(800) 741-8607
Tech Fax:(801) 566-3787
Web Page: http://www.asante.com
email: support@asante.com
BBS:(408) 432-1416

Ascend .... **(660) 463-1412**
Herndon, VA 22070

Tech:(800) 272-3634
Tech Fax:(816) 463-7958
Web Page: http://www.intercon.com
email: stfsupport@stfinc.com
BBS:(816) 463-1131

Ascend Communications ....(510) 769-6001
1701 Harbor Bay Pkwy
Alameda, CA 94502
Tech:(800) 272-3634 Toll Free:(800) 621-9578
Fax:(510) 814-2300
Web Page: http://www.ascend.com
email: info@ascend.com

Ascend Communications ....(978) 692-2600
5 Carlisle Road
Westford, MA 01886
Fax:(978) 692-9214
Web Page: http://www.casc.com

Ascent Solutions Inc ....(937) 847-2374
9009 Springboro Pike
Miamisburg, OH 45342
Web Page: http://www.asizip.com

ASCII Group Inc, The ....(301) 718-2600
7101 Wisconsin Ave, Suite 1000
Bethesda, MD 20814
Fax:(301) 718-0435
Web Page: http://www.ascii.com
email: info@ascii.com

ASD Software Inc ....(909) 624-2594
4650 Arrow Highway, Suite E-6
Montclair, CA 91763-1223
Tech:(909) 624-2594
Fax:(909) 624-9574
Web Page: http://www.asdsoft.com
email: tech@asdsoft.com
Compuserve: GO ASD SOFTWARE 102404,3630
America Online: ASDSOFT

Ashlar Inc ....(408) 615-6840
100 Saratoga Ave, Suite 200
Santa Clara, CA 95051
Tech:(800) 877-2745 Toll Free:(800) 877-2745
Fax:(408) 487-9815
Web Page: http://www.ashlar.com
email: support@ashlar.com
Compuserve: 7133,1060

Ashton-Tate (see Inprise Corp)

ASIC Northwest Inc ....(541) 923-3755
14108 S. Mountain View Drive
Powell Butte, OR 97753
Fax:(541) 923-8752
Web Page: http://www.asicnw.com
email: john@asicnw.com

askSam Systems ....(850) 584-6590
P O Box 1428
Perry, FL 32348
Tech:(850) 584-6590 Toll Free:(800) 800-1997
Fax:(850) 584-7481
Web Page: http://www.asksam.com
email: tech@askSam.com
Compuserve: 74774,352 or GO ASKSAM

ASP Computer Products ....(408) 746-2965
Tech:(408) 746-2965
Fax:(408) 456-8946
Web Page: http://www.asp.net
email: asptech@asp.net
BBS:(408) 456-9139

Aspect Software Engineering (see Microsoft)

Association for Computing Machinery ..(212) 869-7440
1515 Broadway, 17th Floor
New York, NY 10036-5701
Toll Free:(800) 342-6626
Fax:(212) 944-1318
Web Page: http://www.acm.org
email: acmhelp@acm.org

AST Computer ....(714) 727-4141
16215 Alton Parkway
Irvine, CA 92718-3618
Tech:(800) 727-1278 Toll Free:(800) 876-4278
Fax:(714) 727-9355 Fax on Demand:(800) 926-1278
Web Page: http://www.ast.com
email: web.support@ast.com
Compuserve: GO AST
America Online: keyword: AST
BBS:(714) 727-4132

Astec America Inc. ....(760) 930-4600
6339 Paseo Del Lago
Carlsdad, CA 92009
Tech:(760) 757-1880
Fax:(760) 930-4700

Web Page: http://www.astec.com

Astec Standard Power (see Astec America Inc)

Astound Inc .... (408) 720-0337
710 Lakeway Dr, Suite 230
Sunnyvale, CA 94086
Tech:(905) 602-5292 Toll Free:(800) 982-9888
Fax:(408) 720-1011
tech support:(905) 602-0395
Web Page: http://www.astoundinc.com
email: listserve@astound.com
Compuserve: 75300,3433 or GO ASTOUND
America Online: AstoundInc

Asus Computer International .... (510) 739-3777
6737 Mowry Business Center, Bldg. 2
Newark, CA 94560
Fax:(510) 608-4555
RMA Fax:(510) 608-4511
Web Page: http://www.asus.com
email: tsd@support.asus.com
BBS:(510) 739-3774

Asymetrix Corp (see Asymetrix Learning Sys)

Asymetrix Learning Systems .... (603) 883-0220
20 Trafalgar Square, Suite 300
Nashua, NH 03063
Tech:(800) 801-2884 Toll Free:(800) 289-2884
Fax:(603) 883-5582
Web Page: http://www.aimtech.com
email: support@aimtech.com
BBS:(603) 598-8402

Asymetrix Learning Systems Inc. .... (425) 637-1500
110 110th Ave NE, Suite 700
Bellevue, WA 98004-5840
Tech:(425) 637-1600 Toll Free:(800) 448-6523
Fax:(425) 455-3071
Web Page: http://www.asymetrix.com
BBS:(425) 451-1173

AT&T Global Info. Solutions (NCR) .... (800) 746-4722
1700 S. Patterson Blvd
Dayton, OH 45479
Tech:(800) 831-4314 Toll Free:(800) 746-4722
Web Page: http://www.ncr.com

AT&T National Parts Sales Center .... (800) 222-7278
4925 Nome St

Denver, CO 80239
Tech:(800) 628-2888 Toll Free:(800) 222-7278
Fax:(800) 527-4360 Fax on Demand:(800) 527-4360
Web Page: http://www.lucent.com

Atcom/Info ....(619) 699-4000
308 G Street
San Diego, CA 92101
Toll Free:(888) 552-8266
Fax:(619) 699-4040
Web Page: http://www.atcominfo.com
email: help@atcominfo.com

ATI Technologies Inc ....(905) 882-2600
33 Commerce Valley Drive East
Thorn Hill, ON L3T 7N6 Canada
Tech:(905) 882-2626
Fax:(905) 882-2620 Fax on Demand:(905) 882-2600
Web Page: http://www.atitech.com
email: support@atitech.com
Compuserve: 74740,667 or GO ATITECH
BBS:(905) 764-9404

Atronics International Inc ....(510) 656-8400
44700-B Industrial Dr
Fremont, CA 94538-6431
Toll Free:(800) 488-7776
Fax:(510) 656-8560
Web Page: http://www.atronics.com

Attachmate Corp ....(425) 644-4010
3617 131st Ave SE
Bellevue, WA 98006-1332
Tech:(800) 388-3270 Toll Free:(800) 426-6283
Fax:(425) 649-6461 Fax on Demand:(425) 649-6595
Tech Support:(425) 957-4607
Other Address: http://www.atm.com
Web Page: http://www.attachmate.com
Compuserve: GO ATTACHMATE
BBS:(425) 649-6660

Attain (Out of Business)

Attar Software ....(978) 456-3946
Two Deerfoot Trail On Partridge Hill
Harvard, MA 01451
Toll Free:(800) 456-3966
Fax:(978) 456-8383
Web Page: http://www.attar.com
email: info@attar.com

Compuserve: 70400,2035

AudioNet ....**(214) 748-6660**
2914 Taylor Street
Dallas, TX 75226
Fax:(214) 748-6657
Web Page: http://www.audionet.com

Aurora Technologies ....**(781) 290-4800**
176 Second Ave
Waltham, MA 02154
Tech:(781) 290-4800
Fax:(781) 290-4844
Tech Fax:(781) 290-5358
Web Page: http://www.auroratech.com
email: support@auroratech.com

Auspex Systems Inc. ....**(408) 986-2000**
2300 Central Expressway
Santa Clara, CA 95050
Tech:(408) 986-2000 Toll Free:(800) 735-3177
Fax:(408) 986-2020
Web Page: http://www.auspex.com
email: info@auspex.com

Autodesk ....**(816) 891-1040**
10725 Ambassador Dr
Kansas City, MO 64153
Tech:(816) 891-8418 Toll Free:(800) 231-8574
Fax:(816) 891-8018
Pro Product Support:(816) 891-0195
Web Page: http://www.softdesk.com
email: ag3@softdesk.com
Compuserve: 76004,1602

Autodesk Inc ....**(415) 507-5000**
111 McInnis Pkwy
San Rafael, CA 94903
Tech:(206) 487-2934 Toll Free:(800) 538-6401
Fax:(415) 507-5100 Fax on Demand:(415) 446-1919
Web Page: http://www.autodesk.com
Compuserve: GO ADESK

Automap (see Microsoft)

Automatic Answer, The ....**(714) 661-2660**
27121 Calle Arroyo, Bldg 2200
San Juan Capistrano, CA 92675
Tech:(800) 800-9822 Toll Free:(800) 262-0291
Fax:(714) 661-0778
Tech Fax:(203) 744-4560

Web Page: http://www.taa.com
email: Customer.Support@taa.com

AVA Instrumentation Inc ························(408) 336-2281
19 Horseshoe Court
Scotts Valley, CA 95066
Tech:(408) 461-1685
Fax:(408) 461-1883
Web Page: http://www.aimnet.com/~avasales/
email: avasales@aimnet.com

Avalan Technology Inc ···························(508) 429-6482
PO Box 6888
Holliston, MA 01746
Fax:(508) 429-3179
Web Page: http://www.avalan.com
email: saleinfo@avalon.com

Avalon Hill Game (see Monarch Ava.)····(410) 254-9200
4517 Hartford Rd
Baltimore, MD 21214
Tech:(410) 426-9600 Toll Free:(800) 999-3222
Fax:(410) 254-0991
Web Page: http://www.avalonhill.com
Compuserve: 72662,1207
America Online: ahgames@aol.com

Avance Technology Inc ···························(310) 446-1339
1800 Century Park East, Suite 600
Los Angeles, CA 90067
Fax:(310) 446-4759
Web Page: http://www.avance-technologies.com
email: sales@avance-technologies.com

Avantos Performance Systems ··············(510) 654-4600
5900 Hollis St, Suite A
Emeryville, CA 94608-2006
Tech:(510) 235-7593
Fax:(510) 654-5199
Tech Fax:(510) 654-4542
Web Page: http://www.avantos.com
email: avantos@aol.com
Compuserve: 71333,2201

Avatar/DCA (see Attachmate)

Avax International ·······························(519) 833-2900
8 Thompson Crescent
Erin, ON N0B 1T0 Canada
Toll Free:(800) 443-4542
Fax:(519) 833-7469

Web Page: http://www.avax.com
email: sales@avax.com

Aveo .... **(408) 486-7900**
2901 Tasman Drive, Suite 208
Santa Clara, CA 95054
Tech:(408) 486-0444
Fax:(408) 486-7952 Fax on Demand:(408) 486-7920
Web Page: http://www.cypressr.com
email: info@aveo.com

Avery Dennison International .... **(800) 462-8379**
150 North Orange Grove Blvd.
Pasadena, CA 91103
Toll Free:(800) 462-8379
Fax:(800) 831-2496
Web Page: http://www.avery.com
email: feedback@averydennison.com

Avery Label .... **(714) 674-8500**
777 East Foothill Boulevard
Azusa, CA 91702
Tech:(972) 389-3699 Toll Free:(800) 252-8379
Fax:(800) 848-7741 Fax on Demand:(626) 584-1681
Web Page: http://www.avery.com
email: softwaresupport@averydennison.com

Avnet Inc .... **(516) 466-7000**
80 Cutter Mill Road
Great Neck, NY 11021
Fax:(516) 466-1203
Web Page: http://www.avnet.com
email: www-marcom@avnet.com

Award Software International Inc .... **(650) 237-6800**
777 E. Middlefield Road
Mountain View, CA 94043-4023
Tech:(650) 237-6800
Fax:(650) 968-0274
Web Page: http://www.award.com
email: sales@award.com
BBS:(650) 968-0249

Axent Technologies .... **(617) 487-7700**
Tech:(781) 890-6944 Toll Free:(888) 727-8671
Tech Support:(617) 977-2001
Web Page: http://www.raptor.com
email: support@raptor.com

Axis Communications Inc .... **(781) 938-1188**
4 Constitution Way, Suite G

Woburn, MA 01801-1030
Tech:(800) 444-2947 Toll Free:(800) 444-2947
Fax:(781) 938-6161
Tech Fax:(781) 938-0774
Web Page: http://www.axis.com
email: tech-support@axis.com
BBS:(781) 932-3363

Az-Tech Software ....(816) 776-2700
201 E. Franklin St, Suite 11
Richmond, MO 64085-1883
Tech:(816) 776-2700
Fax:(816) 776-8398
Web Page: http://www.az-tech.com
email: tsg@az-tech.com

Azerty Inc ....(716) 662-0200
13 Centre Dr
Orchard Park, NY 14127
Tech:(716) 662-7616 Toll Free:(800) 888-8080
Fax:(716) 662-7616
Web Page: http://www.azerty.com

Aztech New Media ....(416) 449-4787
1 Scarsdale Rd
Don Mills, ON M3B 2R2
Fax:(416) 449-1058
Web Page: http://www.aztech.com
email: support@aztech.com

Azure Technologies (see GN Nettest Group)

B & L Associates Inc ....(781) 444-1404
56 Kearney Road
Needham Heights, MA 02194
Fax:(781) 444-5805
Web Page: http://www.bandl.com
email: info@bandl.com

B. C. Software Inc ....(310) 636-4711
Tech:(310) 636-4711 Toll Free:(800) 231-4055
Web Page: http://www.bcsoftware.com
email: info@bcsoftware.com

Baker & Taylor Entertainment ....(847) 965-8060
8140 Lehigh Ave
Morton Grove, IL 60053
Tech:(847) 965-8060 Toll Free:(800) 775-4200
Fax:(847) 470-7860
Web Page: http://www.btent.com
email: info@btent.com

Balboa Software ....(800) 763-8542
5845 Yonge Street, P O Box 69539
Willowdale, ON M2M 4K3 Canada
Tech:(416) 730-8980 Toll Free:(800) 763-8542
Fax:(416) 730-9715
Web Page: http://www.balboa-software.com
email: support@balboa-software.com

Baler Software Corp (see Tech Tools)

Balt Inc ....(817) 697-4953
201 N. Crockett
Cameron, TX 76520
Tech:(817) 697-4953 Toll Free:(800) 749-2258
Fax:(800) 529-7577
Fax:(800) 697-6258
Web Page: http://www.baltinc.com
email: furniture@baltinc.com

Banner Blue Software (see Broderbund)

Banyan Systems Inc ....(508) 898-1000
120 Flanders Rd, PO Box 5013
Westboro, MA 01581
Tech:(508) 898-1000 Toll Free:(800) 222-6926
Fax:(508) 898-1755
Web Page: http://www.banyan.com
email: webmaster@banyan.com
Compuserve: GO BAN FORUM
BBS:(508) 836-1834

Barbey Electronics ....(610) 376-7451
333 N. 4th
Reading, PA 19603
Tech:(610) 376-7451 Toll Free:(800) 822-2251
Fax:(610) 372-8622
Web Page: http://www.barbeyele.com
email: info@barbeyele.com

BASF Magnetics (See Emtech Data Store Media)

Basic Needs Inc ....(760) 738-7020
118 State Place, Suite 202
Escondido, CA 92029
Tech:(800) 633-3703 Toll Free:(800) 633-3703
Fax:(760) 738-0515
Web Page: http://www.basicneeds.com
email: sales@basicneeds.com

Bason Computer, Inc. ....(818) 727-9054
20130 Plummer St

Chatsworth, CA 91311 USA
Toll Free:(800) 238-4453
Fax:(818) 727-9064
Web Page: http://www.basoncomputer.com

Bate Tech Software Inc. (see Ixchange Inc.)

Battery Express (see Fuller's Wholesale Ele)

Battery Technology Inc (BTI) ....(213) 728-7874
5700 Bandini Blvd
Commerce, CA 90040
Tech:(800) 982-8284 Toll Free:(800) 982-8284
Fax:(213) 728-7996
Web Page: http://www.batterytech.com
email: techsupport@batterytech.com

Bay Networks Inc (Nortel)..........................(408) 988-2400
4401 Great America Pkwy
Santa Clara, CA 95054
Tech:(800) 252-6926 Toll Free:(800) 822-9638
Fax:(408) 988-5525 Fax on Demand:(800) 786-3228
Public Relations:(408) 764-7548
Web Page: http://www.baynetworks.com
email: infosvc@baynetworks.com
Compuserve: GO BAYNET

Bayer Corp. (AGFA Division) ....(201) 440-2500
100 Challenger Road
Ridgefield Park, NJ 07660
Tech:(800) 879-2432 Toll Free:(800) 424-8973
Fax:(201) 440-5733 Fax on Demand:(800) 879-2432
Web Page: http://www.agfa.com

BayWare Inc (see Transparent Language)

BBN Inc. (see GTE Internetworking)

BCAM Int. Inc. (see HumanCAD Systems Inc.)

BDM International Inc ....(888) 842-9237
1501 BDM Way
McLean, VA 22102
Toll Free:(888) 842-9237
Web Page: http://www.cybershield.com
email: info@www.cybershield.com

BE Inc ....(650) 462-4100
800 El Camino Real, Suite 300
Menlo Park, CA 94025
Fax:(650) 462-4129
Web Page: http://www.be.com
email: custsupport@be.com

Beckman Industrial (see Wavetek)

BEI Corp ....(425) 644-6000
200 Kelsey Creek Center, 15015 Main St
Bellevue, WA 98007
Fax:(425) 644-8222
Web Page: http://www.ultrabac.com
email: support@ultrabac.com

Belden Wire And Cable ....(765) 983-5200
2200 US Highway 27 South, PO Box 1980
Richmond, IN 47345
Tech:(765) 983-5200 Toll Free:(800) 235-3361
Fax:(765) 983-5737
Web Page: http://www.belden.com
email: info@belden.com

Belkin Components...................................(310) 898-1100
501 W. Walnut Street
Compton, CA 90220
Tech:(800) 223-5546 Toll Free:(800) 223-5546
Fax:(310) 898-1111
Web Page: http://www.belkin.com
email: support@belkin.com

Belmont Distributing (see Almo Corp)

Benefit Software Inc ....(805) 568-0240
212 Cottage Grove Ave
Santa Barbara, CA 93101
Toll Free:(800) 533-1388
Fax:(805) 568-0239
Web Page: http://www.bsiweb.com
email: info@bsiweb.com

Bentley Systems Inc ....(610) 458-5000
690 Pennsylvania Drive
Exton, PA 19341
Toll Free:(800) 236-8539
Fax:(610) 458-1060
Sales:(610) 458-1059
Web Page: http://www.bentley.com
email: select.online@bentley.com

Berkeley Software Design Inc ....(719) 593-9445
5575 Tech Center Drive, # 110
Colorado Springs, CO 80918
Tech:(800) 487-2738 Toll Free:(800) 800-4273
Fax:(719) 598-4238
toll free order:(800) 776-2734
Web Page: http://www.bsdi.com

email: info@bsdi.com

Berkeley Systems Inc ··························(510) 540-5535
2095 Rose Street
Berkeley, CA 94709
Tech:(510) 549-2300 Toll Free:(800) 757-7707
Fax:(510) 849-9426
Web Page: http://www.berksys.com
email: support@sierra.com
Compuserve: GO BERKFORUM 75300,1375

Berkshire Products ··························(770) 271-0088
PO Box 1015
Suwanee, GA 30024
Fax:(770) 932-0082
Web Page: http://www.berkprod.com
email: email@berkprod.com

Best Data Products Inc ··························(818) 773-9600
21800 Nordhoff Street
Chatsworth, CA 91311-3943
Tech:(818) 773-9600
Fax:(818) 773-9619
Tech Fax:(818) 717-1721
Web Page: http://www.bestdata.com
email: service@bestdata.com
BBS:(818) 773-9627

Best Power ··························(608) 565-7200
N 9246 Hwy 80, PO Box 280
Necedah, WI 54646
Tech:(800) 356-5737 Toll Free:(800) 356-5794
Fax:(608) 565-2221 Fax on Demand:(800) 487-6813
Fax:(608) 565-7675
Web Page: http://www.bestpower.com
email: service@bestpower.com
Compuserve: 71333,2001
BBS:(608) 565-2901

Best Programs Inc. (see Best Software Inc.)

Best Software Inc ··························(703) 709-5200
11413 Isaac Newton Square
Reston, VA 20190
Tech:(800) 331-8514 Toll Free:(800) 368-2405
Fax:(703) 318-0499
Fax:(703) 709-9359
Web Page: http://www.bestsoftware.com
email: VACustServ@bestprograms.com

Best! Software .... **(813) 579-1111**
888 Executive Center Dr W, Suite 300
St. Petersburg, FL 33702
Tech:(813) 579-1111 Toll Free:(800) 424-9392
Fax:(813) 578-2178
Web Page: http://www.abra.com
email: custserv@bestsoftware.com

BestWare Inc .... **(973) 586-2200**
300 Roundhill Drive
Rockaway, NJ 07866
Tech:(800) 322-6962 Toll Free:(800) 322-6962
Fax:(973) 586-8885 Fax on Demand:(973) 586-1553
Marketing Fax on Demand:(973) 586-2200
Web Page: http://www.bestware.com
Compuserve: GO MYOB

Bethesda Softworks .... **(301) 926-8300**
1370 Piccard Drive, Suite 120
Rockville, MD 20850-4304
Tech:(301) 963-2002 Toll Free:(800) 677-0700
Fax:(301) 926-8010
Hints:(900) 884-1687
Web Page: http://www.bethsoft.com
email: tech@bethsoft.com

Beverly Hills Software .... **(310) 358-8311**
8845 W. Olympic Blvd, Suite 200
Beverly Hills, CA 90211
Fax:(310) 358-0326
Web Page: http://www.bhs.com
email: webmaster@bhs.com

Bible Research Systems .... **(512) 251-7541**
2013 Wells Branch Pkwy, # 304
Austin, TX 78728
Tech:(512) 251-7541 Toll Free:(800) 423-1228
Fax:(512) 251-4401
Web Page: http://www.brs-inc.com/bible.html
email: bible@brs-inc.com

Biblesoft .... **(206) 824-8360**
22014 7th Ave South, Suite 104/205
Seattle, WA 98198-6235
Tech:(206) 870-1463 Toll Free:(800) 877-0778
Fax:(206) 824-1828
Web Page: http://www.biblesoft.com
email: techsupp@biblesoft.com

Bindview..(713) 881-9100
3355 West Alabama, 12th Floor
Houston, TX 77098
Tech:(800) 749-8439 Toll Free:(800) 749-8439
Fax:(713) 881-9200
Web Page: http://www.bindview.com
email: webmaster@bindview.com
Compuserve: GO BINDVIEW

BitShop..(301) 345-6789
4716 Pontiac St, Suite 310-A
College Park, MD 20740
Fax:(301) 345-6745
Web Page: http://www.bitshop.com
email: sales@bitshop.com

Bitstream Inc..(617) 497-6222
215 First Street
Cambridge, MA 02142
Tech:(617) 497-7514 Toll Free:(800) 522-3668
Fax:(617) 868-0784
Tech support Fax:(617) 354-7954
Web Page: http://www.bitstream.com
email: info@bitstream.com
Compuserve: 71221,2145

Biz Base Inc (see ACE Contact Manager)

Black Belt Systems ....(406) 228-8945
398 Johnson Road, Building 2
Glasgow, MT 59230
Tech:(406) 228-8944
Fax:(406) 228-8943
Web Page: http://www.blackbelt.com
email: support@blackbelt.com
Compuserve: 74774,3106

Black Box Corp..(724) 746-5500
1000 Park Drive
Lawrence, PA 15055-1018
Tech:(724) 746-5565
Fax:(724) 746-0746
Fax:(800) 321-0746
Web Page: http://www.blackbox.com
email: info@blackbox.com

Black Ice Software Inc ....(603) 673-1019
292 Route 101
Amherst, NH 03031
Tech:(603) 673-1019

Fax:(603) 672-4112
email: blackice@mv.mv.com
Compuserve: GO BLACKICE

Blackstar Publishing Company ....(212) 679-3288
116 East 27th Street
New York, NY 10016
Fax:(212) 889-2052
Web Page: http://www.blackstar.com
email: sales@blackstar.com

Blastronix ....(209) 795-0738
PO Box 6255
Arnold, CA 95223
Fax:(209) 795-0646
Web Page: http://www.blastronix.com
email: dbarnes@blastronix.com

Blue Sky Software ....(619) 459-6365
7777 Fay Ave, Suite 201
La Jolla, CA 92037
Tech:(619) 551-5680 Toll Free:(800) 793-0364
Fax:(619) 459-6366
Tech Fax:(619) 551-2486
Web Page: http://www.blue-sky.com
email: support@blue-sky.com

Blue Squirrel ....(801) 523-1063
Palo Alto, CA 94306

Blue Willow Inc ....(303) 932-1600
8241 S. Carr Street
Littleton, CO 80128
Toll Free:(800) 932-1600
Fax:(303) 932-1800
Web Page: http://www.bluewillow.com
email: glover@bluewillow.com

Bluestone Inc ....(609) 727-4600
1000 Briggs Road
Mount Laurel, NJ 08054
Tech:(609) 778-7900
Fax:(609) 727-5077
Tech Fax:(609) 234-2877
Web Page: http://www.bluestone.com
email: support@bluestone.com

BMDP Statistical Software Inc (see SPSS)
Web Page: http://www.spss.com

Boardwatch Magazine ....(303) 973-6038
13949 W. Colfax Ave., Suite 250
Golden, CO 80401
Toll Free:(800) 933-6038
Fax:(303) 973-3731
Web Page: http://www.boardwatch.com
email: subscriptions@boardwatch.com

Boca Research ....(561) 997-6227
1377 Clint More Road
Boca Raton, FL 33487
Tech:(561) 241-8088
Fax:(561) 995-9456 Fax on Demand:(561) 995-9456
Web Page: http://www.bocaresearch.com
email: support@bocaresearch.com
Compuserve: GO BOCA
BBS:(561) 241-1601

Boffin Limited ....(612) 894-0595
2500 W. County Road 42, Suite 5
Burnsville, MN 55337
Toll Free:(800) 248-5328
Fax:(612) 894-6175
Web Page: http://www.boffin.com
email: sales@boffin.com

Borland International (see Inprise Corp)

Boston Computer Exchange ....(617) 542-4414
210 South Street
Boston, MA 02111
Tech:(617) 542-4414 Toll Free:(800) 262-6399
Fax:(617) 542-8849

Bottom Line Industries Inc ....(818) 700-1922
9556 Cozycroft Ave
Chatsworth, CA 91311
Tech:(818) 700-1922 Toll Free:(800) 334-6044
Fax:(818) 700-4549
Web Page: http://bottomline-ind.com
email: sales@bottomline-ind.com

Boundless Technologies ....(516) 342-7887
Tech:(800) 231-5445 Toll Free:(800) 231-5445
Web Page: http://www.boundless.com
email: sales@boundless.com

Bourbaki Inc ....(208) 342-5849
PO Box 2867
Boise, ID 83701
Fax:(208) 342-5823

Web Page: http://www.primenet.com/cnsbbk/
email: cnsbbk@bourbaki.com

Box Hill Systems Corp ....(212) 989-4455
161 Avenue Of The Americas
New York, NY 10013
Tech:(800) 727-3863 Toll Free:(800) 727-3863
Fax:(212) 989-6817
Web Page: http://www.boxhill.com
email: tech-support@boxhill.com

Boxer Software ....(602) 485-1635
PO Box 14545
Scottsdale, AZ 85267-4545
Toll Free:(800) 982-6937
Fax:(602) 485-1636
Web Page: http:/www.boxersoftware.com/users/dham
email: dhamel@boxersoftware.com
Compuserve: 74777,170; GO BOXER

Brain-Storm Technologies Inc ....(818) 760-7974
11440 Chandler Blvd, Suite 200
North Hollywood, CA 91601
Tech:(818) 760-7974 Toll Free:(800) 829-7974
Customer Service:(800) 979-8321
Web Page: http://www.mindtech.com
email: brainstorm@nwc.com

Brick Computer Company, The...............(978) 535-7510
One Intercontinental Way
Peabody, MA 01960
Tech:(800) 633-1922 Toll Free:(800) 495-3746
Fax:(978) 535-7512 Fax on Demand:(800) 723-0778
Web Page: http://www.ergo-computing.com
email: support@ergo-computing.com
Compuserve: GO ERGO
BBS:(978) 535-7228

Brightwork Developement (see McAfee East)

Brilliance Labs Inc...................................(407) 306-9554
10600 Bloomfield Drive, Suite 1528
Orlando, FL 32825
Fax:(407) 306-9554
Web Page: http://www.brlabs.com
email: info@brlabs.com

BroadVision ....(650) 261-5100
585 Broadway
Redwood City, CA 94063
Tech:(888) 825-5121 Toll Free:(800) 269-9375

Fax:(650) 261-5900
Web Page: http://www.broadvision.com
email: bvhelp@broadvision.com

Broderbund Software Inc ....(415) 382-4400
500 Redwood Blvd
Novato, CA 94947
Tech:(415) 382-4700 Toll Free:(800) 521-6263
Fax:(415) 382-4419 Fax on Demand:(800) 474-8840
Other:(800) 548-1798
Other Address: http://www.broder.com
Web Page: http://www.broderbund.com
Compuserve: GO BB
America Online: keyword: Broderbund

Active Mind Series..(415) 382-4740
Toll Free:(800) 548-1798
Fax on Demand:(800) 474-8840

Carmen Sandiego..(415) 382-4747
Toll Free:(800) 548-1798
Fax on Demand:(800) 474-8840

Living Books..(415) 382-4710
Toll Free:(800) 548-1798
Fax on Demand:(800) 474-8840

Print Shop..(415) 382-4750
Toll Free:(800) 548-1798
Fax on Demand:(800) 474-8840

Red Orb Entertainment Center......................(415) 382-4799
Toll Free:(800) 548-1798
Fax on Demand:(800) 474-8840

T/Maker products..(415) 382-4775
Toll Free:(800) 548-1798
Fax on Demand:(800) 474-8840

Brooktrout Technology...........................(781) 449-4100
410 First Ave
Needham, MA 02494-2722
Fax:(781) 449-9009
Web Page: http://www.brooktrout.com
email: info@brooktrout.com

Brother International................................(908) 356-8880
200 Cottontail Lane
Somerset, NJ 08875-6714
Tech:(901) 373-6256 Toll Free:(800) 284-4357
Fax:(800) 947-1445 Fax on Demand:(800) 521-2846
Web Page: http://www.brother.com
BBS:(714) 859-2610

Dealer parts..(901) 373-6371

Fax Service .... (800) 284-4329
100 Somerset Corporate Blvd
Bridgewater, NJ 08807-0911
Fax:(908) 575-8790
Accessory:(888) 879-3232
BBS:(888) 298-3616

Printer service .... (800) 276-7746
15 Musick
Irvine, CA 92618
Fax:(714) 859-2272 Fax on Demand:(800) 521-2846
Accessories:(888) 879-3232
Web Page: http://www.brother.com
BBS:(888) 298-3616

Word Processor Service .... (901) 373-6256
Fax:(901) 373-6213

BTC (Behavior Tech Computer Corp) ....(510) 657-3956
4180 Business Center Dr
Freemont, CA 94538
Fax:(510) 657-3965
Web Page: http://www.behavior.com/index.htm
email: service@btc.com.tw

BTG Inc .... (703) 383-8000
3877 Fairfax Ridge Road
Fairfax, VA 22030-7448
Tech:(800) 899-6200 Toll Free:(800) 899-6200
Fax:(703) 383-8999
Web Page: http://www.btg.com

Buerg Software And Computers .... (707) 769-5477
850 Petaluma Blvd N
Petaluma, CA 94952
Toll Free:(800) 442-8374
Fax:(707) 769-5479
Web Page: http://www.buerg.com
email: buerg@buerg.com

Buffalo Creek Software .... (515) 225-9552
913 39th Street
West Des Moines, IA 50265
BBS:(515) 225-8496

Buffalo Inc .... (512) 349-1580
2805 19th St SE
Salem, OR 97302-1520
Tech:(512) 349-1580 Toll Free:(800) 345-2356
Fax:(512) 794-8520
Web Page: http://www.buffinc.com
email: tech@buffinc.com

Compuserve: 72350,2252

Bulldog Computer Products ....(800) 438-6039
851 Commerce Court
Buffalo Grove, IL 60089

Bungie Software Products Corp ....(312) 397-0500
350 W. Onterio 7th Floor
Chicago, IL 60601
Toll Free:(800) 295-0060
Web Page: http://www.bungie.com
email: pcsupport@bungie.com

Bureau of Electronic Publishing (see Thynx)

Burr-Brown Corp ....(520) 746-1111
PO Box 11400
Tucson, AZ 85734-1400
Tech:(800) 548-6132 Toll Free:(800) 227-3947
Fax:(520) 746-7401
Web Page: http://www.burr-brown.com

Business Resource Software Inc............(512) 251-7541
2013 Wells Branch Pkwy, # 305
Austin, TX 78728
Tech:(512) 251-7541 Toll Free:(800) 423-1228
Fax:(512) 251-4401
Web Page: http://www.brs-inc.com
email: brs@brs-inc.com

BusLogic Inc (see Mylex Corp)

Button Ware Inc (see Buttonware Software)

Buttonware Software Inc ....(972) 238-2535
13800 Montfort Drive, Suite 100
Dallas, TX 75240
Tech:(972) 734-3306
Fax:(972) 734-2957
Sales Only:(888) 734-3824
Web Page: http://www.outlooksoftware.com
email: support@buttonware.com
Compuserve: GO PCVENA
BBS:(214) 713-5122

Cable Connection.....................................(408) 395-6700
102 Cooper Court
Los Gatos, CA 95030
Fax:(408) 354-3980
Web Page: http://www.cable-connection.com
email: cables4u@cable-connection.com

Cables To Go (CTG) .... **(937) 224-8646**
1501 Webster Street
Dayton, OH 45404
Tech:(513) 275-0886 Toll Free:(800) 826-7904
Fax:(800) 331-2841
Other:(800) 225-8646
Web Page: http://www.cablestogo.com
email: expert@cablestogo.com

Cabletron Systems Inc .... **(603) 332-9400**
35 Industrial Way
Rochester, NH 03866-5005
Tech:(603) 332-9400
Fax:(603) 337-2211 Fax on Demand:(603) 337-2444
Web Page: http://www.ctron.com
email: support@cabletron.com
BBS:(603) 335-3358

Cactus Development Company Inc .... **(512) 453-2244**
7113 Burnet Road, Suite 214
Austin, TX 78757-2216
Tech:(512) 453-2244 Toll Free:(800) 336-9444
Fax:(512) 453-3757
Web Page: http://www.cactusdevelopment.com
Compuserve: 72662,1356 or GO CACTUS

Cadence Design Systems Inc .... **(408) 733-1595**
555 N. Mathilda Ave
Sunnyvale, CA 94086
Web Page: http://www.cadence.com/alta

Cadix International Inc .... **(770) 804-9951**
1200 Ashwood Pkwy, Suite 135
Atlanta, GA 30338
Tech:(770) 804-9951 Toll Free:(800) 876-4605
Web Page: http://www.cadix.com
email: sales@cadix.com

CADRE Technologies (see Cayenne Software)

Caere Corp .... **(408) 395-7000**
100 Cooper Court
Los Gatos, CA 95032
Tech:(408) 395-8319 Toll Free:(800) 535-7226
Fax:(408) 354-2743 Fax on Demand:(408) 354-8471
Other:(408) 395-8319
Web Page: http://www.caere.com
email: support@caere.com
BBS:(408) 395-1631

CakeWalk Music Software ....(617) 441-7870
5 Cambridge Center
Cambridge, MA 02142
Tech:(617) 924-6275 Toll Free:(888) 225-3925
Fax:(617) 441-7887
Web Page: http://www.cakewalk.com
email: support@cakewalk.com
Compuserve: 74774,1773

Cal Comp ....(714) 821-2000
2411 West La Palma Ave
Anaheim, CA 92801-2689
Tech:(800) 225-2667 Toll Free:(800) 225-2667
Fax:(714) 821-2832
Other Address: ftp.summagraphics.com
Web Page: http://www.calcomp.com
email: websupportitd@calcomp.com

CalComp Inc ....(714) 821-2000
2411 W Lapalma Ave
Anaheim, CA 92801
Tech:(800) 225-2667 Toll Free:(800) 225-2667
Fax:(714) 821-2832 Fax on Demand:(714) 821-2914
Web Page: http://www.calcomp.com

Calculus Inc ....(650) 854-3130
325 Sharon Park Drive, Suite M
Menlo Park, CA 94025
Fax:(650) 854-1248

Caldera Inc ....(801) 377-7687
240 W. Center Street
Orem, UT 84057
Tech:(801) 377-7687 Toll Free:(800) 850-7779
Fax:(801) 377-8752
Web Page: http://www.caldera.com
email: info@caldera.com

Calera Recognition Systems (see Caere)

Caligari Corporation ....(650) 390-9600
1959 Landings Drive
Mountain View, CA 95043
Tech:(650) 390-9600 Toll Free:(800) 351-7620
Fax:(650) 390-9755
Web Page: http://www.caligari.com
email: support@caligari.com
Compuserve: GO GUGRPA
America Online: Keyword: Caligari

Caliper Corp .... **(617) 527-4700**
1172 Beacon Street
Newton, MA 02161-9926
Tech:(617) 527-4700
Fax:(617) 527-5113
Web Page: http://www.caliper.com
email: support@caliper.com

Callware Technologies .... **(801) 984-1100**
8911 South Sandy Parkway
Sandy, UT 84070
Tech:(801) 984-6230 Toll Free:(800) 888-4226
Fax:(801) 984-1120
Web Page: http://www.callware.com
email: info@callware.com

Cambrix Publishing .... **(800) 992-8781**
9304 Deering Ave
Chatsworth, CA 91311
Tech:(818) 993-4274 Toll Free:(800) 992-8781
Fax:(818) 992-8781
Tech Fax:(818) 993-6201
Web Page: http://www.cambrix.com
email: cambrix@earthlink.net

Camelot Corporation .... **(972) 733-3005**
17770 Preston Road
Dallas, TX 75252
Tech:(972) 733-3005 Toll Free:(800) 528-7822
Fax:(972) 733-0574
Web Page: http://www.camelotcorp.com

Camintonn Z-Ram .... **(949) 454-1500**
22 Morgan Ave
Irvine, CA 92618-2022
Tech:(949) 454-1500 Toll Free:(800) 368-4726
Fax:(949) 830-4726
Web Page: http://www.camintonn.com
email: techsupp@z-ram.com

Campbell Services Inc .... **(248) 559-5955**
21700 Northwestern Hwy, 10th Floor
Southfield, MI 48075
Tech:(900) 454-8324 Toll Free:(800) 559-5955
Fax:(248) 559-1034
Web Page: http://www.ontime.com
email: support@ontime.com
Compuserve: GO ONTIME
BBS:(248) 559-6434

Canary Communications .... **(408) 365-7100**
5884 Eden Park Place
San Jose, CA 95138
Toll Free:(800) 883-9201
Fax:(408) 365-1600
Web Page: http://www.canarycom.com
email: info@canarycom.com

Canon Business Machines Inc .... **(714) 556-4700**
3191 Red Hill Avenue
Costa Mesa, CA 92626
Web Page: http://www.canon.com

Canon Computer Systems Inc .... **(714) 438-3000**
2995 Red Hill Ave
Costa Mesa, CA 92626-5923
Tech:(800) 423-2366
Toll Free:(800) 423-2366
Fax:(714) 438-3099
Fax on Demand:(800) 526-4345
Web Page: http://www.ccsi.canon.com
BBS:(757) 420-2000

Canon Information Systems Inc .... **(714) 438-7100**
110 Innovation Dr
Irvine, CA 92612
Web Page: http://www.canon.com

Canon Research Center America Inc ..... **(650) 354-1200**
4009 Miranda Avenue
Palo Alto, CA 94304
Web Page: http://www.canon.com

Canon Software Inc .... **(510) 327-2270**
1 Canon Plaza
New Hyde Park, NY 11042-1119

Canon USA Inc .... **(516) 488-6700**
One Canon Plaza
Lake Success, NY 11042-1113
Tech:(800) 423-2366
Fax on Demand:(800) 526-4345
Other Address: http://www.usa.canon.com -2nd web addr
Web Page: http://www.usa.canon.com

Canon USA Inc .... **(714) 753-4000**
15955 Alton Parkway
Irvine, CA 92718
Broadcast Equipment .... (201) 816-2900
400 Sylvan Ave
Edgewood Cliffs, NJ 07632
Web Page: http://www.canon.com

Custom Integrated Technology Inc··············(757) 881-6300
120 Enterprise Drive
Newport News, VA 23603
Fax:(757) 881-6400
Web Page: http://www.canon.com

Industrial Resource Technologies Inc·········(804) 695-7000
6000 Industrial Drive
Gloucester, VA 23061
Fax:(804) 695-7099
Web Page: http://www.canon.com

MCS Business Solutions Inc·························(212) 850-1000
633 Third Ave
New York, NY 10017
Web Page: http://www.canon.com

Canon USA Inc (E)····································(732) 521-7000
100 Jamesburg Road
Jamesburg, NJ 08831
Tech:(732) 521-7000 Toll Free:(800) 221-3333
Web Page: http://www.canon.com

Canon USA Inc (E)····································(703) 807-3400
2110 Washington Blvd, Suite 150
Arlington, VA 22204-5799
Tech:(703) 807-3400
Fax on Demand:(703) 807-3400
Web Page: http://www.canon.com

Canon USA Inc (Hawaii)·····························(808) 522-5930
210 Ward Avenue, Suite 200
Honolulu, HI 96814
Tech:(808) 522-5930
Web Page: http://www.canon.com

Canon USA Inc (MW)··································(630) 250-6200
100 Park Boulevard
Itasca, IL 60143
Tech:(630) 250-6200
Fax:(630) 250-1572
Web Page: http://www.canon.com

Canon USA Inc (S)·····································(770) 849-7700
5625 Oakbrook Parkway
Norcross, GA 30093
Tech:(770) 849-7700
Web Page: http://www.canon.com

Canon USA Inc (SW)··································(214) 830-9600
3200 Regent Boulevard
Irving, TX 75063

Tech:(214) 830-9600
Web Page: http://www.canon.com

Canon USA Inc (W)......................................**(714) 753-4000**
15955 Alton Parkway
Irvine, CA 92618
Tech:(714) 753-4000
Web Page: http://www.canon.com

Canon USA Inc (W)......................................**(408) 982-5200**
2051 Mission COllege Blvd
Santa Clara, CA 95054
Tech:(408) 982-5200
Web Page: http://www.canon.com

Affiliated Business Solutions Inc..................(609) 387-8700
300 Commerce Square Blvd
Burlington, NJ 08016
Web Page: http://www.canon.com

Ambassador Business Solutions Inc...........(847) 706-3400
425 N Martingale Road, Suite 1400
Schaumburg, IL 60173
Web Page: http://www.canon.com

Astro Business Solutions Inc........................(310) 217-3000
110 W Walnut
Gardena, CA 90248
Web Page: http://www.canon.com

C S Polymer Inc...(804) 249-5500
11900 Canon Blvd
Newport News, VA 23606
Web Page: http://www.canon.com

South Tech Inc...(804) 443-8000
PO Box 126
Tappahannock, VA 22560
Web Page: http://www.canon.com

Canon Virginia Inc......................................**(804) 881-6000**
12000 Canon Blvd
Newport News, VA 23606-4299
Tech:(804) 881-6000 Toll Free:(800) 423-2366
Web Page: http://www.canon.com

Canyon Software......................................**(415) 453-9779**
1537 4th Street, Suite 131
San Rafael, CA 94901
Tech:(415) 453-9779 Toll Free:(800) 280-3691
Fax:(415) 453-6195
Web Page: http://www.canyonsw.com
email: support@canyonsw.com
Compuserve: 74774,554

BBS:(415) 453-4289

CAP Automation ······································(817) 560-7007
3737 Ramona Ave
Fort Worth, TX 76116
Tech:(817) 560-7007 Toll Free:(800) 826-5009
Fax:(817) 560-8249
Web Page: http://www.capauto.com
email: tech_support@capauto.com
BBS:(817) 560-1296

Capital Computing Services ·····················(919) 828-7770
19 West Hargett St, Suite 200
Raleigh, NC 27602-0367
Fax:(919) 833-8975
Web Page: http://www.capitalcomputing.com/
email: CharlesBrown@CapitalComputing.com

Capsoft Development Corporation ·········(801) 354-8000
2222 South 950 East
Provo, UT 84606
Tech:(801) 354-8080 Toll Free:(800) 500-3627
Fax:(801) 354-8099
Tech Fax:(800) 533-1645
Web Page: http://www.capsoft.com
email: technical.support@bender.com

Capstone (Intracorp Entertainment) ·······(305) 373-7700
501 Brickll Key Drive, 6th Floor
Miami, FL 33131
Tech:(305) 373-3770
Web Page: http://www.intracorp.com

Caravelle Networks Corp ·························(613) 225-1172
210 Colonnade Road South, Suite 301
Nepean, ON K2E 7L5 Canada
Tech:(613) 225-1172 Toll Free:(800) 363-5292
Fax:(613) 225-4777
Web Page: http://www.caravelle.com
email: support@caravelle.com

Cardexpert Technology Inc ·····················(510) 252-1118
47881 Fremont Blvd
Fremont, CA 94538

Cardiff Software Inc ································(760) 752-5200
1782 La Costa Meadows Drive
San Marcos, CA 92069
Tech:(760) 752-5212 Toll Free:(800) 659-8755
Fax:(760) 752-5222
Tech Fax:(760) 752-5221

Web Page: http://www.cardiffsw.com
email: sales@cardiffsw.com
Compuserve: GO CARDIFF

Cardinal Technologies Inc........................(770) 840-2157
5854 Peachtree Corners East
Norcross, GA 30092
Tech:(770) 840-2157
Fax:(770) 729-6513 Fax on Demand:(800) 947-0808
Web Page: http://www.cardtech.com
email: cardinal.support@hayes.com

Cardoza Entertainment............................(619) 296-0595
591 Camino De La Reina, Suite 728
San Diego, CA 92108
Web Page: http://www.cardozaent.com
email: support@CardozaEnt.com

Carleton Corp..(612) 828-0300
7275 Flying Cloud Drive
Eden Prairie, MN 55344
Toll Free:(800) 328-3998
Fax:(612) 828-0773
Fax:(612) 828-0454
Web Page: http://www.apertus.com
email: info@carleton.com

Cartesia Software.....................................(609) 397-1611
80 Lambert Lane, Suite 100
Lambertville, NJ 08530
Tech:(609) 397-1611 Toll Free:(800) 334-4291
Fax:(609) 397-5724
Web Page: http:/www.map-art.com
email: cartesia@map-art.com

Casady & Greene Inc...............................(408) 484-9228
22734 Portola Dr
Salinas, CA 93908-1119
Tech:(408) 484-9228 Toll Free:(800) 359-4920
Fax:(408) 484-9218
Fax:(800) 359-4264
Web Page: http://www.casadyg.com
email: tech@casadyg.com
America Online: casadygree@aol.com

Cascade (see Ascend Communications)

Casio Inc..(973) 361-5400
570 Mt Pleasant Ave
Dover, NJ 07801
Tech:(800) 962-2746 Toll Free:(800) 634-1895

Fax:(973) 361-3819
Cassiopeia:(888) 204-7765
Web Page: http://www.casio.com

Castelle..**(408) 496-0474**
3255-3 Scott Blvd
Santa Clara, CA 95054
Tech:(408) 496-6966 Toll Free:(800) 289-7555
Fax:(408) 492-1964 Fax on Demand:(800) 289-9998
Web Page: http://www.castelle.com
email: support@ibex.com
Compuserve: GO CASTELLE

Castelle Ibex Technologies Inc................**(916) 939-8888**
4921 R J Matthews Pkwy
El Dorado Hills, CA 95762
Tech:(916) 939-8888 Toll Free:(800) 975-4239
Fax:(916) 939-8899 Fax on Demand:(800) 289-9998
Fax Back:(916) 939-8875
Other Address: http://www.castelle.com
Web Page: http://www.ibex.com
email: info@ibex.com

Cayenne Software Inc..............................**(781) 273-9003**
8 New England Executive Park
Burlington, MA 01803
Tech:(800) 356-2224 Toll Free:(800) 528-2388
Fax:(781) 229-9904
Fax:(781) 280-6000
Web Page: http://www.cayennesoft.com
email: support@cayennesoft.com

Cayman Systems.....................................**(781) 279-1101**
100 Maple Street
Stoneham, MA 02180
Tech:(781) 279-1101 Toll Free:(800) 473-4776
Fax:(781) 438-5560
Web Page: http://www.cayman.com
email: support@cayman.com

CCOM Information Systems.....................**(732) 603-7750**
120 Wood Ave South
Iselin, NJ 08830
Tech:(732) 603-7750
Fax:(732) 603-7751
Web Page: http://www.ccom-infosys.com
email: info@ccom-infosys.com

CD Concepts Inc....................................**(317) 651-9848**
229 S. Baldwin

Marion, IN 46952
Tech:(317) 651-9848

CD International Inc ....(207) 985-6370
128 York St, Suite 2
Kennebunk, ME 04043
Fax:(207) 985-6467
Web Page: http://www.cdcomputer.com
email: cdi@ime.net

CD Technologies ....(408) 752-8500
762 San Aleso Ave
Sunnyvale, CA 94086
Web Page: http://www.cdtech.com
Compuserve: 73647,2163

CD-Rom Strategies Inc. (see DV Studio Inc)

CDB Systems Inc ....(303) 444-7071
1035 Coffman Street
Longmont, CO 80501
Tech:(303) 444-7071
Fax:(303) 444-0035
Web Page: http://www.cdbsystems.com
email: cdb@cdbsystems.com

CE Software ....(515) 221-1801
1801 Industrial Circle, PO Box 65580
West Des Moines, IA 50265
Tech:(515) 221-1803 Toll Free:(800) 523-7638
Fax:(515) 221-1806
Tech Fax:(515) 221-2169
Web Page: http://www.cesoft.com
Compuserve: ce_support@cesoft.com

Cedar Software ....(802) 888-5275
RR1 Box 4495
Wolcott, VT 05680
Fax:(802) 888-3009
Web Page: http://www.netletter.com
email: helper@netletter.com
Compuserve: 72330.1574@compuserv.com
America Online: DickOliver@aol.com

Cedar Technologies ....(612) 830-1993
5250 W. 74th St, Ste. 8C
Edina, MN 55439
Fax:(612) 830-1039
Web Page: http://www.cedar-tech.com

Centerline .... **(617) 498-3000**
10 Fawcett Street
Cambridge, MA 02138-1110
Toll Free:(800) 669-2687
Fax:(617) 868-6655
Fax:(617) 868-5004
Web Page: http://www.centerline.com
email: customerline@centerline.com

Centigram Communications Corp .... **(408) 944-0250**
91 East Tasman Drive
San Jose, CA 95134
Fax:(408) 428-3732
Web Page: http://www.centigram.com

Centon Electronics Inc .... **(949) 855-9111**
20 Morgan
Irvine, CA 92618
Toll Free:(800) 234-9292
Fax:(949) 855-6035
Web Page: http://www.centon.com
email: techsupport@centon.com

Central Data Corp .... **(217) 359-8010**
1602 Newton Drive
Champaign, IL 61822-1098
Toll Free:(800) 482-0315
Fax:(217) 359-6904
Web Page: http://www.cd.com
email: support@cd.com

Central Point Software (see Symantec)

Centron Software Inc .... **(910) 215-5708**
300 American Legion Lane
Pinehurst, NC 28374
Tech:(910) 215-5708 Toll Free:(800) 848-2424
Fax:(910) 295-8908
Web Page: http://www.centronsoftware.com
email: centron@ac.net

Centrum Business Systems Inc .... **(303) 939-8888**
1050 Walnut St, Ste. 320
Boulder, CO 80302
Web Page: http://www.centrumpro.com

Centura Software Corp .... **(650) 596-3400**
975 Island Drive
Redwood Shores, CA 94065
Tech:(888) 523-6887 Toll Free:(800) 444-8782
Fax:(650) 596-4900 Fax on Demand:(650) 596-4600

Web Page: http://www.centurasoft.com
email: info_usa@centurasoft.com
Compuserve: GO GUPTA
BBS:(650) 321-0549

Century Microelectronics Inc ··················(408) 748-7788
4800 Great America Parkway
Santa Clara, CA 95054
Fax:(408) 748-8688
Web Page: http://www.century-micro.com
email: info@century-micro.com

Century Software Inc ·······························(801) 268-3088
5284 S. Commerce Drive, Suite C-134
Salt Lake City, UT 84107
Tech:(800) 877-3088 Toll Free:(800) 877-3088
Fax:(801) 268-2772 Fax on Demand:(800) 329-2384
Web Page: http://www.censoft.com
email: support@censoft.com
BBS:(801) 266-0330

Certus (see Symantec)

CH Products···(760) 598-2518
970 Park Center Drive
Vista, CA 92083
Tech:(760) 598-7833
Fax:(760) 598-2524
Web Page: http://www.chproducts.com
email: sales@chproducts.com
Compuserve: GO GAMEDPUB

Chaco Communications Inc (see Likeminds Inc)

Chain Store Guide Info Serv (CSGIS) ·····(813) 664-6800
3922 Coconut Palm Drive
Tampa, FL 33619-8321
Tech:(800) 252-9793 Toll Free:(800) 778-9794
Fax:(813) 664-6882
Web Page: http://www.d-net.com/csgis

Champion Business Systems Inc ···········(303) 792-3606
6726 S. Revere Pkwy
Englewood, CO 80112
Tech:(800) 243-2626 Toll Free:(800) 243-2626
Fax:(303) 792-0255
Web Page: http://www.champbiz.com
email: champion@champbiz.com

Changeling Inc···(512) 419-7085
2507 Albata Avenue

Austin, TX 78757-2102
Tech:(512) 419-7085 Toll Free:(800) 769-2768
Fax:(512) 419-7288
Other Address: E-World: changelin
Web Page: http://www.changeling.com
email: change@changeling.com
Compuserve: 72662,1547
America Online: changelingS

Chaplet Systems USA Inc........................**(408) 732-7950**
252 N Wolfe Road
Sunnyvale, CA 94086
Fax:(408) 732-6050

Chase Advanced Technologies..............**(860) 526-2400**
500 Main Street
Deep River, CT 06417
Web Page: http://www.chase-at.com/chase/
email: sales@chase-at.com
Compuserve: 100434,3006@compuserve.com

Chatsworth Products Inc..........................**(818) 735-6100**
31425 Agoura Road
Westlake Village, CA 91361-4614
Fax:(818) 735-6199
Web Page: http://www.chatsworth.com
email: techsupport@chatsworth.com

CheckFree..**(614) 825-3000**
8275 North High Street
Columbus, OH 43235
Tech:(614) 825-3000 Toll Free:(800) 882-5280
Web Page: http://www.checkfree.com

CheckMark Software..................................**(970) 225-0522**
724 Whalers Way, Building H
Fort Collins, CO 80525
Tech:(970) 225-0387 Toll Free:(800) 444-9922
Fax:(970) 225-0611
Web Page: http://www.checkmark.com
email: rgilmore@checkmark.com

Chemtronics Inc..**(770) 424-4888**
8125 Cobb Center Dr
Kennesaw, GA 30152
Tech:(800) 424-9300 Toll Free:(800) 645-5244
Fax:(770) 424-4267 Fax on Demand:(770) 426-5447
Web Page: http://www.chemtronics.com
email: chem-admin@chemtronics.com

Cherry Electrical Products ....(847) 662-9200
3600 Sunset Avenue
Waukegan, IL 60087-3298
Fax:(847) 662-2990
Web Page: http://www.cherrycorp.com
email: khuebner@cherrycorp.com

Cheyenne Software Inc ....(516) 465-5000
3 Expressway Plaza
Rosalyn Heights, NY 11577
Tech:(800) 243-9832 Toll Free:(800) 243-9462
Fax:(516) 630-2026 Fax on Demand:(516) 465-5979
Web Page: http://www.cheyenne.com
email: sales@cheyenne.com
Compuserve: GO CHEYENNE

Chicago Case Co ....(773) 927-1600
4446 S Ashland Ave
Chicago, IL 60609
Tech:(773) 927-1600 Toll Free:(800) 927-2602
Fax:(800) 333-8172

Chicago-Soft Ltd ....(603) 643-4002
45 Lyme Road, Suite 307
Hanover, NH 03755
Fax:(603) 643-4571
Other Address: http://www.chicago-soft.com
Web Page: http://www.quickref.com
email: support@quickref.com

Chinon America (See Tech Media?)

Chipcom Corp (3Com Corp)

Chips And Technologies Inc ....(408) 434-0600
2950 Zanker Road
San Jose, CA 95134
Toll Free:(800) 944-6284
Fax:(408) 894-2082
Web Page: http://www.chips.com
email: sales@chips.com

ChipSoft Inc (see Intuit)

Chorus Systems ....(650) 960-1300
1999 South Bascom Ave, Suite 400
Campbell, CA 95008

Chuck Atkinson Programs (see CAP Automation)

Cidco Inc ....(408) 779-1162
220 Cochrane Circle

Morgan Hill, CA 95037
Fax:(408) 779-3106
Web Page: http://www.cidco.com
email: feedback@cidco.com

Cipher Data Products (see Overland Data)

Ciprico Inc ....(612) 551-4035
2800 Campus Drive
Plymouth, MN 55441
Tech:(612) 551-4131 Toll Free:(800) 727-4669
Fax:(612) 551-4002
Web Page: http://www.ciprico.com
email: support@ciprico.com

Cirque Corp ....(801) 467-1100
433 W. Lawndale Drive
Salt Lake City, UT 84115-2916
Tech:(801) 467-1100 Toll Free:(800) 454-3375
Fax:(801) 467-0208 Fax on Demand:(800) 454-3375
Other Address: http://www.glidepoint.com
Web Page: http://www.cirque.com
email: tech@cirque.com
BBS:(801) 467-0128

Cirrus Logic Inc ....(510) 623-8300
3100 W Warren Ave
Fremont, CA 94538
Tech:(510) 623-8300
Fax:(510) 252-6020 Fax on Demand:(800) 359-6414
Web Page: http://www.cirrus.com
email: pc-support@cirrus.com
BBS:(510) 440-9080

Cisco Systems ....(408) 526-4000
170 W. Tasman Drive
San Jose, CA 95134-1706
Tech:(800) 553-2447 Toll Free:(800) 553-6387
Fax:(408) 526-4100
Tech Support:(408) 526-7209
Web Page: http;//www.cisco.com
email: tac@cisco.com

Citadel Computer Systems ....(713) 522-5625
3260 Sul Ross
Houston, TX 77098
Tech:(800) 325-3587 Toll Free:(800) 962-0701
Fax:(713) 522-8965
Tech Support:(214) 520-9292
Web Page: http://www.kentmarsh.com

email: support@kentmarsh.com

Citizen America Corp ....(310) 453-0614
2450 Broadway, Suite 600
Santa Monica, CA 90404
Tech:(310) 453-0614
Fax:(310) 315-1881
Web Page: http://www.citizen-america.com
email: techsupcac@mcimail.com
BBS:(310) 828-9729

Citizen CBM America Corp ....(310) 781-1460
365 Van Ness Way, Suite 510
Torrance, CA 90501
Tech:(800) 843-8270 Toll Free:(800) 218-9045
Fax:(310) 781-9152
Web Page: http://www.cbma.com

Citrix Systems Inc ....(954) 267-3000
6400 NW 6th Way
Ft. Lauderdale, FL 33309
Tech:(800) 424-8749 Toll Free:(800) 437-7503
Fax:(954) 267-9319 Fax on Demand:(800) 437-7503
Tech Fax:(954) 267-9342
Web Page: http://www.citrix.com

CLARiiON ....(508) 480-7350
Coslin Drive
Southboro, MA 01772
Tech:(800) 344-1314 Toll Free:(800) 672-7729
Fax:(508) 480-7950
Web Page: http://www.clariion.com
email: service@clariion.com

Clarion Software (see Top Speed Corp)

Claris Corp (see Filemaker Inc)

Clarity Software ....(408) 557-6725
2700 Garcia Ave, Suite 200
Mountain View, CA 94043
Fax:(650) 964-4383
Web Page: http://www.clarity.com
email: info@clarity.com

Clark Development Comp. (out of business?) ....

Clary Corp ....(626) 359-4486
1960 S. Walker Ave
Monrovia, CA 91016
Tech:(800) 551-6111 Toll Free:(800) 442-5279
Fax:(626) 305-0254

Web Page: http://www.clary.com
email: clary@clary.com
Compuserve: 73420,1166

Classic PIO Partners ··································(626) 564-8106
87 E. Green Street, Suite 309
Pasadena, CA 91105
Tech:(626) 564-8106 Toll Free:(800) 370-2746
Fax:(626) 564-8554
email: classicpio@aol.com

Classic Software Inc ·································(513) 232-6764
7665 Athenia Drive
Cincinnati, OH 45244
Tech:(513) 232-6764 Toll Free:(800) 677-2952
Fax:(513) 232-9844
Web Page: http://www.classicsoftware.com
email: sales@classicsoftware.com
Compuserve: 72103,1513

CLEAR Software Inc··································(617) 965-6755
199 Wells Ave
Newton, MA 02159
Tech:(617) 965-5019 Toll Free:(800) 338-1759
Fax:(617) 965-5310
Web Page: http://www.clearsoft.com

Cleo Communications (see Interface Systems)

Clickable Software (see Virtual Publisher)
Web Page: http://www.virtualpublisher.com

Client/Server Connection ·························(914) 921-0800
14 Elm Place, Suite 200
Rye, NY 10580
Fax:(914) 921-3276
Web Page: http://www.cscl.com
email: webmaster@cscl.com

Clipper Products ·······································(513) 528-7011
675 Cinti-Batavia Pike
Cincinnati, OH 45245
Fax:(513) 528-7676
Web Page: http://www.clipperproducts.com

CMD Technology Inc··································(949) 454-0800
19 Morgan
Irvine, CA 92618
Tech:(949) 870-2553
Fax:(949) 455-1656
Tech Fax:(949) 454-8314

Web Page: http://www.cmd.com
email: tech-support@cmd.com

CMH Software ....**(406) 293-3616**
PO Box 469
Libby, MT 59923
Toll Free:(800) 680-7638
Fax:(406) 293-5075
Web Page: http://www.libby.org/cmh/
email: cmhsftwr@libby.org
BBS:(406) 293-4240

CMS Enhancements ....**(714) 424-5520**
3095 Redhill Ave
Costa Mesa, CA 92626
Tech:(800) 555-1671 Toll Free:(800) 327-5773
Fax:(714) 435-9504
Tech Fax:(714) 435-9483
Web Page: http://www.cmsenhancements.com
email: service@cmsperipheralsinc.com

CMS Enhancements (see AmeriQuest Tech.)

CNet Technology Inc ....**(408) 934-0800**
1455 McCandless Drive
Milpitas, CA 95035
Tech:(408) 934-0800 Toll Free:(800) 486-2638
Fax:(408) 934-0900
Web Page: http://www.cnetusa.com
email: vickyc@hq.cnetusa.com

Cobalt Microserver Inc ....**(650) 930-2500**
440 Clyde Ave, Building B
Mountain View, CA 94043
Fax:(650) 930-2501
Web Page: http://www.cobaltmicro.com
email: support@cobaltnet.com

Coconut Computing (see ITU Engineering)

Codenoll Technology Corp ....**(914) 965-6300**
200 Corporate Blvd South
Yonkers, NY 10701
Tech:(914) 965-6300 Toll Free:(888) 263-6655
Fax:(914) 965-9811
Web Page: http://www.codenoll.com
email: aracelis_jimenez@codenell.com
BBS:(914) 965-1972

Cogent Data Tech Inc (Adaptec) ....**(425) 603-0333**
15375 SE 30th Place, Suite 310

Seattle, WA 98007
Tech:(408) 934-7274 Toll Free:(800) 426-4368
Fax:(425) 603-9223
Web Page: http://www.adaptec.com/networking
email: sales@adaptec.com

CogniTech Corp ....(770) 395-3000
Atlanta, GA 30350
Tech:(770) 395-3090 Toll Free:(800) 947-5075
Fax on Demand:(770) 640-3102
network support:(770) 518-5010
Web Page: http://www.sharkware.com
email: sharkware@sharkware.com
Compuserve: 72662,3417
BBS:(770) 518-7617

Cognitronix ....(619) 549-8955
12642 Poway Road, Suite 16-120
Poway, CA 92064
Tech:(619) 549-8955 Toll Free:(800) 217-0932
Fax:(619) 549-8327
Web Page: http://www.cognitronix.com
email: support@cognitronix.com
Compuserve: 76306,402

Cognos Corporation ....(781) 229-6600
67 S. Bedford St, Suite 200W
Burlington, MA 01803-5164
Tech:(613) 228-7900 Toll Free:(800) 426-4667
Fax:(781) 229-9844
Tech Fax:(613) 228-3066
Web Page: http://www.cognos.com
email: support@cognos.com

Coleman Research Corp..................(407) 244-3700
201 South Orange Ave, Suite 1300
Orlando, FL 32801
Web Page: http://www.crc.com

Collabra Software Inc..................(650) 254-1900
501 E. Middlefield Road
Mountain View, CA 94043
Tech:(800) 639-0939 Toll Free:(800) 639-0939
Fax:(650) 528-4124
Web Page: http://www.collabra.com

Colorado Memory Systems (HP)..................(970) 669-8000
800 S. Taft Ave
Loveland, CO 80537
Tech:(970) 635-1501

Fax:(970) 667-0997 Fax on Demand:(800) 368-9373
Web Page: http://www.hp.com
email: colorado_support@hp-loveland-om1o.om.hp.
Compuserve: 72662,2165

ColorAge Inc (see Splash Technology)

Colorspan Corp ······························(612) 944-9330
7090 Shady Oak Road
Eden Prairie, MN 55347
Tech:(800) 925-0563 Toll Free:(800) 477-7714
Fax:(612) 944-0522
Web Page: http://www.colorspan.com

Columbia Data Products Inc ··················(407) 869-6700
1070-B Rainer Drive, PO Box 163088
Altamonte Springs, FL 32714
Tech:(407) 869-6700 Toll Free:(800) 613-6288
Fax:(407) 862-4725
Web Page: http://www.cdp.com
email: support@cdp.com
BBS:(407) 862-4724

Columbia Power & Data (see Computer Sys&Edu)

Com-Kyle Inc ·································(408) 734-9660
1366 Borregas Ave
Sunnyvale, CA 94089
Toll Free:(800) 722-1123
Fax:(408) 744-1650

Comark Inc ···································(630) 924-6700
444 Scott Drive
Bloomingdale, IL 60108
Tech:(800) 955-1488 Toll Free:(800) 888-5390
Fax:(630) 351-7497
Web Page: http://www.comark.com

Comfy ··(800) 992-6639
1054 Saratoga-Sunnyvale, Suite 102
San Jose, CA 95129
Tech:(617) 746-2929 Toll Free:(800) 992-6639
Web Page: http://www.comfyland.com
email: info@comfyland.com

Command Communications Inc ···········(303) 751-7000
10800 E. Bethany Drive
Aurora, CO 80014
Tech:(800) 288-6794 Toll Free:(800) 288-3491
Fax:(303) 752-1903
email: 76323.3036@CompuServ.com

Compuserve: 76323,3036

Command Software Systems Inc············(561) 575-3200
1061 E. Indiantown Road, Suite 500
Jupiter, FL 33477
Tech:(561) 575-3200 Toll Free:(800) 423-9147
Fax:(561) 575-3026
Web Page: http://www.commandcom.com
email: support@commandcom.com
Compuserve: 75300,3645 or GO FPROT
BBS:(561) 575-1281

Common Ground Software······················(650) 917-2360
480 San Antonio Road, Suite 200
Mountain View, CA 94040
Tech:(800) 598-3821 Toll Free:(800) 598-3821
Fax:(650) 917-2369
Other Address: http://www.commonground.com
Web Page: http://www.hummingbird.com

CommTouch Software Inc ······················(408) 245-8682
298 S. Sunnyvale Ave, Suite 209
Sunnyvale, CA 94086
Fax:(408) 245-3466
Web Page: http://www.commtouch.com
email: sales@commtouch.com

Compaq Computer Corp··························(281) 370-0670
20555 SH-249
Houston, TX 77070-2698
Tech:(800) 652-6672 Toll Free:(800) 888-5858
Fax:(281) 514-1740 Fax on Demand:(800) 231-9977
Web Page: http://www.compaq.com
email: support@compaq.com
BBS:(713) 518-1418

Compatible Systems Corp······················(303) 444-9532
PO Box 17220
Boulder, CO 80308
Tech:(800) 356-0283 Toll Free:(800) 356-0283
Fax:(303) 444-9595
Web Page: http://www.compatible.com
email: support@compatible.com
BBS:(303) 443-0845

Compex Inc ···(714) 630-7302
4051 E LaPalma
Anaheim, CA 92807
Tech:(714) 630-5451 Toll Free:(800) 279-8891
Fax:(714) 630-6521

Web Page: http://www.cpx.com
email: support@cpx.com
BBS:(714) 630-2570

Compex Technology Inc (see Kenpax)

Compix Media Inc ....(213) 487-8222
3250 Wilshire Blvd, Suite 1505
Los Angeles, CA 90010
Tech:(213) 487-3215
Fax:(213) 487-9251
email: compixla@aol.com

Complete PC, The (see Boca Research)

Compsee Inc ....(407) 724-4321
400 N. Main Street, PO Box 1457
Mount Gilead, NC 27306
Tech:(407) 724-4321
Fax:(407) 723-2895
Web Page: http://www.compsee.com
email: sales@compsee.com

Compton's NewMedia ....(800) 862-2206
2320 Camino Vida Roble
Carlsbad, CA 92009
Tech:(800) 893-5458 Toll Free:(800) 862-2206
Fax:(716) 871-7591 Fax on Demand:(716) 871-7337
Web Page: http://www.comptons.com
email: cust_servAlearningco.com
Compuserve: GO LEARNING
America Online: Keyword: Comptons

Compu-Teach ....(425) 885-0517
16541 Redmond Way, #137-C
Redmond, WA 95052-4482
Tech:(425) 867-0767 Toll Free:(800) 448-3224
Fax:(425) 883-9169
Web Page: http://www.wolfenet/~cmpteach
email: support@compu-teach.com

CompuCover Inc ....(850) 862-4448
2104 Lewis Turner Blvd
Fort Walton Beach, FL 32547-1349
Toll Free:(800) 874-6391
Fax:(850) 863-2200
Web Page: http://www.compucover.com
email: info@compucover.com

CompuLink Laserfiche Inc ....(310) 212-5465
370 S Crenshaw Blvd, Suite E-106

Torrance, CA 90503
Tech:(310) 212-5465
Fax:(310) 212-5064 Fax on Demand:(310) 212-5064
Another BBS:(310) 212-5045
Web Page: http://www.laserfiche.com

CompuMart ....**(512) 992-6400**
1101 Santa Fe
Corpus Christi, TX 78404
Toll Free:(800) 864-1155
Fax:(512) 888-6073
Web Page: http://www.compumart.com
email: cmart@intcomm.net

CompUSA Inc ....**(800) 266-7872**
14951 N. Dallas Parkway
Dallas, TX 75240
Tech:(800) 266-7872 Toll Free:(800) 266-7872
Fax:(800) 669-8329
Alaska toll free:(800) 998-9967
Web Page: http://www.compusa.com
email: customer_service@compusa.com

CompuServe Inc ....**(614) 457-8600**
5000 Arlington Centre Blvd
Columbus, OH 43220-5439
Tech:(800) 944-9871 Toll Free:(800) 848-8990
Fax:(614) 761-7040
Fax:(800) 343-8913
Web Page: http://www.compuserve.com
email: 70006.101@compuserve.com

Computational Mechanics Inc ....**(512) 467-0618**
7800 Shoal Creek Blvd, Suite 290E
Austin, TX 78757-1024
Fax:(512) 467-1382
Web Page: http://www.comco.com
email: info@comco.com

Computer Associates International ....**(516) 342-5224**
One Computer Associates Plaza
Islandia, NY 11788-7000
Tech:(516) 342-4100 Toll Free:(800) 225-5224
Fax:(516) 342-5329
Tech Support:(800) 645-3042
Web Page: http://www.cai.com
email: info@cai.com
Compuserve: GO CAI

Computer Discount Warehouse (CDW)··**(847) 465-6000**
1020 East Lake Cook Road
Buffalo Grove, IL 60089
Tech:(800) 383-4239 Toll Free:(800) 840-4239
Fax:(847) 465-7700
Web Page: http://www.cdw.com

Computer Friends Inc ································**(503) 626-2291**
13865 NW Cornell Road
Portland, OR 97229
Tech:(503) 626-2291 Toll Free:(800) 547-3303
Fax:(503) 643-5379
Web Page: http://www.cfriends.com
email: cfi@cfriends.com

Computer Industry Almanac ·····················**(702) 749-5053**
PO Box 600
Glenbrook, NV 89413-0600
Tech:(702) 749-5053 Toll Free:(800) 377-6810
Fax:(702) 749-5864 Fax on Demand:(702) 749-5053
Web Page: http://www.c-i-a.com
email: info@c-i-a.com

Computer Intelligence InfoCorp··············**(619) 450-1667**
3344 N. Torrey Pines Ct
La Jolla, CA 92037
Toll Free:(800) 645-5795
Fax:(619) 452-7491
Web Page: http://www.ci.zd.com
email: cisupport@zd.com

Computer Knacks Inc ·······························**(732) 530-0262**
621 Shrewsbury Ave
Shrewsbury, NJ 07702
Tech:(732) 530-0262 Toll Free:(800) 551-1433
Fax:(732) 741-0972
Compuserve: GO KNACKS

Computer Library (Information Acc.) ·····**(212) 503-4400**
1 Park Avenue
New York, NY 10016
Tech:(212) 503-4444
Fax:(212) 503-4414

Computer Parts Outlet Inc························**(561) 265-1206**
33 SE 1st Ave
Delray Beach, FL 33444
Tech:(561) 265-1655 Toll Free:(800) 475-1655
Fax:(561) 265-1209
Web Page: http://www.cpoinc.com

email: cpodan@flinet.com

Computer Parts Unlimited (see CPU Mart)

Computer Peripherals International ·······(949) 454-2441
7 Whatney
Irvine, CA 92618
Tech:(949) 454-2441 Toll Free:(800) 854-7600
Fax:(949) 454-8527
Web Page: http://www.cpinternational.com
email: support@CPInternational.com
BBS:(949) 470-1759

Computer Products Plus (see Road Warrior)

Computer Review ·····································(978) 283-2100
19 Pleasant Street
Gloucester, MA 01930
Fax:(978) 281-3125
Web Page: http://www.computerreview.com
email: info@computerreview.com
Compuserve: 74264,371

Computer Shopper Magazine···················(212) 503-3800
One Park Avenue
New York, NY 10016
Toll Free:(800) 274-6384
Web Page: http://www.cshopper.com

Computer Support Corp ·························(972) 661-8960
15926 Midway Road
Dallas, TX 75244-2196
Tech:(972) 661-8960 Toll Free:(800) 752-9057
Fax:(972) 661-5429
Web Page: http://www.arts-letters.com
email: support@arts-letters.com
BBS:(972) 404-8652

Computer Systems And Education ········(360) 693-6165
8808 NE 21st Avenue
Vancouver, WA 98665
Tech:(800) 791-1181
Fax:(360) 693-6109
Web Page: http://www.nwbiz.com

Computer Teaching Corp ························(217) 352-6363
1713 South State Street
Champaign, IL 61820
Fax:(217) 352-3104
Web Page: http://www.tencore.com
email: sales@tencore.com

Computer Technology Review ....(310) 376-9500
420 N. Camden Dr
Beverly Hills, CA 90210
Fax:(310) 246-1405
Web Page: http://www.ctreview.com
email: cheryl_smith@wwpi.com

Computer Tyme Software Lab ....(417) 866-1222
309 N. Jefferson, Suite 220
Springfield, MO 65806
Toll Free:(800) 548-5353
Fax:(417) 866-1665
Web Page: http://www.ctyme.com
email: marc@ctyme.com
BBS:(417) 866-1665

ComputerPREP Inc ....(602) 275-7700
410 N 44th St, Suite 600
Phoenix, AZ 85008
Toll Free:(800) 228-1027
Fax:(602) 275-1603
Web Page: http://www.computerprep.com
email: computerprep@computerprep.com

ComputerTrend Systems (see Premio Computer)

Computone Corp ....(770) 475-2725
1060 Windward Ridge Pkwy, Suite 100
Alpharetta, GA 30005
Tech:(770) 475-2725 Toll Free:(800) 241-3946
Fax:(770) 664-1510
another BBS:(770) 664-1210
Web Page: http://www.computone.com
email: support@computone.com
BBS:(770) 343-9737

Computron Software Inc ....(201) 935-3400
301 Route 17 North
Rutherford, NJ 07070
Toll Free:(800) 828-7660
Fax:(201) 935-7678
Other Address: http://www.computronsoftware.com
Web Page: http://www.ctronsoft.com

Compuware Corp ....(248) 737-7300
31440 Northwestern Highway
Farmington Hills, MI 48334-2564
Fax:(248) 737-7119
Web Page: http://www.compuware.com

Comstor .... **(303) 442-4747**
5718 Central Avenue
Boulder, CO 80301
Tech:(800) 543-6098 Toll Free:(800) 543-6090
Fax:(303) 442-7985
Other:(800) 266-7867
Web Page: http://www.comstor.com
email: techsupport@comstor.com

Comstor .... **(972) 407-0222**
3200 Commander Drive
Carrollton, TX 75006
Tech:(800) 644-6311 Toll Free:(800) 266-7867
Fax:(972) 407-9732
Web Page: http://www.comstor.com
email: techsupport@comstor.com

Comtech Publishing .... **(702) 825-9000**
PO Box 12340
Reno, NV 89510-2340
Tech:(702) 825-9000
Fax:(702) 825-1818
Web Page: www.accutek.com/comtech
Compuserve: 70724,561

Comtech Research .... **(419) 278-6790**
5220 Milton Road
Custar, OH 43511-9766
Fax:(419) 278-7744
Web Page: http://www.comtech-pcs.com
email: mwaters@comtech-pcs.com

Comtrol Corp .... **(612) 631-7654**
900 Long Lake Road, Suite 210
Saint Paul, MN 55112
Tech:(800) 926-6876 Toll Free:(800) 926-6876
Fax:(612) 631-8117
Web Page: http://www.comtrol.com
email: info@comtrol.com

Concentric Data .... **(508) 366-1122**
110 Turnpike Rd
Westborough, MA 01581
Tech:(800) 325-9035 Toll Free:(800) 325-9035
Fax:(508) 366-2954
support contract:(800) 487-8622
Web Page: http://www.walldata.com
Compuserve: 71601,1465

Concept Software ································ **(407) 786-4457**
1017 Woodall Dr
Altamonte Springs, FL 32714-7225
Fax:(407) 786-4458
Web Page: http://www.consoft.com
email: support@consoft.com
Compuserve: 71441,3045

Concord Communications ····················· **(508) 460-4646**
33 Boston Post Road West
Marlboro, MA 01752
Tech:(888) 832-4340 Toll Free:(800) 851-8725
Fax:(508) 481-9772
Web Page: http://www.concord.com
email: info@concord.com

Concurrent ·· **(954) 974-1700**
2101 W. Cypress Creek Road
Fort Lauderdale, FL 33309
Toll Free:(800) 666-4544
Fax:(954) 977-5580
Customer Service:(800) 245-6453
Web Page: http://www.ccur.com
email: ccurevents@mail.ccur.com

Connectix Corp ································· **(650) 571-5100**
2655 Campus Dr, Suite 100
San Mateo, CA 94403
Tech:(800) 839-3627 Toll Free:(800) 950-5880
Fax:(650) 571-0850 Fax on Demand:(800) 571-7558
Tech Fax:(650) 571-5195
Other Address: AppleLink
Web Page: http://www.connectix.com
email: info@connectix.com

ConnectSoft Inc ································ **(425) 827-6467**
11130 NE 33rd Place, Suite 250
Bellevue, WA 98004-1448
Fax:(425) 822-9095
Web Page: http://www.connectsoft.com
email: stereo@connectsoft.com

Conner Peripherals Inc (see Seagate)

Conner Tape Products ·························· **(714) 641-1230**
1650 Sunflower Ave
Costa Mesa, CA 92626
Tech:(800) 426-6637
Fax:(714) 966-5534 Fax on Demand:(408) 456-4903
BBS:(408) 456-4415

Contact East .... **(978) 682-2000**
335 Willow Street South
North Andover, MA 01845-5995
Toll Free:(888) 925-2960
Fax:(800) 225-5317
Fax:(978) 688-7829
Web Page: http://www.contacteast.com
email: support@contacteast.com

Contact Software International Inc .... **(214) 418-1866**
1625 West Crosby Road
Carrollton, TX 75006-6654
Tech:(214) 484-4349

Contour Design Inc .... **(603) 893-4556**
354 N. Broadway
Salem, NH 03079
Tech:(603) 893-4556 Toll Free:(800) 462-6678
Fax:(603) 893-4558
Other Address: http://www.contourdesign.com
Web Page: http://www.contourdes.com
email: info@contourdes.com

Control Data Systems (CDC) .... **(612) 415-2999**
4201 Lexington Avenue North
Arden Hills, MN 55126-6198
Tech:(800) 257-6736 Toll Free:(888) 742-5864
Fax:(612) 415-3000
Tech Support:(612) 482-6736
Web Page: http://www.cdc.com
email: info@cdc.com

Copia International Ltd .... **(630) 778-8898**
1220 Iroquois Dr
Naperville, IL 60563
Tech:(630) 778-8864 Toll Free:(800) 689-8898
Fax:(630) 665-9841 Fax on Demand:(888) 332-6742
Fax:(630) 778-8848
Web Page: http://www.copia.com
email: copia@copia.com

Core International .... **(561) 997-6055**
6500 E. Rogers Circle
Boca Raton, FL 33487
Tech:(561) 997-6033
Fax:(561) 997-6202
Core Engineering:(561) 998-3800
BBS:(407) 241-2929

Corel Corporation..**(613) 728-3733**
1600 Carling Avenue
Ottawa, ON K1Z 8R7 Canada
Tech:(613) 728-7070 Toll Free:(800) 772-6735
Fax:(613) 761-9176 Fax on Demand:(613) 728-0826
Fax:(613) 761-1295
Web Page: http://www.corel.com
Compuserve: GO COREL
Microsoft: Keyword: COREL
BBS:(613) 728-4752

Corel Corporation..**(801) 765-4010**
567 East Timpanogos Parkway
Orem, UT 84097-6209
Fax:(801) 222-4379
Artshow (all versions)..............................(613) 728-6173
CD Office Companion..............................(613) 728-6173
CD+ for A&M CD's
Corel CADD 3D..............................(613) 728-6418
Toll Free:(800) 291-4434
Corel CD Home..............................(613) 728-1010
Corel Click & Create..............................(613) 728-1010
Toll Free:(800) 754-8209
Corel Family Tree Suite..............................(613) 728-6891
Corel GALLERY all versions..............................(613) 728-6173
Corel Graphics Pack..............................(613) 728-6891
Toll Free:(800) 205-4295
Corel Learning Series..............................(613) 728-6548
Corel MEGA GALLERY all versions..............................(613) 728-6173
Corel Office Pro. 7 Win 95..............................(716) 871-2317
Canada:(613) 728-2822
Corel Office Pro. Suite (Win 3.1)..............................(716) 871-2315
Canada:(613) 728-2650
Corel Paradox 8..............................(613) 728-4657
BBS:(613) 728-4752
Corel PHOTO-PAINT all versions..............................(613) 728-6398
Toll Free:(800) 792-6735
Corel Print House all versions..............................(613) 728-6891
Corel VisualCADD..............................(613) 728-6418
Toll Free:(800) 291-4434
Corel WEB.DATA..............................(613) 728-6625
Toll Free:(800) 856-5650
Corel WEB.DESIGNER..............................(613) 728-6625
Toll Free:(800) 856-5650
Corel WEB.GALLERY..............................(613) 728-6173
Corel WEB.GRAPHICS Suite..............................(613) 728-6625
Toll Free:(800) 856-5650

Corel WebMaster Suite .... (613) 728-6625
Toll Free:(800) 856-5650
Corel Word Perfect Suite 7 Win 95 .... (716) 871-2316
Toll Free:(800) 757-2133
Canada:(613) 728-2702
Corel Word Perfect Suite 8 Win 95 .... (716) 871-2325
Toll Free:(800) 757-2133
Canada:(613) 728-5324
Corel WordPerfect Ste 8 Legal Ed. .... (801) 765-4042
Corel WordPerfect Suite 7 Win 3.1 .... (716) 871-2313
Toll Free:(800) 757-2133
Canada:(613) 728-2311
Corel WordPerfect Suite 8 - Dragon .... (801) 765-4081
CorelDRAW 3, 4, 5 .... (613) 728-6641
Toll Free:(800) 582-6735
CorelDRAW 6, 7, 8 .... (613) 728-7070
Toll Free:(800) 205-4295
CorelDRAW for OS/2 .... (613) 728-6641
Toll Free:(800) 582-6735
CorelDRAW for UNIX .... (613) 728-7070
CorelFLOW 2, 3 .... (613) 728-6173
Toll Free:(800) 856-5650
CorelSCSI .... (613) 728-1010
CorelVENTURA 5, 7, 8 .... (613) 728-6398
Toll Free:(800) 792-6735
CorelXARA .... (613) 728-6891
Toll Free:(800) 992-6735
Envoy/Quickfinder 7 Win 95 .... (716) 871-2323
Toll Free:(800) 861-2310
InfoCentral 1.1, 7.0 (Win 3.1) .... (716) 871-2300
IVAN .... (613) 728-1937
Language Modules .... (716) 871-2301
Canada:(613) 728-0234
Make-It-Perfect .... (613) 728-1645
Novell GroupWise Support .... (801) 431-3400
Toll Free:(800) 858-4000
Web Page: http://support.novell.com
Paradox 7, 8 .... (613) 728-4657
Tech:(613) 728-5258 Toll Free:(888) 761-6909
PerfectWorks Support (Arkose) .... (801) 228-9936
Photo CD .... (613) 728-6173
Presentations 2.1 (DOS) .... (801) 765-4031
Toll Free:(800) 861-2410
Presentations 3.0 (Win 3.1) .... (716) 871-2304
Canada:(613) 728-0380
Presentations 7 Win 95 .... (716) 871-2322
Canada:(613) 728-4187

Product Orders ·········· (800) 772-6735
Quattro Pro 5.6 ·········· (716) 871-2305
Canada:(613) 728-0432
Quattro Pro 6.0 (Win 3.1) ·········· (716) 871-2310
Canada:(613) 728-1243
Quattro Pro 7 Win 95 ·········· (716) 871-2321
Canada:(613) 728-3751
Service Sales ·········· (800) 861-2160
SGML ·········· (716) 871-2301
SGML 7 ·········· (801) 765-4059
Toll Free:(800) 861-2310
UNIX (Priority Service) ·········· (801) 765-4019
WordPerfect 5.1+ (DOS) ·········· (716) 871-2307
Canada:(613) 728-0750
WordPerfect 5.2 and higher (UNIX) ·········· (801) 765-4019
WordPerfect 6.1 (Win 3.1) ·········· (716) 871-2309
Canada:(613) 728-1198
WordPerfect 6.1, 6.2 (DOS) ·········· (716) 871-2308
Canada:(613) 728-0883
WordPerfect 7 Win 95 ·········· (716) 871-2320
Canada:(613) 728-3413
WordPerfect French Canadian Support ·········· (613) 728-9035

Coriolis Group Inc, The ·········· (602) 483-0192
14455 N. Hayden Road, Suite 220
Scottsdale, AZ 85260
Toll Free:(800) 410-0192
Fax:(602) 483-0193
Web Page: http://www.coriolis.com
email: webmaster@coriolis.com
Compuserve: 74242,1304
America Online: Corilois1

Cornerstone Imaging ·········· (408) 435-8900
1710 Fortune Dr
San Jose, CA 95131-1744
Tech:(408) 435-8900 Toll Free:(800) 562-2552
Fax:(408) 435-8998 Fax on Demand:(408) 325-3300
Web Page: http://www.corimage.com
email: support@corimage.com
BBS:(408) 435-8943

Cornerstone Solutions Group Inc ·········· (314) 469-9910
12400 Olive Blvd, Ste. 505
St. Louis, MO 63141-5439
Tech:(314) 469-6927
Web Page: http://www.csgsolutions.com
email: webmaster@csgstl.com

Cornerstone Training ································(732) 251-6300
5 Jillian Court
Spotswood, NJ 08884

Corollary Inc ··(949) 250-4040
2802 Kelvin Ave
Irvine, CA 92614
Fax:(949) 250-4043
Web Page: http://www.corollary.com

Corporate Express Promotional Marketing ···········(800) 854-1199
1400 North Price Road
St Louis, MO 63132-2308
Toll Free:(800) 854-1199
Fax:(314) 432-1818

CoStar Corp ···(203) 661-9700
599 West Putnam Avenue
Greenwich, CT 06830-6092
Tech:(203) 661-9700
Toll Free:(800) 426-7827
Fax:(203) 661-1540
Fax on Demand:(800) 350-6280
Tech Fax:(203) 661-6534
Web Page: http://www.costar.com
email: support@costar.com
Compuserve: 75300,2225

Cougar Mountain Software Inc ················(208) 375-4455
7180 Potomac Drive, Suite D
Boise, ID 83704
Tech:(800) 727-0656
Toll Free:(800) 388-3038
Fax:(208) 375-4460
Web Page: http://www.cougarmtn.com
BBS:(208) 323-9011

Covey Leadership Center (see Franklin Covey)

Cox Recorders/Energy Reserve Inc ·······(704) 825-8146
69 McAdenville Road
Belmont, NC 28012
Fax:(704) 825-4498
Sales:(909) 946-4441
Web Page: http://www.cx-en.com/cox.htm
email: cxen@loclnet.com

CPU Mart ···(805) 532-2500
5069 Maureen Lane
Moorpark, CA 93021
Toll Free:(800) 644-4494
Fax:(805) 532-2599
Other Address: http://www.compparts.com

Web Page: http://www.cpumart.com
email: custserv@cpumart.com

Cray Research Inc .... **(612) 683-3800**
655 A Lone Oak Drive
Eagan, MN 55121
Fax:(612) 683-3599
Web Page: http://www.cray.com
email: crayinfo@cray.com

Creative Assistance Software .... **(704) 544-0001**
9431 Kings Falls Drive
Charlotte, NC 28210
Tech:(704) 544-0001
Fax:(704) 544-8031
Web Page: http://www.vnet.net/casoft/homepage.ht
email: support@casoft.com
Compuserve: 70741,3451 or GO OS2 AVEN
BBS:(704) 544-2515

Creative Labs Inc .... **(408) 428-6600**
1901 McCarthy Blvd
Milpitas, CA 95035
Tech:(405) 742-6622 Toll Free:(800) 998-1000
Fax:(408) 428-6611 Fax on Demand:(405) 372-5227
Tech Fax:(405) 742-6633
Other Address: http://www.creaf.com
Web Page: http://www.creativelabs.com
Compuserve: GO BLASTER
BBS:(405) 742-6660

Creative Multimedia Corp .... **(503) 241-4351**
225 SW Broadway, Suite 600
Portland, OR 97205
Tech:(503) 241-1530
Fax:(503) 241-4370

Crescent Software .... **(781) 280-3000**
14 Oak Park
Bedford, MA 01730
Tech:(781) 280-3000 Toll Free:(800) 352-2742
Fax:(781) 280-4025
Web Page: http://crescent.progress.com
email: crescent@progress.com
Compuserve: 70662,2605

Cross Pen Computing Group .... **(401) 333-1200**
1 Albion Rd
Lincoln, RI 02865
Web Page: http://www.cross-pcg.com

Crosstalk Communication .... **(770) 442-4000**
1000 Alderman Drive
Alpharetta, GA 30202
Tech:(425) 957-7764 Toll Free:(800) 426-6283
BBS:(425) 649-6660

Crosswise Corp .... **(408) 459-9060**
105 Locust Street, Suite 301
Santa Cruz, CA 95060
Tech:(408) 459-9060
Fax:(408) 426-3859
Other Address: AppleLink:CROSSWISE
Web Page: http://www.crosswise.com
email: support@crosswise.com
Compuserve: GO CROSSWISE

Crystal Services (Seagate) .... **(604) 681-3435**
1095 W Pender Street, 4th Floor
Vancouver, BC V6E 3M6 Canada
Tech:(604) 669-8379 Toll Free:(800) 877-2340
Fax:(604) 681-2934 Fax on Demand:(604) 681-3450
Web Page: http://www.crystalinc.com
email: crystal@crystalinc.com
Compuserve: 72223,2632 or GO REPORTS

CS Electronics .... **(714) 475-9100**
17500 Gillette Ave
Irvine, CA 92614
Fax:(714) 475-9119
Web Page: http://www.scsi-cables.com
email: sales@scsi-cables.com

CTX International Inc .... **(626) 839-0500**
748 Epperson Drive
City Of Industry, CA 91748
Tech:(800) 282-2205 Toll Free:(800) 888-2012
Fax:(626) 810-6703
Web Page: http://www.ctxintl.com
email: techsup@ctxintl.com
BBS:(909) 595-3870

Desktop .... (800) 742-5289
Toll Free:(800) 285-1889
Web Page: http://www.ctxintl.com
email: tech_dt@ctxintl.com

Monitors .... (800) 888-2012
Toll Free:(800) 888-2120
Fax:(909) 598-8294
Web Page: http://www.ctxintl.com
email: tech_mon@ctxintl.com

Notebook .... (800) 888-9052
Tech:(800) 281-1052 Toll Free:(800) 888-2017
Fax:(909) 598-8294
Parts:(800) 289-8808
Web Page: http://www.ctxintl.com
email: tech_nb@ctxintl.com

Cubix Corp .... (702) 888-1000
2800 Lockheed Way
Carson City, NV 89706-0719
Toll Free:(800) 829-0550
Fax:(702) 888-1001
Web Page: http://www.cubix.com
email: dianew@cubix.com
BBS:(702) 888-1003

CUC International Inc .... (425) 649-9800
SE Bellevue, WA 98015-8506
Tech:(425) 644-4343 Toll Free:(800) 757-7707
Fax:(425) 644-7697
Tech Fax:(402) 393-3224
Other Address: ftp bbs.sierra.com
Web Page: http://www.sierra.com
email: support@sierra.com
Compuserve: GO SIERRA
America Online: sierra@aol.com

Curtis Manufacturing Co (Rolodex) .... (800) 727-7656
245 Secaucus Road
Secaucus, NJ 07096
Tech:(800) 955-5544 Toll Free:(800) 955-5544
Fax:(201) 348-0239

Curtis Mathes (see uniView Technologies)

Curtis/Qualtec .... (319) 263-8144
2210 Second Ave, PO Box 599
Muscatine, IA 52761
Tech:(800) 553-9647 Toll Free:(800) 272-2366
Fax:(800) 272-2382
Web Page: http://www.ringking.com

Cutting Edge .... (619) 667-7888
8191 Center St
La Mesa, CA 91942
Toll Free:(800) 257-1666
Fax:(619) 667-7890
Web Page: http://www.cuttedge.com
email: info@cuttedge.com

Cway Software .... **(215) 368-9494**
30 E. Lincoln Ave
Hatfield, PA 19440
Tech:(215) 368-9494
Fax:(215) 368-7233
Web Page: http://www.cway.com
email: cway@ix.netcom.com

CyberMax Computer inc .... **(888) 403-1515**
133 N. 5th St
Allentown, PA 18102
Toll Free:(888) 566-1313
Fax:(800) 218-4883
Web Page: http://www.cybmax.com
email: cmxtech@cybmax.com

CyberMedia Inc .... **(310) 581-4700**
3000 Ocean Park Blvd, Suite 2001
Santa Monica, CA 90405
Tech:(310) 581-4710 Toll Free:(800) 721-7824
Fax:(310) 581-4720
Tech Fax:(310) 581-4737
Web Page: http://www.cybermedia.com
email: support@cybermedia.com
Compuserve: 74777,3470 or GO CyberMedia
America Online: cymediatek@aol.com

Cybersound .... **(650) 812-7380**
Palo Alto, CA 94303-9750
Tech:(650) 812-7380
Fax:(650) 961-3395
Web Page: http://www.cybersound.com
email: support@cybersound.com

Cybex Computer Products Corp .... **(205) 430-4000**
4912 Research Drive
Huntsville, AL 35805
Tech:(205) 430-4000 Toll Free:(800) 793-3758
Fax:(205) 430-4030 Fax on Demand:(800) 462-9239
Web Page: http://www.cybex.com
email: tech-support@cybex.com

CyLink Corp .... **(408) 735-5800**
910 Hermosa Court
Sunnyvale, CA 94086
Tech:(800) 545-6608 Toll Free:(800) 533-3958
Fax:(408) 735-6643
Web Page: http://www.cylink.com
email: support@cylink.com

Cyma Systems Inc .... **(602) 303-2962**
2330 W. University Drive, Suite 7
Tempe, AZ 85281
Tech:(602) 303-2962 Toll Free:(800) 292-2962
Fax:(602) 303-2969
Web Page: http://www.cyma-systems.com
email: support@cyma-systems.com

Cypress Research (see Aveo)

Cypress Semiconductor Corp .... **(408) 943-2600**
3901 N. First Street
San Jose, CA 95134
Tech:(800) 858-1810
Fax:(408) 943-2843 Fax on Demand:(800) 858-1810
Tech Support:(408) 943-2821
Web Page: http://www.cypress.com
email: cyapps@cypress.com

Cyrix Corp .... **(972) 968-8388**
PO Box 850118
Richardson, TX 75085-0118
Tech:(800) 462-9749 Toll Free:(800) 462-9749
Fax:(972) 699-9857 Fax on Demand:(800) 462-9749
Web Page: http://www.cyrix.com
email: tech_support@cyrix.com
BBS:(972) 968-8610

D-Link Systems Inc .... **(714) 455-1688**
5 Musick
Irvine, CA 92618
Tech:(714) 598-8150 Toll Free:(800) 326-1688
Tech Support:(714) 788-0805
Web Page: http://www.dlink.com
email: tech@dlink.com

DacEasy Inc .... **(972) 248-0305**
17950 Preston Rd, Suite 800
Dallas, TX 75252
Tech:(972) 248-0205 Toll Free:(800) 322-3279
Fax:(972) 713-6331
Web Page: http://www.daceasy.com

Dalco Electronics .... **(513) 743-8042**
425 S. Pioneer Blvd, PO Box 550
Springboro, OH 45066
Tech:(800) 543-2526 Toll Free:(800) 445-5342
Fax:(513) 743-9251
Web Page: http://www.dalco.com
email: custserv@dalco.com

Compuserve: GO DA
BBS:(513) 743-2244

Dallas Semiconductor..............................**(972) 788-2197**
4100 Spring Valley Road, Suite 302
Dallas, TX 75244
Tech:(972) 371-4167
Fax:(972) 980-4290 Fax on Demand:(972) 371-4441
Web Page: http://www.dalsemi.com
email: Paul.Clark@dalsemi.com

Damark International Inc ....**(612) 531-4500**
7101 Winnetka Avenue North, PO Box 9437
Minneapolis, MN 55440-9437
Tech:(800) 729-9000 Toll Free:(800) 729-9000
Fax:(612) 531-0281
Web Page: http://www.damark.com
email: customerservice@damark.com

Danmere USA Inc ....**(408) 399-4520**
Tech:(408) 399-4520
Web Page: http://www.danmere.com

Danpex Corp ....**(408) 434-1688**
2114 Ringwood Ave
San Jose, CA 95131-1715
Tech:(408) 434-1688 Toll Free:(800) 452-1551
Fax:(408) 434-1699
Other:(888) 432-6739
Web Page: http://www.danpex.com
email: support@danpex.com

Dantz Development Corp..........................**(925) 253-3000**
4 Orinda Way, Building C
Orinda, CA 94563
Tech:(925) 253-3050
Fax:(925) 253-9099
Upgrades:(800) 225-4880
Web Page: http://www.dantz.com
email: customer_service@dantz.com

Dariana Software (see E-Ware)

Data Access Corp.....................................**(305) 238-0012**
14000 SW 119 Avenue
Miami, FL 33186
Tech:(305) 232-3142 Toll Free:(800) 451-3539
Fax:(305) 238-0017
Web Page: http://www.daccess.com
email: doug-g@dataaccess.com
Compuserve: GO DACCESS

BBS:(305) 238-0640

Data Assist Inc .... **(614) 888-8088**
651 Lakeview Plaza Blvd, Suite G
Worthington, OH 43085
Tech:(800) 326-8088 Toll Free:(800) 326-8088
Fax:(614) 888-8072
Web Page: http://www.data.assist.com
email: sales@data.assist.com
BBS:(614) 888-8056

Data Code Inc .... **(407) 352-5215**
7380 Sand Lake Road, Suite 500
Orlando, FL 32819
Toll Free:(800) 762-1480
Fax:(407) 352-5294
Web Page: http://www.datacode.com
email: datacode@datacode.com

Data Conversion Laboratory .... **(718) 357-8700**
184-13 Horace Harding Expressway
Fresh Meadows, NY 11365
Fax:(718) 357-8776
Web Page: http://www.dclab.com
email: convert@dclab.com

Data Direct Networks .... **(310) 247-0006**
9201 Oakdale Ave
Chatsworth, CA 91311
Tech:(818) 700-7676 Toll Free:(800) 322-4744
Fax:(818) 700-7601
Tech Fax:(818) 700-7677
Web Page: http://www.megadrive.com
email: support@datadirectnetworks.com
BBS:(818) 700-4060

Data Fellows .... **(408) 938-6700**
675 N. First Street, 8th Floor
San Jose, CA 95112
Tech:(408) 938-6700
Fax:(408) 938-6701
Web Page: http://www.datafellows.com
email: info@DataFellows.com

Data General Corp .... **(508) 898-5000**
4400 Computer Drive
Westborough, MA 01580
Tech:(800) 344-3577 Toll Free:(800) 328-2436
Fax:(508) 366-1319
Web Page: http://www.dg.com

Data I/O Corp ....**(425) 881-6444**
PO Box 97046, 10525 Willows Rd NE
Redmond, WA 98073-9746
Tech:(800) 247-5700 Toll Free:(800) 426-1045
Fax:(425) 882-1043
Web Page: http://www.data-io.com
email: ptst@data-io.com
BBS:(425) 882-3211

Data Pro Accounting Software................**(813) 885-9459**
5439 Beaumont Center Blvd, Suite 1050
Tampa, FL 33634
Tech:(813) 888-5847
Fax:(813) 882-8143
Web Page: http://www.dpro.com
email: support@dpro.com
BBS:(813) 888-8892

Data Race ....**(210) 263-2000**
12400 Network Blvd
San Antonio, TX 78249
Tech:(210) 263-2010 Toll Free:(800) 329-7223
Fax:(210) 263-2075
Web Page: http://www.datarace.com
email: tss@datarace.com
BBS:(210) 263-2111

Data Storage Marketing (see Comstor)

Data Technology Corp (DTC) ....**(408) 942-4000**
1515 Centre Pointe Drive
Milpitas, CA 95035
Tech:(408) 262-7700
Fax:(408) 942-4027 Fax on Demand:(408) 942-4005
RMA:(408) 942-4045
Web Page: http://www.datatechnology.com
email: support@datatechnology.com
BBS:(408) 942-4010

Data Watch Corp ....**(978) 988-9700**
3235 Satellite Blvd, Bldg 400 Suite 300
Duluth, GA 30096
Tech:(978) 658-0040 Toll Free:(800) 445-3311
Fax:(978) 988-2040
Tech Fax:(978) 988-0697
Web Page: http://www.datawatch.com

Data Watch Corp ....**(978) 988-9700**
234 Ballardvale Street
Willmington, MA 01887

Tech:(978) 988-9700 Toll Free:(800) 988-4739
Fax:(978) 988-2040
Tech Fax:(978) 988-0697
Web Page: http://www.datawatch.com

Database America ....(201) 476-2000
100 Paragon Drive
Montvale, NJ 07645-0416
Tech:(201) 476-2000 Toll Free:(888) 362-2533
Fax:(201) 476-2419
Web Page: http://www.databaseamerica.com
email: reception@databaseamerica.com

DataCal Corp ....(602) 813-3100
531 E. Elliot Rd
Chandler, AZ 85225
Tech:(602) 545-8089 Toll Free:(800) 223-0123
Fax:(602) 545-8090
Tech Support:(602) 813-3260
Web Page: http://www.datacal.com
email: info@datacal.com

DataEase International ....(203) 374-8000
7 Cambridge Dr
Trumbull, CT 06611
Tech:(203) 374-2825
Fax:(203) 740-4005
Web Page: http://www.multi-ware.com
email: 110257,2456@compuserve.com
Compuserve: GO DATAEASE

Datalight Inc ....(360) 435-8086
18810 59th Ave NE
Arlington, WA 98223
Toll Free:(800) 221-6630
Fax:(360) 435-0253
Web Page: http://www.datalight.com
email: jamief@datalight.com

Datalux Corp ....(540) 662-1500
155 Aviation Dr
Winchester, VA 22602
Toll Free:(800) 328-2589
Fax:(540) 662-1682
Web Page: http://www.datalux.com
email: support@datalux.com

Dataproducts Corporation ....(805) 578-4000
1757 Tapo Canyon Road, Suite 200
Simi Valley, CA 93063-3393

Tech:(805) 578-4455 Toll Free:(800) 887-8848
Fax:(805) 578-4001 Fax on Demand:(805) 578-9255
Web Page: http://www.dpc.com
BBS:(805) 578-9251

Dataquest Interactive (see Gartner Group)

Datashield Unison (see Tripp Lite)

Datasouth Computer Corp ....(704) 523-8500
4216 Stuart Andrew Blvd
Charlotte, NC 28217
Tech:(800) 476-2450 Toll Free:(800) 476-2120
Fax:(704) 523-9298
Web Page: http://www.datasouth.com
email: service@datasouth.com

DataSpec (see ORA Electronics)

Datastor ....(714) 833-8000
1815 E. Carnegie Ave
Santa Anna, CA 92705
Tech:(714) 833-8001 Toll Free:(800) 777-6621
Fax:(714) 833-9600
Web Page: http://www.dstor.com
email: tech@dstor.com

Datastorm Tech (Quarterdeck Corp) ......(573) 443-3282
3212 Lemone Industrial Blvd
Columbia, MO 65201
Tech:(573) 875-0530 Toll Free:(800) 354-3222
Fax:(800) 354-3329 Fax on Demand:(800) 371-4566
Web Page: http://www.datastorm.com
email: info@quarterdeck.com
Compuserve: 73707,1100 or GO DATASTORM
BBS:(573) 875-0503

DataViews Corp ....(413) 586-4144
47 Pleasant Street
Northampton, MA 01060
Tech:(413) 586-4144
Fax:(413) 586-3805
Web Page: http://www.dvcorp.com
email: info@dvcorp.com

DataViz Inc ....(203) 268-0030
55 Corporate Drive
Trumbull, CT 06611
Tech:(203) 268-0030 Toll Free:(800) 733-0030
Fax:(203) 268-4345
Other Address: d0248@AppleLink.apple.com

Web Page: http://www.dataviz.com
email: info@list.dataviz.com

Dataware Technologies Inc ·····················(617) 621-0820
222 Third Street, Suite 3300
Cambridge, MA 02142
Tech:(617) 621-0820
Fax:(617) 494-0740
Web Page: http://www.dataware.com
email: Adl@dataware.com

Datum Inc ···(949) 598-7500
3 Parker
Irvine, CA 92618
Fax:(949) 598-7524
Web Page: http://www.datum.com
email: sales@datum.com

Dauphin Technology Inc·························(847) 358-4406
800 East Northwest Highway, Suite 950
Palatine, IL 60067
Fax:(847) 358-4407
Web Page: http://www.dauphintech.com
email: support@dauphintech.com

David Systems Inc (see 3 Com Corp)

Davidson & Associates (CUC Intl) ··········(310) 793-0600
19840 Pioneer Ave
Torrance, CA 90503
Tech:(800) 556-6141 Toll Free:(800) 545-7677
Fax:(310) 793-0601
Web Page: http://www.davd.com
email: support@davd.com
America Online: keyword: Davidson

Day Runner ···(714) 680-3500
15295 Alton Parkway, PO Box 57027
Irvine, CA 92619-7027
Toll Free:(800) 232-9786
Fax:(714) 441-4848
Fax:(714) 441-4840
Web Page: http://www.dayrunner.com
email: tunger@dayrunner.com

Dayna Communications ·························(801) 269-7200
849 W Levoy Drive
Salt Lake City, UT 84123
Tech:(801) 569-7200 Toll Free:(800) 531-0600
Fax:(801) 269-7363
Web Page: http://www.dayna.com

America Online: DAYNACOM

Db-Tech Inc (see WebSci Technologies)

DCA/IRMA (see Attachmate)

Deadly Games ··(516) 537-6060
PO Box 676
Bridgehampton, NY 11932
Tech:(516) 537-6060
Fax:(516) 537-3299
Other Address: E-world:DeadlyGame
Web Page: http://www.deadlygames.com
email: tech@deadlygames.com
Compuserve: 74431,2470
America Online: Deadly G

DEC (see Digital Equipment Corp)

Decisive Technology Corp ······················(650) 528-4300
1991 Landings Drive
Mountain View, CA 94043
Toll Free:(800) 987-9995
Fax:(650) 528-4321
Web Page: http://www.decisive.com
email: info@decisive.com

Deep River Publishing Inc ······················(207) 871-1684
565 Congress Street, PO Box 9715-975
Portland, ME 04104
Tech:(207) 871-1684 Toll Free:(800) 643-5630
Fax:(207) 871-1683
Web Page: http://www.deepriver.com
email: webmaster@deepriver.com

Dell Computer Corp ·································(512) 338-4400
2214 W Braker Lane, Suite D
Austin, TX 78758
Tech:(800) 624-9896 Toll Free:(800) 289-3355
Fax:(800) 727-8320 Fax on Demand:(800) 950-1329
Web Page: http://www.dell.com
Compuserve: GO DELL
America Online: Keyword:Dell
BBS:(512) 728-8528

DeLorme Mapping Corp ···························(207) 846-7000
Two DeLorme Drive, PO Box 298
Yarmouth, ME 04096
Tech:(207) 846-8900 Toll Free:(800) 452-5931
Fax:(800) 575-2244 Fax on Demand:(207) 846-7058
Tech Suppport Fax:(207) 846-7050

Web Page: http://www.delorme.com
email: support@delorme.com
Compuserve: 72030,2146
BBS:(207) 846-7059

Delphi (see News Corp/MCI Online Ventures)

Delphi Internet Service ....(617) 441-4801
1030 Massachusetts Ave
Cambridge, MA 02138
Tech:(617) 441-4801 Toll Free:(800) 695-4005
Fax:(617) 441-4902
Web Page: http://www.delphi.com
email: service@delphi.com

Delrina Software (Symantec)....................(416) 441-3676
895 Don Mils Rd, 500-2 Park Center
Toronto, ON M3C1W3 Canada
Tech:(800) 268-6082 Toll Free:(800) 268-6082
Fax:(416) 441-0333 Fax on Demand:(416) 443-1614
Web Page: http://www.delrina.com
Compuserve: GO DELRINA

Delta Software Systems Inc ....(901) 758-0123
146 Timber Creek Drive, Suite 200
Memphis, TN 38018
Tech:(901) 758-0123
Fax:(901) 758-0211
Web Page: http://www.deltasoftware.com
email: dtbarton@deltasoftware.com

Deltec (see Excide Electronics)

Deneba Software ....(305) 596-5644
7400 SW 87th Ave
Miami, FL 33173
Tech:(305) 596-5644 Toll Free:(800) 733-6322
Fax:(305) 273-9069
Web Page: http://www.deneba.com
email: support@deneba.com
Compuserve: 76004,2154 or GO MACBVEN
America Online: Keyword: Deneba

Derby and Associates ....(303) 979-6054
Littleton, CO
Fax:(303) 972-8043

DeScribe Inc (Out of Business)

DesignCAD..(918) 825-7555
One American Way
Pryor, OK 74361

Tech:(918) 825-4844 Toll Free:(800) 233-3223
Fax:(918) 825-6359
Training Products:(800) 842-4723
Other Address: http://www.viagrafix.com
Web Page: http://www.designcad.com
email: support@ViaGrafix.com
Compuserve: GO DESIGNCAD
Microsoft: Go To: ViaGrafix

Develcon Electronics Ltd ....(306) 933-3300
856 51st Street East
Saskatoon, SK S7K 5C7 Canada
Toll Free:(800) 667-9333
Fax:(306) 931-1370
Web Page: http://www.develcon.com
email: support@develcon.com

DFI Inc ....(916) 568-1234
135 Main Ave
Sacramento, CA 95838
Tech:(916) 568-1234
Fax:(916) 568-1233
Web Page: http://www.dfiusa.com
email: support@dfiusa
BBS:(732) 390-4820

Di-USA ....(813) 832-6439
Tampa, FL 33687
Toll Free:(800) 477-3483
Fax:(813) 832-5426
Web Page: http://www.di-usa.com
email: support@di-usa.com

Dia-Nielsen ....(609) 829-9441
2615 River Road, Suite 1
Cinnaminson, NJ 08077-1624
Tech:(609) 829-9381 Toll Free:(800) 893-6361
Fax:(609) 829-8814

Diagnostic Technologies Inc
6511C Mississauga Road
Mississauga, ON L5N1A6 Canada
Tech:(905) 347-0486
Fax:(905) 542-8458
Web Page: http://www.diagtec.com
email: diagnost@io.org

DiagSoft Inc ....(813) 207-7000
6200 Courtney Campbell Cswy, Suite 320
Tampa, FL 33607

Toll Free:(800) 342-4763
Fax:(813) 207-7001
Web Page: http://www.diagsoft.com
email: support@diagsoft.com

Dialogic GammaLink ....(973) 993-3000
1515 Route Ten
Parsippany, NJ 07054
Toll Free:(800) 329-4727
Fax:(973) 993-3093 Fax on Demand:(800) 755-5599
Fax Back:(973) 993-1063
Web Page: http://www.dialogic.com
email: techs@gammalink.com

Dialogic Sunnyvale ....(408) 744-1400
1314 Chesapeake Terrace
Sunnyvale, CA 94089
Toll Free:(800) 755-4444
Web Page: http://www.gammalink.com

Diamond Computer (see Diamond Multimedia)

Diamond Flower Electric Inst (see DFI Inc)

Diamond Multimedia ....(541) 967-2400
7101 Supra Dr SW
Albany, OR 97321
Tech:(541) 967-2450 Toll Free:(800) 468-5846
Fax:(541) 967-2401 Fax on Demand:(800) 380-0030
Tech FAx:(541) 967-2401
Web Page: http://www.diamondmm.com
email: techsupt@diamondmm.com
Compuserve: GO DIAMOND
America Online: SupraCorp2

Diamond Multimedia Systems ....(408) 325-7000
2880 Junction Ave
San Jose, CA 95134-1922
Tech:(541) 967-2450 Toll Free:(800) 468-5846
Fax:(408) 325-7070 Fax on Demand:(800) 380-0030
Tech Fax:(541) 967-2401
Web Page: http://www.diamondmm.com
email: techsupt@diamondmm.com
Compuserve: GO DIAMOND

Digi International ....(612) 912-3444
11001 Bren Road E.
Minnetonka, MN 55343
Toll Free:(800) 344-4273
Fax:(612) 912-4952 Fax on Demand:(612) 912-4990
Tech Fax:(612) 912-4958

Web Page: http://www.digibd.com
email: info@dgii.com
BBS:(612) 912-4800

Digi Lan Connect ....(408) 752-2770
1299 Orleans Drive
Sunnyvale, CA 94089
Tech:(408) 744-2751 Toll Free:(800) 466-4526
Fax:(408) 744-2793
Tech Support Fax:(408) 744-2771
Web Page: http://www.milan.com
email: support@MiLan.com

Digi-Data Corp ....(301) 498-0200
8580 Dorsey Run Road
Jessup, MD 20794
Tech:(301) 498-0200
Fax:(301) 498-0771
Web Page: http://www.digidata.com
email: support@digidata.com
BBS:(301) 604-9357

Digi-Key Corporation ....(218) 681-6674
701 Brooks Ave South
Thief River Falls, MN 56701-2757
Toll Free:(800) 344-4539
Fax:(218) 681-3380
Web Page: http://www.digikey.com
email: webmaster@digikey.com

Digiboard Inc (see Digi International)

Digicom Systems Inc ....(408) 262-1277
188 Topaz Street
Milpitas, CA 95035
Tech:(408) 934-1601 Toll Free:(800) 833-8900
Fax:(408) 262-1390
Web Page: http://www.digicomsys.com
email: support@digicomsys.com

Digimarc Corp...(503) 223-0118
One Centerpoint, Suite 500
Portland, OR 97035
Tech:(503) 626-8811 Toll Free:(800) 344-4627
Fax:(503) 223-6015 Fax on Demand:(800) 344-4627
Other:(503) 968-2908
Web Page: http://www.digimarc.com
email: info@digimarc.com

Digit Head Inc ....(703) 524-0101
2420 Wilson Blvd

Arlington, VA 22201
Tech:(703) 524-0101
Fax:(703) 524-0102
Web Page: http://www.digithead.com
email: support@digithead.com

Digital Design Inc ....(973) 857-0901
67 Sand Park Road
Cedar Grove, NJ 07009
Toll Free:(800) 469-2205
Fax:(973) 857-0607
Web Page: http://www.omniprint.com
email: ddi@omniprint.com

Digital Dynamics ....(408) 438-4444
5274 Scotts Valley Drive
Scotts Valley, CA 95066
Tech:(408) 438-4444
Fax:(408) 438-6825
Web Page: http://www.digitaldynamics.com
email: ddi@digitaldynamics.com
BBS:(714) 529-5313

Digital Equipment Corp ....(508) 841-3111
334 South Street
Shrewsbury, MA 01545-4182
Toll Free:(800) 354-9000
Fax:(508) 841-6100
Web Page: http://www.digital.com
Computer Systems Division ....(800) 722-9332
111 Powdermill Road
Maynard, MA 01754-1499
Tech:(800) 354-9000
Web Page: http://www.digital.com
Digital Components and Peripherals ....(800) 354-9000
Toll Free:(800) 365-0696
Digital Semiconductor ....(508) 568-6872
Digital Storage Information ....(800) 786-7967
Internet Business Group ....(800) 344-4825
Mobile Software Business ....(508) 486-2111

Digital Impact Inc ....(918) 742-2022
6506 S. Lewis Ave, Suite 275
Tulsa, OK 74136-1047
Tech:(918) 742-2022 Toll Free:(800) 775-4232
Fax:(918) 742-8176
Web Page: http://www.digitalimpact.com
email: sales@digitalimpact.com

Digital Vision .... **(781) 329-5400**
270 Bridge St, Suite 201
Dedham, MA 02026
Tech:(781) 329-5400 Toll Free:(800) 346-0090
Fax:(781) 329-6286
Web Page: http://www.digvis.com
email: support@digvis.com
Compuserve: GO DIGVIS
America Online: Keyword: DIGITAL VISION

Dimension X Inc (see Microsoft)
Web Page: http://www.microsoft.com/dimensionx/
email: info@dimensionx.com

Direct Network Services .... **(978) 772-9978**
543 Great Road
Littleton, MA 01460
Tech:(978) 772-9978
Fax:(978) 772-9984
Web Page: http://www.windata.com
email: dkirkland@windata.com

Disc Distributing Corp .... **(310) 322-6700**
19430 S. VanNess Ave
Torrance, CA 90501
Toll Free:(800) 688-4545
Fax:(310) 322-6711
Web Page: http://www.discmart.com

Disc Tec .... **(407) 671-5500**
925 S. Semoran Blvd, Suite 114
Winter Park, FL 32792
Fax:(407) 671-6606
Web Page: http://www.disctec.com

Discis Knowledge Research .... **(416) 250-6537**
90 Shppard Ave East, 7th Floor
Toronto, ON M2N3A1 Canada
Tech:(416) 250-6537
Web Page: http://www.goodmedia.com
Compuserve: 75474,1415

Disney Interactive .... **(818) 841-3326**
500 S. Buena Vista St
Burbank, CA 91521-8432
Tech:(800) 228-0988 Toll Free:(800) 900-9234
Fax:(818) 846-0454 Fax on Demand:(800) 965-5360
Web Page: http://www.disney.com
email: sysop@disneysoft.com
Compuserve: 71333,14

America Online: Disneysoft
BBS:(818) 567-4027

Distinct Corp ··(408) 366-8933
12900 Saratoga Ave, PO Box 3410
Saratoga, CA 95070-1410
Tech:(408) 342-3216
Fax:(408) 366-0153 Fax on Demand:(408) 366-2101
Tech Fax:(408) 366-0149
Web Page: http://www.distinct.com
email: tech@distinct.com

Distributed Processing Tech····················(407) 830-5522
140 Candace Drive
Maitland, FL 32751
Tech:(407) 830-5522 Toll Free:(800) 322-4378
Fax:(407) 260-5366
Tech Fax:(407) 830-4793
Web Page: http://www.dpt.com
email: support@dpt.com
BBS:(407) 831-6432

Diversified Technology····························(601) 856-4121
112 E. State St
Ridgeland, MS 39157
Toll Free:(800) 443-2667
Fax:(601) 856-2888

DMA (see Symantec)

DocuMagix Inc ··(650) 324-0600
1378 Willow Road
Menlo Park, CA 94025-1430
Tech:(650) 470-6990 Toll Free:(800) 362-8624
Fax:(650) 470-6955
Web Page: http://www.documagix.com
email: support@documagix.com

Dorling Kindersley Publishing Inc··········(212) 213-4800
95 Madison Ave
New York, NY 10016
Tech:(800) 356-6575 Toll Free:(888) 342-5357
Fax:(212) 213-5240
Web Page: http://www.dkonline.com/dkcom
email: support@dk.com

DPS Software Group ·····························(905) 944-4000
11 Spiral Dr, Suite 10
Florence, KY 41042
Toll Free:(800) 775-3314
Fax:(606) 371-3729

Web Page: http://www.dps.com
email: support.us@dps.com

Dr. Dobb's Journal (Miller Freeman) ......(650) 358-9500
411 Borel Ave, Suite 100
San Mateo, CA 94402
Toll Free:(800) 444-4881
Fax:(650) 358-9749
Web Page: http://www.ddj.com
email: editors@ddj.com

Dr. Solomon's Software ....(781) 273-7400
1 New England Executive Park
Burlington, MA 01803
Tech:(781) 273-7400 Toll Free:(888) 377-6566
Fax:(781) 273-7474
Tech Support:(800) 595-9175
Web Page: http://www.drsolomon.com
email: techhelp@us.drsolomon.com
Compuserve: Go DRSOLOMON

Dr. T's Music Software
124 Crescent Road
Needham, MA 02194
Tech:(770) 428-0008
Fax:(617) 272-9097
Compuserve: 71154,346

Dragon Systems Inc ....(617) 965-5200
320 Nevada Street
Newton, MA 02160
Tech:(617) 965-7670 Toll Free:(800) 825-5897
Fax:(617) 527-0372 Fax on Demand:(617) 965-9454
Tech Fax:(617) 527-4576
Web Page: http://www.dragonsys.com
email: support@dragonsys.com
Compuserve: GO DRAGON
BBS:(617) 332-7371

Dream Theater ....(818) 773-4979
21630 Marilla St
Chatsworth, CA 91311
Tech:(818) 773-4979
Fax:(818) 773-8314
Web Page: http://www.dreamtheater.com
email: info@dreamtheater.com

Dresselhaus Computer Products ....(909) 937-1137
305 N. Sacramento Place
Ontario, CA 91764-4423

Tech:(909) 937-1137 Toll Free:(800) 368-7737
Fax:(909) 937-1150 Fax on Demand:(800) 373-3600
Web Page: http://www.dresselhaus.com
email: productsupport@dresselhaus.com

DS Design .... (919) 319-1770
1157 Executive Circle, Suite D
Cary, NC 27511
Toll Free:(800) 745-4037
Fax:(919) 460-5983
Web Page: http://www.dsdesign.com
email: sales@dsdesign.com

DSP Group Inc .... (408) 986-4300
3120 Scott Blvd
Santa Clara, CA 95054-3317
Fax:(408) 986-4323
Web Page: http://www.dspg.com

DSP Solutions Inc .... (650) 919-4000
1157 San Antonio Road
Mountain View, CA 94043
Tech:(650) 919-4100
Fax:(650) 919-4040
Web Page: http://www.dsps.com
email: support@dspnet.dsps.com
BBS:(650) 919-4199

DTK Computer Inc .... (626) 810-0098
770 Epperson Drive
City Of Industry, CA 91748
Tech:(626) 810-0098 Toll Free:(800) 289-2385
Fax:(626) 810-0090 Fax on Demand:(800) 806-1385
Web Page: http://www.dtkcomputer.com
email: tech-sup@dtkcomputer.com
BBS:(626) 854-0797

Dukane Corporation .... (630) 584-2300
2900 Dukane Drive
Saint Charles, IL 60174
Fax:(630) 584-5156
Web Page: http://www.dukane.com
email: corporate@dukane.com

Durand Communications Inc .... (805) 961-8700
147 Castilian Drive
Santa Barbara, CA 93117
Fax:(805) 961-8701
Web Page: http://www.durand.com
email: support@durand.com

DV Studio Inc .... **(714) 453-1702**
6 Venture, Suite 208
Irvine, CA 92618
Tech:(714) 453-1702 Toll Free:(800) 454-1702
Fax:(714) 453-1311
Web Page: http://www.dv-studio.com
email: tech_sup@dv-studio.com

Dynacomp Inc .... **(716) 346-9788**
2935 E. Lake Road
Livonia, NY 14487
Tech:(716) 242-0908
Other Address: http://www.dynacompsoftware.com
Web Page: http://www.dynacomp.com
email: info@dynacompsoftware.com

Dynalink Technologies .... **(514) 489-3007**
PO Box 593
Beaconsfield, QC H9W5V3 Canada
Tech:(514) 489-3007
Fax:(514) 486-2901
Web Page: http://www.dynalinktech.com
email: webmaster@dynalinktech.com
Compuserve: 72220,2276

Dynatech Computer Power (see SL Waiber)

Dynatran .... **(800) 423-7650**
5150 SW Griffith Drive
Beaverton, OR 97005
Tech:(888) 350-4100 Toll Free:(800) 423-7650
Web Page: http://www.dynatran.com
email: support@dynatran.com

E-mu Systems Inc .... **(408) 438-1921**
1600 Green Hills Road, PO Box 660015
Scottsvalley, CA 95067
Tech:(408) 438-1921
Fax:(408) 439-8612 Fax on Demand:(408) 438-1921
Web Page: http://www.emu.com
email: service@emu.com

E-Tech Research Inc .... **(510) 438-6700**
47400 Seabridge Drive
Fremont, CA 94538
Tech:(888) 413-7433 Toll Free:(888) 609-8885
Fax:(408) 438-6701 Fax on Demand:(510) 438-6700
Web Page: http://www.e-tech.com
email: techsupt@e-tech.com
BBS:(408) 988-3663

E-Ware Systems Inc .... **(212) 581-5858**
331 West 57th Street, Suite 267
New York, NY 10019
Toll Free:(888) 600-4934
Fax:(310) 260-0075
Web Page: http://www.soundbiz.com
email: support@soundbiz.com

Eagle Data Protection Inc .... **(801) 363-7300**
350 S 400 E, Suite 101
Salt Lake City, UT 84111
Tech:(801) 363-7300
Fax:(801) 538-0200
Web Page: http://www.eagledata.com
Compuserve: GO PIRACY

Eagle Point Software Corp .... **(319) 556-8392**
Dubuque, IA
Web Page: http://www.netins.net/showcase/eagleww

Eagle Technology .... **(414) 241-3845**
10500 N Port Washington Rd
Mequon, WI 53092
Tech:(414) 241-3845 Toll Free:(800) 388-3268
Fax:(414) 241-5248
Web Page: http://www.eagleone.com
email: eagle@execpc.com

Eastman Kodak Co .... **(716) 724-9977**
343 State St
Rochester, NY 14650
Tech:(800) 235-6325 Toll Free:(800) 242-2424
Fax:(716) 724-3282
Help Line:(716) 781-5224
Web Page: http://www.kodak.com
Compuserve: GO KODAK
America Online: Keyword: Kodak

Easy Software Products .... **(301) 373-9603**
44145 Airport View Drive, Suite 204
Hollywood, MD 20636-3111
Fax:(301) 373-9604
Web Page: http://www.easysw.com
email: info@easysw.com

EBM Corporation .... **(800) 815-5719**
2249 S. Grout Road
Gladwin, MI 48624
Toll Free:(800) 815-5719
Web Page: http://www.thecollective.com/ebm/

Eccentric Software .... **(206) 628-2687**
PO Box 2777
Seattle, WA 98111-2777
Tech:(206) 628-2687 Toll Free:(800) 436-6758
Fax:(206) 628-2681
Web Page: http://www.eccentricsoftware.com
email: xcentric@aol.com

Echo Speech Corporation .... **(805) 684-4593**
6460 Via Real
Carpinteria, CA 93013
Tech:(805) 684-4593
Web Page: http://www.echospeech.com
email: help@echospeech.com

Eclipse Technologies Inc .... **(408) 523-5700**
547 Oakmead Parkway
Sunnyvale, CA 94086
Web Page: http://www.eclipse-technologies.com/
email: techsupport@eclipse-technologies.com

Edmark Corp .... **(425) 556-8400**
6727 185th Ave NE, P O Box 97021
Redmond, WA 98073-9721
Tech:(425) 556-8480 Toll Free:(800) 691-2986
Fax:(425) 556-8430
automated tech support:(800) 320-8381
Web Page: http://www.edmark.com
email: pctech@edmark.com
America Online: keyword: EDMARK

EDS Internet New Media .... **(972) 604-7445**
5400 Legacy Dr
Plano, TX 75024-3199
Toll Free:(800) 890-1841
Web Page: http://www.eds.com
email: info@eds.com

EDS Unigraphics .... **(314) 344-5900**
13736 Riverport Drive
Maryland Heights, MO 63043
Fax:(314) 344-4180
Web Page: http://www.ug.eds.com

EDUCORP Multimedia .... **(619) 536-9999**
7434 Trade Street
San Diego, CA 92121-2410
Tech:(619) 693-4030 Toll Free:(800) 843-9497
Fax:(619) 536-2345
Web Page: http://www.educorp.com

email: sales@educorp.com

EFA Corp of America ....(408) 987-5400
3040 Oakmead Village Dr
Santa Clara, CA 95051
Web Page: http://www.efacorp.com

EFI Electronics Corp ....(801) 977-9009
2415 South 2300 West
Salt Lake City, UT 84119
Tech:(800) 877-1174 Toll Free:(800) 877-1174
Fax:(801) 977-0200
Web Page: http://www.efinet.com
email: efi@efinet.com

Egghead Software ....(425) 391-0800
PO Box 7004
Issaquah, WA 98027
Toll Free:(800) 344-4323
Fax:(425) 391-0880
Web Page: http://www.egghead.com

Eicon Technology Corp ....(972) 490-3270
14755 Preston Road, Suite 620
Dallas, TX 75240
Tech:(972) 490-3270 Toll Free:(800) 342-6660
Fax:(972) 239-8069
Web Page: http://www.eicon.com
email: sales@eicon.com

Eidos Interactive...(415) 547-1200
Tech:(415) 547-1244
Web Page: http://www.eidosinteractive.com
email: techsupp@eidos.com

Eigentech Inc ....(609) 985-9185
115 Church Road
Marlton, NJ 08053-9410
Toll Free:(800) 676-8689
Web Page: http://www.webpress.net/eigentech
email: bobmar@ix.netcom.com

Eizo Nanao Technologies Inc.....................(562) 431-5011
5710 Warland Dr
Cypress, CA 90630
Tech:(800) 800-5202 Toll Free:(800) 800-5202
Fax:(310) 431-4811 Fax on Demand:(800) 416-3539
Other Address: http://www.nanao.com
Web Page: http://www.eizo.com
email: support@eizo.com

Elan Computer Group .... **(650) 964-2200**
888 Villa Street, Suite 300
Mountain View, CA 94041
Toll Free:(800) 536-3526
Fax:(650) 964-8588
Web Page: http://www.elan.com
email: info@elan.com

Elan Software Corp (see Goldmine Software)

Electronic Arts .... **(650) 571-7171**
1450 Fashion Island Blvd
San Mateo, CA 94404
Tech:(650) 572-2787 Toll Free:(800) 448-8822
Fax:(650) 571-7995
Tech Fax:(650) 286-5080
Web Page: http://www.ea.com

Electronic City .... **(818) 842-5275**
4001 W. Burbank Blvd
Burbank, CA 91505
Fax:(818) 842-0419
Web Page: http://www.electroniccity.com
email: howardp@electroniccity.com

Electronic Data Systems Corp (EDS) ..... **(972) 605-6000**
5400 Legacy Drive
Plano, TX 75024-3199
Tech:(972) 605-6000 Toll Free:(800) 566-9337
Fax:(972) 604-3562
Web Page: http://www.eds.com
email: info@eds.com

Electronic Energy Control Inc .... **(614) 464-4470**
380 South Fifth Street, Suite 604
Columbus, OH 43215-5491
Tech:(614) 464-4470 Toll Free:(800) 842-7714
Fax:(614) 464-9656
Web Page: http://www.eeci.com
email: eeci1@ibm.net

Electronic Press Services Group .... **(617) 225-9023**
101 Rogers Street, Suite 201
Cambridge, MA 02142-1049
Tech:(800) 680-6856 Toll Free:(800) 680-6856
Fax:(617) 225-7983
Web Page: http://www.epsg.com
email: info@epsg.com

Electronics Of Salina .... **(913) 827-7377**
235 North Santa Fe

Salina, KS 67401
Toll Free:(800) 874-8204
Fax:(913) 827-7611
Web Page: http://users.aol.com/eossln/page1.html
email: eossln@aol.com

Electronix Corp
1 Herald Square
Fairborn, OH 45324-5036
Fax:(937) 878-1972
Web Page: http://www.electronix.com
email: tech@electronix.com

Elgin Interactive Software ....(847) 697-9654
104 Joslyn Drive
Elgin, IL 60120
Fax:(847) 697-9689
Web Page: http://www.mcs.net/~elgin/
email: elgin@mcs.com

Elite Products ....(800) 929-1747
704 A Nursery Rd
Linthicum, MD 21090
Toll Free:(800) 929-1747
Fax:(410) 636-5037

Elitegroup Computer System (ECS) ....(510) 226-7333
45225 N. Port Ct
Fremont, CA 94538
Tech:(510) 226-7333
Fax on Demand:(510) 226-7333
BBS:(510) 683-0928

Elo TouchSystems Inc ....(510) 651-2340
6500 Kaiser Drive
Fremont, CA 94555
Tech:(615) 220-4299 Toll Free:(800) 356-8682
Fax:(510) 651-3511 Fax on Demand:(888) 329-6335
Web Page: http://www.elotouch.com
email: eloinfo@elotouch.com
Compuserve: GO ELO TOUCH
BBS:(615) 482-9840

Elsa Inc ....(408) 919-9100
2231 Calle De Luna
Santa Clara, CA 95054
Toll Free:(800) 272-3572
Fax:(408) 919-9120
Web Page: http://www.elsa.com
email: sup-us@elsa.com

Emblem Interactive ....(800) 232-8324
1400 SW First St
Miami, FL 33135
Tech:(305) 541-4331 Toll Free:(800) 323-8324
Fax:(305) 541-0074
Compuserve: 71270,2642

Emerald Systems (see NCE Storage Solutions)

EMPaC International Corp ....(510) 683-8800
47490 Seabridge Drive
Fremont, CA 94538
Fax:(510) 683-8662
Fax:(510) 683-0777
Web Page: http://www.empac.com
email: info@empac.com

Empower Pro ....(408) 879-7900
1999 S. Bascom Ave, Suite 700
Campbell, CA 95008
Tech:(408) 879-7911 Toll Free:(800) 806-2462
Fax:(408) 879-7979
Web Page: http://www.magna1.com
email: support@empowerpro.com

Empress Software Inc ....(301) 220-1919
6401 Golden Triangle Drive, Suite 220
Greenbelt, MD 20770
Tech:(301) 220-1919
Fax:(301) 220-1997
Web Page: http://www.empress.com
email: supportus@empress.com

Emtech Data Store Media ....(781) 271-4000
9 Oak Park Drive
Bedford, MA 01730
Tech:(800) 225-3326 Toll Free:(800) 343-4600
Fax:(781) 275-2708
Tech Support:(800) 708-6334
Web Page: http://www.datastoremedia.com
email: sanzp@datastoremedia.com

Emulex ....(714) 662-5600
3535 Harbor Blvd
Costa Mesa, CA 92626-1437
Tech:(714) 513-8270 Toll Free:(800) 854-7112
Fax:(714) 241-0792 Fax on Demand:(714) 513-8277
Web Page: http://www.emulex.com
email: support@emulex.com
BBS:(714) 662-1445

Enable Software Inc
Northway 10 Executive Park, 313 Ushers Road
Ballston Lake, NY 12019
Fax:(518) 877-3337

Encore Computer Corp ···························(954) 587-2900
6901 West Sunrise Blvd
Fort Lauderdale, FL 33313-4499
Tech:(800) 936-2673
Fax:(954) 797-5793
Web Page: http://www.encore.com

Encore Software Inc ·································(800) 507-1375
Tech:(800) 507-1375

Endl Publications ····································(408) 867-6642
14426 Black Walnut Court
Saratoga, CA 95070
Fax:(408) 867-2115 Fax on Demand:(408) 741-1600

Enhance 3000 ···(818) 343-3066
18730 Oxnard St, Suite 201
Tarzana, CA 91356
Tech:(818) 343-3066 Toll Free:(800) 343-0100
Fax:(818) 343-1436
Web Page: http://www.enhancememory.com
email: support@enhance3000.com

Enhance Memory Products (see Enhance 3000)

Enhanced Software Technologies Inc····(602) 470-1115
4014 E. Broadway, Ste. 405
Phoenix, AZ 85040-8822
Toll Free:(800) 998-8649
Fax:(602) 470-1116
Web Page: http://www.estinc.com
email: support@estinc.com

Enlight Corp ··(562) 693-8885
1227 Slauson Ave
Whittier, CA 90606
Fax:(562) 693-9455
Web Page: http://www.enlightcorp.com
email: brian@enlightcorp.com

ENSONIQ Corp ···(610) 647-3930
155 Great Valley Pkwy, PO Box 3035
Malvern, PA 19355-0735
Fax:(610) 647-8908 Fax on Demand:(800) 257-1439
Web Page: http://www.ensoniq.com
email: multimedia@ensoniq.com

Compuserve: GO MIENSONIQ

Envirogen International ....(714) 574-1440
5319 University Drive, Suite 344
Irvine, CA 92715
Tech:(714) 574-1440 Toll Free:(800) 228-8839
Fax:(714) 574-1432
Web Page: http://www.softspot.com

Environmental Systems Research Inst ..(909) 793-2853
380 New York St
Redlands, CA 92373-8100
Tech:(909) 793-2853 Toll Free:(800) 447-9778
Fax:(909) 793-5953
Tech Fax:(909) 792-0960
Web Page: http://www.esri.com
email: service@esri.com

EO (see AT&T)

Epilogue Technology Corp......................(408) 542-1500
201 Moffett Park Drive
Sunnyvale, CA 94089
Tech:(781) 245-0804
Fax:(408) 542-1961
Tech Fax:(781) 245-8122
Web Page: http://www.epilogue.com
email: support@epilogue.com

Epox International Inc ....(714) 990-8858
499 Nibus St, Ste A
Brea, CA 92821
Fax:(714) 990-8972
Web Page: http://www.epox.com
email: epox@epox.com

Epson America Inc ....(310) 782-0770
20770 Madrona Ave
Torrance, CA 90503
Tech:(800) 922-8911 Toll Free:(800) 289-3776
Fax:(310) 782-5220 Fax on Demand:(800) 922-8911
Download Service:(800) 442-2007
Web Page: http://www.epson.com
email: epso@com.s
BBS:(310) 782-4531

Equilibrium Inc ....(415) 332-4343
3 Harbor Drive, Suite 111
Sausalito, CA 94965
Tech:(415) 332-4343 Toll Free:(800) 524-8651
Fax:(415) 332-4433

Other Address: Applelink: EQUILIBRIUM
Web Page: http://www.equilibrium.com
email: info@equil.com
Compuserve: 76420,320
America Online: EQUILIBRIU
BBS:(415) 332-6152

Equinox Systems Inc ....(954) 746-9000
1 Equinox Way
Sunrise, FL 33351
Tech:(954) 746-9000 Toll Free:(800) 275-3500
Fax:(954) 746-9101
Web Page: http://www.equinox.com
email: info@equinox.com
BBS:(954) 746-0282

Erdas Inc ....(404) 248-9000
2801 Buford Hwy NE, Ste. 300
Atlanta, GA 30329-2137
Tech:(404) 248-9777
Fax:(404) 320-8458
Web Page: http://www.erdas.com
email: support@erdas.com

Ergo Computing (see Brick Computer Co., The)

eSoft Inc. ....(303) 699-6565
5335-C Sterling Drive
Boulder, CO 80301
Tech:(303) 444-1600
Fax:(303) 444-1640
Web Page: http://www.esoft.com
email: ipad-support@esoft.com
BBS:(303) 444-4232

EST (Engineering Services & Tech) ....(603) 673-9907
17 Old Nashua Road, # 6
Amherst, NH 03031-2839
Fax:(603) 673-9913
Web Page: http://www.stgrp.com/est.htm
email: est@stgrp.com

Etak Inc..(650) 328-3825
1430 O'brien Dr
Menlo Park, CA 94025
Toll Free:(800) 765-0555
Fax:(650) 328-3148
Web Page: http;//www.etak.com
email: info@etak.com

eTek International ···································· **(303) 627-4737**
15200 E. Girard Ave, Suite 3500
Aurora, CO 80014-5040
Toll Free:(800) 888-6894
Fax:(303) 627-4738
Web Page: http://www.mtxi.com
email: cpomeroy@mtxi.com

Europa Software (see WebCo Int)

Evans & Sutherland ···································· **(801) 588-1000**
600 Komas Drive
Salt Lake City, UT 84108
Fax:(801) 588-4500
Web Page: http://www.es.com

Everex Systems ···································· **(510) 498-1111**
5020 Brandin Court
Fremont, CA 94538
Tech:(510) 498-4411 Toll Free:(800) 383-7391
Fax:(510) 683-2186 Fax on Demand:(510) 683-2800
Tech Support:(800) 262-3312
Web Page: http://www.everex.com
email: support@everex.com
BBS:(510) 226-9694

Evergreen Technologies Inc ···················· **(541) 757-0934**
808 NW Buchanan Ave
Corvallis, OR 97330-6218
Tech:(541) 757-7341
Fax:(541) 757-7350
Tech Fax:(541) 752-9851
Web Page: http://www.evertech.com
email: techsupport@evertech.com

Evolution Computing ···································· **(602) 967-8633**
437 S. 48th St, Suite 106
Tempe, AZ 85281
Tech:(800) 874-4028 Toll Free:(800) 874-4028
Fax:(602) 968-4325
Web Page: http://www.fastcad.com
email: support@fastcad.com

Ex Machina Inc (see Air Media)

Exabyte Corp ···································· **(303) 417-7792**
1685 38th St
Boulder, CO 80301
Tech:(800) 445-7736 Toll Free:(800) 392-2983
Fax:(303) 417-7160 Fax on Demand:(303) 417-7792
Web Page: http://www.exabyte.com

email: exasoft@exabyte.com
BBS:(303) 417-7100

Excalibur Communications Inc .... **(918) 488-9801**
2504 East 71st Street, Suite E
Tulsa, OK 74136
Tech:(918) 488-9801 Toll Free:(800) 392-2522
Fax:(918) 491-0033
Web Page: http://www.excalbbs.com
email: sales@excalbbs.com

Excide Electronics .... **(619) 291-4211**
2727 Kurtz Street
San Diego, CA 92110
Tech:(800) 848-4734 Toll Free:(800) 854-2658
Fax:(619) 296-8039
Web Page: http://www.deltecpower.com
email: info@deltecpower.com

Excite Inc .... **(650) 568-6000**
555 Broadway
Redwood City, CA 94063
Fax:(650) 568-6030
Web Page: http://www.architext.com

Executive Software .... **(818) 547-2050**
701 N. Brand Blvd, 6th Floor
Glendale, CA 91203
Toll Free:(800) 829-6468
Fax:(818) 545-9241
Web Page: http://www.execsoft.com
email: tech_support@executive.com

Exide Electronics Group Inc .... **(919) 872-3020**
8609 Six Forks Rd
Raleigh, NC 27615
Toll Free:(800) 554-3448
Fax:(800) 753-9433
Web Page: http://www.exide.com
email: techsupt@email.exide.com

Expert Software .... **(305) 567-9990**
800 Douglas Road, Suite 600
Coral Gables, FL 33134-3160
Tech:(305) 567-9990 Toll Free:(800) 759-2562
Fax:(305) 443-0786 Fax on Demand:(800) 772-5706
Tech Fax:(305) 569-1350
Web Page: http://www.expertsoftware.com

ExperVision Inc .... **(510) 623-7071**
48065 Fremont Blvd

Fremont, CA 94538
Tech:(800) 732-3897 Toll Free:(800) 732-3897
Fax:(510) 623-9290
Web Page: http://www.expervision.com
email: marketing@expervision.com

Exponent Corp .... **(973) 808-9424**
I-80 & New Maple Avenue, PO Box 104
Pine Brook, NJ 07058
Tech:(973) 808-9423 Toll Free:(800) 772-7077
Fax:(973) 808-9419

Express Systems Inc (see WRQ)

Extended Systems Inc .... **(208) 322-7800**
5777 N. Meeker
Boise, ID 83713
Tech:(800) 235-7576 Toll Free:(800) 235-7576
Fax:(406) 587-9170 Fax on Demand:(800) 251-2612
Tech Fax:(406) 585-3606
Web Page: http://www.extendsys.com
email: support@extendsys.com
BBS:(208) 327-5020

Extensis .... **(503) 274-2020**
1800 SW First Ave, Suite 500
Portland, OR 97201
Tech:(503) 274-7030 Toll Free:(800) 796-9798
Fax:(503) 274-0530
Web Page: http://www.extensis.com
1800 SW First Ave, Ste 500 .... (503) 274-2020
Portland, OR 97201 USA
email: http://www.extensis.com

EZI America Corp .... **(805) 639-0385**
5040 Shennandoah St
Ventura, CA 93003
Fax:(805) 639-0386

Facet Corp .... **(972) 985-9901**
4031 West Plano Pkwy
Plano, TX 75093
Tech:(972) 985-9901 Toll Free:(800) 235-9901
Fax:(972) 612-2035
Fax:(800) 982-9901
Web Page: http://www.facetcorp.com
email: support@facetcorp.com

Fairhaven Software .... **(508) 994-6400**
295 Phillips Avenue
New Bedford, MA 02746

Tech:(508) 994-6464 Toll Free:(800) 582-4747
Fax:(508) 994-6465
Web Page: http://www.fairsoft.com
email: office@fairsoft.com

Farallon Computing (see Netopia)

Fargo Electronics Inc ···························· (612) 941-9470
7901 Flying Cloud Dr
Eden Prairie, MN 55344
Tech:(612) 941-0050 Toll Free:(800) 327-4622
Fax:(612) 941-7836
Tech Support Fax:(612) 941-1852
Web Page: http://www.fargo.com
email: id.support@fargo.com

Fast Color.Com ···························· (800) 899-2595
801 Commerce St
Sinking Springs, PA 19608
Tech:(610) 678-8035 Toll Free:(800) 899-2595
Fax:(610) 678-8152
Web Page: http://www.fastcolor.com

FastComm Communications Corp ········· (703) 318-7750
45472 Holiday Drive
Sterling, VA 22091
Tech:(703) 318-4350 Toll Free:(800) 521-2496
Fax:(703) 787-4625
Web Page: http://www.fastcomm.com
email: support@fastcomm.com

Faulkner Information Services ················ (609) 662-2070
114 Cooper Center, 7905 Browning Road
Pennsauken, NJ 08109-4319
Toll Free:(800) 843-0460
Fax:(609) 662-3380
Web Page: http://www.faulkner.com
email: systems@faulkner.com

FaxBack Inc ···························· (503) 645-1114
1100 NW Compton Drive, Suite 200
Beaverton, OR 97006-1900
Tech:(503) 614-5350 Toll Free:(800) 329-2225
Fax:(503) 690-6399 Fax on Demand:(503) 614-5390
Web Page: http://www.faxback.com
email: support@faxback.com

Fedco Electronics Inc (Energy+) ············ (920) 922-6490
1363 Capital Dr, PO Box 1403
Fond Du Lac, WI 54936-1403
Tech:(800) 542-9761 Toll Free:(800) 542-9761

Fax:(920) 922-6750

FedWorld Info Net ....(703) 487-4650
Springfield, VA 22161
Web Page: http://www.fedworld.gov

Fessenden Technologies ....(417) 485-2501
116 N. 3rd Street
Ozark, MO 65721
Fax:(417) 485-3133
Web Page: http://www.oznet.com/fessenden
email: fessenden@oznet.com
Compuserve: 76660,1035

FGS (see Symantec)

Fibermux (see ADC Kentrox)

Ficus Systems (Out of Business)

Fidelity International Technologies ....(732) 417-2230
215 Campus Drive
Edison, NJ 08837-3939
Tech:(732) 417-2230
Fax:(732) 417-5994
Web Page: http://www.fitusa.com
email: tech@fitusa.com

Fifth Generation Sys (see Symantec)

Filemaker Inc ....(408) 987-7000
5201 Patrick Henry Drive
Santa Clara, CA 95054
Tech:(408) 727-9004 Toll Free:(800) 325-2747
Fax:(408) 987-7447 Fax on Demand:(800) 800-8954
Other:(408) 727-8227
Web Page: http://www.claris.com
Compuserve: GO CLARIS
America Online: claris@aol.com

FileNet Corp ....(425) 646-1066
10900 NE Eighth St, Suite 700 Plaza Center
Bellevue, WA 98004
Tech:(425) 450-1500 Toll Free:(800) 345-3638
Fax:(425) 462-0879
Web Page: http://www.saros.com

FileNET Corp...(714) 966-3400
3565 Harbor Blvd
Costa Mesa, CA 92626
Toll Free:(800) 345-3638
Fax:(714) 966-2490
Web Page: http://www.filenet.com

email: product_info@filenet.com

Financial Navigator Int'l ....(650) 962-0300
254 Polaris Ave
Mountain View, CA 94043
Tech:(650) 962-8510 Toll Free:(800) 468-3636
Fax:(650) 962-0730
Web Page: http://www.finnav.com
email: support@finnav.com

Firefox Inc (see FTP Software Inc)

First Floor Inc (Calico Tech) ....(650) 968-1101
444 Castro Street, Suite 200
Mountain View, CA 94041
Tech:(650) 968-1101 Toll Free:(800) 639-6387
Fax:(650) 968-1193
Other Address: http://www.calicotech.com
Web Page: http://www.firstfloor.com
email: lobby@firstfloor.com
BBS:(650) 968-0428

First Things First ....(503) 246-6200
1820 SW Vermont, Suite A
Portland, OR 97219
Tech:(503) 246-6200
email: ftfsoft@aol.com
America Online: keyword:visionary

Fitnesoft Inc ....(801) 221-7777
11 E. 200 N., Suite 204
Orem, UT 84057
Tech:(801) 221-7708 Toll Free:(800) 607-7637
Fax:(801) 221-7707
Web Page: http://www.fitnesoft.com
email: lifeform@fitnesoft.com
Compuserve: GO LFORM

Flagship Systems Inc...............................(972) 458-8828
4601 Langland Road, Suite 106
Dallas, TX 75244
Tech:(972) 458-8828
Fax:(972) 458-8728
Web Page: http://www.flagsys.com
email: info@flagsys.com

Flambeaux Software Inc ....(800) 833-7355
1147 E. Broadway, Suite 56
Glendale, CA 91205
Tech:(818) 957-0097 Toll Free:(800) 833-7355
Fax:(818) 957-0194

Fluke Corporation .... **(425) 347-6100**
PO Box 9090
Everett, WA 98206-9090
Tech:(800) 443-5853 Toll Free:(800) 443-5853
Fax:(425) 356-5116 Fax on Demand:(800) 358-5332
Web Page: http://www.fluke.com
email: fluke-info@tc.fluke.com

Focus Enhancements Inc .... **(978) 371-2000**
142 North Road
Sudbury, MA 01776
Tech:(978) 371-8500 Toll Free:(800) 538-8865
Fax:(781) 938-7741
Customer Service:(800) 538-8862
Web Page: http://www.focusinfo.com
email: tech@focusinfo.com
America Online: focustech @ aol.com

Foley Hi-Tech Systems .... **(510) 597-1621**
5337 College Ave, Suite 308
Oakland, CA 94618
Tech:(510) 597-1621
Fax:(510) 595-0862
Web Page: http://www.fht.com
email: info@fht.com
Compuserve: 71333,3657

Folio Corp .... **(801) 229-6700**
5072 N. 300 West
Provo, UT 84604
Tech:(801) 229-6650 Toll Free:(800) 543-6546
Fax:(801) 229-6787
Tech Fax:(801) 229-6791
Web Page: http://www.folio.com
email: support@folio.com
BBS:(801) 229-6668

Fore Front Direct .... **(813) 724-8994**
25400 US Highway 19 N, Suite 285
Clearwater, FL 33763
Tech:(813) 725-2755 Toll Free:(800) 653-4933
Fax:(813) 726-6922
Web Page: http://www.ffg.com
email: productinfo@ffg.com

Fore Systems Inc .... **(724) 742-4444**
1000 Fore Drive
Pittsburgh, PA 15086-7502
Tech:(724) 772-6600 Toll Free:(888) 404-0444
Fax:(724) 742-7777

Web Page: http://www.fore.com

Forefront .... **(713) 961-1101**
1360 Post Oak Blvd, Suite 2050
Houston, TX 77056
Tech:(800) 475-5831 Toll Free:(800) 475-5831
Fax:(713) 961-1149
Web Page: http://www.ffg.com

Foresight Resources Corp (see Autodesk)

FormGen Inc .... **(602) 443-4109**
15649 Greenway-Hayden Loop
Scottsdale, AZ 85260-1750
Tech:(602) 443-4109
Fax:(602) 951-6810
Web Page: http://www.formgen.com

Forte Inc .... **(760) 431-6460**
2141 Palomar Airport Road, Suite 200
Carlsbad, CA 92009
Fax:(760) 431-6465
Web Page: http://www.forteinc.com

Four Star Systems LC .... **(713) 790-0333**
8825 Knight Rd
Houston, TX 77054
Toll Free:(800) 474-3947
Fax:(713) 790-2499
Web Page: http://www.fourstarsystems.com
email: sales@fourstarsystems.com

Fractal Design Corp (MetaCreations) ····· **(408) 430-4100**
PO Box 66959
Scotts Valley, CA 95067-6959
Tech:(408) 430-4200 Toll Free:(800) 846-0111
Fax:(408) 438-9673
Tech Fax:(408) 438-9672
Other Address: http://www.metacreations.com
Web Page: http://www.fractal.com
Compuserve: GO GUGRPA
America Online: fractal

Frame Technology (see Adobe Systems)

Franklin Covey .... **(801) 975-1776**
2200 W. Parkway Blvd
Salt Lake City, UT 84119
Tech:(801) 975-9999 Toll Free:(800) 654-1776
Fax:(800) 446-1492
Web Page: http://www.covey.com

Frederick Engineering Inc ....(410) 290-9000
10200 Old Columbia Rd
Columbia, MD 21046
Tech:(410) 290-9000 Toll Free:(888) 866-9008
Fax:(410) 381-7180
Other:(888) 866-9001
Web Page: http://www.fe-engr.com
email: rich@fetest.com

FreeSoft Co ....(724) 846-2700
150 Hickory Drive
Beaver Falls, PA 15010
email: freesoft@ccia.com

Fry's Electronics ....(650) 496-6000
382 Portage Avenue
Palo Alto, CA 94306

Frye Computer (see Seagate EMS)

FTP Software Inc ....(978) 685-4000
2 High St
North Andover, MA 01845
Tech:(978) 685-3600 Toll Free:(800) 282-4387
Fax:(978) 794-4488 Fax on Demand:(978) 794-4477
Tech Support:(800) 832-4387
Web Page: http://www.ftp.com
email: support@ftp.com
Compuserve: GO PCVENJ
BBS:(978) 684-6240

Fujitsu America Inc ....(408) 432-1300
3055 Orchard Drive
San Jose, CA 95134-2022
Fax:(408) 432-1318
Web Page: http://www.fujitsu.com

Fujitsu Business Communications ....(800) 654-0715
3190 Miraloma Ave
Anaheim, CA 92806
Fax:(714) 764-2573

Fujitsu Computer Products of Amer ......(408) 432-6333
2904 Orchard Pkwy
San Jose, CA 95134-2009
Toll Free:(800) 626-4686
Fax:(408) 894-1709 Fax on Demand:(408) 428-0456
Web Page: http://www.fcpa.com
email: info@fcpa.fujitsu.com
Compuserve: GO FCPA

Fujitsu HAL ....(800) 425-0329

Fujitsu ICL ....(800) 538-8716

Toll Free:(800) 345-0845

Fujitsu Interactive ....(415) 538-2900

128 Spear St Second Floor
San Francisco, CA 94105

Toll Free:(888) 992-5433
Fax:(415) 538-2990
Web Page: http://www.fujitsu-interactive.com

Fujitsu PC Corp ....(408) 935-8800

598 Gibralter Dr
Milpitas, CA 95035

Toll Free:(800) 838-5487
Web Page: http://www.fujitsu-pc.com

Fujitsu Personal Systems Inc ....(408) 982-9500

5200 Patrick Henry Dive
Santa Clara, CA 95054

Tech:(408) 982-9500 Toll Free:(800) 831-3183
Fax:(408) 496-0609
Web Page: http://www.fpsi.com
email: CustServ@fpsi.fujitsu.com

Fujitsu ROSS ....(800) 767-7937

Full Armor ....(781) 641-1500

955 Massachusetts Ave, Suite 365
Cambridge, MA 02139

Tech:(781) 641-2017 Toll Free:(800) 653-1783
Fax:(781) 641-1973
Web Page: http://www.micah.com
email: info@micah.com

Full Circle Technologies ....(650) 691-1091

Mountain View, CA 94040-0316

Fax:(650) 691-1108
Web Page: http://www.virusweb.com
email: info@fcircle.com

Full Time Software ....(650) 572-0200

177 Bovet Rd, 2nd Floor
San Mateo, CA 94402

Tech:(800) 245-8649 Toll Free:(800) 233-4559
Fax:(650) 572-1300
Qualix Direct:(800) 455-9273
Web Page: http://www.qualix.com
email: info@qualix

Fuller's Wholesale Electronics Inc. ········**(304) 428-2296**
713 Gladstone Street
Parkersburg, WV 26101-5661
Tech:(304) 428-2296 Toll Free:(800) 666-2296
Fax:(304) 428-2297
Fax:(800) 727-2297
Web Page: http://www.batteryking.com
email: fullers@citynet.net

Funk Software Inc ······································**(617) 497-6339**
222 Third Street
Cambridge, MA 02142
Tech:(617) 497-6339 Toll Free:(800) 828-4146
Fax:(617) 547-1031
Web Page: http://www.funk.com
email: support@funk.com

Future Domain Corp (see Adaptec)

Future Thinking (Precision Powerh..)·····**(612) 333-9111**
911 Second Street South
Minneapolis, MN 55415
Tech:(612) 332-9262
Fax:(612) 332-9200
Web Page: http://www.future-think.com
email: profrom@future-think.com

FutureSoft Engineering Inc ······················**(281) 496-9400**
12012 Wickchester Lane, Suite 600
Houston, TX 77079-1222
Tech:(281) 588-6868 Toll Free:(800) 989-8908
Fax:(281) 496-1090
another BBS:(281) 588-6805
Other Address: http://www.futuresoft.com
Web Page: http://www.fse.com
email: info@futuresoft.com
Compuserve: 76702,755
BBS:(281) 588-6870

FutureTense Inc···**(978) 635-3600**
43 Nagog Park
Acton, MA 01720
Tech:(978) 635-3144
Fax:(978) 635-3610
Tech Fax:(978) 635-3616
Web Page: http://www.futuretense.com
email: support@futuretense.com

Futurus Corp (see Novell)

FWB Software LLC ································ **(650) 482-4800**
2750 El Camino Real
Redwood City, CA 94061-3911
Tech:(650) 482-4800 Toll Free:(800) 581-4392
Fax:(650) 482-4858
Upgrade Info:(408) 848-2409
Web Page: http://www.fwb.com
email: info@fwb.com

G.V.C. ································ **(905) 738-9300**
485 Millway Ave, Building G
Concord, ON L4K 3V4 Canada
Tech:(905) 738-5736
Fax:(905) 738-5563
Web Page: http://www.gvc.ca/
email: eli@gvc.ca
BBS:(905) 738-7931

Galacticomm Inc ································ **(954) 583-5990**
4101 SW 47th Ave, Suite 101
Fort Lauderdale, FL 33314
Tech:(800) 625-6782 Toll Free:(800) 328-1128
Fax:(954) 583-7846
Web Page: http://www.gcomm.com
email: service@gcomm.com
BBS:(954) 583-7808

Gametek ································ **(415) 289-0220**
3 Harbor Drive, Suite 110
Sausalito, CA 94965
Web Page: http://www.gametek.com
email: webmaster@sf.gametek.com

Gammalink (see Dialogic Sunnyvale)

Gandalf Systems Corp ································ **(425) 462-8996**
606 122 Ave NE
Bellevue, WA 98005
Fax:(425) 462-8984
Web Page: http://www.gandalf.com
email: megda@ix.netcom.com

Gap Development ································ **(949) 496-3774**
24242 Porto Fino
Monarch Beach, CA 92629
Tech:(949) 496-3774
Fax:(949) 496-3774
Web Page: http://www.gapdev.com
email: support@gapdev.com
Compuserve: 76054,210

BBS:(949) 493-3819

Gartner Group/Dataquest ....**(408) 468-8000**
251 River Oaks Parkway
San Jose, CA 95134-1913
Tech:(408) 748-1111 Toll Free:(800) 419-3282
Fax:(408) 954-1780
Web Page: http://www.dataquest.com
email: help@gartner.com

Gates Arrow ....**(510) 489-5371**
1502 Crocker Avenue
Hayward, CA 94544
Tech:(800) 332-2315 Toll Free:(800) 332-2222
Fax:(510) 489-9393
Customer service:(800) 332-2299
Web Page: http://www.gatesarrow.com

Gateway 2000 Inc ....**(605) 232-2000**
610 Gateway Drive, PO Box 2000
North Sioux City, SD 57049
Tech:(800) 846-2301 Toll Free:(800) 846-2000
Fax:(605) 232-2023 Fax on Demand:(800) 846-4526
Fax Back:(605) 232-2561
Other Address: http://www.gateway.com
Web Page: http://www.gw2k.com
email: custserv@gw2kbbs.com
BBS:(800) 846-7562

Gateway Electronics ....**(314) 427-6116**
8123 Page Blvd
Saint Louis, MO 63130
Tech:(800) 669-5810 Toll Free:(800) 669-5810
Fax:(314) 427-3147
Web Page: http://www.gatewayelex.com
email: gateway@mvp.net

Gateway Electronics ....**(303) 458-5444**
2525 N Federal Blvd
Denver, CO 80211
Tech:(800) 669-5810 Toll Free:(800) 669-5810
Fax:(303) 458-6988
Web Page: http://www.gatewayelex.com
email: gateway@mvp.net

Gateway Electronics ....**(619) 279-6802**
9222 Chesapeake Drive
San Diego, CA 92123
Tech:(800) 669-5810 Toll Free:(800) 669-5810
Fax:(619) 279-7294

Web Page: http://www.gatewayelex.com
email: gateway@mvp.net

Gazelle Systems (see GTM Software)

GBC Technologies ··(609) 767-2500
100 GBC Corporation
Berlin, NJ 08009

GCC Technologies ··(781) 275-5800
209 Burlington Road
Bedford, MA 01730-9143
Tech:(781) 276-8620 Toll Free:(800) 422-7777
Fax:(781) 275-1115
Web Page: http://www.gcctech.com
email: support@gcctech.com

General Computer Engineering ··············(714) 999-2894
1501 N. Raymond Ave, Suite J
Anaheim, CA 92801
Fax:(714) 999-2793
Web Page: http://www.gcei.com
email: general@gcei.com

General DataComm Inc ······························(203) 574-1118
1579 Straits Turnpike
Middlebury, CT 06762-1299
Fax:(203) 758-8507
Other Address: http://www.voiceofatm.com
Web Page: http://www.gdc.com

General Interactive Inc ·····························(617) 354-8585
66 Church St
Cambridge, MA 02138
Toll Free:(888) 354-8585
Fax:(617) 354-8899
Web Page: http://www.icybernetics.com
email: info@interactive.com

General Magic Inc ···(408) 774-4000
420 N. Mary Avenue
Sunnyvale, CA 94086
Toll Free:(800) 468-4342
Fax:(408) 774-4010
Web Page: http://www.genmagic.com
email: webmaster@generalmagic.com

General Signal Networks (Telenex) ········(609) 234-7900
13000 Midlantic Drive
Mount Laurel, NJ 08054
Toll Free:(800) 222-0187

Fax:(609) 778-8700
Web Page: http://www.gsnetworks.com

General Software ······························(425) 454-5755
12737 Bel-Red Road, Suite 100
Bellevue, WA 98005
Toll Free:(800) 850-5755
Fax:(425) 454-5744
Web Page: http://www.gensw.com
email: salesgsi@gensw.com

Generic Software (see AutoDesk)

Genicom ······························(703) 802-9200
14800 Conference Center Drive, Suite 400
Chantilly, VA 20151-3820
Tech:(703) 802-9200 Toll Free:(800) 436-4266
Fax:(703) 802-9039
Tech Fax:(540) 949-1505
Web Page: http://www.genicom.com
email: techsupport@genicom.com
BBS:(540) 949-1576

Genoa Systems Corp ······························(510) 668-4688
3358 Gateway Blvd
Fremont, CA 94538
Tech:(510) 668-4677
Fax:(510) 668-4699
Web Page: http://www.genoasys.com
email: support@genoasys.com
Compuserve: GO GENOA
BBS:(510) 624-4999

Genovation Inc ······························(714) 833-3355
17741 Mitchell North
Irvine, CA 92614
Tech:(714) 833-3355 Toll Free:(800) 822-4333
Fax:(714) 833-0322
Web Page: http://www.genovation.com
email: mail@genovation.com

GEO Interactive Publishing ······················(818) 703-8436
21110 Oxnard St
Woodland Hills, CA 91367
Tech:(503) 968-2280 Toll Free:(800) 576-7751
Fax:(818) 703-8654
Web Page: http://www.emblaze.com

Geographic Data Technologies Inc ········(603) 643-2815
11 Lafayette St
Lebanon, NH 03766-1445

Toll Free:(800) 331-7881
Fax:(603) 643-6808
Web Page: http://www.geographic.com
email: support@gdt1.com

GeoWorks .... **(510) 814-1660**
960 Atlantic Ave
Alameda, CA 94501
Toll Free:(800) 436-7735
Fax:(510) 814-4250
Web Page: http://www.geoworks.com

GFI Fax & Voice .... **(716) 265-1380**
26 E. Main St
Webster, NY 14580
Tech:(716) 265-1397 Toll Free:(888) 243-4329
Fax:(716) 265-1016
Web Page: http://www.gfifax.com
email: sales@gfifax.com

Gibson Research .... **(949) 348-7100**
27071 Cabot Road, Suite 105
Laguna Hills, CA 92653
Toll Free:(800) 736-0637
Fax:(949) 348-7110
Web Page: http://www.grc.com
email: offices@grc.com

Giga-Byte Technology Co Ltd .... **(626) 854-9334**
18325 Valley Blvd, Unit E
LaPuente, CA 91744
Tech:(626) 854-9334
Fax:(626) 854-9339
Web Page: http://www.giga-byte.com
email: info-gbt@giga-byte.com
BBS:(626) 854-9340

GigaTrend Inc .... **(760) 931-9122**
2234 Rutherford Road, Ste 100
Carlsbad, CA 92008
Tech:(760) 931-9122 Toll Free:(800) 743-4442
Fax:(760) 929-0846
Web Page: http://www.gigatrend.com
BBS:(760) 931-9469

GITI .... **(212) 344-0400**
17 Battery Pl, 26th Floor
New York, NY 10004
Web Page: http://www.wbt.giti.com

Global Computer Supply ....(800) 845-6225
2318 E. Delano Blvd
Compton, CA 90220
Toll Free:(800) 845-6225
Fax:(310) 637-6191
Web Page: http://www.globalcomputer.com
email: custsvc@globalcomputer.com

Global Engineering Documents ....(303) 792-2181
15 Inverness Way East
Englewood, CO 80112-5704
Toll Free:(800) 854-7179
Fax:(303) 397-2740
Web Page: http://global.ihs.com
email: global@ihs.com

Global Payment Systems ....(404) 235-4400
4 Corporate Square
Atlanta, GA 30329
Tech:(404) 235-4400
Web Page: http://www.globalpay.com

Global Village Communications ....(408) 523-1000
1144 E. Arques Avenue
Sunnyvale, CA 94086
Tech:(408) 523-1040 Toll Free:(800) 736-4821
Fax:(408) 523-2407 Fax on Demand:(408) 523-2402
Another Fax on Demand:(800) 890-4562
Web Page: http://www.globalvillage.com
email: pcsupport@globalvillage.com
Compuserve: 75300,3473 or GO GLOBAL
America Online: keyword: Global
BBS:(800) 335-6003

Globalink Inc ....(703) 273-5600
9302 Lee Highway, 12th Floor
Fairfax, VA 22031
Tech:(703) 934-2734 Toll Free:(800) 255-5660
Fax:(703) 273-3866
Tech Fax:(703) 273-6098
Web Page: http://www.globalink.com
email: techsupp@globalink.com
Compuserve: 75352.635@compuserve.com

Globe Manufacturing Inc ....(908) 232-7301
1159 US Route 22
Mountainside, NJ 07092
Fax:(908) 232-4729
Web Page: http://www.akstamping.com
email: truth@akstamping.com

Globecomm Systems .... **(516) 231-9800**
45 Oser Ave
Hauppauge, NY 11788
Fax:(516) 231-1557
Web Page: http://www.worldcomm.com
email: info@globecommsystems.com

Globelle Corporation .... **(612) 947-1000**
6100 West 110th Street
Minneapolis, MN 55438-8866
Web Page: http://www.globelle.com
email: plus@usa.globelle.com

GN Nettest Group .... **(800) 233-3800**
63 South Street
Hopkinton, MA 01748
Toll Free:(800) 233-3800
Fax:(508) 435-0448
Web Page: http://www.nettestca.gn.com
email: sales@azure-tech.com

Go Ahead Software Inc .... **(425) 882-1900**
8652 154th Ave NE
Redmond, WA 98052
Fax:(425) 882-1117
Web Page: http://www.goahead.com
email: support@goahead.com

Gold Disk Inc (see Astound Inc)

Gold Standard Multimedia Inc .... **(813) 287-1775**
3825 Henderson Blvd, Suite 200
Tampa, FL 33629-5002
Tech:(813) 287-1775 Toll Free:(800) 375-0943
Fax:(813) 287-1810
Web Page: http://www.gsm.com
email: support@gsm.com

Golden Bow Systems .... **(619) 298-9349**
PO Box 3039
San Diego, CA 92163
Tech:(800) 284-3269 Toll Free:(800) 284-3269
Fax:(619) 298-9950
Web Page: http://www.goldenbow.com
email: hi@goldenbow.com

Golden Coast Information Systems .... **(619) 268-8447**
9323 Chesapeake Drive, Suite E
San Diego, CA 92123
Tech:(619) 278-0948
Fax:(619) 278-0948

Golden Imaging ....(303) 443-6966
3075 N. 75th Street
Boulder, CO 80301
Fax:(303) 443-1660

Golden Ribbon (see Golden Imaging)

Golden Software ....(303) 279-1021
809 14th Stret
Golden, CO 80401-1866
Tech:(303) 279-1021 Toll Free:(800) 972-1021
Fax:(303) 279-0909
Web Page: http://www.goldensoftware.com
email: info@goldensoftware.com
BBS:(303) 279-0909

Goldmine Software Corp ....(310) 454-6800
17383 Sunset Blvd, Suite 101
Pacific Palisades, CA 90272
Tech:(310) 459-1222 Toll Free:(800) 654-3526
Fax:(310) 454-4848 Fax on Demand:(310) 459-1222
Tech Fax:(310) 459-8222
Web Page: http://www.goldminesw.com
email: support@goldminesw.com
Compuserve: GO GOLDMINE
BBS:(310) 459-3443

GoldStar USA Inc (see LG Electronics USA)

Good Software (see Outlook Software)

GRACE Electronic Materials ....(781) 861-6600
77 Dragon Court
Middlesex Essex Gmf, MA 01888
Tech:(800) 832-4929 Toll Free:(800) 832-4929
Fax:(781) 933-4318

Gradient Technologies Inc ....(508) 624-9600
2 Mount Royal Ave
Marlborough, MA 01752-1995
Tech:(508) 229-0239 Toll Free:(800) 525-4343
Fax:(508) 229-0338
Web Page: http://www.gradient.com
email: support@gradient.com

Grand Junction Network (see Cisco Systems)

Granite Communications Inc ....(603) 881-8666
9 Townsend West, Suite 1
Nashua, NH 03063
Fax:(603) 881-4042
Web Page: http://www.gci.com

email: sales@gci.com

Graphic Utilities Inc··········(408) 577-0334
2149 O'Toole Ave, Suite L
San Jose, CA 95131
Tech:(800) 669-4723 Toll Free:(800) 400-5253
Fax:(408) 577-0348
Tech Support:(207) 473-7587
Web Page: http://www.inklink.com

Graphix Zone··········(714) 833-3838
42 Corporate Pk, Suite 200
Irvine, CA 92714
Tech:(812) 829-1007
Fax:(800) 828-3838
Other Address: http://www.ignited.com
Web Page: http://www.gzone.com

GraphOn Corp··········(408) 370-4080
150 Harrison Ave
Campbell, CA 95008
Fax:(408) 370-5047
Web Page: http://www.graphon.com
email: support@graphon.com

GraphPad Software··········(619) 457-3909
10855 Sorrento Valley Road, # 203
San Diego, CA 92121
Toll Free:(800) 388-4723
Fax:(619) 457-8141
Web Page: http://www.graphpad.com
email: support@graphpad.com

Graphsoft Inc··········(410) 290-5114
10270 Old Columbia Road
Columbia, MD 21046-1751
Tech:(410) 290-5114
Fax:(410) 290-8050
Web Page: http://www.graphsoft.com
email: tech@graphsoft.com
Compuserve: 72662,1320
America Online: MCadtech

Graymark··········(800) 854-7393
PO Box 2015
Tustin, CA 92681

Great Falls Computer (see Microtec)

Great Plains Software··········(701) 281-0555
1701 38th St SW

Fargo, ND 58103
Tech:(701) 281-0550 Toll Free:(800) 456-0025
Fax:(701) 281-3328 Fax on Demand:(800) 456-0025
Fax:(701) 281-3752
Web Page: http://www.GPS.com

Great Wave Software ....(408) 438-1990
5353 Scotts Valley Dr
Scotts Valley, CA 95066
Tech:(800) 423-1144 Toll Free:(800) 423-1144
Fax:(408) 438-7171
Web Page: http://www.greatwave.com
email: support@greatwave.com

Greentree Technologies ....(516) 271-6995
33 Walt Whitman Rd, Suite 236
Huntington Station, NY 11746-3627
Tech:(800) 257-7708 Toll Free:(800) 257-7708
Fax:(516) 271-8067
Web Page: http://www.green-tree.com
email: support@green-tree.com

Greenview Data ....(734) 996-1300
PO Box 1586
Ann Arbor, MI 48106-1586
Tech:(800) 458-3348
Fax:(734) 996-1308
Web Page: http://www.vedit.com
email: support@vedit.com
Compuserve: 71333,3656

Greenwich Mean Time - UTA ....(703) 908-6600
Tech:(703) 908-6600
Web Page: http://www.gmt-2000.com

Grolier Interactive Inc ....(203) 797-3530
90 Sherman Turnpike
Danbury, CT 06816
Tech:(203) 796-2536 Toll Free:(800) 621-1115
Fax:(203) 797-3130
Tech Fax:(203) 797-3835
Web Page: http://www.grolier.com
email: techsup@grolier.com
BBS:(203) 797-6872

Group 1 Software ....(301) 731-2300
4200 Parliament Place, Suite 600
Lanham, MD 20706-1844
Tech:(301) 731-2300 Toll Free:(800) 368-5806
Fax:(301) 731-0360 Fax on Demand:(301) 918-0781

PC support:(800) 578-8324
Web Page: http://www.g1.com
email: support@g1.com

Gruber Industries Inc ....(602) 863-2655
21439 N. Second Ave
Phoenix, AZ 85027
Tech:(602) 581-1697 Toll Free:(800) 658-5883
Fax:(602) 257-4313
Web Page: http://www.gruber.com
email: support@gruber.com

Gryphon Software Corp ....(619) 536-8815
7220 Trade Street
San Diego, CA 92121-2325
Tech:(619) 536-8815 Toll Free:(800) 795-0981
Fax:(619) 536-8932
Web Page: http://www.gryphonsw.com
email: sales@gryphonsw.com

GSI Inc ....(949) 261-7949
PO Box 17118
Irvine, CA 92623-7118
Tech:(949) 261-9744 Toll Free:(800) 486-7800
Fax:(949) 757-1778
Web Page: http://www.gsi-inc.com

GT Interactive Software Corp ....(212) 726-6500
New York, NY
Toll Free:(800) 610-4847
Web Page: http://www.gtinteractive.com

GTCO Corp ....(410) 381-6688
7125 Riverwood Drive
Columbia, MD 21046
Tech:(410) 381-6688 Toll Free:(800) 344-4723
Fax:(410) 290-9065 Fax on Demand:(410) 381-6688
Web Page: http://www.gtco.com
email: techsup@gtco.com

GTE Internetworking ....(617) 873-2000
150 Cambridgepark Drive
Cambridge, MA 02140
Toll Free:(800) 472-4565
Fax:(617) 873-5011
Other:(800) 632-7638
Web Page: http://www.bbn.com
email: ops@bbnplanet.com

GTEK Inc ....(800) 282-4835
PO Box 2310

Bay Saint Louis, MS 39521-2310
Tech:(601) 467-8048 Toll Free:(800) 282-4835
Fax:(601) 467-0935
Web Page: http://www.gtek.com
email: support@gtek.com

GTM Software ....(801) 235-7000
305 North 500 West
Provo, UT 84601-2644
Tech:(801) 235-7000
Fax:(801) 235-7099
Compuserve: 76702,265

Guillemot International ....(800) 967-0863
625 Third St, 3rd Floor
San Francisco, CA 94107
Tech:(888) 893-2648 Toll Free:(800) 967-0863
Fax:(514) 490-0027
Web Page: http://www.guillemot.com
email: support@guillemot.com

Gupta Corp (see Centura Software Corp)

Guru Software Corp ....(617) 576-9100
1 Kendall Square, Ste. 2200
Cambridge, MA 02139
Fax:(617) 576-9101
Web Page: http://www.gurucorp.com
email: mail@g1news.com

GVC Technologies Inc (see MaxTech GVC)

GW Instruments Inc ....(617) 625-4096
35 Medford Street
Somerville, MA 02143-4237
Fax:(617) 625-1322
Web Page: http://www.gwinst.com
email: info@gwinst.com

H45 Technology Inc ....(650) 961-9114
620B Clyde Ave
Mountain View, CA 94043
Tech:(800) 220-6346 Toll Free:(800) 220-6346
Fax:(650) 964-2426
Web Page: http://www.h45.com
email: h45tech@h45.com

HAHT Software Inc ....(919) 786-5100
4200 Six Forks Road, Suite 200
Raleigh, NC 27609
Tech:(919) 786-5200 Toll Free:(888) 438-4248

Fax:(919) 786-5250
Web Page: http://www.haht.com
email: support@haht.com

Hal Computer Systems ....(408) 379-7000
1315 Dell Ave
Campbell, CA 95008
Tech:(800) 425-9111 Toll Free:(800) 425-0329
Fax:(408) 341-5401
Web Page: http://www.hal.com
email: action@hal.com

Hansol Electronics Inc ....(714) 562-5151
6 Centerpointe Dr, Ste 220
La Palma, CA 90623
Tech:(888) 426-7651 Toll Free:(888) 426-7651
Web Page: http://www.hansol-us.com

Harbinger ....(925) 602-2000
1000 Burnett Ave, Suite 2000
Concord, CA 94520
Tech:(800) 578-4334 Toll Free:(800) 426-3836
Fax:(925) 688-2895
Web Page: http://www.harbinger.com
email: info@harbinger.com

Harbinger Corp ....(404) 467-3000
1055 Lenox Park Blvd
Atlanta, GA 30319
Tech:(404) 841-4334 Toll Free:(800) 578-4334
Web Page: http://www.harbinger.com
email: support@harbinger.com

Harbor Electronics ....(408) 988-6544
805 Aldo Ave, Unit 101
Santa Clara, CA 95054
Fax:(408) 988-2948
Web Page: http://www.harbor-electronics.com
email: info@harbor-electronics.com

Harlequin Incorporated ....(617) 374-2400
One Cambridge Center, 8th Floor
Cambridge, MA 02142
Fax:(617) 252-6505
Web Page: http://www.harlequin.com
email: web@harlequin.com

Harris Computer Systems (see Concurrent)

Hauppauge Computer Works Inc ....(516) 434-1600
91 Cabot Court

Hauppauge, NY 11788
Tech:(516) 434-3197 Toll Free:(800) 443-6284
Fax:(516) 434-3198
Web Page: http://www.hauppauge.com
email: techsupport@hauppauge.com
BBS:(516) 434-8454

HavenTree Software Ltd .... (613) 544-6035
110 Railway Street
Kingston, ON K7K 2L9 Canada
Tech:(613) 544-6035 Toll Free:(800) 267-0668
Fax:(613) 544-9632
Web Page: http://www.haventree.com
email: Tech-sup@haventree.com
Compuserve: 76366,512

Hayes Corp .... (301) 921-8600
1300 Quince Orchard Blvd
Gaithersburg, MD 20878
Tech:(800) 456-7844 Toll Free:(800) 456-7844
Fax:(301) 921-8376 Fax on Demand:(301) 921-8600
Tech Support Fax:(301) 840-1528
Web Page: http://www.accessbeyond.com
email: support@accessbeyond.com

Hayes Microcomputer Products Inc .... (770) 840-9200
5854 Peachtree Corners East
Norcross, GA 30092
Tech:(770) 441-1617 Toll Free:(800) 377-4377
Fax:(770) 441-1213 Fax on Demand:(800) 429-3739
Tech Support Fax:(770) 449-0087
Web Page: http://www.hayes.com
email: hayes.support@hayes.com
Compuserve: GO HAYES
America Online: keyword Hayes
BBS:(770) 446-6336

Haystack Labs .... (512) 918-3555
Tech:(512) 918-3555
Web Page: http://www.haystack.com

HDC Computer (see WRQ)

HDS Network Systems (see Neoware)

Heathkit Educational Systems .... (616) 925-6000
455 Riverview Drive
Benton Harbor, MI 49022
Tech:(616) 925-6000 Toll Free:(800) 253-0570
Fax:(616) 925-2898
Web Page: http://www.heathkit.com

email: techsupport@heathkit.com

Helix Software Co. (Network Assoc) ······(718) 392-3100
47-09 30th Street
Long Island City, NY 11101
Tech:(718) 392-3735 Toll Free:(800) 451-0551
Fax:(718) 392-4212 Fax on Demand:(718) 392-3100
Web Page: http://www.helix.com
email: info@helix.com
Compuserve: GO HELIX

Hercules Computer Technology Inc·······(510) 623-6030
3839 Spinnaker Court
Fremont, CA 94538-6524
Tech:(510) 623-6050 Toll Free:(800) 323-0601
Fax:(510) 623-1112 Fax on Demand:(800) 711-4372
Tech Support Fax:(510) 623-4215
Web Page: http://www.hercules.com
email: support@hercules.com
Compuserve: GO HERCULES or 71333,2532
America Online: support@hercules.com

Herne Data Systems Ltd ··························(416) 364-9955
31 Adelaide St E, PO Box 357
Toronto, ON M5C 2J4 Canada
Fax:(416) 364-9955
Web Page: http://www.herne.com
email: herne@herne.com
Compuserve: 72060,1153

Heurikon Corp (see Artesyn Technologies)

Hewlett-Packard ··(301) 670-4300
Building 51, PO Box 58059
Santa Clara, CA 95051-8059
Tech:(208) 323-2551 Toll Free:(800) 752-0900
Fax on Demand:(800) 333-1917
Web Page: http://www.hp.com
email: ftp-boi.external.hp.com
Compuserve: GO HP
America Online: keyword "HP"

Communication Semiconductor Div.···········(800) 235-0312
350 W. Trimble Road
San Jose, CA 95131
Web Page: http://www.hp.com/go/ir

DeskJet/DeskWriter··································(208) 344-4131
Toll Free:(877) 283-4684

Disk Memory Division·······························(208) 396-6000
11311 Chinden Blvd
Boise, ID 83714

Tech:(208) 323-2551
Fax:(208) 333-3182
Fax Back...............(208) 344-4809
Toll Free:(800) 333-1917
Handheld Products...............(970) 392-1001
Tech:(970) 635-1000 Toll Free:(800) 443-1254
Information Storage Group...............(970) 635-1500
800 S Taft Avenue
Loveland, CO 80537
Tech:(970) 635-1000
Fax on Demand:(800) 333-1917
Web Page: http://www.hp.com/go/storage_support
email: ftp-boi.external.hp.com
Compuserve: GO HP
America Online: keyword: HP
JetDirect...............(208) 323-2551
Toll Free:(877) 283-4684
Mass Storage Division
Tech:(970) 635-1000 Toll Free:(800) 826-4111
Net Server Products...............(800) 322-4772
Toll Free:(800) 533-1333
Web Page: http://www.hp.com/go/netserver
Network Connectivity Products...............(970) 635-1000
Toll Free:(800) 533-1333
Web Page: http://www.hp.com/go/network_city
Omnibook Notebook PC's...............(970) 635-1000
Tech:(970) 346-8682 Toll Free:(800) 443-1254
Web Page: http://www.hp.com/omnibook
Pavilion PC...............(208) 323-4663
Toll Free:(877) 283-4684
Peripheral Group...............(208) 323-2551
19091 Pruneridge Avenue, MS #46UL
Cupertino, CA 95014
Personal Computer Products...............(800) 752-0900
Personal Information Products Group
5301 Stevens Creek Boulevard
Santa Clara, CA 95052
Fax on Demand:(800) 333-1917
Web Page: http://www.hp.com
email: ftp-boi.external.hp.com
Compuserve: GO HP
America Online: keyword "HP"
PhotoSmart...............(208) 376-3686
Toll Free:(877) 283-4684
RISC Systems...............(800) 752-0900
Software Distribution Services...............(303) 739-4009
Greeley, CO 80632

Fax:(303) 739-4143
Unix Workstation & Server·····(800) 637-7740
Vectra Desktop PC's·····(800) 322-4772
Web Page: http://www.hp.com/go/vectra
Windows Client·····(800) 752-0900

Hi-Image ·····(408) 232-8726
1820 Gateway Drive, Suite 370
San Mateo, CA 94404
Toll Free:(800) 345-3540
Web Page: http://www.hi-image.com
Compuserve: GO IMAGEIN; Library 13, Image A
BBS:(415) 358-9795

Hi-Val ·····(714) 648-0220
Santa Ana, CA
Tech:(714) 648-0220
Fax:(714) 543-0802
Web Page: http://www.hival.com
email: tech@hival.com
BBS:(714) 543-7129

Hilbert Computing ·····(913) 780-5051
13632 S. Sycamore Drive
Olathe, KS 66062
Fax:(913) 829-2450
Web Page: http://www.hilbertinc.com
email: postmaster@hilbertinc.com
Compuserve: 73457,365
BBS:(913) 829-2450

Hilgraeve Inc ·····(734) 243-0576
111 Conant Ave, Suite A
Monroe, MI 48161
Tech:(734) 243-0576 Toll Free:(800) 826-2760
Fax:(734) 243-0645
Web Page: http://www.hilgraeve.com
email: support@hilgraeve.com
Compuserve: GO HILGRAEVE
BBS:(734) 243-5915

Hitachi America ·····(914) 332-5800
50 Prospect Ave
Tarrytown, NY 10591
Tech:(800) 448-2244 Toll Free:(800) 448-2244
Fax:(914) 332-5555 Fax on Demand:(800) 448-3291
Web Page: http://www.hitachi.com
Computer Div/Storage Products Group·····(650) 589-8300
2000 Sierra Point Parkway
Brisbane, CA 94005-1835

Toll Free:(800) 448-2244
Fax:(650) 583-4207 Fax on Demand:(800) 448-3291
Web Page: http://www.hitachi.com

Home Electronics ········· (770) 279-5600
3890 Steve Reynolds Blvd
Norcross, GA 30093
Tech:(800) 241-6558 Toll Free:(800) 323-9712
Other Address: http://www.mpegcam.com
Web Page: http://www.hitachi.com

Hitachi PC Corp ········· (408) 546-8000
1565 Barber Lane
Milpitas, CA 95035
Tech:(800) 555-6820 Toll Free:(800) 943-7607
Fax:(800) 555-4625 Fax on Demand:(800) 555-9621
Fax:(408) 546-8218
Web Page: http://www.hitachipc.com
BBS:(408) 546-8173

HockWare Inc ········· (919) 380-0616
315 N. Academy Street, PO Box 336
Cary, NC 27512-0336
Tech:(919) 380-0616
Fax:(919) 380-0757
Web Page: http://www.vispro.com
email: support@vispro.com
Compuserve: 71333,3226

Hollywood Interactive Digital Ent. ········· (818) 897-2020
12420 Montague St, Suite B
Arleta, CA 91331
Tech:(818) 897-2020 Toll Free:(800) 423-7779
Fax:(818) 897-1878
email: hide@aol.com

Hopkins Technology ········· (612) 931-9376
421 Hazel Lane
Hopkins, MN 55343-7116
Tech:(800) 397-9211 Toll Free:(800) 397-9211
Fax:(612) 931-9377

Horizons Technology Inc ········· (619) 292-8331
3990 Ruffin Road
San Diego, CA 92123-1826
Tech:(619) 292-8320 Toll Free:(800) 828-3808
Fax:(619) 292-9439
Another Fax:(619) 292-7321
Web Page: http://www.horizons.com
BBS:(619) 268-0380

Hot Wire Data Security Inc ························(610) 435-7700
1227 Liberty Street
Allentown, PA 18102
Toll Free:(888) 468-9473
Fax:(610) 435-6449
Web Page: http://www.hotwire.net
email: hotwire@hotwire.net

Houghton Mifflin Interactive ······················(617) 503-4800
120 Beacon St
Somerville, MA 02143
Toll Free:(800) 210-0241
Fax:(617) 503-4900
Web Page: http://www.hminet.com
email: hmi@hmco.com

Houston Instrument (see Summagraphics)

Howard W. Sams ·······································(317) 298-5400
2647 Waterfront Pkwy E Drive
Indianapolis, IN 46214
Toll Free:(800) 428-7267
Fax:(317) 298-5604
Web Page: http://www.a1.com/hwsams
email: csmgr@in.net

Howling Dog Systems Inc ························(613) 376-3868
RR1 3833 Daley Road
Sydenham, ON K0H 2T0 Canada
Tech:(613) 376-3868 Toll Free:(800) 267-4695
Fax:(613) 376-3584
Web Page: http://www.howlingdog.com
email: howling.dog@howlingdog.com
Compuserve: 71333,2166 or GO HOWLING

HP (see Hewlett-Packard)

HPS Simulations···(408) 554-8381
PO Box 3245
Santa Clara, CA 95055-3245
Tech:(408) 554-8381
Fax:(408) 241-6886
Web Page: http://www.hpssims.com
email: hpssims@compuserve.com
Compuserve: 74774,771
BBS:(770) 474-7304

HSC Software (see Metacreations)

Hughes Network Systems ·························(301) 428-5500
11717 Exploration Lane

Germantown, MD 20876
Tech:(301) 428-5500
Fax:(301) 428-1868
Web Page: http://www.hns.com
email: info@hns.com

HumanCAD Systems Inc ....(905) 761-7681
210-3100 Steeles Ave West
Concord, ON L4K 3R1
Tech:(516) 752-3507 Toll Free:(800) 248-3746
Fax:(905) 761-7682
Other:(888) 730-3746
Web Page: http://www.mqpro.com
email: support@humancad.com
Compuserve: 74103,2413@compuserv.com

Hummingbird Communications Ltd ....(416) 496-2200
1 Sparks Avenue
North York, ON M2H 2W1 Canada
Tech:(416) 496-2200
Fax:(416) 496-2207
Web Page: http://www.hummingbird.com
email: support@hummingbird.com
BBS:(416) 496-9233

HyperGlot Software ....(615) 558-8270
314 Erins Dr
Knoxville, TN 37919
Tech:(800) 726-5087 Toll Free:(800) 726-5087

Hyperion Software...................................(203) 703-3000
900 Long Ridge Road
Stamford, CT 06902
Tech:(203) 703-3000
Fax:(203) 329-4541
Web Page: http://www.hysoft.com
email: info@hyperion.com

Hyundai Electronics America.................(408) 232-8000
3101 North First Street
San Jose, CA 95134
Tech:(408) 232-8191 Toll Free:(800) 568-0060
Fax:(408) 232-8121 Fax on Demand:(800) 501-4986
Web Page: http://www.hea.com

I/O Software Inc ....(909) 222-7600
1533 Spruce St
Riverside, CA 92507
Fax:(909) 222-7601
Web Page: http://www.iosoftware.com

email: support@iosoftware.com

IBC ....(805) 527-8792
2685C Park Center Dr
Simi Valley, CA 93065
Fax:(805) 527-6362
Web Page: http://www.ibc.com
email: webmaster@IBC.com

Ibex Technologies (see Castelle Ibex Tech)

IBM Corporation ....(800) 426-3333
11400 Burnet Road
Austin, TX 78758
Tech:(800) 772-2227 Toll Free:(800) 456-0550
Web Page: http://www.ibm.com
BBS:(919) 517-0001

IBM Corporation ....(404) 238-7000
4111 Northside Pkwy
Atlanta, GA 30327
Web Page: http://www.ibm.com
BBS:(919) 517-0001

AIX Systems Support Center ....(800) 237-5511
Tech:(817) 962-7379
Fax:(512) 823-7634
Ambra Technical Support ....(800) 363-0066
Anti-Virus Services ....(541) 465-8520
ARTIC Technical Support ....(800) 237-5511
Authorized Dealer Locator ....(800) 447-4700
Automated Fax System ....(800) 426-3395
Bulletin Board System ....(919) 517-0001
Canada BBS:(905) 316-4244
CAD Assistance ....(800) 964-6432
Catalog Solutions Center ....(800) 426-2255
Customer Support Center ....(800) 967-7882
Desktop Software Support Hotline ....(800) 336-5430
Direct (Supplies, Orders, Price Info) ....(800) 426-2468
DisplayWrite End-User Support ....(800) 336-5430
Easy Options Technical Support ....(800) 933-7573
End User Support ....(800) 772-2227
General Info ....(800) 426-3333
IBM Teach ....(800) 426-8322
Canada:(800) 661-2131
Web Page: http://www.can.ibm.com/edu
Independence Series Info, TDD ....(800) 426-4833
Independence Series Product Info ....(800) 426-4832
Industrial PC Support Line ....(800) 526-6602

Information Center .... (914) 288-3000
Old Orchard Road
Armonk, NY 10504
Fax on Demand:(800) 426-4329
Web Page: http://www.ibm.com
IPDS Tech Support, Developer .... (800) 553-1623
IPDS Tech Support, End Users .... (800) 241-1620
ISSS Tech Support, Developer .... (800) 553-1623
ISSS Tech Support, End users .... (800) 241-1620
Link Customer Support Center .... (800) 543-3912
Maintenance Agreements Dept .... (800) 624-6875
Materials Safety Information .... (800) 426-4333
Microelectronics(PowerPC,Blue Lgtng) .... (800) 426-0181
Multi-Media Technical Support .... (800) 241-1620
NSD Hardware Service PC Repair .... (800) 426-7378
OEM Division .... (520) 574-4600
1133 Westchester Ave
White Plains, NY 10604
Toll Free:(800) 426-3333
email: ibm_direct@vnet.ibm.com
OS/2 Hardware Testing/Certification .... (407) 443-4014
OS/2 ServicePak Defect/Missing Disk .... (800) 897-2755
OS/2 Support BBS (Montreal, Canada) .... (514) 938-3022
BBS:(514) 938-3022
OS/2 Support BBS (Toronto, Canada) .... (416) 492-1823
BBS:(416) 492-1823
OS/2 Support BBS (Toronto/Markham) .... (416) 946-4255
BBS:(416) 946-4255
OS/2 Support BBS (Vancouver, Can.) .... (604) 664-6466
BBS:(604) 664-6466
OS/2 Support Center .... (800) 992-4777
PartnerLink (CSS/RICS) Dealer Supt .... (800) 426-3325
Parts Order Center .... (800) 388-7080
PC Company Bulletin Board System .... (919) 517-0001
BBS:(919) 517-0001
PC Company Product Info Faxback .... (800) 426-4329
Fax on Demand:(800) 426-4329
PC Company Tech Support Faxback .... (800) 426-3395
Fax on Demand:(800) 426-3395
PC Help Center .... (800) 772-2227
PenAssist Developer's Program .... (404) 238-2200
Personal Software Solutions Ctr .... (800) 992-4777
Personal Systems Card Repair Service .... (800) 759-6995
Personal Systems HelpCenter .... (800) 772-2227
Toll Free:(800) 772-2227
Fax:(800) 426-3395
BBS:(919) 517-0001

Platinum Accounting Software Support ·······(800) 333-5242
PS/1 Bulletin Board System ·······(404) 835-8230
PS/1 Dealer Locator ·······(800) 426-3377
PS/1 Help Line ·······(800) 765-4747
PSP Support Center ·······(800) 992-4777
RISC System/6000 ·······(800) 426-7378
Software Manufacturing & Delivery Ctr ·······(800) 879-2755
Software Support Line ·······(800) 237-5511
Storage Systems Division ·······(507) 286-4200

3605 Highway 52N
Rochester, MN 55901
Tech:(507) 253-4110
Fax:(507) 253-4111 Fax on Demand:(415) 903-0955
email: hddtech@vnet.ibm.com
BBS:(507) 253-4112

Supplies Technical Hotline ·······(800) 426-1484
Systems Storage Division (Ad/Star) ·······(408) 284-6039
SystemXtra for Personal Systems ·······(800) 547-1283
ThinkPad Helpdesk for NBA Coaches ·······(800) 622-8465
VoiceType Inquires (Dragon Systems) ·······(800) 825-5897
VoiceType Tech Support, End Users ·······(800) 241-1620

IBM Desktop Software (Talklink Info) ·····(800) 547-1283

IBM PC Company ·······(800) 772-2227

3039 Cornwallis Drive, Building 203
Research Triangle Park, NC 27709
Tech:(800) 426-7378 Toll Free:(800) 772-2227
Fax on Demand:(800) 426-3395

IC Systems (see IC Verify)

IC Verify ·······(510) 553-7500

473 Roland Way
Oakland, CA 94621-2014
Tech:(800) 811-1371 Toll Free:(800) 900-6133
Fax:(510) 553-7553 Fax on Demand:(510) 553-7555
Web Page: http://www.icverify.com
email: support@icverify.com
Compuserve: GO ICVERIFY
BBS:(510) 553-7554

ICA (Int Communications Assoc) ·······(972) 620-7020

2735 Villa Creek Drive, #200
Dallas, TX 75234
Toll Free:(800) 422-4636
Fax:(972) 488-9985
Web Page: http://www.icanet.com
email: intlcoma@onramp.net

Iceberg Software LLC .... **(703) 435-3427**
11654 Plaza America, Suite 187
Reston, VA 20190
Fax:(703) 435-9049
Web Page: http://www.icebergsoftware.com
email: Help@IcebergSoftware.com

ICG Netcom .... **(408) 881-5000**
2 North Second Street, Plaza A
San Jose, CA 95113
Tech:(408) 881-1810 Toll Free:(888) 316-1122
Fax:(408) 325-6479 Fax on Demand:(800) 638-6383
Express Support:(800) 638-6383
Web Page: http://www.netcom.com
email: businessinfo@netcom.com

Icon CMT Corp .... **(201) 601-2000**
1200 Harbor Blvd, 8th Floor
Weehawken, NJ 07087
Tech:(800) 275-4266 Toll Free:(800) 572-4266
Fax:(201) 601-2018
Web Page: http://www.icon.com
email: contact@icon.com

Iconovex Corp .... **(612) 930-4675**
7900 Xerxes Ave S., Suite 550
Bloomington, MN 55431
Tech:(612) 930-4675
Fax:(612) 896-5101
Web Page: http://www.iconovex.com
email: fyi@iconovex.com
Compuserve: 74064,440

IDG Books Worldwide Inc .... **(416) 293-8464**
919 E. Hillsdale Blvd, Suite 400
Foster City, CA 94404-2112
Tech:(800) 434-3422 Toll Free:(800) 762-2974
Canada Toll Free:(800) 667-1115
Web Page: http://www.idgbooks.com

iDream Software LLC .... **(425) 486-3646**
18939 120th Ave NE, Ste. 111
Bothell, WA 98011
Web Page: http://www.i-dream.com

IDT .... **(201) 928-1000**
294 State St
Hackensack, NJ 07601
Toll Free:(800) 573-9438
Fax:(201) 928-1057

Web Page: http://www.idt.net

IEC Electronics Corp. .... **(205) 931-8000**
350 11th Street S.W.
Arab, AL 35016
Fax:(205) 931-8251
Web Page: http://www.iec-electronics.com
email: avery@accutek-iec.com

IEEE Computer Society .... **(714) 821-8380**
10662 Los Vaqueros Circle, PO Box 3014
Los Alamitos, CA 90720-1264
Toll Free:(800) 272-6657
Fax:(714) 821-4010
Web Page: http://www.computer.org
email: technical@computer.org

Ikon .... **(602) 266-9029**
2700 N. Central, Suite 900
Phoenix, AZ 85004
Toll Free:(800) 264-9029
Fax:(602) 266-6252
Web Page: http://www.midak.com
email: info@midak.com

Ikon Office Solutions .... **(520) 577-2661**
1010 N. Finance Center Dr
Tucson, AZ 85710
Tech:(520) 918-4045 Toll Free:(800) 916-2075
Fax:(520) 918-4502
Web Page: http://www.ikon.net
email: support@ikon.net

Illinois Lock Co .... **(847) 537-1800**
301 West Hintz Road
Wheeling, IL 60090-5754
Tech:(800) 733-3907 Toll Free:(800) 733-3907
Fax:(847) 537-1881
Web Page: http://www.illinoislock.com
email: illock@aol.com

Illustra Information Technologies .... **(510) 652-8000**
1111 Broadway, Suite 2000
Oakland, CA 94607
Tech:(510) 652-8000
Fax:(510) 869-6388 Fax on Demand:(510) 873-6299
Web Page: http://www.illustra.com

Image Club Graphics Inc .... **(800) 387-9193**
833 Fourth Avenue Southwest, Suite 800
Calgary, AB T2P 3T5 Canada

Tech:(403) 262-8008 Toll Free:(800) 387-9193
Fax:(403) 261-7013 Fax on Demand:(206) 628-5737
Tech Fax:(800) 814-7783
Other Address: Apple: CDA0573
Web Page: http://www.imageclub.com
email: support@adobestudios.com
Compuserve: 72560,2323
America Online: imageclub@aol

Image Control Corp ....(416) 694-7509
1396 Kingston Road
Toronto, ON M1N 1R3 Canada
Tech:(416) 694-7747
Fax:(416) 694-7929
Web Page: http://www.image-control.com

Image Smith ....(408) 460-9155
Santa Cruz, CA 95061-1724
Tech:(408) 457-0854 Toll Free:(800) 876-6679
Fax:(408) 460-9154
Web Page: http://www.imagesmith.com
email: image@imagesmith.com

Image-In (see Hi Image)

Imageline Inc ....(804) 798-8156
1051 E. Cary Street, Suite 204
Richmond, VA 23219
Tech:(804) 798-8156
Fax:(804) 644-0769
Web Page: http://www.imagelineinc.com
email: info@imagelineinc.com

Imagine Media ....(415) 468-4684
150 N. Hill Drive
Brisbane, CA 94005
Tech:(415) 468-4684
Fax:(415) 468-4686
Other Address: http://www.imaginemedia.com
Web Page: http://www.imagine-inc.com

Imagine Publishing Inc (see Imagine Media)

IMAJA ....(510) 526-4621
PO Box 6386
Albany, CA 94706
Fax:(510) 559-9571
Web Page: http://www.imaja.com
email: software@imaja.com
Compuserve: 72767,1101
America Online: IMAJA

Imation Corp .... **(888) 466-3456**
1 Imation Pl
Oakdale, MN 55128-3414
Fax:(650) 596-4434
Web Page: http://www.imation.com

IMC Networks Corp .... **(949) 724-1070**
16931 Millikan Avenue
Irvine, CA 92606
Tech:(800) 624-1070 Toll Free:(800) 624-1070
Fax:(949) 724-1020
Web Page: http://www.imcnetworks.com
email: techsupport@imcnetworks.com

Impediment Inc .... **(781) 834-3800**
541 Plain Street
Marshfield, MA 02050-2713
Fax:(781) 834-3666
Web Page: http://www.impediment.com
email: sales@impediment.com

IMSI Software .... **(415) 257-3000**
1895 Fransisco Blvd East
San Rafael, CA 94901-5506
Tech:(415) 257-3000 Toll Free:(800) 833-8082
Fax:(415) 257-3565
Web Page: http://www.imsisoft.com
BBS:(415) 257-8468

IMT Systems Inc .... **(713) 690-2990**
7240 Brittmoore Rd, Suite 106
Houston, TX 77041
Web Page: http://mfginfo.com/comp/

In Focus Systems Inc .... **(503) 685-8888**
27700B SW Parkway Ave
Wilsonville, OR 97070-9215
Tech:(800) 799-9911 Toll Free:(800) 294-6400
Fax:(503) 685-8887
After Hours Support:(888) 592-6800
Web Page: http://www.infs.com
email: webmaster@infocus.com

InContext Systems Inc .... **(905) 819-1173**
6733 Mississauga Road, 7th Floor
Mississauga, ON L5N 6J5 Canada
Tech:(905) 819-1173 Toll Free:(888) 819-2500
Fax:(905) 819-9245
Web Page: http://www.incontext.com
email: support@incontext.com

Indiana Cash Drawer Co ....(317) 398-6643
1315 S. Miller St, PO Box 236
Shelbyville, IN 46176
Tech:(800) 227-4832 Toll Free:(800) 227-4379
Fax:(317) 392-0958
Tech Fax:(317) 392-6726
Web Page: http://www.icdpos.com
email: cowens@icdpos.com

Individual Software Inc ....(510) 734-6767
4255 Hopyard Rd, #2
Pleasanton, CA 94588-9900
Tech:(800) 331-3313 Toll Free:(800) 822-3522
Fax:(510) 734-8337
Web Page: http://www.individualsoftware.com
email: techsupport@individualsoftware.com

Inference Corp, The ....(415) 893-7200
100 Rowland Way
Novato, CA 94945
Toll Free:(800) 332-9923
Fax:(415) 899-9080
Web Page: http://www.inference.com
email: support@inference.com

Infinite Technologies ....(410) 363-1097
11433 Cronridge Drive
Owings Mills, MD 21117
Tech:(410) 363-9453 Toll Free:(800) 678-1097
Fax:(410) 363-0846
Fax:(410) 363-3779
Other Address: http://www.ilink.com
Web Page: http://www.ihub.com
email: support@infiniteMail.com
Compuserve: library@infinite

InfiniText Software ....(714) 651-0640
105 Eastshore
Irvine, CA 92714
Fax:(714) 651-0640

Info Access Inc ....(425) 201-1915
15821 NE 8th Street
Bellevue, WA 98008
Tech:(425) 201-1916 Toll Free:(800) 344-9737
Fax:(425) 201-1922
Web Page: http://www.infoaccess.com
email: techsupp@infoaccess.com

InfoGold American Multisystems ....(408) 945-2296
1830 Houret Court
Milpitas, CA 95035
Fax:(408) 945-2299
Web Page: http://www.infogold.com
email: infogold@infogold.com

InfoMagic Inc ....(520) 526-9565
11950 N. Hwy 89, PO Box 30370
Flagstaff, AZ 86004
Tech:(520) 526-9565 Toll Free:(800) 800-6613
Fax:(520) 526-9573
Web Page: http://www.infomagic.com
email: support@infomagic.com

Infonet Communications Inc ....(209) 435-1561
2109 W. Bullard, Suite 145
Fresno, CA 93711

Information Builders Inc ....(212) 736-4433
Two Penn Plaza
New York, NY 10121-2898
Tech:(800) 736-6130 Toll Free:(800) 969-4636
Fax:(212) 967-6406
Customer Support:(800) 736-6130
Web Page: http://www.ibi.com
email: askinfo@ibi.com

Information Cybernetics (see General Inter.)

Informative Graphics Corp ....(602) 971-6061
706 East Bell Road, Suite 207
Phoenix, AZ 85022
Tech:(602) 971-6061
Fax:(602) 971-1714
Web Page: http://www.infograph.com
email: info@infograph.com
Compuserve: 72662,3001 or GO IGCORP

Informix Software Inc ....(650) 926-6300
4100 Bohannon Drive
Menlo Park, CA 94025
Tech:(800) 274-8184 Toll Free:(800) 331-1763
Web Page: http://www.informix.com

Inforonics Inc ....(978) 486-8976
550 Newtown Road, PO Box 458
Littleton, MA 01460-0458
Fax:(978) 486-0027
Web Page: http://www.infor.com
email: info@inforonics.com

InfoVision Technologies Inc ···················· **(508) 366-3660**
PO Box 989
Westboro, MA 01581
Tech:(508) 366-3660
Fax:(508) 366-2544
Web Page: http://www.ivtinc.com
email: info@ivtinc.com

Infoworld ·· **(650) 572-7341**
155 Bovet Road
San Mateo, CA 94402
Toll Free:(800) 227-8365
Fax:(650) 312-0580
Web Page: http://www.infoworld.com/lionsden

Ingram Micro ······································ **(714) 566-1000**
1600 E. St Andrew Place, PO Box 25125
Santa Ana, CA 92799-5125
Tech:(800) 445-5066 Toll Free:(800) 274-4800
Fax:(714) 566-7720 Fax on Demand:(714) 566-1900
Orders:(800) 456-8000
Web Page: http://www.ingrammicro.com

Initio Corp ·· **(408) 577-1919**
2188-B Del Franco St
San Jose, CA 95131-1575
Tech:(408) 577-1919
Fax:(408) 577-0640
Web Page: http://www.initio.com
email: support@initio.com
BBS:(408) 577-0431

Inline Inc ·· **(714) 921-4100**
22860 Savi Ranch Parkway
Yorba Linda, CA 92887
Tech:(800) 882-7117 Toll Free:(800) 882-7117
Fax:(714) 921-4160
Web Page: http://www.inlineinc.com
email: inlinemail@aol.com

Inmagic Inc ······································· **(781) 938-4442**
800 W. Cummings Park
Woburn, MA 01801-6357
Tech:(781) 938-4442 Toll Free:(800) 229-8398
Fax:(781) 938-6393
Web Page: http://www.inmagic.com
email: inmagic@inmagic.com
Compuserve: 71333,105

Inmark Development **(see Rogue Wave Software)**

Innovative Electronics Corp (IEC) ....**(303) 288-5000**
6185 E. 56th Ave, Ste A
Commerce City, CO 80022-3927
Toll Free:(800) 765-4432
Fax:(303) 288-5099
Web Page: http://www.iec.net
email: info@iec.net

Innovative Quality Software ....**(702) 435-9077**
4680 S Eastern Avenue, Suite D
Las Vegas, NV 89119-6192
Tech:(702) 435-9077 Toll Free:(800) 844-1554
Fax:(702) 435-9106
Web Page: http://www.iqsoft.com
email: tech@iqsoft.com
Compuserve: GO IQSSUPPORT

Inprise Corp ....**(408) 431-1000**
100 Enterprise Way
Scotts Valley, CA 95066-3249
Tech:(800) 782-5558 Toll Free:(800) 331-0877
Fax on Demand:(800) 822-4269
Tech Support:(800) 446-3565
Other Address: BIX:JOIN BORLAND
Web Page: http://www.inprise.com
Compuserve: GO BORLAND

Assist ....(408) 431-1064
Toll Free:(800) 523-7070
C++ Installation...(408) 461-9133
Customer Service.......................................(800) 331-0877
Scotts Valley, CA 95067-0001
Tech:(800) 437-8884
Fax:(408) 431-4122
credit card orders:(800) 932-9994
email: customer-service@borland.com
D-Base Installation....................................(408) 461-9110
DataGateway Installation............................(408) 461-9123
Delphi Installation ....(408) 461-9195
Installation Assist......................................(408) 461-9144
Interbase Software Corp ....(888) 345-2015
JBuilder..(408) 461-9144
Per Incident Support ....(888) 456-2003
Tech:(800) 446-3565 Toll Free:(888) 683-2378
ReportSmith Installation.............................(408) 461-9150

Inset Systems Inc (see Quarterdeck Office)

Insight Development Corp.......................**(510) 244-2000**
2420 Camino Ramone, Suite 202

San Ramone, CA 94583
Tech:(510) 244-2000 Toll Free:(800) 825-4115
Fax:(510) 244-2020
Other Address: http://www.hotofftheweb.com
Web Page: http://www.insightdev.com
email: info@insightdev.com

Insight Software Solutions ....(801) 295-1890
Bountiful, UT 84011-0354
Fax:(801) 299-1781
Web Page: http://www.smartcode.com/iss
email: info@wintools.com

Insignia Solutions ....(510) 360-3700
41300 Christy St
Fremont, CA 94538-3115
Tech:(408) 327-6500 Toll Free:(800) 848-7677
Fax:(408) 327-6105 Fax on Demand:(800) 876-3872
Web Page: http://www.insignia.com
email: unixtech@isinc.insignia.com

INSO Corporation ....(312) 692-5100
330 N. Wabash, 15th Floor
Chicago, IL 60611
Tech:(312) 527-4357 Toll Free:(800) 333-1395
Fax:(312) 670-0820
Web Page: http://www.inso.com
email: supportchi@inso.com

Inspiration Software ....(503) 297-3004
7412 SW Beaverton Hillsdale Hwy, Suite 102
Portland, OR 97225-2167
Tech:(800) 877-4292 Toll Free:(800) 877-4292
Fax:(503) 297-4676
Web Page: http://www.inspiration.com

Int'l Electronic Research (IERC) ....(818) 842-7277
135 W. Magnolia Blvd
Burbank, CA 91502
Fax:(818) 848-8872
Web Page: http://www.iercdya.com/index.html

Integral Peripherals Inc ....(303) 449-8009
5775 Flatiron Parkway, Suite 100
Boulder, CO 80301
Tech:(303) 449-8009 Toll Free:(800) 333-8009
Fax:(303) 449-8089
Web Page: http://www.integralnet.com
email: klowe@integralnet.com

Integrated Data Systems Inc ....(912) 236-4374
6001 Chatham Center Dr, Suite 300
Savannah, GA 31405
Tech:(912) 236-4374
Fax:(912) 236-6792
Web Page: http://www.ids-net.com
email: techinfo@ids-net.com

Integrated Electronics Corp ....(303) 292-5537
420 E. 58th Ave
Denver, CO 80216
Fax:(303) 292-0114

Integrated Info. Tech (see 8 X 8)

Integrated Systems Inc ....(408) 542-1500
201 Moffett Park Drive
Sunnyvale, CA 94089
Tech:(800) 458-7767 Toll Free:(800) 543-7767
Fax:(408) 542-1950
Other:(800) 770-3338
Web Page: http://www.isi.com
email: psos-support@isi.com

Intel Corp ....(408) 765-8080
2200 Mission College Blvd
Santa Clara, CA 95052-8119
Tech:(800) 321-4044 Toll Free:(800) 238-0486
Fax:(408) 765-9904 Fax on Demand:(800) 525-3019
Web Page: http://www.intel.com
BBS:(503) 264-7999

PC Enhancement Division ....(503) 696-8080
2111 NE 25th Street
Hillsboro, OR 97124
Tech:(503) 264-7000 Toll Free:(800) 538-3373
Web Page: http://www.intel.com
Compuserve: GO INTEL
BBS:(503) 264-7999

Intelitool Inc ....(630) 406-1041
117 S. Prairie St, PO Box 459
Batavia, IL 60510
Tech:(630) 406-1041 Toll Free:(800) 227-3805
Fax:(630) 406-1079
Web Page: http://www.com/intelitool/index.html
email: support@intelitool.com

Intellicom Inc ....(818) 407-3900
20415 Nordhoff Street
Chatsworth, CA 91311

Tech:(818) 407-3900
Fax:(818) 882-2404
Web Page: http://www.intellicom.com

Interactive Magic Inc .... (919) 461-0722
RTP, NC 27709
Tech:(919) 461-0948
Fax:(919) 461-0723
Web Page: http://www.imagicgames.com

InterCon Systems (see Ascend)

Interex Inc .... (316) 636-5544
8447 E. 35th St. N
Wichita, KS 67226
Toll Free:(800) 513-9744
Web Page: http://www.interex.com
email: tech@interexinc.com

Interface Group, The .... (781) 449-6600
800 First Ave
Needham Heights, MA 02194
Fax:(781) 449-2674
Web Page: http://www.interfacegroup.com

Interface Systems .... (313) 769-5900
5855 Interface Drive
Ann Arbor, MI 48103
Tech:(800) 544-4072 Toll Free:(800) 544-4072
Fax:(313) 769-1047
Web Page: http://www.interfacesystems.com
BBS:(313) 769-1358

Intergraph Corp .... (205) 730-5441
1 Madison Industrial Park
Huntsville, AL 35894-0001
Tech:(800) 633-7248 Toll Free:(800) 345-4856
Fax:(205) 730-9441
Hardware Products:(800) 763-0242
Web Page: http://www.intergraph.com
email: info@intergraph.com

Interleaf Inc .... (781) 290-0710
62 Fourth Avenue
Waltham, MA 02154
Tech:(800) 688-5151 Toll Free:(800) 955-5323
Fax:(781) 290-4943
Web Page: http://www.ileaf.com
email: support@interleaf.com

International Transware .... **(650) 903-2300**
1503 Grant Road, Suite 155
Mountain View, CA 94040
Tech:(650) 903-2300 Toll Free:(800) 999-6387
Fax:(650) 903-9544
Web Page: http://www.transware.com
email: info@transware.com

Internet Factory, The .... **(650) 559-9000**
4546 El Camino Real, Suite 222
Los Altos, CA 94022
Web Page: http://www.ifact.com
email: customer_service@ifact.com

Internet Solutions Group .... **(508) 416-1000**
600 Worcester Road
Framingham, MA 01702
Tech:(508) 879-9000 Toll Free:(800) 828-2608
Fax:(508) 626-8515
Web Page: http://www.microsys.com
email: info@microsys.com
BBS:(508) 875-8009

Internex Information Services .... **(408) 327-2355**
2306 Walsh Ave
Santa Clara, CA 95051
Tech:(408) 327-2200
Fax:(408) 496-5485
Sales:(408) 327-2388
Web Page: http://www.internex.net
email: sales@internex.net
BBS:(408) 496-5480

Interphase Corp .... **(214) 654-5000**
13800 Senlac
Dallas, TX 75234
Toll Free:(800) 327-8638
Fax:(214) 654-5500
Customer Service:(214) 654-5555
Web Page: http://www.iphase.com
email: intouch@iphase.com

Interplay Productions .... **(949) 553-6655**
16815 Von Karman Ave
Irvine, CA 92606
Tech:(949) 553-6678 Toll Free:(800) 468-3775
Fax:(949) 252-2820
Other Address: http://www.ipoem.com
Web Page: http://www.interplay.com
email: support@interplay.com

BBS:(949) 252-2822

Interse Corp (see Microsoft)

Intershop Communications ·····················(415) 229-0100
600townsend St, Top Floor W
San Francisco, CA 94103
Toll Free:(800) 736-5197
Fax:(415) 229-0555
Web Page: http://www.intershop.com

Intersolv ···(301) 838-5000
9420 Key West Ave
Rockville, MD 20850
Tech:(800) 443-1601 Toll Free:(800) 547-4000
Fax:(919) 461-4526
Fax:(301) 838-5064
Web Page: http://www.intersolv.com
email: info@intersolv.com

Intex Solutions Inc ································(781) 449-6222
35 Highland Circle
Needham, MA 02194
Tech:(781) 449-6222
Fax:(781) 444-2318
Web Page: http://www.intex.com
email: desk@intex.com

IntraServer Technology Inc ·····················(508) 429-0425
7 October Hill Road
Holliston, MA 01746
Fax:(508) 429-0430
Web Page: http://www.intraserver.com

Intuit Inc ··(800) 446-8848
Mountain View, CA 94043
Web Page: http://www.intuit.com
Compuserve: GO INTUIT

Intuitive Manufacturing Systems Inc······(425) 821-0740
12006 98th Ave NE, Ste 702
Kirkland, WA 98034

Invisible Software (Out of Business)······(415) 570-5967
Web Page: http://www.invisiblesoft.com

IOMEGA Corp ··(801) 778-1000
1821 West Iomega Way
Roy, UT 84067
Tech:(800) 456-5522 Toll Free:(888) 446-6342
Fax:(801) 778-3460 Fax on Demand:(801) 778-5763
Web Page: http://www.iomega.com

email: info@iomega.com
Compuserve: GO PCVENE or GO MACCVEN
Microsoft: GoWord: Iomega
America Online: keyword:IOMEGA
BBS:(801) 392-9819

Ipswitch Inc ··(781) 676-5700
81 Hartwell Ave
Lexington, MA 02421
Tech:(781) 676-5784
Fax:(781) 676-5710
Web Page: http://www.ipswitch.com
email: support@ipswitch.com

IQ Software Corp ··································(770) 446-8880
3295 River Exchange Drive, Suite 550
Norcross, GA 30092
Tech:(800) 458-0386 Toll Free:(800) 458-0386
Fax:(770) 448-4088
Tech Fax:(770) 446-2481
Other Address: http://www.iqsc.com
Web Page: http://www.iqsoftware.com
email: support@iqsc.com
BBS:(206) 821-5486

IQ Technologies Inc (see Smart Cable)

Irwin Magnetic Systems (see Conner)

ISDN*tek ··(650) 712-3000
PO Box 3000
San Gregorio, CA 94074
Fax:(650) 712-3003
Web Page: http://www.isdntek.com
email: info@isdntek.com

Island Software ··(916) 985-0511
90 Digital Drive
Nevato, CA 94949
Tech:(800) 595-4907 Toll Free:(800) 255-4499
Web Page: http://www.islandsoft.com
email: help@islandsoft.com

Isys/Odyssey Development Inc ··············(303) 689-9998
8775 E. Orchard Road, Suite 811
Greenwood Village, CO 80111
Fax:(303) 689-9997
Web Page: http://www.isysdev.com
email: support@isysdev.com

IT Bridge Int ....(408) 757-7878
2900 Corvin Dr
Santa Clara, CA 95051
Toll Free:(800) 444-7300
Fax:(408) 737-0808
Web Page: http://www.itbridge.com
email: info@itbridge.com

IT Designs USA Inc ....(408) 342-0435
10430 S. DeAnza Blvd, Suite 130
Cupertino, CA 95014
Toll Free:(800) 437-7339
Web Page: http://www.itdesign.ie/
email: info@itdesign.com

ITAC Systems Inc ....(972) 494-3073
3113 Benton Street
Garland, TX 75042
Tech:(800) 533-4822 Toll Free:(800) 533-4822
Fax:(972) 494-4159
Web Page: http://www.mousetrak.com
email: support@mouse-trak.com

ITK Telecommunications ....(978) 441-2181
One Executive Drive
Chelmsford, MA 01824
Tech:(514) 381-7155 Toll Free:(888) 485-4685
Fax:(978) 441-9060
Tech Fax:(603) 647-5414
Web Page: http://www.itk.com
email: support@itk-tele.com
Compuserve: 75300,2170:also-GO TELEBIT
America Online: keyword:telebit TS

ITT Pomona Electronics ....(909) 469-2900
1500 East 9th Street
Pomona, CA 91766-3835
Tech:(909) 469-2900
Fax:(909) 629-3317 Fax on Demand:(800) 444-6785
Web Page: http://www.ittpomona.com

ITU Engineering ....(619) 456-2002
565 Pearl Street, Suite 300
Lajolla, CA 92037
Tech:(619) 456-2002
email: info@coconut.com
BBS:(619) 456-0815

IVI Publishing (see On Health Network)

Ixchange Inc .... **(303) 763-8333**
7550 W. Yale Ave, Suite B-130
Denver, CO 80227
Toll Free:(800) 743-6238
Fax:(303) 763-2783
Web Page: http://www.ixchange-inc.com
email: sales@ixchange-inc.com

Ixla USA Inc .... **(203) 730-8805**
17 Jansen St
Danbury, CT 06810
Fax:(203) 730-8802
Web Page: http://www.ixla.com
email: support@ixla.com

iXMicro .... **(408) 369-8282**
2085 Hamilton Ave, Third Floor
San Jose, CA 95125
Toll Free:(888) 467-8282
Fax:(408) 369-0128
Web Page: http://www.ixmicro.com
email: support@ixmicro.com

J-Mark Computer Corp .... **(909) 305-8800**
13111 Brooks Dr, Suite A
Baldwin Park, CA 91706
Tech:(909) 305-8800
Fax:(909) 305-1168
Web Page: http://www.j-mark.com
email: tech@j-mark.com

J. D. Edwards .... **(303) 488-4000**
One Technology Way
Denver, CO 80237
Tech:(800) 289-2999 Toll Free:(800) 727-5333
Fax:(303) 334-4141
Web Page: http://www.jdedwards.com
email: denver_customer_support@jdedwards.com

J. River Inc .... **(612) 339-2521**
125 North First Street
Minneapolis, MN 55401
Fax:(612) 339-4445
Web Page: http://www.jriver.com
email: support@jriver.com

Jade Computer .... **(310) 370-7474**
18511 Hawthorne Blvd
Torrence, CA 90504
Tech:(800) 421-5500

email: jade-support@jadecomputer.com

Jameco Electronics .... **(650) 592-8097**
1355 Shoreway Road
Belmont, CA 94002
Tech:(650) 592-8097 Toll Free:(800) 831-4242
Fax:(650) 592-2503
Fax:(800) 237-6948
Web Page: http://www.jameco.com
email: tech@jameco.com

Janna Systems Inc .... **(416) 483-7711**
3080 Yonge St, Ste 6020
Toronto, ON M4N 3N1
Toll Free:(800) 268-6107
Fax:(416) 483-3220
Web Page: http://www.janna.com
email: info@janna.com

JASC Software Inc .... **(612) 930-9800**
11011 Smetana Rd
Minnetonka, MN 55343
Tech:(612) 930-9171 Toll Free:(800) 622-2793
Fax:(612) 930-9172
Web Page: http://www.jasc.com
email: techsup@jasc.com
Compuserve: 74774,570

Jazz Multimedia Inc .... **(408) 727-8900**
1040 Richard Ave
Santa Clara, CA 95050
Tech:(408) 727-8900 Toll Free:(888) 568-3676
Fax:(408) 727-9092
Web Page: http://www.jazzmm.com
email: techsupp@JazzMM.com

JDR Microdevices .... **(408) 494-1400**
1850 S. 10th St
San Jose, CA 95112-4108
Tech:(800) 538-5002 Toll Free:(800) 538-5000
Fax:(800) 538-5005
Fax:(408) 494-1420
Web Page: http://www.jdr.com/jdr
email: custserv@jdr.com
BBS:(408) 494-1430

Jensen Tools Inc .... **(602) 968-6231**
7815 S. 46th St
Phoenix, AZ 85044-5399
Tech:(602) 968-6241 Toll Free:(800) 426-1194

Fax:(602) 438-1690 Fax on Demand:(602) 968-6241
Web Page: http://www.jensentools.com
email: techsupport@jensentools.com

JES Hardware Solutions Inc .... **(305) 597-3980**
Toll Free:(800) 482-1866
Fax:(305) 594-4443
Web Page: http://www.jescdrom.com

JETFAX Inc .... **(650) 324-0600**
1378 Willow Road
Menlo Park, CA 94025
Tech:(650) 324-0600 Toll Free:(800) 753-8329
Fax:(650) 326-6003
Web Page: http://www.jetfax.com
email: techsupp@jetfax.com

JetForm Corp .... **(613) 230-3676**
560 Rochester Street
Ottawa, ON K1S 5K2 Canada
Tech:(613) 230-4700 Toll Free:(800) 538-3676
Fax:(613) 751-4804
Web Page: http://www.jetform.com

JIAN .... **(650) 254-5600**
1975 W El Camino Real, Suite 301
Mountain View, CA 94040
Tech:(650) 254-5600 Toll Free:(800) 346-5426
Fax:(650) 254-5640
Web Page: http://www.jianusa.com
email: tech@jianusa.com

JL Chatcom Inc .... **(818) 709-1778**
9600 Topanga Canyon Blvd
Chatsworth, CA 91311
Toll Free:(800) 456-1333
Fax:(818) 882-9134
Web Page: http://www.jlchatcom.com
email: sales@jlchatcom.com

JL Cooper Electronics .... **(310) 322-9990**
142 Arena Street
El Segundo, CA 90245
Tech:(310) 322-9990
Fax:(310) 335-0110
Web Page: http://www.jlcooper.com
email: 75300.1373@compuserve.com
Compuserve: 75300,1373
America Online: JLC Rick@aol.com

Johnson-Grace Co (See America Online)

Joseph Electronics ....(847) 297-4200
8830 N. Milwaukee Ave
Niles, IL 60714
Fax:(847) 297-6923

Jostens Home Learning ....(619) 587-0087
9920 Pacific Heights Blvd, Suite 500
San Diego, CA 92121
Tech:(800) 548-8372 Toll Free:(800) 548-8372
Fax:(619) 587-1629

Jovian Logic Corp ....(510) 651-4823
47929 Fremont Blvd
Fremont, CA 94538
Tech:(510) 651-4823
Fax:(510) 651-1343
Web Page: http://www.jovianlogic.com
email: info@jovianlogic.com
Compuserve: 75300,221

JTS Corporation ....(888) 587-0945
166 Baypointe Parkway
San Jose, CA 95134
Tech:(888) 587-0945
Fax:(408) 468-1619
Web Page: http://www.jtscorp.com
email: info@jtscorp.com

Just Logic Technologies Inc ....(514) 943-3749
PO Box 63050, 40 Commerce Street
Nuns Island, QC H3E 1V6 Canada
Toll Free:(800) 267-6887
Web Page: http://www.justlogic.com
email: sales@justlogic.com

JVC (Victor Company Of Japan Ltd)
17811 Mitchell Avenue
Irvine, CA 92714
Tech:(714) 816-6500
Fax:(714) 261-9690
Web Page: http://www.jvc.com

Kaetron Software ....(281) 298-1500
26119 Oak Ridge Dr, Suite 1024
Spring, TX 77380
Tech:(281) 298-1547 Toll Free:(800) 938-8900
Fax:(281) 298-2520
Web Page: http://www.kaetron.com
email: support@kaetron.com

Kalok Corp (see JTS)

Kalpana (see Cisco Systems)

Kasco Technologies Inc ....(516) 692-6363
220 5th Avenue
New York, NY 10001
Fax:(212) 725-8062

Katz and Associates Inc ....(908) 464-7048
219 South Street, Suite 202
Murray Hill, NJ 07974
Toll Free:(800) 348-3774
Fax:(908) 464-4636
Web Page: http://www.katzassoc.com
email: info@katzassoc.com

KDS USA (Korea Data Systems) ....(714) 379-5599
12300 Edison Way
Garden Grove, CA 92841
Tech:(800) 283-1311 Toll Free:(800) 237-9988
Fax:(714) 379-5591
Fax:(714) 891-2661
Other Address: http://www.orchestra.com
Web Page: http://www.kdsusa.com
email: tech_support@kdsusa.com

Kennsco Inc ....(612) 559-5100
15755 32nd Avenue N.
Plymouth, MN 55447
Fax:(612) 559-5548

Kenpax ....(626) 855-7988
15334 E. Valley Blvd
La Puente, CA 91746
Fax:(626) 855-7980

Kensington Technology Group...............(650) 572-2700
2855 Campus Dr
San Mateo, CA 94403
Tech:(800) 535-4242 Toll Free:(800) 280-8318
Fax:(650) 572-9675
Tech Fax:(610) 231-1039
Other Address: Applelink: Kensington
Web Page: http://www.kensington.com
email: info@kensington.com
Compuserve: 76077,231
America Online: keyword: KENSINGTON

Kent Marsh Ltd (see Citadel Computer)

Kerr Publications....................................(406) 356-2126
PO Box 976

Forsyth, MT 59327
Web Page: http://www.mcn.net/~kerrlaw/
email: kerrlaw@mcn.net

Key Tronic Corp ....(509) 928-8000
Spokane, WA 99214-0687
Tech:(800) 262-6006 Toll Free:(800) 262-6006
Fax:(509) 927-5248 Fax on Demand:(800) 262-6006
Tech Support Fax:(509) 927-5252
Web Page: http://www.keytronic.com
email: info@keytronic.com
BBS:(509) 927-5288

Keyfile Corp ....(603) 883-3800
22 Cotton Road
Nashua, NH 03063
Toll Free:(800) 453-9345
Fax:(603) 889-9259
International Fax:(603) 598-8284
Web Page: http://www.keyfile.com
email: northeast@keyfile.com
BBS:(603) 883-5968

Keystone Learning Systems ....(801) 375-8680
2241 Larsen Pkwy
Provo, UT 84606
Tech:(888) 557-2677 Toll Free:(800) 748-4838
Fax:(801) 373-6872
Web Page: http://www.klscorp.com

Kidasa Software Inc ....(512) 328-0168
1114 Lost Creek Blvd, Suite 300
Austin, TX 78746
Tech:(800) 765-0167 Toll Free:(800) 765-0167
Fax:(512) 328-0247
Web Page: http://www.kidasa.com
email: support@kidasa.com
Compuserve: 76702,1305 or GO WINAPB
America Online: kidasa1

KidSoft L.L.C. ....(408) 255-3434
10275 De Anza Blvd
Cupertino, CA 95014
Tech:(408) 255-1328 Toll Free:(800) 354-6150
Fax:(408) 342-3500
Web Page: http://www.kidsoft.com
email: corpcom@kidsoft.com

Kinetix ....(415) 547-2000
642 Harrison Street

San Francisco, CA 94107
Web Page: http://www.ktx.com

Kingston Technology Corp ....(714) 435-2600
17600 Newhope St
Fountain Valley, CA 92708-4298
Tech:(800) 435-0640 Toll Free:(800) 337-8410
Fax:(714) 438-1820 Fax on Demand:(800) 435-0056
Tech Support Fax:(714) 437-3310
Web Page: http://www.kingston.com
email: tech_support@kingston.com
Compuserve: GO KINGSTON
BBS:(714) 435-2636

Kiss Software Corp ....(714) 979-5477
5000 Birch St, Suite 4000 West Tower
Newport Beach, CA 92660
Tech:(503) 598-8710
Fax:(714) 832-7805
Tech Fax:(503) 598-8516
Web Page: http://www.kissco.com
email: support@kissco.com

KL Group ....(416) 594-1026
260 King Street East
Toronto, ON m5A 1K3 canada
Toll Free:(800) 663-4723
Fax:(416) 594-1919
Web Page: http://www.klg.com
email: info@klg.com

Knowledge Adventure ....(818) 246-4400
1311 Grand Central Avenue
Glendale, CA 91201
Tech:(818) 246-4811 Toll Free:(800) 469-3466
Fax:(818) 246-8412
Tech Fax:(818) 246-5604
Web Page: http://www.adventure.com
email: info@adventure.com

Knowledge Based Systems Inc ....(409) 260-5274
1500 University Drive E., One KBSI Place
College Station, TX 77840
Toll Free:(800) 808-5274
Fax:(409) 260-1965
Web Page: http://www.kbsi.com
email: products@kbsi.com

Knowledge Garden Inc ....(561) 432-9550
3861 Wry Rd

Lake Worth, FL 33467
Fax:(561) 432-9139
Web Page: http://www.kgarden.com
email: support@kgarden.com
Compuserve: 76004,1603

Knowledge Media Inc ....(530) 872-7487
436 Nunneley Road, Suite B
Paradise, CA 95969
Tech:(530) 872-7487 Toll Free:(800) 782-3766
Fax:(530) 872-3826
Web Page: http://www.km-cd.com
email: pbenson@km-cd.com

Knowledge Quest ....(714) 376-8150
301 Forest Ave
Laguna Beach, CA 92651
Tech:(714) 376-8150
Other Address: http://www.kq.com
Web Page: http://www.knowledgequest.com
email: mark.dawson@kq.com

KnowledgePoint Software ....(707) 762-0333
1129 Industrial Avenue
Petaluma, CA 94952
Tech:(707) 762-0333 Toll Free:(800) 727-1133
Fax:(707) 762-0802
Web Page: http://www.knowledgepoint.com
email: kp@knowledgepoint.com

Kodak (see Eastman Kodak Co)

Konami Of America Inc ....(847) 215-5100
900 Deerfield Parkway
Buffalo Grove, IL 60089-4510
Web Page: http://www.konami.com
email: cmunoz@konami.com

Konexx Unlimited Systems Corp ....(619) 622-1400
5550 Oberlin Dr
San Diego, CA 92121
Toll Free:(800) 275-6354
Fax:(619) 550-7330
Web Page: http://www.konexx.com
email: support@konexx.com

Korenthal Associates Inc ....(212) 242-1790
511 Avenue Of The Americas, Suite 400
New York, NY 10011
Tech:(212) 242-1790 Toll Free:(800) 527-7647
Fax:(212) 242-2599

Compuserve: GO KORENTHAL

KorTeam International Inc ········ **(408) 733-7888**
777 Palomar Ave
Sunnyvale, CA 94086
Tech:(408) 523-4757 Toll Free:(800) 763-1688
Fax:(408) 733-9888
Web Page: http://www.korteam.com
email: korteam@korteam.com

Koss Corp ········ **(414) 964-5000**
4129 N. Port Washington Ave
Milwaukee, WI 53212
Tech:(800) 558-8305 Toll Free:(800) 872-5677
Fax:(414) 964-8615
Canada:(800) 263-9607
Web Page: http://www.koss.com

Kurta ········ **(301) 572-2555**
12210 Plum Orchard Dr
Silver Spring, MD 20904
Tech:(301) 572-2555
Fax:(301) 572-2510
Web Page: http://www.kurta.com
email: kurta@clark.net

Kyocera Electronics Inc ········ **(732) 560-3400**
2301-300 Cottontail Lane, PO Box 6727
Somerset, NJ 08873
Tech:(732) 560-3400 Toll Free:(800) 232-6797
Fax:(732) 560-8380 Fax on Demand:(800) 459-6329
Web Page: http://www.kyocera.com
email: techsupport@kyocera.com

LA Computer ········ **(310) 533-7177**
2230 Amapola Court, Unit #1
Torrance, CA 90501
Fax:(310) 533-6955
Web Page: http://www.lacomp.com
email: lacom@ix.netcom.com

LAB Tech ········ **(978) 470-0099**
2 Dundee Park
Andover, MA 01810
Tech:(800) 879-5228 Toll Free:(800) 879-5228
Fax:(978) 470-3338
Other:(800) 899-1609
Web Page: http://www.labtech.com
email: info@labtech.com

Labtec Enterprises Inc ································(360) 896-2000
1499 SE Tech Center Place, Suite 350
Vancouver, WA 98683
Tech:(360) 896-2000
Fax:(360) 896-2020
Web Page: http://www.labtec.com
email: feedback@labtec.com

LaCie Limited ··(503) 844-4500
22985 NW Evergreen Parkway
Hillsboro, OR 97124
Tech:(503) 844-4503 Toll Free:(800) 999-1179
Fax:(503) 844-4508
Tech Fax:(503) 844-4501
Web Page: http://www.lacie.com
email: support@lacie.com

LAN Times ··(650) 513-6800
1900 O'Farrell Street, Suite 200
San Mateo, CA 94403
Fax:(650) 513-6985
Web Page: http://www.lantimes.com

LANart Corp ··(781) 444-1994
145 Rosemary Street, Ste D
Needham, MA 02194
Tech:(800) 292-1994 Toll Free:(800) 292-1994
Fax:(781) 444-3692
Web Page: http://www.lanart.com
email: mbrown@lanart.com

Landmark Research (see Quarterdeck Select)

LANshark Systems Inc ··································(614) 751-1111
784 Morrison Road
Colombus, OH 43230
Tech:(614) 751-1111
Fax:(614) 751-1112
Web Page: http://www.lanshark.com
email: support@lanshark.com
Compuserve: 72567,1151
BBS:(614) 751-1113

LANsource Technologies Inc ························(416) 535-3555
221 Duffrin Street, Suite 310A
Toronto, ON M6K 3J2 Canada
Tech:(416) 535-2668 Toll Free:(800) 677-2727
Fax:(416) 535-6225
Web Page: http://www.lansource.com
email: tech_support@lansource.com

Compuserve: GO LANSOURCE
BBS:(416) 535-5878

Lantec ....(801) 375-7050
3549 N. University Ave, Suite 325
Provo, UT 84604
Tech:(800) 352-6832 Toll Free:(800) 352-6832
Fax:(801) 375-7043

Lantronix ....(949) 453-3990
15353 Barranca Parkway
Irvine, CA 92618
Tech:(800) 422-7044 Toll Free:(800) 422-7055
Fax:(949) 453-3995
International tech support:(949) 453-7198
Web Page: http://www.lantronix.com
email: support@lantronix.com
BBS:(949) 367-1051

Lapis Technologies (see Focus Enhancements)

Laser Age (Cartridges USA) ....(888) 866-3787
6905 Oslo Circle, Suite D
Buena Park, CA 90621
Toll Free:(888) 866-3787
Fax:(714) 994-8030
Web Page: http://www.cartridgesusa.com
email: info@cartridgesusa.com

Laser Card Systems Corp ....(650) 969-4428
2644 Bayshore Pkwy
Mountain View, CA 94043
Fax:(650) 967-6524
Web Page: http://www.lasercard.com
email: techsupport@lasercard.com

Laser Mag. Stor.Int.(see Philips Laser Mag)

Laser Master Tech (see Colorspan Corp)

Laser Printer Accessories (see PCPI)

Laser Publishing Group..........................(510) 222-0199
64 Green View Lane
Richmond, CA 94803
Fax:(510) 222-4748
Web Page: http://www.lp-group.com
email: lp-group@lp-group.com

LaserGo Inc...(619) 578-3100
9715 Carroll Center Road, Suite 107
San Diego, CA 92126-6506
Tech:(619) 578-3100

Fax:(619) 578-4502
Web Page: http://www.lasergo.com
email: support@lasergo.com
BBS:(619) 578-3818

LaserMedia·····(416) 977-2001
11 Charlotte St
Toronto, ON M5V 2H5
Tech:(416) 977-2001 Toll Free:(888) 639-0628
Fax:(416) 977-7353
Web Page: http://www.lmcommunications.com
email: techsupport@lasermedia.com

LaserTools Corp·····(510) 704-7430
1250 45th Street, # 340
Emeryville, CA 94608
Tech:(510) 704-7433
Fax:(510) 420-1150
Web Page: http://www.lasertoolscorp.com
email: support@lasertoolscorp.com

Lasonic Electronics Corp·····(626) 281-3957
1827 W. Valley Blvd
Alhambra, CA 91803
Tech:(626) 281-3957
Fax:(626) 576-7314
Web Page: http://www.lasonic.com

Lattice Incorporated·····(630) 769-4060
3020 Woodcreek Drive, Suite D
Downers Grove, IL 60515
Tech:(630) 769-4060 Toll Free:(800) 444-4309
Fax:(630) 769-4083
BBS:(630) 769-4084

Lava Computer Mfg. Inc·····(416) 674-5942
28A Dansk Ct
Rexdale, ON M9W 5V8
Toll Free:(800) 241-5282
Fax:(416) 674-8262
Web Page: http://www.lavalink.com
email: sales@lavalink.com

Lazer Impact·····(512) 832-9151
10435 Burnet Road, Suite 114
Austin, TX 78758
Toll Free:(800) 777-4323
Fax:(512) 832-9321

Lead Technologies Inc·····(704) 332-5532
900 Baxter Street, Suite 103

Charlotte, NC 28204
Tech:(704) 372-9681 Toll Free:(800) 637-4699
Fax:(704) 372-8161
Tech Fax:(704) 332-5868
Web Page: http://www.leadtools.com
email: sales@leadtools.com
Compuserve: GO LEADTECH

Leader Technologies ....(714) 757-1787
4590 MacArthur Blvd, Suite 500
Newport Beach, CA 92660
Tech:(505) 822-0700 Toll Free:(800) 922-1787
Other Address: AppleLink:leadertechnologies
Web Page: http://www.leadertech.com
email: leader@leadertech.com
America Online: leadertechnologies

Learned-Mahn Inc (see Global Payment System)

Learning Company ....(617) 494-1200
1 Athenaeum Street
Cambridge, MA 02142
Tech:(423) 670-2020 Toll Free:(800) 227-5609
Fax:(617) 494-5898 Fax on Demand:(423) 670-2022
Web Page: http://www.softkey.com
email: cust_serv@learningco.com
Compuserve: GO SOFTKEY

Learning Company, The ....(510) 792-2101
9715 Parkside Drive
Knoxville, TN 37922
Tech:(423) 670-2020 Toll Free:(800) 852-2255
Fax:(423) 670-2021 Fax on Demand:(423) 670-2022
Web Page: http://www.learningco.com
email: support@learningco.com
Compuserve: GO SOFTKEY
BBS:(423) 670-2023

Learning Company, The ....(612) 569-1500
6160 Summit Dr N.
Minneapolis, MN 55430-4003
Tech:(423) 670-2020 Toll Free:(800) 685-6322
Fax:(612) 569-1551
Order Fax:(612) 569-1755
Web Page: http://www.mecc.com

LearnIT Corp ....(352) 375-6655
2233 NW 41st Street, Suite 200
Gainesville, FL 32606
Tech:(800) 352-4806 Toll Free:(888) 532-7648

Fax:(352) 376-0022
Fax:(800) 594-8436
Web Page: http://www.learnitcorp.com
email: learnit@learnitcorp.com

LearnKey Inc ....(801) 674-9733
1845 W. Sunset Blvd
St George, UT 84770
Toll Free:(800) 865-0165
Fax:(801) 674-9734
Web Page: http://www.learnkey.com
email: techsupport@learnkey.com

Legacy Storage Systems Corp ....(905) 475-1077
43 Rivera Dr
Markham, ON L3R 5J6
Toll Free:(800) 361-5685
Fax:(905) 475-1088
Web Page: http://www.legacy.ca

Legato Systems ....(650) 812-6000
3210 Porter Drive
Palo Alto, CA 94034
Tech:(650) 812-6100 Toll Free:(888) 853-4286
Fax:(650) 812-6032 Fax on Demand:(650) 812-6156
Web Page: http://www.legato.com
email: service@legato.com

Legi-tech ....(916) 447-1886
1029 J Street, Suite 450
Sacramento, CA 95814
Fax:(916) 447-1109
Web Page: http://www.legitech.com

Lenel Systems International Inc ....(716) 248-9720
290 Woodcliffe Office Park
Fairport, NY 14450
Tech:(716) 248-9720
Fax:(716) 248-9185
Web Page: http://www.lenel.com
Compuserve: 71333,622

Leverage Technologists Inc ....(301) 309-8783
6701 Democracy Blvd, Suite 324
Bethesda, MD 20817
Web Page: http://stout.levtech.com/home.html
email: info@levtech.com

Lexington Technology Inc ....(714) 903-2435
15175 Springdale St
Huntington Beach, CA 92649

Tech:(714) 903-5856 Toll Free:(888) 432-3683
Fax:(714) 903-1316
Web Page: http://www.lexingtontech.com
email: support@lexingtontech.com

Lexmark International Inc ························(606) 232-2000
740 New Circle Rd NW, Bldg 004-1
Lexington, KY 40511
Tech:(800) 453-9872 Toll Free:(800) 258-8575
Fax:(606) 232-5179 Fax on Demand:(800) 453-9329
Electronic Support:(800) 553-9457
Web Page: http://www.lexmark.com
BBS:(606) 232-5238
Customer Support ······························(800) 539-6275
Hardware Service Support ······················(800) 253-9778
Medley ···(800) 236-1751
Notebook Computers ····························(800) 554-3202
Printer Technical Support ······················(606) 232-3000
Toll Free:(800) 358-5835

LG Electronics USA Inc ··························(201) 816-2000
1000 Sylvan Ave
Englewood Cliffs, NJ 07632
Tech:(800) 243-0000 Toll Free:(800) 243-0000
Fax:(201) 816-0636
Web Page: http://www.lgeus.com
email: techsupport@lge.co.kr

Liant Software Corp ·····························(508) 872-8700
354 Waverly St
Framingham, MA 01702
Fax:(508) 626-2221
Web Page: http://www.liant.com
Product Division ······························(800) 349-9222
3006 Long Horn Blvd, Suite 107
Austin, TX 78758
Fax:(512) 345-8010
Web Page: http://www.liant.com
R. M. Division ································(512) 343-1010
8911 N Capitol Of Texas Highway
Austin, TX 78759
Tech:(512) 343-1010 Toll Free:(800) 349-9222
Fax:(512) 343-9487
Web Page: http://www.liant.com/
Software Serv ·································(512) 371-7028
8711 Burnet Road, Building C
Austin, TX 78757
Fax:(512) 371-7609
Web Page: http://www.liant.com

Libra Corp ··(800) 453-3827
4001 South 700 East, Suite 301
Salt Lake City, UT 84107-2177

Lifeboat Assoc (Programmers Paradise) ···············(800) 445-7899
1163 Shrewsbury Ave
Shrewsbury, NJ 07702
Toll Free:(800) 445-7899

Lifestyle Software Group ·························(904) 794-7070
2155 Old Moultrie Road
St. Augustine, FL 32086
Tech:(904) 794-7955 Toll Free:(800) 289-1157
Fax:(904) 825-0223
Web Page: http://www.lifeware.com
email: support@lifeware.com

Light Source Computer Images Inc ········(415) 446-4200
4040 Civic Center Drive, 4th Floor
San Rafael, CA 94903
Tech:(415) 499-9390 Toll Free:(800) 231-7226
Fax:(415) 492-8011 Fax on Demand:(415) 499-1520
Web Page: http://www.ls.com
email: ts@ls.com
Compuserve: 74774,666 of GO LIGHTSOURCE

Lighten Inc ··(510) 528-4376
2124 Kittredge Street, # 775
Berkeley, CA 94704
Tech:(510) 528-4376 Toll Free:(800) 398-4545
Fax:(510) 528-0246
Web Page: http://www.lighten.com
email: info@lighten.com

Likeminds Inc. ···································(415) 284-6965
457 Bryant St.
San Francisco, CA 94107
Tech:(415) 284-0300
Fax:(415) 284-6969
Web Page: http://www.likeminds.com

Lilly Software Associates Inc ··················(603) 926-9696
500 Lafayette Rd
Hampton, NH 03842
Fax:(603) 926-9698
Other Address: http://www.visualmfg.com
Web Page: http://mfginfo.com/cadcam/visual/
email: info@visualmfg.com

Lind Electronics ·································· **(612) 927-6303**
6414 Cambridge Street
Minneapolis, MN 55426
Tech:(800) 697-3702 Toll Free:(800) 659-5956
Fax on Demand:(612) 927-4671
Web Page: http://www.lindelectronics.com
email: lrlind@lindelectronics.com

Linguist's Software ······························ **(425) 775-1130**
Edmunds, WA 98020-0580
Tech:(425) 775-1130
Fax:(425) 771-5911
Web Page: http://www.linguistsoftware.com
email: techsupport@linguistsoftware.com

Link Instruments Inc ····························· **(973) 808-8990**
369 Passaic Ave, # 100
Fairfield, NJ 07004
Fax:(973) 808-8786
Web Page: http://www.linkinstruments.com
email: siteadministrator@LinkInstruments.com

Link Technologies (see Wyse Technology)

Linksys Group Inc ································ **(949) 261-1288**
17401 Armstrong Ave
Irvine, CA 92714
Tech:(949) 261-1288 Toll Free:(800) 546-5797
Fax:(949) 261-8868
Web Page: http://www.linksys.com
email: info@linksys.com
BBS:(714) 261-2888

Lino Color ··· **(888) 546-6265**
425 Oser Ave
Hauppauge, NY 11788
Tech:(703) 903-0254
Web Page: http://www.linocolor.com
email: info@linocolor.com

Lite-On Technology ······························· **(408) 946-4873**
720 South Hillview Drive
Milipitas, CA 95035 USA
Fax:(408) 946-1751

Liuski International Inc ························· **(404) 447-9454**
6585 Crescent Drive
Norcross, GA 30072
Tech:(800) 347-5454 Toll Free:(800) 454-8754
Web Page: http://www.liuski.com
email: techsupt@liuski.com

BBS:(404) 840-1080

Live Picture .... **(408) 558-4301**
910 E. Hamilton Ave, Suite 300
Campbell, CA 95008
Tech:(408) 371-4455 Toll Free:(800) 724-7900
Web Page: http://www.livepicture.com
email: service@livepicture.com

Live Pix Company, The .... **(415) 908-1067**
531 Howard St
San Francisco, CA 94105
Tech:(425) 889-7041 Toll Free:(800) 727-1621
Fax:(415) 908-1058
Web Page: http://www.livepix.com
email: info@livepix.com

LM Soft Inc .... **(514) 948-1000**
1280 Bernard St. W, Ste. 401
Outremont, PQ H2V 1V9
Web Page: http://www.lmsoft.ca

Locus Computing Corp (see Platinum Tech)

Logical Connection Inc (see Buffalo Inc)

Logitech Inc .... **(510) 795-8500**
6505 Kaiser Drive
Fremont, CA 94555
Tech:(702) 269-3457 Toll Free:(800) 231-7717
Fax:(510) 792-8901 Fax on Demand:(800) 245-0000
Web Page: http://www.logitech.com
email: logi_sales@logitech.com

Lotus .... **(617) 577-8500**
55 Cambridge Pkwy
Cambridge, MA 02142
Tech:(978) 988-2500 Toll Free:(800) 343-5414
Fax:(617) 693-4551 Fax on Demand:(800) 346-9508
Web Page: http://www.lotus.com

Lotus Development Corp .... **(770) 391-0011**
55 Cambridge Pkwy
Cambridge, MA 02142
Tech:(800) 553-4270 Toll Free:(800) 346-1305
Other:(800) 343-5414
Web Page: http://www.lotus.com
BBS:(617) 693-7000

Academic .... (800) 343-5414
Tech:(800) 343-5414
Fax on Demand:(800) 346-3508

Business Sales and Service .... (800) 343-5414
Tech:(800) 343-5414
Fax on Demand:(800) 346-3508

CC:Mail .... (415) 961-8800
800 El Camino Real West, 1st Floor
Mountain View, CA 94040
Tech:(800) 448-2500 Toll Free:(800) 448-2500
Fax on Demand:(415) 966-4951
BBS:(415) 691-0401

Notes .... (800) 437-6391
Tech:(800) 346-1305

Passport .... (800) 266-8720
Tech:(800) 553-4270

Word Processing .... (770) 391-0011
1000 Abernathy Rd, Bldg 400, Suite 1700
Atlanta, GA 30328
Tech:(508) 988-2500 Toll Free:(800) 343-5414
Fax:(770) 698-7659
Web Page: http://www.lotus.com

LSI Logic .... (719) 533-7000
4420 Arrows West Dr
Colorado Springs, CO 80907
Tech:(719) 533-7230 Toll Free:(800) 856-3093
Fax:(719) 536-3301
Tech Fax:(719) 533-7271
Web Page: http://www.symbios.com
email: support@symbios.com
BBS:(719) 533-7235

LSI Logic Corp .... (408) 433-8000
1551 McCarthy Blvd
Milpitas, CA 95035
Toll Free:(800) 433-8778
Fax:(408) 433-8989 Fax on Demand:(800) 574-4286
Web Page: http://www.lsilogic.com
email: feedback@lsilogic.com

LucasArts Entertainment .... (415) 507-0400
PO Box 10307
San Rafael, CA 94912
Tech:(415) 507-4545
Fax:(818) 587-6629
Tech Fax:(415) 507-0300
Web Page: http://www.lucasarts.com
email: webjedi@lucasarts.com
Compuserve: GO GAMEPUB
America Online: Lucasart1

Lucent Technologies .... **(908) 582-7472**
600 Mountain Ave
Murray Hill, NJ 07974
Toll Free:(800) 243-7883
Other Address: http://www.lucent-inferno.com
Web Page: http://www.lucent.com

Lucid Corp .... **(972) 480-9600**
PO Box 864527
Plano, TX 75086-4527
Tech:(972) 480-9600
Fax:(972) 480-9610
Order Fax:(972) 480-9610
email: support@lucidcorp.com
Compuserve: 72662,305

Lynx Real-Time Systems Inc .... **(408) 879-3900**
2239 Samaritan Drive
San Jose, CA 95124
Toll Free:(800) 255-5969
Fax:(408) 879-3920
Web Page: http://www.lynx.com
email: sales@lynx.com

Lytec Systems Inc .... **(801) 562-1568**
7050 Union Park Center, Suite 390
Midvale, UT 84047
Tech:(800) 895-6700 Toll Free:(800) 735-1991
Fax:(801) 562-0256
Web Page: http://www.lytec.com
email: products@lytec.com

M-USA Business Systems .... **(972) 386-6100**
15806 Midway Road
Dallas, TX 75244-2195
Tech:(972) 407-9059
Fax:(972) 404-1957
Compuserve: GO PCVENC

MA Laboratories Inc .... **(408) 954-9388**
1972 Concourse Drive
San Jose, CA 95131
Tech:(408) 941-0808
Web Page: http://www.malabs.com
email: webmaster@malabs.com

Mackie Designs Inc .... **(425) 487-4333**
16220 Wood Red Road NE
Woodinville, WA 98072
Tech:(800) 898-3211 Toll Free:(800) 258-6883

Fax:(425) 487-4337
Web Page: http://www.mackie.com
email: mackie@mackie.com
Compuserve: GO MACKIE
America Online: GO MACKIE
BBS:(206) 488-4586

Macmillan New Media (see Elect Press)

Macola Software ....(740) 382-5999
333 E. Center St, PO Box 1824
Marion, OH 43301
Toll Free:(800) 468-0834
Fax:(740) 382-0239
Web Page: http://www.macola.com

Macromedia Inc ....(415) 252-2000
600 Townsend Street, Suite 310W
San Francisco, CA 94103
Tech:(415) 252-9080 Toll Free:(800) 888-9335
Fax:(415) 626-0554
Other:(800) 326-2128
Web Page: http://www.macromedia.com

Madge Networks ....(408) 955-0700
2314 N. 1st Street
San Jose, CA 95131-1011
Tech:(800) 876-2343 Toll Free:(800) 876-2343
Fax:(408) 955-0970
Web Page: http://www.madge.com
Compuserve: go madge

MaeDae Enterprises ....(888) 683-3860
16615 Cathys Loop
Peyton, CO 80831
Tech:(719) 683-3860
Fax:(719) 683-5199
Fax:(719) 683-3549
Web Page: http://www.maedae.com
email: support@maedae.com

MAG InnoVision Inc ....(949) 855-4930
20 Goodyear
Irvine, CA 92618-1813
Toll Free:(800) 827-3998
Fax:(949) 855-4535 Fax on Demand:(714) 751-0166
Web Page: http://www.maginnovision.com

Magee Enterprises Inc ....(770) 446-6611
2909 Langford Road Suite A-600, P O Box 1587
Norcross, GA 30091

Tech:(770) 446-6611
Fax:(770) 368-0719
Web Page: http://www.magee.com
email: support@magee.com

Magic Solutions (see Network Associates)

Magna (see Empower Pro)

Magnavox (see Philips Consumer Electronics)

Magnetic Music
PO Box 1899
Aptos, CA 95003
Fax:(408) 662-8967
Compuserve: 76711,463@compuserve.com

Mailer's Software ....(949) 492-7000
970 Calle Negocio
San Clemente, CA 92673-6201
Tech:(949) 492-3900 Toll Free:(800) 800-6245
Fax:(949) 492-7086
Web Page: http://www.800mail.com
email: tech@800mail.com

Mainstay ....(805) 484-9400
591-A Constitution Ave
Camarillo, CA 93012
Tech:(805) 484-9400 Toll Free:(800) 484-9817
Fax:(805) 484-9428
Web Page: http://www.mstay.com
email: info@mstay.com
Compuserve: 76004,1525
America Online: keyword: Mainstay

Maintenance Troubleshooting ....(302) 738-0532
273 Polly Drummond Road
Newark, DE 19711
Fax:(302) 738-3028 Fax on Demand:(800) 886-0532
Web Page: http://www.maint_troubleshooting.com

Mannesmann Tally ....(425) 251-5500
8301 S. 180th St
Kent, WA 98064-9718
Tech:(425) 251-5532 Toll Free:(800) 843-1347
Fax:(425) 251-5520 Fax on Demand:(425) 251-5511
Web Page: http://www.tally.com
email: support@tally.com

Mansfield Software Group Inc.................(860) 429-8402
PO Box 532
Storrs Mansfield, CT 06268

Tech:(860) 429-8402
Fax:(860) 487-1185
Web Page: http://www.kedit.com
email: support@kedit.com
Compuserve: GO PCVENA
BBS:(860) 429-3784

ManTech Systems/InSync ....(703) 218-6000
12015 Lee Jackson Memorial Highway
Fairfax, VA 22033
Tech:(703) 218-6000
Web Page: http://www.mantech.com

Manugistics Inc ....(301) 984-5000
2115 E. Jefferson Street
Rockville, MD 20852-4999
Tech:(301) 984-5489 Toll Free:(800) 592-0050
Fax:(301) 984-5370
Web Page: http://www.manugistics.com
email: answer@manu.com

MapInfo..(518) 285-6000
One Global View
Troy, NY 12180-8399
Tech:(518) 285-7283 Toll Free:(800) 327-8627
Fax:(518) 285-6060 Fax on Demand:(800) 552-2511
Tech Support:(800) 552-2511
Web Page: http://www.mapinfo.com
email: techsupport@mapinfo.com

MapLinx Corp ....(800) 352-3414
5720 LBJ Freeway, Suite 180
Dallas, TX 75240-6328
Fax:(972) 248-2690
Web Page: http://www.maplinx.com
email: rickr@maplinx.com
Compuserve: 72662.157@compuserve.com

Mark IV Industries Inc ....(716) 689-4972
501 John James Audubon Pkwy, PO Box 810
Amherst, NY 14226
Tech:(716) 689-4972
Fax:(716) 689-1529
Web Page: http://www.mark-iv.com

Mark Of The Unicorn Inc ....(617) 576-2760
1280 Massachusetts Ave
Cambridge, MA 02138
Tech:(617) 576-3066
Fax:(617) 576-3609

Tech Support Fax:(617) 354-3068
Web Page: http://www.motu.com
email: techsupport@motu.com
Compuserve: 71333,3666
America Online: Keyword MOTU

MarketArts (see Window on Wallstreet)

MarketForce ....(817) 277-3000
PO Box 120279
Arlington, TX 76012-9925
Tech:(817) 277-3000 Toll Free:(800) 766-7355
Fax:(817) 274-6700
Web Page: http://www.marketforce-inc.com
email: sales@marketforce-inc.com

Marlin P. Jones & Assoc Inc ....(561) 848-8236
PO Box 12685
Lake Park, FL 33403-0685
Tech:(561) 848-8236
Fax:(561) 844-8764 Fax on Demand:(561) 848-1125
Fax:(800) 432-9937
Web Page: http://www.mpja.com
email: mpja@mpja.com

Marshall Industries.................................(626) 307-6000
9320 Telstar Ave
El Monte, CA 91731-2895
Tech:(626) 307-6033 Toll Free:(800) 877-9839
Fax:(626) 307-6187
Web Page: http://www.marshall.com
Compuserve: www.marshall.com

Masque Publishing.................................(303) 290-9853
7200 E. Drycreek Road, PO Box 5223
Englewood, CO 80155
Tech:(303) 290-9853 Toll Free:(800) 765-4223
Fax:(303) 290-6303
Web Page: http://www.masque.com
email: support@masque.com
Compuserve: 71333,1547 or GO GAMECPUB

Mass Micro Systems Mega Tape ....(408) 946-9207
1507 Centre Point Drive
Melipitas, CA 95035
Toll Free:(800) 950-9025
Fax:(408) 946-4746
Other Address: AppleLink:d0817

Masterclips Graphics (see IMSI Software)

MasterSoft Inc (see Adobe Systems)

MathSoft Inc ······························(617) 577-1017
101 Main Street
Cambridge, MA 02142-1521
Tech:(617) 577-1778 Toll Free:(800) 628-4223
Fax:(617) 577-8829 Fax on Demand:(617) 577-1778
Web Page: http://www.mathsoft.com
email: support@mathsoft.com

MathWorks Inc, The ······························(508) 647-7000
24 Prime Park Way
Natick, MA 01760-1500
Tech:(508) 647-7000
Fax:(508) 647-7001
Web Page: http://www.mathworks.com
email: service@mathworks.com

Matrox Graphics Inc ······························(514) 969-6320
1055 St. Regis Boulevard
Dorval, QB H9P 2T4 Canada
Tech:(514) 685-0270 Toll Free:(800) 361-1408
Fax:(514) 969-6363 Fax on Demand:(514) 685-0174
Web Page: http://www.matrox.com
email: graphics.techsupport@matrox.com
Compuserve: GO MATROX
BBS:(514) 685-6008

Maxell Corp Of America ······························(201) 794-5900
22-08 Rte 208 S.
Fair Lawn, NJ 07410
Tech:(201) 795-5900 Toll Free:(800) 533-2836
Fax:(201) 796-8790
tech support:(800) 377-5887
Web Page: http://www.maxell.com
email: feedback@maxell.com

Maxi Switch Inc ······························(520) 294-5450
2901 E. Elvira Road
Tucson, AZ 85706
Tech:(520) 746-9378
Fax:(520) 294-6890
Web Page: http://www.maxiswitch.com
email: maxiswitch@maxiswitch.com

Maximized Software ······························(510) 433-1903
55 Santa Clara Ave, Suite 250
Oakland, CA 94610
Toll Free:(888) 629-7638
Fax:(510) 433-1904

Web Page: http://www.maximized.com
email: info@maximized.com

Maximum Strategy Inc ....(408) 383-1600
801 Buckeye Court
Milpitas, CA 95035-7408
Fax:(408) 383-1616
Web Page: http://www.maxstrat.com
email: info@maxstrat.com

Maxis Software ....(510) 933-5630
2121 North California Boulevard, Suite 600
Walnut Creek, CA 94596-3572
Tech:(510) 927-3905 Toll Free:(800) 336-2947
Fax:(510) 927-3736
Web Page: http://www.maxis.com
email: support@maxis.com
Compuserve: GO GAMEPUB
America Online: Maxis or SimCity

Maxoptix Corp ....(510) 353-9700
3342 Gateway Blvd
Fremont, CA 94538
Tech:(800) 848-3092 Toll Free:(800) 848-3092
Fax:(510) 353-1845
Web Page: http://www.maxoptix.com
email: maxhelp@maxoptix.com
BBS:(510) 353-1448

MaxTech Corporation ....(562) 921-1698
13915 Cerritos Corporate Drive
Cerritos, CA 90703
Tech:(562) 921-4438 Toll Free:(800) 289-4821
Fax:(562) 802-9605 Fax on Demand:(562) 921-9540
Other Address: http://www.maxtech.com
Web Page: http://www.maxcorp.com
email: support@maxcorp.com
Compuserve: 71333,44
America Online: Maxtech
BBS:(562) 921-7180

MaxTech GVC ....(973) 586-3008
400 Commons Way
Rockaway, NJ 07866
Toll Free:(800) 936-7629
Fax:(973) 586-3308 Fax on Demand:(562) 921-9540
Tech Support Fax:(973) 586-2264
Web Page: http://www.maxcorp.com

Maxtor Corp .... **(303) 678-2700**
2190 Miller Drive, Mail Stop 1900-1
Longmont, CO 80501
Tech:(303) 678-2045 Toll Free:(800) 262-9867
Fax:(303) 678-2146 Fax on Demand:(800) 262-9867
Tech Fax:(303) 260-2260
Web Page: http://www.maxtor.com
email: technical_assistantance@maxtor.com

Maxtor Corp .... **(408) 432-1700**
2191 Zanker Rd
San Jose, CA 95131
Tech:(800) 262-9867 Toll Free:(800) 262-9867
Fax:(408) 922-2085 Fax on Demand:(800) 262-9867
Tech Fax:(408) 922-2050
Web Page: http://www.maxtor.com

Maxus Group .... **(216) 687-1000**
28601 Chagrin Blvd, Suite 500
Woodmere, OH 44122
Fax:(216) 687-1009
Web Page: http://www.maxusgroup.com
email: mis@maxusgroup.com

Maynard Electronic (see Seagate)

McAfee Associates (see Network Associates)

McAfee East .... **(732) 530-0440**
766 Shrewsbury Ave, Jerral Center W
Tinton Falls, NJ 07724-3298
Toll Free:(800) 552-9876
Fax:(732) 530-0622
Web Page: http://www.mcafee.com
email: custcare@mcafee.com
BBS:(408) 988-4004

MCI Communications Corp .... **(202) 872-1600**
1801 Pennsylvania Ave NW
Washington, DC 20006
Tech:(800) 444-8722 Toll Free:(800) 444-3333
Web Page: http://www.mci.com

MCM Electronics .... **(937) 434-0031**
650 Congress Park Drive
Dayton, OH 45459
Tech:(800) 824-8324 Toll Free:(800) 543-4330
Fax:(937) 434-6959
Fax:(800) 765-6960
Web Page: http://www.mcmelectronics.com
email: tech@mcmelectronics.com

MCS Products (see Micro Computer Systems)

MD & I Corp .... (626) 442-8899
9440 Telstar Ave, Ste 1
El Monte, CA 91731
Tech:(626) 442-6872 Toll Free:(800) 619-8899
Fax:(626) 442-8710
Web Page: http://www.mdinotebook.com
email: spprt@mdinotebook.com

MDL Corp .... (425) 861-6700
14940 NE 95th St
Redmond, WA 98052
Toll Free:(800) 800-3766
Fax:(425) 861-6767
Web Page: http://www.mdlcorp.com
email: help@mdlcorp.com

MECA Software LLC .... (800) 288-6322
115 Corporate Drive
Trumbull, CT 06611
Tech:(888) 808-6322
Fax on Demand:(203) 268-6160
Web Page: http://www.mymnet.com
Compuserve: GO MECA
America Online: PCapplications
BBS:(203) 268-1800

MECC (see Learning Company, The)

Media Graphics Int. Inc .... (800) 598-2037
8175-A Sheridan Blvd, Ste. 355
Arvada, CO 80003
Tech:(303) 427-8808 Toll Free:(800) 679-6730
Web Page: http://www.media-graphics.net
email: tech@media-graphics.net

Media Synergy .... (416) 369-1100
260 King St. E., Bldg C
Toronto, ON M5A 1K3
Fax:(416) 369-9037
Web Page: http://www.mediasynergy.com
email: askme@mediasynergy.com

MediaForm Inc .... (610) 458-9200
400 Eagleview Blvd, Suite 104
Exton, PA 19341
Tech:(610) 458-9200 Toll Free:(800) 220-1215
Fax:(610) 458-9554
Web Page: http://www.mediaform.com
email: support@mediaform.com

Mediamagic (see IPC Technology)

Mediatrix Peripherals Inc ....(819) 829-8749
4229 Garlock
Sherbrooke, PQ J1L 2C8 canada
Tech:(819) 829-8749
Fax:(819) 829-5100
Web Page: http;//www.mediatrix.com
email: techsupp@mediatrix.com
Compuserve: GO MEDIATRIX 74774, 1335
BBS:(819) 829-5101

Mega Drive Systems (see Data Direct Network)

Megahertz Corp (see 3Com)

MegaImage Inc ....(714) 522-8500
6902 Aragon Circle
Buena Park, CA 90620
Tech:(800) 555-4736 Toll Free:(800) 250-1876
Fax:(714) 522-2890
Web Page: http://www.megai.com
email: megasale@megai.com

Megamedia Corp ....(510) 623-1100
47381 Bayside Parkway
Fremont, CA 94538
Toll Free:(800) 634-2633
Fax:(510) 440-9924
Web Page: http://www.megamedia.com
email: info@megamed.com

Megatech Software ....(310) 320-8287
PO Box 11333
Torrence, CA 90510
Tech:(310) 320-8287
Fax:(310) 320-8286
Web Page: http://www.megatech-software.com
Compuserve: 74431,2473
BBS:(310) 539-7739

Memorex ....(800) 636-8352
Web Page: http://www.memorex.com

Memorex Telex Corp ....(972) 444-3500
545 E. John Carpenter Freeway
Irving, TX 75062
Toll Free:(800) 944-4455
Fax:(972) 444-3501
Web Page: http://www.mtc.com

Memtek Products Inc ....(562) 906-2800

Web Page: http://www.memtek.com

Mentat Inc .... **(310) 208-2650**
1145 Gayley Ave, Suite 315
Los Angeles, CA 90024
Fax:(310) 208-3724
Web Page: http://www.mentat.com
email: info@mentat.com

Mentor Electronics (see Qualtek Electronics)

Mentor Networks .... **(902) 421-5100**
1959 Upper Water St, Ste. 600
Halifax, NS B3J 3N2
Web Page: http://www.mentor.ca

Mercury Interactive Corp .... **(408) 822-5200**
470 Potrero Ave
Sunnyvale, CA 94086
Toll Free:(800) 837-8911
Fax:(408) 523-9911
Web Page: http://www.merc-int.com

Mergent International (see Utimaco Safeware)

Meridian Data Inc .... **(408) 438-3100**
5615 Scotts Valley Drive
Scotts Valley, CA 95066
Tech:(800) 755-8324 Toll Free:(800) 342-1129
Fax:(408) 438-6816
Tech Fax:(408) 438-8001
Web Page: http://www.meridian-data.com
email: support@meridian-data.com

Meridian Software Inc .... **(919) 518-1070**
12204 Old Creedmoor Road
Raleigh, NC 27613
Fax:(919) 518-1170
Web Page: http://www.meridian-software.com
email: merecb@meridian-software.com

Merisel .... **(310) 615-3080**
200 Continental Blvd
El Segundo, CA 90245
Tech:(800) 832-4003 Toll Free:(800) 637-4735
Other:(800) 462-5241
Web Page: http://www.merisel.com

Merit Software .... **(972) 385-2353**
13707 Gamma Road
Dallas, TX 75244
Tech:(972) 385-2957

Fax:(972) 385-8205
Web Page: http://www.softdisk.com/comp/merit/ind
email: 76711.247@compuserve.com
Compuserve: 76711,247
BBS:(972) 702-8641

Merit Studios Inc (see Merit Software)

Meritec ···························· (440) 354-3148
1359 West Jackson St, PO Box 8003
Painesville, OH 44077
Toll Free:(888) 637-4832
Fax:(440) 354-0509
Web Page: http://www.meritec.com
email: info@meritec.com

Merritt Computer Products Inc ··············· (214) 339-0753
5565 Red Bird Center Drive, Suite 150
Dallas, TX 75237
Tech:(214) 339-0753
Fax:(214) 339-1313

Metacard Corp ···························· (303) 447-3936
4710 Shoup Pl
Boulder, CO 80303
Fax:(303) 499-9855
Web Page: http://www.metacard.com
email: www@metacard.com

MetaCreations Inc ···························· (805) 566-6200
6303 Carpinteria Ave
Carpinteria, CA 93013
Tech:(805) 566-6239 Toll Free:(800) 472-9025
Fax:(805) 566-6385
Tech Support:(888) 707-6382
Other Address: http://www.metacreations.com
Web Page: http://www.metatools.com
email: support@metatools.com

MetaTools Inc (see MetaCreations)

Methode Electronics Inc ························ (708) 867-9600
7444 W. Wilson Ave
Chicago, IL 60656
Tech:(708) 867-9600
Fax:(708) 867-9130
Web Page: http://www.methode.com

Metrics Technology Inc ························ (505) 761-9630
3830 Commons Ave NE
Albuquerque, NM 87109

Toll Free:(800) 398-1490
Fax:(505) 761-9641
Web Page: http://www.alltech.com
email: info@metrictech.com

Metrowerks Corp .... **(512) 873-4700**
9801 Metric Blvd, Suite 100
Austin, TX 78758
Toll Free:(800) 377-5416
Fax:(512) 873-4900
Web Page: http://www.metrowerks.com
email: sales@metrowerks.com

Metz Software Inc .... **(425) 641-4525**
PO Box 6699
Bellevue, WA 98008-0699
Tech:(425) 641-4525 Toll Free:(800) 447-1712
Fax:(425) 644-6026
Web Page: http://www.metz.com
email: metz@metz.com
Compuserve: GO METZ
America Online: METZSoft

MGE UPS Systems .... **(714) 557-1636**
1660 Scenic Ave
Costa Mesa, CA 92626
Toll Free:(800) 523-0142
Fax:(714) 557-9788
Tech Fax:(714) 434-7652
Web Page: http://www.mgeups.com

MGI Software Corp .... **(905) 764-7000**
40 W. Wilmot St
Richmond Hill, ON L4B 1H8 Canada
Web Page: http://www.mgisoft.com
email: support@mgisoft.com

Micah Development Corp (see Full Armor)

Micro 2000 Inc .... **(818) 547-0125**
1100 E. Broadway, Suite 301
Glendale, CA 91205
Toll Free:(800) 864-8008
Fax:(818) 547-0397
Web Page: http://www.micro2000.com
email: techsupp@micro2000.com

Micro Accessories Inc .... **(510) 226-6310**
6086 Stewart Ave
Fremont, CA 94538-3152
Tech:(510) 226-6310 Toll Free:(800) 777-6687

Fax:(510) 226-6316
Other Address: AppleLink:MICROA
Web Page: http://www.micro-a.com
email: n_chernoff@micro-a.com

Micro Computer Cable Company............(313) 946-9700
12200 Delta Drive
Taylor, MI 48180
Fax:(313) 946-9645
Web Page: http://www.microccc.com

Micro Computer Systems Inc..................(972) 659-1514
2300 Valley View Lane, Suite 800
Irving, TX 75062
Fax:(972) 659-1624
Web Page: http://www.mcsdallas.com
email: info@mcsdallas.com

Micro Design International Inc.................(407) 677-8333
6985 University Blvd
Winter Park, FL 32792
Toll Free:(800) 920-8205
Fax:(407) 677-8365
Tech Fax:(407) 677-0221
Web Page: http://www.mdi.com
email: support@mdi.com

Micro Exchange Corp ....(973) 872-1200
20 Haines Dr
Wayne, NJ 07470
Web Page: http://www.microexch.com

Micro Firmware Inc....................................(405) 321-8333
330 West Gray St, Suite 120
Norman, OK 73069-7111
Tech:(405) 321-8333 Toll Free:(800) 767-5465
Fax:(405) 573-5535
Other:(888) 472-2467
Web Page: http://www.firmware.com
email: support@firmware.com
Compuserve: PCVENDOR D Section 8
BBS:(405) 573-5538

Micro Focus ....(650) 938-3700
701 E. Middlefield Rd
Mountain View, CA 94043
Tech:(610) 263-3550
Fax:(650) 856-6134
Tech Fax:(610) 263-3555
Web Page: http://www.microfocus.com

email: mfsupp@mfltd.co.uk
Compuserve: GO MICROFOCUS
BBS:(610) 263-3569

Micro House International Inc ··················(303) 443-3388
2477 N. 55th Street, Suite 101
Boulder, CO 80301
Tech:(303) 443-3389 Toll Free:(800) 926-8299
Fax:(303) 443-3323
Web Page: http://www.microhouse.com
email: support@microhouse.com
BBS:(303) 443-9957

Micro Solutions ···(815) 756-3411
132 W. Lincoln Highway
De Kalb, IL 60115
Tech:(815) 754-4500 Toll Free:(800) 890-7227
Fax:(815) 756-2928 Fax on Demand:(815) 754-4600
Tech Support Fax:(815) 756-4986
Web Page: http://www.micro-solutions.com
BBS:(815) 756-9100

Micro Star Software·································(760) 931-4949
2245 Camino Vida Roble, Suite 100
Carlsbad, CA 92009
Tech:(760) 931-4955 Toll Free:(800) 444-1343
Fax:(760) 931-4950
Tech Support Fax:(760) 931-4944
Web Page: http://www.microstar-usa.com
email: tech@microstar-usa.com

MicroBiz Corp ··(201) 512-0900
777 Corporate Drive
Mahwah, NJ 07430
Tech:(201) 512-0900 Toll Free:(800) 637-8268
Fax:(201) 512-1919
Tech Support Fax:(201) 512-5919
Web Page: http://www.microbiz.com

Microchip Technology Inc ························(602) 786-7200
2355 West Chandler Blvd
Chandler, AZ 85224-6199
Web Page: http://www.microchip.com
BBS:(800) 848-8980

Microcom Corp ···(740) 548-6262
8333-A Green Meadows Drive N
Westerville, OH 43081
Toll Free:(800) 642-7626
Fax:(740) 548-6556

Web Page: http://www.microcomcorp.com
email: msales@microcomcorp.com

Microcom Inc ....(781) 551-1000
500 River Ridge Drive
Norwood, MA 02062-5028
Tech:(781) 551-1414 Toll Free:(800) 822-8224
Fax:(781) 551-1006 Fax on Demand:(800) 285-2802
Web Page: http://www.microcom.com
BBS:(781) 762-5134

MicroData Corp ....(813) 573-5900
2727 Ulmerton Rd, Suite 300
Clearwater, FL 33762
Tech:(408) 261-7090 Toll Free:(800) 539-0123
Fax:(813) 572-5085
Web Page: http://www.quicktech.com
email: quicktech@quicktech.com

Microdyne Corp ....(352) 687-4633
491 Oak Road
Ocala, FL 34472-0213
Fax:(352) 687-3392
Other Address: http://www.mcdy.com
Web Page: http://www.microdyne.com

Microdyne Corp ....(703) 329-3700
3601 Eisenhower Ave, Suite 300
Alexandria, VA 22304-6495
Other Address: http://www.microdyne.com
Web Page: http://www.mcdy.com
email: mwebmaster@microdyne.com

Micrografx Inc ....(972) 234-1769
1303 E. Arapaho Road
Richardson, TX 75081
Tech:(972) 234-2694 Toll Free:(800) 671-0144
Fax:(972) 234-2410
Tech Support Fax:(972) 644-3688
Web Page: http://www.micrografx.com
Compuserve: GO MICROGRAFX
America Online: go max

MicroHelp Inc ....(757) 873-6707
728 Thimble Shoals Blvd #E
Newport News, VA 23606
Fax:(757) 873-0322

Microid Research Inc ....(978) 686-6468
1538 Turnpike Street
North Andover, MA 01845

Fax:(978) 683-1630
Web Page: http://www.mrbios.com
email: mrbios@mrbios.com

Microleague Int Sftr(Out of Business)

Microleague Multimedia............................(717) 872-6567
1001 Millersville Road, PO Box 4547
Lancaster, PA 17601
Tech:(717) 872-2442 Toll Free:(800) 545-9009
Fax:(717) 871-9959
Customer Service:(800) 467-3550
Web Page: http://www.mmi.com
email: techsupp@mmi.com

MicroLogic Software Inc..........................(510) 652-5464
1351 Ocean Ave
Emeryville, CA 94608
Tech:(510) 652-5464 Toll Free:(800) 888-9078

MicroMedium Inc.....................................(800) 561-2098
1434 Farrington Road
Apex, NC 27502
Fax:(919) 303-6011
Web Page: http://www.micromedium.com
email: support@micromedium.com
Compuserve: GO MICROMEDIUM

Micron Electronics Inc..............................(888) 742-4331
900 E. Karcher Rd
Nampa, ID 83687
Toll Free:(800) 209-9686
Fax:(208) 893-7390
Fax:(800) 270-1232
Web Page: http://www.micronpc.com
email: techsupport.meic@micronpc.com
BBS:(800) 270-1207

Micron Technology Inc.............................(208) 368-4000
8000 S. Federal Way, PO Box 6
Boise, ID 83707-0006
Tech:(888) 349-6972 Toll Free:(800) 964-2766
Fax:(208) 368-4431
Web Page: http://www.micron.com
email: username@micron.com

MicroNet Technology Inc..........................(949) 453-6000
80 Technology
Irvine, CA 92618
Tech:(949) 453-6060 Toll Free:(800) 800-3475
Fax:(949) 453-6101

Web Page: http://www.micronet.com
email: tech@micronet.com
BBS:(949) 453-6063

Micronetics Design Corp ....(301) 258-2605
1375 Piccard Drive, Suite 300
Rockville, MD 20850
Toll Free:(800) 433-7581
Fax:(301) 840-8943
Web Page: http://www.micronetics.com
email: support@us.micronetics.com
Compuserve: GO MUMPS

Micronics Computers Inc ....(510) 651-2300
45365 Northport Loop West
Fremont, CA 94538
Tech:(510) 661-3000 Toll Free:(800) 577-0977
Fax:(510) 651-5612 Fax on Demand:(510) 661-3199
Tech Fax:(510) 651-6982
Web Page: http://www.micronics.com
email: service@micronics.com
BBS:(510) 651-6837

Microplex Systems Ltd ....(604) 444-4232
8525 Commerce Court
Burnaby, BC V5A 4N3 Canada
Toll Free:(800) 665-7798
Fax:(604) 444-4239
Web Page: http://www.microplex.com
email: support@microplex.com

Micropolis Corp (Out of Business) ....(818) 225-1370

MicroProcessors Unlimited ....(918) 267-4961
24000 S. Peoria Ave
Beggs, OK 74421
Tech:(918) 267-4961

Microprose Software ....(510) 522-1164
2490 Mariner Square Loop
Alameda, CA 94501
Tech:(510) 864-4550 Toll Free:(800) 695-4263
Fax:(510) 522-9357 Fax on Demand:(800) 832-4958
Other Address: http://www.microprose.com
Web Page: http://www.holobyte.com
email: support@microprose.com
Compuserve: 76004,2223
America Online: MicroProse
BBS:(510) 522-8909

MicroRidge Systems Inc .... **(541) 593-1656**
PO Box 3249
Sunriver, OR 97707-0249
Tech:(541) 689-3265
Fax:(541) 593-5652
Web Page: http://www.microridge.com
email: info@microridge.com

Microsoft Corporation .... **(425) 882-8080**
1 Microsoft Way
Redmond, WA 98052-6399
Tech:(800) 322-1233 Toll Free:(800) 426-9400
Fax:(425) 936-7329
Other:(800) 227-4679
Web Page: http://www.microsoft.com
email: gopher.microsoft.com
BBS:(425) 936-6735

Access .... (425) 635-7050
Canada:(905) 568-2294
Authorized Support Centers .... (800) 936-3500
Web Page: http://www.microsoft.com/support
Basic (discontinued support) .... (425) 635-7053
Bob (see Works)
Bulletin Board System .... (425) 936-6735
Canada BBS:(905) 507-3022
CD-ROM Installation .... (425) 635-7033
Delta (see Source Safe)
Developer Network .... (800) 759-5474
Canada:(800) 759-5474
Web Page: http://www.microsoft.com/devnews
Download Service-USA .... (425) 936-6735
Web Page: http://www.microsoft.com/msdownload
Excel for Windows .... (425) 635-7070
Canada:(905) 568-2294
Excel SDK .... (425) 635-7048
Fast Tips, Business Systems .... (800) 936-4400
Canada:(800) 936-4400
Fast Tips, Desktop Applications .... (800) 936-4100
Canada:(800) 936-4100
Fast Tips, Desktop Systems .... (800) 936-4200
Fast Tips, Development Tools .... (800) 936-4300
Canada:(800) 936-4300
Fast Tips, Home Products .... (800) 936-4100
Canada:(800) 936-4100
Fox Development .... (425) 635-7191
Hardware Products .... (425) 635-7040
Kids Products .... (425) 635-7140

Canada:(905) 568-3503

Microsoft Automap .... (425) 635-7146
Canada:(905) 568-3503

Microsoft C/C++ .... (425) 635-7007
Canada:(905) 568-3503

Microsoft FrontPage .... (425) 635-7088
Canada:(905) 568-3503

Microsoft Internet Explorer .... (425) 635-7123
Canada:(905) 568-4494
Web Page: http://www.microsoft.com/ie

Microsoft Macro Assembler (MASM) .... (425) 646-5109

Microsoft Money .... (425) 635-7131
Canada:(905) 568-3503

Microsoft MS-DOS .... (425) 646-5104
Canada:(905) 568-4494

Microsoft Plus! .... (425) 635-7122
Canada:(905) 568-4494

Microsoft Press .... (800) 677-7377
Tech:(800) 426-9400 Toll Free:(800) 426-9400
BBS:(425) 936-6735

Microsoft Project .... (425) 635-7155
Canada:(905) 568-3503

Microsoft Publisher .... (425) 635-7140
Canada:(905) 568-3503

Microsoft SourceSafe .... (425) 635-7014

Microsoft Visual Basic .... (425) 646-5105
Canada:(905) 568-3503

Microsoft Visual J++ .... (425) 635-7011

Microsoft Windows .... (425) 635-3329

Microsoft Windows 95 .... (425) 635-7000
Canada:(905) 568-4494

Multimedia Products .... (425) 635-7172
Canada:(905) 568-3503

Office - Switch Line .... (425) 635-7041
Canada:(905) 568-2294

Office for Windows .... (425) 635-7056
Canada:(905) 568-2294
Web Page: http://www.microsoft.com/office

Outlook .... (425) 635-7031
Canada:(905) 568-3503

PowerPoint .... (425) 635-7145
Canada:(905) 568-3503

Priority Comprehensive .... (800) 936-5900
Toll Free:(800) 936-4600

Priority Desktop Applications .... (800) 936-5700
Canada:(800) 668-7975

Priority Development Issues .... (800) 936-5500

Priority Development Products ·····(800) 936-5800
Priority Home Products ·····(800) 936-5600
Profiler
Profit ·····(800) 723-3333
QuickBasic (see Visual Basic)
QuickC ·····(425) 635-7010
Scenes and Games ·····(425) 637-9308
Canada:(905) 568-3503
Schedule + ·····(425) 635-7049
Canada:(905) 568-2294
TechNet ·····(800) 344-2121
Canada:(800) 344-2121
TT/TDD (Text Telephone) ·····(425) 635-4948
Canada:(905) 568-9641
Visual InterDev ·····(425) 635-7016
Visual Test (Rational Software) ·····(703) 761-4400
Toll Free:(800) 728-1212
Windows Entertainment Products ·····(425) 637-9308
Canada:(905) 568-3503
Windows NT Workstation ·····(425) 635-7018
Canada:(905) 568-4494
Windows/Windows for Workgroups ·····(425) 637-7098
Canada:(905) 568-4494
Word for MS-DOS ·····(425) 635-7210
Canada:(905) 568-2294
Word for Windows ·····(425) 462-9673
Canada:(905) 568-2294
Works for MS-DOS ·····(425) 635-7150
Canada:(905) 568-3503
Works for Windows ·····(425) 635-7130
Canada:(905) 568-3503

Microspeed Inc ·····(510) 259-1270
2495 Industrial Pkwy West
Hayward, CA 94545-5007
Tech:(800) 232-7888 Toll Free:(800) 232-7888
Fax:(510) 259-1291
Web Page: http://www.microspeed.com
email: support@microspeed.com

Microspot USA Inc ·····(408) 253-2000
12380 Saratoga Sunnyvale Road, Suite 6
Saratoga, CA 95070
Tech:(408) 257-4000 Toll Free:(800) 622-7568
Fax:(408) 253-2055
Other Address: AppleLink:microspot
Web Page: http://www.microspot.com
email: support@microspot.com

Microstar Laboratories Inc .... **(425) 453-2345**
2265 116th Avenue NE
Bellevue, WA 98004
Fax:(425) 453-3199
Web Page: http://www.mstarlabs.com
email: appeng@mstarlabs.com

Microstar Software Ltd .... **(613) 596-2233**
3775 Richmond Road
Nepean, ON K2H 5B7 Canada
Toll Free:(800) 267-9975
Fax:(613) 596-5934
Web Page: http://www.microstar.com
email: info@microstar.com

MicroSupply .... **(425) 885-5420**
12368 Northrup Way
Bellevue, WA 98005
Tech:(425) 885-5420
Fax:(425) 885-9181
Web Page: http://bellevue.microsupply.com
email: bellevue@microsupply.com

Arizona .... (602) 829-1258
1407 W. 10th Place, Suite 105
Tempe, AZ 85281
Tech:(602) 829-1258
Fax:(602) 829-1966
Web Page: http://tempe.microsupply.com
email: tempe@microsupply.com

Colorado .... (303) 792-5474
7399 S. Tuscon Way, Suite A-3
Englewood, CO 80112
Tech:(303) 792-5474
Fax:(303) 792-5667
Web Page: http://englewood.microsupply.com
email: englewood@microsupply.com

Nevada .... (702) 739-3393
5525 S. Valley View Blvd, Suite 9
Las Vegas, NV 89118
Tech:(702) 739-3393
Fax:(702) 798-9897
Web Page: http://las_vegas.microsupply.com
email: las_vegas@microsupply.com

New Mexico .... (505) 345-5272
5445 Edith Ave NE, Suite K
Albuquerque, NM 87107
Fax:(505) 345-5291
email: etapang@microsupply.com

Oregon ....(503) 627-0359
9970 SW Arctic Drive
Beaverton, OR 97005
Tech:(503) 627-0359
Fax:(503) 627-0360
Web Page: http://beaverton.microsupply.com
email: beaverton@microsupply.com

Utah ....(801) 972-3680
2550 S. 2300 West, Suite 1
West Valley City, UT 84119
Tech:(801) 972-3680
Fax:(801) 972-3808
Web Page: http://west_valley.microsupply.com
email: west_valley@microsupply.com

Washington ....(253) 922-1127
1407 Willow Road East, Suite B
Tacoma, WA 98424
Tech:(253) 922-1127
Fax:(253) 922-1224
Web Page: http://tacoma.microsupply.com
email: tacoma@microsupply.com

Washington ....(509) 533-9280
109 S. Scott Dr, Suite #D1
Spokane, WA 99202
Fax:(509) 535-1690
email: shorty@microsupply.com

Microsystems (see Internet Solutions Group)

MicroSystems Development Tech ....(408) 296-4000
4100 Moorpark Ave, Suite 104
San Jose, CA 95117
Tech:(408) 296-4200
Fax:(408) 296-5877
Web Page: http://www.msd.com
email: info@msd.com

MicroTac Software (see Globalink Inc)

Microtec...(408) 487-7125
880 Ridder Park Drive
San Jose, CA 95131
Fax:(408) 487-7001
Web Page: http://www.microtec.com
email: info@microtec.com

Microtech Corp ....(540) 937-3298
685 Battle Mountain Road
Amissville, VA 20106
Toll Free:(800) 223-3693

Fax:(540) 937-3299
Web Page: http://www.microtech.com

Microtech International ....(203) 468-6223
158 Commerce Street
New Haven, CT 06512
Tech:(800) 666-9689 Toll Free:(800) 220-9489
Fax:(203) 469-3926
Web Page: http://www.microtechint.com
email: tech@microtechint.com

Microtek Lab Inc ....(310) 297-5000
3715 Doolittle Drive
Redondo Beach, CA 90278-1226
Tech:(310) 297-5100 Toll Free:(800) 654-4160
Fax:(310) 297-5050 Fax on Demand:(310) 297-5101
Web Page: http://www.microtekusa.com
email: pcsupport@microtek.com
Compuserve: GRAPHSUPc
America Online: Keyword MGR
BBS:(310) 538-4032

Microtest Inc ....(602) 952-6400
4747 N. 22nd St
Phoenix, AZ 85016-4708
Tech:(800) 638-3497 Toll Free:(800) 526-9675
Fax:(602) 952-6401 Fax on Demand:(602) 952-6450
Tech Fax:(800) 419-8991
Web Page: http://www.microtest.com
email: support@microtest.com
Compuserve: GO MICROTEST
BBS:(602) 957-7716

MicroTouch Systems Inc ....(978) 659-9000
300 Griffin Brook Park Drive
Methuen, MA 01844
Tech:(978) 659-9200 Toll Free:(800) 642-7686
Fax:(978) 659-9100
Web Page: http://www.microtouch.com
email: support@microtouch.com
BBS:(978) 659-9250

Microware Education (see IT Bridge)

MicroWay Inc ....(508) 746-7341
Research Park, PO Box 79
Kingston, MA 02364
Tech:(508) 746-7341
Fax:(508) 746-4678
Web Page: http://www.microway.com

email: tech@microway.com
BBS:(508) 746-7946

Midak (see Ikon)

MIDI Solutions Inc ....(604) 794-3013
PO Box 3010
Vancouver, BC V6B3X5 Canada
Tech:(604) 794-3013 Toll Free:(800) 561-6434
Fax:(604) 794-3396
Web Page: http://www.midisolutions.com
email: info@midisolutions.com
Compuserve: 72662,140

Midisoft Corporation ....(425) 391-3610
1605 NW Sammamish Road Suite 205
Issaquah, WA 98027
Tech:(425) 313-3495 Toll Free:(800) 776-6434
Fax:(425) 391-3422
Tech Support Fax:(425) 313-3491
Web Page: http://www.midisoft.com
email: techsup@midisoft.com

Milan Technology (see Digi Lan Connect)

Miles Spicer ....(800) 923-8463
1660 South Highway 100, Suite 210
Minneapolis, MN 55416
Toll Free:(800) 923-8463
Fax:(800) 923-8463
Web Page: http://www.opendoors.com
email: support@opendoors.com

Miles Tek ....(940) 455-7444
1 Lake Trail Drive
Argyle, TX 76226
Toll Free:(800) 524-7444
Fax:(940) 484-9402
email: milestek@milestek.com

Miller Freeman Inc ....(415) 905-2200
600 Harrison Street
San Francisco, CA 94107
Tech:(415) 905-2200 Toll Free:(800) 227-4675
Fax:(415) 905-2232
Web Page: http://www.mfi.com

Mindscape ....(415) 897-9900
88 Rowland Way
Novato, CA 94947
Tech:(423) 670-2040 Toll Free:(800) 234-3088

Fax:(415) 897-8286 Fax on Demand:(800) 409-1497
Tech Support Fax:(415) 897-5186
Web Page: http://www.mindscape.com
email: msales@mindscape.com

Ministor Peripherals (Out of Business)

Minolta Corp .... (201) 825-4000
101 Williams Dr
Ramsey, NJ 07446
Web Page: http://www.minoltausa.com
Printers
Tech:(800) 459-3250 Toll Free:(888) 264-6658
Web Page: http://www.minoltaprinters.com

Minuteman UPS .... (972) 446-7363
1455 LeMay Dr
Carrollton, TX 75007
Toll Free:(800) 238-7272
Fax:(972) 446-9011 Fax on Demand:(972) 664-3833
Fax Back:(800) 263-3933
Web Page: http://www.minuteman-ups.com

Miramar Systems .... (805) 966-2432
121 Gray Ave, Suite 200
Santa Barbara, CA 93101
Tech:(805) 965-5161 Toll Free:(800) 862-2526
Fax:(805) 965-1824
Web Page: http://www.miramarsys.com
email: support@miramarsys.com

Miros Inc .... (781) 235-0330
572 Washington St, Suite 18
Wellesley, MA 02482
Fax:(781) 235-0720
Web Page: http://www.miros.com
email: sales@miros.com

Misco Power Up .... (732) 264-8200
One Misco Plaza
Holmdel, NJ 07733
Tech:(800) 876-4726 Toll Free:(800) 876-4726
Fax:(732) 264-5955 Fax on Demand:(732) 264-5955

Mitac Industrial Corp .... (510) 656-5288
42001 Christy Street
Fremont, CA 94538
Toll Free:(800) 648-2295
Fax:(510) 656-2669
Web Page: http://www.mitacinds.com
email: webmaster@mitacinds.com

Mitsubishi Electronics America .... **(714) 220-2500**
5665 Plaza Drive, PO Box 6007
Cypress, CA 90630-6007
Tech:(800) 843-2515 Toll Free:(800) 344-6352
Fax:(714) 229-3854 Fax on Demand:(800) 937-2094
Other Address: http://www.mitsubishi.com
Web Page: http://www.mela-itg.com

Mitsumi Electronics Corp .... **(516) 752-7730**
35 Pinelawn Rd
Melville, NY 11747
Tech:(800) 801-7927 Toll Free:(800) 648-7864
Fax:(516) 752-7490 Fax on Demand:(650) 691-4460
Other Address: http://www.eciusa.com
Web Page: http://www.mitsumi.com
email: support@ecicorp.com
BBS:(515) 288-2608

Mitsumi Electronics Corp .... **(972) 550-7300**
5808 W. Campus Circle Dr
Irving, TX 75063
Toll Free:(800) 801-7927
Fax:(972) 550-7424
Web Page: http://www.mitsumi.com

MKS .... **(519) 884-2251**
185 Columbia Street West
Waterloo, ON N2L 5Z5 Canada
Tech:(519) 884-2270 Toll Free:(800) 265-2797
Fax:(519) 884-8861
Web Page: http://www.mks.com
email: support@mks.com

MMB Development Corporation .... **(310) 318-1322**
904 Manhattan Ave
Manhattan Beach, CA 90266
Toll Free:(800) 832-6022
Fax:(310) 318-2162
Web Page: http://www.mmb.com
email: info@mmb.com
BBS:(310) 318-5302

MMF Industries .... **(847) 537-7890**
370 Alice St
Wheeling, IL 60090
Tech:(800) 445-8293 Toll Free:(800) 323-8181
Fax:(847) 537-1120
Web Page: http://www.mmfind.com

Mobility Electronics .... **(602) 596-0061**
7955 E. Redfield Rd
Scottsdale, AZ 85260
Fax:(602) 596-0349
Web Page: http://www.mobilityelectronics.com
email: techsupport@mobilityelectronics.com

Mobius Computer Corp .... **(510) 556-1500**
6529 Sierra Lane
Dublin, CA 94568
Web Page: http://www.mobius.com
email: info@mobius.com

Molloy Group Inc, The .... **(973) 540-1212**
4 Century Dr
Parsippany, NJ 07054
Fax:(973) 292-9407
Web Page: http://www.molloy.com
email: info@molloy.com

Monotype Typography Inc .... **(847) 718-0400**
985 Busse Road
Elk Grove Village, IL 60007-2400
Tech:(800) 666-6897 Toll Free:(800) 803-6964
Fax:(847) 718-0500
Web Page: http://www.monotype.com
email: tech@monotype.com

Monster Cable .... **(415) 840-2000**
274 Wattis Way
San Francisco, CA 94080
Fax:(415) 468-0202
Web Page: http://www.monstercable.com
email: n_zachlod@monstercable.com

Moon Valley Software .... **(805) 781-3890**
1880 Santa Barbara Street, Suite D
San Luis Obispo, CA 93401
Tech:(800) 473-5509 Toll Free:(800) 473-5509
Fax:(805) 781-3898
Web Page: http://www.moonvalley.com
email: info@moonvalley.com

Mortice Kern Systems Inc (see MKS)

Most Significant Bits Inc .... **(440) 934-1397**
37207 Colorado Ave
Avon, OH 44011
Tech:(440) 934-1397 Toll Free:(800) 755-4619
Fax:(440) 934-1386
Web Page: http://www.msbcd.com

email: support@msbcd.com

Motion Works Group .... **(604) 685-9975**
1020 Mainland St, Suite 130
Vancouver, BC V6B2T4 Canada
Fax:(604) 685-6105
Web Page: http://www.mwg.com
email: info@mwg.com

Motorola Inc .... **(708) 576-5000**
1303 E. Algonquin Road
Schaumburg, IL 60196
Tech:(800) 311-6456
Fax:(708) 576-7653
Web Page: http://www.mot.com

Semiconductor Products Sector .... (512) 895-2000
6501 William Cannon Drive W
Austin, TX 78735
Fax:(512) 891-2652
Web Page: http://www.mot.com

Motorola ISG .... **(205) 430-8000**
5000 Bradford Dr
Huntsville, AL 35805
Tech:(205) 726-0798 Toll Free:(800) 446-0144
Web Page: http://www.mot.com or www.motorola.com

Motorola ISG .... **(508) 261-4307**
20 Cabot Blvd
Mansfield, MA 02048-1193
Tech:(508) 261-0366 Toll Free:(800) 544-0062
Fax:(508) 339-1105
Web Page: http://www.mot.com

MountainGate Datasystems .... **(702) 851-9393**
9393 Gateway Drive
Reno, NV 89511-8910
Tech:(800) 447-8302 Toll Free:(800) 556-0222
Fax:(702) 851-5533 Fax on Demand:(702) 851-6610
Tech Support Fax:(800) 447-8303
Web Page: http://www.mountaingate.com
email: tech_center@mountaingate.lmco.com

Mouse Systems Corp .... **(510) 656-1117**
41660 Boscell Rd
Fremont, CA 94538
Tech:(510) 656-1117 Toll Free:(800) 886-6423
Fax:(510) 770-1924 Fax on Demand:(510) 683-0720
Fax:(510) 656-4409
Web Page: http://www.mousesystems.com

email: support@mousesystems.com
BBS:(510) 683-0617

Mouser Electronics ....(800) 346-6873
958 N. Main Street
Mansfield, TX 76063
Toll Free:(800) 346-6873
Fax:(817) 483-0931
Web Page: http://www.mouser.com
email: tech@mouser.com

MPI Media Group ....(708) 460-0555
16101 S. 108th Ave
Orland Park, IL 60467
Tech:(708) 460-0555 Toll Free:(800) 777-2223
Fax:(708) 460-0175
Web Page: http://www.mpimedia.com
email: webmaster@mpimedia.com

Muddy Shoes Software LLC ....(847) 381-7758
342 Lageschulte St
Barrington, IL 60010
Tech:(847) 381-3695
Fax:(847) 381-7762
Web Page: http://www.muddyshoes.com
email: sales@muddyshoes.com

Mueller Technical Research ....(708) 726-0709
21718 Mayfield Lane
Barrington, IL 60010-9733
Fax:(708) 726-0710
email: 73145.1566@CompuServ.com
Compuserve: 73145,1566

Multi-Ad Services ....(309) 692-1530
1720 West Detweiller Drive
Peoria, IL 61615
Fax:(309) 692-6566
Web Page: http://www.multi-ad.com
email: corporate@multi-ad.com

Multi-Net Communications ....(541) 579-7006
15702 Black Bear Court
Klamath Falls, OR 97601
Toll Free:(800) 235-7789
Fax:(503) 883-7879
Web Page: http://www.multinet.com
email: support@multinet.com
BBS:(503) 883-8197

Multi-Tech Systems Inc ....(612) 785-3500
2205 Wooddale Drive
Mounds View, MN 55112
Tech:(800) 972-2439 Toll Free:(800) 328-9717
Fax:(612) 785-9874 Fax on Demand:(612) 717-5888
BBS:(612) 785-3702
Web Page: http://www.multitech.com
email: info@multitech.com
BBS:(800) 392-2432

Multimedia 2000 ....(206) 622-5530
1100 Olive Way, 12th Floor
Seattle, WA 98101
Tech:(206) 343-5934 Toll Free:(800) 850-7272
Fax:(206) 622-4380
Web Page: http://www.multicom.com
email: techsupport@m-2k.com

Multimedia Integrated ....(650) 872-7100
379 Oyster Point Blvd, Suite 7
South San Francisco, CA 94080
Tech:(650) 872-7120
Fax:(650) 872-7133

Multimedia Learning Inc ....(972) 869-8282
5215 N. O'Connor Blvd, Suite 760
Irving, TX 75039
Tech:(972) 869-8282 Toll Free:(800) 870-6608
Fax:(972) 869-8280
Web Page: http://www.media.com
email: webmaster@media.com

Music Quest Inc ....(214) 881-7408
1700 Alma Drive, Suite 108
Plano, TX 75075
Tech:(214) 881-7408 Toll Free:(800) 876-1376
Compuserve: GO MUSICQUEST
BBS:(214) 881-7311

Musicator ....(530) 759-9424
PO Box 73793
Davis, CA 95617
Tech:(530) 756-9807 Toll Free:(800) 551-4050
Fax:(530) 759-8852
Web Page: http://www.musicator.com
email: musicator@musicator.com
Compuserve: GO MCATOR

Musicware Inc ....(425) 881-9797
8654 154th Avenue NE

Redmond, WA 98052
Tech:(425) 881-1419 Toll Free:(800) 997-4266
Fax:(425) 881-9664
Web Page: http://www.musicwareinc.com
email: tech@musicwareinc.com
Compuserve: 75162,433

Musitek .... (805) 646-8051
410 Bryant Circle, Suite K
Ojai, CA 93023
Tech:(805) 646-5841 Toll Free:(800) 676-8055
Fax:(805) 646-8099
Web Page: http://www.musitek.com
email: tech@musitek.com
Compuserve: 71333,3723

Mustang Software Inc .... (805) 873-2500
6200 Lake Ming Road, PO Box 2264
Bakersfield, CA 93306
Tech:(805) 873-2550 Toll Free:(800) 999-9619
Fax:(805) 873-2599
Web Page: http://www.mustang.com
email: support@mustang.com
BBS:(805) 873-2400

Mustek Inc .... (714) 790-3800
121 Water Works Way, #100
Irvine, CA 92618
Tech:(949) 788-3600 Toll Free:(800) 468-7835
Fax:(949) 788-3670
Web Page: http://www.mustek.com
email: webmaster@mustek.com

Mutoh America .... (602) 276-5533
3007 E. Chambers
Phoenix, AZ 85040-3796
Tech:(800) 445-8782
Fax:(602) 276-7823
Tech Fax:(602) 276-9007
Web Page: http://www.mutoh.com
email: support@mutoh.com
BBS:(602) 243-9440

My Software Company .... (650) 473-3620
2197 E. Bayshore Rd
Palo Alto, CA 94303
Tech:(970) 522-3000
Fax:(650) 325-0873
Web Page: http://www.mysoftware.com
email: info@mysoftware.com

Mylex Corp .... **(510) 796-6100**
34551 Ardenwood Blvd
Fremont, CA 94555-3607
Tech:(510) 608-2400 Toll Free:(800) 776-9539
Fax:(510) 745-7654
Tech Fax:(510) 745-7715
Web Page: http://www.mylex.com
email: tsup@mylex.com
BBS:(510) 793-3491

Mystic River Software .... **(978) 371-1100**
142 North Rd
Sudbury, MA 01776
Toll Free:(800) 298-3500
Fax:(978) 371-1818
Web Page: http://www.mysticriver.com
email: info@mysticriver.com

Nanao USA Corp (see EIZO Nanao Tech Inc)

Narrative Communications .... **(781) 290-5300**
1601 Trapelo Road
Waltham, MA 02154
Toll Free:(800) 978-8670
Fax:(781) 290-5312
Web Page: http://www.narrative.com
email: support@narrative.com

National Assoc. Of Service Managers .... **(619) 562-7004**
PO Box 712500
Santee, CA 92071
Toll Free:(888) 562-7004
Fax:(619) 562-7153
Web Page: http://www.nasm.com
email: nasm@starnetink.com

National Computer Dist (see AmeriQuest)

National Computer Systems Inc .... **(612) 829-3000**
11000 Prairie Lakes Drive
Eden Prairie, MN 55344
Toll Free:(800) 431-1421
Web Page: http://www.ncs.com
email: info@ncs.com

National Instruments .... **(512) 794-0100**
11500 N. Mopal Expwy
Austin, TX 78759-3504
Toll Free:(800) 329-7177
Fax:(512) 683-8411 Fax on Demand:(512) 418-1111
Web Page: http://www.natinst.com

email: support@natinst.com

National Semiconductor ·························(408) 721-5000
2900 Semiconductor Drive, PO Box 58090
Santa Clara, CA 95051
Toll Free:(800) 272-9959
Fax:(408) 721-7662
Web Page: http://www.national.com

National Technical Info Service ·············(800) 553-6847
5285 Port Royal Road
Springfield, VA 22161
Fax:(703) 321-8547
Web Page: http://www.csce.ca/epub/softdir

Natural Intelligence Inc ···························(617) 876-4876
725 Concord Ave, 2nd Floor
Cambridge, MA 02138-1052
Tech:(617) 876-7680
Fax:(617) 492-7425
Web Page: http://www.natural.com
email: support@natural.com

Navision Software US Inc ·······················(770) 798-8300
500 Pinnacle Ct, Ste. 510
Norcross, GA 30071
Toll Free:(800) 552-8478
Web Page: http://www.navision-us.com
email: sales@navision-us.com

NavPress Software
1934 Rutland Drive, Suite 500
Austin, TX 78758
Fax:(512) 834-1888

NCE Storage Solutions ····························(619) 452-7974
9717 Pacific Heights Blvd
San Diego, CA 92121-3719
Tech:(619) 658-9720 Toll Free:(800) 446-6456
Fax:(619) 452-3271
Sales Fax:(619) 658-9736
Web Page: http://www.ncegroup.com
email: support@ncegroup.com
Compuserve: GO EMERALD
BBS:(619) 626-8550

NCR Microelectronics (see LSI Logic)

NDC Communications Inc ······················(408) 730-0888
265 Santa Ana Court
Sunnyvale, CA 94086

Toll Free:(800) 632-1118
Fax:(408) 730-0889
Web Page: http://www.ndclan.com
email: sales@ndclan.com

Neamco ....(617) 269-7600
510 E. Second Street
Boston, MA 02127
Tech:(617) 269-7600 Toll Free:(800) 937-1300
Fax:(617) 268-0473
Web Page: http://www.neamco.com

NEBS Software (One-Write Plus) ....(800) 225-9550
20 Industrial Park Drive
Nashua, NH 03062
Toll Free:(800) 225-9550

NEC Systems Laboratory Inc ....(408) 433-1358
110 Rio Robles
San Jose, CA 95134
Toll Free:(888) 223-4385
Fax:(408) 433-1448
Web Page: http://netmc.neclab.com
email: info@auraline.com

NEC Technologies Inc ....(978) 264-8000
1414 Massachusetts Ave
Acton, MA 01719
Tech:(800) 388-8888 Toll Free:(800) 366-3632
Fax on Demand:(800) 366-0476
CDROM Main Line:(415) 528-6000
Other Address: http://www.nec-computer.com
Web Page: http://www.nec.com
email: tech-support@nectech.com
Compuserve: GO NECTECH
America Online: Keyword:NECTECH
BBS:(978) 742-8706

Neoware ....(610) 277-8300
400 Feheley Drive
King Of Prussia, PA 19406
Tech:(800) 437-1551 Toll Free:(800) 437-1551
Fax:(610) 275-5739
Web Page: http://www.hds.com
email: info@hds.com

Net Manage ....(703) 506-0500
1604 Springhill Road, Suite 400
Vienna, VA 22182-7509
Tech:(703) 902-8700 Toll Free:(800) 795-8674

Fax:(703) 506-0510
Web Page: http://www.relay.com
BBS:(703) 902-8720

Net Phone Inc ....(508) 787-1000
313 Boston Post Road W
Marlborough, MA 01752
Fax:(508) 787-1030
Web Page: http://www.netphone.com
email: support@NetPhone.com

Net USA Software ....(650) 948-6200
201 San Antonio Circle, Suite C250
Mountain View, CA 94040
Tech:(650) 948-6200 Toll Free:(800) 628-3475
Fax:(650) 948-6296
Web Page: http://www.netusa.com/pacmicro
email: webmaster@netusa.com

net.Genesis Corp ....(617) 577-9800
215 First Street
Cambridge, MA 02142-1119
Tech:(617) 577-9800
Fax:(617) 577-9850
Web Page: http://www.netgen.com
email: support@netgen.com

Net2Net Corp (see Visual Networks Broadband)

Netaccess ....(603) 898-1800
18 Keewaydin Dr
Salem, NH 03079
Tech:(800) 435-7926 Toll Free:(800) 435-7926
Fax:(603) 894-4545
Web Page: http://www.netacc.com
email: info@netacc.com

Netcom Online Comm. (see ICG Netcom)

NetIQ Corp ....(408) 556-0888
5410 Betsy Ross Dr, Ste 260
Santa Clara, CA 95054
Tech:(408) 330-7000
Fax:(408) 330-0979
Web Page: http://www.netiq.com
email: support@netiq.com

NetManage Inc ....(408) 973-7171
10725 N. De Anza Blvd
Cupertino, CA 95014-2030
Tech:(408) 973-8181

Fax:(408) 257-6405
Web Page: http://www.netmanage.com
email: support@netmanage.com
Compuserve: GO ECCO

Netopia .... **(510) 814-5100**
2470 Mariner Square Loop
Alameda, CA 94501
Tech:(510) 814-5000
Fax:(510) 814-5023 Fax on Demand:(510) 814-5040
Other Address: Applelink:FARALLON /// eWorld:FARALLO5
Web Page: http://www.netopia.com
email: ask_netopia@netopia.com
Compuserve: 75410,2702
America Online: FARALLON
BBS:(510) 865-1321

Netrix .... **(703) 742-6000**
13595 Dulles Technology Drive
Herndon, VA 20171-3424
Tech:(800) 776-1477
Fax:(703) 742-4049
Fax:(703) 742-4048
Web Page: http://www.netrix.com
email: marketing@netrix.com
BBS:(703) 793-1233

Netscape Communications Corp .... **(650) 254-1900**
501 E. Middlefield Road
Mountain View, CA 94043
Tech:(650) 937-2555
Fax:(650) 528-4124
Web Page: http://www.netscape.com
email: info@netscape.com

netViz Corp .... **(301) 258-5087**
9210 Corporate Blvd, Suite 150
Rockville, MD 20850-4608
Tech:(301) 258-5087 Toll Free:(800) 827-1856
Fax:(301) 258-5088
Web Page: http://www.quyen.com
email: techsupport@netviz.com
BBS:(301) 990-3979

Netwave Technologies Inc .... **(510) 737-1600**
6663 Owens Dr
Pleasanton, CA 94588
Tech:(510) 737-1613
Fax:(510) 847-8744
Web Page: http://www.netwave-wireless.com

email: info@netwave-wireless.com

Network 1 Software & Technology Inc ···(212) 293-3068
909 Third Avenue, 9th Floor
New York, NY 10022
Tech:(972) 606-8200 Toll Free:(800) 638-9751
Fax:(212) 293-3090
Web Page: http://www.network-1.com
email: support@network-1.com

Network Appliance (NetApp)····················(408) 367-3000
2770 San Tomas Expressway
Santa Clara, CA 95051
Tech:(888) 463-8277
Fax:(408) 367-3151
Web Page: http://now.netapp.com
email: support@netapp.com

Network Associates ·································(408) 988-3832
2805 Bowers Ave
Santa Clara, CA 95051
Tech:(408) 988-3832 Toll Free:(888) 847-8766
Fax:(408) 970-9727 Fax on Demand:(408) 988-3034
Web Page: http://www.mcafee.com
email: custcare@mcafee.com
Compuserve: 76702,1714 or GO MCAFEE
America Online: keyword: mcafee
BBS:(408) 988-4004

Network Associates ·································(201) 587-1515
10 Forest Ave
Paramus, NJ 07652
Tech:(800) 966-9695 Toll Free:(800) 966-9695
Fax:(201) 587-8005
Web Page: http://www.magicsolutions.com

Network Associates ·································(408) 988-3832
4200 Bohannon Drive
Menlo Park, CA 94025
Toll Free:(800) 764-3337
Web Page: http://www.ngc.com
email: sales@ngc.com

Network Associates ·································(408) 988-3832
3965 Freedom Circle
Santa Clara, CA 95054
Tech:(408) 988-3832
Fax:(408) 970-9727
Web Page: http://www.networkassociates.com
BBS:(408) 988-4004

Network Computing Devices··················**(650) 694-0650**
350 North Bernardo Ave
Mountain View, CA 94043-5207
Tech:(650) 691-7445 Toll Free:(800) 800-9599
Fax:(650) 961-7711
Web Page: http://www.ncd.com
email: info@ncd.com

Network General Corp (see Network Associates

Network Peripherals································**(408) 321-7300**
1371 McCarthy Blvd
Milpitas, CA 95035
Tech:(408) 321-7375 Toll Free:(800) 674-8855
Fax:(408) 321-9218 Fax on Demand:(800) 674-8855
Web Page: http://www.npix.com
email: support@npix.com

Networth Inc (Netelligent)·······················**(214) 929-1700**
8404 Esters Blvd
Irving, TX 75063
Tech:(214) 929-6984 Toll Free:(800) 544-5255
Fax:(214) 929-1720
Web Page: http://www.compaq.com/productinfo/syst
BBS:(214) 929-4882

New Horizons Computer Learning Ctr ···**(714) 556-1220**
1231 E. Dyer Road, Suite 140
Santa Ana, CA 92705
Toll Free:(800) 811-2530
Fax:(714) 438-9499
Fax:(714) 556-4612
Web Page: http://www.newhorizons.com
email: 75411.447@compuserve.com
Compuserve: GO NEW HORIZONS

New Media Corp ·····································**(949) 453-0100**
1 Technology, Building A
Irvine, CA 92618
Tech:(888) 595-2195 Toll Free:(800) 227-3748
Fax:(949) 453-0114
Tech Support Fax:(949) 453-0614
Web Page: http://www.newmediacorp.com
email: support@newmediacorp.com
BBS:(949) 453-0214

New Vision Technology Inc·····················**(613) 727-8184**
38 Auriga Drive, Unit 13
Nepean, ON K2E 8A5 Canada
Tech:(613) 727-0884

Fax:(613) 727-8190

New World Computing Inc (see 3DO)

Newbridge Microsystems ························(613) 591-3600
Tech:(703) 834-5300
Web Page: http://www.newbridge.com

Newbridge Networks Corp ·····················(703) 834-3600
593 Herdon Parkway
Herndon, VA 20170
Fax:(703) 471-7080
Web Page: http://www.newbridge.com
email: webmaster@newbridge.com

NewCom, Inc. ···(818) 597-3200
31166 Via Colinas
Westlake Village, CA 91362-4500 USA
Fax:(818) 597-3210
Web Page: http://www.newcominc.com

Newer Technology Inc ····························(316) 943-0222
4848 Irving Street
Wichita, KS 67209
Tech:(888) 656-8324 Toll Free:(800) 678-3726
Fax:(316) 943-0555
Fax:(316) 943-4515
Web Page: http://www.newertech.com
email: techsupport@newertech.com

NewGen Imaging Systems Corp ············(714) 641-8600
3545 Cadillac Avenue
Costa Mesa, CA 92626
Tech:(714) 436-5150 Toll Free:(800) 756-0556
Fax:(714) 641-2800 Fax on Demand:(800) 888-1689
Fax:(714) 436-5189
Web Page: http://www.newgen.com
email: sales@newgen.com

Newport Systems (see Cisco Systems)

News Corp/MCI Online (see Delphi Internet)

NewTek ···(210) 370-8000
8200 IH-10 W, Ste. 900
San Antonio, TX 78230
Toll Free:(800) 862-7837
Fax:(210) 370-8001
Tech Fax:(210) 370-8002
Web Page: http://www.newtek.com
email: Customer_service@newtek.com

NexGen Inc (see Advanced Micro Devices)

NHC Communications ....**(514) 735-2741**
5450 Cote De Liesse
Mount Royal, QC H4P1A5 Canada
Tech:(514) 734-4361 Toll Free:(800) 361-1965
Fax:(514) 735-8057
Web Page: http://www.nhc.com
email: support@nhc.com
BBS:(514) 735-8006

Nikon, Inc. ....**(516) 547-4200**
1300 Walt Whitman Rd
Melville, NY
Toll Free:(800) 526-4566
Fax:(516) 547-0314
Web Page: http://www.nikonusa.com

Nimax Inc ....**(619) 452-2220**
9275 Carroll Park Drive
San Diego, CA 92121
Tech:(619) 452-2220 Toll Free:(800) 876-4629
Fax:(619) 452-6669
Tech Fax:(619) 452-6341
Web Page: http://www.nimax.com

Nirvana Systems Inc ....**(512) 345-2545**
3415 Greystone Dr, Suite 205
Austin, TX 78731
Tech:(512) 345-2592 Toll Free:(800) 880-0338
Fax:(512) 345-4225
Web Page: http://www.nirv.com
email: support@nirv.com
Compuserve: 72662,2166 or GO PCVENJ

Nisus Software Inc ....**(619) 481-1477**
107 S. Cedrous Ave, PO Box 1300
Solana Beach, CA 92075
Tech:(619) 481-1477 Toll Free:(800) 922-2993
Fax:(619) 481-6154 Fax on Demand:(619) 481-4366
Other Address: AppleLink:nisus.tech
Web Page: http://www.nisus.com
email: support@nisus.com
Compuserve: 75300,1243
America Online: nisus@aol.com

Nolo Press..**(510) 549-1976**
950 Parker Street
Berkeley, CA 94710
Tech:(510) 549-4660 Toll Free:(800) 728-3555
Fax:(800) 645-0895
Other:(800) 992-6656

Web Page: http://www.nolo.com
email: NoloTec@nolo.com

Nomai ....(408) 542-5900
592 Weddell Drive, Suite 5 & 6
Sunnyvale, CA 94089
Tech:(408) 542-5900
Fax:(408) 542-5925
Web Page: http://www.nomai.com

Nombas Inc ....(781) 391-6595
34 Salem Street
Medford, MA 02155
Fax:(781) 391-3842
Web Page: http://www.nombas.com
email: nombas@nombas.com

Nortel (Bay Networks)................................(972) 684-1000
2211 Lakeside Blvd
Richardson, TX 75081
Tech:(972) 684-1335 Toll Free:(800) 667-8437
Fax:(972) 684-3866 Fax on Demand:(800) 667-8437
Other Address: http://www.ntcr.com
Web Page: http://www.nortel.com
email: support@listmail.ntcr.com

North Edge Software (see Timeslips Corp)

Norton-Lambert Corp.................................(805) 964-6767
PO Box 4085
Santa Barbara, CA 93140-4085
Tech:(805) 964-6767
Fax:(805) 683-5679
Web Page: http://www.norton-lambert.com
email: techsupport@norton-lambert.com
Compuserve: GO CLOSEUP
BBS:(805) 683-2249

Novastor Corporation ....(805) 579-6700
80-B W. Cochran Street
Simi Valley, CA 93065
Tech:(805) 579-6700
Fax:(805) 579-6710
Web Page: http://www.novastor.com
email: support@novastor.com

NovaWeb Technologies Inc.......................(510) 249-9500
48511 Warm Springs Blvd, Suite 208
Fremont, CA 94539
Toll Free:(888) 722-6932
Fax:(510) 249-9380

Web Page: http://www.novatech.com
email: infonova@novawebtech.com

Novell Corporation .... **(800) 526-7937**
122 East 1700 South
Provo, UT 84606
Tech:(800) 858-4000 Toll Free:(800) 638-9273
Fax on Demand:(801) 861-5363
Fax Back:(801) 429-3030
Web Page: http://www.novell.com

Novell Corporation .... **(801) 429-7000**
1555 N. Technology Way
Orem, UT 84097-2399
Toll Free:(800) 453-1267
Web Page: http://wp.novell.com/groupwar/groupwar
BBS:(801) 221-5197
Applications
ConvertPerfect
DataPerfect
Desktop Systems .... (408) 434-2300
Toll Free:(800) 274-4374
Edutainment
Envoy .... (800) 861-2401
Hearing Impaired (TDD) .... (801) 765-4032
InfoCentral
Piracy Hotline .... (800) 747-2837
Piracy-BSA .... (800) 688-2721
Support Connection .... (800) 861-2507
Tech:(800) 274-4374 Toll Free:(800) 858-4000
Tech Support .... (800) 861-2146

Now Software .... **(503) 274-2800**
921 SW Washington Street, Suite 500
Portland, OR 97205-2823
Tech:(503) 274-2815 Toll Free:(800) 237-2078
Fax:(503) 274-0670
Web Page: http://www.nowsoft.com
email: help@nowsoft.com
America Online: NOW@AOL.COM

nStor Corp Inc .... **(407) 829-3500**
450 Technology Park
Lake Mary, FL 32746
Toll Free:(800) 724-3511
Fax:(407) 829-3555
Web Page: http://www.nstor.com
email: support@nstor.com

NTP Software .... **(603) 622-4400**
1750 Elm St
Manchester, NH 03104
Fax:(603) 641-6934
Web Page: http://www.ntp.com
email: sales@ntp.com

NTT Software Corp .... **(650) 688-1100**
350 Cambridge Ave, Ste. 300
Palo Alto, CA 94306
Fax:(650) 688-1121
Web Page: http://www.ntts.com
email: interspace@ntts.com

NuKote International Inc .... **(615) 794-9000**
17950 Preston Road, Suite 690 LB21
Dallas, TX 75252
Toll Free:(800) 448-1422
Web Page: http://www.nukote.com

Number Nine Visual Technology .... **(781) 674-0009**
18 Hartwell Ave
Lexington, MA 02173-3103
Tech:(781) 869-7214 Toll Free:(800) 438-6463
Fax:(617) 674-2919 Fax on Demand:(800) 438-6463
Tech Support Fax:(617) 273-0899
Web Page: http://www.nine.com
email: dom-sales@nine.com

Numega Labs .... **(617) 267-9743**
321 Columbus Ave
Boston, MA 02116
Tech:(888) 686-3427 Toll Free:(800) 468-6342
Fax:(617) 424-1839
Other Address: http://www.numega.com
Web Page: http://www.uw.com
email: info@uw.com
Compuserve: GO UNDERWARE

NuMega Technologies .... **(603) 578-8400**
9 Townsend W
Nashua, NH 03063
Toll Free:(800) 468-6342
Fax:(603) 578-8401
Web Page: http://www.numega.com
email: tech@numega.com

Numera Software Corp .... **(206) 622-2233**
1501 4th Ave, Suite 2880
Seattle, WA 98101

Web Page: http://www.numera.com
Compuserve: GO NUMERA

NVIDIA Corp .... **(408) 617-4000**
1226 Tiros Way
Sunnyvale, CA 94086
Tech:(408) 617-4000
Fax:(408) 617-4100
Web Page: http://www.nvidia.com
email: info@nvidia.com

O'Reilly and Associates Inc .... **(707) 829-0515**
101 Morris Street
Sebastopol, CA 95472
Toll Free:(800) 998-9938
Fax:(707) 829-0104
toll free:(800) 889-8969
Web Page: http://www.ora.com
email: booktech@oreilly.com

O.R. Technologies, Inc. .... **(303) 473-9170**
42 W. Campbell Ave
Campbell, CA 95008
Web Page: http://www.ortechnology.com
email: mktg@ortechnology.com

Oak Technology .... **(408) 737-0888**
139 Kifer Court
Sunnyvale, CA 94086
Fax:(408) 737-3838 Fax on Demand:(800) 239-0319
Web Page: http://www.oaktech.com
email: tech@oaktech.com

Oberon Software .... **(617) 494-0990**
215 First Street, Suite 3100
Cambridge, MA 02142
Tech:(800) 654-1364 Toll Free:(800) 654-1215
Fax:(617) 494-0414
Web Page: http://www.oberon.com
email: info@oberon.com

Object Design Inc .... **(617) 674-5000**
25 Mall Road
Burlington, MA 01803
Tech:(617) 674-5040 Toll Free:(800) 962-9620
Fax:(617) 674-5010
Web Page: http://www.odi.com
email: support@odi.com
Compuserve: GO ODIFORUM

Object Share ....(949) 833-1122
16811 Hale Ave, Suite A
Irvine, CA 92614
Tech:(800) 727-2555 Toll Free:(800) 759-7272
Fax:(949) 833-0209
Tech:(408) 773-7474
Web Page: http://www.objectshare.com
email: support@objectshare.com

Ocean Information Systems Inc ....(626) 339-8888
688 Arrow Grand Circle
Covina, CA 91722
Tech:(626) 339-8888 Toll Free:(800) 325-2496
Fax:(626) 859-7668
Web Page: http://www.ocean-usa.com/ocean
email: tech@ocean-usa.com
BBS:(626) 859-7639

Ocean Isle Soft (see Stac Electronics)

Ocean Office Automation ....(626) 339-8888
688 Arrow Grand Circle
Covina, CA 91722
Web Page: http://www.ocean-usa.com
email: tech@ocean-usa.com

OCLI (Optical Coating Laboratory) ....(707) 545-6440
2789 Northpoint Parkway
Santa Rosa, CA 95407-7397
Tech:(800) 545-6254 Toll Free:(800) 545-6254
Fax:(707) 525-7410
Web Page: http://www.ocli.com

Octel Communications Inc ....(408) 321-2000
1001 Murphy Ranch Road
Milpitas, CA 95035-7912
Fax:(408) 324-2702
Web Page: http://www.octel.com

Odyssey Computing Inc ....(619) 675-3660
16981 Via Tazon, Suite D
San Diego, CA 92127
Toll Free:(800) 965-7224
Fax:(619) 675-1130
Other Address: support@odysseyinc.com
Web Page: http://www.odysseyinc.com
email: info@odysseyinc.com
Compuserve: GO ODYSSEY

Oki Semiconductor ....(408) 720-1900
785 N. Mary Ave

Sunnyvale, CA 94086
Tech:(800) 832-6654 Toll Free:(800) 832-6654
Fax:(408) 720-1918 Fax on Demand:(800) 832-6654
Web Page: http://www.okisemi.com

Oki Telecom ....(770) 995-9800
437 Old Peachtree Road
Suwanee, GA 30174
Toll Free:(800) 554-3112
Fax:(770) 822-2681
Web Page: http://www.oki.com

Okidata ....(609) 235-2600
532 Fellowship Road
Mt. Laurel, NJ 08054-3405
Tech:(800) 634-0089 Toll Free:(800) 654-3282
Fax:(609) 778-4184 Fax on Demand:(800) 654-6651
Fax:(609) 222-5320
Web Page: http://www.okidata.com
email: comments@okidata.com
BBS:(609) 234-5344

Okna Corp (Out of Business)
Web Page: http://www.okna.com

Olicom USA..(800) 265-4266
900 E. Park Blvd, Suite 250
Plano, TX 75074
Tech:(800) 654-2661 Toll Free:(800) 265-4266
Web Page: http://www.olicom.com
email: support@olicom.com
Compuserve: GO OLICOM

Olivr Corp
4 Militia Drive, Suite 12
Lexington, MA 02173
Fax:(781) 863-6155
Web Page: http://www.olivr.com
email: info@olivr.com

Omni Data Systems..............................(314) 273-6800
132 Crestmont Circle
Wildwood, MO 63040
Tech:(800) 766-2449 Toll Free:(800) 766-2449
Fax:(314) 405-1177
Web Page: http://www.omni1.com
email: omni1@stinet.com

Omni Development Inc ....(206) 523-4152
2707 NE Blakeley Street
Seattle, WA 98105-3118

Toll Free:(800) 315-6664
Fax:(206) 523-5896
Web Page: http://www.omnigroup.com
email: info@omnigroup.com

Omnicomp Graphics Corp........................(713) 464-2990
1734 W. Sam Houston Pkwy N.
Houston, TX 77043
Tech:(713) 464-2990 Toll Free:(800) 995-6664
Fax:(713) 827-7540
Web Page: http://www.omnicomp.com
email: techs@omnicomp.com

OmniData International Inc........................(801) 753-7760
124 South 600 West
Logan, UT 84321
Fax:(801) 753-6756
Web Page: http://www.omnidatai.com
email: info@omnidatai.com

Omniprint Inc (see Digital Design Inc)

Omnitech Gencorp Inc...............................(305) 599-9898
1900 NW 97th Ave
Miami, FL 33172
Tech:(305) 599-9898 Toll Free:(800) 222-9618
Fax:(305) 594-2997
Web Page: http://www.omnitechgc.com

Omnitrend ....(860) 673-8910
201 Johnnycake Mountain Road
Burlington, CT 06013-2011
Tech:(860) 673-8910
Fax:(860) 673-3023
Web Page: http://www.omnitrend.com
email: support@omnitrend.com
Compuserve: 72662,455 GO WIRELESS

Omron Advanced Systems Inc ....(408) 727-6644
3945 Freedom Circle, # 700
Santa Clara, CA 95054-1224
Tech:(408) 727-1444
Fax:(408) 727-5540
Web Page: http://www.oas.omron.com
email: sales@oas.omron.com
Compuserve: 75070,2275

On Health Network ....(206) 583-0100
7500 Flying Cloud Dr, Suite 400
Eden Prairie, MN 55344-3739
Toll Free:(800) 952-4773

Web Page: http://www.ivi.com

On Technology Corp ....(617) 374-1400
1 Cambridge Center, 6th Floor
Cambridge, MA 02142-9773
Tech:(800) 767-6683 Toll Free:(800) 767-6683
Fax:(617) 374-1433
Web Page: http://www.on.com
email: info@on.com

One World Systems ....(408) 523-1100
1144 E. Arques Ave
Sunnyvale, CA 94086
Toll Free:(877) 697-2537
Fax:(408) 523-2408
Web Page: http://www.oneworldsystems.com
email: sales@oneworldsystems.com

Ontrack Computer Sys. (see Ontrack Data Rec)

Ontrack Data Recovery ....(612) 937-5161
6321 Bury Drive, Suite 1519
Eden Prairie, MN 55346
Tech:(612) 937-2121 Toll Free:(800) 872-2599
Fax:(612) 937-5815
Web Page: http://www.ontrack.com
email: tech@ontrack.com
Compuserve: 72662,33 or GO ONTRACK
BBS:(612) 937-0860

OnTrack Media Corp ....(415) 331-1692
150-B Shoreline Highway, Suite 22
Mill Valley, CA 94941
Tech:(415) 331-1693 Toll Free:(800) 520-5627
Fax:(415) 331-1695
Web Page: http://www.ontrackmedia.com
email: careerpath@ontrackmedia.com

OnWord Press ....(505) 474-5130
1580 Center Drive
Santa Fe, NM 87505
Tech:(800) 466-9673 Toll Free:(888) 763-8786
Fax:(505) 474-5020
Fax:(505) 474-5030
Web Page: http://www.onwordpress.com
email: orders@hmp.com

Opcode Music Quest ....(650) 856-3333
3950 Fabian Way, Suite 100
Palo Alto, CA 94303
Tech:(650) 812-3205

Fax:(650) 856-3332 Fax on Demand:(415) 812-3207
Tech Support:(650) 856-3331
Web Page: http://www.opcode.com

Open Environment ....(617) 562-1111
25 Travis Street
Boston, MA 02134
Web Page: http://www.oec.com

Open Group, The ....(617) 621-8700
11 Cambridge Center
Cambridge, MA 02142
Fax:(617) 621-0631
Web Page: http://www.opengroup.org
email: direct@opengroup.org

Open Route Networks Inc..........................(508) 898-2800
9 Technology Drive
Westborough, MA 01581-1799
Tech:(508) 898-3100 Toll Free:(800) 545-7464
Fax:(508) 836-5346 Fax on Demand:(800) 545-7464
RMA:(508) 898-2800
Web Page: http://www.openroute.com
email: prohelp_cs@proteon.com
BBS:(508) 366-7827

Open Systems Inc ....(612) 496-2465
1157 Valley Park Dr, Suite 105
Shakopee, MN 55379
Tech:(800) 582-5000 Toll Free:(800) 328-2276
Web Page: http://www.osas.com
email: info@osas.com

OPTi Inc ....(408) 486-8000
1440 McCarthy Blvd
Milpitas, CA 95035
Fax:(408) 486-8001
Web Page: http://www.opti.com
email: syslogic@opti.com

Optibase Inc ....(408) 260-6760
3031 Tisch Way Plaza West, Suite 1
San Jose, CA 95128
Toll Free:(800) 451-5101
Fax:(408) 244-0545
Web Page: http://www.optibase.com
email: sales@optibase.co.il

Optical Access International ....(617) 937-3910
36 Commerce Way
Woburn, MA 01801

Toll Free:(800) 433-5133
Fax:(616) 935-7673

Optical Data Systems Inc ....(972) 234-6400
1101 E. Arapahoe Rd
Richardson, TX 75081-2399
Fax:(972) 234-1467
Web Page: http://www.ods.com
email: ocorporate@ods.com

Optics Storage
85, Defu Lane 10, #04-00, Mode Circle Building
Singapore, 539218 Japan
Web Page: http://www.optics-storage.com.sg

Optima Technology Corp ....(714) 476-0515
17062 Murphy Ave
Irvine, CA 92614
Tech:(714) 476-0515 Toll Free:(800) 411-4237
Fax:(714) 476-0613
Web Page: http://www.optimatech.com
email: techsupport@optimatech.com
BBS:(714) 476-0626

Optio Software ....(770) 283-8500
4800 River Green Pkwy
Duluth, GA 30096
Fax:(770) 283-8699
Web Page: http://www.optiosoftware.com
email: info@optiosoftware.com

ORA Electronics ....(818) 772-2700
9410 Owensmouth Ave
Chatsworth, CA 91311
Tech:(818) 772-2700 Toll Free:(800) 877-7448
Fax:(818) 718-8626
Other Fax:(818) 718-8667
Web Page: http://www.orausa.com
email: info@orausa.com

Oracle Corp ....(650) 506-7000
500 Oracle Pkwy
Redwood Shores, CA 94065
Tech:(650) 506-1500 Toll Free:(800) 542-1170
Fax:(650) 506-7200 Fax on Demand:(650) 506-6985
Tech Support:(800) 392-2999
Web Page: http://www.oracle.com

Orange Cherry/New Media Schoolh.. .....(914) 764-4387
69 Westchester Ave
Pound Ridge, NY 10576

Tech:(914) 764-4104
Fax:(914) 764-0104
Web Page: http://www.nmsh.com
email: nmsh@cloud9.net

Orange Micro Inc ....**(714) 779-2772**
1400 N. Lakeview Ave
Anaheim, CA 92807
Fax:(714) 779-9332
Web Page: http://www.orangemicro.com
email: support@orangemicro.com

Orchid Technology....................................**(510) 651-2300**
45365 Northport Loop West
Fremont, CA 94538
Tech:(510) 661-3000 Toll Free:(800) 577-0977
Fax:(510) 651-6982 Fax on Demand:(510) 661-3199
Tech Support Fax:(510) 651-6982
Web Page: http://www.orchid.com
email: tech@orchid.com
BBS:(510) 651-6837

Origin Systems Inc ....**(512) 434-4263**
5918 W. Courtyard Drive
Austin, TX 78730
Tech:(512) 434-4357 Toll Free:(800) 245-4525
Fax:(512) 794-8959
Tech Fax:(512) 795-8014
Web Page: http://www.origin.ea.com
email: support@origin.ea.com
Compuserve: GO GAMEAPUB
America Online: keyword:origin

Ornetix Network Products ....**(650) 354-0250**
2465 E. Bayshore Rd, Suite 301
Palo Alto, CA 94303
Toll Free:(888) 676-3849
Fax:(650) 354-0253
Web Page: http://www.ornetix.com

OSC (see Macromedia)

Osicom ....**(888) 675-2668**
7402 Hollister Ave
Santa Barbara, CA 93117-2590
Tech:(800) 262-2290
Web Page: http://www.rns.com
email: support@rns.com

Osicom Technologies ....**(781) 647-1234**
411 Waverly Oaks Road, Suite 227

Waltham, MA 02154
Tech:(800) 984-9004 Toll Free:(800) 243-2333
Fax:(781) 647-4474 Fax on Demand:(781) 398-4950
Other Address: http://www.osicom.com
Web Page: http://www.digprod.com
email: techsup@digprod.com
Compuserve: 76366,2245
BBS:(781) 647-5959

OTC (see Output Tech Corp)

Other 90% Technologies Inc, The··········(415) 460-9710
PO Box 2669
San Rafael, CA 94912-2669
Tech:(415) 460-9710
Fax:(415) 460-1919
Web Page: http://www.other90.com
email: support@other90.com

Outlook Software (see Buttonware Software)

Output Technology Corp (OTC)·············(509) 536-0468
2310 N. Fancher Road
Spokane, WA 99212-1381
Tech:(509) 536-0468 Toll Free:(800) 468-8788
Fax:(509) 533-1290
Tech Support Fax:(509) 533-1295
Web Page: http://www.output.com
Compuserve: www.iea.com:80/~output/
BBS:(509) 533-1217

Overland Data Inc·····································(619) 571-5555
8975 Balboa Ave
San Diego, CA 92123-1599
Tech:(619) 571-5555 Toll Free:(800) 729-8725
Fax:(619) 571-0982
Web Page: http://www.overlanddata.com
email: techsupport@overlanddata.com

P.A.C.E. ···(801) 753-1067
PO Box 3986
Logan, UT 84341
Toll Free:(800) 359-6670

Pacific HiTech Inc·····································(801) 501-0866
3855 S. 500 West, Suite M
Salt Lake City, UT 84115
Toll Free:(800) 765-8369
Fax:(801) 261-0310
Web Page: http://www.pht.com
email: sstone@pht.com

Pacific Image Communications Inc ....(626) 457-8880
2121 W. Mission Rd, Ste 301
Alhambra, CA 91803
Tech:(626) 457-9684
Fax:(626) 457-8881
Web Page: http://www.supervoice.com
email: support@supervoice.com

Pacific Magtron Inc ....(408) 956-8888
1600 California Circle
Milpitas, CA 95035
Tech:(408) 956-8888
Fax:(408) 956-8488
Web Page: http://www.pacificmagtron.com
email: sales@pacmag.com

Pacific Micro Data Inc ....(714) 955-9090
16751 Millikan Ave, Building 140
Irvine, CA 92714
Tech:(714) 955-9090 Toll Free:(800) 933-7575
Fax:(714) 955-9490
Web Page: http://www.pmdraid.com
email: sales@pmicro.com

Pacific Microelectronics (see Net USA)

Packard Bell ....(888) 211-4159
31717 Latienda Dr
Westlake Village, CA 91362
Tech:(800) 733-4433 Toll Free:(800) 733-4411
Fax:(801) 579-0093
Web Page: http://www.packardbell.com
email: support@packardbell.com
BBS:(916) 386-9899

Palindrome Corp (see Seagate Software)

Panacea (see Spacetec IMC)

Panamax ....(415) 499-3900
150 Mitchell Blvd
San Rafael, CA 94903-2057
Tech:(800) 472-5555 Toll Free:(800) 472-5555
Fax:(415) 472-5540
Web Page: http://www.panamax.com
email: custrelations@panamax.com

Panasonic Comm & Systems ....(201) 348-7000
One Panasonic Way
Secaucus, NJ 07094
Tech:(800) 222-0584 Toll Free:(800) 726-2797

Service Locator:(800) 447-4700
Web Page: http://www.panasonic.com
BBS:(201) 863-7845

Panasonic Office Automation ....(714) 373-7412
6550 Jatella Avenue
Cypress, CA 90630
Tech:(800) 726-2797 Toll Free:(800) 726-2797
Fax on Demand:(800) 222-0854
Sales:(800) 662-3537
Web Page: http://www.panasonic.com
BBS:(201) 863-7845
Broadcast & Digital Systems ....(800) 526-6610
Computer & System Support Hot Line ....(800) 726-2797
Web Page: http://www.panasonic.com
Computer & Systems Support ....(800) 726-2797
Toll Free:(800) 742-8086
Computer Peripherals Support Line ....(800) 346-4768
Tech:(800) 993-2333 Toll Free:(800) 222-0584
Web Page: http://www.panasonic.com
Consumer Products ....(800) 435-7329
Laptop Computer Information ....(800) 527-8675
Toll Free:(800) 662-3537
Web Page: http://www.panasonic.com
Manuals and Repair Parts ....(800) 833-9626
Web Page: http://www.panasonic.com
Printers ....(800) 742-8086
Toll Free:(800) 222-0584

Panasonic Services Company ....(800) 833-9626
Fax:(800) 237-9080
Web Page: http://www.panasonic.com

Pantheon ....(650) 429-4400
444 Castro St, Suite 101
Mountain View, CA 94041
Fax:(650) 429-4500
Web Page: http://www.pantheoninc.com
email: pantheoninfo@zip2.com

PaperClip Software Inc ....(201) 329-6300
3 University Plaza, Suite 600
Hackensack, NJ 07601
Tech:(201) 329-6300
Fax:(201) 487-0613
Web Page: http://www.paperclip.com
email: support@paperclip.com

Paperless Corp ....(972) 235-4008
1750 N. Collins, Suite 104

Richardson, TX 75080
Toll Free:(800) 658-6486
Fax:(972) 680-2566
Web Page: http://www.paperlesscorp.com
email: paper@onramp.net

Paracel Online Systems Inc ···················(626) 744-2000
12770 Coit Road, Suite 450
Dallas, TX 75251
Tech:(626) 744-2000 Toll Free:(888) 727-2235
Fax:(626) 744-2001
Web Page: http://www.online.paracel.com
email: info@paracel.com

ParaCom Corp ·······································(978) 470-8686
10 Tower Office Park, Suite 301
Woburn, MA 01801

Paradigm Software (see Work Wise Software)

Paradise (see Western Digital Corp)

Paragraph International ··························(408) 364-7700
2011 N. Shoreline Blvd, Bldg 10 MS571
Mountain View, CA 94043
Web Page: http://www.paragraph.com
email: info@paragraph.com

Parallax Inc ···(916) 624-8333
3805 Atherton Road, Suite 102
Rocklin, CA 95765
Tech:(916) 624-8333 Toll Free:(888) 512-1024
Fax:(916) 624-8003 Fax on Demand:(916) 624-1869
Web Page: http://www.parallaxinc.com
email: info@parallaxinc.com

Parana Supplies Corp ····························(310) 793-1325
3625 Del Amo Blvd, Suite 260
Torrance, CA 90503
Tech:(800) 472-7262 Toll Free:(800) 472-7262
Fax:(310) 793-1343

Parc Place Digitalk (see Object Share)

Parsons Technology ······························(319) 395-9626
1 Martha's Way, PO Box 100
Hiawatha, IA 52233-0100
Tech:(319) 395-7314 Toll Free:(800) 779-6000
Fax:(319) 395-0102
Automated Help:(888) 830-4357
Web Page: http://www.parsonstech.com
Compuserve: GO PA

Microsoft: Parsons
America Online: keyword PARSONS

Parts Now Inc ·· **(608) 276-8688**
3517 W. Beltline Hwy
Madison, WI 53713
Toll Free:(800) 886-6688
Fax:(608) 276-9593
Web Page: http://www.partsnowinc.com
email: ibuy@partsnow.com

Patton & Patton Software Corp ················ **(408) 778-6557**
16890 Church Street, Building 16
Morgan Hill, CA 95037
Tech:(408) 778-6557 Toll Free:(800) 525-0082
Fax:(408) 778-9972
Web Page: http://www.patton-patton.com
email: support@patton-patton.com

Paul Mace Software Inc ···························· **(541) 488-2322**
400 Williamson Way
Ashland, OR 97520
Tech:(541) 488-0224 Toll Free:(800) 944-0191
Fax:(541) 488-1549
Web Page: http://www.pmace.com
email: tech@pmace.com
Compuserve: 76704,17

PC & MAC Connection ······························ **(603) 446-7721**
6 Mill Street
Marlow, NH 03456
Tech:(800) 800-0024 Toll Free:(800) 800-5555
Fax:(603) 446-7791
Web Page: http://www.pcconnection.com
email: corporate@pcconnection.com

PC Checks & Supplies Inc ························ **(205) 969-0024**
371 Summit Drive
Birmingham, AL 35243
Toll Free:(800) 322-5317
Fax:(800) 322-5318
Web Page: http://www.pcchecks.com
email: pcchecks@pcchecks.com

PC DOCS ·· **(781) 273-3800**
25 Burlington Mall Road, 4th Floor
Burlington, MA 01803
Tech:(850) 942-5000 Toll Free:(800) 933-3627
Fax:(850) 656-5559
Web Page: http://www.pcdocs.com

email: support@pcdocs.com

PC Dynamics Inc ....(818) 889-1741
31332 Via Colinas, Suite 102
Westlake Village, CA 91362
Tech:(818) 889-1742 Toll Free:(800) 888-1741
Fax:(818) 889-1014
Web Page: http://www.pcdynamics.com
email: sales@pcdynamics.com

PC Guardian..(415) 459-0190
1133 E. Francisco Blvd, Suite D
San Rafael, CA 94901
Tech:(415) 459-0190 Toll Free:(800) 288-8126
Fax:(415) 459-1162
Web Page: http://www.pcguardian.com
email: pbasaran@pcguardian.com

PC Power & Cooling Inc ....(760) 931-5700
5995 Avenida Encinas
Carlsbad, CA 92008
Toll Free:(800) 722-6555
Fax:(760) 931-6988
Web Page: http://www.pcpowercooling.com
email: PCPower@ix.netcom.com

PC Service Source ....(972) 406-8583
2350 ValleyView Lane
Dallas, TX 75234
Tech:(972) 481-4000 Toll Free:(800) 727-2787
Fax:(972) 406-9081
Web Page: http://www.pcservice.com

PC-Kwik Corp (see Micro Design)

PC-Sig/Spectra Pub (see CD World)

PC411 Inc ....(310) 645-1114
9800 La Cienega Blvd, Suite 411
Inglewood, CA 90301-4440
Tech:(310) 645-1114
Fax:(310) 645-1112
Web Page: http://www.pc411.com
email: info@pc411.com
America Online: tspc411@aol

PCMCIA ....(408) 433-2273
2635 N. 1st Street, Suite 209
San Jose, CA 95134-3209
Fax:(408) 433-9558
Web Page: http://www.pc-card.com

email: office@pcmcia.org

PCPI Technologies .... **(619) 485-8411**
11031 Via Frontera
San Diego, CA 92127
Tech:(619) 485-8411
Fax:(619) 487-5809
Web Page: http://www.pcpi-itc.com
email: submit@pcpi-itc.com
BBS:(619) 674-2196

Peachtree Software .... **(770) 724-4000**
1505 Pavillion Place
Norcross, GA 30093
Tech:(770) 492-6311 Toll Free:(800) 247-3224
DOS support:(770) 492-6312
Web Page: http://www.peachtree.com
email: sales@peachtree.com

Pegasus Disk Technologies .... **(925) 938-5340**
1600 S. Main St, Suite 210
Walnut Creek, CA 94596
Tech:(925) 938-5340
Fax:(925) 938-5341
Web Page: http://www.pegasus-ofs.com
email: support@pegasus-ofs.com

Pegasus Imaging Corp .... **(813) 875-7575**
4522 Spruce St, Suite 200
Tampa, FL 33607
Toll Free:(800) 875-7009
Fax:(813) 875-7705
Web Page: http://www.jpg.com
email: support@jpg.com

Pelikan Inc (see NuKote International)

Pen Magic Software (see Pivotal Graphics)

Penril Datability Networks (see Hayes Corp)

Pentafour Software .... **(562) 467-1141**
12750 Center Court Dr, Ste 410
Cerritos, CA 90703
Web Page: http://www.pentafour.com
email: ibd@pentafour.com

Pentax Technologies Corp .... **(303) 460-1600**
100 Technology Dr
Broomfield, CO 80021
Tech:(303) 460-1820 Toll Free:(800) 543-6144
Fax:(303) 460-1628

Web Page: http://www.pentaxtech.com
email: comments@pentaxtech.com

Penton Overseas .... **(760) 431-0060**
2470 Impala Dr
Carlsbad, CA 92008
Tech:(800) 748-5804
Fax:(760) 431-8110
Web Page: http://www.pentonoverseas.com
email: info@pentonoverseas.com

PeopleSoft .... **(925) 225-3000**
4440 Rosewood Drive, Bldg 4 1st Floor
Pleasanton, CA 94588-3031
Tech:(800) 477-5738 Toll Free:(888) 773-8277
Fax:(925) 225-3100
Tech Support:(925) 467-7239
Web Page: http://www.peoplesoft.com
email: info@peoplesoft.com

Perceptive Solutions Inc .... **(214) 954-1774**
2700 Flora St
Dallas, TX 75201
Tech:(214) 954-1774 Toll Free:(800) 486-3278
Fax:(214) 953-1774
Web Page: http://www.psidisk.com
email: tech@psidisk.com
BBS:(214) 954-1856

PerfectData Corp .... **(805) 581-4000**
110 W. Easy Street
Simi Valley, CA 93065
Toll Free:(800) 973-7332
Fax:(805) 522-5788
Web Page: http://www.perfectdata.com

Persoft Inc .... **(608) 273-6000**
465 Science Dr, PO Box 44953
Madison, WI 53744-4953
Tech:(608) 273-4357 Toll Free:(888) 657-3776
Fax:(608) 273-8227
Other:(800) 368-5283
Web Page: http://www.persoft.com
email: support@persoft.com
BBS:(608) 273-6595

Persona Technologies **(see Monster Cable)**

Personal Training Systems .... **(650) 614-5950**
1005 Hamilton Ct
Menlo Park, CA 94025

Tech:(800) 832-2499 Toll Free:(800) 832-2499
Fax:(650) 463-2522
Web Page: http://www.ptst.com

Personics Corp (see Data Watch Corp)

PHD - Professional Help Desk················(203) 356-7700
800 Summer St, 5th Floor
Stamford, CT 06901
Tech:(203) 356-7700 Toll Free:(800) 474-3725
Fax:(203) 356-7900
Web Page: http://www.prohelpdesk.com
email: phdhelp@phd.com

Philips Consumer Electronics ················(423) 521-4316
One Philips Drive, PO Box 14810
Knoxville, TN 37914-1810
Tech:(423) 475-8869 Toll Free:(800) 531-0039
Fax:(423) 521-4586
Tech Support:(423) 475-6186
Other Address: http://www.philips.com
Web Page: http://www.magnavox.com

Philips Laser Magnetic Storage ··············(719) 593-7900
4425 Arrowswest Dr
Colorado Springs, CO 80907
Tech:(719) 593-4393 Toll Free:(800) 777-5674
Fax:(719) 593-4597
Web Page: http://www.philipslms.com
email: techsupport@lmsmail.lms.com

Phoenix Technologies ····························(781) 551-5000
320 Norwood Park South
Norwood, MA 02062-3950
Tech:(781) 551-4000 Toll Free:(800) 677-7300
Fax:(781) 551-3750
Fax:(781) 551-5001
Web Page: http://www.phoenix.com

Phoenix Technologies ····························(408) 570-1000
411 E. Plumeria
San Jose, CA 95134
Tech:(312) 541-0262 Toll Free:(800) 677-7305
Fax:(408) 570-1001
Web Page: http://www.phoenix.com

PhotoDisc Inc ···(206) 441-9355
2013 Fourth Ave, 4th Floor
Seattle, WA 98121-2460
Tech:(206) 441-9355 Toll Free:(800) 528-3472
Fax:(206) 441-9379 Fax on Demand:(800) 528-3472

Fax:(206) 441-4961
Web Page: http://www.photodisc.com
email: sales@photodisc.com

Physician Micro Systems Inc ....(206) 441-8490
2033 Sixth Avenue, Suite 707
Seattle, WA 98121
Fax:(206) 441-8915
Web Page: http://www.pmsi.com
email: info@pmsi.com
BBS:(206) 443-1730

Piiceon ....(408) 362-0266
3610 Shell Ave
San Jose, CA 95136
Tech:(800) 366-2983 Toll Free:(800) 366-2983
Fax:(408) 943-1309
Fax:(408) 362-0267
Web Page: http://www.piiceon.com
email: tgil@piiceon.com

Pinnacle Data Systems Inc ....(614) 487-1150
2155 Dublin Road
Columbus, OH 43228
Toll Free:(800) 882-8282
Fax:(614) 487-8568
Web Page: http://www.pinnacle.com
email: info@pinnacle.com

Pinnacle Micro Inc ....(714) 789-3000
140 Technology Dr, Suite 500
Irvine, CA 92618-2334
Tech:(714) 595-2185 Toll Free:(800) 553-7070
Fax:(714) 789-3150
Tech Support:(888) 805-3588
Web Page: http://www.pinnaclemicro.com
email: fasteddie@codenet.com
BBS:(714) 789-3048

Pinnacle Publishing ....(770) 565-1763
PO Box 7225
Marietta, GA 30007-2255
Tech:(206) 251-3513 Toll Free:(800) 788-1900
Fax:(770) 565-8232
Web Page: http://www.pinpub.com
email: custserve@pinpub.com
Compuserve: GO PINNACLE
BBS:(206) 251-6217

Pinnacle Software ....**(514) 345-9578**
CP386
Mont Royal, PQ H3P 3C6 Canada
Tech:(514) 345-9578 Toll Free:(800) 242-4775
Web Page: http://users.aol.com/psoftinfo
email: psoftinfo@aol.com
Compuserve: 70154,1577

Pinnacle Systems ....**(650) 526-1600**
280 N. Bernardo Ave
Mountain View, CA 94043
Tech:(650) 237-1800
Fax:(650) 526-1601
Web Page: http://www.pinnaclesys.com
email: info@pinnaclesys.com

Pinpoint Publishing ....**(707) 523-0400**
PO Box 1359
Glen Ellen, CA 95442
Toll Free:(800) 788-5236
Fax:(707) 523-0469
Web Page: http://www.pinpointpub.com
email: pinpoint@pinpointpub.com
Compuserve: 74774,572
BBS:(707) 523-0468

Pioneer New Media Technologies ....**(310) 952-2111**
2265 E. 220th Street
Long Beach, CA 90810-1643
Tech:(800) 872-4159 Toll Free:(800) 444-6784
Fax:(310) 952-2990 Fax on Demand:(310) 952-2309
Web Page: http://www.pioneerusa.com
BBS:(310) 835-7980

Pivotal Graphics Inc ....**(408) 451-0201**
1435 Koll Circle, Suite 111
San Jose, CA 95112-4610
Tech:(408) 954-2700
Fax:(408) 954-0118
Fax:(408) 451-0214
Web Page: http://www.pivotalusa.com
email: sales@pivotalusa.com
Compuserve: 102016,435

Pivotal Software ....**(604) 988-9982**
224 W. Esplande, Suite 300
North Vancouver, BC V7M 3M6 Canada
Tech:(604) 988-9982 Toll Free:(888) 275-7486
Fax:(604) 988-0035
Web Page: http://www.pivotal.com

email: info@pivotal.com

Pixar Animation Studios ....(510) 236-4000
1001 W. Cutting Blvd
Richmond, CA 94804
Tech:(800) 937-3179 Toll Free:(800) 888-9856
Fax:(510) 236-0388
Web Page: http://www.pixar.com
email: webmaster@pixar.com

Pixar Interactive (see Pixar Animation)

PKware Inc ....(414) 354-8699
9025 N. Deerwood Drive
Brown Deer, WI 53223-2480
Tech:(414) 354-8699
Fax:(414) 354-8559 Fax on Demand:(414) 354-8699
Web Page: http://www.pkware.com
email: support@pkware.com
Compuserve: 75300,730:GO PKWARE
BBS:(414) 354-8670

PlainTree Systems ....(617) 965-5811
150 Wells Ave
Newton, MA 02159
Tech:(800) 831-1095 Toll Free:(800) 370-2724
Fax:(617) 965-2466
Tech Support Fax:(613) 831-6120
Web Page: http://www.plaintree.com
email: sales@plaintree.com
BBS:(613) 831-8312

Plasmon Data, Inc. ....(612) 946-4100
9625 West 76th Street
Minneapolis, MN 55344 USA
Fax:(612) 946-4141
Web Page: http://www.plasmon.com
email: support@plasmon.com
BBS:(612) 946-4130

Platinum Software Corp.........................(949) 453-4000
195 Technology Drive
Irvine, CA 92618
Tech:(800) 309-9990 Toll Free:(800) 999-1809
Fax:(949) 453-4091
Tech Support:(949) 727-1282
Web Page: http://www.platsoft.com

Platinum Technology Inc ....(630) 620-5000
1815 S. Meyers Road
Oakbrook Terrace, IL 60181-5241

Tech:(800) 833-7528 Toll Free:(800) 442-6861
Fax:(630) 691-0718
Tech Fax:(630) 691-0708
Web Page: http://www.platinum.com
email: info@platinum.com

Play Inc ....(916) 851-0800
2890 Kilgore Road
Rancho Cordova, CA 95670-6133
Tech:(916) 851-0900 Toll Free:(800) 306-7529
Fax:(916) 851-0801
Tech Fax:(916) 853-9831
Web Page: http://www.play.com
email: customerservice@play.com

Play Pro Software Inc.................................(408) 969-0800
3350 Scott Blvd
Santa Clara, CA 95054
Web Page: http://www.playprosoft.com

Plextor ....(408) 980-1838
4255 Burton Drive
Santa Clara, CA 95054
Tech:(800) 886-3935 Toll Free:(800) 475-3986
Fax:(408) 986-1010
another BBS:(408) 986-1474
Web Page: http://www.plextor.com
email: support@plextor.com
BBS:(408) 986-1569

Plus and Minus Software Corp ....(713) 984-7626
908 Town & Country Blvd, Suite 120
Houston, TX 77024
Fax:(713) 984-7576
Web Page: http://www.talyon.com
email: +and-@plusandminus.com

Plus Development Corp (see Quantum Corp)

PNY Electronics...(973) 515-9700
299 Webro Road
Parsippany, NJ 07054
Tech:(973) 438-6300 Toll Free:(800) 234-4597
Fax:(973) 560-5590
Tech support:(800) 234-4597
Web Page: http://www.pny.com

Poet Software ....(650) 286-4640
999 Baker Way, Ste 100
San Mateo, CA 94404
Toll Free:(800) 950-8845

Fax:(650) 286-4630
Web Page: http://www.poet.com
email: support@poet.com

PointCast Inc ....(408) 990-7000
501 Macara Ave
Sunnyvale, CA 94086
Toll Free:(800) 548-2203
Fax:(408) 990-0080
Web Page: http://www.pointcast.com

Polaris Software ....(760) 747-1528
1928 Don Lee Place
Escondido, CA 92029
Toll Free:(800) 338-5943
Fax:(760) 738-0113
Fax:(760) 489-8243
Web Page: http://www.polarissoftware.com
email: support@PolarisSoftware.com
Compuserve: GO POLARIS

Polaroid Corporation ....(781) 386-2000
549 Technology Square
Cambridge, MA 02139
Toll Free:(800) 343-5000
Fax:(781) 386-3263
Digital Products:(800) 432-5355
Web Page: http://www.polaroid.com

Polygon Inc ....(314) 432-4142
1350 Baur Blvd, PO Box 8470
Saint Louis, MO 63132
Tech:(314) 432-4142
Fax:(314) 997-9696
Web Page: http://www.polygon.com
email: support@polygon.com

Port Inc ....(203) 852-1102
66 Fort Point Street
Norwalk, CT 06855
Tech:(800) 350-7678
Toll Free:(800) 242-3133
Fax:(203) 866-0221
Fax on Demand:(800) 322-7678
Fax:(203) 899-8469
Web Page: http://www.port.com/
email: TechSupport@port.com

Portable Graphics (see Template Graphics)

Portrait Displays Inc.................................(925) 227-2700
5117 Johnson Dr
Pleasanton, CA 94588

Tech:(925) 227-2716
Fax:(925) 227-2705
Web Page: http://www.portrait.com
email: mestrin@portrait.com

POSitive Software Co .... (509) 735-9194
1300 Columbia Center Blvd
Richland, WA 99352
Tech:(509) 735-9194 Toll Free:(800) 735-6860
Fax:(509) 735-6299
Web Page: http://www.pointofsale.com
email: postech@pointofsale.com
BBS:(509) 736-9544

Power BBS Computing .... (516) 938-0506
35 Fox Ct
Hicksville, NY 11801
Fax:(516) 681-3226
Web Page: http://www.powwwerworkgroup.com
BBS:(516) 822-7396

Powercom America Inc .... (714) 632-8889
1040A S. Melrose Street
Placentia, CA 92870-7119
Tech:(800) 666-8931 Toll Free:(800) 666-8931
Fax:(714) 632-8868
Web Page: http://www.powercom-usa.com

Powercore Inc (see CE Software)

PowerProduction Software .... (408) 358-2358
432 Los Gatos Blvd
Los Gatos, CA 95032
Fax:(408) 358-1186
Web Page: http://www.powerproduction.com
email: techsupport@powerproduction.com

PowerQuest Corp .... (801) 437-8900
1359 N. Research Way, Building K
Orem, UT 84097
Tech:(801) 226-6834 Toll Free:(800) 379-2566
Fax:(801) 226-8941 Fax on Demand:(800) 379-2566
Web Page: http://www.powerquest.com
email: support@powerquest.com

Powersoft Corp (see Sybase Inc)

Practical Peripherals .... (770) 441-0896
5853 Peachtree Industrial Blvd, PO Box 921789
Norcross, GA 30092-3405
Tech:(770) 840-9966 Toll Free:(800) 934-2937

Fax:(770) 734-4615 Fax on Demand:(800) 225-4774
Tech Fax:(770) 734-4601
Web Page: http://www.practinet.com
email: practical.support@hayes.com
Compuserve: GO PPIFORUM
America Online: Keyword PPI
BBS:(770) 734-4600

Prairie Group ....(515) 225-3720
PO Box 65820
West Des Moines, IA 50265
Tech:(515) 225-4122
Fax:(515) 225-2422
Orders only:(800) 346-5392
Web Page: http://www.prgrsoft.com
email: support@prgrsoft.com
Compuserve: 72662,131
America Online: Prairiesft

Precision Digital Images Corp ....(425) 882-0218
8520 154th Ave NE
Redmond, WA 98052
Toll Free:(800) 678-6505
Fax:(425) 867-9177
Web Page: http://www.precisionimages.com
email: techsupport@precisionimages.com

Premenos Corp (see Harbinger)

Premio Computer ....(626) 333-5121
938 Radecki Court
City Of Industry, CA 91748
Tech:(626) 839-3100 Toll Free:(800) 677-6477
Fax:(626) 369-6803
Fax:(626) 839-3111
Web Page: http://www.premiopc.com
email: support@premiopc.com
BBS:(626) 330-9749

Prescience (see Waterloo Maple Software)

Priam Systems (see AST Research)

Primavera Systems Inc ....(610) 667-8600
2 Bala Plaza
Bala Cynwyd, PA 19004
Tech:(610) 668-3030 Toll Free:(800) 423-0245
Fax:(610) 667-7894
Sure Track Tech:(610) 667-7100
Web Page: http://www.primavera.com
email: usatech@primavera.com

Compuserve: GO PRIMAVERA
BBS:(610) 660-5833

Princeton ....(714) 751-8405
2801 S. Yale St
Santa Ana, CA 92704
Toll Free:(800) 747-6249
Fax:(714) 751-5736
Web Page: http://www.prgr.com
email: sales@prgr.com

Printronix Inc ....(949) 221-2515
17500 Cartwright Rd, PO Box 19559
Irvine, CA 92623-9559
Tech:(949) 863-1900 Toll Free:(800) 826-3874
Fax:(949) 660-8682
Web Page: http://www.printronix.com
email: support@printronix.com

Pro CD Inc ....(402) 537-6180
222 Rosewood Drive
Danvers, MA 01923-4520
Tech:(402) 537-6181
Fax:(402) 537-6182 Fax on Demand:(800) 397-7623
Web Page: http://www.procd.com
email: technical.support@procd.com

Pro-C Limited ....(519) 742-9521
709 Glasgow Street
Kitchener, OT N2M 2N7 Canada
Fax:(519) 742-1099
Web Page: http://www.pro-c.com
email: support@pro-c.com
Compuserve: GO PROC

Process Software Corp ....(508) 879-6994
959 Concord Street
Framingham, MA 01701
Tech:(508) 879-6994 Toll Free:(800) 722-7770
Fax:(508) 879-0042
Web Page: http://www.process.com
email: Sales@Process.com

Procom Technology ....(949) 852-1000
2181 Dupont Drive
Irvine, CA 92612
Tech:(800) 800-8600 Toll Free:(800) 800-8600
Fax:(949) 852-1221
Tech Support Fax:(949) 261-6452
Web Page: http://www.procom.com

email: support@procom.com
Compuserve: 75300,2312

Prodigy Services Company ······················(914) 448-8000
44 S. Broadway
White Plains, NY 10601
Tech:(800) 284-5933 Toll Free:(800) 776-3449
Service Information:(800) 776-0845
Web Page: http://www.prodigy.com

Programmer's Paradise Inc ······················(732) 389-8950
1163 Shrewsbury Ave
Shrewsbury, NJ 07702-4321
Tech:(732) 389-9229 Toll Free:(800) 445-7899
Web Page: http://www.pparadise.com

Programmer's Super Shop ······················(617) 740-2510
90 Industrial Park Road
Hingham, MA 02043
Toll Free:(800) 421-8006
Fax:(617) 740-2728
Web Page: http://computing.supershops.com

Programmers Warehouse (see Breakthrough) ··············

Progress Software Corp ··························(781) 280-4000
14 Oak Park
Bedford, MA 01730
Tech:(781) 280-4999 Toll Free:(800) 477-6473
Fax:(781) 275-4543
Web Page: http://www.progress.com
email: support@progress.com

Progressive Networks (see Real Networks)

Prometheus Products (Out of Business)
Web Page: http://www.prometheusproducts.com

Promise Technology Inc ··························(408) 452-0948
1460 Koll Circle, Suite 103
San Jose, CA 95112
Tech:(408) 452-1180 Toll Free:(800) 888-0245
Fax:(408) 452-1534 Fax on Demand:(408) 452-9160
Tech Fax:(408) 452-9163
Web Page: http://www.promise.com
email: support@promise.com
BBS:(408) 452-1267

Prostar Interactive MediaWorks··············(604) 273-4099
13880 Mayfield Place
Richmond, BC V6V2E4 Canada
Tech:(604) 273-4099

Fax:(604) 273-4046
Web Page: http://www.minicat.com

Protec Microsystems Inc ·························(514) 630-5832
297 Labrosse
Pointe-Claire, QU H9R 1A3
Tech:(514) 630-5832 Toll Free:(800) 363-8156
Fax:(514) 630-2987
Tech Fax:(514) 694-6973
Web Page: http://www.protec.ca
email: techsupp@protec.ca

Proteon Inc (see Open Route Networks)

Proteq Technologies Pte Ltd
413 Whitree Lane
Chesterfield, MO 63017
Fax:(314) 434-1993

Provantage Corp ···································(330) 494-8715
7249 Whipple Ave NW
North Canton, OH 44720-7143
Tech:(330) 494-8715 Toll Free:(800) 336-1166
Fax:(330) 494-5260
Web Page: http://www.provantage.com
email: sales@provantage.com

ProVUE Development ····························(714) 841-7779
18411 Gothard St, Unit A
Huntington Beach, CA 92648
Tech:(714) 841-8779 Toll Free:(800) 966-7878
Fax:(714) 841-1479
Web Page: http://www.provue.com
email: support@provue.com

Proxim Inc ··(650) 960-1630
295 N. Bernardo Ave
Mountain View, CA 94043
Tech:(650) 526-3640 Toll Free:(800) 229-1630
Fax:(650) 960-1984
Tech Fax:(650) 960-1106
Web Page: http://www.proxim.com
email: support@proxim.com
BBS:(650) 960-2519

Proxima Corp ···(619) 457-5500
9440 Carroll Park Drive
San Diego, CA 92121-2298
Tech:(800) 447-7694 Toll Free:(800) 447-7692
Fax:(619) 457-9647 Fax on Demand:(800) 285-6613
Web Page: http://www.proxima.com

PSI Integration (see Supra Corp)

Psion Inc ....(978) 371-0310
150 Baker Ave
Concord, MA 01742-2727
Toll Free:(800) 997-7466
Fax:(978) 371-9611 Fax on Demand:(978) 369-4337
Web Page: http://www.psioninc.com
email: usa-support@psion.com

Psygnosis Ltd (see Sony Interactive)

Public Software Library ....(713) 524-6394
PO Box 35705-F
Houston, TX 77235
Toll Free:(800) 242-4775
Fax:(713) 524-6398
Web Page: http://www.pslweb.com
email: feedback@pslweb.com
Compuserve: 71355,470

Pulse Systems ....(516) 329-4222
190 Willow Avenue, 2nd Floor
Bronx, NY 10454-3596
Tech:(800) 933-6003 Toll Free:(800) 444-1585
Fax:(718) 402-9603
Web Page: http://www.willow.com
email: peripherals@willow.com

PureData..(905) 731-6444
9225 Leslie St
Richmond Hill, ON L4B 3H6 Canada
Toll Free:(800) 661-8210
Fax:(905) 731-7017
Web Page: http://www.puredata.com
email: sales@puredata.com
Compuserve: GO PUREDATA
BBS:(905) 731-4679

QLogic ....(714) 438-2200
3545 Harbor Blvd
Costa Mesa, CA 92626
Tech:(800) 737-6524 Toll Free:(800) 662-4471
Fax:(714) 668-5090
Web Page: http://www.qlc.com
email: a_dunbar@qlc.com

QMS Inc ....(334) 633-4300
One Magnum Pass
Mobile, AL 36618
Tech:(334) 633-4500 Toll Free:(800) 523-2696

Fax:(334) 633-4866 Fax on Demand:(800) 633-7213
Web Page: http://www.qms.com
email: info@qms.com
Compuserve: GO QMSPRINT
BBS:(334) 633-3632

QNX Software Systems Ltd ····················(613) 591-0931
175 Terence Matthews Crescent
Kanata, ON K2M 1W8 Canada
Toll Free:(800) 676-0566
Fax:(613) 591-3579
Web Page: http://www.qnx.com
email: info@qnx.com

Quadralay Corp ·································(512) 719-3399
9101 Burnet Road, Suite 105
Austin, TX 78758
Fax:(512) 719-3606
Web Page: http://www.quadralay.com
email: info@quadralay.com

Quadtel Corp (see Phoenix Technologies)

Qualcomm Inc·································(619) 587-1121
6455 Lusk Blvd
San Diego, CA 92121-2779
Fax:(619) 658-2100
Web Page: http://www.qualcomm.com

Qualitas ···(301) 578-8400
8601 Georgia Ave, #908
Silver Spring, MD 20910
Tech:(301) 907-7400 Toll Free:(800) 733-1377
Fax:(301) 589-8872
Web Page: http://www.qualitas.com
email: tech@qualitas.com

Qualix Group (see Full Time Software)

Qualtec Data Products Inc·····················(319) 263-8144
47767 Warm Springs Blvd
Fremont, CA 94539
Tech:(800) 628-4413 Toll Free:(800) 628-4413
Fax:(510) 490-8471
Web Page: http://www.pcsecurity.com

Qualtek Electronic Corp ·······················(440) 951-1884
7675 Jenther Dr, Suite E
Mentor, OH 44060
Tech:(440) 951-3300
Fax:(440) 951-0107

Fax:(440) 951-7252
Web Page: http://www.qualtekusa.com
email: qualtek@ix.netcom.com

Quantum Corp ··(408) 894-4000
500 McCarthy Blvd
Milpitas, CA 95035
Tech:(800) 826-8022 Toll Free:(800) 624-5545
Fax:(408) 894-3282 Fax on Demand:(800) 434-7532
Other:(800) 345-3377
Web Page: http://www.quantum.com
email: qsupport@qntm.com
BBS:(408) 894-3214

Quark Inc ··(303) 894-8888
1800 Grant Street
Denver, CO 80203
Tech:(303) 894-8899 Toll Free:(800) 676-4575
Fax:(303) 894-3399
Tech Fax:(303) 894-3398
Web Page: http://www.quark.com
email: cservice@quark.com

Quarter-Inch Cartridge Dr Stds ················(805) 963-3853
311 East Carrillo Street
Santa Barbara, CA 93101
Fax:(805) 962-1541
Web Page: http://www.qic.org
email: ray@qic.org

Quarterdeck Corp ··(310) 309-3700
13160 Mindanao Way
Marina Del Rey, CA 90292-9705
Tech:(573) 875-0530 Toll Free:(800) 225-8148
Fax:(813) 523-2331 Fax on Demand:(800) 762-6832
Tech Support Fax:(573) 443-3282
Web Page: http://www.qdeck.com
email: info@quarterdeck.com
Compuserve: GO QUARTERDECK
BBS:(573) 875-0503

Customer Service ··(573) 443-3282
Toll Free:(800) 354-3222
Fax:(800) 354-3329
Automated Tech Support:(800) 762-6832
email: info@quarterdeck.com

Internet/Communications Products ··············(573) 875-0530
Toll Free:(800) 339-1136
IWare Connect:(573) 499-4558
Web Page: http://support.quarterdeck.com

Utilities Products .... (573) 875-0932
Toll Free:(800) 339-1136
Web Page: http://support.quarterdeck.com

Quarterdeck Select .... **(813) 523-9700**
5770 Rosevelt Blvd, Suite 400
Clearwater, FL 34620
Tech:(800) 683-0854 Toll Free:(800) 683-6696
Fax:(813) 532-4222
Fax:(813) 523-2391
Web Page: http://www.quarterdeck.com/
email: info@quarterdeck.com
Compuserve: GO QUARTERDECK

Quercus Systems .... **(408) 372-7399**
PO Box 51218
Pacific Grove, CA 93950
Toll Free:(800) 440-5944
Fax:(408) 372-5776
Web Page: http://www.quercus-sys.com
email: info@quercus-sys.com
Compuserve: 75300,2450 or GO QUERCUS

Qume (see Wyse Technology)

Qume Corp (see Data Technology)

Quyen Systems (see netViz Corp)

Rabbit Software (see Tangram)

Racal Data Group .... **(954) 846-1601**
1601 N. Harrison Parkway
Sunrise, FL 33323-2899
Toll Free:(800) 722-2555
Fax:(954) 846-4942
Web Page: http://www.racal.com/rdg

RAD Data Communications .... **(201) 529-1100**
900 Corporate Drive
Mahwah, NJ 07430
Tech:(201) 529-1100 Toll Free:(800) 332-9225
Fax:(201) 529-5777
Web Page: http://www.rad.co.il
email: amnon@radusa.com

Radio Shack .... **(817) 390-3200**
1800 One Tandy Center
Fort Worth, TX 76102
Tech:(800) 843-7422 Toll Free:(800) 843-7422
Fax:(817) 390-3240
Web Page: http://www.tandy.com

email: support@tandy.com

Radius Inc ....(650) 404-6000
215 Moffett Park Drive
Sunnyvale, CA 94089-1374
Tech:(408) 541-5700 Toll Free:(800) 227-2795
Fax:(408) 541-6150 Fax on Demand:(800) 966-7360
Automated Tech support:(800) 332-9225
Other Address: http://research.radius.com
Web Page: http://www.radius.com
Compuserve: 76004,2155
America Online: TadiusTS

RAG Electronics (see Test Equity)

Rail Systems Center ....(412) 751-8470
2013 Country Club Drive
Mount Vernon, PA 15135-3040
Fax:(412) 754-0176
Web Page: http://www.lm.com/~urichard/rsc_1.htm
email: urichard@lm.com

Raima Corp ....(206) 515-9477
4800 Columbia Center, 701 Fifth Ave
Seattle, WA 98104
Tech:(206) 557-5333 Toll Free:(800) 327-2462
Fax:(206) 748-5200
Other:(800) 275-4724
Web Page: http://www.raima.com
email: sales@raima.com

Rainbow Technologies ....(949) 450-7300
50 Technology Dr
Irvine, CA 92618
Tech:(800) 852-8569 Toll Free:(888) 667-4728
Fax:(949) 450-7447
Other Address: http://www.rainbow.com
Web Page: http://isg.rainbow.com
email: cryptoswift@rainbow.com

Rambus inc ....(650) 944-8000
2465 Latham St
Mountain View, CA 94040
Fax:(650) 944-8080
Web Page: http://www.rambus.com
email: info@rambus.com

Rancho Technology Inc. ....(909) 987-3966
10783 Bell Court
Rancho Cucamonga, CA 91730
Fax:(909) 989-2365

Web Page: http://www.rancho.com
email: support@rancho.com

Rand McNally & Company .... **(800) 333-0136**
Tech:(847) 329-6968
Web Page: http://www.randmcnally.com

Rand Software Corp .... **(802) 362-0663**
12 Bonnet St
Manchester Ctr, VT 05255
Toll Free:(888) 726-3763
Web Page: http://www.randsoft.com
email: support@randsoft.com

Raosoft Inc .... **(206) 525-4025**
6645 NE Windermere Road
Seattle, WA 98115-7942
Tech:(206) 525-4025
Fax:(206) 525-4947
Web Page: http://www.raosoft.com
email: raosoft@raosoft.com

Raster OPS (see True Vision Raster OPS)

Ray Dream (MetaCreations) .... **(831) 430-4100**
1804 N. Shoreline Blvd
Mountain View, CA 94043
Tech:(831) 430-4200 Toll Free:(800) 846-0111
Fax:(650) 960-1198
Corporate:(805) 566-6200
Web Page: http://www.raydream.com

Rayovac Corp .... **(608) 275-4694**
601 Rayovac Drive
Madison, WI 53711-2497
Toll Free:(800) 237-7000
Fax:(608) 275-4577
Web Page: http://www.rayovac.com

Raytheon .... **(508) 470-9393**
Web Page: http://www.raytheon.com

Reach Software
872 Hermosa Drive
Sunnyvale, CA 94086
Fax:(408) 733-9265
Web Page: http://www.reachsoft.com
email: support@reachsoft.com
Compuserve: GO REACH

ReadMe.DOC .... **(717) 264-0843**
975 Progress Road

Chambersburg, PA 17201
Toll Free:(800) 678-1473
Fax:(717) 264-8614
Web Page: http://www.readmedotdoc.com
email: readme@cvn.net

Ready-To-Run Software Inc ....(978) 692-9922
4 Pleasant Street, PO Box 2038
Forge Village, MA 01886
Toll Free:(800) 743-1723
Fax:(978) 692-9990
Web Page: http://www.rtr.com
email: support@rtr.com

Real Networks Inc ....(206) 674-2700
1111 3rd Ave, Suite 2900
Seattle, WA 98101
Tech:(206) 674-2650 Toll Free:(888) 768-3248
Fax:(206) 674-2699
Web Page: http://www.realaudio.com
email: customer_service@prognet.com

Real Time Integration Inc..........................(425) 576-0822
733 7th Ave, Suite 214
Kirkland, WA 98033
Toll Free:(888) 675-1122
Fax:(888) 670-1122
Fax:(206) 374-2418
Web Page: http://www.realtimeint.com
email: info@realtimeint.com

Reality Online Inc ....(800) 346-2024
1000 Madison Ave
Norristown, PA 19403
Tech:(800) 777-7424
Web Page: http://www.moneynet.com
email: websupport@moneynet.com

Reality Tech (see Reality Online Inc)

RealWorld Corp ....(603) 641-0200
670 Commercial Street, PO Box 9516
Manchester, NH 03108-9516
Tech:(603) 288-3433 Toll Free:(800) 678-6336
Fax:(603) 641-0230
Web Page: http://www.megamart.com/megascore
email: sales@realworldcorp.com

Red Hat Software Inc ....(919) 547-0012
4201 Research Commons, Ste 100
Research Triangle Park, NC 27709

Web Page: http://www.redhat.com

Red Wing Business Systems Inc············(612) 388-1106
491 Highway 19, PO Box 19
Red Wing, MN 55066
Toll Free:(800) 732-9464
Fax:(612) 388-7950
Web Page: http://www.pressenter.com/~rwbs
email: info@redwingsoftware.com

Relay Point, Inc.·····································(323) 663-9127
1950 Rodney Drive, Suite 205
Los Angeles, CA 90027
Toll Free:(800) 432-8233
Fax:(323) 663-9137
Web Page: http://www.datadepot.com
email: sales@relaypoint.net

Relay Technology (see Net Manage)

Relialogic Corp··(510) 770-3990
48006 Fremont Blvd
Fremont, CA 94538
Tech:(510) 770-3990 Toll Free:(800) 998-3966
Fax:(510) 770-3994
BBS:(510) 226-0632

Relisys···(408) 945-9000
320 S. Milpitas Blvd
Milpitas, CA 95035 USA
Fax:(408) 945-1499
Web Page: http://www.relisys.com

Remco Software Inc·································(701) 225-8336
26 West Villard
Dickinson, ND 58601

Remote Control Intl (see Telemagic)

Reply Corp (see Radius Inc)

Research Information Systems···············(760) 438-5526
2355 Camino Vida Roble
Carlsbad, CA 92009
Tech:(800) 722-1227 Toll Free:(800) 722-1227
Fax:(760) 438-5573
Fax:(760) 438-5266
Web Page: http://www.risinc.com
email: tech@risinc.com

Reseller Management······························(617) 558-4723
800 South St, Ste 305
Waltham, MA 02154

Fax:(617) 558-4757
Web Page: http://www.resellermgmt.com
email: dbarrett@resellermgmt.com

Responsive Software ....(415) 945-3876
1901 Tunnel Road
Berkeley, CA 94705
Toll Free:(800) 669-4611
Fax:(510) 644-1013
Web Page: http://www.responsivesoftware.com
email: westing3@aol.com

Retix (see Sonoma Systems)

Revelation Software Inc ....(617) 577-0300
201 Broadway
Cambridge, MA 02139
Tech:(800) 262-4747 Toll Free:(800) 262-4747
Fax:(617) 494-0008
Web Page: http://www.revelation.com
email: info@revelation.com

Rexon Data Storage (see Tecmar Technologies)

RGB Spectrum ....(510) 814-7000
950 Marina Village Parkway
Alameda, CA 94501
Fax:(510) 814-7026
Web Page: http://www.rgb.com
email: sales@rgb.com

Rhode Island Soft Systems Inc ....(401) 767-3106
342 Park Ave, PO Box 748
Woonsocket, RI 02895-0784
Tech:(401) 767-3106 Toll Free:(800) 959-7477
Fax:(401) 767-3108
Web Page: http://www.risoftsystems.com
email: info@risoftsystems.com
Compuserve: GO RISS,SEC/LIB 5
BBS:(401) 767-3931

Ricoh Corp ....(973) 882-2000
5 Dedrick Place
West Caldwell, NJ 07006
Fax:(973) 882-5840
Web Page: http://www.ricohcorp.com

Consumer Products ....(702) 352-1600
475 Lillard Dr
Sparks, NV 89434
Toll Free:(800) 225-1899
Fax:(702) 352-1615

Web Page: http://www.ricohcorp.com
email: tech_support@ricohcpg.com

Peripheral Products .... (800) 955-3453
3001 Orchard Pkwy
San Jose, CA 95134-2088
Tech:(800) 955-3453 Toll Free:(800) 955-3453
Fax:(408) 944-3312
Web Page: http://www.ricohcorp.com
email: tech@ricohdms.com

Scanners .... (714) 259-1310
1231 Warner Ave
Tustin, CA 92680
Tech:(210) 520-0951 Toll Free:(800) 955-3453
Fax:(714) 556-3505
Web Page: http://www.ricohcorp.com

Rinda Technologies Inc .... (312) 736-6633
4563 N. Elston Avenue
Chicago, IL 60630
Fax:(312) 736-2950
email: rinda technologies.com

Ring King Visibles (see Curtis/Qualtec)

Ring Zero Systems Inc .... (650) 349-9664
1650 S. Amphlett Blvd, Ste 300
San Mateo, CA 94402
Tech:(650) 349-0600
Web Page: http;//www.rzs.com
email: support@rzs.com

River Run Software Group Inc .... (203) 861-0090
8 Greenwich Office Park
Greenwich, CT 06831
Tech:(919) 941-0722
Fax:(203) 861-0096
Tech Fax:(919) 941-0527
Web Page: http://www.riverrun.com
email: support@riverrun.com

RKS Software Inc .... (703) 534-1726
3820 North Dittmar Road
Arlington, VA 22207-4565
Fax:(703) 534-4358
Web Page: http://www.rks-software.com
email: info@rks-software.com

RNS Inc (see Osicom)

Road Warrior International .... (714) 418-1400
3201 S. Shannon

Santa Ana, CA 92704
Tech:(714) 434-8600 Toll Free:(800) 274-4277
Fax:(714) 839-6282
Fax:(714) 434-8601
Web Page: http://www.warrior.com
email: command@warrior.com

Rocket Science Games Inc......................(415) 442-5000
139 Townsend Street, Suite 100
San Francisco, CA 94107
Fax:(415) 442-5001
Tech Support Fax:(415) 442-5002
Web Page: http://www.rocketsci.com

Rockwell Semiconductor Systems.........(714) 221-4600
4311 Jamboree Rd, PO Box C
Newport Beach, CA 92658-8902
Fax:(714) 221-6375
Web Page: http://www.nb.rockwell.com

Rockwell Software Inc ....(440) 646-7800
2424 S. 102nd St
West Allis, WI 53227
Tech:(440) 646-7800
Fax:(414) 321-2211 Fax on Demand:(440) 646-7777
Tech Fax:(440) 646-7801
Web Page: http://www.software.rockwell.com
email: info@software.rockwell.com

Rogue Wave Software..............................(650) 691-9000
1871 Landings Drive, 2nd Floor
Mountain View, CA 94043
Toll Free:(888) 442-9641
Fax:(650) 691-9099
Web Page: http://www.roguewave.com
Compuserve: 70550,2570
BBS:(650) 691-9990

Roland Corp US..(323) 685-5141
7200 Dominion Circle
Los Angeles, CA 90040-3696
Fax:(323) 722-0911 Fax on Demand:(323) 685-5141
Web Page: http://www.rolandus.com

Roland Digital Group ....(714) 727-2100
15271 Barranca Parkway
Irvine, CA 92618-2201
Tech:(800) 542-2307 Toll Free:(800) 542-2307
Fax:(714) 727-2112 Fax on Demand:(800) 542-2307
Web Page: http://www.rolanddga.com

Compuserve: 102121,2763

Ross Systems .... **(770) 351-9600**
Two Concourse Pkwy, Suite 800
Atlanta, GA 30328
Fax:(770) 351-0036
Web Page: http://www.rossinc.com
email: clientsupport@rossinc.com

Ross Technology Inc .... **(512) 349-3108**
5316 Hwy 290 West
Austin, TX 78735
Tech:(512) 436-2061 Toll Free:(800) 767-7937
Fax:(512) 349-3101
Web Page: http://www.ross.com
email: techsupport@ross.com

RSA Data Security .... **(650) 295-7600**
2955 Campus Drive, Suite 400
San Mateo, CA 94403-2507
Fax:(650) 295-7700
Web Page: http://www.rsa.com
email: tech-support@rsa.com

Rupp Technology Corporation .... **(602) 941-4789**
2240 N. Scottsdale Road
Tempe, AZ 85281
Tech:(602) 941-5602 Toll Free:(800) 844-7775
Fax:(602) 941-5505 Fax on Demand:(602) 675-7290
Web Page: http://www.rupp.com
email: sglasgow@rupp.com
Compuserve: 75300,1232

Rybs Electronics .... **(303) 444-6073**
351 W. Arapahoe, PO Box 4521
Boulder, CO 80306-4521
Tech:(303) 444-7927
Web Page: http://www.rybs.com
email: ryoung@rybs.com

S&K Computers, Inc. .... **(303) 699-2220**
7174 S. Sedlia St
Aurora, CO 80016 USA
Fax:(303) 672-1694
Web Page: http://www.skcomputer.com

S3 Incorporated .... **(408) 980-5400**
2841 Mission College Blvd, PO Box 58058
Santa Clara, CA 95052-8058
Tech:(408) 588-8585
Fax:(408) 980-5444

Other:(408) 588-8000
Web Page: http://www.s3.com
email: support@s3.com

Saber Software (see Network Associates)

Sage U.S. Inc .... (972) 818-3900
17950 Preston Road, Suite 800
Dallas, TX 75252
Tech:(978) 768-7490 Toll Free:(800) 285-0999
Fax:(972) 248-9245 Fax on Demand:(978) 768-6100
Tech Fax:(978) 768-7532
Web Page: http://www.timeslips.com
Compuserve: GO TIMESLIPS
America Online: Timeslips

SAI Inc (see Microleague Multimedia)........................

Sampo Technology Inc .... (770) 449-6220
5550 Peachtree Industrial Blvd
Norcross, GA 30071
Tech:(770) 449-6220
Fax:(770) 447-1109

Samsung America Inc .... (562) 802-2211
14251 E. Firestone Blvd, Suite 201
La Mirada, CA 90638
Toll Free:(800) 933-4110
Fax:(562) 802-3011 Fax on Demand:(800) 229-2239
Web Page: http://www.samsungla.com

Samsung Electronics America .... (201) 229-7000
105 Challenger Road
Ridgefield Park, NJ 07660-0511
Tech:(201) 229-4000 Toll Free:(800) 726-7864
Fax:(201) 229-7030 Fax on Demand:(800) 229-2239
Other Address: http://www.samsung.com
Web Page: http://www.sosimple.com

Sandalwood Software Inc .... (801) 379-0789
555 S. State St, Ste 234
Orem, UT 84058
Web Page: http://www.sandalwood.com
email: info@sandalwood.com

SanDisk Corp .... (408) 542-0500
140 Caspian Court
Sunnyvale, CA 94089
Tech:(408) 542-0730
Fax:(408) 542-0503
Web Page: http://www.sandisk.com

email: support@sandisk.com
Compuserve: GO SUNDISK
BBS:(408) 986-1186

Santa Cruz Operations··························(408) 425-7222
425 Encinal Street, PO Box 1900
Santa Cruz, CA 95061-1900
Tech:(800) 347-4381 Toll Free:(800) 726-8649
Fax:(408) 458-4227 Fax on Demand:(408) 427-6800
Tech Support:(408) 425-4726
Web Page: http://www.sco.com
email: info@sco.com
BBS:(408) 426-9495

Santa Fe Software (see ACE Contact Manager)

SAP America Inc··································(610) 355-2500
701 Lee Road
Wayne, PA 19087
Fax:(610) 725-4555
Web Page: http://www.sap.com
email: info@sap-ag.de

Saros Corp (see Filenet Corp)

SAS Institute Inc··································(919) 677-8000
SAS Campus Drive, PO Box 8000
Cary, NC 27513-2414
Fax:(919) 677-8123
Fax:(919) 677-4444
Web Page: http://www.sas.com
email: software@sas.com

Savin Corp ···(203) 967-5000
333 Ludlow Street, PO Box 10270
Stamford, CT 06904-2270
Fax:(203) 967-5014
Web Page: http://www.savin.com
email: jsahm@savin.com

Sayett Technology Inc ···························(716) 264-1290
7 Norton Street
Honeoye Falls, NY 14472
Tech:(800) 836-7730 Toll Free:(800) 678-7469
Fax:(716) 624-6080
Web Page: http://www.eznet.net/sayett
email: sayett@eznet.net

SBT Accounting Systems······················(415) 444-9900
1401 Los Gamos Drive
San Rafael, CA 94903

Tech:(415) 444-9700 Toll Free:(800) 944-1000
Fax:(415) 444-9901
Web Page: http://www.sbt.com
email: info@sbt.com
Compuserve: 76557,653
BBS:(415) 444-9939

Scala Inc ....(818) 673-1300
2323 Horse Pen Road, Suite 300
Herndon, VA 20171
Toll Free:(888) 967-2252
Web Page: http://www.scala.com

SciNet Corp ....(408) 328-0160
268 Santa Ana Court
Sunnyvale, CA 94086
Fax:(408) 328-0168
Web Page: http://www.scinetcorp.com
email: info@scinetcorp.com

SciTech Software Inc ....(530) 894-8400
505 Wall Street
Chico, CA 95928-5624
Fax:(530) 894-9069
Web Page: http://www.scitechsoft.com
email: support@scitechsoft.com
Compuserve: GO SCITECH
America Online: KEYWORD SCITECH
BBS:(530) 894-9047

Scitor Corp ....(650) 462-4200
333 Middlefield Road, 2nd Floor
Menlow Park, CA 94025
Tech:(650) 462-4350
Fax:(650) 462-4201
Web Page: http://www.scitor.com
email: pcsupport@scitor.com
Compuserve: 72662,261

SCO (see Santa Cruz Operations)

Scopus Technology (see Siebel)

Seagate NSMG ....(800) 327-2232
Tech:(407) 531-7600
Web Page: http://www.nsmg.seagatesoftware.com

Seagate Software ....(407) 531-7500
400 International Pkwy
Heathrow, FL 32746
Toll Free:(800) 327-2232

Fax on Demand:(800) 327-2232
Fax Back:(407) 333-7767
Web Page: http://www.sems.com

Seagate Technology ···························(408) 438-6550
920 Disc Drive
Scotts Valley, CA 95066-4544
Tech:(800) 732-4283 Toll Free:(800) 468-3472
Fax:(405) 936-1685 Fax on Demand:(405) 936-1620
Web Page: http://www.seagate.com
email: ftp.seagate:com
Compuserve: GO SEAGATE

Disc Drive Products·································(405) 936-1200
Tech:(405) 936-1234 Toll Free:(800) 468-3472
Fax:(405) 936-1685 Fax on Demand:(405) 936-1620
Automated Help:(405) 936-1234
email: discsupport@seagate.com
BBS:(405) 936-1630

Seagate Express·································(972) 481-4246
Toll Free:(800) 531-0968
Fax:(972) 481-4812
Other:(800) 656-2794

Tape Drive Products (IDE/SCSI) ···················(405) 936-1400
Tech:(405) 936-1410 Toll Free:(800) 732-4283
Fax:(405) 936-1683 Fax on Demand:(405) 936-1640
Automated Help:(405) 936-1234
BBS:(405) 936-1630

Tape Drive Products (Internal/External)········(405) 936-1300
Tech:(800) 732-4283 Toll Free:(800) 468-3472
Fax:(405) 936-1683
Automated Help:(405) 936-1234
BBS:(405) 936-1630

TSSI···(805) 778-1773
Fax on Demand:(405) 936-1640

Sealevel Systems ·································(864) 843-4343
155 Technology Place, PO Box 830
Liberty, SC 29657
Tech:(864) 843-4343
Fax:(864) 843-3067
Web Page: http://www.sealevel.com
email: support@sealevel.com
Compuserve: GO SEALEVEL

Seanix Technology Corp ·······················(604) 303-2900
140-6631 Elmbridge Way
Richmond, BC V7C 4N1 Canada
Tech:(888) 252-1197 Toll Free:(888) 584-4121
Fax:(604) 303-2932

Web Page: http://www.seanix.com
email: techsprt@seanix.com

Searchlight Software..............................**(216) 631-9290**
6516 Detroit
Cleveland, OH 44102
Tech:(800) 988-9290 Toll Free:(800) 988-5483
Fax:(216) 631-9289
Web Page: http://www.searchlight.com
email: info@searchlight.com
BBS:(216) 631-9285

Seattle Lab..............................**(425) 402-6003**
9606 Northeast 180th Street
Bothell, WA 98011
Tech:(425) 481-7619
Fax:(425) 828-9011
Fax:(425) 486-2766
Web Page: http://www.seattlelab.com
email: support@seattlelab.com

Sega Entertainment..............................**(510) 371-3131**
Tech:(510) 371-3131
Web Page: http://www.sega.com

Segasoft Inc
Tech:(888) 734-2763
Web Page: http://www.segasoft.com

Seiko Instruments USA..............................**(408) 922-5806**
1130 Ringwood Court
San Jose, CA 95131-1726
Tech:(800) 688-0817 Toll Free:(800) 553-5312
Fax:(408) 922-5840
Tech Fax:(408) 922-5867
Web Page: http://www.seiko-usa.com

Compact Convenience Products..............................(408) 922-5900
Tech:(800) 757-1011 Toll Free:(800) 688-0817
Fax:(716) 873-0906
email: info@seikosmart.com

Consumer Products..............................(310) 517-7810
Tech:(310) 517-8121 Toll Free:(800) 873-4508
Fax:(310) 517-7793
email: cpd.info@seiko-la.com

Digital Imaging..............................(800) 553-5312
Tech:(800) 553-5312 Toll Free:(800) 888-0817
Fax:(408) 922-5838
Tech Fax:(408) 922-5867
email: techsupp@seikosj.com

Electronic Components..............................(310) 517-7822

Tech:(310) 517-8113
Fax:(310) 517-8131
Other:(310) 517-7792
email: ecd.info@seiko-la.com

Factory Automation ....(310) 517-7850
Tech:(310) 517-7842
Fax:(310) 517-8158
email: info@seikorobots.com

Micro Printers ....(310) 517-7778
Fax:(310) 517-8154
email: mpd@seiko-la.com

SemWare Corp ....**(770) 641-9002**
4343 Shallowford Road, Suite C3A
Marietta, GA 30062-5022
Tech:(770) 641-9002 Toll Free:(800) 467-3692
Fax:(770) 640-6213
Web Page: http://www.semware.com
email: tech.support@semware.com
Compuserve: 75300,2710 or GO SEMWARE
BBS:(770) 641-8968

Sequel Inc ....**(408) 987-1000**
2777 San Tomas Expressway, PO Box 4972
Santa Clara, CA 95054-4972
Tech:(408) 987-1417 Toll Free:(800) 848-5837
Web Page: http://www.sequel-inc.com
email: gen@sequel-inc.com

Sequent Computer Systems Inc ....**(503) 626-5700**
15450 SW Koll Parkway
Beaverton, OR 97006-6063
Toll Free:(800) 257-9044
Product or literature info:(800) 257-9044
Web Page: http://www.sequent.com

Sequoia Publishing Inc ....**(303) 972-4167**
9350 W Cross Drive, Suite 101
Littleton, CO 80123
Tech:(303) 972-4167 Toll Free:(800) 873-7126
Fax:(303) 972-0158
Web Page: http://www.sequoiapub.com
email: tglover@sequoiapub.com

Serome Technology Inc ....**(408) 453-9771**
2025 Gateway Pl, Ste 312
San Jose, CA 95110
Fax:(408) 453-9799
Web Page: http://www.serome.com
email: dkim@serome.com

Server Technology ·························(408) 745-0300
521 E. Weddell Drive, Suite 120
Sunnyvale, CA 94089-2232
Tech:(800) 835-1515 Toll Free:(800) 835-1515
Fax:(408) 745-0392
Web Page: http://www.servertech.com
email: webmaster@servertech.com

Service 2000 (see Kennsco Inc)

Service News Magazine ·························(207) 856-0600
38 Lafayette St, PO Box 995
Yarmouth, ME 04096
Fax:(207) 846-0657
Web Page: http://www.servicenews.com
email: klipp@servicenews.com

SES (Scientific & Engin Software) ··········(512) 328-5544
4301 WestBank Drive, Building A
Austin, TX 78746-6564
Tech:(512) 328-3377 Toll Free:(800) 759-6333
Fax:(512) 327-6646
Web Page: http://www.ses.com
email: info@ses.com

Set Enterprises Inc ·························(602) 837-3628
15402 E. Verbena Drive
Fountain Hills, AZ 85268
Tech:(602) 837-3628 Toll Free:(800) 351-7765
Fax:(602) 837-5644
Web Page: http://www.setgame.com
email: setgame@setgame.com

Severn Companies ·························(301) 794-9680
4640 Forbes Boulevard
Lanham, MD 20706 USA
Fax:(301) 459-3272
Web Page: http://www.severn.com

SGS-Thomson Microelectronics ············(602) 485-6201
1000 E. Bell Road
Phoenix, AZ 85022
Fax:(602) 485-6330
Web Page: http://www.st.com

Shaffstall Corp ·························(317) 842-2077
7901 E. 88th Street
Indianapolis, IN 46256
Tech:(317) 842-2077 Toll Free:(800) 248-3475
Fax:(317) 842-8294
Web Page: http://www.shaffstall.com

email: support@shaffstall.com

Shapeware (Out of Business)

Sharp Electronics Corp ....(201) 529-8200
Sharp Plaza
Mahwah, NJ 07430-2135
Tech:(800) 237-4277 Toll Free:(800) 237-4277
Fax:(201) 529-8425
Web Page: http://www.sharp-usa.com

Sherwood Terminals ....(510) 266-5600
21056 Forbes St
Hayward, CA 94545
Tech:(800) 777-8755 Toll Free:(800) 777-8755
Fax:(510) 623-8945
Fax:(510) 266-5627
Web Page: http://www.sherwoodterm.com
email: sherwood@sherwoodterm.com

Shiva Corporation ....(781) 687-1000
28 Crosby Drive
Bedford, MA 01730
Tech:(781) 270-8400 Toll Free:(800) 458-3550
Fax:(781) 270-8599 Fax on Demand:(800) 370-6917
Other:(800) 997-4482
Web Page: http://www.shiva.com
email: sales@shiva.com
Compuserve: GO SHIVA

ShowCase Corporation............................(507) 288-5922
4131 Highway 52 North, Suite G-111
Rochester, MN 55901-3144
Tech:(507) 288-5922 Toll Free:(800) 829-3555
Fax:(507) 287-2803 Fax on Demand:(800) 988-0014
Web Page: http://www.showcasecorp.com
email: richardw@showcasecorp.com
Compuserve: GO SHOWCASE

Shugart Corporation ....(520) 294-0898
26611 Kappat Road
Mission Viejo, CA 92653
Fax:(714) 367-8843

Shuttle Computer Group ....(909) 595-5060
3555 Lemon Ave
Walnut, CA 91789 USA
Fax:(909) 595-2667
Web Page: http://www.shuttlela.com

Siebel .... **(510) 597-5800**
1900 Powell Street
Emeryville, CA 94608
Fax:(510) 597-8600
Web Page: http://www.siebel.com
email: info@scopus.com

Sierra On-Line (see CUC International)

Sigma Data .... **(603) 526-6909**
26 Newport Road
New London, NH 03257-4565
Tech:(603) 526-7100 Toll Free:(800) 446-4525
Fax:(603) 526-6915
Web Page: http://www.sigmadata.com
email: tsupport@sigmadata.com

Sigma Designs Inc .... **(510) 770-0100**
46501 Landing Parkway
Fremont, CA 94538
Tech:(970) 339-7120 Toll Free:(800) 845-8086
Fax:(510) 770-2640
Web Page: http://www.sigmadesigns.com
email: tech_support@sdesigns.com
BBS:(510) 770-0111

Silicon Graphics Inc .... **(650) 960-1980**
2011 N. Shoreline Blvd
Mountain View, CA 94043
Tech:(800) 800-4744 Toll Free:(800) 800-7441
Fax:(650) 960-0197
Web Page: http://www.sgi.com

Silicon Valley Technical Staff .... **(408) 749-8989**
1095 E. Duane Ave.
Sunnyvale, CA 94086
Web Page: http://www.svts.com
Compuserve: GO MEDIAVISION

Sir-Tech Software Inc .... **(315) 393-6451**
PO Box 245
Ogdensburg, NY 13669
Tech:(315) 393-6644 Toll Free:(800) 447-1230
Fax:(315) 393-1525
Web Page: http://www.sir-tech.com
email: tech@sir-tech.com

Sirius Publishing .... **(602) 951-3288**
7320 E. Butherus Drive, Suite 100
Scottsdale, AZ 85260-2438
Tech:(602) 951-8405 Toll Free:(800) 247-0307

Fax:(602) 951-3884
Other Address: http://www.treasurequest.com
Web Page: http://www.siriuspub.com
email: techsupport@siriuspub.com

Skill Dynamics (see IBM Tech)

SkiSoft ··························(781) 863-1876
1644 Massachusetts Ave, Suite 79
Lexington, MA 02173
Fax:(781) 861-0086
Web Page: http://www.skisoft.com

SL Waber Inc ··························(609) 866-8888
520 Fellowship Road, Suite 306C
Mount Laurel, NJ 08054
Tech:(800) 257-8384 Toll Free:(800) 634-1485
Fax:(609) 866-1945
Tech Fax:(215) 752-4859
Web Page: http://www.waber.com
email: waber@voicenet.com

Smart and Friendly Inc ··························(818) 772-8001
20520 Nordhoff St
Chatsworth, CA 91311
Tech:(818) 772-2888
Web Page: http://www.smartandfriendly.com
email: webmaster@smartandfriendly.com

Smart Cable ··························(253) 474-9967
7403 Lakewood Drive, # 14
Tacoma, WA 98467
Tech:(206) 823-2273 Toll Free:(800) 752-6526
Fax:(253) 474-9940
Web Page: http://www.nrsnet.com/smartcable
email: SmartCable@nrsnet.com
BBS:(206) 821-5486

Smart Modular Technologies ··························(510) 623-1231
4305 Cushing Parkway
Fremont, CA 94538
Tech:(510) 249-1605 Toll Free:(800) 841-2739
Fax:(510) 249-1600
Tech Fax:(510) 249-1604
Other Address: http://www.smartm.com
Web Page: http://www.apexdata.com
email: support@smartm.com
BBS:(510) 249-1601

Smartronics Inc ··························(603) 437-1975
East Derry, NH 03041-0310

Fax:(603) 434-5470
Web Page: http://www.smartronics.com
email: info@smartronics.com

SMC (Standard Microsystems Corp)······(516) 273-3100
80 Arkay Drive
Hauppauge, NY 11788-3774
Tech:(800) 992-4762 Toll Free:(800) 762-4968
Fax:(516) 273-1803 Fax on Demand:(800) 762-8329
Fax Back:(516) 435-6107
Web Page: http://www.smc.com
email: techsupport@smc.com
Compuserve: GO SMC
BBS:(516) 434-3162

Smith Micro Software Inc ························(949) 362-5800
51 Columbia
Aliso Viejo, CA 92656
Tech:(949) 362-5810 Toll Free:(800) 964-7674
Fax:(949) 362-2300 Fax on Demand:(949) 362-2396
Other:(949) 362-5811
Web Page: http://www.smithmicro.com
email: sales@smithmicro.com
Compuserve: GO SMITHMICRO
BBS:(949) 362-5822

SMS Data Products Group Inc ················(703) 709-9898
1501 Farm Credit Dr
McLean, VA 22102-5004
Fax:(703) 356-5167
Web Page: http://www.sms.com
email: bdoody@sms.com

SMS Technology (see Televideo Inc)

Snow Software·······································(813) 784-8899
2360 Congress Ave
Clearwater, FL 34623
Fax:(813) 786-5904
Fax Orders:(813) 787-1904
Web Page: http://www.snowsoft.com

SNX ···(718) 499-6293
692 10th Street
Brooklyn, NY 11215-4502
Tech:(718) 369-2944 Toll Free:(800) 447-9639
Fax:(718) 768-3997
Other:(800) 619-0299
Web Page: http://www.snx.com
email: snx@snx.com

Socket Communications ....(510) 744-2700
37400 Central Court
Newark, CA 94560
Tech:(510) 744-2720 Toll Free:(800) 552-3300
Fax:(510) 744-2727 Fax on Demand:(800) 503-3853
Web Page: http://www.socketcom.com
email: info@socketcom.com
Compuserve: GO SOCKET
BBS:(510) 744-2820

SoftArc Inc ....(905) 415-7000
100 Allstate Parkway
Markham, ON L3R 6H3 Canada
Tech:(905) 415-7000 Toll Free:(800) 763-8272
Fax:(905) 415-7151
Web Page: http://www.softarc.com
email: support@softarc.com
BBS:(905) 415-7070

Softbank Comdex (see ZD Comdex)

Softbite International ....(630) 833-0006
1101 31st St, Suite 180
Downers Grove, IL 60515
Tech:(630) 493-0009
Fax:(630) 833-0584 Fax on Demand:(630) 833-9122
Fax:(630) 493-1710
Web Page: http://www.softbite.com
email: access@softbite.com

SoftBooks Inc ....(714) 225-1463
17975 Sky Park Circle, Suite G
Irvine, CA 92614-6317
Web Page: http://www.softbooks.com

SoftCad International ....(510) 376-0117
1620 School Street, Suite 101
Moraga, CA 94556
Tech:(800) 763-8223 Toll Free:(800) 763-8223
Fax:(510) 376-0118
Web Page: http://www.softcad.com
email: info@softcad.com

SoftCraft Inc ....(608) 232-1859
16 N. Carroll Street, Suite 220
Madison, WI 53703
Toll Free:(800) 351-0500
Fax:(608) 231-0363
email: softcraft@compuserve.com
Compuserve: 76702,1304

Softdesk Retail Products (see Autodesk)

Softkey International (see Learning Co)

Softklone .... (850) 878-8564
327 Office Plaza Drive, Suite 100
Tallahassee, FL 32301-2776
Tech:(850) 878-8564 Toll Free:(800) 634-8670
Fax:(850) 877-9763 Fax on Demand:(800) 634-8670
Web Page: http://www.softklone.com
email: support@softklone.com
Compuserve: 76224,134
BBS:(850) 878-9884

SofTouch Systems Inc .... (405) 947-8080
1300 S. Meridian, Suite 600
Oklahoma City, OK 73108-1751
Tech:(800) 944-3028 Toll Free:(800) 944-3036
Fax:(405) 947-8169
Web Page: http://www.softouch.com
email: ssgtech@softouch.com

SoftQuad Inc .... (416) 544-9000
20 Eglinton Ave. West 13th Floor, PO Box 2025
Toronto, ON M4R 1K8 Canada
Tech:(416) 544-8879 Toll Free:(800) 387-2777
Fax:(416) 544-0300
Web Page: http://www.softquad.com
email: support@sq.com

Softronics Inc .... (719) 593-9540
5085 List Drive
Colorado Springs, CO 80919
Tech:(719) 593-9550
Fax:(719) 548-1878
Web Page: http://www.softronics.com
email: support@softronics.com
BBS:(719) 593-9295

SoftTalk Inc (see Callware Tech)

Software Business Technologies (see SBT)

Software Directions Inc .... (408) 763-0606
935 Metivers Way
La Selva Beach, CA 95076
Fax:(408) 763-1608
Web Page: http://www.softwaredirections.com
email: info@softwaredirections.com

Software Made Simple .... (800) 846-9726
PO Box 897

San Ramon, CA 94583-9985
Tech:(520) 670-7100
Fax:(510) 328-1117
Web Page: http://www.aristosoft.com
email: WFS@PACBELL.NET
Compuserve: 75300,3507@compuserve.com

Software Marketing (see Softkey Intl)

Software Publishers Assoc ····················(202) 452-1600
1730 M Street NW, Suite 700
Washington, DC 20036-4510
Toll Free:(800) 388-7478
Fax:(202) 223-8756 Fax on Demand:(800) 637-6823
Web Page: http://www.spa.org
email: webmaster@spa.org

Software Publishing Corp ······················(408) 537-3000
111 North Market St
San Jose, CA 95113
Tech:(408) 988-6005 Toll Free:(800) 336-8360
Fax:(408) 537-3500
Web Page: http://www.spco.com
email: support@spco.com

Software Support Inc ·····························(800) 873-4357
300 International Pkwy, Suite 320
Heathrow, FL 32746
Tech:(800) 873-4357 Toll Free:(800) 756-4463
Fax:(407) 333-9080
Web Page: http://www.ssisupport.com

Software Toolworks (see Mindscape)

Software Ventures ·································(800) 336-6477
2907 Claremont Blvd
Berkeley, CA 94705
Tech:(510) 644-1325

Sola Electric ··(708) 439-2800
1717 Busse Road
Elk Grove Village, IL 60007
Tech:(800) 289-7652 Toll Free:(800) 289-7652
Fax:(708) 439-1160
Web Page: http://www.solaelex.com

Solectek Accessories ·····························(619) 450-1220
6370 Nancy Ridge Drive, Suite 109
San Diego, CA 92121-3212
Tech:(800) 437-1518 Toll Free:(800) 437-1518
Fax:(619) 457-2681

Tech Fax:(619) 642-2793
Web Page: http://www.solectek.com
email: rlittle@solectek.com
BBS:(619) 450-6537

Solidex (see Unitech Industries)

Solomon Software ....(419) 424-1422
200 E. Hardin Street, PO Box 414
Findlay, OH 45840
Tech:(419) 424-0422 Toll Free:(800) 476-5666
Fax:(419) 422-2044
Fax:(419) 424-3400
Web Page: http://www.solomon.com
email: clients@solomon.com

Solsource Computers ....(760) 929-7800
2075 Corte Del Nogal, Suite D
Carlsbad, CA 92009
Toll Free:(800) 858-4405
Fax:(760) 929-7810
Web Page: http://www.solsource.com
email: tech@solsource.com

Sonera Technologies ....(732) 747-6886
PO Box 565
Rumson, NJ 07760
Tech:(732) 747-6886 Toll Free:(800) 932-6323
Fax:(732) 747-4523
Web Page: http://www.displaymate.com
email: info@sonera.com

Sonic (see Fast Color.Com)

Sonic Foundry ....(608) 256-3133
754 Williamson St, Suite 204
Madison, WI 53703
Tech:(608) 256-5555 Toll Free:(800) 577-6642
Fax:(608) 256-7300
Other Address: http://www.sonicfoundry.com
Web Page: http://www.sfoundry.com
email: support@sonicfoundry.com
Compuserve: 74774,1340

Sonic Systems ....(408) 736-1900
575 N. Pastoria Ave
Sunnyvale, CA 94086
Tech:(408) 736-1900 Toll Free:(888) 222-6563
Fax:(408) 736-7228
Other Address: AppleLink
Web Page: http://www.sonicsys.com

email: tech@sonicsys.com

Sonoma Systems .... **(310) 827-8000**
4640 Admiralty Way, 6th Floor
Marina Del Rey, CA 90292-6695
Tech:(800) 997-3849 Toll Free:(800) 255-2333
Fax:(310) 828-2255
Fax:(310) 305-2525
Web Page: http://www.sonoma-systems.com
email: support@sonoma-systems.com

Sony Corp Of America .... **(201) 930-1000**
1 Sony Drive
Park Ridge, NJ 07645
Tech:(941) 768-7669 Toll Free:(800) 222-7669
Fax on Demand:(888) 476-6972
Web Page: http://www.sony.com
BBS:(408) 955-5107

Sony Electronics .... **(408) 432-1600**
3300 Zanker Road
San Jose, CA 95134
Tech:(800) 326-9551 Toll Free:(800) 352-7669
Fax:(408) 955-5111 Fax on Demand:(800) 883-7669
Web Page: http://www.sony.com/storagebysony
BBS:(408) 955-5107

CD-Rom Discman Support .... (800) 766-9236
Web Page: http://www.sony.com

Computer Peripheral .... (800) 326-9551
Web Page: http://www.sony.com

Direct Response Center .... (800) 222-7669
Web Page: http://www.sony.com

Media Support .... (800) 766-9328
Web Page: http://www.sony.com

Monitors .... (708) 773-7579
1 Sony Drive
Parkridge, NJ 07656
Toll Free:(800) 222-7669
Fax on Demand:(800) 282-2848
Web Page: http://www.sony.com

PlayStation Support .... (800) 345-7669
Web Page: http://www.sony.com

Service & Parts .... (800) 488-7669
Web Page: http://www.sony.com

Service Center Locations .... (800) 282-2848
Web Page: http://www.sony.com

Storage Devices .... (408) 922-0699
Web Page: http://www.sony.com

Sony Interactive Studios .... **(415) 655-8000**
919 E. Hillsdale Blvd, 2nd Floor
Foster City, CA 94404
Tech:(415) 655-5683
Fax:(415) 655-8001
Web Page: http://www.sony.com
Compuserve: helpline@interactive.sony.com

Computer Products .... (888) 476-6972
Tech:(800) 326-9551
Fax on Demand:(800) 883-7669
BBS:(408) 955-5107

Sophisticated Circuits .... **(425) 485-7979**
19017 120th Ave NE, Suite 106
Bothell, WA 98041-0727
Tech:(425) 485-7979 Toll Free:(800) 827-4669
Fax:(425) 485-7172
Other:(800) 769-3773
Web Page: http://www.sophisticated.com
email: support@sophisticated.com
Compuserve: 74431,335
America Online: sophcir

Sophos .... **(781) 932-0222**
18 Commerce Way
Woburn, MA 01801
Fax:(781) 932-0251
Web Page: http://www.sophos.com
email: support@sophos.com

Sound Source Unlimited .... **(818) 878-0505**
26115 Mureau Road, Suite B
Calabasas, CA 91302-3126
Tech:(818) 878-0505 Toll Free:(800) 877-4778
Fax:(818) 878-0007
Fax:(818) 871-1972
Web Page: http://www.soundsourceinteractive.com
email: soundsource@ssiimail.com
Compuserve: 75361,1544
America Online: ssi online

SourceMate Info (see AccountMate Software)

Southpeak Interactive .... **(919) 677-4499**
Web Page: http://www.southpeak.com
email: southpeak_support@aqinc.com

Spacetec IMC .... **(978) 275-6128**
The Boott Mill, 100 Foot Of John Street
Lowell, MA 01852-1126

Tech:(978) 970-0440 Toll Free:(800) 788-9994
Fax:(978) 275-6200
Web Page: http://www.panacea.com
email: support@spacetech.com

Spalding Software ....(770) 449-0594
154 Technology Parkway, Suite 250
Norcross, GA 30092
Tech:(770) 449-0594
Fax:(770) 449-0052
Web Page: http://www.spaldingsoft.com
email: info@spaldingsoft.com
Compuserve: GO SPALDING or 74431,240

SPARC International Inc ....(408) 748-9111
3333 Bowers Ave, Suite 280
Santa Clara, CA 95054-2913
Fax:(408) 748-9777
Web Page: http://www.sparc.com
email: info@sparc.com

Specialix ....(408) 378-7919
745 Camden Avenue
Campbell, CA 95008-4146
Tech:(800) 423-5364 Toll Free:(800) 423-5364
Fax:(408) 378-0786
Web Page: http://www.specialix.com
email: support@specialix.com
BBS:(408) 378-4766

Specialized Products Co ....(214) 550-1923
3131 Premier Dr
Irving, TX 75063
Tech:(800) 527-5018 Toll Free:(800) 866-5353
Fax:(214) 550-1386
Web Page: http://www.specializedproducts.com
email: spcompany@aol.com

Spectragraphics ....(619) 450-0611
9707 Waples Street
San Diego, CA 92121
Tech:(900) 934-3200 Toll Free:(800) 821-4822
Fax:(619) 450-0218
order:(619) 450-3213
Web Page: http://www.spectra.com

Spectrum HoloByte (see Microprose Software)

Spider Island Software.............................(714) 508-9366
4790 Irvine Blvd, Suite 105-347
Irvine, CA 92620

Fax:(714) 508-9233
Web Page: http://www.spiderisland.com

Spinnaker Software (Learning Co)

Splash Technology ....(978) 667-8585
900 Middlesex Turnpike, Building 8
Billerica, MA 01821
Tech:(978) 663-8213 Toll Free:(800) 437-3336
Fax:(978) 667-8821 Fax on Demand:(800) 260-7797
Web Page: http://www.colorage.com
email: techsupport@splashtech.com
BBS:(978) 663-2704

Sprague Magnetics ....(818) 364-1800
12806 Bradley Ave
Sylmar, CA 91342
Tech:(800) 553-8712 Toll Free:(800) 553-8712
Fax:(818) 364-1810
Web Page: http://www.sprague-magnetics.com
email: smitech@sprague-magnetics.com

Spry Net..(425) 957-8000
3535 128th Ave SE
Bellevue, WA 98006
Fax:(425) 957-6000
Web Page: http://www.spry.com
email: info@spry.com

SPSS Inc..(312) 329-2400
444 N. Michigan Ave
Chicago, IL 60611
Tech:(800) 543-2185 Toll Free:(800) 521-1337
Fax:(312) 329-3668
Web Page: http://www.spss.com
email: suggest@spss.com

Spyglass..(630) 505-1010
1240 E. Diehl Road
Naperville, IL 60563
Toll Free:(888) 677-9452
Fax:(630) 505-4944
Other Address: http://www.surfwatch.com
Web Page: http://www.spyglass.com
email: needs@spyglass.com

SRW Computer Components ....(714) 963-5500
1402 Morgan Circle
Tustin, CA 92680
Tech:(800) 547-7766 Toll Free:(800) 547-7766
Fax:(714) 259-8037

Stac Electronics .... **(619) 794-4300**
12636 High Bluff Dr, Suite 400
San Diego, CA 92130-2093
Tech:(619) 794-3700 Toll Free:(800) 305-7822
Fax:(619) 794-3717 Fax on Demand:(619) 794-3710
Other:(800) 522-7822
Web Page: http://www.stac.com
BBS:(619) 794-3711

Stallion Technologies Inc .... **(408) 477-0440**
2880 Research Park Drive, Suite 160
Soquel, CA 95073
Tech:(800) 729-2342 Toll Free:(800) 347-7979
Fax:(408) 477-0444
Web Page: http://www.stallion.com
email: support@stallion.com
BBS:(408) 477-4352

Stampede Technologies Inc .... **(937) 291-5035**
65 Rhoads Center Drive
Dayton, OH 45458
Toll Free:(800) 763-3423
Fax:(937) 291-5040
Web Page: http://www.stampede.com

Standard Microsystems Corp (see SMC)

Star Media Systems (see DPS Software)

Star Micronics America .... **(732) 572-9512**
70 Ethel Road West
Piscataway, NJ 08854
Tech:(732) 572-3300 Toll Free:(800) 506-7827
Fax:(732) 572-5095 Fax on Demand:(732) 572-4004
Tech Fax:(732) 572-5995
Web Page: http://www.starmicronics.com
email: Userhelp@Starus.com
BBS:(732) 572-5010

Starfish Software .... **(408) 461-5800**
1700 Green Hills Road
Scotts Valley, CA 95066
Tech:(970) 522-4610 Toll Free:(888) 782-7347
Fax:(408) 461-5900 Fax on Demand:(800) 503-3847
Questions about an order:(800) 765-7839
Web Page: http://www.starfishsoftware.com
email: cs@starfish.com
Compuserve: GO STARFISH
Microsoft: GO STARFISH
America Online: KEYWORD STARFISH

BBS:(408) 461-5930

Starquest Connectivity Software ··········(510) 704-2000
2150 Shattuck Ave, Suite 600
Berkeley, CA 94704
Tech:(510) 704-2570 To[illegible]ree:(800) 763-0050
Fax:(510) 704-2001
Web Page: http://www.starquest.com
email: sales@starquest.com

State Of The Art ··········(916) 791-7730
8211 Sierra College Blvd, Suite 440
Roseville, CA 95661
Tech:(800) 447-5700 Toll Free:(800) 447-5700
Fax:(916) 791-5525 Fax on Demand:(800) 527-6587
Other Address: http://www.sota.com
Web Page: http://www.stateoftheart.com
BBS:(916) 791-2061

StatView Software ··········(415) 623-2032
1918 Bonita Ave
Berkley, CA 94704-1014
Tech:(919) 677-8008 Toll Free:(800) 666-7828
Fax:(510) 540-0260 Fax on Demand:(415) 623-2083
Web Page: http://www.abacus.com
email: techsupport@abacus.com

STB Systems Inc ··········(972) 234-8750
1651 N. Glenville, Suite 210
Richardson, TX 75081
Tech:(972) 669-0989 Toll Free:(800) 234-4334
Fax:(972) 234-1306
Tech Support Fax:(972) 669-1326
Web Page: http://www.stb.com
email: sales@stb.com
BBS:(972) 437-9615

Steinberg/Jones ··········(818) 993-4091
9312 Deering
Chatsworth, CA 91311
Tech:(818) 993-4161
Fax:(818) 701-7452 Fax on Demand:(800) 888-7510
Compuserve: 71333,2447

Sterling Commerce ··········(770) 804-8100
5 Concourse Pkwy, Suite 850
Atlanta, GA 30328
Toll Free:(800) 322-3366
Fax:(770) 804-8102
Web Page: http://www.xcellenet.com

email: info@xcellenet.com

Storage Dimensions ································ **(408) 954-0710**
1656 McCarthy Boulevard
Milpitas, CA 95035
Tech:(408) 894-1325 Toll Free:(800) 765-7895
Fax:(408) 944-1200
Other Address: techconnect@xstor.com
Web Page: http://www.storagedimensions.com
email: info@xstor.com
BBS:(408) 944-1221

Storage Technology Corporation ··········· **(303) 673-5151**
2270 S. 88th St
Louisville, CO 80028
Tech:(719) 536-4055 Toll Free:(800) 786-7835
Fax:(303) 673-7577
Fax:(719) 536-4053
Web Page: http://www.stortek.com

Storm Technology ································ **(650) 691-6600**
1395 Charleston Road
Mountain View, CA 94043
Tech:(650) 969-9555 Toll Free:(800) 275-5734
Fax:(650) 691-9825
Other:(888) 438-3279
Web Page: http://www.stormtech.com
email: support@stormtech.com
Compuserve: 73060,3227
America Online: stormsoft

Strata Distributing Inc ························· **(510) 656-9848**
5405 Randall Place
Fremont, CA 94538
Tech:(510) 656-9848
Fax:(510) 656-9891
Web Page: http://www.stratadist.com
email: dax@str.uucp.netcom.com

Strata Inc ·· **(435) 628-5218**
2 W. St. George Blvd
Saint George, UT 84770-2858
Tech:(435) 628-9751 Toll Free:(800) 678-7282
Fax:(435) 628-9756
Tech Fax:(435) 652-5408
Web Page: http://www.strata3d.com
email: support@strata3d.com

Strategic Mapping (see Software Support)

Strategic Networks ....(617) 912-8300
66 B St
Needham, MA 02194
Toll Free:(800) 999-7621
Fax:(781) 433-2897
Web Page: http://www.snci.com
email: snci@snci.com

Strategic Simulations Inc (see Mindscape)

Strategic Studies Group ....(850) 469-8880
3186 Hyde Park Place, PO Box 30085
Pensacola, FL 32503-1085
Tech:(850) 469-8880 Toll Free:(888) 522-5919
Fax:(850) 469-8885
Fax:(850) 983-7304
Web Page: http://www.ssgus.com
email: pctech@ssgus.com

Street Electronics (see Echo Speech Corp)

Streetwise Software ....(310) 829-7827
2118 Wilshire Blvd, Suite #836
Santa Monica, CA 90403
Tech:(310) 998-3361 Toll Free:(800) 743-6765
Fax:(310) 828-8258
Web Page: http://www.swsoftware.com
email: tech.support@swsoftware.com
Compuserve: 7323,2304
BBS:(310) 998-3373

Structured Software (see Facet Corp)

Structured Software Services Group .....(972) 960-7555
14651 Dallas Pkwy, Suite 230
Dallas, TX 75240
Toll Free:(800) 767-6547
Fax:(972) 991-3244
Web Page: http://www.sssg.com
email: admin@sssg.com

Summagraphics Corp (see Cal Comp)

Summit Software Co ....(315) 445-9000
4933 Jamesville Rd
Jamesville, NY 13078-9428
Fax:(315) 445-9567
Web Page: http://www.summsoft.com
email: info@summsoft.com

Sun Microsystems Computer Co ....(650) 960-1300
2550 Garcia Ave

Mountain View, CA 94043-1100
Tech:(800) 872-4786 Toll Free:(800) 821-4643
Fax on Demand:(800) 329-7869
Other:(888) 786-3463
Web Page: http://www.sun.com

SunSoft (see Sun Microsystems Computer)

Superbase Inc ··(516) 244-1500
80 Orville Drive
Bohemia, NY 11716
Toll Free:(800) 315-7944
Fax:(516) 244-0250
Compuserve: GO SUPERBASE; library 21

Supermac Technology (see Radius)

Supra Corp ··(360) 604-1400
312 SE Stonemill Dr, Suite 150
Vancouver, WA 98684 Canada
Tech:(541) 967-2450 Toll Free:(800) 727-8772
Fax:(360) 604-1401 Fax on Demand:(800) 380-0030
Tech Support Fax:(541) 967-2401
Web Page: http://www.supra.com
email: sales@supra.com
Compuserve: GO SUPRA

Supra Corp (see Diamond Multimedia)

Surflogic LLC ···(415) 731-2732
293 Downey Street
San Francisco, CA 94117
Fax:(415) 731-0584
Web Page: http://www.surflogic.com
email: sales@surflogic.com

SusTeen Inc ···(310) 787-1589
22301 S. Western Ave, Suite 107
Torrance, CA 90501
Tech:(310) 787-1589
Fax:(310) 787-1590
Web Page: http://www.susteen.com

Swan Technologies Corp ·························(800) 533-1131
313 Boston Post Road, Suite 200
Marldorough, MA 01752
Tech:(800) 468-7926 Toll Free:(800) 446-2499
Sales Fax:(814) 237-4450
Web Page: http://www.swantech.com
email: http://www.tisco.com/swan
Compuserve: GO SWAN

Swfte International (see Expert Software)

Sybase Inc..**(978) 287-1500**
561 Virginia Road
Concord, MA 01742-2732
Tech:(978) 287-1500 Toll Free:(800) 395-3525
Fax on Demand:(508) 287-1600
Tech Support:(800) 879-2273
Web Page: http://www.sybase.com/products/tools
email: csfeedback@powersoft.com

Sybase Inc..**(510) 922-3555**
6475 Christie Ave
Emeryville, CA 94608-1050
Tech:(510) 922-3500 Toll Free:(800) 879-2273
Fax:(510) 658-9441
Web Page: http://www.sybase.com
email: webmaster@sybase.com

Sybex Inc..**(510) 523-8233**
1151 Marina Village Parkway
Alameda, CA 94501
Tech:(800) 227-2346 Toll Free:(800) 227-2346
Fax:(510) 523-6840
Order Fax:(510) 523-2373
Web Page: http://www.sybex.com
email: support@sybex.com

Symantec Corp....................................**(408) 253-9600**
10201 Torre Ave
Cupertino, CA 95014-2132
Tech:(541) 465-8645 Toll Free:(800) 441-7234
Fax:(408) 253-3968 Fax on Demand:(800) 554-4403
Customer Service:(800) 685-2349
Web Page: http://www.symantec.com
Compuserve: GO SYMANTEC
America Online: Keyword: Symantec
BBS:(541) 484-6669

ACT! for Windows and MAC..........................(541) 465-8645
Tech:(541) 465-8645
Tech support per incident fee:(800) 927-3989
Web Page: http://www.symantec.com

Norton..(310) 453-4600
2500 Broadway, Suite 200
Santa Monica, CA 90404
Tech:(541) 465-8484 Toll Free:(800) 927-4017
Fax:(310) 453-0636
Web Page: http://www.norton.com

Priority Support.......................................(541) 334-6054

Tech:(800) 927-4012 Toll Free:(800) 927-3991
Web Page: http://www.norton.com

Standard Care Tech Support .... (541) 465-8420
Web Page: http://www.symantec.com

Symantec C++ .... (408) 253-9600
10201 Torre Ave
Cupertino, CA 95014
Tech:(541) 465-8470
Tech support per incident fee:(800) 927-4014
Web Page: http://www.symantec.com

Customer Operations .... (541) 345-3322
175 W. Broadway
Eugene, OR 97401
Tech:(541) 465-8430 Toll Free:(800) 441-7234
Fax:(541) 334-7473 Fax on Demand:(541) 984-2490
Fax Back:(800) 554-4403
Web Page: http://www.symantec.com
Compuserve: GO SYMANTEC
America Online: keyword: symantec

Symbios Logic (see LSI Logic)

Symbol Technologies Inc .... (516) 738-2400
One Symbol Plaza
Holtsville, NY 11742-1300
Tech:(516) 738-5200 Toll Free:(800) 722-6234
Fax:(516) 738-5990
Web Page: http://www.symbol.com

Synchronics .... (901) 761-1166
6584 Poplar Avenue, Suite 200
Memphis, TN 38138
Tech:(800) 852-8755 Toll Free:(800) 852-5852
Fax:(901) 683-8303
Web Page: http://www.sync-link.com
email: support@sync-link.com

Synergy Interactive Corp
444 De Haro Street, Suite 123
San Francisco, CA 94107
Fax:(415) 431-3684

Synergy Software .... (610) 779-0522
2457 Perkiomen Ave
Reading, PA 19606
Tech:(610) 779-0522 Toll Free:(800) 876-8376
Fax:(610) 370-0548 Fax on Demand:(610) 779-9315
Web Page: http://www.synergy.com
email: support@synergy.com

Synergy Solutions (see Artisoft)

Synex (see SNX)

SynOptics Commun (see Bay Network)

SyQuest Technology..............................(510) 226-4000
47071 Bayside Parkway
Fremont, CA 94538-6517
Tech:(510) 226-5400 Toll Free:(800) 245-2278
Fax:(510) 226-4100 Fax on Demand:(510) 226-4120
Web Page: http://www.syquest.com
email: support@syquest.com
BBS:(510) 656-0473

Syracuse Language Systems..................(315) 449-4500
5790 Widewaters Parkway
Syracuse, NY 13214
Tech:(800) 688-1937 Toll Free:(800) 797-5264
Fax:(315) 449-9886
Web Page: http://www.syrlang.com
email: customerservice@syrlang.com

SysKonnect..(408) 437-3800
1922 Zanker Road
San Jose, CA 95112
Tech:(408) 437-3857 Toll Free:(800) 752-3334
Fax:(408) 437-3866
Web Page: http://www.syskonnect.com
email: support@syskonnect.com
BBS:(408) 437-3869

Systems Compatibility (see Inso)

Systems Plus Inc.....................................(650) 969-7047
500 Clyde Ave
Mountain View, CA 94043
Tech:(650) 969-7066 Toll Free:(800) 222-7701
Fax:(650) 969-8936
Web Page: http://www.systemsplus.com
BBS:(650) 969-5287

SystemSoft Corp ....(508) 651-0088
One Innovation Dr
Natick, MA 01760-2059
Tech:(508) 651-0088 Toll Free:(800) 449-7973
Fax:(508) 651-8188
Web Page: http://www.systemsoft.com
email: support@systemsoft.com

Systran Software Inc................................(619) 459-6700
7855 Fay Ave, Ste 300

La Jolla, CA 92037
Fax:(619) 459-8487
Web Page: http;//www.systransoft.com
email: techsupport@systransoft.com

Sytron Corp (see Arcada)

T/Maker Company ....(650) 962-0195
1390 Villa St
Mountain View, CA 94041-1105
Tech:(650) 962-0195 Toll Free:(800) 986-2537
Fax:(650) 962-0201
Web Page: http://www.clickart.com
email: clickart-info@tmaker.com

TAC Systems Inc ....(205) 721-1976
Toll Free:(800) 659-4440
Web Page: http://www.tacsystems.com
email: tech-support@tacsystems.com

Tadiran Electronic Industries ....(516) 621-4980
2 Seaview Blvd
Port Washington, NY 11050
Toll Free:(800) 537-1368
Fax:(516) 621-4517
Web Page: http://www.tadiranbat.com
email: Sales@tadiranbat.com

Tadpole Technology Inc ....(512) 219-2200
12012 Technology Blvd
Austin, TX 78727
Tech:(800) 232-1881 Toll Free:(800) 232-6656
Fax:(512) 219-2222
Sales:(800) 232-6656
Web Page: http://www.tadpole.com
email: sales@tadpole.com

Tallgrass Technologies (see Exabyte)

Talyon Software Corp (see Plus and Minus)

Tandem Computers Inc. ....(408) 285-6000
19333 Vallco Parkway
Cupertino, CA 95014-2599
Tech:(800) 255-5010 Toll Free:(800) 482-6336
Web Page: http://www.tandem.com

Tandon (TSL Holdings Inc)
609 Science Drive
Moorpark, CA 93021

Tandy Corp (see Radio Shack)

Tangent Computers Inc .... (800) 342-9388
197 Airport Blvd
Burlingame, CA 94010
Tech:(800) 399-8324 Toll Free:(888) 826-4368
Fax:(650) 342-9380
Other:(650) 342-9388
Web Page: http://www.tangent.com
email: support@tangent.com

Tangram Enterprise Solutions .... (919) 462-9096
11000 Regency Parkway, Suite 401
Cary, NC 27511-8504
Tech:(800) 722-2482 Toll Free:(800) 482-6472
Fax:(919) 851-6004
Tech Support:(919) 653-6002
Web Page: http://www.tesi.com
email: info@tesi.com

Targus .... (714) 523-5429
Cerritos, CA 90703-5039
Tech:(714) 523-5429
Fax:(714) 523-0153
Web Page: http://www.targus.com
email: info@targus.com

Tatung Company Of America .... (310) 637-2105
2850 El Presidio St
Long Beach, CA 90810
Tech:(800) 827-2850 Toll Free:(800) 827-2850
Fax:(310) 637-8484
Web Page: http://www.tatungusa.com
email: tatung@pacbell.net

Taylored Graphics .... (800) 346-3629
PO Box 1729
Taylor, MI 48180
Other Address: AppleLink: D2588

TDA/IPC .... (425) 402-7000
1518 Seattle Hill Road
Bothell, WA 98012
Tech:(425) 402-7000 Toll Free:(800) 624-2101
Fax:(425) 402-1900

TDA/WINK Data Products (see TDA/IPC)

TDK Electronics Corp .... (516) 625-0100
12 Harbor Park Dr
Port Washington, NY 11050
Tech:(800) 835-8273 Toll Free:(800) 835-8273
Fax:(516) 625-0651

Fax:(516) 625-0171
Web Page: http://www.tdk.com
email: customer.support@tec.tdk.com

TEAC America Inc ······························(213) 726-0303
7733 Telegraph Road
Montebello, CA 90640
Fax:(213) 727-7656 Fax on Demand:(213) 727-7629
Web Page: http://www.teac.com
email: airborne@teac.com
BBS:(213) 727-7660

Tech Data Corp ······························(813) 539-7429
5350 Tech Data Dr
Clearwater, FL 34620
Tech:(800) 553-7976 Toll Free:(800) 237-8931
Fax:(813) 538-7816 Fax on Demand:(813) 538-7876
Tech Fax:(813) 532-6024
Web Page: http://www.techdata.com
email: sales@techdata.com

Tech Media Computer Systems Corp·····(800) 379-0077
7301 Orange Wood Avenue
Garden Grove, CA 92841
Fax:(714) 379-9069

Tech Tools ······························(908) 813-2400
43 Newburg Rd
Hackettstown, NJ 07840
Tech:(603) 424-2886 Toll Free:(800) 252-2748
Fax:(603) 888-8413
Tech Fax:(603) 429-3833
Web Page: http://www.techtools.com
email: support@techtools.com
Compuserve: 72730,1770

Techni-Tool Inc ······························(610) 941-2400
5 Apollo Rd, PO Box 368
Plymouth Meeting, PA 19462
Tech:(610) 941-2400 Toll Free:(800) 832-4866
Fax:(610) 828-5623
Web Page: http://www.techni-tool.com
email: techtool@interserv.com

Technical Communications Corp ··········(303) 693-2408
5503 South Malta Street
Aurora, CO 80015
Web Page: http://www.techcomm.com

Technology Concepts (see Prometheus)

Technology Group Inc, The ....(410) 576-2040
36 S. Charles St, Suite 2200
Baltimore, MD 21201
Tech:(410) 576-2040
Fax:(410) 576-1968
Web Page: http://www.technologygroup.com
Compuserve: 74603,761
BBS:(410) 576-8806

Technology Works ....(512) 794-8533
4030 W. Braker Lane, Suite 500
Austin, TX 78759-5319
Tech:(800) 688-7466 Toll Free:(800) 814-3306
Fax:(512) 794-8520 Fax on Demand:(800) 285-3084
Other:(800) 753-8360
Other Address: E-world:TechWorks
Web Page: http://www.techworks.com
email: techsupport@techworks.com
America Online: TechWorks
BBS:(512) 329-6327

TechSmith Corp ....(517) 333-2100
3001 Coolidge Rd, Ste 400
East Lansing, MI 48823-6320
Fax:(517) 333-1888
Web Page: http://www.techsmith.com
email: support@techsmith.com

TechWorks ....(512) 794-8533
4030 W. Braker Lane, Suite 350
Austin, TX 78759
Tech:(800) 933-6113 Toll Free:(800) 753-8360
Fax:(512) 794-8520
Web Page: http://www.techworks.com
email: techsupport@techworks.com

Tecmar Technologies ....(303) 682-3700
1900 Pike Road, Building E
Longmont, CO 80501
Toll Free:(800) 422-2587
Fax:(303) 776-7706 Fax on Demand:(303) 776-1085
Web Page: http://www.tecmar.com
email: tech.support@tecmar.com

Tecra Tool ....(303) 338-9224
2452 S. Trenton Way
Denver, CO 80231
Tech:(800) 284-0808 Toll Free:(800) 284-0808
Fax:(303) 338-9289
Web Page: http://www.tecratools.com

email: info@tecratools.com

Tekram Technology .... **(512) 833-6550**
11500 Metric Blvd, Suite 190
Austin, TX 78758
Tech:(512) 833-8158 Toll Free:(800) 556-6218
Fax:(512) 833-7276
Web Page: http://www.tekram.com
email: support@tekram.com
BBS:(512) 833-7985

TekSoft Inc .... **(602) 942-4982**
3320 W. Cheryl Drive, Suite 240
Phoenix, AZ 85051
Tech:(602) 942-4982
Fax:(602) 866-9016
Web Page: http://www.teksoft.com
email: info@teksoft.com

Tektronix .... **(800) 835-9433**
26600 SW Parkway Ave, PO Box 1000
Wilsonville, OR 97070-1000
Tech:(800) 547-8949
Fax:(503) 682-2980
Web Page: http://www.tek.com
email: xpresssupport@tek.com
BBS:(503) 685-4504

Teldar Corp .... **(602) 814-8400**
1110 S. Alma School Road, Suite 5-250
Mesa, AZ 85210
Fax:(602) 892-9877
Web Page: http://www.teldar.com
email: info@teldar.com

Telebit Corp (see ITK Telecommunications)

Telemagic .... **(800) 835-6244**
5973 Avienda Encina, Suite 101
Carlsbad, CA 92008
Tech:(972) 713-0259
Fax:(760) 736-7038
Web Page: http://www.telemagic.com
email: bagheri@telemagic.com
Compuserve: go tmagic.com

Televideo Inc .... **(408) 954-8333**
2345 Harris Way, PO Box 49048
San Jose, CA 95131
Tech:(408) 955-7711 Toll Free:(800) 345-6050
Fax:(408) 954-0623

Web Page: http://www.televideoinc.com
email: ts_support@televideoinc.com
BBS:(408) 954-8231

Teltone Corp ....(425) 487-1515
22121-20th Ave SE
Bothell, WA 98021-4408
Toll Free:(800) 426-3926
Fax:(425) 487-2288
Web Page: http://www.teltone.com
email: info@teltone.com

Template Graphics Software....................(512) 719-8000
3006 Longhorn Blvd, Suite 105
Austin, TX 78758
Toll Free:(800) 544-4847
Fax:(512) 832-0752
Tech Support Fax:(801) 588-4540
Web Page: http://www.tgs.com
email: support@tgs.com

Template Graphics Software....................(619) 457-5359
9920 Pacific Heights Blvd, Suite 200
San Diego, CA 92121-4331
Toll Free:(800) 544-4847
Fax:(619) 452-2547
Web Page: http://www.tgs.com
email: support@tgs.com

Ten X Technology ....(512) 918-9182
13091 Pond Springs Rd, B200
Austin, TX 78729
Toll Free:(800) 922-9050
Fax:(512) 918-9495
Web Page: http://www.tenx.com
email: jwuest@tenx.com

Teradyne ....(617) 482-2700
321 Harrison Ave
Boston, MA 02118
Fax:(617) 422-2910
Web Page: http://www.teradyne.com
email: tom.newman@teradyne.com

Test Equity ....(805) 498-9933
2450 Turquoise Circle
Thousand Oaks, CA 91320
Toll Free:(800) 950-3457
Fax:(805) 498-3733
Fax:(800) 272-4329

Web Page: http://www.testequity.com
email: sales@testequity.com

Texas Instruments Inc ....(214) 995-6611
13500 N. Central Expressway
Dallas, TX 75222
Tech:(800) 848-3927 Toll Free:(800) 848-3927
Fax:(800) 443-2984
Web Page: http://www.ti.com

Texas Memory Systems Inc ....(713) 266-3200
11200 Westheimer, Suite 1000
Houston, TX 77042
Fax:(713) 266-0332
Web Page: http://www.texmemsys.com
email: info@texmemsys.com

Texas Micro..(713) 541-8200
5959 Corporate Drive
Houston, TX 77036
Tech:(713) 541-8200 Toll Free:(800) 627-8700
Fax:(713) 541-8226
Web Page: http://www.texasmicro.com
email: support@texasmicro.com

Text 100 Corp...(206) 443-8004
2003 Western Ave, Suite 315
Seattle, WA 98121
Fax:(206) 443-5838
Web Page: http://www.text100.com
email: info@text100.com

Thermalloy Inc ....(972) 243-4321
2021 W. Valley View Lane, PO Box 810839
Dallas, TX 75234
Tech:(972) 243-4321
Fax:(972) 241-4656
Web Page: http://www.thermalloy.com

Thomas Computer Corporation ....(407) 855-2020
7101 President Drive, Suite 200
Orlando, FL 32809
Toll Free:(800) 621-3906
Fax:(407) 851-9700
Web Page: http://www.thomascomputer.com
email: polk02@aol.com

Thomas-Conrad Corp (see Compaq)

Three D Graphics ....(310) 553-3313
1801 Ave Of The Stars, Ste 600

Los Angeles, CA 90067
Toll Free:(800) 913-0008
Fax:(310) 788-8975
Web Page: http://www.threedgraphics.com
email: info@threedgraphics.com

ThrustMaster Inc ....(503) 615-3200
7175 NW Evergreen Parkway, Suite 400
Hillsboro, OR 97124
Tech:(503) 615-3200
Fax:(503) 615-3300
Web Page: http://www.thrustmaster.com
email: csweb@thrustmaster.com

Tiara Computer Sys (see Internex Info Serv)

TigerSoftware ....(305) 229-1119
8700 West Flagler Street, Suite 400
Miami, FL 33174
Toll Free:(800) 879-1597
Fax:(305) 228-3400
Web Page: http://www.tigerdirect.com
email: webmaster@tigerdirect.com

Timberline Software ....(503) 626-6775
9600 SW Nimbus
Beaverton, OR 97008
Tech:(800) 858-7098
Fax:(503) 641-7498
Web Page: http://www.timberline.com
email: webmaster@timberline.com

Time Line Solutions Inc ....(415) 898-1919
7599 Redwood Blvd, Suite 209
Novato, CA 94945
Tech:(415) 898-8737
Fax:(415) 898-0177
Web Page: http://www.tlsolutions.com

Time Motion Tools...(619) 679-0303
12778 Brookprinter Place
Poway, CA 92064
Tech:(619) 679-0303 Toll Free:(800) 779-8170
Fax:(800) 779-8171
Web Page: http://www.timemotion.com
email: info@timemotion.com

Timeslips Corp (see Sage U.S. Inc)

Tivoli Systems ....(512) 794-9070
9442 Capital Of Texas Hwy N, Arboretum Plaza One Suite 500

Austin, TX 78759
Tech:(512) 794-9070 Toll Free:(800) 284-8654
Fax:(512) 794-0623
Web Page: http://www.tivoli.com
email: support@tivoli.com

Tivoli Systems ....(408) 988-2800
5101 Patrick Henry Drive
Santa Clara, CA 95054
Tech:(408) 988-2800 Toll Free:(800) 848-6548
Fax:(408) 988-2236
Web Page: http://www.unison.com
email: info@unison.com

TMS Sequoia ....(405) 377-0880
206 W. 6th Ave
Stillwater, OK 74074
Toll Free:(800) 944-7654
Fax:(405) 372-9288
Tech Fax:(405) 742-1707
Web Page: http://www.tmsinc.com
email: contact_us@tmsinc.com

Todd Enterprises ....(516) 777-8633
65 E. Bethpage Rd
Plainview, NY 11803
Toll Free:(800) 445-8633
Fax:(516) 777-2750
Web Page: http://www.toddent.com

Tool Kit Specialists (see Com-Kyle)

Top Speed Corp..(954) 785-4555
150 E. Sample Road
Pompano Beach, FL 33064
Tech:(954) 785-4556 Toll Free:(800) 354-5444
Fax:(954) 946-1650 Fax on Demand:(954) 785-4555
Web Page: http://www.topspeed.com
email: sales@topspeed.com

Toray Industries ....(650) 341-7152
1875 S. Grant Street, Suite 720
San Mateo, CA 94402
Tech:(650) 341-7152 Toll Free:(800) 867-2973
Fax:(650) 341-0845
Web Page: http://www.toray.com
email: info@toray.com

Toshiba America ....(212) 596-0600
1251 Avenue Of The Americas, 41st Floor
New York, NY 10020

Tech:(800) 457-7777 Toll Free:(800) 457-7777
Fax:(212) 593-3875 Fax on Demand:(714) 583-3800
Fax Back:(888) 598-7802
Web Page: http://www.toshiba.com
BBS:(714) 837-4408

CD Rom Support .... (714) 583-3000
Toll Free:(800) 457-7777

Consumer Products .... (201) 628-8000
82 Totowa Road
Buffalo Grove, IL 60089-6900
Tech:(800) 999-4273 Toll Free:(800) 631-3811
Fax:(201) 628-1875 Fax on Demand:(888) 598-7802
Fax Back:(714) 583-3800
Other Address: http://www.tais.com
Web Page: http://www.toshiba.com/tacp
BBS:(714) 837-4408

Disk Products Division .... (714) 457-0777
9740 Irvine Blvd
Irvine, CA 92713-9724
Tech:(714) 455-0407 Toll Free:(800) 457-7777
Fax:(714) 587-6144
Tech Fax:(714) 587-6382
Web Page: http://www.toshiba.com
BBS:(714) 837-8864

Electronic Components .... (714) 455-2000
Tech:(714) 583-3000
Web Page: http://www.toshiba.com/taec

Electronic Imaging .... (800) 468-6744
Tech:(800) 950-4373 Toll Free:(800) 927-3133
Fax:(714) 583-3573
BBS:(714) 581-7600

Industrial Electronics .... (800) 231-1412

Information Systems .... (714) 583-3000
9740 Irvine Blvd
Irvine, CA 92618
Tech:(714) 455-0407 Toll Free:(800) 457-7777
Fax:(800) 950-4373 Fax on Demand:(714) 583-3800
Web Page: http://www.toshiba.com or www.tais.com
Compuserve: GO TOSHIBA
BBS:(714) 837-4408

Laptop Support .... (800) 999-4273

Magnia Servers .... (888) 882-8247

Out of Warranty .... (800) 999-4273

PC Support .... (800) 400-4172

Tosoh USA Inc .... (650) 286-2385
383 E. Grand Ave, Suite E
San Francisco, CA 94080

Tech:(650) 286-2385 Toll Free:(800) 238-6764
Fax:(650) 286-2392

Total Computer Supplies..........................**(810) 673-5000**
2524 Airport Road
Waterford, MI 48329

Total Management Inc..............................**(812) 476-5049**
2340 Ashington Ave
Evansville, IN 47714
Tech:(800) 553-5783 Toll Free:(888) 948-6825
Fax:(812) 476-5145
Web Page: http://www.totalmanagement.com
email: tcs@totalmanagement.com

Totally Hip Software Inc............................**(604) 685-6525**
201-1040 Hamilton Street
Vancouver, BC V6R 2R9 Canada
Tech:(604) 685-0984
Fax:(604) 685-4057
Web Page: http://www.totallyhip.com
email: support@totallyhip.com

TouchStone Software Corp.......................**(714) 969-7746**
2124 Main Street, 2nd Floor
Huntington Beach, CA 92648-9809
Tech:(714) 374-2801
Fax:(714) 969-4444 Fax on Demand:(714) 969-7746
Fax:(714) 960-1886
Web Page: http://www.checkit.com
email: tschelp@touchstone-sc.com
Compuserve: GO TOUCHSTONE
BBS:(714) 969-0688

Trade'Ex Electronic Commerce Sys.......**(813) 222-2050**
501 E. Kennedy Blvd, Suite 750
Tampa, FL 33602
Fax:(813) 222-5658
Web Page: http://www.tradeex.com
email: support@tradeex.com

Transcend Information Inc......................**(714) 921-2000**
1645 N. Brian Street
Orange, CA 92867
Tech:(714) 921-2000 Toll Free:(800) 886-5590
Fax:(714) 921-2111
Web Page: http://www.transcendusa.com
email: techsupport@transcendusa.com

Transparent Language Inc......................**(603) 465-2230**
22 Proctor Hill Road, PO Box 575

Hollis, NH 03049-0575
Tech:(800) 752-1767 Toll Free:(800) 752-1767
Fax:(603) 465-2779
Web Page: http://www.transparent.com
email: support@transparent.com

Trantor Systems Ltd (see Adaptec)

Traquair Data Systems Inc ....(607) 266-6000
114 Sheldon Road
Ithaca, NY 14850
Fax:(607) 266-8221
Web Page: http://www.traquair.com
email: traquair@traquair.com

Traveling Software ....(425) 483-8088
18702 N. Creek Pkwy
Bothell, WA 98011
Tech:(425) 487-8803 Toll Free:(800) 343-8080
Fax:(425) 485-6786 Fax on Demand:(425) 487-5410
Tech Fax:(425) 487-5440
Web Page: http://www.travsoft.com
email: sales@travsoft.com
BBS:(425) 485-1736

Trellix Corp ....(781) 788-9436
51 Sawyer Rd
Waltham, MD 02453
Toll Free:(888) 873-5549
Fax:(503) 684-4647
Web Page: http://www.trellix.com
email: support@trellix.com

Trend Micro Inc...(408) 257-1500
10101 N. Deanza Blvd, Suite 400
Cupertino, CA 95014
Tech:(310) 936-1188 Toll Free:(800) 586-7803
Fax:(408) 257-2003
Other:(800) 228-5651
Other Address: http://www.antivirus.com
Web Page: http://www.trendmicro.com
email: info@trendmicro.com
Compuserve: 72662,432
BBS:(310) 936-1192

Tri-Mark Engineering ....(615) 966-3667
406 Monitor Lande
Knoxville, TN 37922
Tech:(615) 966-3667
Fax:(615) 675-3458

email: mdudley@brbbs.brbbs.com
BBS:(615) 966-3574

Tri-Star Computer ....(602) 707-6450
3832 E. Watkins St
Phoenix, AZ 85034
Toll Free:(800) 844-2993
Web Page: http://www.tristar.com
email: support@tristar.com

Tribe Computer Works (see Zoom Telephonics)

Trident Microsystems Inc ....(650) 691-9211
189 N. Bernardo Ave
Mountain View, CA 94043-5203
Tech:(650) 934-2123
Fax:(650) 691-9260
Other Address: http://www.tridentmicro.com
Web Page: http://www.trid.com
email: techsupport@tridmicr.com
BBS:(650) 691-1165

TriniTech Inc ....(813) 442-8882
1430 Court St, Suite 3
Clearwater, FL 34616-6147
Tech:(813) 442-8882
Fax:(813) 581-4411
Web Page: http://www.pcdiagnostics.com
email: ttimart@aol.com

Trio Information Systems ....(919) 846-4990
8601 Six Forks Road, Suite 105
Raleigh, NC 27615
Tech:(919) 846-4985 Toll Free:(800) 880-4400
Fax:(919) 846-4997
Web Page: http://www.trio.com
email: suppquestions@trio.com
Compuserve: GO TRIO

Tripp Lite Worldwide ....(773) 869-1111
1111 W. 35th St
Chicago, IL 60609
Tech:(773) 869-1234
Fax:(773) 869-1329 Fax on Demand:(773) 869-1877
Web Page: http://www.tripplite.com
email: techsupport@tripplite.com

TriTech Microelectronics ....(408) 941-1300
1440 McCandless Dr
Milpitas, CA 95035
Web Page: http://www.tritechmicro.com

Trius Inc .... (978) 794-9377
North Andover, MA 01845-0249
Tech:(978) 794-0140
Fax:(978) 688-6312
Web Page: http://www.triusinc.com
email: info@triusinc.com
Compuserve: 71333,103
BBS:(978) 794-0762

Truevision .... (317) 841-0332
2500 Walsh Ave
Santa Clara, CA 95051
Tech:(317) 577-8788 Toll Free:(800) 522-8783
Fax:(317) 576-7770 Fax on Demand:(800) 522-8783
Web Page: http://www.truevision.com
email: support@truevision.com
Compuserve: GO TRUEVISION

Tseng Laboratories Inc .... (610) 313-9388
6 Terry Drive
Newtown, PA 18940
Web Page: http://www.tseng.com
email: prodsupp@tseng.com

TSSI (Tech Support & Service Inc) .... (805) 778-1773
375 Conejo Ridge Ave
Thousand Oaks, CA 91361
Tech:(800) 992-9916 Toll Free:(800) 286-0651
Fax:(805) 373-3000
Web Page: http://www.tssi4u.com
email: tssi4u@sprynet.com
BBS:(805) 582-3620

Tucker Electronics .... (214) 348-8800
1717 Reserve Street
Garland, TX 75042
Toll Free:(800) 527-4642
Fax:(214) 348-0367
Web Page: http://www.tucker.com
email: sales@tucker.com

Tulin Technology (see Computer Review)

Turbopower Software Company .... (719) 260-9136
4775 Centennial Blvd, Suite 114
Colorado Springs, CO 80919
Tech:(719) 260-6641 Toll Free:(800) 333-4160
Fax:(719) 260-7151
Web Page: http://www.turbopower.com
email: info@turbopower.com

Compuserve: 76004,2611

Turtle Beach Systems .... **(914) 966-0600**
5 Odell Plaza
Yonkers, NY 10701-1406
Tech:(914) 966-2150 Toll Free:(800) 233-9377
Fax:(914) 966-1102 Fax on Demand:(914) 966-0600
Web Page: http://www.tbeach.com
email: support@tbeach.com
Compuserve: GO TURTLE
America Online: tbsupport
BBS:(914) 966-1216

Tutankhamon Electric .... **(925) 682-6510**
2495 Estand Way, Suite 213
Pleasant Hill, CA 94523-3911
Tech:(800) 998-4888 Toll Free:(800) 998-4888
Fax:(925) 682-4125
Web Page: http://www.tutsys.com
email: info@tutsys.com

Twelve Tone Systems (see Cake Walk Music)

TwinBridge Software Corp .... **(323) 263-3926**
1055 Corporate Center Drive, Suite 400
Monterey Park, CA 91754
Tech:(323) 263-5931 Toll Free:(800) 894-6114
Fax:(323) 263-8126
Web Page: http://www.twinbridge.com
email: techsup@twinbridge.com
Compuserve: 70630,1010

Twinhead Corp .... **(510) 492-0828**
48295 Fremont Blvd
Fremont, CA 94538
Tech:(510) 492-0828 Toll Free:(800) 995-8946
Fax:(510) 492-0832
East:(800) 552-8946
Web Page: http://www.twinhead.com
email: technical_support@twinhead.com
Compuserve: GO LAPTOP
BBS:(510) 492-0835

Tyan Computer Corp .... **(510) 651-8868**
3288 Laurelview Ct
Fremont, CA 94538
Tech:(510) 440-8808
Fax:(510) 651-7688
Tech Fax:(510) 659-8288
Web Page: http://www.tyan.com

email: techsupport@tyan.com

Typhoon Software ....(805) 966-7633
1303A State St
Santa Barbara, CA 93101
Tech:(800) 728-8318 Toll Free:(800) 933-6520
Fax:(805) 962-6811
Web Page: http://www.typhoon.com
email: talktech@typhoon.com

U.S. Robotics Inc (3Com) ....(847) 982-5010
8100 N. McCormick Boulevard
Skokie, IL 60076
Tech:(800) 982-5151 Toll Free:(800) 550-7800
Fax:(847) 982-0823 Fax on Demand:(800) 762-6163
Web Page: http://www.usr.com
email: support@usr.com
Compuserve: GO USROBOTICS
BBS:(847) 982-5092

UB Networks Inc ....(408) 496-0111
Santa Clara, CA
Web Page: http://www.ub.com
email: kpagano@ub.com

Ubisoft Inc ....(415) 547-4000
625 Third St, 3rd Floor
San Francisco, CA 94107
Tech:(415) 547-4028
Fax:(415) 547-4001
Web Page: http://www.ubisoft.com

Ulead Systems Inc ....(310) 523-9393
970 W. 190th Street, Suite 520
Torrance, CA 90502
Tech:(310) 523-9391
Fax:(310) 523-9399
Web Page: http://www.ulead.com
email: support@ulead.com
Compuserve: GO ULEAD

Ultra-X Inc ....(408) 261-7090
PO Box 730010
San Jose, CA 95173-0010
Tech:(408) 261-7090 Toll Free:(800) 722-3789
Fax:(408) 261-7077
Web Page: http://www.ultra-x.com
email: sales@ultra-x.com

UltraCoach ....(909) 625-0463
9635 Monte Vista Ave, Suite 201

Montclair, CA 91763
Tech:(909) 398-1870 Toll Free:(800) 400-1390
Fax:(909) 625-4504
Web Page: http://www.ultracch.com
email: uctech@ultracch.com

UMAX Technologies Inc ....(510) 651-4000
3561 Gateway Blvd
Fremont, CA 94538
Tech:(510) 651-8883 Toll Free:(800) 562-0311
Fax:(510) 651-2610 Fax on Demand:(510) 651-3710
Notebook Support:(888) 815-8629
Web Page: http://www.umax.com
BBS:(510) 651-2550

Underware (see Numega Labs)

Unimark Inc ....(913) 649-2424
9400 Reeds Road
Overland Park, KS 66207
Tech:(913) 649-2232 Toll Free:(800) 255-6356
Fax:(913) 649-5795
Tech Fax:(913) 438-8952
Web Page: http://www.unimark.com
email: service@unimark.com

Unison Software Inc (see Tivoli Systems)

Unisys Corp ....(716) 924-0480
1100 Corporate Drive
Farmington, NY 14425
Tech:(800) 328-0440 Toll Free:(800) 448-1424
Fax:(716) 742-6671
Web Page: http://www.unisys.com

Unitech Industries ....(602) 303-9853
15035 N. 75th Street, Suite A
Scottsdale, AZ 85260
Toll Free:(800) 328-6483
Web Page: http://www.unitech-industries.com

Univel (see Novell)

Universal Software ....(310) 866-1274
PO Box 3683
Lakewood, CA 90711-3683

University Research & Development ......(412) 363-0990
1811 Jancey Street, Suite 200 (Rear)
Pittsburgh, PA 15206-1065
Toll Free:(800) 338-0517
Web Page: http://www.urda.com

uniView Technologies Corp ····················(214) 503-8880
10911 Petal St
Dallas, TX 75234
Tech:(214) 553-5586 Toll Free:(888) 864-8439
Web Page: http://curtismathes.com
email: support@uniview.com

Unixware (see Novell)

USA Flex (see Comark Inc)

Utimaco Safeware ····································(860) 688-4454
7 Waterside Crossing
Windsor, CT 06095
Tech:(800) 688-3227 Toll Free:(800) 688-1199
Fax:(860) 688-4496
Web Page: http://www.utimaco.com

UUNet Technologies ································(703) 206-5600
3060 Williams Dr
Fairfax, VA 22031-4648
Tech:(800) 900-0361 Toll Free:(800) 488-6384
Fax:(703) 206-5601
Web Page: http://www.us.uu.net
email: info@uu.net

V Communications Inc·····························(408) 965-4000
2290 North First Street, Suite 101
San Jose, CA 95131
Tech:(408) 965-4018 Toll Free:(800) 648-8266
Fax:(408) 965-4014
Web Page: http://www.v-com.com
email: support@v-com.com
Compuserve: 75031,3042

V-One Corp ···(301) 515-5200
20250 Century Blvd, Suite 300
German Town, MD 20874
Tech:(301) 515-5200
Fax:(301) 515-5280
Web Page: http://www.v-one.com
email: support@v-one.com

ValueStor Inc ··(408) 437-2300
1609-B Regatta Lane
San Jose, CA 95112
Tech:(408) 437-2310 Toll Free:(800) 873-8258
Fax:(408) 437-9333
Web Page: http://users.aol.com/valustorcw
email: ValuStorCW@aol.com
America Online: valuestor@aol.com

BBS:(408) 437-1616

ValueWare Software .... **(423) 675-7958**
142 West End Ave
Knoxville, TN 37922
Toll Free:(800) 441-7604
Fax:(423) 675-0657
Web Page: http://www.value-ware.com
email: value@1stresource.com

Varta Batteries .... **(914) 592-2500**
300 Executive Blvd
Elmsford, NY 10523
Toll Free:(800) 468-2782
Fax:(914) 592-2667
Web Page: http://www.varta.com

Vartech Inc (Out of Business)

VDONet Corp
170 Knowles Dr, Suite 206
Los Gatos, CA 95035
Web Page: http://www.vdo.net
email: sales@vdo.net

Velocity Inc .... **(415) 274-8840**
4 Embarcadero Center, Suite 3100
San Francisco, CA 94111-4106
Tech:(415) 274-8840 Toll Free:(800) 856-2489
Fax:(415) 776-8099
Web Page: http://www.velocitygames.com
email: info@velo.com
Compuserve: 766670,2202

Ven-Tel Inc .... **(408) 436-7400**
2121 Zanker Road
San Jose, CA 95131
Tech:(800) 538-5121 Toll Free:(800) 538-5121
Fax on Demand:(800) 841-4699
Web Page: http://www.zoom.com/ventel

Ventura Software (see Corel)

VenturCom Inc .... **(617) 661-1230**
215 First Street
Cambridge, MA 02142
Toll Free:(800) 334-8649
Fax:(617) 577-1607
Web Page: http://www.vci.com
email: info@vci.com

Verbatim Corp .... **(408) 773-3000**
445 Indio Way
Sunnyvale, CA 94086
Tech:(800) 538-8589 Toll Free:(888) 837-2284
Fax:(408) 746-3877
Other:(800) 259-6222
Web Page: http://www.verbatimcorp.com
email: info@verbatimcorp.com

Verbatim Corp .... **(704) 547-6500**
1200 WT Harris Blvd
Charlotte, NC 28262
Fax:(704) 547-6565
Web Page: http://www.verbatim.com

Verbex Voice Systems Inc .... **(732) 225-5225**
275 Raritan Center Parkway
Edison, NJ 08837-3613
Tech:(888) 483-7239 Toll Free:(888) 483-7239
Fax:(732) 225-7764
Web Page: http://www.verbex.com
email: support@listen.verbex.com

VeriFone .... **(408) 496-0444**
4988 Great America Parkway
Santa Clara, CA 95054-1200
Toll Free:(800) 654-1674
Fax:(408) 919-1405
Web Page: http://www.verifone.com
email: webmaster@verifone.com

VeriSign Inc .... **(650) 961-7500**
1390 Shorebird Way
Mountain View, CA 94043
Tech:(650) 429-3400
Fax:(650) 961-7300
Web Page: http://www.verisign.com
email: info@verisign.com

Verity Inc .... **(408) 541-1500**
892 Ross Drive
Sunnyvale, CA 94089
Tech:(408) 542-2222
Fax:(408) 541-1600
Tech Support:(408) 542-2152
Web Page: http://www.verity.com
email: tech-support@verity.com

Vermont Microsystems .... **(802) 655-2860**
11 Tigan Street

Winooski, VT 05404
Tech:(800) 354-0055 Toll Free:(800) 354-0055
Compuserve: 75410,2627

Versant Object Technology ....(510) 789-1500
6539 Dumbarton Circle
Fremont, CA 94555
Toll Free:(800) 837-7268
Fax:(510) 789-1515
Web Page: http://www.versant.com
email: info@versant.com

Vertex Industries ....(201) 503-1919
23 Carol Street, PO Box 996
Clifton, NJ 07014
Tech:(201) 777-3500
Fax:(201) 472-0814
Web Page: http://www.vetx.com
email: KREIFER@vertexindustries.com

Vertisoft Systems ....(803) 295-5875
153-B Grace Drive
Easley, SC 29640
Tech:(803) 269-9969 Toll Free:(800) 466-5875
Fax:(800) 466-4719
Web Page: http://www.vertisoftsys.com
email: info@vertisoftsys.com
Compuserve: GO VERTISOFT

Vertisoft Systems (Corporate) ....(415) 956-5999
Four Embarcadero Center, #3470
San Fransisco, CA 94111
Fax:(415) 956-5355
Web Page: http://www.vertisoftsys.com
email: info@vertisoftsys.com

Viacom New Media C/O Star Pak ....(303) 339-7114
237 22nd Street
Greeley, CO 80631
Tech:(303) 339-7114 Toll Free:(800) 469-2539
Fax:(303) 339-7022
Compuserve: 76702,1604:GO CDROM
America Online: VNM SUPPORT

ViaGrafix Corp (see DesignCAD)

Victory Enterprises Tech ....(512) 450-0801
223 W. Anderson Lane, Suite B300
Austin, TX 78752
Toll Free:(800) 727-3475
Fax:(512) 450-0869

Web Page: http://www.victoryent.com
email: info@victoryent.com

Video Electronic Standards Assn ··········(408) 435-0333
2150 North 1st Street, Suite 440
San Jose, CA 95131-2029
Fax:(408) 435-8225
Web Page: http://www.vesa.org
email: webmaster@vesa.org

Videodiscovery Inc ································(206) 285-5400
1700 Westlake Avenue N., Suite 600
Seattle, WA 98109-3012
Tech:(800) 548-3472 Toll Free:(800) 548-3472
Fax:(206) 285-9245
Web Page: http://www.videodiscovery.com/vdyweb
email: techsupp@videodiscovery.com

Viewpoint DataLabs ·····························(801) 229-3000
625 S. State St
Orem, UT 84058
Tech:(800) 552-4669 Toll Free:(800) 328-2738
Fax:(801) 229-3300
Tech Support:(801) 229-3372
Web Page: http://www.viewpoint.com
email: support@viewpoint.com

ViewSonic Corp ·································(909) 869-7976
20480 Business Pkwy
Walnut, CA 91789-0708
Tech:(909) 468-5800 Toll Free:(800) 888-8583
Fax:(909) 869-7958 Fax on Demand:(909) 869-7318
Tech Fax:(909) 468-1202
Web Page: http://www.viewsonic.com
email: vstech@viewsonic.com

Vinca Corp ···(801) 223-3100
1815 S. State St, Ste 2000
Orem, UT 84097-8068
Tech:(801) 223-3104 Toll Free:(888) 808-4622
Fax:(801) 223-3107
Web Page: http://www.vinca.com
email: support@vinca.com

Vireo Software Inc ·····························(978) 369-3380
30 Monument Square
Concord, MA 01742
Fax:(978) 318-6946
Web Page: http://www.vireo.com
email: info@vireo.com

Virgil Corp .... **(415) 433-9025**
290 Green Street, Suite 1
San Francisco, CA 94133
Tech:(415) 433-9025 Toll Free:(800) 662-8256
Fax:(415) 433-8411
Web Page: http://www.stockcenter.com
Compuserve: GO VIRGIL

Virgin Interactive Entertainment .... **(714) 833-8710**
18061 Fitch
Irvine, CA 92614
Tech:(714) 833-1999 Toll Free:(800) 874-4607
Fax:(714) 833-8717
Web Page: http://www.vie.com
BBS:(714) 833-2001

Virtual Comtech International Inc .... **(616) 399-8934**
977 Butternut Drive, Suite 181
Holland, MI 49424
Tech:(616) 399-8934
another BBS:(616) 399-8791
Web Page: http://virtualc.com
email: rdgraaf@virtualc.com
BBS:(616) 399-4818

Virtual Publisher .... **(702) 833-0622**
Incline Village, NV 89450
Web Page: http://www.virtualpublisher.com
email: support@virtualpublisher.com

Virtual Reality Laboratories .... **(805) 545-8515**
3534 A Empleo
San Luis Obispo, CA 93401
Tech:(805) 545-8515 Toll Free:(800) 829-8754
Web Page: http://www.vrli.com/vrli
email: vrli@aol.com
BBS:(805) 781-2257

Virtual Technologies (see Virtual Comtech)

Virtual Vegas .... **(310) 581-3636**
1223 Wilshire Blvd, Suite 808
Santa Monica, CA 90403
Toll Free:(800) 958-3427
Fax:(310) 581-3645
Web Page: http://www.virtualvegas.com
email: webmaster@virtualvegas.com

Virtus Corp .... **(919) 467-9700**
114 MacKenan Dr, Suite 100
Cary, NC 27511

Tech:(919) 467-9599 Toll Free:(800) 847-88
Fax:(919) 460-4530
Other:(888) 847-8875
Web Page: http://www.virtus.com
email: technical-support@virtus.com
Compuserve: 75300,3251

Visio Corp .... **(206) 521-4500**
520 Pike Street, Suite 1800
Seattle, WA 98101-4001
Tech:(541) 882-8687 Toll Free:(800) 248-4746
Fax:(206) 521-4501 Fax on Demand:(206) 521-4550
Tech Fax:(541) 882-8446
Other Address: micronet- go to visio
Web Page: http://www.visio.com
Compuserve: GO VISIO

Vision Computers Inc .... **(770) 840-0015**
5865 Jimmy Carter Blvd, Suite 125
Norcross, GA 30071
Tech:(770) 840-9249 Toll Free:(800) 886-4466
Fax:(770) 840-7115
Web Page: http://www.visioncomputers.com
email: sales@visioncomputers.com

Vision Imaging (see Advanced Media)

Vision Research Inc .... **(973) 696-4500**
190 Parish Drive
Wayne, NJ 07470
Toll Free:(800) 737-6588
Fax:(973) 696-0560
Web Page: http://www.visiblesolutions.com
email: phantom@visiblesolutions.com

Visionary Software (see First Things First)

Visioneer .... **(510) 608-0300**
34800 Campus Drive
Fremont, CA 94555
Tech:(888) 368-9633 Toll Free:(800) 787-7007
Fax:(716) 871-2138 Fax on Demand:(800) 505-0175
Web Page: http://www.visioneer.com

Visual Business Systems .... **(508) 263-9900**
1740 Massachusetts Ave
Acton, MA 01719-2209
Fax:(508) 263-9957
Web Page: http://www.visbussys.com/bbs
email: vbs@visbussys.com
BBS:(508) 266-0076

ual Networks Broadband Tech Grp···(978) 568-0600
1 Coolidge Street
udson, MA 01749
Tech:(978) 568-0600 Toll Free:(888) 638-2638
Fax:(978) 568-8858
Web Page: http://www.net2net.com
email: support@net2net.com

Visual Numerics ·····································(713) 784-3131
9990 Richmond Ave, Suite 400
Houston, TX 77042-4548
Toll Free:(800) 222-4675
Fax:(713) 781-9260
Web Page: http://www.vni.com
email: info@houston.vni.com

Viziflex Seels···(201) 487-8080
16 E. Lafayette St
Hackensack, NJ 07601-6895
Toll Free:(800) 307-3357
Fax:(201) 487-6637
Web Page: http://www.viziflex.com
email: seels@viziflex.com

VMark Software (see Ardent Software Inc)

VocalTec Inc ··(201) 768-9400
35 Industrial Parkway
Northvale, NJ 07647
Tech:(201) 768-9400
Fax:(201) 768-8893
Web Page: http://www.vocaltec.com
email: info@vocaltec.com

Vorton Technologies·································(613) 721-1107
1100-303 Moodie Dr
Nepean, ON K2H 9C4
Fax:(613) 721-0116
Web Page: http://www.vorton.com
email: service@vorton.com

Voxware Inc ··(609) 514-4100
305 College Road East
Princeton, NJ 08540
Fax:(609) 514-4101
Web Page: http://www.voxware.com
email: info@voxware.com

Voyager Company, The ····························(212) 219-2522
578 Broadway, Suite 1106
New York, NY 10012

Tech:(212) 219-2522
Fax:(212) 431-5799
sales:(888) 292-5584
Web Page: http://www.voyagerco.com
email: techsupport@voyagerco.com

Voyetra Technologies .... **(914) 966-0600**
5 Odell Plaza
Yonkers, NY 10701-1406
Tech:(914) 966-0600 Toll Free:(800) 233-9377
Fax:(914) 966-1102 Fax on Demand:(914) 966-0600
Web Page: http://www.voyetra.com
email: support@voyetra.com
Compuserve: 76702,2037
America Online: voyetra
BBS:(914) 966-1216

VST Power Systems .... **(508) 287-4600**
1620 Sudbury Road, Suite 3
Concord, MA 01742
Tech:(508) 287-4600
Fax:(508) 287-4068

Wacom Technology Corp .... **(360) 896-9833**
1311 SE Cardinal Ct, Suite 300
Vancouver, WA 98661
Tech:(800) 922-6613 Toll Free:(800) 922-9348
Fax:(360) 896-9724 Fax on Demand:(800) 922-9348
Web Page: http://www.wacom.com
email: support@wacom.com
Compuserve: go wacom
America Online: wacom
BBS:(360) 896-9714

Walker Richer & Quinn Inc (see WRQ)

Wall Data Inc .... **(425) 814-9255**
11332 NE 122nd Way
Kirkland, WA 98034-6931
Tech:(425) 814-3403 Toll Free:(800) 915-9255
Fax:(650) 856-9265 Fax on Demand:(425) 814-4362
Other:(888) 786-2268
Web Page: http://www.walldata.com
Compuserve: GO WALLDATA
BBS:(425) 814-4361

Walnut Creek CDROM .... **(925) 674-0783**
4041 Pike Lane, Suite E
Concord, CA 94520
Tech:(925) 603-1234 Toll Free:(800) 786-9907

Fax:(925) 674-0821
Web Page: http://www.cdrom.com
email: support@cdrom.com

Wang Laboratories Inc..............................**(978) 967-5000**
600 Technology Park Drive
Billerica, MA 01821-4130
Tech:(800) 247-9264 Toll Free:(800) 225-0654
Fax:(978) 967-0829
Web Page: http://www.wang.com
email: webteam@wang.com

Wangtek/WangDAT (see Tecmar Tech)

Warever Corp..**(801) 572-2555**
112 W. Business Park Dr
Draper, UT 84020
Tech:(801) 572-8923 Toll Free:(800) 766-7229
Fax:(801) 572-2444
Web Page: http://www.actionplus.com
email: support@actionplus.com

Watergate Software Inc..............................**(510) 596-2080**
2000 Powell St, Suite 1200
Emeryville, CA 94608
Fax:(510) 596-2092
Web Page: http://www.ws.com
email: support@ws.com

Waterloo Maple Software..........................**(519) 747-2373**
57 Erb St
Waterloo, ON N2L 6C2 Canada
Tech:(519) 747-2505 Toll Free:(800) 267-6583
Fax:(519) 747-5284
Web Page: http://www.maplesoft.com
email: support@maplesoft.com

Watermark Business Unit..........................**(617) 229-2600**
15 3rd Ave
Burlington, MA 01803
Tech:(714) 850-7680
Fax:(617) 229-2989
Web Page: http://www.watermark.com
BBS:(617) 229-5789

Wave Technologies....................................**(314) 995-5767**
Tech:(800) 826-4640 Toll Free:(800) 714-9468
Other:(888) 204-6143
Web Page: http://www.wavetech.com

Wavefront Communications·····················**(612) 638-9594**
1450 Energy Park Drive, Suite 110-J
Saint Paul, MN 55108
Fax:(612) 603-0269
Web Page: http://www.wavefront.com
email: training@wavefront.com

WaveMetrics Inc ·····································**(503) 620-3001**
PO Box 2088
Lake Oswego, OR 97035
Fax:(503) 620-6754
Web Page: http://www.wavemetrics.com
email: support@wavemetrics.com

Wavetek Corp ···**(619) 279-2200**
9045 Balboa
San Diego, CA 92123
Tech:(619) 279-2200 Toll Free:(800) 854-2708
Fax:(619) 627-0130
Web Page: http://www.wavetek.com
email: webteam@wavetek.com

Wayzata Technology Inc (Out of Business)

WebCo International Inc. ··························**(503) 417-2900**
506 SW Sixth Ave, Suite 602
Portland, OR 97204
Fax:(503) 227-7344
Web Page: http://www.europasoftware.com
email: support@webquick.com

WebManage Technologies Inc ················**(603) 594-9226**
547 Amherst St
Nashua, NH 03063
Fax:(603) 594-9227
Web Page: http://www.webmanage.com
email: info@webmanage.com

WebMaster Inc ··**(408) 345-1800**
1601 Civic Center Drive, Suite 200
Santa Clara, CA 95050
Fax:(408) 247-9372
Web Page: http://www.webmaster.com
email: support@webmaster.com

WebSci Technologies ·······························**(732) 329-9000**
4214 US Route 1
Monmouth Junction, NJ 08852
Toll Free:(800) 234-4500
Fax:(732) 329-0066
Web Page: http://www.dbpower.com

email: info@dbpower.com

WeiSheng Enterprise (Compucase) ·······(310) 464-2646
8319 Allport Ave
Santa Fe Springs, CA 90670
Fax:(310) 464-2648
Web Page: http://www.tata.com.tw/compucase
email: dwu@ksts.seed.net.tw

Westbrook Technologies Inc ··················(203) 483-6666
22 Summit Place
Branford, CT 06405
Tech:(203) 483-6666 Toll Free:(800) 949-3453
Fax:(203) 483-3350
Web Page: http://www.filemagic.com
email: webinfo@filemagic.com
BBS:(203) 483-3348

Westech Corp ··(973) 729-6584
57 Sparta Ave
Sparta, NJ 07871
Tech:(800) 745-4378 Toll Free:(800) 829-4767
Fax on Demand:(973) 729-9468
Tech Support:(973) 729-4378
Web Page: http://www.westech.com
email: support@westech.com
Compuserve: GO PCVENJ

Western Digital Corp ································(714) 932-5000
8105 Irvine Center Drive
Irvine, CA 92618
Tech:(800) 275-4932 Toll Free:(800) 832-4778
Fax:(714) 932-6294 Fax on Demand:(714) 932-4300
Web Page: http://www.wdc.com
email: ftp.wdc.com
Microsoft: WDC
America Online: search: westerndigital
BBS:(714) 753-1234

Western Micro Technology Inc ···············(408) 379-0177
254 E. Hacienda Ave
Campbell, CA 95008
Toll Free:(800) 338-1600
Fax:(408) 341-4762
Web Page: http://www.westernmicro.com
email: info@westernmicro.com

Western Scientific Inc ·····························(619) 565-6699
9445 Farnham Street
San Diego, CA 92123

Toll Free:(800) 443-6699
Fax:(619) 565-6938
Web Page: http://www.wsm.com
email: support@wsm.com

Western Telematic Inc ....(949) 586-9950
5 Sterling
Irvine, CA 92718-2517
Tech:(800) 854-7226 Toll Free:(800) 854-7226
Fax:(949) 583-9514
Web Page: http://www.westtel.com
email: info@westtel.com

Westwood Studios (see Virgin Interactive)

White Pine Software Inc............................(603) 886-9050
542 Amherst Street
Nashua, NH 03063
Toll Free:(800) 241-7463
Fax:(603) 886-9051
Web Page: http://www.wpine.com
email: cuservice@wpine.com

Whittaker Xyplex (see Xyplex)

Wholesale Computer Exchange..............(203) 459-8222
525 Fan Hill Road, #3
Monroe, CT 06468
Fax:(203) 459-8022
Web Page: http://www.wholesalecomputer.com
email: TomQ@snet.net

WildCard Technologies Inc (see PureData)

Willies Computer Software.......................(281) 360-4232
6215 Longflower Lane
Kingwood, TX 77345
Tech:(281) 360-3187 Toll Free:(800) 966-4832
Fax:(281) 360-3231
Web Page: http://www.wcscnet.com
email: support@wcscnet.com
Compuserve: 75300,3427

Willow Peripherals (see Pulse Systems)

Wind2 Software Inc ....(970) 482-7145
1901 Sharp Point Dr, Ste A
Fort Collins, CO 80525
Toll Free:(800) 779-4632
Web Page: http://www.wind2.com

Windata (see Direct Network Services)

ndow on Wallstreet ....(972) 235-9594
320 N. Glenville Drive, Suite 100
Richardson, TX 75081
Tech:(972) 783-6793 Toll Free:(800) 998-8439
Fax:(972) 783-6798
Web Page: http://www.wallstreet.net
email: support@wallstreet.net
BBS:(972) 414-7982

Windsoft International Inc ....(407) 240-2300
PO Box 590006
Orlando, FL 32859
Tech:(407) 240-3350 Toll Free:(800) 542-4455
Fax:(407) 240-2323
Web Page: http://www.windsoft.com
email: sales@windsoft.com
Compuserve: GO WINDSOFT

Windsor Technologies Inc ....(415) 456-2200
130 Alto St
San Rafael, CA 94901-4768
Tech:(415) 456-2200
Fax:(415) 456-2244
Web Page: http://www.windsortech.com
email: sales@windsortech.com

Wingra Technologies Inc ....(608) 238-4454
450 Science Drive, One West
Madison, WI 53711-1056
Tech:(608) 238-4454 Toll Free:(800) 544-5465
Fax:(608) 238-8986
Web Page: http://www.wingra.com
email: csc@wingra.com

WinSoft Corp ....(800) 494-6763
1928 E. Deere Ave, Suite 110
Santa Ana, CA 92705
Web Page: http://www.winsoft.com
email: winsoft@ix.netcom.com
Compuserve: 74534,2561

WinWay Corp ....(916) 965-7878
5431 Auburn Blvd, Suite 398
Sacramento, CA 95841-2801
Tech:(916) 965-7878 Toll Free:(800) 494-6929
Fax:(916) 965-7879
Web Page: http://www.winway.com
email: support@winway.com

Wizardware Multimedia Ltd
918 Delaware Avenue
Bethlehem, PA 18015-7969
Fax:(610) 691-8258

Wizardworks Group Inc ....(800) 229-2714
2300 Berkshire Lane
Minneapolis, MN 55441
Tech:(425) 398-3051
Fax:(612) 577-0631 Fax on Demand:(425) 398-3051
Web Page: http://www.wizworks.com
email: support@wizworks.com
BBS:(612) 559-6197

Wollongong (see Attachmate)

Wonderware Corp ....(714) 727-3200
100 Technology Dr
Irvine, CA 92718
Tech:(714) 727-3299
Fax:(714) 727-3270
Web Page: http://www.wonderware.com
email: support@wonderware.com
Compuserve: GO WONDER
BBS:(714) 727-0726

WordStar Int (see Learning Co)

Wordstar USA (see Learning Co)

Work Wise Software..................................(206) 545-1191
146 Canal Street, Suite 200
Seattle, WA 98103
Tech:(206) 545-8679 Toll Free:(800) 967-5947
Fax:(206) 545-1273
Web Page: http://www.workwise.com
email: techsupport@workwise.com

Working Software Inc...............................(408) 423-5696
303 Potrero Street, Suite 22
Santa Cruz, CA 95060
Toll Free:(800) 229-9675
Fax:(408) 423-5699
Web Page: http://www.webcom.com/~working
email: info@working.com

World Software Corp...............................(201) 444-3228
124 Prospect Street
Ridgewood, NJ 07450
Tech:(201) 444-3290 Toll Free:(800) 962-6360
Fax:(201) 444-9065

Web Page: http://www.worldox.com
email: worldox@worldox.com
Compuserve: 72662,460

Worldcomm Systems (see Globecomm Systems)

Worthington Data Solutions····················(408) 458-9938
3004 Mission St, Suite 220
Santa Cruz, CA 95060
Toll Free:(800) 345-4220
Fax:(408) 458-9964
Web Page: http://www.barcodehq.com
email: wds@barcodehq.com

Wright Strategies·····································(619) 702-0500
600 B St, 18th Floor
San Diego, CA 92101-4590
Web Page: http://www.wrightstrat.com

Wrox Press··(312) 397-1900
1512 N. Fremont Street, Suite 103
Chicago, IL 60622
Tech:(800) 873-9769 Toll Free:(800) 814-4527
Fax:(312) 397-8990
Web Page: http://www.wrox.com
email: support@wrox.com
Compuserve: 100063,2152

WRQ···(206) 217-7500
1500 Dexter Avenue N.
Seattle, WA 98109
Tech:(206) 217-7000 Toll Free:(800) 872-2829
Fax:(206) 217-0293
Other:(206) 217-7100
Web Page: http://www.wrq.com
email: support@wrq.com
BBS:(206) 217-0145

Wyatt River Software ·······························(408) 458-2600
Tech:(408) 458-2600
Web Page: http://www.wyattriver.com

Wyse Technology·····································(408) 473-1200
3471 N. First St
San Jose, CA 95134-1803
Tech:(800) 800-9973 Toll Free:(800) 438-9973
Fax:(408) 473-2401 Fax on Demand:(800) 800-9973
Web Page: http://www.wyse.com
email: techsupport@wyse.com
BBS:(408) 922-4400

X Consortium (see Open Group, The)

X-10 (USA) Inc..(201) 784-97
91 Ruckman Road, Box 420
Closter, NJ 07624
Tech:(201) 784-1936 Toll Free:(800) 411-2888
Fax:(201) 784-9464
Orders:(800) 675-3044
Web Page: http://www.x10.com
email: support@x10.com
America Online: x-10usa@aol.com

X3 Secretarist (see ICA)

XcelleNet Inc (see Sterling Commerce)

Xebec (Out of Business)

XenoSoft..(510) 644-9366
2210 6th Street
Berkeley, CA 94710

Xerox Corp..(716) 423-5090
Xerox Square
Rochester, NY 14644
Tech:(800) 822-2979
Web Page: http://www.xerox.com
Documentation/Software Services..............(800) 327-9753
Font Services..............................(800) 445-3668
Imaging Systems (Xerox DDS)..................(978) 977-2000
9 Centennial Drive
Peabody, MA 01960
Tech:(800) 248-6550 Toll Free:(800) 248-6550
Fax on Demand:(978) 977-2462
Web Page: http://www.xerox.com
Compuserve: GO XEROX
BBS:(978) 532-0869
Parts..(800) 828-5881
Personal Document Product Division..........(800) 821-2797
East Rochester, NY 14445
Web Page: http://www.xerox.com
Quality Services..............................(800) 438-5077
Scanning/Image Capture......................(888) 362-8462
ScanSoft..(800) 343-0311
Small Office/Home Office......................(800) 832-6979
Supplies..(800) 822-2200

Xerox/X-Soft..(650) 424-0111
3400 Hillview Avenue
Palo Alto, CA 94304
Toll Free:(800) 334-6200

eb Page: http://www.xsoft.com
email: info@XSoft.xerox.com

net Inc .... (510) 845-0555
2560 Ninth Street, Suite 312
Berkeley, CA 94710
Fax:(510) 644-2680
Web Page: http://www.xinet.com
email: www@xinet.com

Xing Technology Corp .... (805) 783-0400
810 Fiero Lane
San Luis Obispo, CA 93401
Toll Free:(800) 298-6448
Fax:(805) 783-4930
Web Page: http://www.xingtech.com

Xircom Inc .... (805) 376-9300
2300 Corporate Center Drive
Thousand Oaks, CA 91320-1420
Tech:(805) 376-9200 Toll Free:(800) 438-4526
Fax:(805) 376-9311 Fax on Demand:(800) 775-0400
Support Fax:(805) 376-9100
Web Page: http://www.xircom.com
email: webmaster@xircom.com
Compuserve: GO XIRCOM
BBS:(805) 376-9130

Xtend Micro Products .... (714) 699-1400
2 Faraday
Irvine, CA 92618
Toll Free:(800) 232-9836
Fax:(714) 699-1434
Web Page: http://www.xmpi.com
email: xtend@xmpi.com

XTree Company (see Symantec Corp)

Xylan Corp .... (818) 878-4507
26707 W. Agoura Rd
Calabasas, CA 91302
Tech:(800) 995-2696 Toll Free:(800) 999-9526
Other:(800) 995-2612
Web Page: http://www.xylan.com
email: switchexpert@xylan.com

Xylogics Inc (see Bay Networks)

Xyplex .... (978) 952-4700
295 Foster Street
Littleton, MA 01460

Tech:(800) 435-7997 Toll Free:(800) 338-5316
Fax:(978) 952-4702
Web Page: http://www.xyplex.com
email: support@xyplex.com

Yahoo! Inc ....(408) 731-3300
3420 Central Expressway, 2nd Floor
Santa Clara, CA 95051
Fax:(408) 731-3301
Web Page: http://www.yahoo.com

Yamaha Corporation Of America ....(714) 522-9011
600 Orange Thorpe Ave, PO Box 6600
Buena Park, CA 90622-6600
Tech:(714) 522-9000 Toll Free:(800) 823-6414
Fax:(714) 527-5782
Web Page: http://www.yamaha.com

Yamaha Systems Technology Inc ....(408) 467-2300
100 Century Center Court, Ste 800
San Jose, CA 95112
Toll Free:(800) 543-7457
Fax:(408) 437-8791
Web Page: http://www.yamahayst.com

YBM Magnex Inc ....(800) 692-5296
110 Terry Drive
Newtown, PA 18940
Tech:(215) 579-0400
Fax:(215) 579-3444

Young Minds Inc......................................(909) 335-1350
1906 Orange Tree Lane, Suite 240
Redlands, CA 92374
Tech:(909) 335-5780 Toll Free:(800) 964-4964
Fax:(909) 798-0488
Tech Support:(800) 496-4237
Web Page: http://www.ymi.com
email: support@ymi.com

Z-Code Software (see Net Manage)
101 Rowland Way, Suite 300
Novato, CA 94945
Web Page: http://www.z-mail.com

Z-Ram (see Camintonn Z-Ram)

Zane Publishing Inc ....(800) 460-2323
Tech:(612) 368-9767
Web Page: http://www.zane.com
email: support@zane.com

ZD comdex & Forums Inc .... **(781) 433-1500**
300 First Ave
Needham, MA 02194-2722
Tech:(650) 578-6941
Fax:(781) 444-4806
Web Page: http://www.comdex.com
email: sci@zd.com

Zebra Technologies .... **(847) 634-6700**
333 Corporate Woods Pkwy
Vernon Hills, IL 60061-3109
Tech:(847) 634-6700 Toll Free:(800) 423-0422
Fax:(847) 913-8766 Fax on Demand:(888) 267-9181
Fax Back:(503) 402-1367
Web Page: http://www.zebra.com

Zedcor .... **(520) 881-8101**
3420 N. Dodge Blvd, Suite Z
Tucson, AZ 85716
Tech:(520) 881-2310 Toll Free:(800) 482-4567
Fax:(520) 881-1841
Web Page: http://www.zedcor.com
email: support@zedcor.com

Zenith Data Systems (Packard Bell) .... **(800) 654-1394**
2455 Horse Pen Road, Suite 100
Herndon, VA 22071
Tech:(800) 227-3360 Toll Free:(800) 654-1394
Fax on Demand:(800) 582-8194
Web Page: http://www.zds.com
BBS:(916) 386-9899

Zenographics Inc .... **(714) 851-6352**
34 Executive Park, Suite 150
Irvine, CA 92614
Tech:(800) 566-7468 Toll Free:(800) 366-7494
Fax:(714) 851-1314 Fax on Demand:(714) 833-7472
Tech Fax:(714) 833-7465
Web Page: http://www.zeno.com
email: eservice@zeno.com
Compuserve: 76702,1351 GO ZENO

Zeos International (Div of Micron) .... **(800) 423-5891**
1301 Industrial Blvd
Minneapolis, MN 55413
Tech:(612) 633-7337 Toll Free:(800) 423-5891
Fax on Demand:(800) 845-2341
Marketing:(208) 465-3434
Web Page: http://www.micron.com
email: support@zeos

Compuserve: GO ZEOS
America Online: keyword: zeos
BBS:(612) 633-0815

Zilog Inc..**(408) 558-8500**
910 E. Hamilton Ave, Suite 110
Campbell, CA 95008-6600
Tech:(800) 880-9665 Toll Free:(800) 880-9665
Fax:(408) 558-8300
Tech Fax:(408) 558-8571
Web Page: http://www.zilog.com
email: csupport@zilog.com

Zoom Telephonics...................................**(617) 423-1072**
2020 Challenger Dr, Suite 101
Alameda, CA 94501
Tech:(617) 423-1076
Fax:(617) 423-5536 Fax on Demand:(617) 423-4651
Other Address: http://www.zoomtel.com
Web Page: http://www.tribe.com
email: tribe-support@zoomtel.com
Compuserve: GO ZOOM
America Online: keyword: ZOOMT

Zoom Telephonics Inc ....**(617) 423-1072**
207 South St
Boston, MA 02111
Tech:(617) 423-1076 Toll Free:(800) 666-6191
Fax:(617) 423-3923 Fax on Demand:(617) 423-4651
Web Page: http://www.zoomtel.com
email: sales@zoomtel.com
Compuserve: GO ZOOM
America Online: keyword: ZOOMT
BBS:(617) 423-3733

ZSoft Corp (see Learning Co)

Zydeco...**(504) 539-9300**
1600 Canal St, 14th Floor
New Orleans, LA 70112
Tech:(504) 539-9300 Toll Free:(800) 932-7515
Fax:(504) 539-9304
Web Page: http://www.itssoftware.com
email: itssupport@itssoftware.com

ZyLAB Corp ....**(301) 590-0900**
9210 Corporate Blvd, Suite 200
Rockville, MD 20850
Tech:(301) 590-0900 Toll Free:(800) 544-6339
Fax:(301) 590-0903

Web Page: http://www.zylab.com
email: support@zylab.com
BBS:(301) 670-1596

ZyPCsom Inc··(510) 783-2501
2301 Industrial Pkwy W., Building 7
Hayward, CA 94545-5029
Tech:(510) 783-2501
Fax:(510) 783-2414

ZyXEL Communications Inc····················(714) 632-0882
4920 East La Palma Ave
Orange, CA 92866
Tech:(714) 693-0808 Toll Free:(800) 255-4101
Fax:(714) 693-8811
Web Page: http://www.zyxel.com
email: support@zyxel.com
Compuserve: 71333,2734

Index

D

A

B

C

D

E

F

G

H

I

J

K

L

M

N

O

P

Q

R

S

T

U

V

W

X

Y

Z

Notes: